P9-CKX-317

Statistical Methods

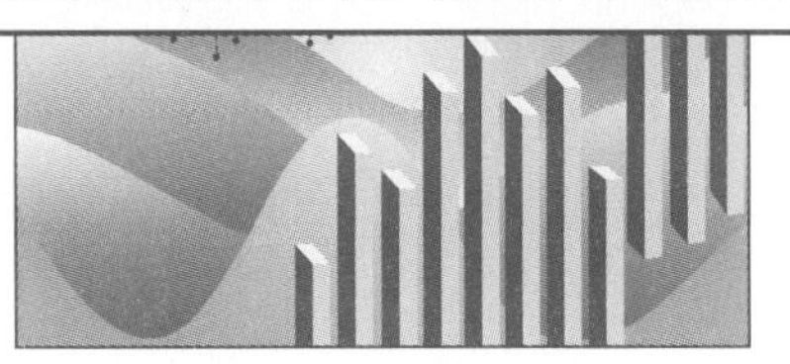

Second Edition

Statistical Methods

Second Edition

Rudolf J. Freund
Texas A&M University

William J. Wilson
University of North Florida

ACADEMIC PRESS
An imprint of Elsevier Science

Amsterdam Boston London New York Oxford Paris
San Diego San Francisco Singapore Sydney Tokyo

Senior Editor, Mathematics	Barbara Holland
Senior Project Manager	Angela Dooley
Editorial Coordinator	Tom Singer
Product Manager	Anne O'Mara
Cover Design	Dick Hannus
Copyeditor	Charles Lauder
Composition	ITC
Printer	Courier

This book is printed on acid-free paper. ∞

Academic Press
An imprint of Elsevier Science
525 B Street, Suite 1900, San Diego, California 92101-4495, USA
http://www.academicpress.com

Academic Press
An imprint of Elsevier Science
84 Theobald's Road, London WC1X 8RR, UK
http://www.academicpressbooks.com

Academic Press
An imprint of Elsevier Science
200 Wheeler Road, Burlington, Massachusetts 01803, USA
http://www.academicpressbooks.com

Library of Congress Control Number: 2002111023
International Standard Book Number: 0-12-267651-3

PRINTED IN THE UNITED STATES OF AMERICA
03 04 05 06 9 8 7 6 5 4 3 2

Contents

Preface

The objective of *Statistical Methods*, Second Edition, is to provide students with a working introduction to statistical methods. Courses using this book are normally taken by advanced undergraduate statistics students and graduate students from various disciplines.

Statistical Methods is an upper-level requirement for undergraduate degrees in disciplines emphasizing quantitative skills, or a requirement for graduate degrees in disciplines where statistics is an important research tool. This book is intended to be used for this type of course. The material in this book provides an overview of a wide range of applications and normally requires two semesters, although a limited knowledge of statistical methods is provided in the first semester. Many students will continue with several additional courses in specialized statistics applications.

Traditionally, textbooks used for statistical methods courses have emphasized plugging numbers into formulas, with computer usage as an afterthought. This approach has led to much mind-numbing drill, which obscures the real issues. The increased usage of computers and availability of comprehensive statistical software packages would seem to imply that statistical methods should now be taught in terms of implementing such software. This approach is likely to make the computer appear as a black box into which one pours data files and automatically receives the correct answers. However, a computer does not know whether it is doing the correct analysis and is capable of a beautifully annotated execution of an incorrect analysis. Also a computer cannot interpret results and write a report.

Guiding Principles

This text provides a reasonable compromise between these two extremes. Our guiding principles are as follows:

- No mathematics beyond algebra is required. However, mathematically oriented students may still find the material in this book challenging, especially if they are also exposed to courses in statistical theory.

- Formulas are presented primarily to show the how and why of a particular statistical analysis. For that reason, there is a minimal number of exercises that plug numbers into formulas.
- The topics in this book are organized in broad categories to facilitate the choice of the best-performance methodology for a specific task and there is considerable cross-referencing to facilitate making this choice.
- All examples containing real data are worked to a logical conclusion, including interpretation of results. Where computer printouts are used, the results are discussed and explained. In general, the emphasis is on conclusions rather than mechanics.
- Throughout the book we stress that certain assumptions about the data must be fulfilled in order for the statistical analyses to be valid, and we emphasize that although the assumptions are often fulfilled, they should be routinely checked.

New to This Edition

- *Friendlier exposition* makes concepts clearer to students without weakening the statistical rigor of the material.
- *New, greater emphasis on graphics* help students to visualize and understand ideas.
- *Examples of contemporary topics*, such as analysis of means, are included at appropriate points in the text.
- *Exercises* or portions of exercises are identified when material is covered from specific sections, allowing students to practice the methods without having to wait until a complete chapter is covered.
- *Examples and exercises* contain both contemporary data and references to additional data on the Internet or in other published works.

Using This Book

Organization

The organization of *Statistical Methods*, Second Edition, follows the "classical" order. The formulas in the book are generally the so-called definitional ones that emphasize concepts rather than computational efficiency. These formulas can be used for a few of the very simplest examples and problems, but we expect that virtually all exercises will be implemented on computers. The first seven chapters, which are normally covered in a first semester, cover data description, probability and sampling distributions, basics of inference for one and two sample problems, the analysis of variance, and one-variable regression. The second portion of the book starts with chapters on multiple regression, factorial experiments, experimental design, and an introduction to general linear models including the analysis of covariance. We have separated factorial experiments and design of experiments, because they are different applications of the same numeric methods.

The last three chapters introduce topics in the analysis of categorical data, nonparametric statistics, and sampling. These chapters provide a brief introduction to these important topics and are intended to round out the statistical education of those who will learn from this book.

Coverage

This book contains more material than can be covered in a two-semester course. We have purposely done this for two reasons:

- Because of the wide variety of audiences for statistical methods, not all instructors will want to cover the same material. For example, courses with heavy enrollments of students from the social and behavioral sciences will want to emphasize nonparametric methods and the analysis of categorical data with less emphasis on experimental design.
- Students who have taken statistical methods courses tend to keep their statistics books for future reference. We recognize that no single book will ever serve as a complete reference, but we hope that the broad coverage in this book will at least lead these students in the proper direction when the occasion demands.

Sequencing

For the most part, topics are arranged so that each new topic builds on previous topics, hence course sequencing should follow the book. There are, however, some exceptions that may appeal to some instructors:

- In some cases it may be preferable to present the material on categorical data at an early stage. Much of the material in Chapter 12 (Categorical Data) can be taught anytime after Chapter 5 (Inferences for Two Populations).
- Some instructors prefer to present nonparametric methods along with parametric methods. Again, any of the material in Chapter 13 (Nonparametric Methods) may be taken at any time after Chapter 3 (Principles of Inference).

Exercises

Properly assigned and executed exercises are an integral part of any course in statistical methods. We have placed all exercises at the ends of chapters to emphasize problem solving rather than mechanics for particular methods. This placement may have the unintended consequence that students may delay starting these exercises until the chapters have been completed, resulting in uneven workloads. To alleviate this potential problem we have placed instructions on initiating work on exercises throughout some of the longer chapters. Students are also encouraged to do all examples. Data files for all exercises and examples are available from the publisher.

Computing

For consistency and convenience and because it is the most widely used single statistical computing package, we have relied heavily on the SAS® System to illustrate examples in this text. However, because student access to computers in general, and the SAS System in particular, is not universal, we have provided generic rather than software specific instructions for performing the analyses for examples and exercises.

Instructional material is available from specific software vendors and an increasing amount of independently published material is becoming available. For those who wish to use the SAS System, data and code for performing the analyses for examples and exercises are available on ASCII files. The data portion of these files can be adapted for use with other software.

Data Sets are available on the Web. Please contact the sales representative or the publisher for further details.

Acknowledgments

We would like to thank the Department of Statistics of Texas A&M University and the Department of Mathematics and Statistics of the University of North Florida for the cooperation and encouragement that made this book possible. We also owe a debt of gratitude to the following reviewers whose comments have made this a much better work:

Erol Pekoz, Boston University; Christine Anderson-Cook, Virginia Tech; Steven Rein, Cal Polytechnic State University; E.D. McCune, Stephen F. Austin State University; Brian Habing, University of South Carolina; Xuming He, Univ. of Illinois Urbana; Pat Goeters, Auburn University; Krzysztof Ostaszewski, Illinois State University; Mark Payton, Oklahoma State University; and Matt Carlton, Cal Polytechnic State University.

We also express our appreciation for the encouragement and guidance provided by Barbara Holland, Senior Editor, Tom Singer, Editorial Coordinator, and Angela Dooley, Senior Project Manager, at Academic Press whose expertise have made this a much more readable book.

We acknowledge Minitab Inc., SAS Institute, SPSS, Inc., and Microsoft Corporation whose software (Minitab®, the SAR® System, SPSS®, and Excel, respectively) are used to illustrate computer output. The SAS System was used to compute the tables for the normal, t, χ^2, and F distributions. We also gratefully acknowledge the Biometric Society, the Trustees of Biometrika, *Journal of Quality Technology*, and American Cyanamid Company for their permission to reproduce tables published under their auspices.

Finally, we owe an undying debt of gratitude to our wives, Marge and Marilyn, who have encouraged our continuing this project despite the often encountered frustrations.

Data and Statistics

1.1 Introduction

To most people the word **statistics** conjures up images of vast tables of confusing numbers, volumes and volumes of figures pertaining to births, deaths, taxes, populations, and so forth, or figures indicating baseball batting averages or football yardage gained flashing across television screens. This is so because in common usage the word *statistics* is synonymous with the word *data*. In a sense this is a reasonably accurate impression because the discipline of statistics deals largely with principles and procedures for collecting, describing, and drawing conclusions from data. Therefore it is appropriate for a text in statistical methods to start by discussing what data are, how data are characterized, and what tools are used to describe a set of data. The purpose of this chapter is to

1. provide the definition of a set of data,
2. define the components of such a data set,
3. present tools that are used to describe a data set, and briefly
4. discuss methods of data collection.

DEFINITION 1.1
A set of **data** is a collection of observed values representing one or more characteristics of some objects or units.

EXAMPLE 1.1

A typical data set Every year, the National Opinion Research Center (NORC) publishes the results of a personal interview survey of U.S. households. This survey is called the General Social Survey (GSS) and is the basis for many studies conducted in the social sciences. In the 1996 GSS, a total of

2904 households were sampled and asked over 70 questions concerning lifestyles, incomes, religious and political beliefs, and opinions on various topics. Table 1.1 lists the data for a sample of 50 respondents on four of the questions asked. This table illustrates a typical mid-sized data set. Each of the rows corresponds to a particular respondent (labeled 1 through 50 in the first column). Each of the columns, starting with column two, are responses to the following four questions:

1. AGE: The respondent's age in years
2. SEX: The respondent's sex coded 1 for male and 2 for female
3. HAPPY: The respondent's general happiness, coded:
   1 for "Not too happy"
   2 for "Pretty happy"
   3 for "Very happy"
4. TVHOURS: The average number of hours the respondent watched TV during a day

This data set obviously contains a lot of information about this sample of 50 respondents. Unfortunately this information is hard to interpret when the data are presented as shown in Table 1.1. There are just too many numbers to make any sense of the data (and we are only looking at 50 respondents!). By summarizing some aspects of this data set, we can obtain much more usuable information and perhaps even answer some specific questions. For example, what can we say about the overall frequency of the various levels of happiness? Do some respondents watch a lot of TV? Is there a relationship between the age of the respondent and his or her general happiness? Is there a relationship between the age of the respondent and the number of hours of TV watched?

We will return to this data set in Section 1.9 after we have explored some methods of summarizing and making sense of data sets like this one. As we develop more sophisticated methods of analysis in later chapters, we will again refer to this data set.[1] ■

DEFINITION 1.2
A **population** is a data set representing the entire entity of interest.

For example, the decennial census of the United States yields a data set containing information about all persons in the country at that time (theoretically all households correctly fill out the census forms). The number of persons per household as listed in the census data constitutes a population of family sizes in the United States. Similarly, the weights of all steers brought to an auction by a particular rancher is a data set that is the population of the weights of that rancher's marketable steers.

Note that elements of a population are really measures rather than individuals. This means that there can be many different definitions of populations that involve the same collection of individuals. For example, the number of

[1]The GSS is discussed on the Web page: http://www.icpsr.umich.edu/GSS/.

Table 1.1

Sample of 50 Responses to the 1996 GSS

Respondent	AGE	SEX	HAPPY	TVHOURS
1	41	1	2	0
2	25	2	1	0
3	43	1	2	4
4	38	1	2	2
5	53	2	3	2
6	43	2	2	5
7	56	2	2	2
8	53	1	2	2
9	31	2	1	0
10	69	1	3	3
11	53	1	2	0
12	47	1	2	2
13	40	1	3	3
14	25	1	2	0
15	60	1	2	2
16	42	1	2	3
17	24	2	2	0
18	70	1	1	0
19	23	2	3	0
20	64	1	1	10
21	54	1	2	6
22	64	2	3	0
23	63	1	3	0
24	33	2	2	4
25	36	2	3	0
26	53	1	1	2
27	26	2	2	0
28	89	2	2	0
29	65	1	1	0
30	45	2	2	3
31	64	2	3	5
32	30	2	2	2
33	75	2	2	0
34	53	2	2	3
35	38	1	2	0
36	26	1	2	2
37	25	2	3	1
38	56	2	3	3
39	26	2	2	1
40	54	2	2	5
41	31	2	2	0
42	44	1	2	0
43	36	2	2	3
44	74	2	2	0
45	74	2	2	3
46	37	2	3	0
47	48	1	2	3
48	42	2	2	6
49	77	2	2	2
50	75	1	3	0

school-age children per household as listed in the census data would constitute a population for another study. As we shall see in discussions about statistical inference, it is important to define the population that we intend to study very carefully.

DEFINITION 1.3
A **sample** is a data set consisting of a portion of a population. Normally a sample is obtained in such a way as to be representative of the population.

The Census Bureau conducts various activities during the years between each decennial census, such as the Current Population Survey. This survey samples a small number of scientifically chosen households to obtain information on changes in employment, living conditions, and other demographics. The data obtained constitute a sample from the population of all households in the country. If two steers were selected from a herd of steers brought to an auction by a rancher, these two steers would be considered a sample from the herd.

Data Sources

Data come from many different sources, depending on the objective of the particular study, the limitations of data collection resources, or any number of other factors. However, in general, data are obtained from two broad categories of sources:

- **Primary** data are collected as part of the study.
- **Secondary** data are obtained from published sources, such as journals, governmental publications, news media, or almanacs.

There are several ways of obtaining primary data. Data are often obtained from simple observation of a process, such as characteristics and prices of homes sold in a particular geographic location, quality of products coming off an assembly line, political opinions of registered voters in the state of Texas, or even a person standing on a street corner and recording how many cars pass each hour during the day. This kind of a study is called an **observational study**. Observational studies are often used to determine whether an association exists between two or more characteristics measured in the study. For example, a study to determine the relationship between high school student performance and the highest educational level of the student's parents would be based on an examination of student performance and a history of the parents' educational experiences. No cause-and-effect relationship could be determined, but a strong association might be the result of such a study. Note that an observational study does not involve any intervention by the researcher.

Much primary data are obtained through the use of **sample surveys** such as Gallup polls or the Nielsen TV ratings. Such surveys normally represent a particular group of individuals and are intended to provide information on the characteristics and/or habits of such a group. Chapter 14 provides some basic principles for planning and conducting sample surveys.

Often data used in studies involving statistics come from **designed experiments**. In a designed experiment researchers impose treatments and controls on the process and then observe the results and take measurements. For example, in a laboratory experiment rats may be subjected to various noise levels and the rapidity of their movements recorded. Designed experiments can be used to help establish causation between two or more characteristics. For example, a study could be designed to determine if high school student performance is affected by a nutritious breakfast. By choosing a proper design and conducting the experiment in a rigorous manner, an actual cause-and-effect relationship might be established. Data from designed experiments are considered a sample. For example, a study relating high school student performance to breakfast may use as few as 25 typical urban high school students. The results of the study would then be inferred to the population of all urban high school students. Chapter 10 provides an introduction to experimental designs.

Using the Computer

Today, comprehensive programs for conducting statistical and data analyses are available in general-use spreadsheet software, graphing calculators, and dedicated statistical software. A person rarely needs to write his or her own programs, since they already exist for almost all aspects of statistics. Because such a large number of such packages are currently available, it is impossible to provide specific instructions for such usage in a single book. Although a few exercises in the beginning of this book, especially those in Chapters 2–5, can be done manually or with the aid of calculators, most exercises even in these chapters, and all exercises in Chapters 8–11, will require the use of a computer. In some examples we have included generic instructions for effective computer usage.

For reasons of consistency and convenience we have used the SAS System almost exclusively for examples in this book. The SAS System is a very comprehensive software package, of which statistical analysis is only a minor portion. Because it is such a large system it may not be optimal for students to have on their personal computers. We assume that additional instructions will be available for the particular software you will be using. In a few instances, especially in the earlier chapters, output from several software packages are used for comparative purposes.

Some general guidelines on using the computer for statistical analyses are, however, useful. There are two types of statistical programs identified by the method in which they accept instructions. Instructions are given to packages either

- by submitting, usually on the computer keyboard, a set of statements that describe the required analysis and options for specific tasks and outputs, or
- by providing menus that describe available analyses and options, which are chosen by pointing with a mouse and clicking the desired analyses and options.

Each of these has advantages and disadvantages. The submitted statements must usually adhere to a specific syntax and are subject to typographical errors that cause error messages and aborted analyses. On the other hand this method of implementing an analysis usually provides more flexibility and a larger number of options. The "point and click" approach is easier to use but often lacks flexibility.

The individual components of these packages are usually very comprehensive in that they can perform a wide variety of tasks and the default output from these components is often exhaustive. For example, this chapter presents various graphical presentations for summarizing data, virtually all of which can be performed by a single such component of most packages. Chapter 6 presents the "one way" analysis of variance for comparing a set of means. Most software not only does this analysis, but also can perform the analyses covered in Chapters 9 and 10 and additional methods beyond the scope of this book. For this reason it is important to be precise in specifying analysis and output options that pertain to a specific problem. Requesting inappropriate options may cause confusing outputs.

Each software package has its own style of output. However, most will contain essentially the same results, although they may appear in a different order and may even have different labels. It is therefore important to study the documentation of any package being used. We should note that most computer outputs in this book have been abbreviated because the full default output often contains information not needed at that particular time, although in a few instances we have presented the full output for illustration purposes.

If a set of data represents an entire population, the techniques presented in this chapter can be used to describe various aspects of that population and a statistical analysis using these descriptors is useful solely for that purpose. However, as is more often the case, the data to be analyzed come from a sample. In this case, the descriptive statistics obtained may subsequently be used as tools for **statistical inference**. A general introduction to the concept of statistical inference is presented in Section 1.8, and most of the remainder of this text is devoted to that subject.

1.2 Observations and Variables

A data set is composed of information from a set of units. Information from a unit is known as an **observation**. An observation consists of one or more pieces of information about the unit; these are called **variables**. Some examples:

- In a study of the effectiveness of a new headache remedy, the units are individual persons, of which 10 are given the new remedy and 10 are given an aspirin. The resulting data set has 20 observations and two variables: the medication used and a score indicating the severity of the headache.
- In a survey for determining TV viewing habits, the units are families. Usually there is one observation for each of thousands of families that have been

contacted to participate in the survey. The variables describe the programs watched as well as descriptions of the characteristics of the families.

- In a study to determine the effectiveness of a college admissions test (e.g., SAT) the units are the freshmen at a university. There is one observation per unit and the variables are the students' scores on the test and their first year's GPA.

Variables that yield nonnumerical information are called **qualitative** variables. Qualitative variables are often referred to as **categorical** variables. Those that yield numerical measurements are called **quantitative** variables. Quantitative variables can be further classified as discrete or continuous. The diagram below summarizes these definitions:

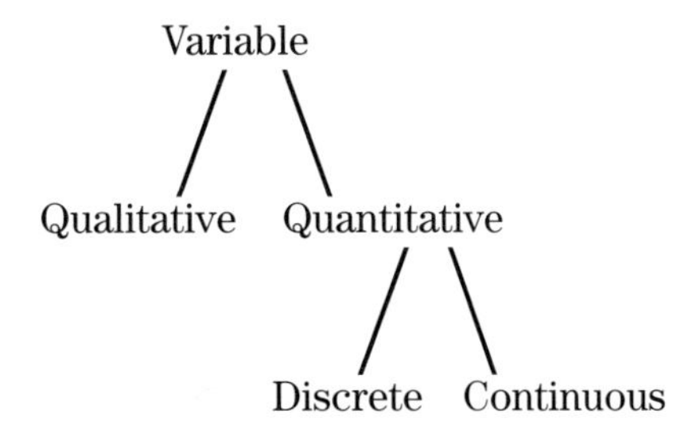

DEFINITION 1.4
A **discrete variable** can assume only a countable number of values. Typically, discrete variables are frequencies of observations having specific characteristics, but all discrete variables are not necessarily frequencies.

DEFINITION 1.5
A **continuous variable** is one that can take any one of an uncountable number of values in an interval. Continuous variables are usually measured on a scale and, although they may appear discrete due to imprecise measurement, they can conceptually take any value in an interval and cannot therefore be enumerated.

In the field of statistical quality control, the term **variable data** is used when referring to data obtained on a continuous variable and **attribute data** when referring to data obtained on a discrete variable (usually the number of defectives or nonconformities observed).

In the preceding examples, the names of the headache remedies and names of TV programs watched are qualitative (categorical) variables. Headache severity scores is a discrete numeric variable, incomes of TV-watching families, and SAT and GPA scores are continuous quantitative variables.

We will use the data set in Example 1.2 to present greater detail on various concepts and definitions regarding observations and variables.

EXAMPLE 1.2

In the fall of 2001, John Mode was offered a new job in a mid-sized city in east Texas. Obviously, the availability and cost of housing will influence his decision to accept, so he and his wife Marsha go to the Internet, find www.realtor.com, and after a few clicks find some 500 single-family residences for sale in that area. In order to make the task of investigating the housing market more manageable, they arbitrarily record the information provided on the first home on each page of six. This information results in a data set that is shown in Table 1.2.

The data set gives information on 69 homes, which comprise the *observations* for this data set. In this example, each property is a **unit**, often called a sample, experimental, or observational unit.[2] The 11 columns of the table provide specific characteristics information for each home and compose the 11 *variables* of this data set. The variable definitions along with brief mnemonic descriptors commonly used in computers are as follows:

- `Obs`[3]: a sequential number assigned to each observation as it is entered into the computer. This is useful for identifying individual observations.
- `zip`: the last digit of the postal service zip code. This variable identifies the area in which the home is located.
- `age`: the age of the home in years.
- `bed`: the number of bedrooms.
- `bath`: the number of bathrooms.
- `size`: the interior area of the home in square feet.
- `lot`: the size of the lot in square feet.
- `exter`: the exterior siding material.
- `garage`: the capacity of the garage; zero means no garage.
- `fp`: the number of fireplaces.
- `price`: the price of the home, in dollars.

The elements of each row define the observed values of the variables. Note that some values are represented by ".". In the SAS System, and other statistical computing packages, this notation specifies a missing value; that is, no information on that variable is available. Such missing values are an unavoidable feature in many date sets and occasionally cause difficulties in analyzing the data.

Brief mnemonic identifiers such as these are used by computer programs to make their outputs easier to interpret and are unique for a given set of data. However, for use in formulas we will follow mathematics convention, where variables are generically identified by single letters taken from the latter part

[2]These different types of units are not always synonymous. For example, an experimental unit may be an animal subjected to a certain diet while the observational units may be several determinations of the weight of the animal at different times. Unless otherwise specified, most of the methods presented in this book are based on the assumption that the three are synonymous and will usually be referred to as experimental units.

[3]The term Obs is used by the SAS System. Other computer software may use other notations.

Table 1.2 Housing Data

Obs	zip	age	bed	bath	size	lot	exter	garage	fp	price
1	3	21	3	3.0	951	64904	Other	0	0	30000
2	3	21	3	2.0	1036	217800	Frame	0	0	39900
3	4	7	1	1.0	676	54450	Other	2	0	46500
4	3	6	3	2.0	1456	51836	Other	0	1	48600
5	1	51	3	1.0	1186	10857	Other	1	0	51500
6	2	19	3	2.0	1456	40075	Frame	0	0	56990
7	3	8	3	2.0	1368	.	Frame	0	0	59900
8	4	27	3	1.0	994	11016	Frame	1	0	62500
9	1	51	2	1.0	1176	6259	Frame	1	1	65500
10	3	1	3	2.0	1216	11348	Other	0	0	69000
11	4	32	3	2.0	1410	25450	Brick	0	0	76900
12	3	2	3	2.0	1344	.	Other	0	1	79000
13	3	25	2	2.0	1064	218671	Other	0	0	79900
14	1	31	3	1.5	1770	19602	Brick	0	1	79950
15	4	29	3	2.0	1524	12720	Brick	2	1	82900
16	3	16	3	2.0	1750	130680	Frame	0	0	84900
17	3	20	3	2.0	1152	104544	Other	2	0	85000
18	3	18	4	2.0	1770	10640	Other	0	0	87900
19	4	28	3	2.0	1624	12700	Brick	2	1	89900
20	2	27	3	2.0	1540	5679	Brick	2	1	89900
21	1	8	3	2.0	1532	6900	Brick	2	1	93500
22	4	19	3	2.0	1647	6900	Brick	2	0	94900
23	2	3	3	2.0	1344	43560	Other	1	0	95800
24	4	5	3	2.0	1550	6575	Brick	2	1	98500
25	4	5	4	2.0	1752	8193	Brick	2	0	99500
26	4	27	3	1.5	1450	11300	Brick	1	1	99900
27	4	33	2	2.0	1312	7150	Brick	0	1	102000
28	1	4	3	2.0	1636	6097	Brick	1	0	106000
29	4	0	3	2.0	1500	.	Brick	2	0	108900
30	2	36	3	2.5	1800	83635	Brick	2	1	109900
31	3	5	4	2.5	1972	7667	Brick	2	0	110000
32	3	0	3	2.0	1387	.	Brick	2	0	112290
33	4	27	4	2.0	2082	13500	Brick	3	1	114900
34	3	15	3	2.0	.	269549	Frame	0	0	119500
35	4	23	4	2.5	2463	10747	Brick	2	1	119900
36	4	25	3	2.0	2572	7090	Brick	2	1	119900
37	4	24	4	2.0	2113	7200	Brick	2	1	122900
38	4	1	3	2.5	2016	9000	Brick	2	1	123938
39	1	34	3	2.0	1852	13500	Brick	2	0	124900
40	4	26	4	2.0	2670	9158	Brick	2	1	126900
41	2	26	3	2.0	2336	5408	Brick	0	1	129900
42	4	31	3	2.0	1980	8325	Brick	2	1	132900
43	2	24	4	2.5	2483	10295	Brick	2	1	134900
44	2	29	5	2.5	2809	15927	Brick	2	1	135900
45	4	21	3	2.0	2036	16910	Brick	2	1	139500
46	3	10	3	2.0	2298	10950	Brick	2	1	139990
47	4	3	3	2.0	2038	7000	Brick	2	0	144900
48	2	9	3	2.5	2370	10796	Brick	2	1	147600
49	2	29	5	3.5	2921	11992	Brick	2	1	149990
50	2	8	3	2.0	2262	.	Brick	2	1	152550

(Continued)

Table 1.2 *(continued)*

Obs	zip	age	bed	bath	size	lot	exter	garage	fp	price
51	4	7	3	3.0	2456	.	Brick	2	1	156900
52	4	1	4	2.0	2436	52000	Brick	2	1	164000
53	3	27	3	2.0	1920	226512	Frame	4	1	167500
54	4	5	3	2.5	2949	11950	Brick	2	1	169900
55	2	32	4	3.5	3310	10500	Brick	2	1	175000
56	4	29	3	3.0	2805	16500	Brick	2	1	179000
57	4	1	3	3.0	2553	8610	Brick	2	1	179900
58	4	1	3	2.0	2510	.	Other	2	1	189500
59	4	33	3	4.0	3627	17760	Brick	3	1	199000
60	2	25	4	2.5	3056	10400	Other	2	1	216000
61	3	16	3	2.5	3045	168576	Brick	3	1	229900
62	4	2	4	4.5	3253	54362	Brick	3	2	285000
63	2	2	4	3.5	4106	44737	Brick	3	1	328900
64	4	0	3	2.5	2993	.	Brick	2	1	313685
65	4	0	3	2.5	2992	14500	Other	3	1	327300
66	4	20	4	3.0	3055	250034	Brick	3	0	349900
67	4	18	5	4.0	3846	23086	Brick	4	3	370000
68	4	3	4	4.5	3314	43734	Brick	3	1	380000
69	4	5	4	3.5	3472	130723	Brick	2	2	395000

of the alphabet. For example the letter Y can be used to represent the variable price. The same lowercase letter, augmented by a subscript identifying the observation number, is used to represent the value of the variable for a particular observation. Using this notation, y_i is the observed price of the ith house. Thus, $y_1 = 30000$, $y_2 = 39900, \ldots, y_{69} = 395000$. The set of observed values of `price` can be symbolically represented as $y_1, y_2, \ldots, y_{69}$, or $y_i, i = 1, 2, \ldots, 69$. The total number of observations is symbolically represented by the letter n; for the data in Table 1.2, $n = 69$. We can generically represent the values of a variable Y, as $y_i, i = 1, 2, \ldots, n$. We will most frequently use Y as the variable and y_i as observations of the variable of interest. ■

1.3 Types of Measurements for Variables

We usually think of data as consisting of numbers, and certainly many data sets do contain numbers. In Example 1.2, for instance, the variable `price` is the asking price of the home, measured in dollars. This measurement indicates a definite metric or scale in the values of the variable `price`. Certainly a $200,000 house costs twice as much as a $100,000 house. As we will see later, not all variables that measure a quantity have this characteristic. However, not all data necessarily consist of numbers. For example, the variable `exter` is observed as either `brick`, `frame`, or `other`, a measurement that does not convey any relative value. Further, variables that are recorded as numbers do not necessarily imply a quantitative measurement. For example, the variable `zip` simply locates the home in some specific area and has no quantitative meaning.

We can classify observations according to a standard measurement scale that goes from "strong" to "weak" depending on the amount or precision of information available in the scale. These measurement scales are discussed at some length in various publications, including Conover (1998). We present the characteristics of these scales in some detail since the nature of the data description and statistical inference is dependent on the type of variable being studied.

DEFINITION 1.6
The **ratio scale** of measurement uses the concept of a unit of distance or measurement and requires a unique definition of a zero value.

Thus, in the ratio scale the difference between any two values can be expressed as some number of these units. Therefore, the ratio scale is considered the "strongest" scale since it provides the most precise information on the value of a variable. It is appropriate for measurements of heights, weights, birth rates, and so on. In the data set in Table 1.2, all variables except `zip` and `exter` are measured in the ratio scale.

DEFINITION 1.7
The **interval scale** of measurement also uses the concept of distance or measurement and requires a "zero" point, but the definition of zero may be arbitrary.

The interval scale is the second "strongest" scale of measurement, because the "zero" is arbitrary. An example of the interval scale is the use of degrees Fahrenheit or Celsius to measure temperature. Both have a unit of measurement (degree) and a zero point, but the zero point does not in either case indicate the absence of temperature. Other popular examples of interval variables are scores on psychological and educational tests, in which a zero score is often not attainable but some other arbitrary value is used as a reference value.

We will see that many statistical methods are applicable to variables of either the ratio or interval scales in exactly the same way. We therefore usually refer to both of these types as **numeric variables**.

DEFINITION 1.8
The **ordinal scale** distinguishes between measurements on the basis of the relative amounts of some characteristic they possess. Usually the ordinal scale refers to measurements that make only "greater," "less," or "equal" comparisons between consecutive measurements.

In other words, the ordinal scale represents a ranking or ordering of a set of observed values. Usually these ranks are assigned integer values starting with "1" for the lowest value, although other representations may be used. The ordinal scale does not provide as much information on the values of a variable and is therefore considered "weaker" than the ratio or interval scale.

Table 1.3

Example of Ordinal Data

Pie	Rank
1	4
2	3
3	1
4	2
5	5

For example, if a person were asked to taste five chocolate pies and rank them according to taste, the result would be a set of observations in the ordinal scale of measurement.

A set of data illustrating an ordinal variable is given in Table 1.3. In this data set, the "1" stands for the most preferred pie while the worst tasting pie receives the rank of "5." The values are used only as a means of arranging the observations in some order. Note that these values would not differ if pie number 3 were clearly superior or only slightly superior to pie number 4.

It is sometimes useful to convert a set of observed ratio or interval values to a set of ordinal values by converting the actual values to ranks. Ranking a set of actual values induces a loss of information, since we are going from a stronger to a weaker scale of measurement. Ranks do contain useful information and, as we will see (especially in Chapter 13), may provide a useful base for statistical analysis.

DEFINITION 1.9
The **nominal scale** identifies observed values by name or classification.

A nominally scaled variable is also often called a categorical or qualitative variable. Although the names of the classifications may be represented by numbers, these are used merely as a means of identifying the classifications and are usually arbitrarily assigned and have no quantitative implications. Examples of nominal variables are sex, breeds of animals, colors, and brand names of products. Because the nominal scale provides no information on differences among the "values" of the variable, it is considered the weakest scale. In the data in Table 1.2, the variable describing the exterior siding material is a nominal variable.

We can convert ratio, interval, or ordinal scale measurements into nominal level variables by arbitrarily assigning "names" to them. For example, we can convert the ratio-scaled variable `size` into a nominal-scaled variable, by defining homes with less than 1000 square feet as "cottages," those with more than 1000 but less than 3000 as "family sized," and those with more than 3000 as "estates."

Note that the classification of scales is not always completely clear-cut. For example, the "scores" assigned by judges for track or gymnastic events are usually treated as possessing the ratio scale but are probably closer to being ordinal in nature.

1.4 Distributions

Very little information about the characteristics of recently sold houses can be acquired by casually looking through Table 1.2. We might be able to conclude that most of the houses have brick exteriors, or that the selling price of houses ranges from $30,000 to $395,000, but a lot more information about this data set can be obtained through the use of some rather simple organizational tools.

Table 1.4

Distribution of `exter`

exter	Frequency
Brick	48
Frame	8
Other	13

To provide more information, we will construct **frequency distributions** by grouping the data into categories and counting the number of observations that fall into each one. Because we want to count each house only once, these categories (called classes) are constructed so they don't overlap. Because we count each observation only once, if we add up the number (called the frequency) of houses in all the classes, we get the total number of houses in the data set. Nominally scaled variables naturally have these classes or categories. For example, the variable `exter` has three values, `Brick`, `Frame`, and `Other`. Handling ordinal, interval, and ratio scale measurements can be a little more complicated, but, as subsequent discussion will show, we can easily handle such data simply by correctly defining the classes.

Once the frequency distribution is constructed, it is usually listed in tabular form. For the variable `exter` from Table 1.2 we get the frequency distribution presented in Table 1.4. Note that one of our first impressions is substantiated by the fact that 48 of the 69 houses are brick while only 8 have frame exteriors. This simple summarization shows how the frequency of the exteriors is distributed over the values of `exter`.

DEFINITION 1.10
A **frequency distribution** is a listing of frequencies of all categories of the observed values of a variable.

We can construct frequency distributions for any variable. For example, Table 1.5 shows the distribution of the variable `zip`, which despite having numeric values, is actually a categorical variable. This frequency distribution is produced by `Proc Freq` of the SAS System where the frequency distribution is shown in the column labeled `Frequency`. Apparently the area represented by zip code 4 has the most homes for sale.

DEFINITION 1.11
A **relative frequency distribution** consists of the **relative** frequencies, or proportions (percentages), of observations belonging to each category.

The relative frequencies expressed as percents are provided in Table 1.5 under the heading `Percent` and are useful for comparing frequencies among categories. These relative frequencies have a useful interpretation: They give the

Table 1.5

Distribution of `zip`

THE FREQ PROCEDURE

zip	Frequency	Percent	Cumulative Frequency	Cumulative Percent
1	6	8.70	6	8.70
2	13	18.84	19	27.54
3	16	23.19	35	50.72
4	34	49.28	69	100.00

Table 1.6

Distribution of Home Prices in Intervals of $50,000

THE FREQ PROCEDURE

Range	Frequency	Percent	Cumulative Frequency	Cumulative Percent
less than 50k	4	5.80	4	5.80
50k to 100k	22	31.88	26	37.68
100k to 150k	23	33.33	49	71.01
150k to 200k	10	14.49	59	85.51
200k to 250k	2	2.90	61	88.41
250k to 300k	1	1.45	62	89.86
300k to 350k	4	5.80	66	95.65
350k to 400k	3	4.35	69	100.00

chance or **probability** of getting an observation from each category in a blind or random draw. Thus if we were to randomly draw an observation from the data in Table 1.5, there is an 18.84% chance that it will be from zip area 2. For this reason a relative frequency distribution is often referred to as an observed or **empirical probability distribution** (Chapter 2).

Constructing a frequency distribution of a numeric variable is a little more complicated. Defining individual values of the variable as categories will usually only produce a listing of the original observations since very few, if any, individual observations will normally have identical values. Therefore, it is customary to define categories as intervals of values, which are called **class** intervals. These intervals must be nonoverlapping and usually each class interval is of equal size with respect to the scale of measurement. A frequency distribution of the variable `price` is shown in Table 1.6. The table is produced by `Proc Freq`, but because SAS does not automatically generate class intervals, it was necessary to write a short program to produce those shown in the table. Clearly the preponderance of homes is in the 50- to 150-thousand-dollar range.

The column labeled `Cumulative Frequency` in Table 1.6 is the **cumulative frequency distribution**, which gives the frequency of observed values less than or equal to the upper limit of that class interval. Thus, for example, 59 of the homes are priced at less than $200,000. The column labeled `Cumulative Percent` is the cumulative relative frequency distribution, which gives the proportion (percentage) of observed values less than the upper limit of that class interval. Thus the 59 homes priced at less than $200,000 represent 85.51% of the number of homes offered. We will see later that cumulative relative frequencies — especially those near 0 and 100% — can be of considerable importance.

Graphical Representation of Distributions

Using the principle that a picture is worth a thousand words (or numbers), the information in a frequency distribution is more easily grasped if it is presented in graphical form. The most common graphical presentation of a frequency distribution for numerical data is a **histogram** while the most common

Figure 1.1

Bar Chart for `exter`

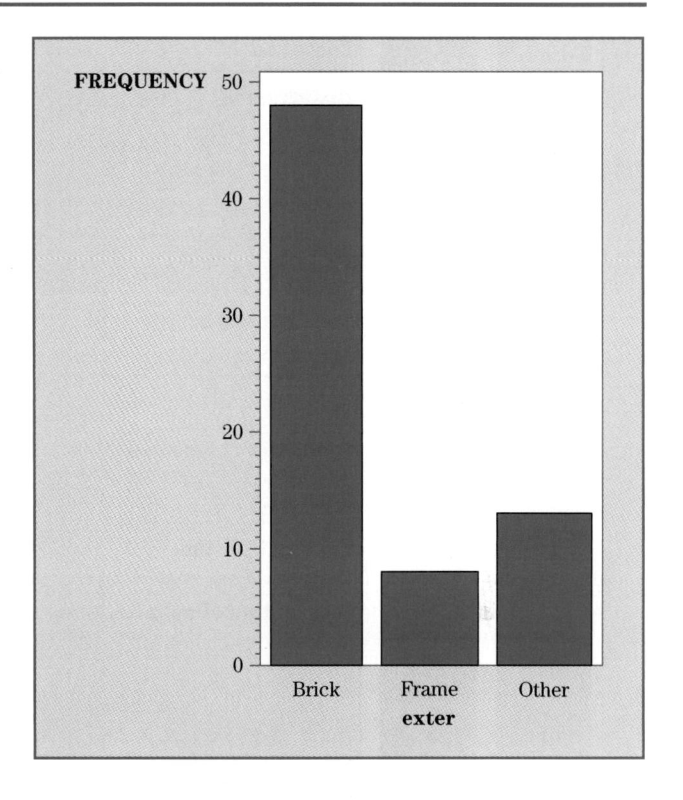

presentation for nominal, categorical, or discrete data is a **bar chart**. Both these graphs are constructed in the same way. Heights of vertical rectangles represent the frequency or the relative frequency. In a histogram, the width of each rectangle represents the size of the class and the rectangles are usually contiguous and of equal width so that the *areas* of the rectangles reflect the relative frequency. In a bar chart the width of the rectangle has no meaning; however, all the rectangles should be the same width to avoid distortion. Figure 1.1 shows a frequency bar chart for `exter` from Table 1.2 which shows the large proportion of brick homes clearly. Figure 1.2 shows a frequency histogram for `price`, clearly showing the preponderance of homes selling from 50 to 150 thousand dollars.

Another presentation of a distribution is provided by a **pie chart** which is simply a circle (pie) divided into a number of slices whose sizes correspond to the frequency or relative frequency of each class. Figure 1.3 shows a pie chart for the variable `zip`. We have produced these graphs with different programs and options to show that, although there may be slight differences in appearances, the basic information remains the same.

The use of graphs and charts is pervasive in the news media, business and economic reports, and governmental reports and publications, mainly due to the ease of storage, retrieval, manipulation, and summary of large sets of data using modern computers. Because of this, it is extremely important to be able to evaluate critically the information contained in a graph or chart.

Figure 1.2

Histogram of `price`

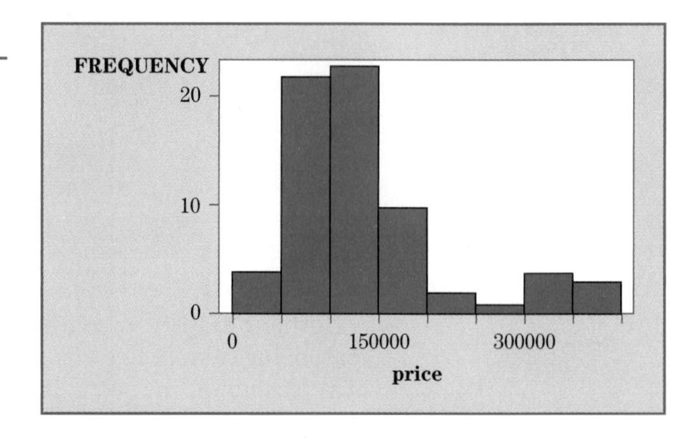

Figure 1.3

Pie Chart for the Relative Frequency Distribution of `zip`

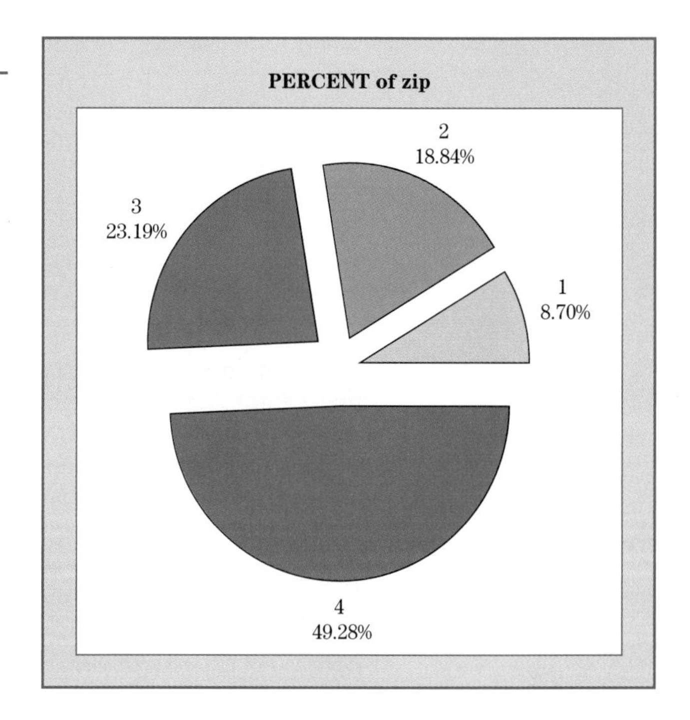

After all, a graphical presentation is simply a visual impression, which is quite easy to distort. In fact, distortion is so easy and commonplace that in 1992 the Canadian Institute of Chartered Accountants deemed it necessary to begin setting guidelines for financial graphics, after a study of hundreds of the annual reports of major corporations reported almost 10% of the reports contained at least one misleading graph that masked unfavorable data.

Whether intentional or by honest mistake, it is very easy to mislead with an incorrectly presented chart or graph. Darrell Huff, in a book entitled *How to Lie with Statistics* (1982) illustrates many such charts and graphs and discusses various issues concerning misleading graphs. In general, a correctly constructed chart or graph should have

1. all axes labeled correctly, with clearly identifiable scales,
2. be captioned correctly,
3. have bars and/or rectangles of equal width to avoid distortion,
4. have sizes of figures properly proportioned, and
5. contain only relevant information.

Histograms of numeric variables provide information on the **shape** of a distribution, a characteristic that we will later see to be of importance when performing statistical analyses. The shape is roughly defined by drawing a reasonably smooth line through the tops of the bars. In such a representation of a distribution, the center is known as the "peak" and the ends as "tails." If the tails are of approximately equal length, the distribution is said to be symmetric. If the distribution has an elongated tail on toward the right side, the distribution is skewed to the right and vice versa. Other features may consist of a sharp peak and long "fat" tails, or a broad peak and short tails. We can see that the distribution of `price` is slightly skewed to the right, which, in this case, is due to a few unusually high prices. We will see later that recognizing the shape of a distribution can be quite important.

We continue the study of shapes of distributions with another example.

EXAMPLE 1.3

The discipline of forest science is a frequent user of statistics. An important activity is to gather data on the physical characteristics of a random sample of trees in a forest. The resulting data may be used to estimate the potential yield of the forest, to obtain information on the genetic composition of a particular species, or to investigate the effect of environmental conditions.

Table 1.7 is a listing of such a set of data. This set consists of measurements of three characteristics of 64 sample trees of a particular species. The researcher would like to summarize this set of data in graphic form to aid in its interpretation.

Solution As we can see from Table 1.7, the data set consists of 64 observations of three ratio variables. The three variables are measurements characterizing each tree and are identified by brief mnemonic identifiers in the column headings as follows:

1. `DFOOT`, the diameter of the tree at one foot above ground level, measured in inches,
2. `HCRN`, the height to the base of the crown measured in feet, and
3. `HT`, the total height of the tree measured in feet.

A histogram for the heights (`HT`) of the 64 trees is shown in Fig. 1.4 as produced by `PROC INSIGHT` of the SAS System. Due to space limitations, not all boundaries of class intervals are shown, but we can deduce that the default option of `PROC INSIGHT` yielded a class interval width of 1.5 feet with the first interval being from 20.25 to 21.75 and the last from 30.75 to 32.25. In this program the user can adjust the size of class intervals by clicking on an

Table 1.7 Data on Tree Measurements

OBS	DFOOT	HCRN	HT	OBS	DFOOT	HCRN	HT	OBS	DFOOT	HCRN	HT
1	4.1	1.5	24.5	23	4.3	2.0	25.6	45	4.7	3.3	29.7
2	3.4	4.7	25.0	24	2.7	3.0	20.4	46	4.6	8.9	26.6
3	4.4	2.8	29.0	25	4.3	2.0	25.0	47	4.8	2.4	28.1
4	3.6	5.1	27.0	26	3.3	1.8	20.6	48	4.5	4.7	28.5
5	4.4	1.6	26.5	27	5.0	1.7	24.6	49	3.9	2.3	26.0
6	3.9	1.9	27.0	28	5.2	1.8	26.9	50	4.4	5.4	28.0
7	3.6	5.3	27.0	29	4.7	1.5	26.7	51	5.0	3.2	30.4
8	4.3	7.6	28.0	30	3.8	3.2	26.3	52	4.6	2.5	30.5
9	4.8	1.1	28.5	31	3.8	2.6	27.6	53	4.1	2.1	26.0
10	3.5	1.2	26.0	32	4.2	1.8	23.5	54	3.9	1.8	29.0
11	4.3	2.3	28.0	33	4.7	2.7	25.0	55	4.9	4.7	29.5
12	4.8	1.7	28.5	34	5.0	3.1	27.3	56	4.9	8.3	29.5
13	4.5	2.0	30.0	35	3.2	2.9	26.2	57	5.1	2.1	28.4
14	4.8	2.0	28.0	36	4.1	1.3	25.8	58	4.4	1.7	29.0
15	2.9	1.1	20.5	37	3.5	3.2	24.0	59	4.2	2.2	28.5
16	5.6	2.2	31.5	38	4.8	1.7	26.5	60	4.6	6.6	28.5
17	4.2	8.0	29.3	39	4.3	6.5	27.0	61	5.1	1.0	26.5
18	3.7	6.3	27.2	40	5.1	1.6	27.0	62	3.8	2.7	28.5
19	4.6	3.0	27.0	41	3.7	1.4	25.9	63	4.8	2.2	27.0
20	4.2	2.4	25.4	42	5.0	3.8	29.5	64	4.0	3.1	26.0
21	4.8	2.9	30.4	43	3.3	2.4	25.8				
22	4.3	1.4	24.5	44	4.3	3.0	25.2				

Figure 1.4

Histogram of Tree Height

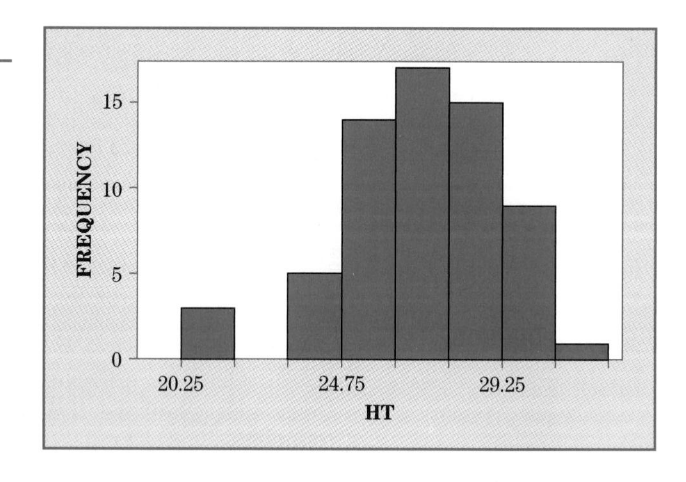

arrow at the lower left (not shown in Fig 1.4) which causes a menu to pop up allowing such changes. For example, by changing the first "tick" to 20, the last to 32, and the "tick interval" to 2, the histogram will have 6 classes instead of the 8 shown. Many graphics programs allow this type of interactive modification. Of course, the basic shape of the distribution is not changed by such modifications. Also note that in these histograms, the legend gives the boundaries of the intervals; other graphic programs may give the midpoints.

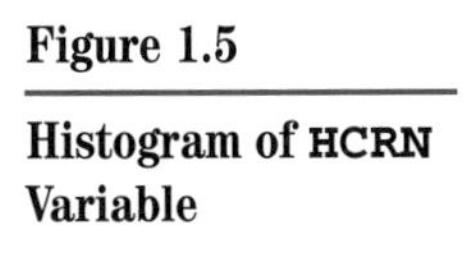

Figure 1.5

Histogram of HCRN Variable

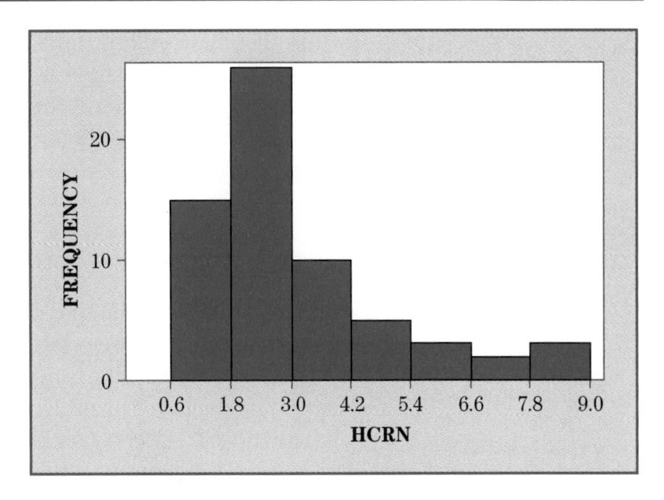

The histogram for the variable HCRN is shown in Fig. 1.5. We can now see that the distribution of HT is slightly skewed to the left while the distribution of HCRN is quite strongly skewed to the right. ■

1.5 Numerical Descriptive Statistics

Although distributions provide useful descriptions of data, they still contain too much detail for some purposes. Assume, for example, that we have collected data on tree dimensions from several forests for the purpose of detecting possible differences in the distribution of tree sizes among these forests. Side-by-side histograms of the distributions would certainly give some indication of such differences, but would not produce measures of the differences that could be used for quantitative comparisons. **Numerical** measures that provide descriptions of the characteristics of the distributions, which can then be used to provide more readily interpretable information on such differences, are needed. Of course, since these are numerical measures, their use is largely restricted to numeric variables, that is, variables measured in the ratio or interval scales (see, however, Chapter 13).

Note that when we first started evaluating the tree measurement data (Table 1.7) we had 64 observations to contend with. As we attempted to summarize the data using a frequency distribution of heights and the accompanying histogram (Fig. 1.4) we represented these data with only eight entries (classes). We can use numerical descriptive statistics to reduce the number of entries describing a set of data even further, typically using only using two numbers. This action of reducing the number of items used to describe the distribution of a set of data is referred to as **data reduction**, which is unfortunately accompanied by a progressive loss of information. In order to minimize the loss of information, we need to determine the most important characteristics of the distribution and find measures to describe these characteristics. The two most important aspects are the **location** and the **dispersion** of the data. In other words, we need to find a number that indicates where the observations are on the measurement scale and another to indicate how widely the observations vary.

Location

The most useful single characteristic of a distribution is some typical, average, or representative value that describes the set of values. Such a value is referred to as a descriptor of **location** or **central tendency**. Several different measures are available to describe this concept. We present two in detail. Other measures not widely used are briefly noted.

The most frequently used measure of location is the arithmetic mean, usually referred to simply as the mean.

DEFINITION 1.12
The **mean** is the sum of all the observed values divided by the number of values.

Denote by y_i, $i = 1, \ldots, n$, an observed value of the variable Y, then the sample mean[4] denoted by $\bar{y}$, is obtained by the formula

$$\bar{y} = \frac{\sum y_i}{n},$$

where the symbol $\sum$ stands for "the sum of." For example, the mean for DFOOT in Table 1.7 is 4.301, which is the mean diameter (at one foot above the ground) of the 64 trees measured. A quick glance at the observed values of DFOOT reveals that this value is indeed representative of the values of that variable.[5]

Another useful measure of location is the median.

DEFINITION 1.13
The **median** of a set of observed values is defined to be the middle value when the measurements are arranged from lowest to highest; that is, 50% of the measurements lie above it and 50% fall below it.

The precise definition of the median depends on whether the number of observations is odd or even as follows:

1. If n is odd, the median is the middle observation; hence, exactly $(n - 1)/2$ values are greater than and $(n - 1)/2$ values are less than the median, respectively.

---

[4]It is also often called the **average**. However, this term is often used as a generic term for any unspecified measure of location and will therefore not be used in this context.

[5]Some small data sets suitable for practicing computations are available in the following as well as in exercises at the end of the chapter.

Figure 1.6

Data for Comparing Mean and Median

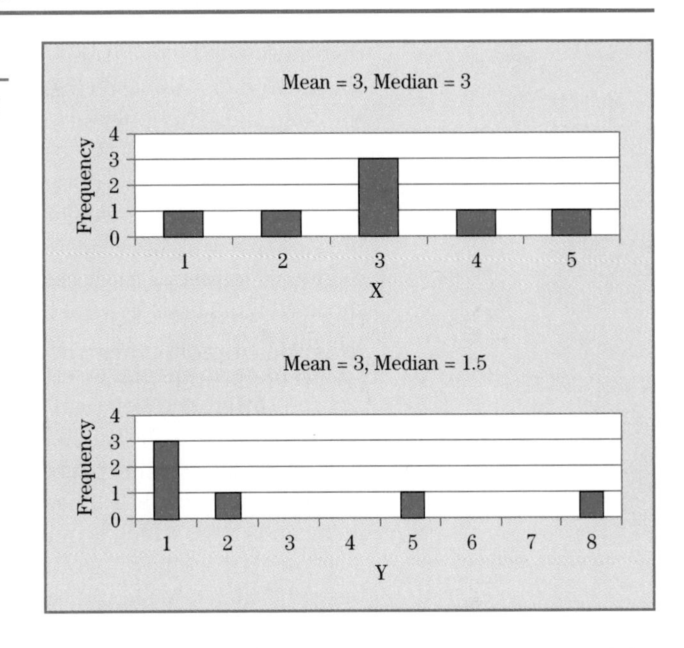

2. If n is even, there are two middle values and the median is the mean of the two middle values and $n/2$ values are greater than and $n/2$ values are less than the median, respectively.[6]

Although both mean and median are measures of central tendency, they do differ in interpretation. For example, consider the following data for two variables, X and Y, given in Table 1.8:

Table 1.8

Data for Comparing Mean and Median

X	Y
1	1
2	1
3	1
3	2
4	5
5	8

We first compute the means

$$\bar{x} = (1/6)(1 + 2 + 3 + 3 + 4 + 5) = 3.0$$

and

$$\bar{y} = (1/6)(1 + 1 + 1 + 2 + 5 + 8) = 3.0.$$

The means are the same for both variables.

Denoting the medians by m_x and m_y, respectively, and noting that there are an even number of observations, we find

$$m_x = (3 + 3)/2 = 3.0$$

and

$$m_y = (1 + 2)/2 = 1.5.$$

The medians are different. The reason for the difference is seen by examining the histograms of the two variables in Fig. 1.6.

The distribution of the variable X is symmetric, while the distribution of the variable Y is skewed to the right. For symmetric or nearly symmetric

[6]If there are some identical values of the variable, the phrase "or equal to" may need to be added to these statements.

distributions, the mean and median will be the same or nearly the same, while for skewed distributions the value of the mean will tend to be "pulled" toward the long tail. This phenomenon can be explained by the fact that the mean can be interpreted as the center of gravity of the distribution. That is, if the observations are viewed as weights placed on a plane, then the mean is the position at which the weights on each side balance. It is a well-known fact of physics that weights placed further from the center of gravity exert a larger degree of influence (also called leverage); hence the mean must shift toward those weights in order to achieve balance. However, the median assigns equal weights to all observations regardless of their actual values; hence the extreme values have no special leverage.

The difference between the mean and median is also illustrated by the tree data (Table 1.7). The heights variable (HT) was seen to have a reasonably symmetric distribution (Fig. 1.4). The mean diameter is 26.96 and its median is 27.0.[7] The variable HCRN has a highly right-skewed distribution (Fig. 1.5) and its mean is 3.04, which is quite a bit larger than its median of 2.4.

Now that we have two measures of location, it is logical to ask which is better? Which one should we use? Note that the mean is calculated using the value of each observation, so all the information available from the data is utilized. This is not so for the median. For the median we only need to know where the "middle" of the data is. Therefore, the mean is the more useful measure and, in most cases, the mean will give a better measure of the location of the data. However, as we have seen, the value of the mean is heavily influenced by extreme values and tends to become a distorted measure of location for a highly skewed distribution. In this case, the median may be more appropriate.

The choice of the measure to be used may depend on its ultimate interpretation and use. For example, monthly rainfall data often contain a few very large values corresponding to rare floods. For this variable, the mean does indicate the total amount of water derived from rain but hardly qualifies as a typical value for monthly rainfall. On the other hand, the median does qualify as a typical value, but certainly does not reflect the total amount of water.

In general, we will use the mean as the single measure of location unless the distribution of the variable is skewed. We will see later (Chapter 4) that variables with highly skewed distributions can be regarded as not fulfilling the assumptions required for methods of statistical analysis that are based on the mean. In Section 1.6 we present some techniques that may be useful for detecting characteristics of distributions that may make the mean an inappropriate measure of location.

Other occasionally used measures of location are as follows:

1. The **mode** is the most frequently occurring value. This measure may not be unique in that two (or more) values may occur with the same greatest

[7] It is customary to give a mean with one more decimal than the observed values. Computer programs usually give all decimal places that the space on the output allows. If a median corresponds to an observed value (n odd), the value is presented as is; if it is the mean of two observations (n even), the extra decimal may be used.

frequency. Also, the mode may not be defined if all values occur only once, which usually happens with continuous numeric variables.

2. The **geometric** mean is the nth root of the product of the values of the n observations. This measure is related to the arithmetic mean of the logarithms of the observed values. The geometric mean cannot exist if there are any values less than or equal to 0.
3. The **midrange** is the mean of the smallest and largest observed values. This measure is not frequently used because it ignores most of the information in the data. (See the following discussion of the range and similar measures.)

Dispersion

Although location is generally considered to be the most important single characteristic of a distribution, the **variability** or **dispersion** of the values is also very important. For example, it is imperative that the diameters of $\frac{1}{4}$-in. nuts and bolts have virtually no variability, or else the nuts may not match the bolts. Thus the mean diameter provides an almost complete description of the size of a set of $\frac{1}{4}$-in. nuts and bolts. However, the mean or median incomes of families in a city provide a very inadequate description of the distribution of that variable since a listing of incomes would include a wide range of values.

Figure 1.7 shows histograms of two small data sets. Both have 10 observations, both have a mean of 5 and, since the distributions are symmetric, both have a median of 5. However, the two distributions are certainly quite different. Data set 2 may be described as having more variability since it has fewer observations near the mean and more observations at the extremes of the distribution.

The simplest and intuitively most obvious measure of variability is the **range**, which is defined as the difference between the largest and smallest observed values. Although conceptually simple, the range has one very serious drawback: It completely ignores any information from all the other values in

Figure 1.7

Illustration of Dispersion

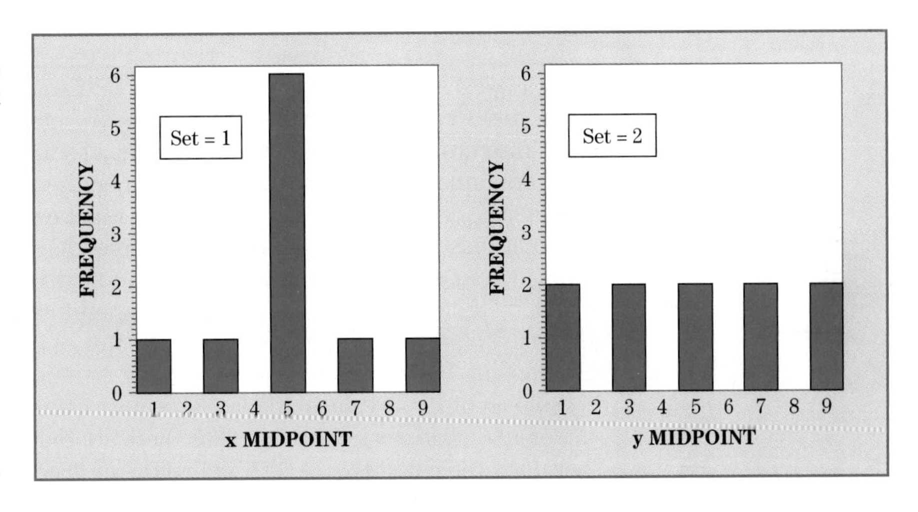

the data. This characteristic is also illustrated by the two data sets in Fig. 1.7. Both of these data sets exhibit the same range (eight), but data set 2 exhibits more variability.

Since greater dispersion means that observations are farther from the center of the distribution, it is logical to consider distances of observations from that center as indication of variability. The preferred measure of variation when the mean is used as the measure of center is based on the set of distances or differences of the observed values (y_i) from the mean ($\bar{y}$). These differences, $(y_i - \bar{y})$, $i = 1, 2, \ldots, n$, are called the **deviations** from the mean. Large magnitudes of deviation imply a high degree of variability, and small magnitudes of deviation imply a low degree of variability. If all deviations are zero, the data set exhibits no variability; that is, all values are identical.

The mean of these deviations would seem to provide a reasonable measure of dispersion. However, a relatively simple exercise in algebra shows that the sum of these deviations, that is, $\sum(y_i - \bar{y})$, is always zero. Therefore, this quantity is not useful. The mean absolute deviation (the mean of deviations ignoring their signs) will certainly be an indicator of variability and is sometimes used for that purpose. However, this measure turns out not to be very useful as the absolute values make theoretical development difficult.

Another way to neutralize the effect of opposite signs is to base the measure of variability on the **squared** deviations. Squaring each deviation gives a nonnegative value and summing the squares of the deviations gives a positive measure of variability. This criterion is the basis for the most frequently used measure of dispersion, the **variance**.

DEFINITION 1.14

The **sample variance**, denoted by s^2, of a set of n observed values having a mean $\bar{y}$ is the sum of the squared deviations divided by $n - 1$:

$$s^2 = \frac{\sum(y_i - \bar{y})^2}{n - 1}.$$

Note that the variance is actually an average or mean of the squared deviations and is often referred to as a **mean square**, a term we will use quite often in later chapters. Note also that we have divided the sum by $(n - 1)$ rather than n. While the reason for using $(n - 1)$ may seem confusing at this time, there is a good reason for it. As we see later in the chapter, one of the uses of the sample variance is to estimate the population variance. Dividing by n tends to underestimate the population variance; therefore by dividing by $(n - 1)$ we get, on average, a more accurate estimate. Recall that we have already noted that the sum of deviations $\sum(y_i - \bar{y}) = 0$; hence, if we know the values of any $(n - 1)$ of these values, the last one must have that value that causes the sum of all deviations to be zero. Thus there are only $(n - 1)$ "free" deviations. Therefore, the quantity $(n - 1)$ is called the **degrees of freedom**.

An equivalent argument is to note that in order to compute s^2, we must first compute $\bar{y}$. Starting with the concept that a set of n observed values of a variable provides n units of information, when we compute s^2 we have already used one piece of information, leaving only $(n - 1)$ "free" units or $(n - 1)$ degrees of freedom.

Computing the variance using the above formula is straightforward but somewhat tedious. First we must compute $\bar{y}$, then the individual deviations $(y_i - \bar{y})$, square these, and then sum. For the two data sets represented by Fig. 1.7 we obtain

Data set 1:

$$s^2 = (1/9)[(1-5)^2 + (3-5)^2 + \cdots + (9-5)^2]$$
$$= (1/9) \cdot 40 = 4.44,$$

Data set 2:

$$s^2 = (1/9)[(1-5)^2 + (1-5)^2 + \cdots + (9-5)^2]$$
$$= (1/9) \cdot 80 = 8.89,$$

showing the expected larger variance for data set 2.

Calculations similar to that for the numerator of the variance are widely used in many statistical analyses and if done as shown in Definition 1.15 are quite tedious. This numerator, called the **sum of squares** and often denoted by SS, is more easily calculated by using the equivalence:

$$\text{SS} = \sum (y_i - \bar{y})^2 = \sum y_i^2 - \left(\sum y_i\right)^2 \Big/ n.$$

The first portion, $\sum y_i^2$, is simply the sum of squares of the original y values. The second part, $(\sum y_i)^2/n$, the square of the sum of the y values divided by the number of observations, is called the **correction factor**, since it "corrects" the sum of squared values to become the sum of squared deviations from the mean. The result, SS, is called the corrected, or centered, sum of squares, or often simply sum of squares. This sum of squares is divided by the degrees of freedom to obtain the mean square, which is the variance. In general, then the variance

$$s^2 = \text{mean square} = (\text{sum of squares})/(\text{degrees of freedom}).$$

For the case of computing a variance from a single set of observed values, the sum of squares is the sum of squared deviations from the mean of those observations, and the degrees of freedom are $(n - 1)$. For more complex situations, which we will encounter in subsequent chapters, we will continue with this general definition of a variance; however, there will be different methods for computing sums of squares and degrees of freedom.

The computations are now quite straightforward, especially since many calculators have single-key operations for obtaining sums and sums of squares.[8]

[8]Many calculators also automatically obtain the variance (or standard deviation). Some even provide options for using either n or $(n - 1)$ for the denominator of the variance estimate! We suggest practice computing a few variances without using this feature.

For the two data sets we have

Data set 1:

$$n = 10, \quad \sum y_i = 50, \quad \sum y_i^2 = 290$$

$$\text{SS} = 290 - 50^2/10 = 40$$

$$s^2 = 40/9 = 4.44$$

Data set 2:

$$n = 10, \quad \sum y_i = 50, \quad \sum y_i^2 = 330$$

$$\text{SS} = 330 - 50^2/10 = 80$$

$$s^2 = 80/9 = 8.89.$$

For purposes of interpretation, the variance has one major drawback: It measures the dispersion in the square of the units of the observed values. In other words, the numeric value is not descriptive of the variability of the observed values. This flaw is remedied by using the square root of the variance, which is called the **standard deviation**.

DEFINITION 1.15
The **standard deviation** of a set of observed values is defined to be the positive square root of the variance.

This measure is denoted by s and does have, as we will see shortly, a very useful interpretation as a measure of dispersion. For the two example data sets, the standard deviations are

$$\text{Data set 1:} \quad s = 2.11,$$

$$\text{Data set 2:} \quad s = 2.98.$$

Usefulness of the Mean and Standard Deviation Although the mean and standard deviation (or variance) are only two descriptive measures, together the two actually provide a great deal of information about the distribution of an observed set of values. This is illustrated by the **empirical rule**: If the shape of the distribution is nearly bell shaped, the following statements hold:

1. The interval $(\bar{y} \pm s)$ contains approximately 68% of the observations.
2. The interval $(\bar{y} \pm 2s)$ contains approximately 95% of the observations.
3. The interval $(\bar{y} \pm 3s)$ contains virtually all of the observations.

Note that for each of these intervals the mean is used to describe the location and the standard deviation is used to describe the dispersion of a given portion of the data. We illustrate the empirical rule with the tree data (Table 1.7). The height (HT) was seen to have a nearly bell-shaped distribution, so the empirical rule should hold as a reasonable approximation. For this variable we compute

$$n = 64, \quad \bar{y} = 26.959, \quad s^2 = 5.163, \quad s = 2.272.$$

According to the empirical rule:

$(\bar{y} \pm s)$, which is 26.959 ± 2.272, which defines the interval 24.687 to 29.231 and should include $(0.68)(64) = 43$ observations,
$(\bar{y} \pm 2s)$, which is 26.959 ± 4.544, which defines the interval from 22.415 to 31.503 and should include $(0.95)(64) = 61$ observations, and
$(\bar{y} \pm 3s)$, which defines the interval from 20.143 to 33.775 and should include all 64 observations.

The effectiveness of the empirical rule is verified using the actual data. This task may be made easier by obtaining an ordered listing of the observed values or using a stem and leaf plot (Section 1.6), which we do not reproduce here. For this variable, 46 values fall between 24.687 and 29.231, 61 fall between 22.415 and 31.503, and all observations fall between 20.143 and 33.775. Thus the empirical rule appears to work reasonably well for this variable.

The empirical rule furnishes us with a quick method of estimating the standard deviation of a bell-shaped distribution. Since at least 95% of the observations fall within 2 standard deviations of the mean in either direction, the range of the data covers about 4 standard deviations. Thus, we can estimate the standard deviation (a crude estimate by the way) by taking the range divided by 4. For example, the range of the data on the HT variable is $31.5 - 20.4 = 11.1$. Divided by 4 we get about 2.77. The actual standard deviation had a value of 2.272, which is approximately "in the ball park," so to speak.

The HCRN variable had a rather skewed distribution (Fig. 1.5); hence the empirical rule should not work as well. The mean is 3.036 and the standard deviation is 1.890. The expected and actual frequencies are given in Table 1.9. As expected, the empirical rule does not work as well. In other words, for a nonsymmetric distribution the mean and standard deviation (or variance) do not provide as complete a description of the distribution as they do for a more nearly bell-shaped one. We may want to include a histogram or general discussion of the shape of the distribution along with the mean and standard deviation when describing data with a highly skewed distribution.

Actually the mean and standard deviation provide useful information about a distribution no matter what the shape. A much more conservative

Table 1.9

The Empirical Rule Applied to a Nonsymmetric Distribution

INTERVAL		NUMBER OF OBSERVATIONS	
Specified	Actual	Should Include	Does Include
$\bar{y} \pm s$	1.146 to 4.926	43	53
$\bar{y} \pm 2s$	−0.744 to 6.816	61	60
$y \pm 3s$	−2.634 to 8.706	64	63

relation between the distribution and its mean and standard deviation is given by Tchebysheff's theorem.

DEFINITION 1.16
Tchebysheff's theorem For any arbitrary constant k, the interval $(\bar{y} \pm ks)$ contains a proportion of the values of at least $[1 - (1/k^2)]$.[9]

Note that Tchebysheff's theorem is more conservative than the empirical rule. This is because the empirical rule describes distributions that are approximately "bell" shaped, whereas Tchebysheff's theorem is applicable for any shaped distribution. For example, for $k = 2$, Tchebysheff's theorem states that the interval $(\bar{y} \pm 2s)$ will contain at least $[1 - (1/4)] = 0.75$ of the data. For the HCRN variable, this interval is from -0.744 to 6.816 (Table 1.9), which actually contains $60/64 = 0.9375$ of the values. Thus we can see that Tchebysheff's theorem provides a guarantee of a proportion in an interval but at the cost of a wider interval.

The empirical rule and Tchebysheff's theorem have been presented not because they are quoted in many statistical analyses but because they demonstrate the power of the mean and standard deviation to describe a set of data. The wider intervals specified by Tchebysheff's theorem also show that this power is diminished if the assumption of a bell-shaped curve is not made.

Other Measures

A measure of dispersion that has uses in some applications is the **coefficient of variation**.

DEFINITION 1.17
The **coefficient of variation** is the ratio of the standard deviation to the mean, expressed in percentage terms.

Usually denoted by CV, it is

$$\text{CV} = \frac{s}{\bar{y}} \cdot 100.$$

That is, the CV gives the standard deviation as a proportion of the mean. For example, a standard deviation of 5 has little meaning unless we can compare it to something. If $\bar{y}$ has a value of 100, then this variation would probably be considered small. If, however, $\bar{y}$ has a value of 1, a standard deviation of 5 would be quite large relative to the mean. If we were evaluating the precision of a laboratory measuring device, the first case, CV = 5%, would probably be acceptable. The second case, CV = 500%, probably would not.

Additional useful descriptive measures are the **percentiles** of a distribution.

[9] Tchebysheff's theorem is usually described in terms of a theoretical distribution rather than for a set of data. This difference is of no concern at this point.

DEFINITION 1.18
The ***p*th percentile** is defined to be that value for which at most $(p)\%$ of the measurements are less and at most $(100 - p)\%$ of the measurements are greater.[10]

For example, the 75th percentile of the diameter variable (DFOOT) corresponds to the 48th ($0.75 \cdot 64 = 48$) ordered observation, which is 4.8. This means that 75% of the trees have diameters of 4.8 in. or less. By definition, cumulative relative frequencies define percentiles.

To illustrate how a computer program calculates percentiles, the Frequency option of SPSS was instructed to find the 30th percentile for the same variable, DFOOT. The program returned the value 4.05. To find this value we note that $0.3 \times 64 = 19.2$. Therefore we want the value of DFOOT for which 19.2 of the observations are smaller and 60.8 are larger. This means that the 30th percentile falls between the 19th observation, 4.00, and the 20th observation, 4.10. The computer program simply took the midpoint between these two values and gave the 30th percentile the value of 4.05.

A special set of percentiles of interest are the **quartiles**, which are the 25th, 50th, and 75th percentiles. The 50th percentile is, of course, the median.

DEFINITION 1.19
The **interquartile range** is the length of the interval between the 25th and 75th percentiles and describes the range of the middle half of the distribution.

For the tree diameters, the 25th and 75th percentiles correspond to 3.9 and 4.8 inches; hence the interquartile range is 0.9 inches. We will use this measure in Section 1.6 when we discuss the box plot. We will see later that we are often interested in the percentiles at the extremes or tails of a distribution, especially the 1, 2.5, 5, 95, 97.5, and 99th percentiles.

Certain measures may be used to describe other aspects of a distribution. For example, a measure of skewness is available to indicate the degree of skewness of a distribution. Similarly, a measure of kurtosis indicates whether a distribution has a narrow "peak" and fat "tails" or a flat peak and skinny tails. Generally, a "fat-tailed" distribution is characterized by having an excessive number of outliers or unusual observations, which is an undesirable characteristic. Although these measures have some theoretical interest, they are not often used in practice. For additional information, see Snedecor and Cochran (1980), Sections 5.13 and 5.14.

[10]Occasionally the percentile desired falls between two of the measurements in the data set. In that case interpolation may be used to obtain the value. To avoid becoming unnecessarily pedantic, most people simply choose the midpoint between the two values involved. Different computer programs may use different interpolation methods.

Computing the Mean and Standard Deviation from a Frequency Distribution

If a data set is presented as a frequency distribution, a good approximation of the mean and variance may be obtained directly from that distribution. Let y_i represent the midpoint and f_i the frequency of the ith class. Then

$$\bar{y} \approx \sum f_i y_i \Big/ \sum f_i$$

and

$$s^2 \approx \sum f_i (y_i - \bar{y})^2 \Big/ \sum f_i$$

or, using the computational form,

$$s^2 \approx \left[\sum f_i y_i^2 - \left(\sum f_i y_i \right)^2 \Big/ \sum f_i \right] \Big/ \sum f_i.$$

Note that these formulas use **weighted** sums of the observed values[11] or squared deviations. That is, each value is weighted by the number of observations it represents. If the y_i are the actual values (rather than midpoints of intervals) of a discrete distribution, these formulas provide exactly the same values as those using the formulas presented previously in this section.

Equivalent formulas may be used for data represented as a relative frequency distribution. Let p_i be the relative frequency of the ith class. Then

$$\bar{y} \approx \sum p_i y_i \quad \text{and} \quad s^2 \approx \sum p_i (y_i - \bar{y})^2$$

or, using the computational form,

$$s^2 \approx \sum p_i y_i^2 - \left(\sum p_i y_i \right)^2.$$

Most data sets are available in their original form and since computers readily perform direct computation of mean and variance these formulas are not often used. We will, however, find these formulas useful in discussions of theoretical probability distributions in Chapter 2.

Change of Scale

Change of scale is often called coding or linear transformation. Most interval and ratio variables arise from measurements on a scale such as inches, grams, or degrees Celsius. The numerical values describing these distributions naturally reflect the scale used. In some circumstances it is useful to change the scale such as, for example, changing from imperial (inches, pounds, etc.) to metric units. Scale changes may take many forms, including a change from ratio to ordinal scales as mentioned in Section 1.3. Other scale changes may involve the use of functions such as logarithms or square roots (see Chapter 6).

A useful form of scaling is the use of a linear transformation. Let Y represent a variable in the observed scale, which is transformed to a rescaled or

[11]These formulas are primarily used for large data sets where $n \approx n - 1$; hence $\sum f_i = n$, rather than $(n - 1)$, is used as the denominator for computing the variance.

transformed variable X by the equation

$$X = a + bY,$$

where a and b are constants. The constant a represents a change in the **origin**, while the constant b represents a change in the unit of measurement, or **scale** identified with a ratio or interval scale variable (Section 1.3). A well-known example of such a transformation is the change from degrees Celsius to degrees Fahrenheit. The formula for the transformation is

$$X = 32 + 1.8Y,$$

where X represents readings in degrees Fahrenheit and Y in degrees Celsius.

Many descriptive measures retain their interpretation through linear transformation. Specifically, for the mean and variance:

$$\bar{x} = a + b\bar{y} \quad \text{and} \quad s_x^2 = b^2 s_y^2.$$

A useful application of a linear transformation is that of reducing round-off errors. For example, consider the following values y_i, $i = 1, 2, \ldots, 6$:

$$10.004 \quad 10.002 \quad 9.997 \quad 10.000 \quad 9.996 \quad 10.001.$$

Using the linear transformation

$$x_i = -10,000 + 1000y_i$$

results in the values of x_i

$$4 \quad 2 \quad -3 \quad 0 \quad -4 \quad 1,$$

from which it is easy to calculate

$$\bar{x} = 0 \quad \text{and} \quad s_x^2 = 9.2.$$

Using the above relationships, we see that $\bar{y} = 10.000$ and $s_y^2 = 0.0000092$.

The use of the originally observed y_i may induce round-off error. Using the original data,

$$\sum y_i = 60.000, \quad \sum y_i^2 = 600.000046, \quad \text{and} \quad \left(\sum y_i\right)^2 \Big/ n = 600.000000.$$

Then

$$\text{SS} = 0.000046 \quad \text{and} \quad s^2 = 0.0000092.$$

If the calculator we are using has only eight digits of precision, then $\sum y^2$ would be truncated to 600.00004, and we would obtain $s^2 = 0.000008$. Admittedly this is a pathological example, but round-off errors in statistical calculations occur quite frequently, especially when the calculations involve many steps as will be required later. Therefore, scaling by a linear transformation is sometimes useful.

1.6 Exploratory Data Analysis

We have seen that the mean and variance (or standard deviation) can do a very good job of describing the characteristics of a frequency distribution. However, we have also seen that these do not work as well when the distribution is skewed and/or includes some extreme or outlying observations. Because the vast majority of statistical analyses make use of the mean and standard deviation, the results of such analyses may prove misleading if the distribution has such features. Therefore, it is imperative that some preliminary checks of the data be performed to see if other methods (see Section 4.5 and Chapter 13) may be more appropriate.

Before the widespread use of automatic data recording equipment and computers, most data were laboriously recorded from laboratory manuals or similar records and then manually entered into calculators where the calculations were usually performed in several stages. During this long and laborious process, it was relatively easy to spot unusual observations and, in general, to get a "feel" for the data and thus recognize the possible need for altering the analysis strategy.

Certainly the automatic recording and computing equipment available today provide greater speed, convenience, and accuracy, as well as more complete and comprehensive analyses. However, these analyses are performed without the help of human intervention and may consequently result in beautifully executed and handsomely annotated computer output of inappropriate analyses on faulty data.

Fortunately, the same computers that can so easily produce inappropriate analyses can just as easily be used to perform preliminary data screening to provide an overview of the nature of the data and thus provide information on unusual distributions and/or data anomalies. A variety of such procedures have been developed and many are available on most popularly used computer software. These procedures are called **exploratory data analysis** techniques or **EDA**, which was first introduced by Tukey (1977). We present here two of the most frequently used EDA tools: the **stem and leaf plot** and the **box plot**.

The Stem and Leaf Plot

The stem and leaf plot is a modification of a histogram for a ratio or interval variable that provides additional information about the distribution of the variable. The first one or two digits specify the class interval, called the "stem," and the next digit (rounded if necessary) is used to construct increments of the bar, which are called the "leaves." Usually in a stem and leaf plot, the bars are arranged horizontally and the leaf values are arranged in ascending order.

We illustrate the construction of a stem and leaf plot using the data on `size` for the 69 homes. To make construction easier, we first arrange the observations from low to high as shown in Table 1.10.

Normally the first or first two digits are used to define stem values, but in this case using one would result in an inadequate five stems, while using

Table 1.10

Home Sizes Measured in Square Feet Arranged from Low to High

.	1550	2456
676	1624	2463
951	1636	2483
994	1647	2510
1036	1750	2553
1064	1752	2572
1152	1770	2670
1176	1770	2805
1186	1800	2809
1216	1852	2921
1312	1920	2949
1344	1972	2992
1344	1980	2993
1368	2016	3045
1387	2036	3055
1410	2038	3056
1450	2082	3253
1456	2113	3310
1456	2262	3314
1500	2298	3472
1524	2336	3627
1532	2370	3846
1540	2436	4106

two would generate an overwhelming 40 stems. A compromise is to use the first two digits, in sets of two, a procedure automatically done by computer programs. In this example, the first stem value (the first "." corresponds to the missing value) is 6, which identifies the range of 600 to 799 square feet. There is one observation in that range, 676, so the leaf value is 8 (76 rounded to 80). The second stem value has two observations, 951 and 994, producing leaf values of 5 and 9. When there are homes represented by both individual stem values, the leaf values for the first precede those for the second. For example, the stem value of 24 represents the range from 2400 to 2599. The first four leaf values 4, 6, and 8, are in the range 2400 to 2499, while the values 1, 5, and 7 are in the range 2500 to 2599. The last stem value is 40 with a leaf value of 1. The resulting plot is shown in Fig. 1.8, produced by `PROC UNIVARIATE` of the SAS System, which automatically also provides the box plot discussed later in this section.[12]

At first glance, the stem and leaf plot looks like a histogram, which it is. However, the stem and leaf plot usually has a larger number of bars (or stems), 18 in this case, which provide greater detail about the nature of the distribution. In this case the stem and leaf chart does not provide any new information on this data set. The leaves provide rather little additional information here, but could, for example, provide evidence of rounding or imprecise measurements by showing an excessive number of zeros and fives. The leaves may

[12]This provides a good illustration of the fact that computer programs do not always provide only what is needed.

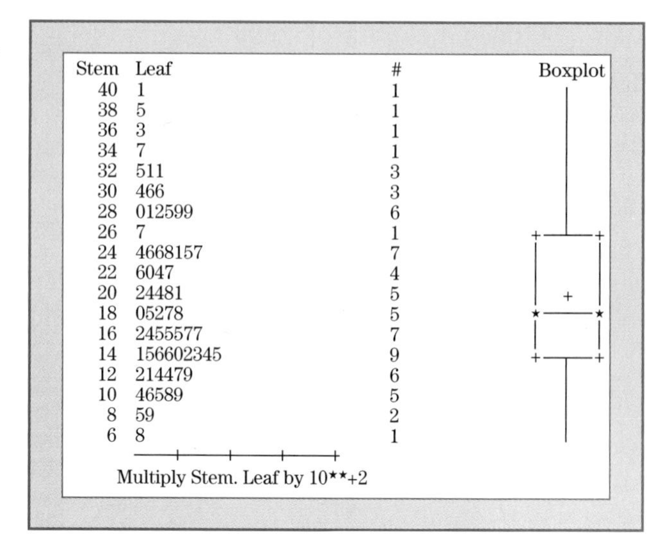

Figure 1.8

Stem and Leaf Plot for `size`

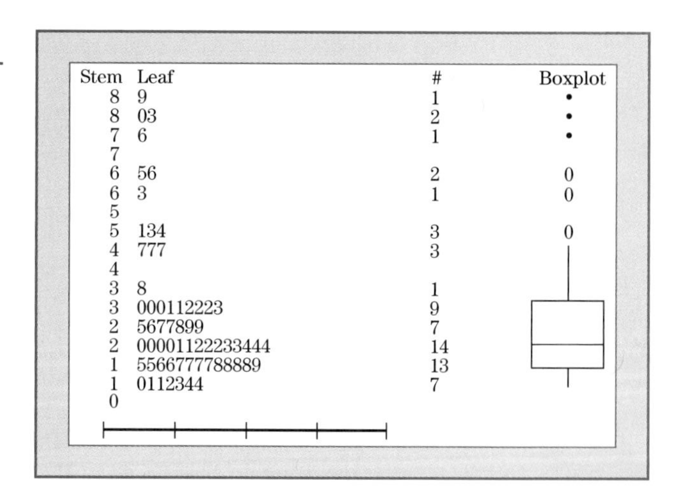

Figure 1.9

Stem and Leaf Plot and Box Plot for `HCRN` Variable

also provide evidence of bunching of specific values within a stem by showing disproportionate frequencies of specific digits.

For some data sets minor modifications may be necessary to provide an informative plot. For example, the first digit of the HCRN variable in the tree data (Table 1.7) provides for only eight stems (classes) while using the first two digits creates too many stems. In such cases it is customary to use two lines for each digit, the first representing leaves with values from 0 through 4, and a second for values from 5 through 9. Most computer programs automatically adjust for such situations. This plot is given in Fig. 1.9 (also produced by PROC UNIVARIATE). The extreme skewness we have previously noted is quite obvious.

Figure 1.10

Typical Box Plot

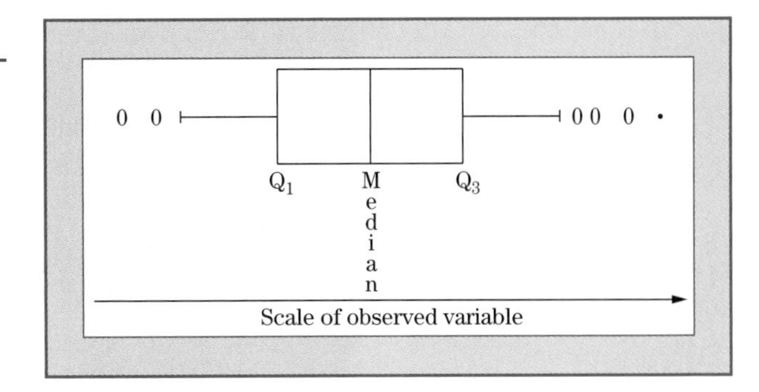

The Box Plot

The box plot[13] is used to show distributional shapes and to detect unusual observations. Figure 1.10 illustrates a typical box plot and the procedure is illustrated in Fig. 1.8 for the `size` variable from the housing data set and in Fig. 1.9 for the `HCRN` variable from the trees data set.

The scale of the plot is that of the observed variable and may be presented horizontally as in Fig. 1.10 or vertically as produced by the SAS System in Figs. 1.8 and 1.9. The features of the plot are as follows:

1. The "box," representing the interquartile range, has a value we denote by R and the endpoints Q_1 and Q_3.
2. A vertical line inside the box indicates the median. If the median is in the center of the box, the middle portion of the distribution is symmetric.
3. Horizontal lines extending from the box represent the range of observed values inside the "inner fences," which are located 1.5 times the value of the interquartile range ($1.5R$) beyond Q_1 to the left and Q_3 on the right. The relative lengths of these lines are an indicator of the skewness of the distribution as a whole.
4. Individual symbols $\circ$ represent "mild" outliers, which are defined as values between the inner and outer fences, which are located $3R$ units beyond Q_1 and Q_3.
5. Individual symbols $\bullet$ represent the location of extreme outliers, which are defined as being beyond the outer fences. Different computer programs may use different symbols for outliers and may provide options for different formats.

Symmetric distributions, which can be readily described by the mean and variance, should have the median line close to the middle of the box and reasonably equal length lines on both sides, a few mild outliers preferably equally distributed on both sides, and virtually no extreme outliers.

An ordered listing of the data or a stem and leaf plot can be used to construct the box plot. We illustrate the procedure for the `HCRN` variable for which the stem and leaf and box plots are shown in Fig. 1.9. Note that the box plot is arranged vertically in that plot. The scale is the same as the stem and leaf plot.

[13] Also referred to as a "box and whisker plot" by Tukey (1977).

on the left. The details of the procedure are as follows:

1. The quartiles Q_1 and Q_3 are found by counting $(n/4) = 16$ leaf values from the top and bottom, respectively. The resulting values of 1.8 and 3.2 define the box. These values also provide the interquartile range: $R = Q_3 - Q_1 = 3.2 - 1.8 = 1.4$. The median of 2.4 defines the line in the box.
2. The inner fences are

$$f_1 = Q_1 - 1.5R = 1.8 - 2.1 = -0.3 \quad \text{and}$$

$$f_2 = Q_3 + 1.5R = 3.2 + 2.1 = 5.3.$$

   The lines extend on each side to the nearest actual values inside the inner fences. In this example the lines extend to 1.0 (the smallest value in the data set) and 5.3, respectively. The much longer line on the high side clearly indicates the skewness.
3. The outer fences are $F_1 = -2.4$ and $F_2 = 7.4$. The fact that the lower fence has a negative value that cannot occur is a clear indicator of a skewed distribution. The four mild outliers lying between the inner and outer fences are 5.4, 6.3, 6.5, and 6.6, and are indicated by the symbol $\circ$. Note that they are all on the high side, again indicating the skewness.
4. The extreme outliers are beyond the outer fences. They are 7.6, 8.0, 8.3, and 8.9, and are indicated by $\bullet$. These are also all on the high side.

Thus we see that the box plot clearly shows the lack of symmetry for the distribution of the HCRN variable. On the other hand, the box plot for the house sizes (Fig. 1.8) shows little lack of symmetry and also has neither mild nor extreme outliers. Obviously the box plot provides a good bit of information on the distribution and outliers, but cannot be considered a complete replacement for the stem and leaf plot in terms of total information about the observations.

Comments

The presence of outliers in a set of data may cause problems in the analysis to be performed. For example, a single outlier (or several in the same direction) usually causes a distribution to be skewed, thereby affecting the mean of the distribution. In the box plot in Fig. 1.9 we see that there are several large values of the HCRN variable identified as outliers. If the mean is to be used for the analysis, it may be larger than is representative of the data due to the presence of these outliers. However, we cannot simply ignore or discard these observations as the trees do exist and to ignore them would be dishonest. A closer examination of the larger trees may reveal that they actually belong to an older grove that represents a different population from that being studied. In that case we could eliminate these observations from the analysis, but note that older trees that belonged to a population not included in the study were present in the data.

Descriptive statistical techniques, and in particular the EDA methods discussed here, are valuable in identifying outliers; however, the techniques very rarely furnish guidance as to what should be done with the outliers. In fact, the concern for "unrepresentative," "rogue," or "outlying" observations in sets of data has been voiced by many people for a long time. There is evidence that concern for outliers predates most of statistical methodology. Treatments of outliers are discussed in many texts, and in fact a book by Barnett and Lewis (1994), entitled *Outliers in Statistical Data*, is completely devoted to the topic. The sheer volume of literature addressing outliers points to the difficulty of adjusting the analysis when outliers are present.

All outliers are not deleterious to the analysis. For example, the experimenter may be tempted in some situations not to reject an outlier but to welcome it as an indication of some unexpectedly useful chemical reaction or surprisingly successful variety of corn. Often it is not necessary to take either of the extreme positions — reject the outlier or include the outlier — but instead to use some form of "robust" analysis that minimizes the effect of the outlier. One such example would be to use the median in the analysis of the variable HCRN in the tree data instead of the mean.

EXAMPLE 1.4

A biochemical assay for a substance we will abbreviate to cytosol is supposed to be an indicator of breast cancer. Masood and Johnson (1987) report on the results of such an assay, which indicates the presence of this material in units per 5 mg of protein on 42 patients. Also reported are the results of another cancer detection method, which are simply reported as "yes" or "no." The data are given in Table 1.11. We would like to summarize the data on the variable CYTOSOL.

Solution All the descriptive measures, stem and leaf plot, and box plot for these observations are given in Fig. 1.11 as provided by the Minitab DESCRIBE, STEM-AND-LEAF, and BOXPLOT commands.

Table 1.11

Cytosol Levels in Cancer Patients

OBS	CYTOSOL	CANCER	OBS	CYTOSOL	CANCER
1	145.00	YES	22	1.00	NO
2	5.00	NO	23	3.00	NO
3	183.00	YES	24	1.00	NO
4	1075.00	YES	25	269.00	YES
5	5.00	NO	26	33.00	YES
6	3.00	NO	27	135.00	YES
7	245.00	YES	28	1.00	NO
8	22.00	YES	29	1.00	NO
9	208.00	YES	30	37.00	YES
10	49.00	YES	31	706.00	YES
11	686.00	YES	32	28.00	YES
12	143.00	YES	33	90.00	YES
13	892.00	YES	34	190.00	YES
14	123.00	YES	35	1.00	YES
15	1.00	NO	36	1.00	NO
16	23.00	YES	37	7.20	NO
17	1.00	NO	38	1.00	NO
18	18.00	NO	39	1.00	NO
19	150.00	YES	40	71.00	YES
20	3.00	NO	41	189.00	YES
21	3.20	YES	42	1.00	NO

Figure 1.11

Descriptive Measures of CYTOSOL

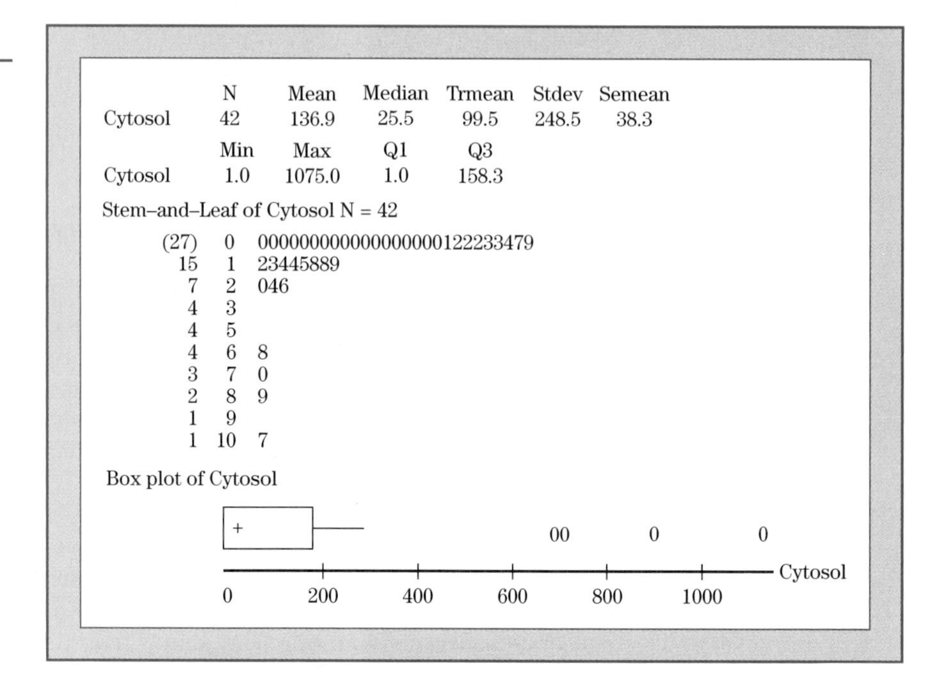

	N	Mean	Median	Trmean	Stdev	Semean
Cytosol	42	136.9	25.5	99.5	248.5	38.3

	Min	Max	Q1	Q3
Cytosol	1.0	1075.0	1.0	158.3

Stem-and-Leaf of Cytosol N = 42

```
(27)  0  000000000000000000122233479
 15   1  23445889
  7   2  046
  4   3
  4   5
  4   6  8
  3   7  0
  2   8  9
  1   9
  1  10  7
```

Box plot of Cytosol

The first portion gives the numerical descriptors. The mean is 136.9 and the standard deviation is 248.5. Note that the standard deviation is greater than the mean. Since the variable (CYTOSOL) cannot be negative, the empirical rule will not be applicable, implying that the distribution is skewed. This conclusion is reinforced by the large difference between the mean and the median. Finally, the first quartile is the same as the minimum value, indicating that at least 25% of the values occur at the minimum. The asymmetry is also evident from the positions of the quartiles, with values of 1.0 and 158.3 respectively. The output also gives the minimum and maximum values, along with two measures (TRMEAN and SEMEAN), which are not discussed in this chapter.

The stem and leaf and box plots reinforce the extremely skewed nature of this distribution. It is of interest to note that in this plot the mild outliers are denoted by * (there are none) and extreme outliers by 0.

A conclusion to be reached here is that the mean and standard deviation are not particularly useful measures for describing the distribution of this variable. Instead, the median should be used along with a brief description of the shape of the distribution. ■

1.7 Bivariate Data

So far we have presented methods for describing the distribution of observed values of a single variable. These methods can be used individually to describe distributions of each of several variables that may occur in a set of data.

However, when there are several variables in one data set, we may also be interested in describing how these variables may be related to or associated with each other. We present in this section some graphic and tabular methods for describing the association between two variables. Numeric descriptors of association are presented in later chapters, especially Chapters 7 and 8.

Specific methods for describing association between two variables depend on whether the variables are measured in a nominal or numerical scale. (Association between variables measured in the ordinal scale is discussed in Chapter 13.) We illustrate these methods by using the variables on home sales given in Table 1.2.

Categorical Variables

Table 1.12 reproduces the home sales data for the two categorical variables sorted in order of `zip` and `exter`. Association between two variables measured in the nominal scale (categorical variables) can be described by a two-way frequency distribution, which is a two-dimensional table showing the frequencies of combinations of the values of the two variables. Table 1.13 is such a table showing the association between the `zip` and `exterior` siding material of the houses. This table has been produced by `PROC FREQ` of the SAS System. The table shows the frequencies of the six combinations of `zip` and `exter`. The headings at the top and left indicate the categories of the two variables. Each of the combinations of the two variables is referred to as

Table 1.12

Home Sales Data for the Categorical Variables

zip	exter	zip	exter	zip	exter
1	Brick	3	Frame	4	Brick
1	Brick	3	Frame	4	Brick
1	Brick	3	Frame	4	Brick
1	Brick	3	Frame	4	Brick
1	Frame	3	Frame	4	Brick
1	Other	3	Other	4	Brick
2	Brick	3	Other	4	Brick
2	Brick	3	Other	4	Brick
2	Brick	3	Other	4	Brick
2	Brick	3	Other	4	Brick
2	Brick	3	Other	4	Brick
2	Brick	3	Other	4	Brick
2	Brick	4	Brick	4	Brick
2	Brick	4	Brick	4	Brick
2	Brick	4	Brick	4	Brick
2	Brick	4	Brick	4	Brick
2	Frame	4	Brick	4	Brick
2	Other	4	Brick	4	Brick
2	Other	4	Brick	4	Brick
3	Brick	4	Brick	4	Frame
3	Brick	4	Brick	4	Other
3	Brick	4	Brick	4	Other
3	Brick	4	Brick	4	Other

Table 1.13

Association between `zip` and `exter`

The FREQ Procedure
Table of zip by exter

ZIP Frequency Row pct	EXTER Brick	Frame	Other	Total
1	4	1	1	6
	66.67	16.67	16.67	
2	10	1	2	13
	76.92	7.69	15.38	
3	4	5	7	16
	25.00	31.25	43.75	
4	30	1	3	34
	88.24	2.94	8.82	
Total	48	8	13	69

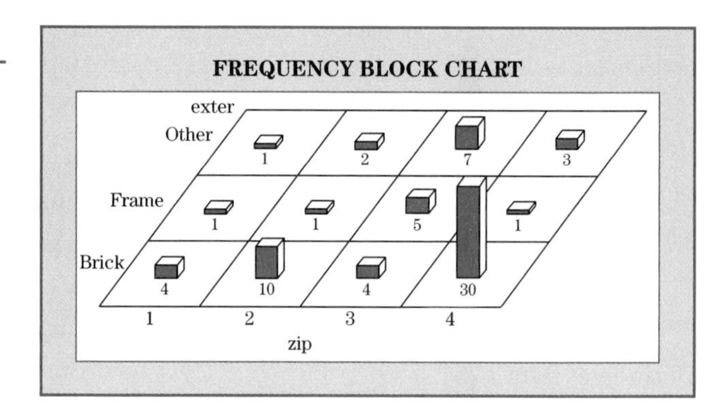

Figure 1.12

Block Chart for `exter` and `zip`

a **cell**. The last row and column (each labeled `Total`) are the individual or marginal frequencies of the two variables. As indicated by the legend at the top left of the table, the first number in each cell is the frequency.

The second number in each cell is the row percentage, that is, the percentage of each row (`zip`) that is brick, frame, or other. We can now see that brick homes predominate in all zip areas except 3, which has a mixture of all types.

The relationship between two categorical variables can also be illustrated with a block chart (a three-dimensional bar chart) with the height of the blocks being proportional to the frequencies. A block chart of the relationship between `zip` and `exter` is given in Fig. 1.12. Numeric descriptors for relationships between categorical variables are presented in Chapter 12.

Categorical and Interval Variables

The relationship between a categorical and interval (or ratio) variable is usually described by computing frequency distributions or numerical descriptors for the interval variables for each value of the nominal variable. For example, the

Figure 1.13

Side-by-Side Box Plots of Home Prices

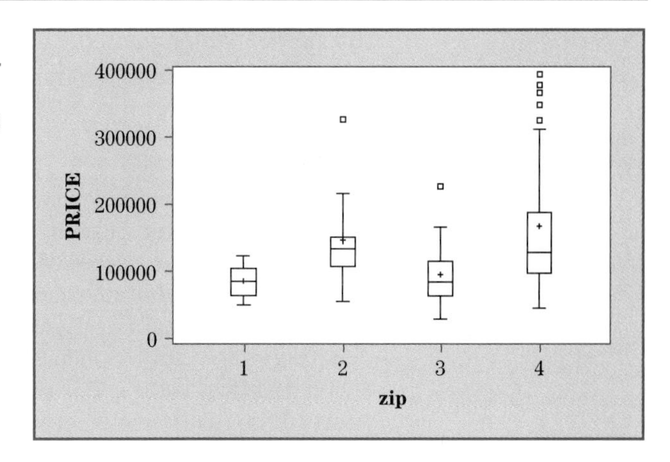

mean and standard deviation of sales prices for the four zip areas are

$$\begin{aligned}
&\texttt{zip}\,\text{area } 1, && \bar{y} = 86,892, && s = 26,877\\
&\texttt{zip}\,\text{area } 2, && \bar{y} = 147,948, && s = 67,443\\
&\texttt{zip}\,\text{area } 3, && \bar{y} = 96,455, && s = 50,746\\
&\texttt{zip}\,\text{area } 4, && \bar{y} = 169,624, && s = 98,929.
\end{aligned}$$

We can now see that `zip` areas 2 and 4 have the higher priced homes. Graphically side-by-side box plots can illustrate this information as shown in Fig. 1.13 for `price` by `zip`. This plot reinforces the information provided by the means and standard deviations, but additionally shows that all of the very-high-priced homes are in `zip` area 4.

Box plots may also be used to illustrate differences among distributions. We illustrate this method with the cancer data, by showing the side-by-side box plots of `CYTOSOL` for the two groups of patients who were diagnosed for cancer by the other method. The results, produced this time with `PROC INSIGHT` of the SAS System in Fig. 1.14, shows that both the location and

Figure 1.14

Side-by-Side Box Plots for Cancer Data

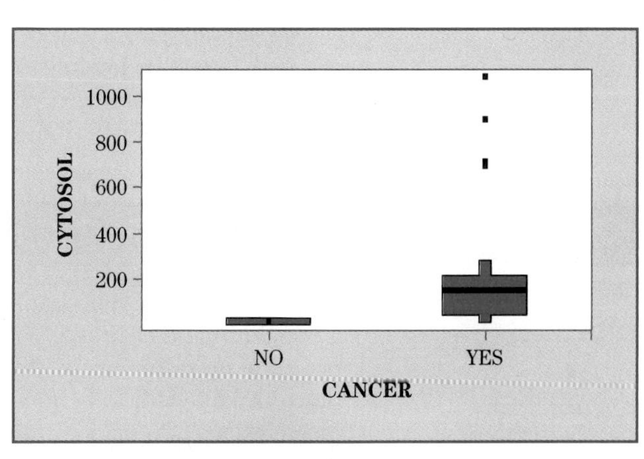

dispersion differ markedly between the two groups. Apparently both methods can detect cancer, although contradictory diagnoses occur for some patients.

Interval Variables

The relationship between two interval variables can be graphically illustrated with a **scatterplot**. A scatterplot has two axes representing the scales of the two variables. The choice of variables for the horizontal or vertical axes is immaterial, although if one variable is considered more important it will usually occupy the vertical axis. Each observation is plotted by a point representing the two variable values. Special symbols may be needed to show multiple points with identical values. The pattern of plotted points is an indicator of the nature of the relationship between the two variables. Figure 1.15 is a scatterplot showing the relationship between `price` and `size` for the data in Table 1.2.

Figure 1.15

Scatter Plot of `price` against `size`

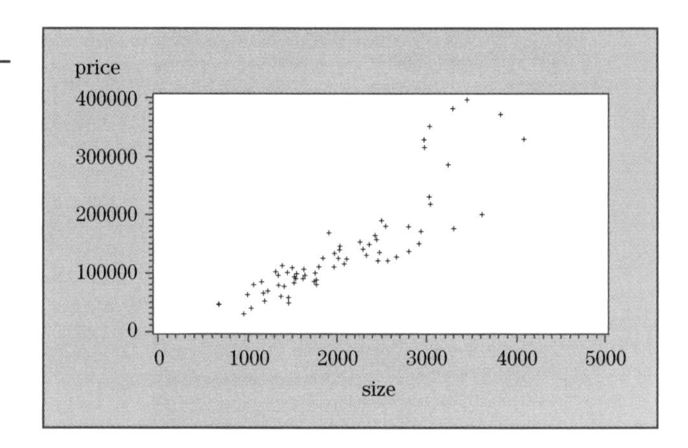

The pattern of the plotted data points shows a rather strong association between `price` and `size`, except for the higher price homes. Apparently these houses have a wider range of other amenities that affect the price. Numeric descriptors for this type of association are introduced in Chapter 7.

We should note at this point that the increased sophistication of computer graphics is rapidly leading to more informative graphs and plots. For example, some software packages provide a scatterplot with box plots on each axis describing the distribution of each of the individual variables.

1.8 Populations, Samples, and Statistical Inference — A Preview

In the beginning of this chapter we noted that a set of data may represent either a population or a sample. Using the terminology developed in this chapter, we can now more precisely define a **population** as the set of values of one or more variables for the entire collection of units relevant to a particular study. Most researchers have at least a conceptual picture of the population for a given study. This population is usually called the **target** population. A target population may be well defined. For example, the trees in Table 1.7 are a sample from a population of trees in a specified forest. On the other hand, a population may be only conceptually defined. For example, an experiment measuring the decrease in blood pressure resulting from a new drug is a sample from a hypothetical population consisting of all sufferers of high blood pressure who are potential users of the drug. A population can, in fact, be infinite. For example, a laboratory experiment can hypothetically be reproduced an infinite number of times.

We are rarely afforded the opportunity of measuring all the elements of an entire population. For this reason, most data are normally some portion or **sample** of the target population. Obviously a sample provides only partial information on the population. In other words, the characteristics of the population cannot be completely known from sample data.

We can, however, draw certain parallels between the sample and the population. Both population and sample may be described by measures such as those presented in this chapter (although we cannot usually calculate them for a population). To differentiate between a sample and the population from which it came, the descriptive measures for a sample are called **statistics** and are calculated and symbolized as presented in this chapter. Specifically, the sample mean is $\bar{y}$ and the sample variance is s^2. Descriptive measures for the population are called **parameters** and are denoted by Greek letters. Specifically, we denote the mean of a population by μ and the variance by σ^2. If the population consists of a finite number of values, $y_1, y_2, \ldots, y_N$, then the mean is calculated by

$$\mu = \sum y_i/N,$$

and the variance is found by

$$\sigma^2 = \frac{\sum(y_i - \mu)^2}{N}.$$

It is logical to assume that the sample statistics provide some information on the values of the population parameters. In other words, the sample statistics may be considered to be **estimates** of the population parameters. However, the statistics from a sample cannot exactly reflect the values of the parameters of the population from which the sample is taken. In fact, two or more individual samples from the same population will invariably exhibit different values of sample estimates. The magnitude of variation among sample estimates is referred to as the **sampling error** of the estimates. Therefore, the magnitude of this sampling error provides an indication of how closely a sample estimate approximates the corresponding population parameter. In other words, if a sample estimate can be shown to have a small sampling error, that estimate is said to provide a good estimate for the corresponding population parameter.

We must emphasize that sampling error is not an error in the sense of making a mistake. It is simply a recognition of the fact that a sample statistic does not exactly represent the value of a population parameter. The recognition and measurement of this sampling error is the cornerstone of statistical inference.

To control as well as to determine the magnitude of the sampling error, we must incorporate in our sampling method as much **randomization** as is physically possible. A random sample is one where "chance" dominates the selection of the units of the population to be included in the sample, in the same sense that chance determines the winners in a properly conducted lottery. That is, the method of randomization results in a sample drawn in such a manner that each possible sample of the specified size has an equal chance of being selected.[14] Actually, the ability of statistical analyses to provide reliable estimates of sampling error is based on the assumption of random samples and is therefore assumed for all statistical methods presented in this book.

The process of drawing a random sample is conceptually simple, but may be difficult to implement in practice. Essentially, a random sample is like drawing for prizes in a lottery: The population consists of all the lottery tickets and the sample of winners is drawn "blindly" from a drum containing all the tickets. The most straightforward method for drawing a random sample is to assign a unique number (usually sequential) to each unit of the population and select for the sample those units that correspond to a set of random numbers that have been picked from a table of random numbers or generated by a computer. This procedure can be used for relatively small finite populations but may not be practical for large finite populations and is an obviously impossible task for infinite populations. Specific instructions for drawing random samples can be found in books on sampling (for example, Scheaffer *et al.*, 1996) or on experimental design (for example, Maxwell and Delaney, 2000). The overriding factor in all types of random sampling is that the actual selection of sample elements not be subject to personal or other bias.

In many cases experimental conditions are such that nonrestricted randomization is impossible; hence the sample is not a random sample. For example, much of the data available for economic research consists of measurements of economic variables over time. For such data the normal sequencing of the data cannot be altered and we cannot really claim to have a random sample of observations. In such situations, however, it is possible to define an appropriate model that contains a random element. Models that incorporate such random elements are introduced in Chapters 6 and 7.

1.9 CHAPTER SUMMARY

Solution to Example 1.1 We now know that the data listed in Table 1.1 consists of 50 observations on four variables from an observational study. Two of the variables (AGE and TVHOURS) are numerical and have the ratio level of

[14] In some special applications the probabilities of selection do not need to be equal, but they must be known and predetermined before the sample is selected.

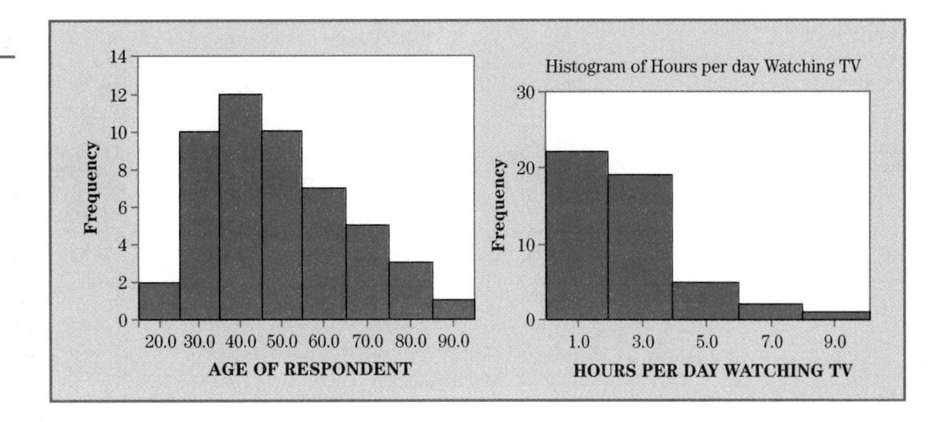

Figure 1.16

Histograms of AGE and TVHOURS

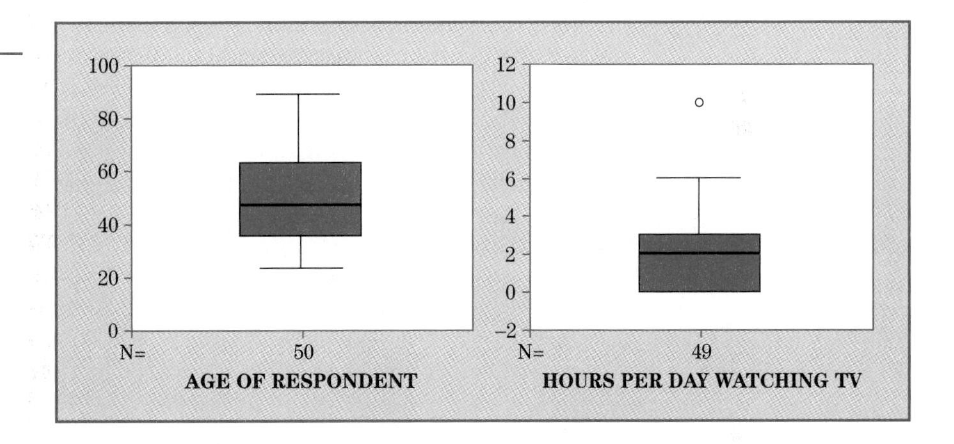

Figure 1.17

Box Plots of AGE and TVHOURS

measurement. The other two are categorical (nominal) level variables. We will explore the nature of these variables and a few of the relationships between them.

We start by using SPSS to construct the frequency histograms of AGE and TVHOURS as shown in Fig. 1.16. From these it appears that the distribution of age is reasonably symmetric while that of TVHOURS is skewed positively.

To further explore the shape of the distributions of the two variables we construct the box plots shown in Fig. 1.17. Note the symmetry of the variable AGE while the obvious positive skewness of TVHOURS is highlighted by the long whisker on the positive side of the boxplot. Also, note that there is one potential outlier identified in the TVHOURS box plot. This is the value 10 corresponding to the 20th respondent in the data set. It is also interesting to see that fully 25% of the respondents reported an average number of hours watching TV as 0 as indicated by the fact that the lower quartile (the lower edge of the box) is at the level "0."

We now examine some of the numerical descriptive statistics for these two measures as seen in Table 1.14.

Table 1.14

Numerical Statistics on AGE and TVHOURS Statistics

	Age of Respondent	Hours per Day Watching TV
N		
Valid	50	49
Missing	0	1
Mean	48.26	1.88
Median	46.00	2.00
Mode	53	0
Std. deviation	17.05	2.14
Variance	290.65	4.60
Minimum	23	0
Maximum	89	10

The first two rows of Table 1.14 tell us that all 50 of our sample respondents answered the question concerning their age while 1 of the respondents did not answer the question about the number of hours per day watching TV. The mean age is 48.26 and the ages of respondents ranges from 23 to 89. The mean number of hours per day watching TV is 1.88 and ranges from 0 to 10. Note that the standard deviation of the number of hours watching TV is actually larger than the mean. This is another indication of the extremely skewed distribution of these values.

Figure 1.18

Bar Chart of HAPPY and Pie Chart of SEX

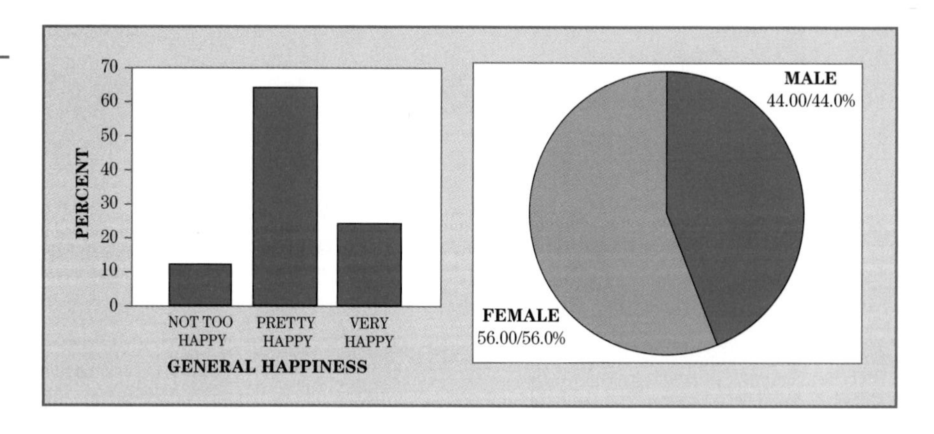

Figure 1.18 shows a relative frequency (percent) bar chart of the variable HAPPY. From this we can see that only about 12% of the respondents considered themselves not happy with their lives. Figure 1.18 also shows a pie chart of the variable SEX. This indicates that 56% of the respondents were female vs 44% male.

To see if there is any noticeable relationship between the variables AGE and TVHOURS, a scatter diagram is constructed. The graph is shown in Fig. 1.19. There does not seem to be a strong relationship between these two variables. There is one respondent who seems to be "separated" from the group, and that is the respondent who watches TV about 10 hours per day.

Figure 1.19

Scatter Diagram of AGE and TVHOURS

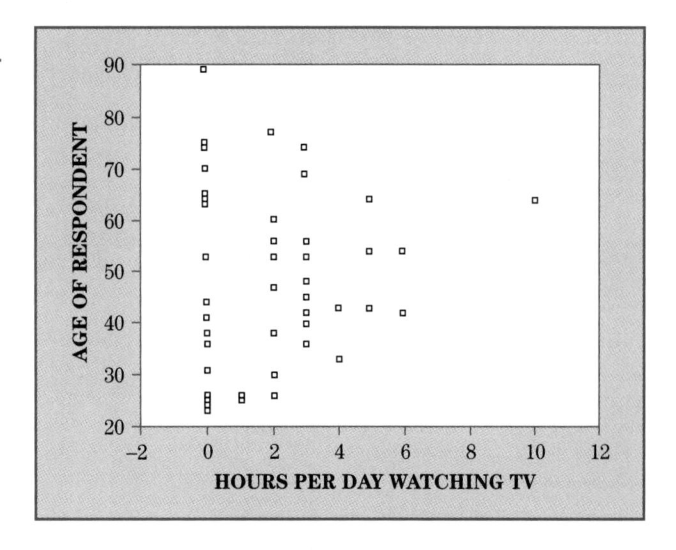

Figure 1.20

Side-by-Side Bar Charts for HAPPY by SEX

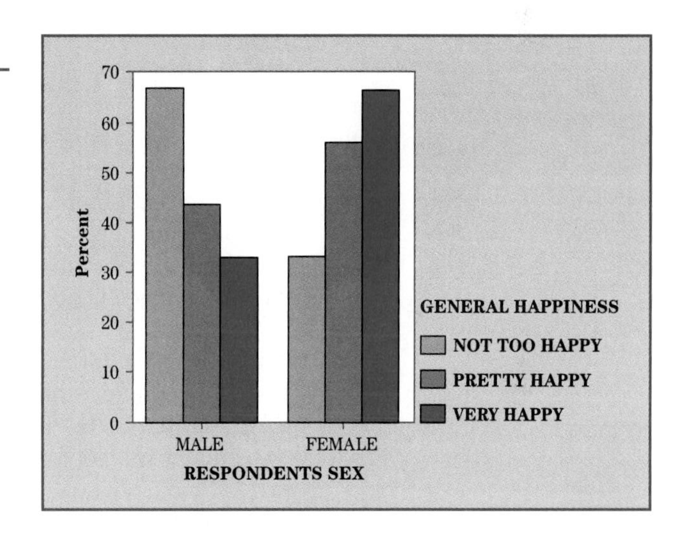

To examine the relationship between the two variables SEX and HAPPY, we will construct side-by-side relative frequency bar charts. These are given in Fig. 1.20. Note that the patterns of "happiness" seem to be opposite for the sexes. For example, of those who identified themselves as being "Very Happy," 67% were female while only 33% were male.

Finally, to see if there is any difference in the relationship between AGE and TVHOURS over the levels of SEX, we construct a scatter diagram identifying points by SEX. This graph is given in Fig. 1.21.

The graph does not indicate any systematic difference in the relationship by sex. The respondent who watches TV about 10 hours per day is male, but other than that nothing can be concluded by examination of this graph. ■

Figure 1.21

AGE vs TVHOURS Identified by SEX

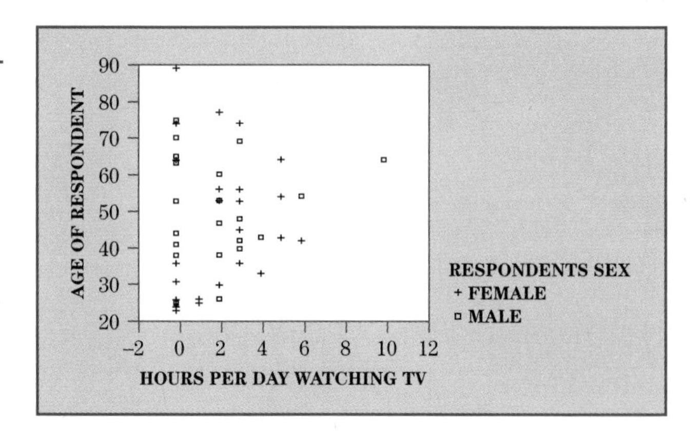

Summary

Statistics is concerned with the analysis of data. A set of data is defined as a set of observations on one or more variables. Variables may be measured on a nominal, ordinal, interval, or ratio scale with the ratio scale providing the most information. Additionally, interval and ratio scale variables, also called numerical variables, may be discrete or continuous. The nature of a statistical analysis is largely dictated by the type of variable being analyzed.

A set of observations on a variable is described by a distribution, which is a listing of frequencies with which different values of the variable occur. A relative frequency distribution shows proportions with which values of a variable occur and is related to a probability distribution, which is extensively used in statistics.

Graphical representation of distributions is extremely useful for investigating various characteristics of distributions, especially their shape and the existence of unusual values. Frequently used graphical representations include bar charts, stem and leaf plots, and box plots.

Numerical measures of various characteristics of distributions provide a manageable set of numeric values that can readily be used for descriptive and comparative purposes. The most frequently used measures are those that describe the location (center) and dispersion (variability) of a distribution. The most frequently used measure of location is the mean, which is the sum of observations divided by the number of observations. Also used is the median, which is the center value.

The most frequently used measure of dispersion is the variance, which is the average of the squared differences between the observations and the mean. The square root of the variance, called the standard deviation, describes dispersion in the original scale of measurement. Other measures of dispersion are the range, which is the difference between the largest and smallest observations, and the mean absolute deviation, which is the average of the absolute values of the differences between the observations and the mean.

Other numeric descriptors of the characteristics of a distribution include the percentiles, of which the quartile and interquartile ranges are special cases.

The importance of the mean and standard deviation is underscored by the empirical rule and Tchebysheff's theorem, which show that these two measures provide a very adequate description of data distributions.

The chapter concludes with brief sections on descriptions of relationships between two variables and a look ahead at the uses of descriptive measures for statistical inference.

1.10 CHAPTER EXERCISES

CONCEPT QUESTIONS

The following multiple choice questions are intended to provide practice in methods and reinforce some of the concepts presented in this chapter.

1. The scores of eight persons on the Stanford–Binet IQ test were:

95 87 96 110 150 104 112 110

The median is:
(1) 107
(2) 110
(3) 112
(4) 104
(5) none of the above.

2. The concentration of DDT, in milligrams per liter, is:
(1) a nominal variable
(2) an ordinal variable
(3) an interval variable
(4) a ratio variable.

3. If the interquartile range is zero, you can conclude that:
(1) the range must also be zero
(2) the mean is also zero
(3) at least 50% of the observations have the same value
(4) all of the observations have the same value
(5) none of the above is correct.

4. The species of each insect found in a plot of cropland is:
(1) a nominal variable
(2) an ordinal variable
(3) an interval variable
(4) a ratio variable.

5. The "average" type of grass used in Texas lawns is best described by
(1) the mean
(2) the median
(3) the mode.

6. A sample of 100 IQ scores produced the following statistics:

mean = 95	lower quartile = 70
median = 100	upper quartile = 120
mode = 75	standard deviation = 30

Which statement(s) is (are) correct?
(1) Half of the scores are less than 95.
(2) The middle 50% of scores are between 100 and 120.
(3) One-quarter of the scores are greater than 120.
(4) The most common score is 95.

7. A sample of 100 IQ scores produced the following statistics:

mean = 100	lower quartile = 70
median = 95	upper quartile = 120
mode = 75	standard deviation = 30

Which statement(s) is (are) correct?
(1) Half of the scores are less than 100.
(2) The middle 50% of scores are between 70 and 120.
(3) One-quarter of the scores are greater than 100.
(4) The most common score is 95.

8. Identify which of the following is a measure of dispersion:
(1) median
(2) 90th percentile
(3) interquartile range
(4) mean.

9. A sample of pounds lost in a given week by individual members of a weight-reducing clinic produced the following statistics:

mean = 5 pounds	first quartile = 2 pounds
median = 7 pounds	third quartile = 8.5 pounds
mode = 4 pounds	standard deviation = 2 pounds

Identify the correct statement:
(1) One-fourth of the members lost less than 2 pounds.
(2) The middle 50% of the members lost between 2 and 8.5 pounds.
(3) The most common weight loss was 4 pounds.
(4) All of the above are correct.
(5) None of the above is correct.

10. A measurable characteristic of a population is:
(1) a parameter
(2) a statistic

(3) a sample
(4) an experiment.

11. What is the primary characteristic of a set of data for which the standard deviation is zero?
(1) All values of the variable appear with equal frequency.
(2) All values of the variable have the same value.
(3) The mean of the values is also zero.
(4) All of the above are correct.
(5) None of the above is correct.

12. Let X be the distance in miles from their present homes to residences when in high school of individuals at a class reunion. Then X is:
(1) a categorical (nominal) variable
(2) a continuous variable
(3) a discrete variable
(4) a parameter
(5) a statistic.

13. A subset of a population is:
(1) a parameter
(2) a population
(3) a statistic
(4) a sample
(5) none of the above.

14. The median is a better measure of central tendency than the mean if:
(1) the variable is discrete
(2) the distribution is skewed
(3) the variable is continuous
(4) the distribution is symmetric
(5) none of the above is correct.

15. A small sample of automobile owners at Texas A & M University produced the following number of parking tickets during a particular year: 4, 0, 3, 2, 5, 1, 2, 1, 0. The mean number of tickets (rounded to the nearest tenth) is:
(1) 1.7
(2) 2.0
(3) 2.5
(4) 3.0
(5) none of the above.

PRACTICE EXERCISES

Most of the exercises in this and subsequent chapters are based on data sets for which computations are most efficiently done with computers. However, manual computations, although admittedly tedious, provide a feel for how various results arise and what they may mean. For this reason, we have included a few exercises with small numbers of simple-valued observations that can be done manually. The solutions to all these exercises are given in the back of the text.

1. A university published the following distribution of students enrolled in the various colleges:

College	Enrollment
Agriculture	1250
Business	3675
Earth sciences	850
Liberal arts	2140
Science	1550
Social sciences	2100

   Construct a bar chart of these data.

2. On ten days, a bank had 18, 15, 13, 12, 8, 3, 7, 14, 16, and 3 bad checks. Find the mean, median, variance, and standard deviation of the number of bad checks.

3. Calculate the mean and standard deviation of the following sample:

$$-1, \quad 4, \quad 5, \quad 0.$$

4. The following is the distribution of ages of students in a graduate course:

Age (years)	Frequency
20–24	11
25–29	24
30–34	30
35–39	18
40–44	11
45–49	5
50–54	1

   (a) Construct a bar chart of the data.
   (b) Calculate the mean and standard deviation of the data.

5. Weekly closing prices of Hewlett–Packard stock from October 1995 to February 1996 are listed below, given in sequential order and rounded to the nearest dollar:

   93, 94, 95, 89, 85, 82, 87, 85, 84, 80, 78, 78, 84, 87, 90.

   (a) Using time as the horizontal axis and closing price as the vertical axis, construct a trend graph showing how the price moved during this period.
   (b) Construct a stem and leaf plot.
   (c) Calculate the mean and median closing price.
   (d) Use the change of scale procedure in Section 1.5 to calculate the standard deviation of the closing price.

EXERCISES

1. Most of the problems in this and other chapters deal with "real" data for which computations are most efficiently performed with computers. Since a little experience in manual computing is healthy, here are 15

observations of a variable having no particular meaning:

12 18 22 17 20 15 19 13 23 8 14 14 19 11 30.

(a) Compute the mean, median, variance, range, and interquartile range for these observations.
(b) Produce a stem and leaf plot.
(c) Write a brief description of this data set.

2. Because waterfowl are an important economic resource, wildlife scientists study how waterfowl abundance is related to various environmental variables. In such a study, the variables shown in Table 1.15 were observed for a sample of 52 ponds.

WATER: the amount of open water in the pond, in acres.
VEG: the amount of aquatic and wetland vegetation present at and round the pond, in acres.
FOWL: the number of waterfowl recorded at the pond during a (random) one-day visit to the pond in January.

The results of some intermediate computations:

$$\text{WATER: } \sum y = 370.5 \qquad \sum y^2 = 25735.9$$
$$\text{VEG: } \sum y = 58.25 \qquad \sum y^2 = 285.938$$
$$\text{FOWL: } \sum y = 3933 \qquad \sum y^2 = 2449535$$

Table 1.15

Waterfowl Data

OBS	WATER	VEG	FOWL	OBS	WATER	VEG	FOWL
1	1.00	0.00	0	27	0.25	0.00	0
2	0.25	0.00	10	28	1.50	0.00	240
3	1.00	0.00	125	29	2.00	1.50	2
4	15.00	3.00	30	30	31.00	0.00	0
5	1.00	0.00	0	31	149.00	9.00	1410
6	33.00	0.00	32	32	1.00	2.75	0
7	0.75	0.00	16	33	0.50	0.00	15
8	0.75	0.00	0	34	1.50	0.00	16
9	2.00	0.00	14	35	0.25	0.00	0
10	1.50	0.00	17	36	0.25	0.25	0
11	1.00	0.00	0	37	0.75	0.00	125
12	16.00	1.00	210	38	0.25	0.00	2
13	0.25	0.00	11	39	1.25	0.00	0
14	5.00	1.00	218	40	6.00	0.00	179
15	10.00	2.00	5	41	2.00	0.00	80
16	1.25	0.50	26	42	5.00	8.00	167
17	0.50	0.00	4	43	2.00	0.00	0
18	16.00	2.00	74	44	0.25	0.00	11
19	2.00	0.00	0	45	5.00	1.00	364
20	1.50	0.00	51	46	7.00	2.25	59
21	0.50	0.00	12	47	9.00	7.00	185
22	0.75	0.00	18	48	0.00	1.25	0
23	0.25	0.00	1	49	0.00	4.00	0
24	17.00	5.25	2	50	7.00	0.00	177
25	3.00	0.75	16	51	4.00	2.00	0
26	1.50	1.75	9	52	1.00	2.00	0

(a) Make a complete summary of one of these variables. (Compute mean, median, and variance, and construct a bar chart or stem and leaf and box plots.) Comment on the nature of the distribution.
(b) Construct a frequency distribution for FOWL, and use the frequency distribution formulas to compute the mean and variance.
(c) Make a scatterplot relating WATER or VEG to FOWL.

3. Someone wants to know whether the direction of price movements of the general stock market, as measured by the New York Stock Exchange (NYSE) Composite Index, can be predicted by directional price movements of the New York Futures Contract for the next month. Data on these variables have been collected for a 46-day period and are presented in Table 1.16. The variables are:

INDEX: the percentage change in the NYSE composite index for a one-day period.

FUTURE: the percentage change in the NYSE futures contract for a one-day period.

Table 1.16

Stock Prices

DAY	INDEX	FUTURE	DAY	INDEX	FUTURE
1	0.58	0.70	24	1.13	0.46
2	0.00	−0.79	25	2.96	1.54
3	0.43	0.85	26	−3.19	−1.08
4	−0.14	−0.16	27	1.04	−0.32
5	−1.15	−0.71	28	−1.51	−0.60
6	0.15	−0.02	29	−2.18	−1.13
7	−1.23	−1.10	30	−0.91	−0.36
8	−0.88	−0.77	31	1.83	−0.02
9	−1.26	−0.78	32	2.86	0.91
10	0.08	−0.35	33	2.22	1.56
11	−0.15	0.26	34	−1.48	−0.22
12	0.23	−0.14	35	−0.47	−0.63
13	−0.97	−0.33	36	2.14	0.91
14	−1.36	−1.17	37	−0.08	−0.02
15	−0.84	−0.46	38	−0.62	−0.41
16	−1.01	−0.52	39	−1.33	−0.81
17	−0.86	−0.28	40	−1.34	−2.43
18	0.87	0.28	41	1.12	−0.34
19	−0.78	−0.20	42	−0.16	−0.13
20	−2.36	−1.55	43	1.35	0.18
21	0.48	−0.09	44	1.33	1.18
22	−0.88	−0.44	45	−0.15	0.67
23	0.08	−0.63	46	−0.46	−0.10

(a) Make a complete summary of one of these variables.
(b) Construct a scatterplot relating these variables. Does the plot help to answer the question posed?

4. The data in Table 1.17 consist of 25 values for four computer-generated variables called Y1, Y2, Y3, and Y4. Each of these is intended to represent a particular distributional shape. Use a stem and leaf and a box plot to ascertain the nature of each distribution and then see whether the empirical rule works for each of these.

Table 1.17

Data for Recognizing Distributional Shapes

Y1	Y2	Y3	Y4
4.0	3.5	1.3	5.0
6.7	6.4	6.7	1.0
6.2	3.3	1.3	0.6
2.4	4.0	2.7	4.5
1.6	3.5	1.3	1.8
5.3	4.8	4.0	0.3
6.8	3.2	1.3	0.1
6.8	6.9	9.4	4.7
2.8	6.5	6.7	2.7
7.3	6.6	6.7	1.1
5.8	4.4	2.7	2.1
6.1	4.2	2.7	2.3
3.1	4.6	2.7	2.5
8.1	4.7	2.7	2.3
6.3	3.3	1.3	0.1
6.9	3.9	2.7	3.9
8.4	5.7	5.4	1.4
3.1	3.3	1.3	2.2
4.5	5.2	4.0	0.9
1.6	4.0	2.7	4.8
1.8	6.7	8.0	1.6
5.3	5.2	4.0	0.1
2.7	5.8	5.4	3.9
3.2	5.9	5.4	0.9
4.2	3.1	0.0	7.4

5. Climatological records provide a rich source of data suitable for description by statistical methods. The data for this example (Table 1.18) are the number of January days in London, England, having rain (Days) and the average January temperature (Temp, in degrees Fahrenheit) for the years 1858 through 1939.

(a) Summarize these two variables.

(b) Draw a scatterplot to see whether the two variables are related.

6. Table 1.19 gives data on population (in thousands) and expenditures on criminal activities (in million $) for the 50 states and the District of Columbia as obtained from the 1988 Statistical Abstract of the United States.

(a) Describe the distribution of states' criminal expenditures with whatever measures appear appropriate. Comment on the features and implications of these data.

(b) Compute the per capita expenditures (`EXPEND/POP`) for these data. Repeat part (a). Discuss any differences in the nature of the distribution you may have stated in part (a).

Table 1.18

Rain Days and Temperatures, London Area, January

Year	Days	Temp	Year	Days	Temp	Year	Days	Temp
1858	6	40.5	1886	23	35.8	1914	12	39.7
1859	10	40.0	1887	13	37.9	1915	19	45.9
1860	21	34.0	1888	9	37.2	1916	14	35.5
1861	7	39.3	1889	10	43.6	1917	18	39.6
1862	19	42.2	1890	21	34.1	1918	18	37.8
1863	15	36.6	1891	14	36.6	1919	22	42.4
1864	8	36.5	1892	13	35.5	1920	21	46.1
1865	13	43.1	1893	17	38.5	1921	20	40.2
1866	23	34.6	1894	25	33.7	1922	20	41.5
1867	17	37.6	1895	16	40.5	1923	15	40.8
1868	19	41.4	1896	9	35.4	1924	18	41.7
1869	15	38.5	1897	21	43.7	1925	11	40.5
1870	17	33.4	1898	9	42.8	1926	18	41.0
1871	17	41.5	1899	19	40.4	1927	17	42.1
1872	22	42.3	1900	21	38.8	1928	21	34.8
1873	18	41.9	1901	12	42.0	1929	12	44.0
1874	17	43.6	1902	11	41.1	1930	17	39.0
1875	23	37.3	1903	17	39.5	1931	20	44.0
1876	11	42.9	1904	22	38.4	1932	13	37.4
1877	25	40.4	1905	8	42.4	1933	14	39.6
1878	15	31.8	1906	18	38.8	1934	18	40.7
1879	12	33.3	1907	8	36.8	1935	13	40.9
1880	5	31.7	1908	10	38.8	1936	21	41.9
1881	8	40.5	1909	13	40.0	1937	23	43.6
1882	7	41.4	1910	14	38.2	1938	21	41.7
1883	21	43.9	1911	12	40.2	1939	22	30.8
1884	16	36.6	1912	17	41.1			
1885	16	36.3	1913	17	38.4			

(c) Make a scatterplot of total and per capita expenditures on the vertical axis against population on the horizontal axis. Which of these plots is more useful?

7. Make scatterplots for all pairwise combinations of the variables from the tree data (Table 1.7). Which pairs of variables have the strongest relationship? Is your conclusion consistent with prior knowledge?

8. The data set in Table 1.20 lists all cases of Down's syndrome in Victoria, Australia, from 1942 through 1957, as well as the number of births classified by the age of the mother (Andrews and Herzberg, 1985).
   (a) Construct a relative frequency histogram for total number of births by age group.
   (b) Construct a relative frequency histogram for number of mothers of Down's syndrome patients by age group.
   (c) Compare the shape of the two histograms. Does the shape of the histogram for Down's syndrome suggest that age alone accounts for number of Down's syndrome patients born?
   (d) Construct a scatter diagram of total number of births versus number of mothers of Down's syndrome. Does the scatter diagram support the conclusion in part (c)?

Table 1.19

Criminal Expenditures

STATE	POP	EXPEND	STATE	POP	EXPEND
AK	525	360	MT	809	123
AL	4083	498	NC	6413	821
AR	2388	219	ND	672	75
AZ	3386	728	NE	1594	206
CA	27663	6539	NH	1057	140
CO	3296	602	NJ	7672	1592
CT	3211	544	NM	1500	296
DC	622	435	NV	1007	256
DE	644	130	NY	17825	5220
FL	12023	2252	OH	10784	1617
GA	6222	835	OK	3272	432
HI	1083	210	OR	2724	463
IA	2834	368	PA	11936	1796
ID	998	120	RI	986	164
IL	11582	2023	SC	3425	427
IN	5531	593	SD	709	79
KS	2476	324	TN	4855	568
KY	3727	417	TX	16789	2313
LA	4461	785	UT	1680	244
MA	5855	1024	VA	5904	914
MD	4535	940	VT	548	74
ME	1187	128	WA	4538	838
MI	9200	1788	WI	4807	863
MN	4246	665	WV	1897	168
MO	5103	660	WY	490	115
MS	2625	245			

Table 1.20

Mongoloid Births in Victoria, Australia[a]

Age Group, Years	Total Number of Births	Number of Mothers of Down's Syndrome Patients
20 or less	35,555	15
20–24	207,931	128
25–29	253,450	208
30–34	170,970	194
35–39	86,046	297
40–44	24,498	240
45 or over	1,707	37

[a] Reprinted with permission from Andrews and Herzberg (1985).

9. Table 1.21 shows the times in days from remission induction to relapse for 51 patients with acute nonlymphoblastic leukemia who were treated on a common protocol at university and private institutions in the Pacific Northwest. This is a portion of a larger study reported by Glucksberg *et al.* (1981).

Since data of this type are notoriously skewed, the distribution of the times can be examined using the following output from PROC UNIVARIATE in SAS as seen in Fig. 1.22.

Table 1.21 Ordered Remission Durations for 51 Patients with Acute Nonlymphoblastic Leukemia (in days)

24	46	57	57	64	65	82	89	90	90	111	117	128	143	148	152
166	171	186	191	197	209	223	230	247	249	254	258	264	269	270	273
284	294	304	304	332	341	393	395	487	510	516	518	518	534	608	642
697	955	1160													

Figure 1.22

Summary Statistics for Remission Data

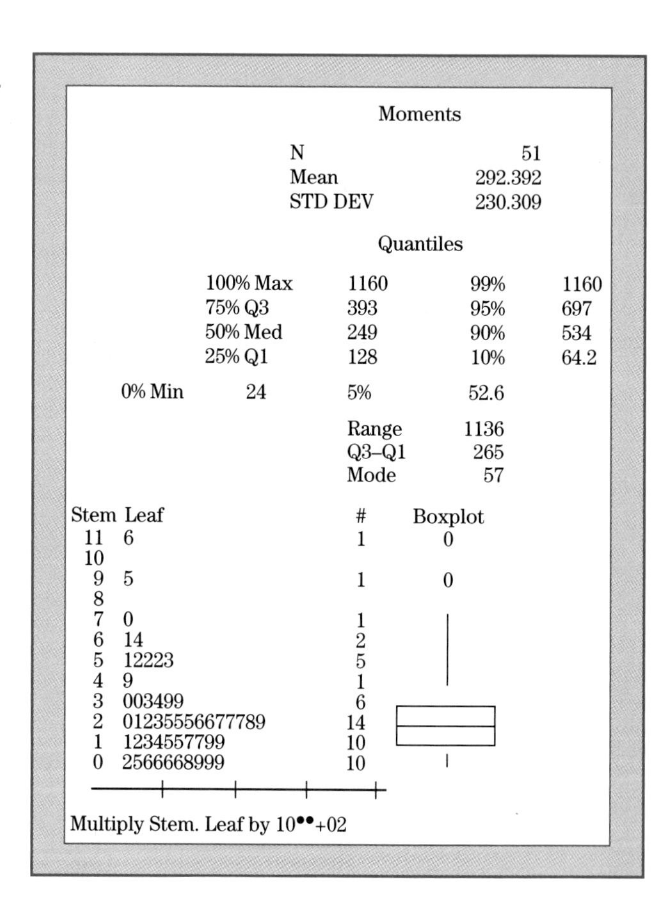

(a) What is the relation between the mean and the median? What does this mean about the shape of the distribution? Do the stem and leaf plot and the box plot support this?

(b) Identify any outliers in this data set. Can you think of any reasons for these outliers? Can we just "throw them away"? Note that the mean time of remission is 292.39 days and the median time is 249.

(c) Approximately what percent of these patients were in remission for less than one year?

10. The use of placement exams in elementary statistics courses has been a controversial topic in recent times. Some researchers think that the use

of a placement exam can help determine whether a student will successfully complete a course (or program). A recent study in a large university resulted in the data listed in Table 1.22. The placement test administered was an inhouse written general mathematics test. The course was Elementary Statistics. The students were told that the test would not affect their course grade. After the semester was over, students were classified according to their status. In Table 1.22 are the students' scores on the placement test (from 0 to 100), and the status of the student (coded as 0 = passed the course, 1 = failed the course, and 2 = dropped out before the semester was over) related?
(a) Construct a frequency histogram for Score. Describe the results.

Table 1.22 Placement Scores for Elementary Statistics

Student	Score	Status	Student	Score	Status	Student	Score	Status
1	90	2	36	85	0	71	97	2
2	65	2	37	99	1	72	90	0
3	30	1	38	45	0	73	30	0
4	55	0	39	90	0	74	1	0
5	1	0	40	10	1	75	1	0
6	5	1	41	56	0	76	70	0
7	95	0	42	55	2	77	90	0
8	99	0	43	50	0	78	70	0
9	40	0	44	1	1	79	75	0
10	95	0	45	45	0	80	75	2
11	1	0	46	50	0	81	70	2
12	55	0	47	85	2	82	85	0
13	85	0	48	95	2	83	45	0
14	95	0	49	15	0	84	50	0
15	15	2	50	35	0	85	55	0
16	95	0	51	85	0	86	15	0
17	15	0	52	85	0	87	55	0
18	65	0	53	50	0	88	20	1
19	55	0	54	10	1	89	1	1
20	75	0	55	60	0	90	75	0
21	15	0	56	45	1	91	45	2
22	35	2	57	90	0	92	70	0
23	90	0	58	1	1	93	70	0
24	10	0	59	80	2	94	45	0
25	10	1	60	45	0	95	90	0
26	20	0	61	90	0	96	65	2
27	25	0	62	45	0	97	75	2
28	15	1	63	20	0	98	70	0
29	40	0	64	35	1	99	65	0
30	15	0	65	40	2	100	55	0
31	50	0	66	40	0	101	55	0
32	80	0	67	60	0	102	40	0
33	50	1	68	15	0	103	56	0
34	50	2	69	45	0	104	85	0
35	97	0	70	45	0	105	80	0

(b) Construct a relative frequency histogram for Score for each value of Status. Describe the differences among these distributions. Are there some surprises?

11. The *1988 Life Insurance Fact Book*, published by the American Council of Life Insurance, gives the net rate of investment income for U.S. life insurance companies from 1968 through 1987 (p. 65). These data are reproduced in Table 1.23.

Table 1.23

Net Rate of Investment Income

Year	Percent	Year	Percent	Year	Percent	Year	Percent
68	4.95	73	5.88	78	7.31	83	8.96
69	5.12	74	6.25	79	7.73	84	9.45
70	5.3	75	6.36	80	8.09	85	9.63
71	5.44	76	6.55	81	8.57	86	9.35
72	5.56	77	6.89	82	8.91	87	9.09

(a) Find the mean rate of investment income and the standard deviation.
(b) What is the median rate of investment? When did the median occur?
(c) Plot the rate of investment income versus the year. What happens prior to 1985? How about after 1985? What would you expect to happen in 1988?

12. A study of characteristics of successful salespersons in a certain industry included a questionnaire given to sales managers of companies in this industry. In this questionnaire the sales manager had to choose a trait that the manager thought was most important for salespersons to have. The results of 120 such responses are given in Table 1.24.

Table 1.24

Traits of Salespersons Considered Most Important by Sales Managers

Trait	Number of Responses
Reliability	44
Enthusiastic/energetic	30
Self-starter	20
Good grooming habits	18
Eloquent	6
Pushy	2

(a) Convert the number of responses to percents of total. What can be said about the first two traits?
(b) Draw a bar chart of the data.

13. A measure of the time a drug stays in the blood system is given by the half-life of the drug. This measure is dependent on the type of drug, the weight of the patient, and the dose administered. To study the half-life of aminoglyco sides in trauma patients, a pharmacy researcher recorded the data in Table 1.25 for patients in a critical care facility. The data consist of measurements of dosage per kilogram of weight of the patient, type of drug, either Amikacin or Gentamicin, and the half-life measured 1 hour after administration.

Table 1.25

Half-Life of Aminoglycosides and Dosage by Drug Type

Patient	Drug	Half-Life	Dosage (mg drug/kg patient)
1	G	1.60	2.10
2	A	2.50	7.90
3	G	1.90	2.00
4	G	2.30	1.60
5	A	2.20	8.00
6	A	1.60	8.30
7	A	1.30	8.10
8	A	1.20	8.60
9	G	1.80	2.00
10	G	2.50	1.90
11	A	1.60	7.60
12	A	2.20	6.50
13	A	2.20	7.60
14	G	1.70	2.86
15	A	2.60	10.00
16	A	1.00	9.88
17	G	2.86	2.89
18	A	1.50	10.00
19	A	3.15	10.29
20	A	1.44	9.76
21	A	1.26	9.69
22	A	1.98	10.00
23	A	1.98	10.00
24	A	1.87	9.87
25	G	2.89	2.96
26	A	2.31	10.00
27	A	1.40	10.00
28	A	2.48	10.50
29	G	1.98	2.86
30	G	1.93	2.86
31	G	1.80	2.86
32	G	1.70	3.00
33	G	1.60	3.00
34	G	2.20	2.86
35	G	2.20	2.86
36	G	2.40	3.00
37	G	1.70	2.86
38	G	2.00	2.86
39	G	1.40	2.82
40	G	1.90	2.93
41	G	2.00	2.95
42	A	2.80	10.00
43	A	0.69	10.00

(a) Draw a scatter diagram of half-life versus dose per kilogram, indexed by drug type (use A's and G's). Does there appear to be a difference in the prescription of initial doses in types of drugs?

(b) Does there appear to be a relation between half-life and dosage? Explain.

(c) Find the mean and standard deviation for half-life for the two types of drugs. Does this seem to support the conclusion in part (a)?

Chapter 2 Probability and Sampling Distributions

EXAMPLE 2.1

A quality control specialist for a manufacturing company that makes complex aircraft parts is concerned about the costs generated by defective screws at two points in the production line. These defective screws must be removed and replaced before the part can be shipped. The two points in the production operate independent of each other, but a single part may have defective screws at one or both of the points. The cost of replacing defective screws at each point, as well as the long-term observed proportion of times defective screws are found at each point, is given in Table 2.1.

Table 2.1

Summary of Defective Screws

Point in the Production Line	Proportion of Parts Having Defective Screws	Cost of Replacing Defective Screws
A	0.008	$0.23
B	0.004	$0.69

On a typical day, 1000 parts are manufactured by this production line. The specialist wants to estimate the total cost involved in replacing the screws. This example illustrates the use of a concept called probability in problem solving. While the main emphasis of this chapter is to develop the use of probability for statistical inference, there are other uses such as that illustrated in this example. The solution is given in Section 2.3 where we discuss discrete probability distributions. ■

2.1 Introduction

Up to now, we have used numerical and graphical techniques to describe and summarize sets of data without differentiating between a sample and a population. In Section 1.8 we introduced the idea of using data from a sample to make inferences to the underlying population, which we called statistical inference, and is the subject of most of the rest of this text. Because inferential statistics involves using information obtained from a sample (usually a small portion of the population) to draw conclusions about the population, we can never be 100% sure that our conclusions are correct. That is, we are constantly drawing conclusions under conditions of uncertainty. Before we can understand the methods and limitations of inferential statistics we need to become familiar with uncertainty. The science of uncertainty is known as **probability** or probability theory. This chapter provides some of the tools used in probability theory as measures of uncertainty, and particularly those tools that allow us to make inferences and evaluate the reliability of such inferences.

Subsequent chapters deal with the specific inferential procedures used for solving various types of problems.

In statistical terms, a population is described by a distribution of one or more variables. These distributions have some unique characteristics that describe their location or shape.

DEFINITION 2.1
A **parameter** is a quantity that describes a particular characteristic of the distribution of a variable. For example, the mean of a variable (denoted by μ) is the arithmetic mean of all the observations in the population.

DEFINITION 2.2
A **statistic** is a quantity calculated from data that describes a particular characteristic of the sample. For example, the sample mean (denoted by $\bar{y}$) is the arithmetic mean of the values of the observations of a sample.

In general, statistical inference is the process of using sample statistics to make deductions about a population probability distribution. If such deductions are made on population parameters, this process is called *parametric* statistical inference. If the deductions are made on the entire probability distribution, without reference to particular parameters, the process is called *nonparametric* statistical inference. The majority of this text concerns itself with parametric statistical inference (with the exception of Chapter 13). Therefore, we will use the following definition:

DEFINITION 2.3
Statistical inference is the process of using sample statistics to make decisions about population parameters.

An example of one form of statistical inference is to estimate the value of the population mean by using the value of the sample mean. Another form of statistical inference is to postulate or hypothesize that the population mean has a certain value, and then use the sample mean to confirm or deny that hypothesis. For example, we take a small sample from a large population with unknown mean, μ, and calculate the sample mean, $\bar{y}$, as 5.87. We use the value 5.87 to estimate the unknown value of the population mean. In all likelihood the population mean is not exactly 5.87 since another sample of the same size from the same population would yield a different value for $\bar{y}$. On the other hand, if we were able to say that the true mean, μ, is between two values, say 5.70 and 6.04 there is a larger likelihood that we are correct. What we need is a way to quantify this likelihood. Alternatively, we may hypothesize that μ actually had the value 6.0 and use the sample mean to test this hypothesis. That is, we ask how likely it is that the sample mean was only 5.87 if the true mean has a value of 6? In order to answer this question, we need to explore a way to actually calculate the probability that $\bar{y}$ is as small as 5.87 if $\mu = 6$. We start the discussion of how to evaluate statistical inferences on the population mean in Section 2.5.

Applications of statistical inferences are numerous, and the results of statistical inferences affect almost all phases of today's world. A few examples follow:

1. The results of a public opinion poll taken from a sample of registered voters. The statistic is the sample proportion of voters favoring a candidate or issue. The parameter to be estimated is the proportion of all registered voters favoring that candidate or issue.
2. Testing light bulbs for longevity. Since such testing destroys the product, only a small sample of a manufacturer's total output of light bulbs can be tested for longevity. The statistic is the mean lifetime as computed from the sample. The parameter is the actual mean lifetime of all light bulbs produced.
3. The yield of corn per acre in response to fertilizer application at a test site. The statistic is the mean yield at the test site. The parameter is the mean yield of corn per acre in response to given amounts of the fertilizer when used by farmers under similar conditions.

It is obvious that a sample can be taken in a variety of ways with a corresponding variety in the reliability of the statistical inference. For example, one way of taking a sample to obtain an estimate of the proportion of voters favoring a certain candidate for public office might be to go to that candidate's campaign office and ask workers there if they will vote for that candidate. Obviously, this sampling procedure will yield less than unbiased results. Another way would be to take a well-chosen sample of registered voters in the state and conduct a carefully controlled telephone poll. (We discussed one method of taking such a sample in Section 1.8, and called it a random sample.) The difference in the credibility of the two estimates is obvious, although voters who do not have a telephone may present a problem. For the most part, we will assume that the data we use have come from a random sample.

The primary purpose of this text is to present procedures for making inferences in a number of different applications and evaluating the reliability of the inferences that go with these procedures. This evaluation will be based on the concepts and principles of probability and will allow us to attach a quantitative measure to the reliability of the statistical inferences we make. Therefore, to understand these procedures for making statistical inferences, some basic principles of probability must be understood.

The subject of probability covers a wide range of topics, from relatively simple ideas to highly sophisticated mathematical concepts. In this chapter we use simple examples to introduce only those topics necessary to provide an understanding of the concept of a sampling distribution which is the fundamental tool for statistical inference. For those who find this topic challenging and want to learn more, there are numerous books on the subject (see Ross, 2002).

In examples and exercises in probability (mainly in this chapter) we assume that the population and its parameters are known and compute the probability of obtaining a particular sample statistic. For example, a typical probability problem might be that we have a population with $\mu = 6$ and we want to know the probability of getting a sample mean of 5.87 if we take a sample of ten items from the population. Starting in Chapter 3 we use the principles developed in this chapter to answer the complement of this question. That is, we want to know what are likely values for the population mean if we get a sample mean of 5.87 from a sample of size 10. Or we ask the question, how likely is it that we get a sample mean of 5.87 if the population mean is actually 6? In other words, in examples and exercises in statistical inference, we know the sample values and ask questions concerning the unknown population parameter.

Chapter Preview

The following short preview outlines our development of the concept of a sampling distribution which provides the foundation for statistical inference. Section 2.2 presents the concept of the **probability** of a simple outcome of an experiment, such as the probability of obtaining a head on a toss of a coin. Rules are then given for obtaining the probability of an event, which may consist of several such outcomes, such as obtaining no heads in the toss of five coins.

In Section 2.3, these rules are used to construct **probability distributions**, which are simply listings of probabilities of all events resulting from an experiment, such as obtaining all possible number of heads in the toss of five coins. In Section 2.4, this concept is generalized to define probability distributions for results of experiments that result in continuous numeric variables. Some of these distributions are derived from purely mathematical concepts and require the use of functions and tables to find probabilities.

Finally, Sections 2.5 and 2.6 present the ultimate goal of this chapter, the concept of a **sampling distribution**, which is a probability distribution that describes how a statistic from a random sample is related to the characteristics of the population from which the sample is drawn.

2.2 Probability

The word **probability** means something to just about everyone, no matter what his or her level of mathematical training. In general, however, most people would be hard pressed to give a rigorous definition of probability. We are not going to attempt such a definition either. Instead, we will use a working definition of probability (Definition 2.7) that defines it as a "long-range relative frequency."

For example, if we proposed to flip a fair coin and asked for the probability that the coin will land head side up, we would probably receive the answer "fifty percent," or maybe "one-half." That is, in the long run we would expect about 50% of the time to get a head, the other 50% a tail, although the 50% may not apply exactly for a small number of flips. This same kind of reasoning can be extended to much more complex situations.

EXAMPLE 2.2

Consider a study in which a city health official is concerned with the incidence of childhood measles in parents of child-bearing age in the city. For each couple she would like to know how likely it is that either the mother or father or both have had childhood measles.

Solution For each person the results are similar to tossing a coin. That is, they have either had measles (a head?) or not (a tail?). However, the probability of an individual having had measles cannot be quite as easily determined as the probability of a head in a single toss of a fair coin. However, we can sometimes obtain this probability by using prior studies or census data. For example, suppose that national health statistics indicate that 20% of adults between the ages of 17 and 35 (regardless of sex) have had childhood measles. The city health official may use 0.20 as the probability that an individual in her city has had childhood measles. Even with this value, the official's work is not finished. Recall that she was interested in determining the likelihood of neither, one, or both individuals in the couple having had measles. To answer this question, we must use some of the basic rules of probability. We will introduce these rules, along with the necessary definitions, and eventually answer the question. ■

Definitions and Concepts

DEFINITION 2.4
An **experiment** is any process that yields an observation.

For example, the toss of a fair coin (gambling activities are popular examples for studying probability) is an experiment.

DEFINITION 2.5
An **outcome** is a specific result of an experiment.

In the toss of a coin, a head would be one outcome, a tail the other. In the measles study, one outcome would be "yes," the other "no."

In Example 2.2, determining whether an individual has had measles is an experiment. The information on outcomes for this experiment may be obtained in a variety of ways, including the use of health certificates, medical records, a questionnaire, or perhaps a blood test.

DEFINITION 2.6
An **event** is a combination of outcomes having some special characteristic of interest.

In the measles study, an event may be defined as "one member of the couple has had measles." This event could occur if the husband has and the wife has not had measles, or if the husband has not and the wife has. An event may also be the result of more than one replicate of an experiment. For example, asking the couple may be considered as a combination of two replicates: (1) asking if the wife has had measles and (2) asking if the husband has had measles.

DEFINITION 2.7
The **probability** of an event is the proportion (relative frequency) of times that the event is expected to occur when an experiment is repeated a large number of times under identical conditions.

We will represent outcomes and events by capital letters. Letting A be the outcome "an individual of childbearing age has had measles," then, based on the national health study, we write the probability of A occurring:

$$P(A) = 0.20.$$

Note that any probability has the property

$$0 \leq P(A) \leq 1.$$

This is, of course, a result of the definition of probability as a relative frequency.

DEFINITION 2.8
If two events cannot occur simultaneously, that is, one "excludes" the other, then the two events are said to be **mutually exclusive**.

Note that two individual observations are mutually exclusive. The sum of the probabilities of all the mutually exclusive events in an experiment must be one.

This is apparent because the sum of all the relative frequencies in a problem must be one.

DEFINITION 2.9
The **complement** of an outcome or event A is the occurrence of any event or outcome that precludes A from happening.

Thus, not having had measles is the complement of having had measles. The complement of outcome A is represented by A'. Because A and A' are mutually exclusive, and because A and A' are all the events that can occur in any experiment, the probabilities of A and A' sum to one:

$$P(A') = 1 - P(A).$$

Thus the probability of an individual not having had measles is

$$P(\text{no measles}) = 1 - 0.2 = 0.8.$$

DEFINITION 2.10
Two events A and B are said to be **independent** if the probability of A occurring is in no way affected by event B having occurred or vice versa.

Rules for Probabilities Involving More Than One Event Consider an experiment with events A and B, and $P(A)$ and $P(B)$ are the respective probabilities of these events. We may be interested in the probability of the event "both A and B occur." If the two events are independent, then

$$P(A \text{ and } B) = P(A) \cdot P(B).$$

If two events are not independent, more complex methods must be used (see, for example, Wackerly *et al.*, 2002).

Suppose that we define an experiment to be two tosses of a fair coin. If we define A to be a head on the first toss and B to be a head on the second toss, these two events would be independent. This is because the outcome of the second toss would not be affected in any way by the outcome of the first toss.

Using this rule, the probability of two heads in a row, $P(A \text{ and } B)$, is (0.5)(0.5) = 0.25. In Example 2.2, any incidence of measles would have occurred prior to the couple getting together, so it is reasonable to assume the occurrence of childhood measles in either individual is independent of the occurrence in the other. Therefore, the probability that both have had measles is

$$(0.2)(0.2) = 0.04.$$

Likewise, the probability that neither has had measles is

$$(0.8)(0.8) = 0.64.$$

We are also interested in the probability of the event "either A or B occurs." If two events are mutually exclusive, then

$$P(A \text{ or } B) = P(A) + P(B).$$

Note that if A and B are mutually exclusive then they both cannot occur at the same time; that is, $P(A \text{ and } B) = 0$.

If two events are not mutually exclusive, then

$$P(A \text{ or } B) = P(A) + P(B) - P(A \text{ and } B).$$

We can now use these rules to find the probability of the event "exactly one member of the couple has had measles." This event consists of two mutually exclusive outcomes:

A: husband has and wife has not had measles.
B: husband has not and wife has had measles.

The probabilities of events A and B are

$$P(A) = (0.2)(0.8) = 0.16$$

$$P(B) = (0.8)(0.2) = 0.16.$$

The event "one has" means either of the above occurred, hence

$$P(\text{one has}) = P(A \text{ or } B) = 0.16 + 0.16 = 0.32.$$

In the experiment of tossing two fair coins, events A (a head on the first toss) and event B (a head on the second) are not mutually exclusive events. The probability of getting at least one head in two tosses of a fair coin would be

$$P(A \text{ or } B) = 0.5 + 0.5 - 0.25 = 0.75.$$

EXAMPLE 2.3

One practical application of probability is in the analysis of screening tests in the medical profession. A recent study of the use of steroid hormone receptors using a fluorescent staining technic (sic) in detecting breast cancer was conducted by the Pathology Department of University Hospital in Jacksonville, Florida (Masood and Johnson 1987). The results of the staining technic were then compared with the commonly performed biochemical assay. The staining technic is quick, inexpensive, and, as the analysis indicates, accurate. Table 2.2 shows the result of 42 cases studied. The probabilities of interest are as follows:

1. The probability of detecting cancer, that is, the probability of a true positive test result. This is referred to as the **sensitivity** of the test.
2. The probability of a true negative, that is, a negative on the test for a patient without cancer. This is known as the **specificity** of the test.

Solution To determine the sensitivity of the test, we notice that the test did identify 23 out of the 25 cases; this probability is $23/25 = 0.92$ or 92%. To determine the specificity of the test, we observe that 15 of the 17 negative

Table 2.2

Staining Technic Result

	STAINING TECHNIC RESULTS		
Biochemical Assay Result	**Positive**	**Negative**	**Total**
Positive	23	2	25
Negative	2	15	17
Total	25	17	42

biochemical results were classified negative by the staining technic. Thus the probability is 15/17 = 0.88 or 88%. Since the biochemical assay itself is almost 100% accurate, these probabilities indicate that the staining technic is both sensitive and specific to breast cancer. However, the test is not completely infallible. ■

System Reliability

An interesting application of probability is found in the study of the reliability of a system consisting of two or more components, such as relays in an electrical system or check valves in a water system. The reliability of a system or component is measured by the probability that the system or component will not fail (or that the system will work). We are interested in knowing the reliability of a system given that we know the reliabilities of the individual components. In practice, reliability is often used to determine which design among those possible for the system meets the required specifications. For example, consider a system with two components, say, component A and component B. If the two components are connected in series, as shown in the diagram, then the system will work only if both components work or, conversely, only if both components do not fail.

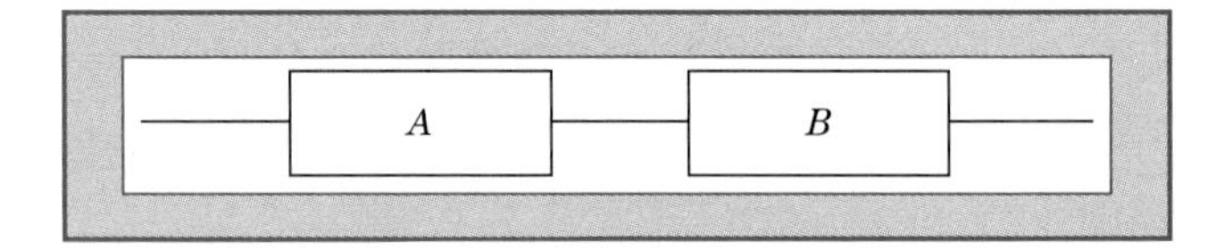

An alternative system that involves two components could be designed as a parallel system. A two-component system with parallel components is shown in the following diagram. In this system, if either of the components fails, the system will still function as long as the other component works. So for the system to fail, both components must fail.

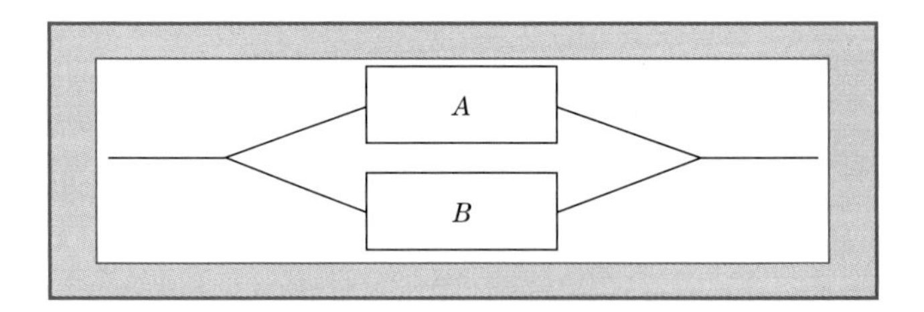

In most practical applications, the probability of failure (often called the failure rate) is known for each component. Then the reliability for each component is 1 – failure rate. Likewise, the reliability of the entire system is 1 – the failure rate of the entire system.

In the series system, if the probability of failure of component A is $P(A)$ and the probability of failure of component B is $P(B)$, then the probability of failure of the system would be $P(\text{system}) = P(A \text{ or } B) = P(A) + P(B) - P(A)P(B)$.

This assumes, of course, that the failure of component A is independent of the failure of component B. The reliability of the system would then be $1 - P(\text{system})$. So, for example, if the probability of component A failing is 0.01 and the probability of component B failing is 0.02, then the probability of the system failing would be $P(\text{system}) = (0.01) + (0.02) - (0.01)(0.02) = 0.0298$. The probability of the system not failing (the reliability) would then be $1 - 0.0298 = 0.9702$.

We could have obtained the same result by considering the probability of each component not failing. Then the probability of the system working would be the probability that both components worked. That is, the probability of the system not failing $= (1 - 0.01)(1 - 0.02) = (0.99)(0.98) = 0.9702$.

In the parallel system, the probability of failure is simply the probability that both components fail, that is, $P(\text{system}) = P(A \text{ and } B) = P(A)P(B)$. The reliability is then $1 - P(A)P(B)$. Assuming the same failure rates, the probability of the system failing is $(0.01)(0.02) = 0.0002$. The probability that the system works (reliability) is $1 - 0.0002 = 0.9998$.

Note that it is more difficult to calculate the reliability of the system by considering the reliability of each component. That is, the probability of the system working is the probability that one or more of the components work. This probability could be calculated by the following:

$$\begin{aligned}
P(\text{system works}) &= P(A \text{ works and } B \text{ fails}) + P(A \text{ fails and } B \text{ works}) \\
&\quad + P(A \text{ and } B \text{ work}) \\
&= [(0.99)(0.02) + (0.01)(0.98) + (0.99)(0.98)] = 0.0198 \\
&\quad + 0.0098 + 0.9702 \\
&= 0.9998.
\end{aligned}$$

Note that this system only needs one component working; the other one is redundant. Hence, systems with this design are often called **redundant systems.** To illustrate the need for redundant systems, consider a space shuttle rocket. It would not be surprising for this rocket to have as many as 1000 components. If these components were all connected in series, then the system reliability might be much lower than would be tolerated. For example, even if the reliability of an individual component was as high as 0.999, the reliability of the entire rocket would be only 0.368! Obviously, more complex arrangements of components can be used, but the same basic principles of probability can be used to evaluate the reliability of the system.

Random Variables

Events of major interest for most statistical inferences are expressed in numerical terms. For example, in Example 2.2 we are primarily interested in the number of adults in a couple that have had measles rather than simply the fact that an adult had measles as a child.

Table 2.3

A Probability Distribution

Y	**Probability**
0	0.64
1	0.32
2	0.04

DEFINITION 2.11
A **random variable** is a rule that assigns a numerical value to an outcome of interest.

This variable is similar to those discussed in Chapter 1, but is not exactly the same. Specifically, a random variable is a number assigned to each outcome of an experiment. In this case, as in many other applications, outcomes are already numerical in nature, and all we have to do is record the value. For others we may have to assign a numerical value to the outcome.

In our measles study we define a random variable Y as the number of parents in a married couple who have had childhood measles. This random variable can take values of 0, 1, and 2. The probability that the random variable takes on a given value can be computed using the rules governing probability. For example, the probability that $Y = 0$ is the same as the probability that neither individual in the married couple has had measles. We have previously determined that to be 0.64. Similarly, we have the probability for each of the possible values for Y. These values are summarized in tabular form in Table 2.3.

DEFINITION 2.12
A **probability distribution** is a definition of the probabilities of the values of a random variable.

The list of probabilities given in Table 2.3 is a probability distribution.

Note the similarity of the probability distribution to the empirical relative frequency distributions of sets of data discussed in Chapter 1. Those distributions were the results of samples from populations and, as noted in Section 1.4, are often called *empirical probability distributions*. On the other hand, the probability distribution we have presented above is an exact picture of the population if the 20% figure is correct. For this reason it is also called a *theoretical probability distribution*. The theoretical distribution is a result of applying mathematical (probability) concepts, while the empirical distribution is computed from data obtained as a result of sampling. If the sampling could be carried out forever, that is, the sample becomes the population, then the empirical distribution would be identical to the theoretical distribution.

In Chapter 1 we found it convenient to use letters and symbols to denote variables. For example, y_i was used to represent the ith observed value of the variable Y in a data set. A random variable is not observed, but is defined for all values in the distribution; however, we use a similar notation for random variables. That is, a random variable is denoted by the capital letter, Y, and specific realizations, such as those shown in Table 2.3, are denoted by the lower case letter, y. A method of notation commonly used to represent the probability that the random variable Y takes on the specific value y is $P(Y = y)$, often written $p(y)$. For example, the random variable describing the number of parents having had measles is denoted by Y, and has values $y = 0$, 1, and 2. Then $p(0) = P(Y = 0) = 0.64$ and so forth. This level of specificity is necessary for our introductory discussion of probability and probability distributions. After Chapter 3 we will relax this specificity and use lower case letters exclusively.

EXAMPLE 2.4

Consider the experiment of tossing a fair coin twice and observing the random variable $Y =$ number of heads showing. Thus Y takes on the values 0, 1, or 2. We are interested in determining the probability distribution of Y.

Solution The probability distribution of Y, the number of heads, is obtained by applying the probability rules, and is seen in Table 2.4. ■

Table 2.4

P (Number of Heads)

y	$p(y)$
0	1/4
1	2/4
2	1/4

Suppose that we wanted to define another random variable that measured the number of times the coin repeated itself. That is, if a head came up on the first toss and a head on the second, the variable would have a value of two. If a head came up on the first and a tail the second, the variable would have a value 1.

Let us define X as the number of times the coin repeats. Then X will have values 1 and 2. The probability distribution of X is shown in Table 2.5. The reader may want to verify the values of $p(x)$.

Table 2.5

P (Number of Repeats)

x	$p(x)$
1	1/2
2	1/2

For our discussion in this text, we classify random variables into two types as defined in the following definitions:

DEFINITION 2.13
A **discrete random variable** is one that can take on only a countable number of values.

DEFINITION 2.14
A **continuous random variable** is one that can take on any value in an interval.

The random variables defined in Examples 2.3 and 2.4 are discrete. Height, weight, and time are examples of continuous random variables.

Probability distributions are also classified as continuous or discrete, depending on the type of random variable the distribution describes.

Before continuing to the subject of sampling distributions, we will examine several examples of discrete and continuous probability distributions with considerable emphasis on the so-called normal distribution, which we will use extensively throughout the book.

Table 2.6

A Discrete Probability Distribution

y	$p(x)$
1	1/6
2	2/6
3	3/6

2.3 Discrete Probability Distributions

A discrete probability distribution displays the probability associated with each value of the random variable Y. This display can be presented as a table, as the previous examples illustrate, as a graph, or as a formula. For example, the probability distribution in Table 2.6 can be expressed in formula form, also called a function, as

$$p(y) = y/6, \qquad y = 1, 2, 3,$$

$$p(y) = 0, \text{ for all other values of } y.$$

It can be displayed in graphic form as shown in Fig. 2.1.

Figure 2.1

Bar Chart of Probability Distribution in Table 2.6

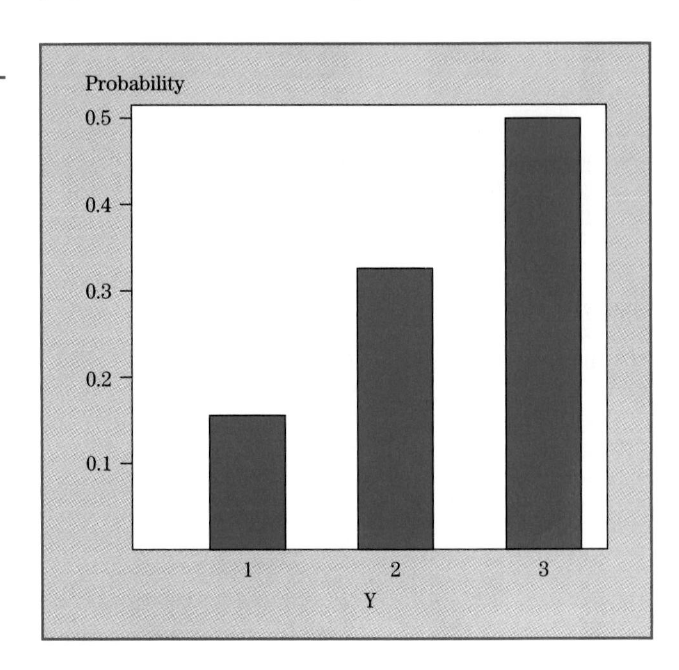

Properties of Discrete Probability Distributions

Any formula $p(y)$ that satisfies the following conditions for discrete values of a variable Y can be considered a probability distribution:

$$0 \leq p(y) \leq 1$$

$$\sum p(y) = 1.$$

All probability distributions presented above are seen to fulfill both conditions.

Descriptive Measures for Probability Distributions

Because empirical and theoretical probability distributions can both be described by similar tables of relative frequencies and/or histograms, it is logical to expect that numerical descriptors of both are the same. Since a theoretical distribution essentially describes a population, the descriptors of such distributions are called *parameters*. For example, we use the Greek letters μ and σ

for the mean and standard deviation of a theoretical probability distribution just as we did for an empirical probability distribution.

Numerically the parameters of a discrete probability distribution are calculated using formulas similar to those used for empirical probability distributions shown in Section 1.5. Specifically,

$$\mu = \sum yp(y),$$

and its variance, which we denote by σ^2, is computed as

$$\sigma^2 = \sum (y - \mu)^2 p(y),$$

where the sums are over all values of Y.

For example, if the 20% figure discussed in the measles example is valid, the mean number of individuals in a couple having had measles calculated from the theoretical probability distribution is

$$\mu = 0(0.64) + 1(0.32) + 2(0.04) = 0.4.$$

That is, the average number of individuals per couple having had measles is 0.4 for the whole city. The variance is

$$\begin{aligned}\sigma^2 &= (0 - 0.4)^2(0.64) + (1 - 0.4)^2(0.32) + (2 - 0.4)^2(0.04)\\ &= 0.1024 + 0.1152 + 0.1024 = 0.320,\end{aligned}$$

and $\sigma = 0.566$.

The mean of a probability distribution is often called the **expected value** of the random variable. For example, the expected number of individuals in a couple having had measles is 0.4. This is a "long-range expectation" in the sense that if we sampled a large number of couples, the expected (average) number of individuals having had measles would be 0.4. Note that the expected value can be (and often is) a value that the random variable may never attain.

Solution to Example 2.1 We can now solve the problem facing the specialist in Example 2.1. The random variable is the cost of replacing screws on a single part for the four outcomes, which we calculate as follows:

Outcome	Probability	Cost
Screw A defective	0.008	\$0.23
Screw B is defective	0.004	\$0.69
Both screws defective	(0.008)(0.004) = 0.000032	\$0.92
Neither screw is defective	1 − 0.008 − 0.004 − 0.000032 = 0.987968	\$0.00

We can now find the expected cost of replacing defective screws on one part:

$$\mu = 0.23(0.008) + 0.69(0.004) + 0.92(0.000032) + 0(0.987968) = 0.00463.$$

There are 1000 parts produced in a day; hence the expected daily cost is 1000(\$0.00463) = \$4.63. ■

The Discrete Uniform Distribution

Suppose the possible values of a random variable from an experiment are a set of integer values occurring with the same frequency. That is, the integers 1 through k occur with equal probability. Then the probability of obtaining any particular integer in that range is $1/k$ and the probability distribution can be written

$$p(y) = 1/k, \quad y = 1, 2, \ldots, k.$$

This is called the **discrete uniform** (or rectangular) distribution, and may be used for all populations of this type, with k depending on the range of existing values of the variable. Note that we are able to represent many different distributions with one function by using a letter (k in this case) to represent an arbitrary value of an important characteristic. This characteristic is the only thing that differs between the distributions, and is called a **parameter** of the distribution. All probability distributions are characterized by one or more parameters, and the descriptive parameters, such as the mean and variance, are known functions of those parameters. For example, for this distribution

$$\mu = (k + 1)/2$$

and

$$\sigma^2 = (k^2 - 1)/12.$$

A simple example of an experiment resulting in a random variable having the discrete uniform distribution consists of tossing a fair die. Let Y be the random variable describing the number of spots on the top face of the die. Then

$$p(y) = 1/6, \quad y = 1, 2, \ldots, 6,$$

which is the discrete uniform distribution with $k = 6$. The mean of Y is

$$\mu = (6 + 1)/2 = 3.5,$$

and the variance is

$$\sigma^2 = (36 - 1)/12 = 2.917.$$

Note that this is an example where the random variable can never take the mean value.

EXAMPLE 2.5

Simulating a Distribution The discrete uniform distribution is frequently used in simulation studies. A simulation study is exactly what it sounds like, a study that uses a computer to simulate a real phenomenon or process as closely as possible. The use of simulation studies can often eliminate the need for costly experiments and is also often used to study problems where actual experimentation is impossible.

When the process being simulated requires the use of a probability distribution to describe it, the technique is often referred to as a Monte Carlo method.

For example, Monte Carlo methods have been used to simulate collisions between photons and electrons, the decay of radioactive isotopes, and the effect of dropping an atomic bomb on a city.

The basic ingredient of a Monte Carlo simulation is the generation of random numbers (see, for example, Owen, 1962). Random numbers can, for example, be generated to consist of single digits having the discrete uniform distribution with $k = 10$. Using the digits 0 through 9, such random digits can be used to simulate the outcomes of Example 2.2. For each simulated interview we generate a random digit. If the value of the digit is 0 or 1, the outcome is "had childhood measles"; otherwise (digits 2 through 9) the outcome is "did not." The outcome "had" then occurs with a probability of 0.2. The result of the experiment involving a single couple is then simulated by using a pair of such integers, one for each individual.

Solution Simulation studies usually involve large numbers of simulated events, but for illustration purposes we use only 10 pairs. Assume that we have obtained the following 10 pairs of random numbers (from a table or generated by a computer):

15 38 68 39 49 54 19 79 38 14

In the first pair (15), the first digit "1" signifies one "has," while the second digit "5" indicates "has not"; hence, for this couple, $y = 1$. For the second pair, $y = 0$, and so forth. The relative frequency distribution for this simulated sample of ten pairs is shown in Table 2.7.

Table 2.7

Simulation of Measles Probabilities

y	$P(y)$
0	0.7
1	0.3
2	0

This result is somewhat different from the theoretical distribution obtained with the use of probability theory because considerable variability is expected in small samples. A sample of 1000 would come much closer but would still not produce the theoretical distribution exactly. ■

The Binomial Distribution

In several examples in this chapter, an outcome has included only two possibilities. That is, an individual had or had not had childhood measles, a coin landed with head or tail up, or a tested specimen did or did not have cancer cells. This dichotomous outcome is quite common in experimental work. For example, questionnaires quite often have questions requiring simple yes or no responses, medical tests have positive or negative results, banks either succeed or fail after the first 5 years, and so forth. In each of these cases, there are two outcomes for which we will arbitrarily adopt the generic labels "success" and "failure." The measles example is such an experiment where each individual in a couple is a "trial," and each trial produces a dichotomous outcome (yes or no).

The **binomial** probability distribution describes the distribution of the random variable Y, the number of successes in n trials, if the experiment satisfies the following conditions:

1. The experiment consists of n identical trials.
2. Each trial results in one of two mutually exclusive outcomes, one labeled a "success," the other a "failure."
3. The probability of a success on a single trial is equal to p. The value of p remains constant throughout the experiment.
4. The trials are independent.

The formula or function for computing the probabilities for the binomial probability distribution is given by

$$p(y) = \frac{n!}{y!(n-y)!}\, p^y(1-p)^{n-y}, \quad \text{for } y = 0, 1, \ldots, n.$$

The notation $n!$, called the factorial of n, is the quantity obtained by multiplying n by every nonzero integer less than n. For example $7! = 7 \cdot 6 \cdot 5 \cdot 4 \cdot 3 \cdot 2 \cdot 1 = 5040$. By definition, $0! = 1$.

Derivation of the Binomial Probability Distribution Function The binomial distribution is one that can be derived with the use of the simple probability rules presented in this chapter. Although memorization of this derivation is not needed, being able to follow it provides an insight into the use of probability rules. The formula for the binomial probability distribution can be developed by first observing that the $p(y)$ is the probability of getting exactly y successes out of n trials. We know that there are n trials so there must be $(n - y)$ failures occurring at the same time. Because the trials are independent, the probability of y successes is the product of the probabilities of the y individual successes, which is p^y and the probability of $(n - y)$ failures is $(1 - p)^{n-y}$. Then the probability of y successes and $(n - y)$ failures is $p^y(1 - p)^{n-y}$.

However, this is the probability of only one of the many sequences of y successes and $(n - y)$ failures and the definition of $p(y)$ is the probability of any sequence of y successes and $(n - y)$ failures. We can count the number of such sequences using a counting rule called **combinations**. This rule says that there are

$$\binom{n}{y} = \frac{n!}{y!(n-y)!}$$

ways that we can get y items from n items. Thus, if we have 5 trials there are

$$\frac{5!}{2!(5-2)!} = \frac{5 \cdot 4 \cdot 3 \cdot 2 \cdot 1}{(2 \cdot 1)(3 \cdot 2 \cdot 1)} = 10$$

ways of arranging 2 successes and 3 failures. (The reader may want to list these and verify that there are ten of them.)

The probability of y successes, then, is obtained by repeated application of the addition rule. That is, the probability of y successes is obtained by multiplying the probability of a sequence by the number of possible sequences, resulting in the above formula.

Note that the measles example satisfies the conditions for a binomial experiment. That is, we label "having had childhood measles" a success, the number of trials is two (a couple is an experiment, and an individual a trial), and $p = 0.2$, using the value from the national health study. We also assume that each individual has the same chance of having had measles as a child, hence p is constant for all trials, and we have previously assumed that the incidence of measles is independent between the individuals. The random variable Y is the number in each couple having had measles. Using the binomial distribution function, we obtain

$$P(Y = 0) = \frac{2!}{0!(2-0)!}(0.2)^0(0.8)^{2-0} = 0.64,$$

$$P(Y = 1) = \frac{2!}{1!(2-1)!}(0.2)^1(0.8)^{2-1} = 0.32,$$

$$P(Y = 2) = \frac{2!}{2!(2-2)!}(0.2)^2(0.8)^{2-2} = 0.04.$$

These probabilities agree exactly with those that were obtained earlier from basic principles, as they should.

Computations involving the binomial distribution can become quite tedious, especially if n is large. Fortunately, a large sample approximation that works well for even moderately large samples is available. The use of this approximation is presented in Section 2.5 and additional applications are presented in subsequent chapters.

The binomial distribution has only one parameter, p (n is usually considered a fixed value). The mean and variance of the binomial distribution are expressed in terms of p as

$$\mu = np,$$

$$\sigma^2 = np(1 - p).$$

For our health study example, $n = 2$ and $p = 0.2$ gives

$$\mu = 2(0.2) = 0.4,$$

$$\sigma^2 = (2)(0.2)(0.8) = 0.32.$$

Again these results are identical to the values previously computed for this example.

The Poisson Distribution

The binomial distribution describes the situation where observations are assigned to one of two categories, and the measurement of interest is the frequency of occurrence of observations in each category. Some data naturally occur as frequencies, but do not necessarily have the category assignment. Examples of such data include the monthly number of fatal automobile accidents in a city, the number of bacteria on a microscope slide, the number of fish caught in a trawl, or the number of telephone calls per day to a switchboard. In a sense such frequencies may be thought of as binomial data without any "failures." The analysis of such data can be addressed using the **Poisson** distribution.

Consider the variable "number of fatal automobile accidents in a given month." Since an accident can occur at any split second of time, there is essentially an infinite number of chances for an accident to occur. If we consider the event "a fatal accident occurs" as a success (!), we have a binomial experiment in which n is infinite. However, the probability of a fatal accident occurring at any given instant is essentially zero. We then have a binomial experiment with a near infinite sample and an almost zero value for p, but np, the number of occurrences, is a finite number. Actually, the formula for the Poisson distribution can be derived by finding the limit of the binomial formula as n approaches infinity and p approaches zero (Wackerly *et al.*, 1996).

The formula for calculating probabilities for the Poisson distribution is

$$P(y) = \frac{\mu^y e^{-\mu}}{y!}, \quad y = 0, 1, 2, \ldots,$$

where y represents the number of occurrences in a fixed time period and μ is the mean number of occurrences in the same time period. The letter e is the Naperian constant, which is approximately equal to 2.71828. For the Poisson distribution both the mean and variance have the value μ.

Use of the formula for calculating probabilities is not too difficult for small y and μ, particularly when using calculators with exponentiation capabilities. Tables for limited ranges of μ are available (for example, Ott, 1993, Appendix Table 7).

EXAMPLE 2.6

Operators of toll roads and bridges need information for staffing tollbooths so as to minimize queues (waiting lines) without using too many operators. Assume that in a specified time period the number of cars per minute approaching a tollbooth has a mean of 10. Traffic engineers are interested in the probability that exactly 11 cars approach the tollbooth in the minute from 12 noon to 12:01.

$$p(11) = \frac{10^{11} e^{-10}}{11!} = 0.114.$$

Thus, there is about an 11% chance that exactly 11 cars would approach the tollbooth the first minute after noon.

Assume that an unacceptable queue will develop when 14 or more cars approach the tollbooth in any minute. The probability of such an event can be computed as the sum of probabilities of 14 or more cars approaching the tollbooth, or more practically by calculating the complement. That is, $P(Y \geq 14) = 1 - P(Y \leq 13)$. We can use the above formula or a computer package with the Poisson option such as Microsoft Excel. Using Excel we find the $P(Y \leq 13) = 0.8645$ or the resulting probability is $1 - 0.8645 = 0.1355$.

2.4 Continuous Probability Distributions

When the random variable of interest can take on any value in an interval, it is called a continuous random variable. Continuous random variables differ from discrete random variables, and consequently continuous probability distributions differ from discrete ones and must be treated separately. For example, every continuous random variable has an infinite, uncountable number of possible values (any value in an interval). Therefore, we must redefine our concept of relative frequency to understand continuous probability distributions. The following list should help in this understanding.

Characteristics of a Continuous Probability Distribution

The characteristics of a continuous probability distribution are as follows:

1. The graph of the distribution (the equivalent of a bar graph for a discrete distribution) is usually a smooth curve. A typical example is seen in Fig. 2.2. The curve is described by an equation or a function that we call $f(y)$. This equation is often called the **probability density** and corresponds to the $p(y)$ we used for discrete variables in the previous section (see additional discussion following).
2. The total area under the curve is one. This corresponds to the sum of the probabilities being equal to 1 in the discrete case.
3. The area between the curve and horizontal axis from the value a to the value b represents the probability of the random variable taking on a value in the interval (a, b). In Fig. 2.2 the area under the curve between the values -1 and 0.5, for example, is the probability of finding a value in this interval.

Figure 2.2

Graph of a Continuous Distribution

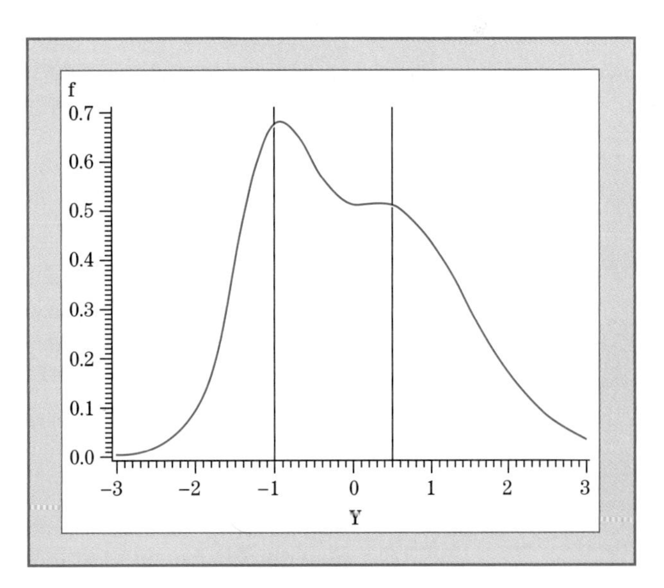

This corresponds to adding probabilities of mutually exclusive outcomes from a discrete probability distribution.

There are similarities but also some important differences between continuous and discrete probability distributions. Some of the most important differences are as follows:

1. The equation $f(y)$ does not give the probability that $Y = y$ as did $p(y)$ in the discrete case. This is because Y can take on an infinite number of values (any value in an interval), and therefore it is impossible to assign a probability value for each y. In fact the value of $f(y)$ is not a probability at all; hence $f(y)$ can take any nonnegative value, including values greater than 1.
2. Since the area under any curve corresponding to a single point is (for practical purposes) zero, the probability of obtaining exactly a specific value is zero. Thus, for a continuous random variable, $P(a \leq Y \leq b)$ and $P(a < Y < b)$ are equivalent, which is certainly not true for discrete distributions.
3. Finding areas under curves representing continuous probability distributions involves the use of calculus and may become quite difficult. For some distributions, areas cannot even be directly computed and require special numerical techniques. For this reason, the areas required to calculate probabilities for the most frequently used distributions have been calculated and appear in tabular form in this and other texts, as well as in books devoted entirely to tables (for example, Pearson and Hartley, 1972). Of course statistical computer programs easily calculate such probabilities.

In some cases, recording limitations may exist that make continuous random variables look as if they are discrete. The round-off of values may result in a continuous variable being represented in a discrete manner. For example, people's weight is almost always recorded to the nearest pound, even though the variable weight is conceptually continuous. Therefore, if the variable is continuous, then the probability distribution describing it is continuous, regardless of the type of recording procedure. As in the case of discrete distributions, several common continuous distributions are used in statistical inference. This section discusses most of the distributions used in this text.

The Continuous Uniform Distribution

A very simple example of a continuous distribution is the continuous uniform or rectangular distribution. Assume a random variable Y has the probability distribution shown in Fig. 2.3. The equation

$$f(y) = 1/(b - a), \quad a \leq y \leq b$$

$$= 0, \quad \text{elsewhere}$$

describes the distribution of such a random variable. Note that this equation describes a straight line, and the area under this line above the horizontal axis

Figure 2.3

The Uniform Distribution

is rectangular in shape as can be seen by the graph in Fig. 2.3. The distribution parameters are a and b, and the graph is a rectangle with width $(b - a)$ and height $1/(b - a)$.

This distribution can be used to describe many processes, including, for example, the error due to rounding. Under the assumption that any real number may occur, rounding to the nearest whole number introduces a round-off error whose value is equally likely between $a = -0.5$ and $b = +0.5$.

The continuous uniform distribution is also extensively used in simulation studies in a manner similar to the discrete uniform distribution. Areas under the curve of the rectangular distribution can be computed using geometry. For example, the total area under the curve is simply the width times the height or

$$\text{area} = \frac{1}{(b-a)} \cdot (b-a) = 1.$$

In a similar manner, other probabilities are computed by finding the area of the desired rectangle. For example, the probability $P(c < Y < d)$, where both c and d are in the interval (a, b), is equal to $(d - c)/(b - a)$.

Principles of calculus are used to derive formulas for the mean and variance of the rectangular distribution in terms of the distribution parameters a and b and are

$$\mu = (a + b)/2$$

and

$$\sigma^2 = (b - a)^2/12.$$

The Normal Distribution

By far the most often used continuous probability distribution is the normal or Gaussian distribution. The normal distribution is described by the equation

$$f(y) = \frac{1}{\sqrt{2\pi}\,\sigma} e^{-(y-\mu)^2/2\sigma^2}, \quad -\infty < y < \infty,$$

where $e \approx 2.71828$, the Naperian constant.

This function is quite complicated and is never directly used to calculate probabilities. However, several interesting features can be determined from

Figure 2.4

Standard Normal Distribution

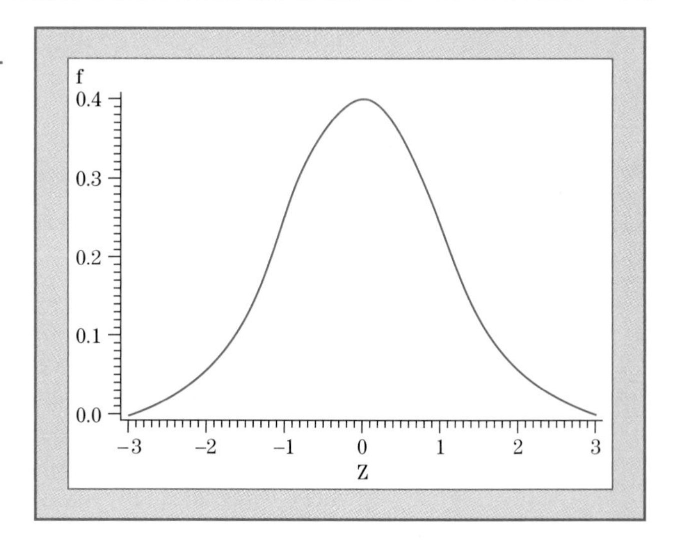

the function without really evaluating it. These features can be summarized as follows:

1. The random variable Y can take on any value from $-\infty$ to $+\infty$.
2. The distribution has only two parameters μ and σ^2 (or σ). These are, in fact, the mean and variance (or standard deviation) of the distribution. Thus, knowing the values of these two parameters completely determines the distribution. The fact that these parameters are also the two most frequently used descriptive measures is a major reason why the normal distribution is so popular.
3. The distribution is bell shaped and symmetric about the mean. This is apparent in the graph of a normal distribution with $\mu = 0$ and $\sigma = 1$, given in Fig. 2.4, and has resulted in the normal distribution being referred to often as the "bell curve."

The primary use of probability distributions is to find probabilities of the occurrence of specified values of the random variable. For example, if it is known that the weights of four-year-old boys can be described by a normal distribution with a mean of 40 lbs and a standard deviation of 3, it may be of interest to determine the probability that a randomly picked four-year-old boy weighs less than 30 lbs. Unfortunately the actual function describing the normal probability distribution (and most other continuous distributions) is much too complicated to easily use to calculate probabilities. Therefore, such probabilities must be obtained by the use of tables or by computer programs which, incidentally, almost always use numerical approximations to the actual distribution functions to calculate probabilities.

Although most of the probabilities associated with various statistical inferences are produced by the computer program that does the analysis, the use of a table for obtaining probabilities of a normally distributed random variable is presented here in some detail. We do this not so much because this method is often used, but rather to help in the interpretation of the probabilities produced by computer outputs.

Since any specific normal distribution is defined by the two parameters, μ and σ, each of which can take on an infinite number of values, it would seem that we need an infinite number of tables. Fortunately normal distributions can easily be *standardized*, which allows us to use a single table for any normal distribution.

All probabilities (areas under the curve) associated with a specific value of the normally distributed variable relate *exactly* to the distance from that value to the mean (μ) as measured in standard deviation (σ) units. For example consider the two normal distributions shown in Figs. 2.5 and 2.6. The one in

Figure 2.5

Area of a Normal Distribution. Area to Right of 20 with μ = 10 and σ = 10

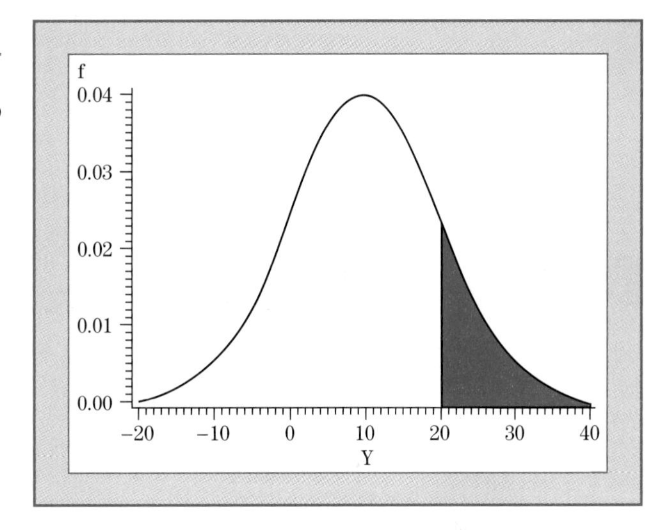

Figure 2.6

Area of a Normal Distribution. Area to Right of 102 with μ = 100 and σ = 2

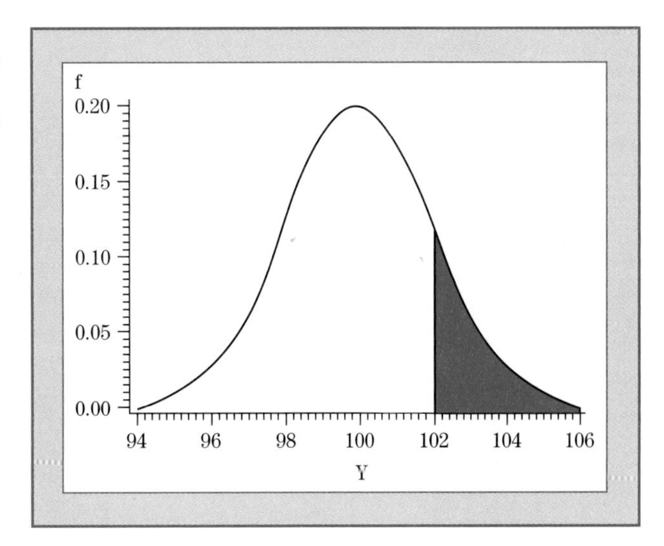

Fig. 2.5 has $\mu = 10$ and $\sigma = 10$, and the one in Fig. 2.6 has $\mu = 100$ and $\sigma = 2$. In both figures, the shaded area is that for $Y > (\mu + \sigma)$; that is, $Y > (10 + 10) = 20$ for Fig. 2.5 and $Y > (100 + 2) = 102$ for Fig. 2.6. The appearance from the plots (supported by mathematical calculations) indicates that both areas are the same. The areas of interest for both variables are those to the right of one standard deviation from the mean. It is this characteristic of the normal distribution that allows the use of a single table to compute probabilities for a normal distribution with any mean and variance. The table used for this purpose is that for $\mu = 0$ and $\sigma = 1$, which is called the **standard normal distribution**. The random variable associated with this distribution is usually denoted by Z. Areas for a normal distribution for a random variable Y with any mean and variance are found by performing a simple transformation of origin and scale. This transformation, called the standardizing transformation, converts the variable Y, which has mean μ and standard deviation σ, to the variable Z, which has the standard normal distribution. This transformation is written

$$Z = \frac{Y - \mu}{\sigma}.$$

Calculating Probabilities Using the Table of the Normal Distribution

The use of the table of probabilities for the normal distribution is given here in some detail. Although you will rarely use these procedures after leaving this chapter, they should help you understand and use tables of probabilities of other distributions as well as appreciate what computer outputs mean.

A table of probabilities for the standard normal distribution is given in Appendix Table A.1. This table gives the area to the right (larger than) of Z for values of z from -3.99 to $+4.00$. Because of the shape of the normal distribution, the area and hence the probability values are almost zero outside this range. Figure 2.7 illustrates the use of the table to obtain standard

Figure 2.7

Area to the Right of 0.9

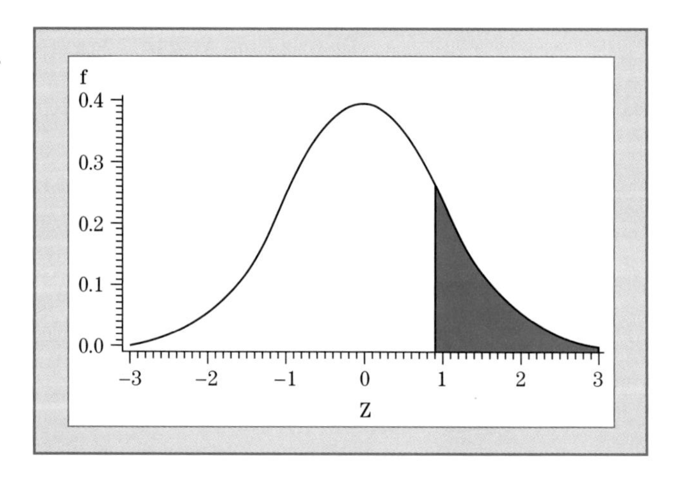

normal probabilities. According to the table, the area to the right of $z = 0.9$ is 0.1841, which is the shaded area in Fig. 2.7.

Obviously we do not always want "areas to the right." The characteristics of the normal distribution allow the following rules to "make the table work":

1. Since the standard normal distribution is symmetric about zero, $P(Z > z) = P(Z < -z)$. This is illustrated later in Fig. 2.11 where the two shaded areas are equal.
2. Since the area under the entire curve is one,

$$P(Z < z) = 1 - P(Z > z).$$

   This is true regardless of the value of z.
3. We may add or subtract areas to get probabilities associated with a combination of values. For example,

$$P(-1 < Z < 1.5) = P(Z > -1) - P(Z > 1.5) = 0.8413 - 0.0668 = 0.7745.$$

This is illustrated in Example 2.9.

With these rules the standard normal table can be used to calculate any desired probability associated with a standard normal distribution, and with the help of the standardization transformation, for any normal distribution with known mean and standard deviation.

EXAMPLE 2.7

Find the area to the right of 2.0; that is, $P(Z \geq 2.0)$.

Solution It helps to draw a picture such as Fig. 2.8. The desired area is the shaded area, which can be directly obtained from the table as 0.0228. Therefore, $P(Z > 2.0) = 0.0228$. ■

Figure 2.8

Area to the Right of 2.0

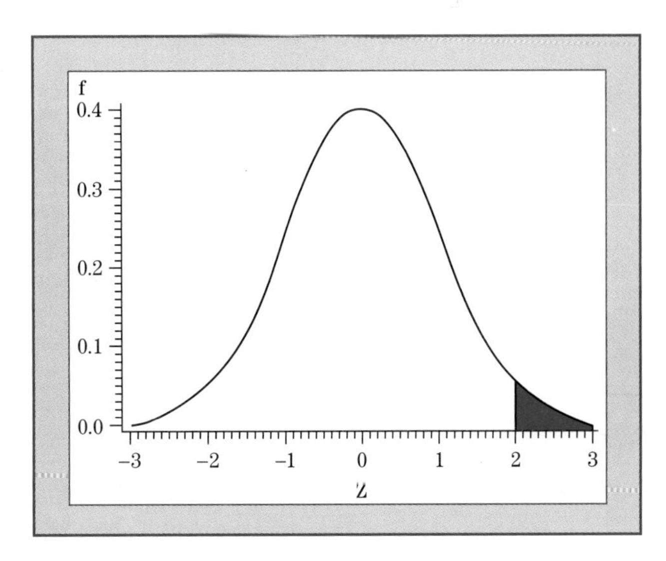

EXAMPLE 2.8

Find the area to the left of −0.5; that is, $P(Z < -0.5)$.

Solution In Fig. 2.9 this is the shaded area. From the table the area to the right of −0.5 is 0.6915. The desired probability is the area to the left; that is, $(1 - 0.6915) = 0.3085$. Alternatively, we can use the symmetry of the normal distribution and find the equivalent area to the right of +0.5. ■

Figure 2.9

Area to the Left of −0.5

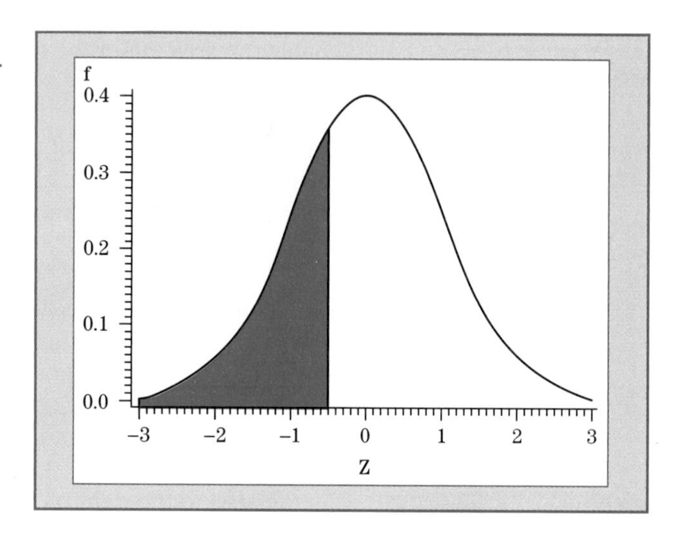

Figure 2.10

Area Between −1.0 and 1.5

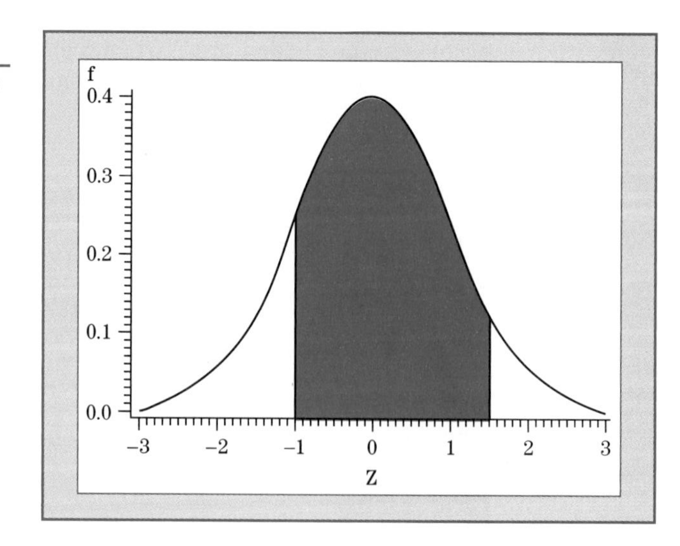

EXAMPLE 2.9

Find $P(-1.0 < Z < 1.5)$.

Solution In Fig. 2.10, the desired area is between −1.0 and 1.5 (shaded). This is obtained by subtracting the area from 1.5 to $+\infty$ from the area from

-1 to $+\infty$. That is,

$$P(-1 < Z < 1.5) = P(Z > -1) - P(Z > 1.5).$$

From the table, the area from 1.5 to ∞ is 0.0668, and the area from -1 to ∞ is 0.8413. Therefore, the desired probability is $0.8413 - 0.0668 = 0.7745$. ■

EXAMPLE 2.10

Sometimes we want to find the value of z associated with a certain probability. For example, we may want to find the value of z that satisfies the requirement $P(|Z| > z) = 0.10$.

Solution Figure 2.11 shows the desired Z values where the total area outside of the vertical lines is 0.10. Due to symmetry the desired value of z satisfies the statement $P(Z > z) = 0.05$. The procedure is to search the table for a value of z such that its value is exceeded with probability 0.05. No area of exactly 0.05 is seen in the table, and the nearest are

$$P(Z > 1.64) = 0.0505,$$

$$P(Z > 1.65) = 0.0495.$$

We can approximate a more exact value by interpolation, which gives $z = 1.645$. ■

Figure 2.11

Symmetry of the Normal Distribution

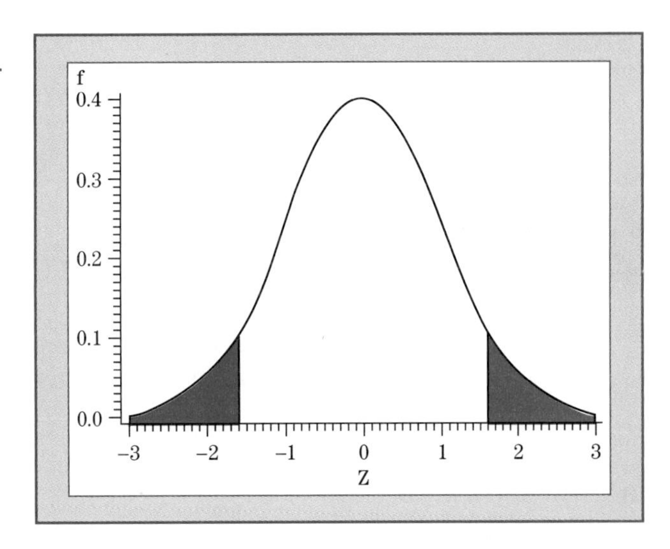

We will often be concerned with finding values of z for given probability values when we start using the normal distribution in statistical inference. To make the writing of formulas easier, we will adopt a form of notation often called the z_α notation. According to this notation, z_α is the value of z such that

$$P(Z > z_\alpha) = \alpha.$$

This definition results in the equivalent statements

$$P(Z < -z_\alpha) = \alpha$$

and, because of the symmetry of the normal distribution,

$$P(-z_{\alpha/2} < Z < z_{\alpha/2}) = 1 - \alpha.$$

Appendix Table A.1A gives a small set of z values for some frequently used probabilities. From this table we can see that the z value exceeded with probability 0.05 (or $z_{0.05}$) is 1.64485.

Finding probabilities associated with a normal distribution other than the standard normal is accomplished in two steps. First use the standardization transformation. As we have noted, this transformation converts a normally distributed random variable having mean μ and variance σ^2 to the standard normal variable having mean zero and variance one. The transformation is

$$Z = \frac{(Y - \mu)}{\sigma},$$

and the resulting Z variable is often called a standard score. The second step is to find the areas as we have already done.

EXAMPLE 2.11

Suppose that Y is normally distributed with $\mu = 10$ and $\sigma^2 = 20$ (or $\sigma = 4.472$).

(a) What is $P(Y > 15)$?
(b) What is $P(5 < Y < 15)$?
(c) What is $P(5 < Y < 10)$?

Solution

(a) **Step 1:** Find the corresponding value of z:

$$z = (15 - 10)/4.472 = 1.12.$$

Step 2: Use the table and find $P(Z > 1.12) = 0.1314$.

(b) **Step 1:** Find the two corresponding values of z:

$$z = (15 - 10)/4.472 = 1.12,$$

$$z = (5 - 10)/4.472 = -1.12.$$

Step 2: From the table, $P(Z > 1.12) = 0.1314$, and $P(Z > -1.12) = 0.8686$, and by subtraction $P(-1.12 < Z < 1.12) = 0.8686 - 0.1314 = 0.7372$.

(c) **Step 1:** $z = (10 - 10)/4.472 = 0$, and
$z = (5 - 10)/4.472 = -1.12$.

Step 2: $P(Z > 0) = 0.5000$, and
$P(Z > -1.12) = 0.8686$, and then
$P(-1.12 < Z < 0) = 0.8686 - 0.5000 = 0.3686$. ■

EXAMPLE 2.12

Let Y be the variable representing the distribution of grades in a statistics course. It can be assumed that these grades are approximately normally distributed with $\mu = 75$ and $\sigma = 10$. If the instructor wants no more than 10% of the class to get an A, what should be the cutoff grade? That is, what is the value of y such that $P(Y > y) = 0.10$?

Solution The two steps are now used in reverse order:

Step 1: Find z from the table so that $P(Z > z) = 0.10$. This is $z = 1.28$ (rounded for convenience).
Step 2: Reverse the transformation. That is, solve for y in the equation $1.28 = (y - 75)/10$. The solution is $y = 87.8$.

Therefore, the instructor should assign an A to those students with grades of 87.8 or higher. Problems of this type can also be solved directly using the formula $y = \mu + z\sigma$, and substituting the given values of μ and σ and the value of z for the desired probability. Specifically, for this example,

$$y = 75 + 1.28(10) = 87.8. \quad \blacksquare$$

2.5 Sampling Distributions

We are now ready to discuss the relationship between probability and statistical inference. Recall that, for purposes of this text, we defined statistical inference as the *process of making inferences on population parameters using sample statistics*. We have two facts that are key to statistical inference. These are: (1) population parameters are fixed numbers whose values are usually unknown and (2) sample statistics are known values for any given sample, but vary from sample to sample taken from the same population. In fact, it is nearly impossible for any two independently drawn samples to produce identical values of a sample statistic.

This variability of sample statistics is always present and must be accounted for in any inferential procedure. Fortunately this variability, which is called *sampling variation*, is readily recognized and is accounted for by identifying probability distributions that describe the variability of sample statistics. In fact, a sample statistic is a random variable as defined in Definition 2.11. And, like any other random variable, a sample statistic has a probability distribution.

DEFINITION 2.15
The **sampling distribution** of a statistic is the probability distribution of that statistic.

This sampling distribution has characteristics that can be related to those of the population from which the sample is drawn. This relationship is usually provided by the parameters of the probability distribution describing the population. The next section presents the sampling distribution of the mean, also referred to as the distribution of the sample mean. In following sections we present sampling distributions of other statistics.

Sampling Distribution of the Mean

Consider drawing a random sample of n observations from a population and computing $\bar{y}$. Repetition of this process a number of times provides a collection of sample means. This collection of values can be summarized by a relative frequency or empirical probability distribution describing the behavior of these means. If this process could be repeated to include all possible samples of size n, then all possible values of $\bar{y}$ would appear in that collection. The relative frequency distribution of these values is defined as the sampling distribution of $\bar{Y}$ for samples of size n and is itself a probability distribution. The next step is to determine how this distribution is related to that of the population from which these samples were drawn.

We illustrate with a very simple population that consists of five identical disks with numbers 1, 2, 3, 4, and 5. The distribution of the numbers can be described by the discrete uniform distribution with $k = 5$; hence

$$\mu = (5 + 1)/2 = 3, \quad \text{and} \quad \sigma^2 = (25 - 1)/12 = 2 \text{ (see Section 2.3).}$$

Blind (random) drawing of these disks, replacing each disk after drawing, simulates random sampling from a discrete uniform distribution having these parameters.

Consider an experiment consisting of drawing two disks, replacing the first before drawing the second, and then computing the mean of the values on the two disks. Table 2.8 lists every possible sample and its mean. Since each of these samples is equally likely to occur, the sampling distribution of these means is, in fact, the relative frequency distribution of the $\bar{y}$ values in the display. This distribution is shown in Table 2.9 and Fig. 2.12. Note that the distribution of the means calculated from a sample of size two more closely resembles a normal distribution than a uniform distribution. Using the

Table 2.8

Samples of Size 2 from Uniform Population

Sample Disks	Mean $\bar{y}$	Sample Disks	Mean $\bar{y}$
(1,1)	1.0	(3,4)	3.5
(1,2)	1.5	(3,5)	4.0
(1,3)	2.0	(4,1)	2.5
(1,4)	2.5	(4,2)	3.0
(1,5)	3.0	(4,3)	3.5
(2,1)	1.5	(4,4)	4.0
(2,2)	2.0	(4,5)	4.5
(2,3)	2.5	(5,1)	3.0
(2,4)	3.0	(5,2)	3.5
(2,5)	3.5	(5,3)	4.0
(3,1)	2.0	(5,4)	4.5
(3,2)	2.5	(5,5)	5.0
(3,3)	3.0		

Table 2.9

Distribution of Sample Means

$\bar{y}$	1.0	1.5	2.0	2.5	3.0	3.5	4.0	4.5	5.0
$P(\bar{y})$	1/25	2/25	3/25	4/25	5/25	4/25	3/25	2/25	1/25

Figure 2.12

Histogram of Sample Means

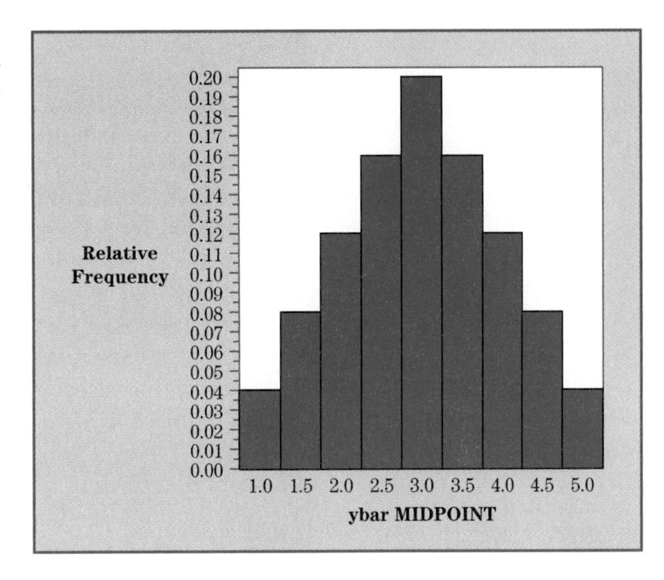

formulas for the mean and variance of a probability distribution given in Section 2.3, we can verify that the mean of the distribution of $\bar{y}$ values is 3 and the variance is 1.

Obviously we cannot draw all possible samples from an infinite population so we must rely on theoretical considerations to characterize the sampling distribution of the mean. A useful theorem, whose proof requires mathematics beyond the scope of this book, states the following:

THEOREM 2.1

Sampling Distribution of the Mean The sampling distribution of $\bar{Y}$ from a random sample of size n drawn from a population with mean μ and variance σ^2 will have mean $= \mu$ and variance $= \sigma^2/n$.

We can now see that the distribution of means from the samples of two disks obeys this theorem:

$$\text{mean} = \mu = 3,$$

and

$$\text{variance} = \sigma^2/2 = 2/2 = 1.$$

A second consideration, called the **central limit theorem**, states that if the sample size n is large, then the following is true:

THEOREM 2.2

> **Central Limit Theorem** If random samples of size n are taken from any distribution with mean μ and variance σ^2, the sample mean $\bar{Y}$ will have a distribution approximately normal with mean μ and variance σ^2/n. The approximation becomes better as n increases.

While the theorem itself is an asymptotic result (being exactly true only if n goes to infinity), the approximation is usually very good for quite moderate values of n. Sample sizes required for the approximation to be useful depend on the nature of the distribution of the population. For populations that resemble the normal, sample sizes of 10 or more are usually sufficient, while sample sizes in excess of 30 are adequate for virtually all populations, unless the distribution is extremely skewed. Finally, if the population is normally distributed, the sampling distribution of the mean is exactly normally distributed regardless of sample size. We can now see why the normal distribution is so important.

We illustrate the characteristics of the sampling distribution of the mean with a simulation study. We instruct a computer to simulate the drawing of random samples from a population described by the continuous uniform distribution with range from 0 to 1 ($a = 0$, $b = 1$, see Section 2.4 on the continuous uniform distribution). We know that for this distribution

$$\mu = 1/2 = 0.5$$

and

$$\sigma^2 = 1/12 = 0.08333.$$

We further instruct the computer to draw 1000 samples of $n = 3$ each, and compute the mean for each of the samples. This provides 1000 observations on $\bar{Y}$ for samples of $n = 3$ from the continuous uniform distribution. The histogram of the distribution of these sample means is shown in Fig. 2.13. This histogram is an empirical probability distribution of $\bar{Y}$ for the 1000 samples. According to theory, the mean and variance of $\bar{Y}$ should be 0.5 and $0.0833/3 = 0.0278$, respectively. From the actual 1000 values of $\bar{y}$ (not reproduced here), we can compute the mean and variance, which are 0.4999 and 0.02759, respectively.

The values from our empirical distribution are not exactly those specified by the theory for the sampling distribution, but the results are quite close. This is, of course, due to the fact that we have not taken all possible samples. Examination of the histogram shows that the distribution of the sample mean looks somewhat like the normal. Further, if the distribution of means is normal, the 5th and 95th percentiles should be

$$0.5 \pm (1.645)(\sqrt{0.0278}), \quad \text{or} \quad 0.2258 \quad \text{and} \quad 0.7742, \text{ respectively.}$$

The corresponding percentiles of the 1000 sample means are 0.2237 and 0.7744, which are certainly close to expected values.

We now repeat the sampling process using samples of size 12. The resulting distribution of sample means is given in Fig. 2.14. The shape of the distribution

Figure 2.13

Means of Sample of Size 3 from a Uniform Population

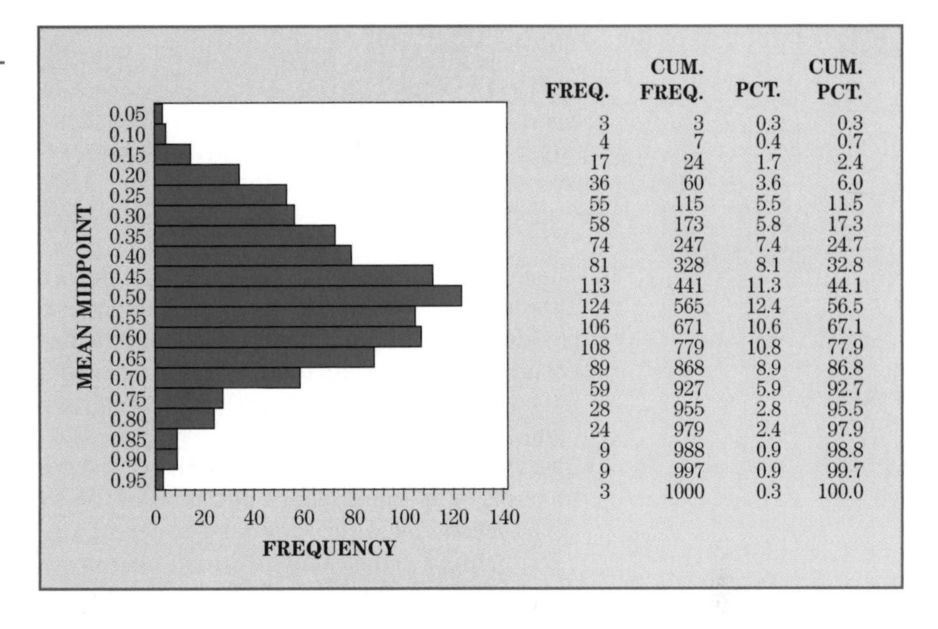

MEAN MIDPOINT	FREQ.	CUM. FREQ.	PCT.	CUM. PCT.
0.05	3	3	0.3	0.3
0.10	4	7	0.4	0.7
0.15	17	24	1.7	2.4
0.20	36	60	3.6	6.0
0.25	55	115	5.5	11.5
0.30	58	173	5.8	17.3
0.35	74	247	7.4	24.7
0.40	81	328	8.1	32.8
0.45	113	441	11.3	44.1
0.50	124	565	12.4	56.5
0.55	106	671	10.6	67.1
0.60	108	779	10.8	77.9
0.65	89	868	8.9	86.8
0.70	59	927	5.9	92.7
0.75	28	955	2.8	95.5
0.80	24	979	2.4	97.9
0.85	9	988	0.9	98.8
0.90	9	997	0.9	99.7
0.95	3	1000	0.3	100.0

Figure 2.14

Means of Samples of Size 12 from a Uniform Population

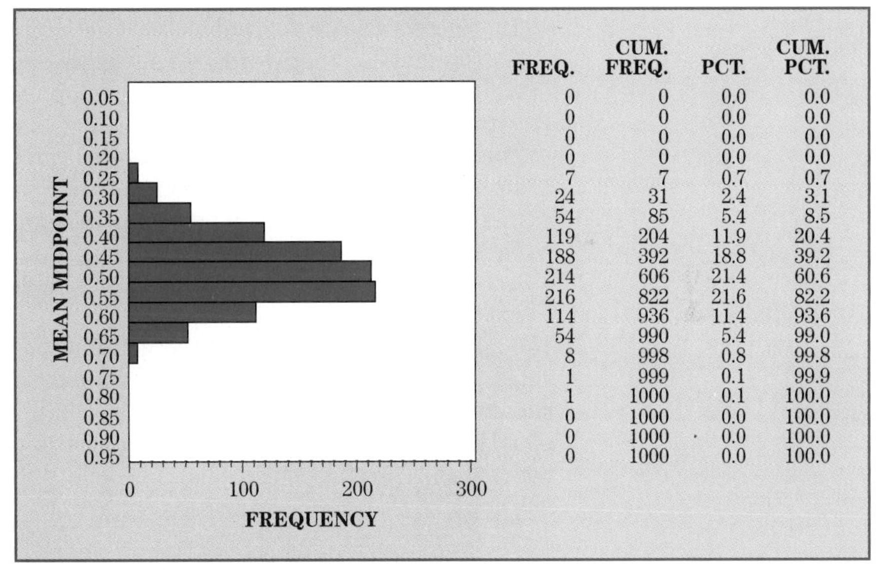

MEAN MIDPOINT	FREQ.	CUM. FREQ.	PCT.	CUM. PCT.
0.05	0	0	0.0	0.0
0.10	0	0	0.0	0.0
0.15	0	0	0.0	0.0
0.20	0	0	0.0	0.0
0.25	7	7	0.7	0.7
0.30	24	31	2.4	3.1
0.35	54	85	5.4	8.5
0.40	119	204	11.9	20.4
0.45	188	392	18.8	39.2
0.50	214	606	21.4	60.6
0.55	216	822	21.6	82.2
0.60	114	936	11.4	93.6
0.65	54	990	5.4	99.0
0.70	8	998	0.8	99.8
0.75	1	999	0.1	99.9
0.80	1	1000	0.1	100.0
0.85	0	1000	0.0	100.0
0.90	0	1000	0.0	100.0
0.95	0	1000	0.0	100.0

of these means is now nearly indistinguishable from the normal, and the mean and variance of the distribution (again computed from the 1000 values not listed) show even more precision, that is, a smaller variance of $\bar{Y}$ than was obtained for samples of three. Specifically, the mean of these 1000 sample means is 0.4987 and the variance is 0.007393, which is quite close to the theoretical values of 0.5 and $0.0833/12 = 0.00694$. Also the actual 5th and 95th percentiles of 0.3515 and 0.6447 agree closely with the values of 0.3586 and 0.6414 based on the additional assumption of normality.

Usefulness of the Sampling Distribution

Note that the sampling distribution provides a bridge that relates what we may expect from a sample to the characteristics of the population. In other words, if we were to know the mean and variance of a population, we can now make probability statements about what results we may get from a sample. The important features of the sampling distribution of the mean are as follows:

1. The mean of the sampling distribution of the mean is the population mean. This implies that "on the average" the sample mean is the same as the population mean. We therefore say that the sample mean is an **unbiased estimate** of the population mean. Most estimates used in this book are unbiased estimates, but not all sample statistics have the property of being unbiased.
2. The variance of the distribution of the sample mean is σ^2/n. Its square root, $\sigma/\sqrt{n}$, is the standard deviation of the sampling distribution of the mean, often called the **standard error of the mean**, and has the same interpretation as the standard deviation of any distribution. The formula for the standard error reveals the two very important features of the sampling distribution:
   - The more variable the population, the more variable is the sampling distribution. In other words, for any given sample size, the sample mean will be a less reliable estimate of the population mean for populations with larger variances.
   - The sampling distribution becomes less variable with increased sample size. We expect larger samples to provide more precise estimates, but this formula specifies by how much: *the standard error decreases with the square root of the sample size.* And if the sample size is infinity, the standard error is zero because then the sample mean is, by definition, the population mean.
3. The approximate normality of the distribution of the sample mean facilitates probability calculations when sampling from populations with unknown distributions. Occasionally, however, the sample is so small or the population distribution is such that the distribution of the sample mean is not normal. The consequences of this occurring are discussed throughout this book.

EXAMPLE 2.13

An aptitude test for high school students is designed so that scores on the test have $\mu = 90$ and $\sigma = 20$. Students in a school are randomly assigned to various sections of a course. In one of these sections of 100 students the mean score is 86. If the assignment of students is indeed random, what is the probability of getting a mean of 86 or lower on that test?

Solution According to the central limit theorem and the sampling distribution of the mean, the sample mean will have approximately the normal distribution with mean 90 and standard error $20/\sqrt{100} = 2$. Standardizing the value of 86, we get

$$z = \frac{(86 - 90)}{2} = -2.$$

Using the standard normal table, we obtain the desired value $P(z < -2) = 0.0228$. Since this is a rather low probability, the actual occurrence of such a result may raise questions about the randomness of student assignments to sections. ■

EXAMPLE 2.14

Quality Control Statistical methods have long been used in industrial situations, such as for process control. Usually production processes will operate in the "in-control" state, producing acceptable products for relatively long periods of time. Occasionally the process will shift to an "out-of-control" state where a proportion of the process output does not conform to requirements. It is important to be able to identify when this shift occurs and take action immediately. One way of monitoring this production process is through the use of a **control chart**. A typical control chart, such as that illustrated in Fig. 2.15, is a graphical display of a quality characteristic that has been measured or computed from a sample plotted against the sample number or time. The chart contains a center line that represents the average value of the characteristic when the process is in control. Two other lines, the upper control limit (UCL) and the lower control limit (LCL), are shown on the control chart. These limits are chosen so that if the process is in control, nearly all of the sample points will fall between them. Therefore, as long as the points plot within these limits the process is considered in control. If a point plots outside the control limits, the process is considered out of control and intervention is necessary. Typically control limits that are three standard deviations of the statistic above and below the average will be established. These are called "3-sigma" control limits. We will use the following simple example to illustrate the use of the sampling distribution of the mean in constructing a control chart.

A manufacturing company uses a machine to punch out parts for a hinge for vent windows to be installed in trucks and vans. This machine produces thousands of these parts each day. To monitor the production of this part and to make sure that it will be acceptable for the next stage of vent window assembly, a sample of 25 parts is taken each hour. The width of a critical area of each part is measured and the mean of each sample is calculated. Thus for each day there are a total of 24 samples of 25 observations each. Listed in Table 2.10 are one day's sampling results. The part will have a mean width of 0.45 in. with a standard deviation of 0.11 in. when the production process is in control.

Solution Using the sampling distribution of the mean, we can determine its standard error as $0.11/\sqrt{25} = 0.022$. Using the control limits of plus or minus 3 standard errors, the control limits on this process are $0.45 + 3(0.022) = 0.516$ and $0.45 - 3(0.022) = 0.384$, respectively. The control chart is shown in

Table 2.10

Data for Control Chart

Sample Number	Mean Width (in.)	Sample Number	Mean Width (in.)
1	0.42	2	0.46
3	0.44	4	0.45
5	0.39	6	0.41
7	0.47	8	0.46
9	0.44	10	0.48
11	0.51	12	0.55
13	0.49	14	0.44
15	0.47	16	0.44
17	0.48	18	0.46
19	0.42	20	0.40
21	0.45	22	0.47
23	0.44	24	0.45

Fig. 2.15. Note that the 12th sample mean has a value of 0.55, which is larger than the upper control limit. This is an indication that the process went "out of control" at that point.

Figure 2.15

Control Chart

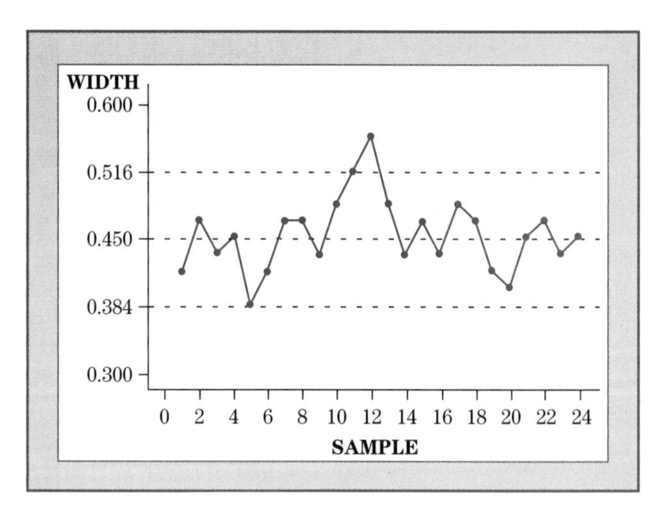

The probability of any sample mean falling outside the control limits can be determined by

$$P(\bar{Y} > 0.516) + P(\bar{Y} < 0.384) = P(Z > 3) + P(Z < -3) = 0.0026.$$

Therefore, the value of 0.55 for the mean is quite extreme if the process is in control. On investigation, the quality manager found out that during that sampling period, there was a thunderstorm in the area, and electric service was erratic, resulting in the machine also becoming erratic. After the storm passed, things returned to normal, as indicated by the subsequent samples. ■

Sampling Distribution of a Proportion

The central limit theorem provides a procedure for approximating probabilities for the binomial distribution presented in Section 2.3. A binomial distribution can be redefined as describing a population of observations, y_i, each having either the value 0 or 1, with the value "1" corresponding to "success" and "0" to "failure." Then each y_i can be described as a random variable from the probability distribution described in Table 2.11.

Table 2.11

Distribution of Binomial Population

y	$p(y)$
0	$1-p$
1	p

Further, the mean and variance of the distribution of the population of y values described in this manner can be shown to be p and $p(1-p)$, respectively (see Section 2.3).

A binomial experiment can be considered a random sample of size n from this population. The total number of successes in the experiment therefore is Σy_i, and the sample proportion of successes is $\bar{y}$, which is usually denoted by $\hat{p}$. Now, according to the central limit theorem, the sample proportion will be an approximately normally distributed random variable with mean p and variance $[p(1-p)]/n$ for sufficiently large n. It is generally accepted that when the smaller of np and $n(1-p)$ is greater than 5, the approximation will be adequate for most purposes. This application of the central limit theorem is known as the large sample approximation to the binomial distribution because it provides the specification of the sampling distribution of the sample proportion $\hat{p}$.

EXAMPLE 2.15

In most elections, a simple majority of voters, that is, a favorable vote of over 50% of voters, will give a candidate a victory. This is equivalent to the statement that the probability that any randomly selected voter votes for that candidate is greater than 0.5. Therefore, if a candidate were to conduct an opinion poll, he or she would hope to be able to substantiate at least 50% support. If such an opinion poll is indeed a random sample from the population of voters, the results of the poll would satisfy the conditions for a binomial experiment given in Section 2.3.

Suppose a random sample of 100 registered voters show 61 with a preference for the candidate. If the election were in fact a toss-up (that is, $p = 0.5$) what is the probability of obtaining that (or a more extreme value)?

Solution Under the assumption $p = 0.5$, the mean and variance of the sampling distribution of $\hat{p}$ are $p = 0.5$ and $p(1-p)/100 = 0.0025$, respectively. Then the standard error of the estimated proportion is 0.05. The probability is obtained by using the z transformation

$$z = (0.61 - 0.5)/0.05 = 2.2,$$

and from the table of the normal distribution the probability of Z being greater than 2.2 is 0.0139. In other words, if the election really is a toss-up, obtaining this large a majority in a sample of 100 will occur with a probability of only 0.0139.

Note that in this section we have been concerned with the proportion of successes, while in previous discussions of the binomial distribution (Section 2.3) we were concerned with the number of successes. Since sample size is fixed, the frequency is simply the proportion multiplied by the sample size, which is a simple linear transformation. Using the rules for means and variances of transformed variables (Section 1.5 on change of scale) we see that the mean and variance of proportions given in this section correspond to the mean and variance of the binomial distribution given in Section 2.3. That is, the mean number of successes is np and the variance is $np(1-p)$. The central limit theorem also holds for both frequency and proportion of successes. Thus, the normal approximation to the binomial can be used for both proportions and frequencies of successes, using the appropriate means and variances, although proportions are more frequently used in practice. ■

EXAMPLE 2.16

Suppose that the process discussed in Example 2.14 also involved the forming of rubber gaskets for the vent windows. When these gaskets are inspected, they are classified as acceptable or nonacceptable based on a number of different characteristics, such as thickness, consistency, and overall size. The process of manufacturing these gaskets is monitored by constructing a control chart using random samples as specified in Example 2.14, where the chart is based on the proportion of nonacceptable gaskets. Such a chart is called an "attribute" chart or simply a p chart.

To monitor the "fraction nonconforming" of gaskets being produced, a sample of 25 gaskets is inspected each hour. The proportion of gaskets not acceptable (nonconforming) is recorded and plotted on a control chart. The center line for this control chart will be the average proportion of nonconforming gaskets when the process is in control. This is found to be $p = 0.10$. The result of a day's sampling, presented in Table 2.12, is to be used to construct a control chart.

Table 2.12

Proportion of Nonconforming Gaskets

Sample	$\hat{p}$	Sample	$\hat{p}$
1	0.17	13	0.09
2	0.12	14	0.10
3	0.15	15	0.07
4	0.10	16	0.09
5	0.09	17	0.05
6	0.11	18	0.04
7	0.14	19	0.06
8	0.13	20	0.08
9	0.08	21	0.05
10	0.09	22	0.04
11	0.11	23	0.03
12	0.10	24	0.04

Figure 2.16

A *p* Chart

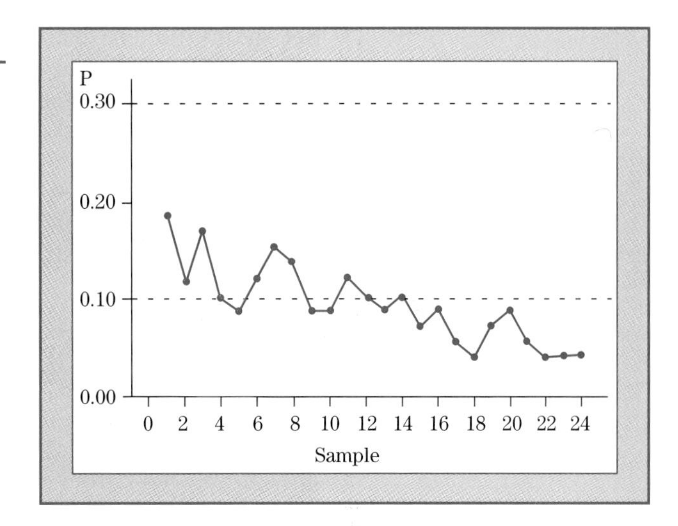

Solution The control limits for the chart are computed by using the sampling distribution of $\hat{p}$ under the assumption that $p = 0.10$. Then the variance of $\hat{p}$ is $(0.10)(0.90)/25 = 0.0036$ and the standard error is 0.06. The upper control limit is $0.10 + 3(0.06) = 0.28$, and the lower control limit is $0.10 - 3(0.06) = -0.08$. For practical purposes, the lower control limit is set at 0, because we cannot have a negative proportion. The chart is illustrated in Fig. 2.16. This chart indicates that the process is in control and seems to remain that way throughout the day. The last 10 samples, as illustrated in the chart, are all below the target value. This seems to indicate a downward "trend." The process does, in fact, appear to be getting better as the control monitoring process continues. This is not unusual, since one way to improve quality is to monitor it. The quality manager may want to test the process to determine whether the process is really getting better. ■

2.6 Other Sampling Distributions

Although the normal distribution is, in fact, used to describe sampling distributions of statistics other than the mean, other statistics have sampling distributions that are quite different. This section gives a brief introduction to three sampling distributions, which are associated with the normal distributions and are used extensively in this text. These distributions are

χ^2: describes the distribution of the sample variance.

t: describes the distribution of a normally distributed random variable standardized by an estimate of the standard deviation.

F: describes the distribution of the ratio of two variances. We will see later that this has applications to inferences on means from several populations.

A brief outline of these distributions is presented here for the purpose of providing an understanding of the interrelationships among these distributions. Applications of these distributions are deferred to the appropriate methods sections in later chapters.

Figure 2.17

χ^2 Distributions for 1, 3, 6, and 10 Degrees of Freedom

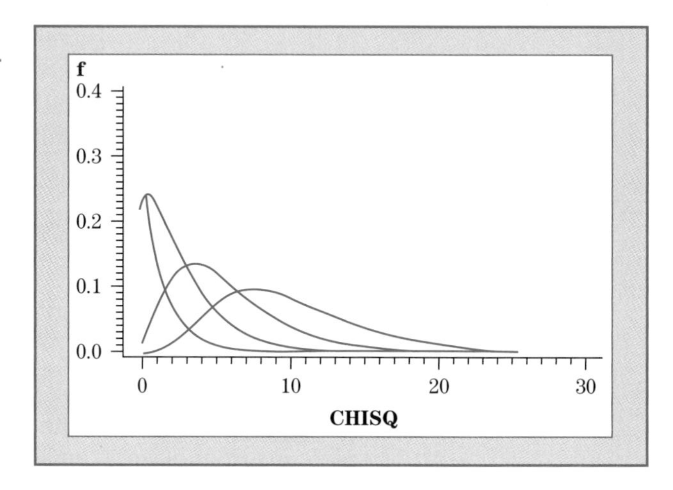

The χ^2 Distribution

Consider n independent random variables with the standard normal distribution. Call these variables Z_i, $i = 1, 2, \ldots, n$. The statistic

$$X^2 = \sum Z_i^2$$

is also a random variable whose distribution we call χ^2 (the Greek lowercase letter chi). The function describing this distribution is rather complicated and is of no use to us at this time, except to observe that this function contains only one parameter. This parameter is called the **degrees of freedom**, and is equal to the number of Z values in the sum of squares. Thus the variable X^2 described above would have a χ^2 distribution with degrees of freedom equal to n. Usually the degrees of freedom are denoted by the Greek letter ν. The distribution is usually denoted by $\chi^2(\nu)$. Graphs of χ^2 distributions for selected values of ν are given in Fig. 2.17.

A few important characteristics of the χ^2 distribution are as follows:

1. χ^2 values cannot be negative since they are sums of squares.
2. The shape of the χ^2 distribution is different for each value of ν; hence, a separate table is needed for each value of ν. For this reason, tables giving probabilities for the χ^2 distribution give values for only a selected set of probabilities similar to the small table for the normal distribution given in Appendix Table A.1A. Appendix Table A.3 gives probabilities for the χ^2 distribution. Values not given in the table may be estimated by interpolation, but such precision is not often required in practice. Computer programs are available for calculating exact values if necessary.

3. The mean of the χ^2 distribution is ν, and the variance is 2ν.
4. For large values of ν (usually greater than 30), the χ^2 distribution may be approximated by the normal, using the mean and variance given in item 3. Thus we may use $Z = (\chi^2 - \nu)/\sqrt{2\nu}$, and find the probability associated with the z value.
5. The ability of the χ^2 distribution to reflect the distribution of $\sum Z^2$ is only moderately affected if the distribution of the Z_i is not exactly normal, although severe departures from normality can affect the nature of the resulting distribution.

Distribution of the Sample Variance

A common use of the χ^2 distribution is to describe the distribution of the sample variance. Let $Y_1, Y_2, \ldots, Y_n$ be a random sample from a normally distributed population with mean $= \mu$ and variance $= \sigma^2$. Then the quantity $(n-1)S^2/\sigma^2$ is a random variable whose distribution is described by a χ^2 distribution with $(n-1)$ degrees of freedom, where S^2 is the usual sample estimate of the population variance given in Section 1.5. That is,

$$S^2 = \sum (Y - \bar{Y})^2/(n-1).$$

In other words the χ^2 distribution is used to describe the sampling distribution of S^2. Since we divide the sum of squares by degrees of freedom to obtain the variance estimate, the expression for the random variable having a χ^2 distribution can be written

$$X^2 = \sum Z^2 = \sum \left(\frac{(Y - \bar{Y})}{\sigma}\right)^2 = \frac{\sum(Y - \bar{Y})^2}{\sigma^2} = \frac{\text{SS}}{\sigma^2} = \frac{(n-1)S^2}{\sigma^2}.$$

EXAMPLE 2.17

In making machined auto parts, the consistency of dimensions, the tolerance as it is called, is an important quality factor. Since the standard deviation (or variance) is a measure of the dispersion of a variable, we can use it as a measure of consistency.

Suppose a sample of 15 such parts shows $s = 0.0125$ mm. If the allowable tolerance of these parts is specified so that the standard deviation may not be larger than 0.01 mm, we would like to know the probability of obtaining that value of S (or larger) if the population standard deviation is 0.01 mm. Specifically, then, we want the probability that $S^2 > (0.0125)^2$ or 0.00015625 when $\sigma^2 = (0.01)^2 = 0.0001$.

Solution The statistic to be compared to the χ^2 distribution has the value

$$X^2 = \frac{(n-1)s^2}{\sigma^2} = \frac{14 \cdot 0.00015625}{0.0001} = 21.875.$$

Figure 2.18 shows the χ^2 distribution for 14 degrees of freedom and the location of the computed value. The desired probability is the area to the right of that value.

Figure 2.18

A χ^2 Distribution for 14 Degrees of Freedom

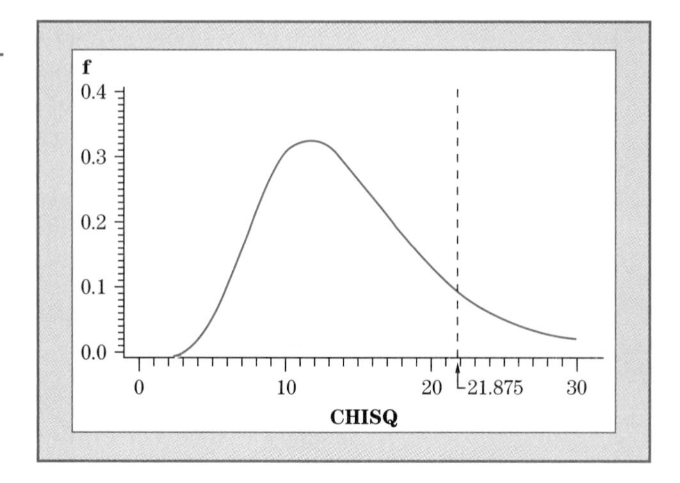

The table of χ^2 probabilities (Appendix Table A.3) gives areas for χ^2 values only for selected probabilities; hence the calculated value does not appear. However, we note that values of $\chi^2 > 21.064$ occur with probability 0.1 and values greater than 23.685 occur with probability 0.05; hence the probability of exceeding the sample value of 21.875 occurs with a probability that lies between 0.05 and 0.1. A computer program provides the exact probability of 0.081. ■

The *t* Distribution

In problems involving the sampling distribution of the mean we have used the fact that

$$Z = \frac{(\bar{Y} - \mu)}{\sigma/\sqrt{n}}$$

is a random variable having the standard normal distribution. In most practical situations σ is not known. The only measure of the standard deviation available may be the sample standard deviation S. It is natural then to substitute S for σ in the above relationship. The problem is that the resulting statistic is not normally distributed.

W. S. Gosset, writing under the pen name "Student," derived the probability distribution for this statistic, which is called the Student t or simply t distribution. The function describing this distribution is quite complex and of little use to us in this text. However, it is of interest that this distribution also has only one parameter, the degrees of freedom; hence the t distribution with ν degrees of freedom is denoted by $t(\nu)$. This distribution is quite similar to the normal in that it is symmetric and bell shaped. However, the t distribution has "fatter" tails than the normal. That is, it has more probability in the extreme or tail areas than does the normal distribution, a characteristic quite apparent for small values of the degrees of freedom, but barely noticeable if the degrees of freedom exceed 30 or so.

Figure 2.19

Student's *t* Distribution

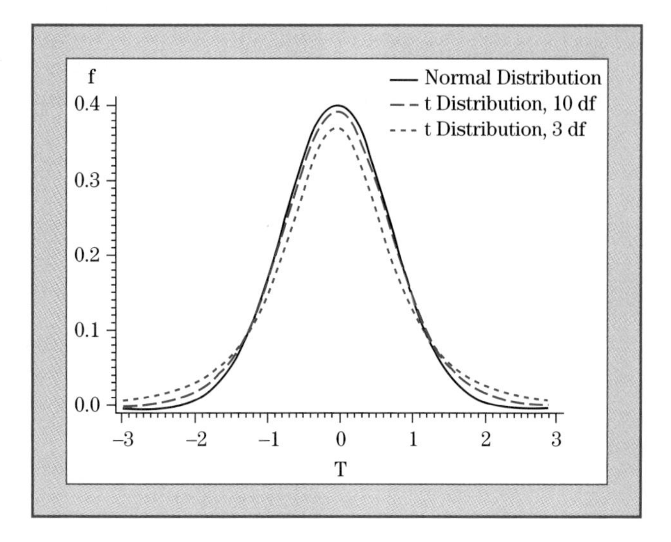

In fact, when the degrees of freedom are ∞, the t distribution is identical to the standard normal distribution as illustrated in Fig. 2.19. A separate table for probabilities from the t distribution is required for each value of the degrees of freedom; hence, as in the table for the χ^2 distribution, only a limited set of probability values is given. Also, since the distribution is symmetric, only the upper tail values are given (see Appendix Table A.2).

The t distribution with ν degrees of freedom actually takes the form

$$t(\nu) = \frac{Z}{\sqrt{\frac{\chi^2(\nu)}{\nu}}},$$

where Z is a standard normal random variable and $\chi^2(\nu)$ is an independent χ^2 random variable with ν degrees of freedom.

Using the *t* Distribution

Using this definition, we can develop the sampling distribution of the sample mean when the population variance, σ^2, is unknown. Recall that

(1) $Z = \frac{\bar{Y}-\mu}{\sigma/\sqrt{n}}$, has the standard normal distribution, and

(2) $\chi^2(n-1) = \text{SS}/\sigma^2 = (n-1)S^2/\sigma^2$ has the χ^2 distribution with $n-1$ degrees of freedom.

These two statistics can be shown to be independent so that

$$T = \frac{\frac{\bar{Y}-\mu}{\sigma/\sqrt{n}}}{\sqrt{\frac{(n-1)S^2/\sigma^2}{n-1}}} = \frac{\bar{Y}-\mu}{S/\sqrt{n}}$$

has the t distribution with $n-1$ degrees of freedom.

EXAMPLE 2.18

Grade point ratios (GPRs) have been recorded for a random sample of 16 from the entering freshman class at a major university. It can be assumed that the distribution of GPR values is approximately normal. The sample yielded a mean, $\bar{y} = 3.1$, and standard deviation, $s = 0.8$. The nationwide mean GPR of entering freshmen is $\mu = 2.7$. We want to know the probability of getting this sample mean (or higher) if the mean GPR of this university is the same as the nationwide population of students. That is, we want the probability of getting a $\bar{Y}$ that is greater than or equal to 3.1 from a population whose mean is 2.7. We compute the value of the statistic as

$$t = \frac{3.1 - 2.7}{0.8/\sqrt{16}} = 2.0.$$

From Appendix Table A.2 we see that for 15 degrees of freedom this value lies between the values 1.7531 for the tail probability 0.05 and 2.1314 for the tail probability 0.025. Therefore, we can say that the probability of obtaining a sample mean this large or larger is between 0.025 and 0.05. As in the case of the χ^2 distribution, more precise values for the probability may be obtained by interpolation or the use of a computer if necessary, which in this example provides the probability as 0.032. We will make extensive use of the t distribution starting in Chapter 4.

The *F* Distribution

A sampling distribution, which occurs frequently in statistical methods, is one that describes the distribution of the ratio of two estimates of σ^2. This is the so-called F distribution, named in honor of Sir Ronald Fisher, who is often called the father of modern statistics. The F distribution is uniquely identified by its set of two degrees of freedom, one called the "numerator degrees of freedom" and the other called the "denominator degrees of freedom." This terminology comes from the fact that the F distribution with ν_1 and ν_2 degrees of freedom, denoted by $F(\nu_1, \nu_2)$, can be written as

$$F(\nu_1, \nu_2) = \frac{\chi_1^2(\nu_1)/\nu_1}{\chi_2^2(\nu_2)/\nu_2}.$$

Where $\chi_1^2(\nu_1)$ is a χ^2 random variable with ν_1 degrees of freedom and $\chi_2^2(\nu_2)$ is an independent χ^2 random variable wtih ν_2 degrees of freedom.

Using of the *F* Distribution

Recall that the quantity $(n-1)S^2/\sigma^2$ has the χ^2 distribution with $n-1$ degrees of freedom. Therefore, if we assume that we have a sample of size n_1 from a population with variance σ_1^2 and an independent sample of size n_2 from another population with variance σ_2^2 then the statistic

$$F = \frac{S_1^2/\sigma_1^2}{S_2^2/\sigma_2^2},$$

where S_1^2 and S_2^2 represent the usual variance estimates of σ_1^2 and σ_2^2, respectively, is a random variable having the F distribution.

The F distribution has two parameters, ν_1 and ν_2. The distribution is denoted by $F(\nu_1, \nu_2)$. If the variances are estimated in the usual manner, the degrees of freedom are $(n_1 - 1)$ and $(n_2 - 1)$, respectively. Also, if both populations have equal variance, that is, $\sigma_1^2 = \sigma_2^2$, the F statistic is simply the ratio S_1^2/S_2^2. The equation describing the distribution of the F statistic is also quite complex and is of little use to us in this text. However, some of the characteristics of the F distribution are of interest:

1. The F distribution is defined only for nonnegative values.
2. The F distribution is not symmetric.
3. A different table is needed for each combination of degrees of freedom. Fortunately, for most practical problems only a relatively few probability values are needed.
4. The choice of which variance estimate to place in the numerator is somewhat arbitrary; hence the table of probabilities of the F distribution always gives the right tail value. That is, it assumes that the larger variance estimate is in the numerator.

Appendix Table A.4 gives values of the F distribution for selected degrees of freedom combinations for right tail areas of 0.1, 0.05, 0.025, 0.01, and 0.005. There is one table for each probability (tail area), and the values in the table correspond to F values for numerator degrees of freedom ν_1 indicated by column headings, and denominator degrees of freedom ν_2 as row headings. Interpolation may be used for values not found in the table, but this is rarely needed in practice.

EXAMPLE 2.19

Two machines, A and B, are supposed to make parts for which a critical dimension must have the same consistency. That is, the parts produced by the two machines must have equal standard deviations. A random sample of 10 parts from machine A has a sample standard deviation of 0.014 and an independently drawn sample of 15 parts from machine B has a sample standard deviation of 0.008. What is the probability of obtaining standard deviations this far apart if the machines are really making parts with equal consistency?

Solution To answer this question we need to calculate probabilities in both tails of the distribution:

$$\text{(A)} \quad P\left[\left(S_A^2/S_B^2\right) > (0.014)^2/(0.008)^2\right] = P\left[\left(S_A^2/S_B^2\right) > 3.06\right],$$

as well as

$$\text{(B)} \quad P\left[\left(S_B^2/S_A^2\right) < (0.008^2)/(0.014)^2\right] = P\left[\left(S_B^2/S_A^2\right) < 0.327\right],$$

assuming $\sigma_A^2 = \sigma_B^2$.

For part (A) we need the probability $P[F(9, 14) > 3.06]$. Because of the limited number of entries in the table of the F distribution, we can find the value 2.65 for $p = 0.05$ and the value 3.21 for $p = 0.01$ for 9 and 14 degrees of freedom. The sample value is between these two; hence we can say that

$$0.025 < P[F(9, 14) > 3.06] < 0.05.$$

For part (B) we need $P[F(14, 9) > 0.327]$, which is the same as $P[F(9, 14) > 1/0.327] = P[F(9, 14) > 3.06]$, which is the same as for part (A). Since we want the probability for both directions, we add the probabilities; hence, the probability of the two samples of parts having standard deviations this far apart is between 0.05 and 0.10. The exact value obtained by a computer is 0.06. ■

Relationships among the Distributions

All of the distributions presented in this section start with normally distributed random variables; hence they are naturally related. The following relationships are not difficult to verify and have implications for many of the methods presented later in this book:

$$t(\infty) = z, \tag{1}$$

$$z^2 = \chi^2(1), \tag{2}$$

$$F(1, v_2) = t^2(v_2), \tag{3}$$

$$F(v_1, \infty) = \chi^2(v_1)/v_1. \tag{4}$$

2.7 CHAPTER SUMMARY

The reliability of statistical inferences is described by probabilities, which are based on sampling distributions. The purpose of this chapter is to develop various concepts and principles leading to the definition and use of sampling distributions.

- A **probability** is defined as the long-term relative frequency of the occurrence of an outcome of an experiment.
- An **event** is defined as a combination of outcomes. Probabilities of the occurrence of a specific event are obtained by the application of rules governing probabilities.
- A **random variable** is defined as a numeric value assigned to an event. Random variables may be discrete or continuous.
- A **probability distribution** is a definition of the probabilities of all possible values of a random variable for an experiment. There are probability distributions for both discrete and continuous random variables. Probability distributions are characterized by parameters.
- The **normal distribution** is the basis for most inferential procedures. Rules are provided for using a table to obtain probabilities associated with normally distributed random variables.

- A **sampling distribution** is a probability distribution of a statistic which relates the statistic to the parameters of the population from which the sample is drawn. The most important of these is the sampling distribution of the mean, but other sampling distributions are presented.

2.8 CHAPTER EXERCISES

CONCEPT QUESTIONS

This section consists of some true/false questions regarding concepts of statistical inference. Indicate if a statement is true or false and, if false, indicate what is required to make the statement true.

1. ______________If two events are mutually exclusive, then $P(A \text{ or } B) = P(A) + P(B)$.

2. ______________If A and B are two events, then $P(A \text{ and } B) = P(A)P(B)$, no matter what the relation between A and B.

3. ______________The probability distribution function of a discrete random variable cannot have a value greater than 1.

4. ______________The probability distribution function of a continuous random variable can take on any value, even negative ones.

5. ______________The probability that a continuous random variable lies in the interval 4 to 7, inclusively, is the sum of $P(4) + P(5) + P(6) + P(7)$.

6. ______________The variance of the number of successes in a binomial experiment of n trials is $\sigma^2 = np(p - 1)$.

7. ______________A normal distribution is characterized by its mean and its degrees of freedom.

8. ______________The standard normal distribution has the mean zero and variance σ^2.

9. ______________The t distribution is used as the sampling distribution of the mean if the sample is small and the population variance is known.

10. ______________The standard error of the mean increases as the sample size increases.

PRACTICE EXERCISES

The following exercises are designed to give the reader practice in using the rules of probability through simple examples. The solutions are given in the back of the text.

1. The weather forecast says there is a 40% chance of rain today and a 30% chance of rain tomorrow.
   (a) What is the chance of rain on both days?
   (b) What is the chance of rain on neither day?
   (c) What is the chance of rain on at least one day?

2. The following is a probability distribution of the number of defects on a given contact lens produced in one shift on a production line:

Number of Defects	0	1	2	3	4
Probability	0.50	0.20	0.15	0.10	0.05

Let A be the event that one defect occurred, and B be the event that 2, 3, or 4 defects occurred. Find:
(a) $P(A)$ and $P(B)$
(b) $P(A \text{ and } B)$
(c) $P(A \text{ or } B)$

3. Using the distribution in Exercise 2, let the random variable Y be the number of defects on a contact lens randomly selected from lenses produced during the shift.
(a) Find the mean and variance of Y for the shift.
(b) Assume that the lenses are produced independently. What is the probability that five lenses drawn randomly from the production line during the shift will be defect-free?

4. Using the distribution in Exercise 2, suppose that the lens can be sold as is if there are no defects for \$20. If there is one defect, it can be reworked at a cost of \$5 and then sold. If there are two defects, it can be reworked at a cost of \$10 and then sold. If there are more than two defects, it must be scrapped. What is the expected revenue generated during the shift if 100 contact lenses are produced?

5. Suppose that Y is a normally distributed random variable with $\mu = 10$ and $\sigma = 2$, and X is an independent random variable, also normally distributed with $\mu = 5$ and $\sigma = 5$. Find:
(a) $P(Y > 12 \text{ and } X > 4)$
(b) $P(Y > 12 \text{ or } X > 4)$
(c) $P(Y > 10 \text{ and } X < 5)$

EXERCISES

1. A lottery that sells 150,000 tickets has the following prize structure:
(1) first prize of \$50,000
(2) 5 second prizes of \$10,000
(3) 25 third prizes of \$1000
(4) 1000 fourth prizes of \$10
   (a) Let Y be the winning amount of a randomly drawn lottery ticket. Describe the probability distribution of Y.
   (b) Compute the mean or expected value of the ticket.
   (c) If the ticket costs \$1.00, is the purchase of the ticket worthwhile? Explain your answer.

(d) Compute the standard deviation of this distribution. Comment on the usefulness of the standard deviation as a measure of dispersion for this distribution.

2. Assume the random variable y has the continuous uniform distribution defined on the interval a to b, that is,

$$f(y) = 1/(b - a), \quad a \leq y \leq b.$$

For this problem let $a = 0$ and $b = 2$.
(a) Find $P(Y < 1)$. (*Hint:* Use a picture.)
(b) Find μ and σ^2 for the distribution.

3. The binomial distribution for $p = 0.2$ and $n = 5$ is:

Value of Y	0	1	2	3	4	5
Probability	0.3277	0.4095	0.2048	0.0512	0.0064	0.0003

(a) Compute μ and σ^2 for this distribution.
(b) Do these values agree with those obtained as a function of the parameter p and sample size n? (See discussion of random variables in Section 2.2.)

4. A system consists of 10 components all arranged in series, each with a failure probability of 0.001. What is the probability that the system will fail? (*Hint:* See Section 2.2.)

5. A system requires two components, A and B, to both work before the system will. Because of the sensitivity of the system, an increased reliability is needed. To obtain this reliability, two duplicate components are to be used. That is, the system will have components $A1$, $A2$, $B1$, and $B2$. An engineer designs the two systems illustrated in the diagram. Assuming independent failure probabilities of 0.01 for each component, compute the probability of failure of each arrangement. Which one gives the more reliable system?

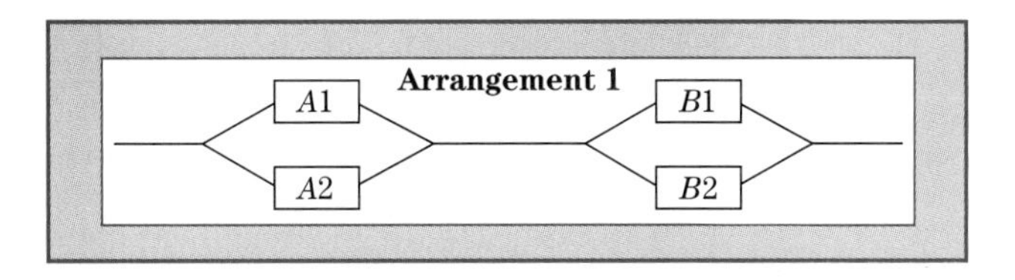

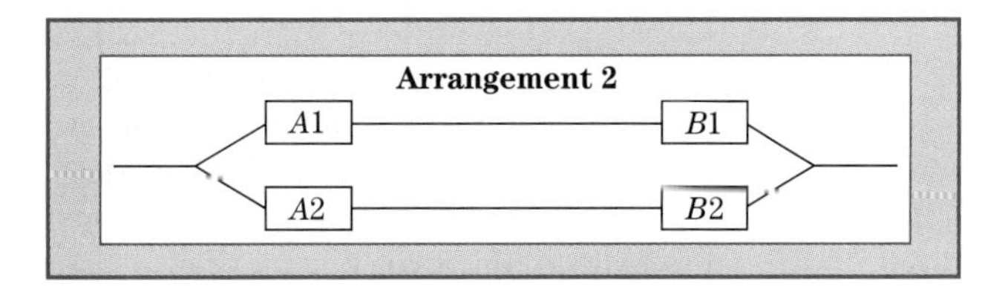

6. Let Z be a standard normal random variable. Use Appendix Table A.1 to find:
(a) $P(Z > 1)$
(b) $P(Z > -1)$
(c) $P(0 < Z < 1)$
(d) $P(Z < -1.5)$
(e) $P(-2.07 < Z < 0.98)$
(f) the value A such that $P(Z < A) = 0.95$
(g) the value C such that $P(-C < Z < C) = 0.95$

7. Let Y be a normally distributed random variable with mean 10 and variance 25. Find:
(a) $P(Y > 15)$
(b) $P(8 < Y < 12)$
(c) the value of C such that $P(Y < C) = 0.90$

8. A teacher finds that the scores of a particularly difficult test were approximately normally distributed with a mean of 76 and standard deviation of 14:
(a) If a score below 60 represents a grade of F (failure), approximately what percent of students failed the test?
(b) If the cutoff for a grade of A is the lowest score of the top 10%, what is that cutoff point?
(c) How many points must be added to the students' scores so that only 5% fail?

9. It is believed that 20% of voters in a certain city favor a tax increase for improved schools. If this percentage is correct, what is the probability that in a sample of 250 voters 60 or more will favor the tax increase? (Use the normal approximation.)

10. The probabilities for a random variable having the Poisson distribution with $\mu = 1$ is given in the following table.

Values of Y	0	1	2	3	4	5	6
Probability	0.368	0.368	0.184	0.061	0.015	0.003	0.001

Note: Probabilities for $Y > 6$ are very small and may be ignored.
(a) Compute the mean and variance of Y.
(b) According to theory, both the mean and the variance of the Poisson distribution are μ. Do the results in part (a) agree with the theory?

11. As μ increases, say, to values greater than 30, the Poisson distribution begins to be very similar to the normal with both a mean and variance of μ. Using this approximation, determine how many telephone operators are needed to ensure at most 5% busy signals if the mean number of phone calls at any given time is 30.

12. The Poisson distribution may also be used to find approximate binomial probabilities when n is large and p is small, by letting μ be np. This method provides for faster calculations of probabilities of rare events such as exotic diseases. For example, assume the incidence rate (proportion in the population) of a certain blood disease is known to be 1%. The probability of getting exactly seven cases in a random sample of 500, where $\mu = np = (0.01)(500) = 5$, is

$$P(Y = 7) = (5^7 e^{-5})/7! = 0.1044.$$

Suppose the incidence of another blood disease is 0.015. What is the probability of getting no occurrences of the disease in a random sample of 200? (Remember that $0! = 1$.)

13. A random sample of 100 is taken from a population with a mean of 140 and standard deviation of 25. What is the probability that the sample mean lies between 138 and 142?

14. A manufacturer wants to state a specific guarantee for the life of a product with a replacement for failed products. The distribution of lifetimes of the product has a mean of 1000 days and standard deviation of 150 days. What life length should be stated in the guarantee so that only 10% of the products need to be replaced?

15. A teacher wants to curve her grades such that 10% are below 60 and 10% above 90. Assuming a normal distribution, what values of μ and σ^2 will provide such a curve?

16. To monitor the production of sheet metal screws by a machine in a large manufacturing company, a sample of 100 screws is examined each hour for three shifts of 8 hours each. Each screw is inspected and designated as conforming or nonconforming according to specifications. Management is willing to accept a proportion of nonconforming screws of 0.05. Use the following result of one day's sampling (Table 2.13) to construct a control chart. Does the process seem to be in control? Explain.

Table 2.13

Data for Exercise 16

Sample	$\hat{p}$	Sample	$\hat{p}$
1	0.04	13	0.09
2	0.07	14	0.10
3	0.05	15	0.09
4	0.03	16	0.11
5	0.04	17	0.10
6	0.06	18	0.12
7	0.05	19	0.13
8	0.03	20	0.09
9	0.05	21	0.14
10	0.07	22	0.11
11	0.09	23	0.15
12	0.10	24	0.16

17. The Florida lottery uses a system of numbers ranging in value from 1 to 49. Every week the lottery commission randomly selects six numbers, and every ticket with those numbers wins a share of the grand prize. Individual numbers appear only once (no repeat values), and the order in which they are chosen does not matter.

(a) What is the probability that a person buying one ticket will win the grand prize? (*Hint:* Use the counting procedure for binomial distributions in Section 2.3.)

(b) The lottery also pays a lesser prize for tickets with five of the six numbers matching. What is the probability that a person buying one ticket will win either the grand prize or the lesser prize?

(c) The lottery also pays smaller prizes for getting three or four numbers matching. What is the probability that a person buying one ticket will win anything? That is, what is the probability of getting six matching numbers, or five matching numbers, or four matching numbers, or three matching numbers?

18. A manufacturer of auto windows uses a thin layer of plastic material between two layers of glass to make safety glass for windshields. The thickness of the layer of this material is important to the quality of the vision through the glass. A constant quality control monitoring scheme is employed by the manufacturer that checks the thickness at 30-minute intervals throughout the manufacturing process by sampling five windshields. The mean thickness is then plotted on a control chart. A perfect windshield will have a thickness of 4 mm. From past experience, it is known that the variance of thickness is about 0.25 mm. The results of one shift's production are given in Table 2.14.

(a) Construct a control chart of these data. (*Hint:* See Example 2.13.) Does the process stay in control throughout the shift?

(b) Does the chart indicate any trends? Explain. Can you think of a reason for this pattern?

Table 2.14

Thickness of Material (in Millimeters)

Sample Number	Thickness
1	4,3,3,4,2
2	5,4,4,4,3
3	3,3,4,4,4,
4	2,3,3,3,5
5	5,5,4,4,5
6	6,4,6,4,5
7	4,4,6,5,4
8	6,5,5,6,5
9	5,5,6,5,5
10	5,4,4,6,4
11	4,6,5,4,4
12	5,5,4,3,3
13	3,3,4,4,5
14	4,4,4,3,4
15	3,3,4,2,4
16	4,3,2,2,3
17	4,5,3,2,2
18	3,4,4,3,4

19. During 1989, a certain trucking company purchased 500 tires from a local dealer. The dealer guaranteed the tires to withstand loads of up to 100,000 pounds at speeds up to 55 mph. The drivers for the trucking company complained that the tires were not living up to this guarantee and were failing the first trip they were used. The trucking company decided to sample the tires and send them to an engineering firm for testing. This testing is expensive and destructive; therefore, the sample size to be tested must be carefully chosen. Construct a sampling plan for the company by doing the following:

(a) Calculate the probability of getting all defective tires in samples of size 25, 30, and 50 for $p = 0.80, 0.90, 0.95$, and 0.99, where p is the probability that an individual tire will fail. (Use the binomial distribution.)

(b) Graph these probabilities against p for the various values of n on the same graph. Use this graph to suggest a sample size.

20. Single-sample acceptance sampling for attributes uses a procedure similar to that of Exercise 19. Suppose that a sampling plan for accepting a lot of size N coming to a manufacturer from a supplier is to be determined by sampling n items from the lot and accepting the entire lot if c of fewer of the items are defective. The lot fraction defective, p, is the true proportion of defective items in the lot. The probability of observing x defective items in a random sample of n items can be calculated using the binomial distribution. The probability of observing c or fewer defective items is the sum of the probabilities from 0 to c and is called the probability of acceptance, P_a. An operating characteristic (OC) curve is then constructed plotting P_a versus p. This OC curve illustrates the performance of a sampling plan using n items and an acceptance value of c. Construct an OC curve for the sampling plan $n = 50$ and $c = 1$. OC curves are further discussed in Section 3.2.

21. A computer slot machine game has a number of different machines. In the simplest machine, there are three "wheels," each of which has four different symbols: three, two, one, or no bars. We will use principles of probability to estimate the mean payout for this machine.

(a) Playing 600 games (it does not cost anything on the computer!) gives the following probabilities of the different symbols for each wheel:

	PROBABILITIES		
Symbols	**Wheel 1**	**Wheel 2**	**Wheel 3**
No bars (blank)	0.487	0.515	0.492
One bar	0.317	0.230	0.095
Two bars	0.163	0.165	0.203
Three bars	0.033	0.090	0.210

The payoff table gives the following information for using $1 coins:

Game Result	**Payoff**
Three bars on each wheel	$100
Two bars on each wheel	$50
One bar on each wheel	$25
Any bar on each wheel	$3

Compute the mean payoff for this machine. Remember that this is only an estimate based on the 600 games.

(b) If you insert five coins, all payoff are five times larger, except that if you get three bars on each wheel, the payoff is $150. Calculate the mean payoff if five coins are used.

(c) With a litte effort, the true makeup of the machine shows that each "wheel" has 30 positions. The number of positions having the different symbols are as follows:

	NUMBER OF POSITIONS		
Symbols	**Wheel 1**	**Wheel 2**	**Wheel 3**
No bars (blank)	15	15	15
One bar	9	7	3
Two bars	5	5	6
Three bars	1	3	6

Compute the mean payoff for this machine. Note that this is close to, but not exactly the same as, the mean payoff using the result of 600 games.

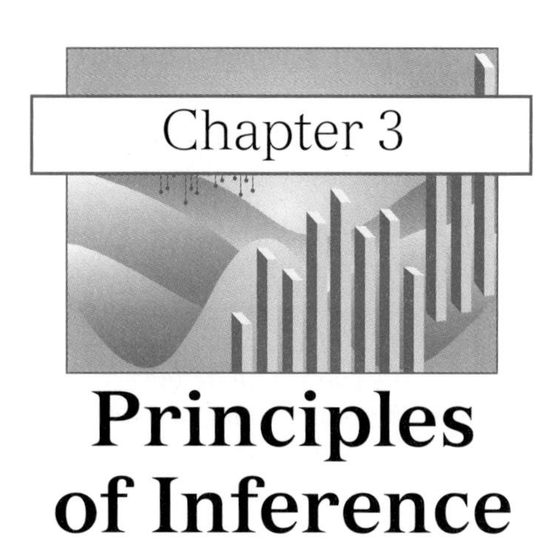

Principles of Inference

EXAMPLE 3.1

Is Office Rent More Expensive in Atlanta? A businessman in Atlanta is considering moving to Jacksonville, Florida, to reduce the office rental costs for his company. In the October 1990 issue of the *Jacksonville Journal*, the mean cost of leasing office space for all downtown buildings in Jacksonville was quoted as being \$12.61 per square foot with a standard deviation of \$4.50. To compare costs with those in Atlanta, the businessman sampled 36 office buildings in Atlanta and found a mean leasing cost of \$13.55 per square foot. Does this mean that leasing office space in Atlanta is really higher? Should the businessman consider moving to Jacksonville to save money on rent (assuming other factors equal)? This chapter presents methodology that can be used to help answer this question. This problem will be solved in Section 3.2. ■

3.1 Introduction

As we have repeatedly noted, one of the primary objectives of a statistical analysis is to use data from a sample to make inferences about the population from which the sample was drawn. In this chapter we present the basic procedures for making such inferences.

As we will see, the sampling distributions discussed in Chapter 2 play a pivotal role in statistical inference. Because inference on an unknown population parameter is usually based solely on a statistic computed from a single sample, we rely on these distributions to determine how reliable this inference is. That is, a statistical inference is composed of two parts:

1. a *statement* about the value of that parameter, and
2. a measure of the *reliability* of that statement, usually expressed as a probability.

Traditionally statistical inference is done with one of two different but related objectives in mind.

1. We conduct tests of hypotheses, in which we hypothesize that one or more parameters have some specific values or relationships, and make our decision about the parameter(s) based on one or more sample statistic. In this type of inference, the reliability of the decision is the probability that the decision is incorrect.
2. We estimate one or more parameters using sample statistics. This estimation is usually done in the form of an interval, and the reliability of this inference is expressed as the level of confidence we have in the interval.

We usually refer to an incorrect decision in a hypothesis test as "making an error" of one kind or another. Making an error in a statistical inference is not the same as making a mistake; the term simply recognizes the fact that the possibility of making an incorrect inference is an inescapable fact of statistical inference. The best we can do is to try to evaluate the reliability of our inference. Fortunately, if the data used to perform a statistical inference are a random sample, we can use sampling distributions to calculate the probability of making an error and therefore quantify the reliability of our inference.

In this chapter we present the basic principles for making these inferences and see how they are related. As you go through this and the next two chapters, you will note that hypothesis testing is presented before estimation. The reason for this is that it is somewhat easier to introduce them in this order, and since they are closely related, once the concept of the hypothesis test is understood, the estimation principles are easily grasped. We want to emphasize that both are equally important and each should be used where appropriate. To avoid extraneous issues, in this chapter we use two extremely simple (and not very interesting) examples that have little practical application. Once we have learned these principles, we can apply them to more interesting and useful applications. That is the subject of the remainder of this book.

3.2 Hypothesis Testing

A hypothesis usually results from speculation concerning observed behavior, natural phenomena, or established theory. If the hypothesis is stated in terms of population parameters such as the mean and variance, the hypothesis is called a **statistical hypothesis**. Data from a sample (which may be an experiment) are used to test the validity of the hypothesis. A procedure that enables us to agree or disagree with the statistical hypothesis using data from a sample is called a **test** of the hypothesis. Some examples of hypothesis tests are:

- A consumer-testing organization determining whether a type of appliance is of standard quality (say, an average lifetime of a prescribed length) would base their test on the examination of a sample of prototypes of the

appliance. The result of the test may be that the appliance is not of acceptable quality and the organization will recommend against its purchase.
- A test of the effect of a diet pill on weight loss would be based on observed weight losses of a sample of healthy adults. If the test concludes the pill is effective, the manufacturer can safely advertise to that effect.
- To determine whether a teaching procedure enhances student performance, a sample of students would be tested before and after exposure to the procedure and the differences in test scores subjected to a statistical hypothesis test. If the test concludes that the method is not effective, it will not be used.

General Considerations

To illustrate the general principles of hypothesis testing, consider the following two simple examples:

EXAMPLE 3.2

There are two identically appearing bowls of jelly beans. Bowl 1 contains 60 red and 40 black jelly beans, and bowl 2 contains 40 red and 60 black jelly beans. Therefore, the proportion of red jelly beans, p, for the two bowls are

$$\text{Bowl 1: } p = 0.6$$

$$\text{Bowl 2: } p = 0.4.$$

One of the bowls is sitting on the table, but you do not know which one it is (you cannot see inside it). You suspect that it is bowl 2, but you are not sure. To test your hypothesis that bowl 2 is on the table you sample five jelly beans.[1] The data from this sample, specifically the number of red jelly beans, is the sample statistic that will be used to test the hypothesis that bowl 2 is on the table. That is, based on this sample, you will decide whether bowl 2 is the one on the table. ■

EXAMPLE 3.3

A company that packages salted peanuts in 8-oz. jars is interested in maintaining control on the amount of peanuts put in jars by one of its machines. Control is defined as averaging 8 oz. per jar and not consistently over- or underfilling the jars. To monitor this control, a sample of 16 jars is taken from the line at random time intervals and their contents weighed. The mean weight of peanuts in these 16 jars will be used to test the hypothesis that the machine is indeed working properly. If it is deemed not to be doing so, a costly adjustment will be needed.[2] ■

These two examples will be used to illustrate the procedures presented in this chapter.

---

[1]To make the necessary probability calculations easier, you replace each jelly bean before selecting a new one; this is called sampling with replacement and allows the use of the binomial probability distribution presented in Section 2.3.

[2]Note the difference between this problem and Example 2.13, the control chart example. In this case, a decision to adjust the machine is to be made on one sample only, while in Example 2.13 it is made by an examination of its performance over time.

The Hypotheses

Statistical hypothesis testing starts by making a set of two statements about the parameter or parameters in question. These are usually in the form of simple mathematical relationships involving the parameters. The two statements are exclusive and exhaustive, which means that one or the other statement must be true, but they cannot both be true. The first statement is called the *null* hypothesis and is denoted by H_0, and the second is called the *alternative* hypothesis and is denoted by H_1.

DEFINITION 3.1
The **null hypothesis** is a statement about the values of one or more parameters. This hypothesis represents the status quo and is usually not rejected unless the sample results strongly imply that it is false.

For Example 3.2, the null hypothesis is

Bowl 2 is on the table.

In bowl 2, since 40 of the 100 jelly beans are red, the statistical hypothesis is stated in terms of a population parameter, $p =$ the proportion of red jelly beans in bowl 2. Thus the null hypothesis is

$$H_0\colon p = 0.4.$$

DEFINITION 3.2
The **alternative hypothesis** is a statement that contradicts the null hypothesis. This hypothesis is declared to be accepted if the null hypothesis is rejected. The alternative hypothesis is often called the research hypothesis because it usually implies that some action is to be performed, some money spent, or some established theory overturned.

In Example 3.2 the alternative hypothesis is

Bowl 1 is on the table,

for which the statistical hypothesis is

$$H_1\colon p = 0.6,$$

since 60 of the 100 jelly beans in bowl 1 are red. Because there are no other choices, the two statements form a set of two exclusive and exhaustive hypotheses. That is, the two statements specify all possible values of parameter p.

For Example 3.3, the hypothesis statements are given in terms of the population parameter μ, the mean weight of peanuts per jar. The null hypothesis is

$$H_0\colon \mu = 8,$$

which is the specification for the machine to be functioning correctly. The alternative hypothesis is

$$H_1: \mu \neq 8,$$

which means the machine is malfunctioning. These statements also form a set of two exclusive and exhaustive hypotheses, even though the alternative hypothesis does not specify a single value as it did for Example 3.2.

Rules for Making Decisions

After stating the hypotheses we specify what sample results will lead to the rejection of the null hypothesis. Intuitively, sample results (summarized as sample statistics) that lead to rejection of the null hypothesis should reflect an apparent contradiction to the null hypothesis. In other words, if the sample statistics have values that are unlikely to occur if the null hypothesis is true, then we decide the null hypothesis is false. The statistical hypothesis testing procedure consists of defining sample results that appear to sufficiently contradict the null hypothesis to justify rejecting it.

In Section 2.5 we showed that a sampling distribution can be used to calculate the probability of getting values of a sample statistic from a given population. If we now define "unlikely" as some small probability, we can use the sampling distribution to determine a range of values of a sample statistic that is unlikely to occur if the null hypothesis is true. The occurrence of values in that range may then be considered grounds for rejecting that hypothesis. Statistical hypothesis testing consists of appropriately defining that region of values.

DEFINITION 3.3
The **rejection region** (also called the **critical region**) is the range of values of a sample statistic that will lead to rejection of the null hypothesis.

In Example 3.2, the null hypothesis specifies the bowl having the lower proportion of red jelly beans; hence observing a large proportion of red jelly beans would tend to contradict the null hypothesis. For now, we will arbitrarily decide that having a sample with all red jelly beans provides sufficient evidence to reject the null hypothesis. If we let Y be the number of red jelly beans, the rejection region is defined as $y = 5$.

In Example 3.3, any sample mean weight $\bar{Y}$ not equal to 8 oz. would seem to contradict the null hypothesis. However, since some variation is expected, we would probably not want to reject the null hypothesis for values reasonably close to 8 oz. For the time being we will arbitrarily decide that a mean weight of below 7.9 or above 8.1 oz. is not "reasonably close," and we will therefore reject the null hypothesis if the mean weight of our sample occurs in this region. Thus, the rejection region for this example contains the values of $\bar{y} < 7.9$ or $\bar{y} > 8.1$.

If the value of the sample statistic falls in the rejection region, we know what decision to make. If it does not fall in the rejection region, we have a choice of decisions. First, we could accept the null hypothesis as being true. As we will see, this decision may not be the best choice. Our other choice would be to "fail to reject" the null hypothesis. As we will see, this is not necessarily the same as accepting the null hypothesis.

Table 3.1

Results of a Hypothesis Test

The Decision	IN THE POPULATION H_0 is True	H_0 is not True
H_0 is not rejected	Decision is correct	A type II error has been committed
H_0 is rejected	A type I error has been committed	Decision is correct

Possible Errors in Hypothesis Testing

In Section 3.1 we emphasized that statistical inferences based on sample data may be subject to what we called errors. Actually, it turns out that results of a hypothesis test may be subject to two distinctly different errors, which are called type I and type II errors. These errors are defined in Definitions 3.4 and 3.5 and illustrated in Table 3.1.

DEFINITION 3.4
A **type I error** occurs when we incorrectly reject H_0, that is, when H_0 is actually true and our sample-based inference procedure rejects it.

DEFINITION 3.5
A **type II error** occurs when we incorrectly fail to reject H_0, that is, when H_0 is actually not true, and our inference procedure fails to detect this fact.

In Example 3.2 the rejection region consisted of finding all five jelly beans in the sample to be red. Hence, the type I error occurs if all five sample jelly beans are red, the null hypothesis is rejected, and we proclaim the bowl to be bowl 1 but, in fact, bowl 2 is actually on the table. Alternatively, a type II error will occur if our sample has four or fewer red jelly beans (or one or more black jelly beans), in which case H_0 is not rejected, and we therefore proclaim that it is bowl 2, but, in fact, bowl 1 is on the table.

In Example 3.3, a type I error will occur if the machine is indeed working properly, but our sample yields a mean weight of over 8.1 or under 7.9 oz., leading to rejection of the null hypothesis and therefore an unnecessary adjustment to the machine. Alternatively, a type II error will occur if the machine is malfunctioning but the sample mean weight falls between 7.9 and 8.1 oz. In this case we fail to reject H_0 and do nothing when the machine really needs to be adjusted.

Obviously we cannot make both types of errors simultaneously, and in fact we may not make either, but the possibility does exist. In fact, we will usually never know whether any error has been committed. The only way to avoid any chance of error is not to make a decision at all, hardly a satisfactory alternative.

Probabilities of Making Errors

If we assume that we have the results of a random sample, we can use the characteristics of sampling distributions presented in Chapter 2 to calculate the **probabilities** of making either a type I or type II error for any specified decision rule.

DEFINITION 3.6

α: denotes the probability of making a type I error
β: denotes the probability of making a type II error

The ability to provide these probabilities is a key element in statistical inference, because they measure the reliability of our decisions. We will now show how to calculate these probabilities for our examples.

Calculating α for Example 3.2 The null hypothesis specifies that the probability of drawing a red jelly bean is 0.4 (bowl 2), and the null hypothesis is to be rejected with the occurrence of five red jelly beans. Then the probability of making a type I error is the probability of getting five red jelly beans in a sample of five from bowl 2. If we let Y be the number of red jelly beans in our sample of five, then

$$\alpha = P(Y = 5 \text{ when } p = 0.4).$$

The use of binomial probability distribution (Section 2.3) provides the result $\alpha = (0.4)^5 = 0.01024$. Thus the probability of incorrectly rejecting a true null hypothesis in this case is 0.01024; that is, there is approximately a 1 in 100 chance that bowl 2 will be mislabeled bowl 1 using the described decision rule.

Calculating α for Example 3.3 For this example, the null hypothesis was to be rejected if the mean weight was less than 7.9 or greater than 8.1 oz. If $\bar{Y}$ is the sample mean weight of 16 jars, the probability of a type I error is

$$\alpha = P(\bar{Y} < 7.9 \text{ or } \bar{Y} > 8.1 \text{ when } \mu = 8).$$

Assume for now that we know[3] that σ, the standard deviation of the population of weights, is 0.2 and that the distribution of weights is approximately normal. If the null hypothesis is true, the sampling distribution of the mean of 16 jars is normal with $\mu = 8$ and $\sigma = 0.2/\sqrt{16} = 0.05$ (see discussion on the normal distribution in Section 2.5). The probability of a type I error corresponds to the shaded area in Fig. 3.1.

---

[3]This is an assumption made here to simplify matters. In Chapter 4 we present the method required if we calculate the standard deviation from the sample data.

Figure 3.1

Rejection Region for Sample Mean

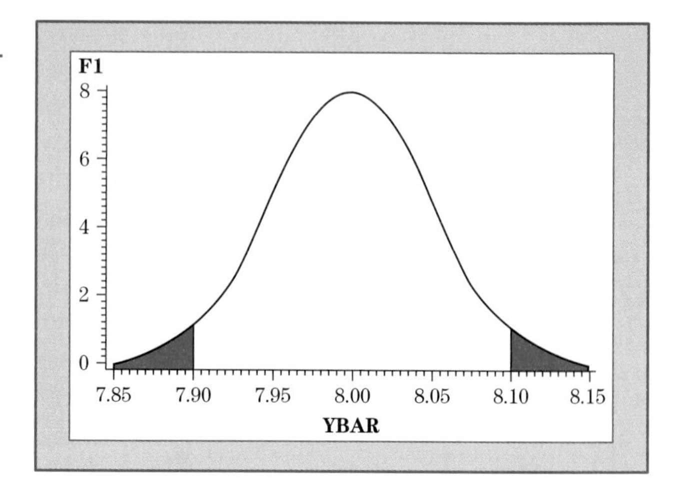

Using the tables of the normal distribution we compute the area for each portion of the rejection region

$$P(\bar{Y} < 7.9) = P\left[Z < \frac{7.9 - 8}{(0.2/\sqrt{16})}\right] = P(Z < -2.0) = 0.0228$$

and

$$P(\bar{Y} > 8.1) = P\left(Z > \frac{8.1 - 8}{0.2/\sqrt{16}}\right) = P(Z > 2.0) = 0.0228.$$

Hence

$$\alpha = 0.0228 + 0.0228 = 0.0456.$$

Thus the probability of adjusting the machine when it does not need it (using the described decision rule) is slightly less than 0.05 (or 5%).

Calculating β for Example 3.2 Having determined α for a specified decision rule, it is of interest to determine β. This probability can be readily calculated for Example 3.2. Recall that the type II error occurs if we fail to reject the null hypothesis when it is not true. For this example, this occurs if bowl 1 is on the table but we did not get the five red jelly beans required to reject the null hypothesis that bowl 2 is on the table. The probability of a type II error, which is denoted by β, is then the probability of getting four or fewer red jelly beans in a size-five sample from bowl 1. If we let Y be the number of red jelly beans in the sample, then

$$\beta = P(Y \leq 4 \text{ when } p = 0.6).$$

Using the probability rules from Section 2.2, we know that

$$P(Y \leq 4) + P(Y = 5) = 1.$$

Since $(Y = 5)$ is the complement of $(Y \leq 4)$,

$$P(Y \leq 4) = 1 - P(Y = 5).$$

Now

$$P(Y = 5) = (0.6)^5,$$

and therefore

$$\beta = 1 - (0.6)^5 = 1 - 0.07776 = 0.92224.$$

That is, the probability of making a type II error in Example 3.2 is over 92%. This value of β is unacceptably large. That is, if on the basis of this test we conclude that bowl 2 is on the table, the probability that we are wrong is 0.92!

Calculating β for Example 3.3 For Example 3.3, H_1 does not specify a single value for μ but instead includes all values of $\mu \neq 8$. Therefore, calculating the probability of the type II error requires that we examine the probability of the sample mean being outside the rejection region for every value of $\mu \neq 8$. These calculations and further discussion of β are presented later in this section where we discuss type II errors.

Choosing between α and β

The probability of making a type II error can be decreased by making rejection easier, which is accomplished by making the rejection region larger. For example, suppose we decide to reject H_0 if either four or five of the jelly beans are red. In this case,

$$\alpha = P(Y \geq 4 \text{ when } p = 0.4) = 0.087$$

and

$$\beta = P(Y < 4 \text{ when } p = 0.6) = 0.663.$$

Note that by changing the rejection region we succeeded in lowering β but we increased α. This will always happen if the sample size is unchanged. In fact, if by changing the rejection region α becomes unacceptably large, no satisfactory testing procedure is available for a sample of five jelly beans, a condition that often occurs when sample sizes are small (see Section 3.4). This relationship between the two types of errors prevents us from constructing a hypothesis test that has a probability of 0 for either error. In fact, the only way to ensure that $\alpha = 0$ is to never reject a hypothesis, while to ensure that $\beta = 0$ the hypothesis should always be rejected, regardless of any sample results.

Five-Step Procedure for Hypothesis Testing

In the above presentation we have shown how to determine the probability of making a type I error for some arbitrarily chosen rejection region. The more frequently used method is to specify an acceptable maximum value for α and then delineate a rejection region for a sample statistic that satisfies this value. A hypothesis test can be formally summarized as a five-step process. Briefly these steps are as follows:

Step 1: Specify H_0, H_1, and an acceptable level of α.
Step 2: Define a sample-based test statistic and the rejection region for the specified H_0.

Step 3: Collect the sample data and calculate the test statistic.
Step 4: Make a decision to either reject or fail to reject H_0. This decision will normally result in a recommendation for action.
Step 5: Interpret the results in the language of the problem. It is imperative that the results be usable by the practitioner.

We now discuss various aspects of these steps.

Step 1 consists of specifying H_0 and H_1 and a choice of a maximum acceptable value of α. This value is based on the seriousness or cost of making a type I error in the problem being considered.

DEFINITION 3.7
The **significance level** of a hypothesis test is the maximum acceptable probability of rejecting a true null hypothesis.[4]

The reason for specifying α (rather than β) for a hypothesis test is based on the premise that the type I error is of prime concern. For this reason the hypothesis statement must be set up in such a manner that the type I error is indeed the more costly. The significance level is then chosen considering the cost of making that error.

In Example 3.2, H_0 was the assertion that the bowl on the table was bowl 2. In this example interchanging H_0 and H_1 would probably not cause any major changes unless there was some extra penalty for one of the errors. Thus, we could just as easily have hypothesized that the bowl was really 1, which would have made H_0: $p = 0.6$ instead of H_0: $p = 0.4$.

In Example 3.3 we stated that the null hypothesis is $\mu = 8$. In this example the choice of the appropriate H_0 is clear: There is a definite cost if we make a type I error since this error may cause an unnecessary adjustment on a properly working machine. Of course, making a type II error is not without cost, but since we have not accepted H_0, we are free to repeat the sampling at another time, and if the machine is indeed malfunctioning, the null hypothesis will eventually be rejected.

Why Do We Focus on the Type I Error?

In general, the null hypothesis is usually constructed to be that of the status quo; that is, it is the hypothesis requiring no action to be taken, no money to be spent, or in general nothing changed. This is the reason for denoting this as the null or nothing hypothesis. Since it is usually costlier to incorrectly reject the status quo than it is to do the reverse, this characterization of the null hypothesis does indeed cause the type I error to be of greater concern. In statistical hypothesis testing, the null hypothesis will invariably be stated in terms of an "equal" condition existing.

---

[4]Because the selection and use of the significance level is fundamental to this procedure, it is often referred to as a significance test. Although some statisticians make a minor distinction between hypothesis and significance testing, we use the two labels interchangeably.

On the other hand, the alternative hypothesis describes conditions for which something will be done. It is the action or research hypothesis. In an experimental or research setting, the alternative hypothesis is that an established (status quo) hypothesis is to be replaced with a new one. Thus, the research hypothesis is the one we actually want to support, which is accomplished by rejecting the null hypothesis with a sufficiently low level of α such that it is unlikely that the new hypothesis will be erroneously pronounced as true.

In Example 3.2, we thought the bowl was 2 (the status quo), and would only change our mind if the sample showed significant evidence that we were wrong. In Example 3.3 the status quo is that the machine is performing correctly; hence the machine would be left alone unless the sample showed so many or so few peanuts so as to provide sufficient evidence to reject H_0.

We can now see that it is quite important to specify an appropriate significance level. Because making the type I error is likely to have the more serious consequences, the value of α is usually chosen to be a relatively small number, and smaller in some cases than in others. That is, α must be selected so that an acceptable level of risk exists that the test will incorrectly reject the null hypothesis. Historically and traditionally, α has been chosen to have values of 0.10, 0.05, or 0.01, with 0.05 being most frequently used. These values are not sacred but do represent convenient numbers and allow the publication of statistical tables for use in hypothesis testing. We shall use these values often throughout the text. (See, however, the discussion of p values later in this section.)

Choosing α

As we saw in Example 3.2, α and β are inversely related. Unless the sample size is increased, we can reduce α only at the price of increasing β. In Example 3.2 there was little difference in the consequences of a type I or type II error; hence, the hypothesis test would probably be designed to have approximately equal levels of α and β. In Example 3.3 making the type I error will cause a costly adjustment to be made to a properly working machine, while if the type II error is committed we do not adjust the machine when needed. This error also entails some cost such as wasted peanuts or unsatisfied customers. Unless the cost of adjusting the machine is extremely high, a reasonable choice here would be to use the "standard" value of 0.05.

Some examples of problems for which one or the other type of error is more serious include the following:

- Malnutrition among young children can have serious consequences. Assume that six-year-old children should average about 10 kg in weight to be considered normal. If a sample of children from a low-income neighborhood is to be tested[5] for subnormal weight, we would probably use H_0: $\mu = 10$ kg

[5]An alternative hypothesis that specifies values in only one direction from the null hypothesis is called a one-sided or one-tailed alternative and requires some modifications in the testing procedure. One-tailed hypothesis tests are discussed later in this section.

and H_1: $\mu < 10$ kg. Rejection of the null hypothesis implies that the children in that neighborhood are of subnormal weight, which may lead to an expanded school lunch program. A type I error would cause the initiation of an expanded school lunch program for children who do not need it, which would be an unnecessary expenditure, but would certainly do no physical harm to the children. Hence the type I error is not very serious. A type II error, on the other hand, would result in no expanded school lunch program being initiated for children who really need it. This error appears to be more serious, and a low level of β would be needed. This, of course, would indicate that a high level of α would be chosen (or a different testing principle, see Section 3.6).

- A chemist working for a major food company has developed a new formulation for instant pudding that he believes tastes better but is more expensive to make. Using a sample of taste testers and a rating scale, he tests H_0: the mean rating of the new formulation is the same as that of the old formulation, against H_1: the mean rating for the new pudding is larger than that of the old. A type I error would result if the hypothesis test concluded that the new pudding tastes better and it really does not. The result of this error would be marketing a product that costs more but does not taste better, probably causing the company to lose a share of the market, which would be a relatively costly error. A type II error would result in failing to market a superior pudding at this time, which could potentially result in some loss of income. Therefore, a low value for α would appear to be appropriate.
- When a drug company tests a new drug, there are two considerations that must be tested: (1) the toxicity (side effects) and (2) the effectiveness. For (1), the null hypothesis would be that the drug is toxic. This is because we would want to "prove" that it is not. For this test we would want a very small α, because a type I error would have extremely serious consequences (a significance level of 0.0001 would not be uncommon). For (2), the null hypothesis would be that the drug is not effective and a type I error would result in the drug being put on the market when it is not effective. The ramifications of this error would depend on the existing competitive drug market and the cost to both the company and society of marketing an ineffective drug.

DEFINITION 3.8

The **test statistic** is a sample statistic whose sampling distribution can be specified for both the null and alternative hypothesis case (although the sampling distribution when the alternative hypothesis is true may often be quite complex). After specifying the appropriate significance level of α, the sampling distribution of this statistic is used to define the rejection region.

DEFINITION 3.9

The **rejection region** comprises the values of the test statistic for which (1) the probability when the null hypothesis is true is less than or equal to the specified α and (2) probabilities when H_1 is true are greater than they are under H_0.

In **Step 2** we define the **test statistic** and the **rejection region**.

For Example 3.3 the appropriate test statistic is the sample mean. The sampling distribution of this statistic has already been used to show that the initially proposed rejection region of $\bar{y} < 7.9$ and $\bar{y} > 8.1$ produces a value of 0.0456 for α. If we had wanted α to be 0.05, this rejection region would appear to have been a very lucky guess! However, in most hypothesis tests it is necessary to specify α first and then use this value to delineate the rejection region. In the discussion of the significance level for Example 3.3 an appropriate level of α was chosen to be 0.05.

Remember, α is defined as

$$P(\bar{Y} \text{ falls in the rejection region when } H_0 \text{ is true}).$$

We define the rejection region by a set of boundary values, often called critical values, that are denoted by $C1$ and $C2$. The probability α is then defined as

$$P(\bar{Y} < C1 \text{ when } \mu = 8) + P(\bar{Y} > C2 \text{ when } \mu = 8).$$

We want to find values of $C1$ and $C2$ so that this probability is 0.05. This is obtained by finding the $C1$ and $C2$ that satisfy the expression

$$\alpha = P\left[Z < \frac{C1 - 8}{0.2/\sqrt{16}}\right] + P\left[Z > \frac{C2 - 8}{0.2/\sqrt{16}}\right] = 0.05,$$

where Z is the standard normal variable. Because of the symmetry of the normal distribution, exactly half of the rejection region is in each tail; hence,

$$P = \left[Z < \frac{C1 - 8}{0.05}\right] = P\left[Z > \frac{C2 - 8}{0.05}\right] = 0.025.$$

The values of $C1$ and $C2$ that satisfy this probability statement are found by using the standard normal table, where we find that the values of $z = -1.96$ and $z = +1.96$ satisfy our probability criteria. We use these values to solve for $C1$ and $C2$ in the equations $[(C1-8)/0.05] = -1.96$ and $[(C2-8)/0.05] = 1.96$. The solution yields $C1 = 7.902$ and $C2 = 8.098$; hence, the rejection region is

$$\bar{y} < 7.902 \quad \text{or} \quad \bar{y} > 8.098,$$

as seen in Fig. 3.2. The rejection region of Fig. 3.2 is given in terms of the test statistic $\bar{Y}$, the sample mean.

It is computationally more convenient to express the rejection region in terms of a test statistic that can be compared directly to a table, such as that

Figure 3.2

Rejection Region for 0.05 Significance

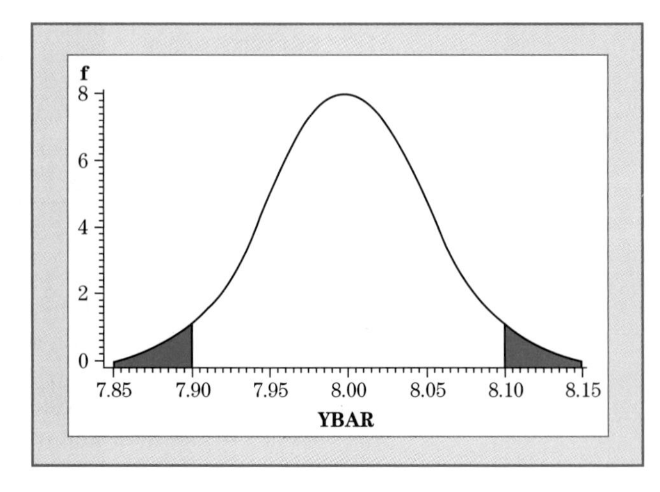

of the normal distribution. In this case the test statistic is

$$Z = \frac{\bar{Y} - \mu}{\sigma/\sqrt{n}}$$

$$= \frac{\bar{Y} - 8}{0.05},$$

which has the standard normal distribution and can be compared directly with the values read from the table. Then the rejection region for this statistic is

$$z < -1.96 \quad \text{or} \quad z > 1.96,$$

which can be more compactly written as $|z| > 1.96$. In other words we reject the null hypothesis if the value we calculate for Z has an absolute value (value ignoring sign) larger than 1.96.

Step 3 of the hypothesis test is to collect the sample data and compute the test statistic. (While this strict order may not be explicitly followed in practice, the sample data should not be used until the first two steps have been completed!) In Example 3.3, suppose our sample of 16 peanut jars yielded a sample mean value $\bar{y} = 7.89$. Then

$$z = (7.89 - 8)/0.05 = -2.20, \quad \text{or} \quad |z| = 2.20.$$

Step 4 compares the value of the test statistic to the rejection region to make the decision. In this case we have observed that the value 2.20 is larger than 1.96 so our decision is to reject H_0. This is often referred to as a "statistically significant" result, which means that the difference between the hypothesized value of $\mu = 8$ and the observed value of $\bar{y} = 7.89$ is large enough to be statistically significant.

In **Step 5** we then conclude that the mean weight of nuts being put into jars is not the desired 8 oz. and the machine should be adjusted.

The Five Steps for Example 3.3

The hypothesis for Example 3.3 is summarized as follows:

Step 1:

$$H_0\colon \mu = 8$$
$$H_1\colon \mu \neq 8$$
$$\alpha = 0.05.$$

Step 2: The test statistic is

$$Z = \frac{\bar{Y} - 8}{0.2/\sqrt{16}}$$

whose sampling distribution is the standard normal. We specify $\alpha = 0.05$; hence we will reject H_0 if $|z| > 1.96$.

Step 3: Sample results: $n = 16$, $\bar{y} = 7.89$, $\sigma = 0.2$ (assumed);

$$z = (7.89 - 8)/[0.2/\sqrt{16}] = -2.20, \quad \text{hence } |z| = 2.20.$$

Step 4: $|z| > 1.96$; hence we reject H_0.
Step 5: We conclude $\mu \neq 8$ and recommend that the machine be adjusted.

Suppose that in our initial setup of the hypothesis test we had chosen α to be 0.01 instead of 0.05. What changes? This test is summarized as follows:

Step 1:

$$H_0\colon \mu = 8$$
$$H_1\colon \mu \neq 8$$
$$\alpha = 0.01.$$

Step 2: Reject H_0 if $|z| > 2.576$.
Step 3: Sample results: $n = 16$, $\sigma = 0.2$, $\bar{y} = 7.89$;

$$z = (7.89 - 8)/0.05 = -2.20.$$

Step 4: $|z| < 2.576$; hence we fail to reject $H_0\colon \mu = 8$.
Step 5: We do not recommend that the machine be readjusted.

We now have a problem. We have failed to reject the null hypothesis and do nothing. However, remember that we have not proved that the machine is working perfectly. In other words, *failing to reject the null hypothesis does not mean the null hypothesis was accepted.* Instead, we are simply saying that this particular test (or experiment) does not provide sufficient evidence to have the machine adjusted at this time. In fact, in a continuing quality control program, the test will be repeated in due time.

P Values

Having to specify a significance level before making a hypothesis test seems unnecessarily restrictive because many users do not have a fixed or definite idea of what constitutes an appropriate value for α. Also it is quite difficult to do when using computers because the user would have to specify an alpha for every test being requested. Another problem with using a specified significance level is that the ultimate conclusion may be affected by very minor changes in sample statistics.

As an illustration, we observed that in Example 3.3 the sample value of 7.89 leads to rejection with $\alpha = 0.05$. However, if the sample mean had been 7.91, certainly a very similar result, the test statistic would be -1.8, and we would not reject H_0. In other words, the decision of whether to reject may depend on minute differences in sample results.

We also noted that with a sample mean of 7.89 we would reject H_0 with $\alpha = 0.05$ but not with $\alpha = 0.01$. The logical question then is this: What about $\alpha = 0.02$, or $\alpha = 0.03$, or ... ? This question leads to a method of reporting the results of a significance test without having to choose an exact level of significance, but instead leaves that decision to the individual who will actually act on the conclusion of the test. This method of reporting results is referred to as reporting the p value.

DEFINITION 3.10

The ***p* value** is the probability of committing a type I error if the actual sample value of the statistic is used as the boundary of the rejection region. It is therefore the smallest level of significance for which we would reject the null hypothesis with that sample. Consequently, the p value is often called the "attained" or the "empirical" significance level. It is also interpreted as an indicator of the weight of evidence against the null hypothesis.

In Example 3.3, the use of the normal table allows us to calculate the p value accurate to about four decimal places. For the sample $\bar{y} = 7.89$, this value is $P(|Z| > 2.20)$. Remembering the symmetry of the normal distribution, this is easily calculated to be $2P(Z > 2.20) = 0.0278$. This means that the management of the peanut-packing establishment can now evaluate the results of this experiment. They would reject the null hypothesis with a level of significance of 0.0278 or higher, and fail to reject it at anything lower.

Using the p value approach, Example 3.3 is summarized as follows:

Step 1:

$$H_0: \mu = 8$$

$$H_1: \mu \neq 8.$$

Step 2: Sample: $n = 16$, $\sigma = 0.2$, $\bar{y} = 7.89$;

$$z = (7.89 - 8)/0.05 = -2.20.$$

Step 3: $p = P(|Z| > 2.20) = 0.0278$; hence the p value is 0.0278. Therefore, we can say that the probability of observing a test statistic at least this extreme if the null hypothesis is true is 0.0278.

One feature of this approach is that the significance level need not be specified by the statistical analyst. In situations where the statistical analyst is not the same person who makes decisions, the analyst provides the p value and the decision maker determines the significance level based on the costs of making the type I error. For these reasons, many research journals now require that the results of such tests be published in this manner.

It is, in fact, actually easier for a computer program to provide p values, which are often given to three or more decimal places. However, when tests are calculated manually we must use tables. And because many tables provide for only a limited set of probabilities, p values can only be approximately determined. For example, we may only be able to state that the p value for the peanut jar example is between 0.01 and 0.05.

Note that the five steps of a significance test require that the significance level α be specified before conducting the test, while the p value is determined after the data have been collected and analyzed. Thus the use of a p value and a significance test are similar, but not strictly identical. It is, however, possible to use the p value in a significance test by specifying α in Step 1 and then altering Step 3 to read: Compute the p value and compare with the desired α. If the p value is smaller than α, reject the null hypothesis; otherwise fail to reject.

ALTERNATE DEFINITION 3.10

A ***p* value** is the probability of observing a value of the test statistic that is at least as contradictory to the null hypothesis as that computed from the sample data.

Thus the p value measures the extent to which the test statistic disagrees with the null hypothesis.

EXAMPLE 3.4

An aptitude test has been used to test the ability of fourth graders to reason quantitatively. The test is constructed so that the scores are normally distributed with a mean of 50 and standard deviation of 10. It is suspected that, with increasing exposure to computer-assisted learning, the test has become obsolete. That is, it is suspected that the mean score is no longer 50, although σ remains the same. This suspicion may be tested based on a sample of students who have been exposed to a certain amount of computer-assisted learning.

Solution The test is summarized as follows:

1.

$$H_0: \mu = 50,$$

$$H_1: \mu \neq 50.$$

2. The test is administered to a random sample of 500 fourth graders. The test statistic is
$$Z = \frac{\bar{Y} - 50}{10/\sqrt{500}}.$$
The sample yields a mean of 51.07. The test statistic has a value of
$$z = \frac{51.07 - 50}{10/\sqrt{500}} = 2.39.$$
3. The p value is computed as $2P(Z > 2.39) = 0.0168$. Because the construction of a new test is quite expensive, it may be determined that the level of significance should be less than 0.01, in which case the null hypothesis will not be rejected. However, the p value of 0.0168 may be considered sufficiently small to justify further investigation, say, by performing another experiment. ■

Type II Error and Power

In presenting the procedures for hypothesis and significance tests we have concentrated exclusively on the control over α, the probability of making the type I error. However, just because that error is the more serious one, we cannot completely ignore the type II error. There are many reasons for ascertaining the probability of that error, for example:

- The probability of making a type II error may be so large that the test may not be useful. This was the case for Example 3.2.
- Because of the trade-off between α and β, we may find that we may need to increase α in order to have a reasonable value for β.
- Sometimes we have a choice of testing procedures where we may get different values of β for a given α.

Unfortunately, calculating β is not always straightforward. Consider Example 3.3. The alternative hypothesis, H_1: $\mu \neq 8$, encompasses all values of μ not equal to 8. Hence there is a sampling distribution of the test statistic for each unique value of μ, each producing a different value for β. Therefore β must be evaluated for all values of μ contained in the alternative hypothesis, that is, all values of μ not equal to 8.

This is not really necessary. For practical purposes it is sufficient to calculate β for a few representative values of μ and use these values to plot a function representing β for all values of μ not equal to 8. A graph of β versus μ is called an "operating characteristic curve" or simply an OC curve.

To construct the OC curve for Example 3.3, we first select a few values of μ and calculate the probability of a type II error at these values. For example, consider $\mu = 7.80, 7.90, 7.95, 8.05, 8.10,$ and 8.20. Recall that for $\alpha = 0.05$ the rejection region is $\bar{y} < 7.902$ or $\bar{y} > 8.098$. The probability of a type II error is then the probability that $\bar{Y}$ does not fall in the rejection region, that is, $P(7.902 \leq \bar{Y} \leq 8.098)$, which is to be calculated for each of the specific values of μ given above.

Figure 3.3

Probability of a Type II Error When the Mean is 7.95

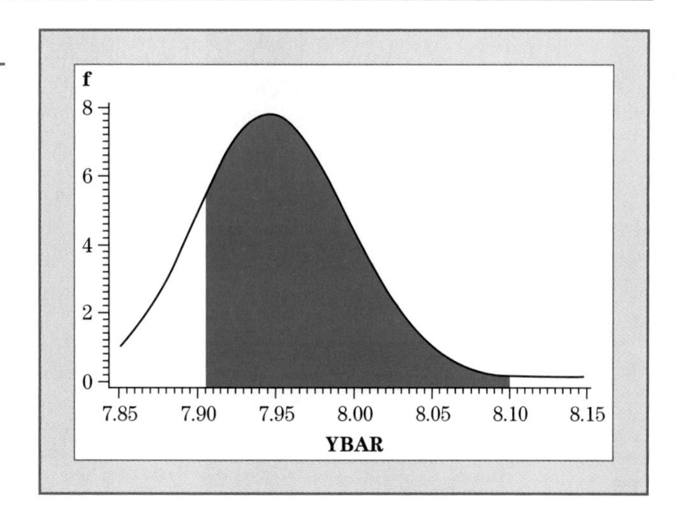

Figure 3.3 shows the sampling distribution for the mean if the population mean is 7.95 as well as the rejection region (nonshaded area) for testing the null hypothesis that $\mu = 8$. The type II error occurs when the sample mean is not in the rejection region. Therefore, as seen in the figure, the probability of a type II error when the true value of μ is 7.95 is

$$\begin{aligned}\beta &= P(7.902 \leq \bar{Y} \leq 8.098 \text{ when } \mu = 7.95)\\ &= P\{[(7.902 - 7.95)/0.05] \leq Z \leq [(8.098 - 7.95)/0.05]\}\\ &= P(-0.96 \leq Z \leq 2.96) = 0.8300,\end{aligned}$$

obtained by using the table of the normal distribution. This probability corresponds to the shaded area in Fig. 3.3.

Similarly, the probability of a type II error when $\mu = 8.05$ is

$$\begin{aligned}\beta &= P(7.902 \leq \bar{Y} \leq 8.098 \text{ when } \mu = 8.05)\\ &= P\{[(7.902 - 8.05)/0.05] \leq Z \leq [(8.098 - 8.05)/0.05]\}\\ &= P(-2.96 \leq Z \leq 0.96) = 0.8300.\end{aligned}$$

These two values of β are the same because of the symmetry of the normal distribution and also because in both cases μ is 0.05 units from the null hypothesis value. The probability of a type II error when $\mu = 7.90$, which is the same as that for $\mu = 8.10$, is calculated as

$$\begin{aligned}\beta &= P(7.902 \leq \bar{Y} \leq 8.098 \text{ when } \mu = 7.90)\\ &= P(0.04 \leq Z \leq 3.96) = 0.4840.\end{aligned}$$

In a similar manner we can obtain β for $\mu = 7.80$ and $\mu = 8.20$, which has the value 0.0207.

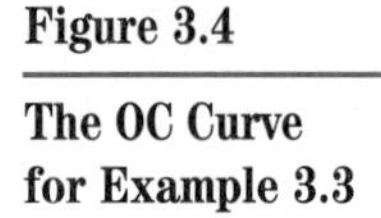

Figure 3.4

The OC Curve for Example 3.3

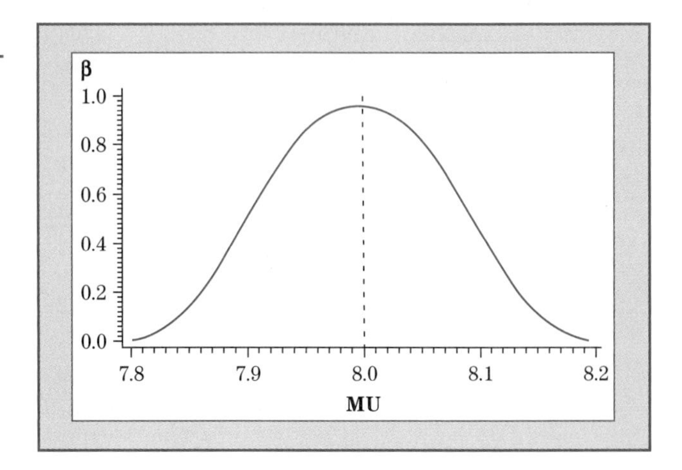

While it is impossible to make a type II error when the true mean is equal to the value specified in the null hypothesis, β approaches $(1 - \alpha)$ as the true value of the parameter approaches that specified in the null hypothesis. The OC curve can now be constructed using these values. Figure 3.4 gives the OC curve for Example 3.3. Note that the curve is indeed symmetric and continuous. Its maximum value is $(1 - \alpha) = 0.95$ at $\mu = 8$, and it approaches zero as the true mean moves further from the H_0 value. From this OC curve we may read (at least approximately) the value of β for any value of μ we desire.

The OC curve shows the logic behind the hypothesis testing procedure as follows:

- We have controlled the probability of making the more serious type I error.
- The OC curve shows that the probability of making the type II error is larger when the difference between the true value of the mean is close to the null hypothesis value, but decreases as that difference becomes greater. In other words, the higher probabilities of failing to reject the null hypothesis occur when the null hypothesis is "almost" true, in which case the type II error may not have serious consequences.

For example, in the peanut jar problem, failing to reject simply means that we continue using the machine but also continue the sampling inspection plan. If the machine is only slightly off, continuing the operation is not likely to have very serious consequences, but since sampling inspection continues, we will have the larger probability of rejection if the machine strays very far from its target.

Power

As a practical matter we are usually more interested in the probability of not making a type II error, that is, the probability of correctly rejecting the null hypothesis when it is false.

Figure 3.5

Power Curve for Example 3.3

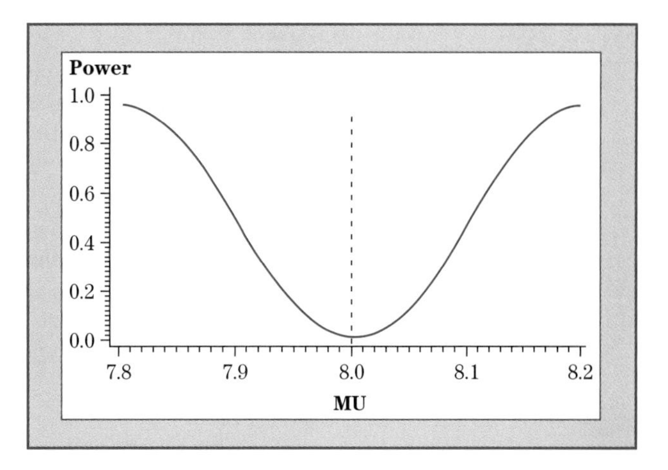

DEFINITION 3.11
The **power** of a test is the probability of correctly rejecting the null hypothesis when it is false.

The power of a test is $(1-\beta)$ and depends on the true value of the parameter μ. The graph of power versus all values of μ is called a **power curve**. The power curve for Example 3.3 is given in Fig. 3.5. Some features of a power curve are as follows:

- The power of the test increases and approaches unity as the true mean gets further from the null hypothesis value. This feature simply confirms that it is easier to deny a hypothesis as it gets further from the truth.
- As the true value of the population parameter approaches that of the null hypothesis, the power approaches α.
- Decreasing α while keeping the sample size fixed will produce a power curve that is everywhere lower. That is, decreasing α decreases the power.
- Increasing the sample size will produce a power curve that has a sharper "trough"; hence (except at the null hypothesis value) the power is higher everywhere. That is, increasing the sample size increases the power.

Uniformly Most Powerful Tests

Obviously high power is a desirable property of a test. If a choice of tests is available, the test with the largest power should be chosen. In certain cases, theory leads us to a test that has the largest possible power for any specified alternative hypothesis, sample size, and level of significance. Such a test is considered to be the best possible test for the hypothesis and is called a "uniformly most powerful" test. The test discussed in Example 3.3 is a uniformly most powerful test for the conditions specified in the example.

The computations involved in the construction of a power curve are not simple, and they become increasingly difficult for the applications in subsequent chapters. Fortunately, the performance of such computations often is not necessary because virtually all of the procedures we will be using provide uniformly the most powerful tests, assuming that basic assumptions are met. We discuss these assumptions in subsequent chapters and provide some information on what the consequences may be of nonfulfillment of assumptions.

Power calculations for more complex applications can be made easier through the use of computer programs. While there is no single program that calculates power for all hypothesis tests, some programs either have the option of calculating power for specific situations or can be adapted to do so. One example using the SAS System can be found in Wright and O'Brien (1988).

One-Tailed Hypothesis Tests

In Examples 3.3 and 3.4 the alternative hypothesis simply stated that μ was not equal to the specified null hypothesis value. That is, the null hypothesis was to be rejected if the evidence showed that the population mean was either larger or smaller than that specified by the null hypothesis. For some applications we may want to reject the null hypothesis only if the value of the parameter is larger or smaller than that specified by the null hypothesis.

Solution to Example 3.1 In the example at the beginning of the chapter, we were interested in determining whether leasing office space in Atlanta costs more than that in Jacksonville. If we let μ be the mean cost per square foot of office space in Atlanta, and if we assume the standard deviation of costs is the same in both cities ($\sigma = 4.50$), we can answer the question by testing the hypothesis[6]

$$H_0\colon \mu = \$12.61,$$

$$H_1\colon \mu > \$12.61.$$

Note that the alternative hypothesis statement is now "greater than." Even though the possibility exists that the cost may be less in Atlanta than in Jacksonville, we really don't care. That is, the decision to move is to be based on the condition that the cost is higher in Atlanta. The businessman will stay in Atlanta if it costs no more to stay. The test statistic is calculated as before: $z = (13.55 - 12.61)/(4.50/6) = 1.25$. However, in this case rejection of H_0 is logical only if the value of $\bar{y}$ is larger than that specified by H_0, which corresponds to positive values for the test statistic z. Thus the entire rejection region is in the upper tail. A test that locates the rejection region only in one tail of the sampling distribution is known as a "one-tailed" (or one-sided) test. For this example, we will let $\alpha = 0.10$, and the rejection value is $z = 1.28$ (the

[6]To be consistent with the specification that the two hypotheses must be exhaustive, some authors will specify the null hypothesis as $\mu \leq 12.61$ for this situation. We will stay with the single-valued null hypothesis statement whether we have a one- or two-tailed alternative. We maintain the exclusive and exhaustive nature of the two hypothesis statements by stating that we do not concern ourselves with values of the parameter in the "other" tail.

Figure 3.6

Power Curve for One- and Two-Tailed Tests

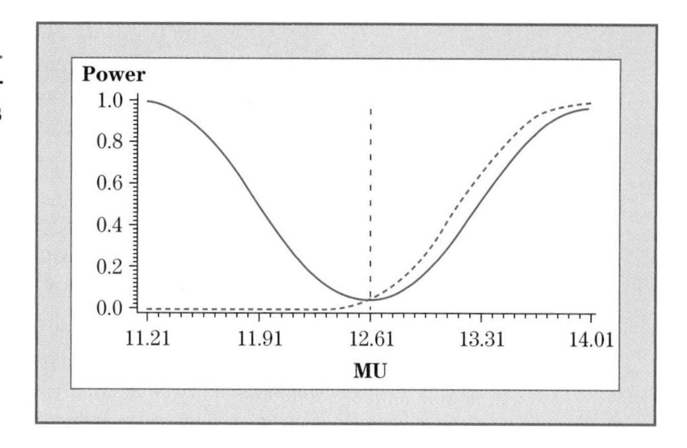

value for the Z distribution exceeded with probability 0.10 from Appendix Table A.1, rounded to two decimals). Since the calculated value of z is 1.25, which does not exceed 1.28, we would not reject the null hypothesis, concluding that there is insufficient evidence that the mean cost in Atlanta is higher than that in Jacksonville. Alternately the p value for this test is obtained directly from the table:

$$p = P(Z > 1.25) = 0.1056.$$

The advantage of a one-tailed test over a two-tailed test is that for a given level of significance, the power is larger when the value of μ is in the range of the alternative hypothesis. This is illustrated by comparing the power curves of a one-tailed (- - -) and two-tailed (—) test as seen in Fig. 3.6. The disadvantage of a one-tailed test is that the power is essentially zero on the "other" side and, in fact, approaches zero as the true value of the parameter moves away from the null hypothesis value. (Obviously we have no interest in the "other side" as shown by the choice of H_1.) In this example, we would not be able to reject the null hypothesis even if $\bar{y}$ were as extreme as \$5, since we do not have a rejection region in that direction. Therefore a one-tailed test should never be used if there is any concern over the true value of the parameter being on the "other" side. ■

The decision on whether to perform a one- or two-tailed test is determined entirely by the problem statement. A one-tailed test is indicated by the alternative or research hypothesis, stating that only larger (or smaller) values of the parameter are of interest. In the absence of such specification, a two-tailed test should be employed.

3.3 Estimation

In many cases we do not necessarily have a hypothesized value for the parameter that we want to test; instead we simply want to make a statement about the value of the parameter. For example, a large business may want to know the mean income of families in a target population near a proposed retail sales outlet. A chemical company may want to know the average amount of a chemical produced in a certain reaction. An animal scientist may want to know the mean yield of marketable meat of animals fed a certain ration. In each of these examples we use data from a sample to estimate the value of a parameter of the population. These are all examples of the inferential procedure called **estimation**.

As we will see, estimation and testing share some common characteristics and are often used in conjunction. For example, assume that we had rejected the hypothesis that the peanut-filling machine was putting 8 oz. of peanuts in the jars. It is then logical to ask, how much is the machine putting in the jars? The answer to this question could be useful in the effort to fix it.

The most obvious estimate of a population parameter is the corresponding sample statistic. This single value is known as a **point estimate**. For example, for estimating the parameter μ, the best point estimate is the sample mean $\bar{y}$. For estimating the parameter p in a binomial experiment, the best point estimate is the sample proportion $\hat{p} = y/n$.

For Example 3.3, the best point estimate of the mean weight of peanuts is the sample mean, which we found to be 7.89. We know that a point estimate will vary among samples from the same population. In fact, the probability that any point estimate exactly equals the true population parameter value is essentially zero for any continuous distribution. This means that if we make an unqualified statement of the form "μ is $\bar{y}$," that statement has almost no probability of being correct.

Thus a point estimate appears to be precise, but the precision is illusory because we have no confidence that the estimate is correct. In other words, it provides no information on the reliability of the estimate. A common practice for avoiding this dilemma is to "hedge," that is, to make a statement of the form "μ is almost certainly between 7.8 and 8." This is an **interval estimate**, and is the idea behind the statistical inference procedure known as the **confidence interval**. Admittedly a confidence interval does not seem as precise as a point estimate, but it has the advantage of having a known (and hopefully high) reliability.

DEFINITION 3.12
A **confidence interval** consists of a range of values together with a percentage that specifies how confident we are that the parameter lies in the interval.

Estimation of parameters with intervals uses the sampling distribution of the point estimate. For example, to construct an interval estimate of μ we use the already established sampling distribution of $\bar{Y}$ (see Section 2.5). Using the characteristics of this distribution we can make the statement

$$P[(\mu - 1.96\sigma/\sqrt{n}) < \bar{Y} < (\mu + 1.96\sigma/\sqrt{n})] = 0.95.$$

An exercise in algebra provides a rearrangement of the inequality inside the parentheses without affecting the probability statement:

$$P[(\bar{Y} - 1.96\sigma/\sqrt{n}) < \mu < (\bar{Y} + 1.96\sigma/\sqrt{n})] = 0.95.$$

In general, using the notation of Chapter 2 we can write the probability statement as

$$P[(\bar{Y} - z_{\alpha/2}\sigma/\sqrt{n}) < \mu < (\bar{Y} + z_{\alpha/2}\sigma/\sqrt{n})] = (1 - \alpha).$$

Then, our interval estimate of μ is

$$(\bar{y} - z_{\alpha/2}\sigma/\sqrt{n}) \text{ to } (\bar{y} + z_{\alpha/2}\sigma/\sqrt{n}).$$

This interval estimate is called a **confidence interval**, and the lower and upper boundary values of the interval are known as **confidence limits**. The probability used to construct the interval is called the **level of confidence** or confidence coefficient. This confidence level is the equivalent of the "almost certainly" alluded to in the preceding introduction. We thus say that we are $(1 - \alpha)$ confident that this interval contains the population mean. The confidence coefficient is often given as a percentage, for example, a 95% confidence interval.

For Example 3.3, a 0.95 confidence interval (or 95% confidence interval) lies between the values

$$7.89 - 1.96(0.2)/\sqrt{16} \text{ and } 7.89 + 1.96(0.2)/\sqrt{16}$$

or

$$7.89 \pm 1.96(0.05), \text{ or } 7.89 \pm 0.098.$$

Hence, we say that we are 95% confident that the true mean weight of peanuts is between 7.792 and 7.988 oz. per jar.

Interpreting the Confidence Coefficient

We must emphasize that the confidence interval statement is not a standard probability statement. That is, we cannot say that with 0.95 probability μ lies between 7.792 and 7.988. Remember that μ is a fixed number, which by definition has no distribution. This true value of the parameter either is or is not in a particular interval, and we will likely never know which event has occurred for a particular sample. We can, however, state that 95% of the intervals constructed in this manner will contain the true value of μ.

DEFINITION 3.13
The **maximum error of estimation**, also called the margin of error, is an indicator of the precision of an estimate and is defined as one-half the width of a confidence interval.

We can write the formula for the confidence limits on μ as $\bar{y} \pm E$, where

$$E = z_{\alpha/2}\sigma/\sqrt{n}$$

is one-half of the width of the $(1 - \alpha)$ confidence interval. The quantity E can also be described as the farthest that μ may be from $\bar{y}$ and still be in the confidence interval. This value is a measure of how "close" our estimate may be to the true value of the parameter. This bound on the error of estimation, E, is most often associated with a 95% confidence interval, but other confidence coefficients may be used. Incidentally, the "margin of error" often quoted in association with opinion polls is indeed E with an unstated 0.95 confidence level.

The formula for E illustrates for us the following relationships among E, α, n, and σ:

1. If the confidence coefficient is increased (α decreased) and the sample size remains constant, the maximum error of estimation will increase (the confidence interval will be wider). In other words, the more confidence we require, the less precise a statement we can make, and vice versa.
2. If the sample size is increased and the confidence coefficient remains constant, the maximum error of estimation will be decreased (the confidence interval narrower). In other words, by increasing the sample size we can increase precision without loss of confidence, or vice versa.
3. Decreasing σ has the same effect as increasing the sample size. This may seem a useless statement, but it turns out that proper experimental design (Chapter 10) can often reduce the standard deviation.

Thus there are trade-offs in interval estimation just as there are in hypothesis testing. In this case we trade precision (narrower interval) for higher confidence. The only way to have more confidence without increasing the width (or vice versa) is to have a larger sample size.

EXAMPLE 3.5

Suppose that a population mean is to be estimated from a sample of size 25 from a normal population with $\sigma = 5.0$. Find the maximum error of estimation with confidence coefficients 0.95 and 0.99. What changes if n is increased to 100 while the confidence coefficient remains at 0.95?

Solution

1. The maximum error of estimation of μ with confidence coefficient 0.95 is

$$E = 1.96(5/\sqrt{25}) = 1.96.$$

2. The maximum error of estimation of μ with confidence coefficient 0.99 is

$$E = 2.576(5/\sqrt{25}) = 2.576.$$

3. If $n = 100$ then the maximum error of estimation of μ with confidence coefficient 0.95 is

$$E = 1.96(5/\sqrt{100}) = 0.98.$$

Note that increasing n fourfold only halved E. The relationship of sample size to confidence intervals is discussed further in Section 3.4. ■

Relationship between Hypothesis Testing and Confidence Intervals

As noted previously there is a direct relationship between hypothesis testing and confidence interval estimation. A confidence interval on μ gives all acceptable values for that parameter with confidence $(1 - \alpha)$. This means that any value of μ not in the interval is not an "acceptable" value for the parameter. The probability of being incorrect in making this statement is, of course, α. Therefore,

> a hypothesis test for H_0: $\mu = \mu_0$ against H_1: $\mu \neq \mu_0$ will be rejected at a significance level of α if μ_0 is not in the $(1 - \alpha)$ confidence interval for μ.

Conversely,

> any value of μ inside the $(1 - \alpha)$ confidence interval will not be rejected by an α-level significance test.

For Example 3.3, the 95% confidence interval is 7.792 to 7.988. The hypothesized value of 8 is not contained in the interval; therefore we would reject the hypothesis H_0:$\mu = 8$ at the 0.05 level of significance. For Example 3.4, a 99% confidence interval on μ is 49.92 to 52.22. The hypothesis H_0:$\mu = 50$ would not be rejected with $\alpha = 0.01$ because the value 50 does lie within the interval. These results are, of course, consistent with results obtained from the hypothesis tests presented previously.

As in hypothesis testing, one-sided confidence intervals can be constructed. In Example 3.1 we used a one-sided alternative hypothesis, H_1: $\mu >$ \$12.61. This corresponds to finding the lower confidence limit so that the confidence statement will indicate that the mean score is at least that amount or higher. For this example, then, the lower $(1 - \alpha)$ limit is

$$\bar{y} - z_\alpha(\sigma/\sqrt{n}),$$

which results in the lower 0.90 confidence limit:

$$13.55 - 1.28(4.50/6) = 12.59.$$

Thus we are 90% confident that the mean cost per square foot of office space in Atlanta is at least \$12.59. This confirms our previous conclusion that there was no evidence that Atlanta cost was higher than the \$12.61 in Jacksonville since 12.61 is in the confidence interval.

3.4 Sample Size

We have noted that in both hypothesis testing and interval estimation, a definite relationship exists between sample size and the precision of our results. In fact, the best possible sample appears to be the one that contains the largest number of observations. This is not necessarily the case. The cost and effort of obtaining the sample and processing and analyzing the data may offset the added precision of the results. Remember that costs often increase linearly with sample size, while precision, in terms of E, decreases only with the square root of the sample size. It is therefore not surprising that the question of sample size is of major concern. Because of the relationship of sample size to the precision of statistical inference, we can answer the question of optimal sample size.

Consider the problem of estimating μ using a sample from a normal population with known standard deviation, σ. We want to find the required sample size, n, for a specified maximum value of E. Using the formula for E,

$$E = \frac{z_{\alpha/2}\sigma}{\sqrt{n}},$$

we can solve for n, resulting in

$$n = \frac{z_{\alpha/2}^2 \sigma^2}{E^2}.$$

Thus, given values for σ and α and a specified maximum E, we can determine the required sample size for the desired precision. For example, suppose that in Example 3.3 we wanted a 99% confidence interval for the mean weight to be no wider than 0.10 oz. This means that $E = 0.05$. The required sample size is

$$\begin{aligned} n &= (2.576)^2(0.2)^2/(0.5)^2 \\ &= 106.2. \end{aligned}$$

We round up to the nearest integer, so the required sample size is 107. This is a large sample, but both the confidence coefficient and the required precision were both quite strict. This example illustrates an often encountered problem: Requirements are often made so strict that unreasonably large sample sizes are required.

Sample size determination must satisfy two prespecified criteria:

1. the value of E, the maximum error of estimation (or, equivalently, the width of the confidence interval), and
2. the required level of confidence (the confidence coefficient, $1 - \alpha$).

In other words, it is not only sufficient to require a certain degree of precision, but it is also necessary to state the degree of confidence. Since the degree of confidence is so often assumed to be 0.95, it is usually not stated, which may give the incorrect impression of 100% confidence! It is, of course, also necessary to have an estimated value for σ^2 if we are estimating μ. In many cases, we have to use rough approximations of the variance. One such approximation can be obtained from the empirical rule discussed in Chapter 1. If we can determine the expected range of values of the results of the experiment, we can use the empirical rule to obtain an estimate of the standard deviation. That is, we could use the range divided by 4 to estimate the standard deviation. This is because the empirical rule states that 95% of the values of a distribution will be plus or minus 2σ from the mean. Thus, 95% of the values will be in the 4σ range.

EXAMPLE 3.6

In a study of the effect of a certain drug on the behavior of laboratory animals, a research psychologist needed to determine the appropriate sample size. The study was to estimate the time necessary for the animal to travel through a maze under the influence of this drug.

Solution Since no previous studies had been conducted on this drug, no independent estimate for the variation of times was available. Using the conventional confidence level of 95%, a bound on the error of estimation of 5 seconds, and an anticipated range of times of from 15 to 60 seconds, what sample size would the psychologist need?

1. First, an estimate of the standard deviation was obtained from the range by dividing by 4:

$$EST(\sigma) = (60 - 15)/4 = 11.25.$$

2. The sample size was determined as $n = [(1.96)^2(11.25)^2]/5^2 = 19.4$.
3. Round up to $n = 20$, so the researcher needs 20 animals in the study.

The formula for the required sample size clearly indicates the trade-off between the interval width (the value of E) and the degree of confidence. In Example 3.6, narrowing the width to 1 would give

$$n = (1.96)^2(11.25)^2/(1)^2 = 487.$$

■

Requirements for being able to detect a specified difference between the null and alternate hypotheses with a given degree of significance can be converted to the desired width of a confidence interval by remembering the equivalence of the two procedures.

In Example 3.4 we may want to be able to detect, at the 0.01 level of significance, a change of one unit in the average test score. According to the equivalence, this requires a 99% confidence interval of plus or minus one unit, hence $E = 1$. The required sample size is

$$n = (2.576)^2(10)^2/(1)^2 = 664.$$

This, of course, may not always be possible, or may not be the best way to approach the problem. What we need is a way to compute directly the required sample size for conducting a hypothesis test, using the constraints usually developed in the process of testing a hypothesis. For example, we might be interested in determining how big a sample we need to have reasonable power against a specified value of μ, say μ_a, in the hypothesis

$$H_0: \mu = \mu_0 \quad \text{vs}$$
$$H_1: \mu > \mu_0.$$

That is, we want to determine what sample size will give us adequate protection against mean values in the alternative (values of μ_a greater than μ_0) that have some negative impact on the process under scrutiny. In this case, however, several prespecified criteria must be considered. We need to satisfy:

1. the required level of significance (α),
2. the difference, called δ (delta), between the hypothesized value and the specified value ($\delta = \mu_a - \mu_0$), and
3. the probability of a type II error (β) when the real mean is at this specified value (or one larger than the specified value).

The value of n that satisfies these criteria can be obtained using the formula

$$n = \frac{\sigma^2(z_\alpha + z_\beta)}{\delta^2},$$

where all the components of this formula have been defined.

Suppose that in Example 3.6 we wanted to test the following set of hypotheses:

$$H_0: \mu = 35 \text{ s} \quad \text{vs}$$
$$H_1: \mu > 35 \text{ s}.$$

We use a level of significance $\alpha = 0.05$, and we decide that we are willing to risk making a type II error of $\beta = 0.10$ if the actual mean time is 37 s. This means that the power of the test at $\mu = 37$ s will be 0.90. The difference between the hypothesized value of the mean and the specified value of the mean is $\delta = 37 - 35 = 2$. In Example 3.6 we estimated the value of the standard deviation as 11.25. We can substitute this value for σ in the formula, obtain the necessary values from Appendix Table A.1A, and calculate n as

$$n = (11.25)^2 \frac{(1.64485 + 1.28155)^2}{(2)^2} = 271.$$

Therefore, if we take a sample of size $n = 271$ we can expect to reject the hypothesis that $\mu = 35$ if the real mean value is 37 or higher with probability 0.90.

The procedure for a hypothesis test with a one-sided alternative in the other direction is almost identical. The only difference is that μ_a will be less than μ_0. To use a two-sided alternative, we use the following formula to calculate the required sample size,

$$n = \frac{\sigma^2(z_{\alpha/2} + z_\beta)^2}{\delta^2},$$

where $\delta = |\mu_a - \mu_0|$.

In Example 3.4 we might want to be more rigorous in our definition of the problem, and rather than saying that we simply want to detect a difference of one unit, say instead that we want to reject the null hypothesis if the deviation from the hypothesized value is one unit or more with probability 99%. That is, we would reject the null hypothesis if it were less than 49 or greater than 51 with power of 0.99. Using the values of $\sigma = 10$, $\alpha = 0.01$, $\beta = 0.01$, and $\delta = 1$, we get

$$n = (10)^2 \frac{(2.57583 + 2.32635)^2}{(1)^2} = 2404.$$

Note that this is larger than the value we obtained using the confidence interval approach; this is because we imposed more rigorous criteria.

These examples of sample size determination are relatively straightforward because of the simplicity of the methods used. If we did not know the standard deviation in a hypothesis test on the mean, or if we were using any of the hypothesis testing procedures discussed in subsequent chapters, we would not have such simple formulas for calculating n. There are, however, tables and charts that enable sample size determination to be done for most hypotheses tests. See, for example, Neter *et al.* (1996).

3.5 Assumptions

In this chapter we have considered inferences on the population mean in situations where it can be assumed that the sampling distribution of the mean is reasonably close to normal. Inference procedures based on the assumption of a normally distributed sample statistic are referred to as normal theory methods.

In Section 2.5 we pointed out that the sampling distribution of the sample mean is normal if the population itself is normal, or if the sample size is large enough to satisfy the central limit theorem. However, normality of the sampling distribution of the mean is not always assured for relatively small samples, especially those from highly skewed distributions or where the observations may be dominated by a few extreme values. In addition, as noted in Chapter 1, some data may be obtained as ordinal values such as ranks, or nominal values such as categorical data. Such data are not readily amenable to analysis by the methods designed for interval data.

When the assumption of normality does not hold, use of methods requiring this assumption may produce misleading inferences. That is, the significance level of a hypothesis test or the confidence level of an estimate may not be as specified by the procedure. For instance, the use of the normal distribution for a test statistic may indicate rejection at the 0.05 significance level, but due to nonfulfillment of the assumptions, the true protection against making a type I error may be as high as 0.10. (Refer to Section 4.5 for ways to determine whether the normality assumption is valid.)

Unfortunately, we cannot know the true value of α in such cases. For this reason alternate procedures have been developed for situations in which normal theory methods are not applicable. Such methods are often described as "robust" methods, because they provide the specified α for virtually all situations. However, this added protection is not free: Most of these robust methods have wider confidence intervals and/or have power curves generally lower than those provided by normal theory methods when the assumption of normality is indeed satisfied.

Various principles are used to develop robust methods. Two often used principles are as follows:

1. Trimming, which consists of discarding a small prespecified portion of the most extreme observations and making appropriate adjustments to the test statistics.
2. Nonparametric methods, which avoid dependence on the sampling distribution by making strictly probabilistic arguments (often referred to as distribution-free methods).

In subsequent chapters we will give examples of situations in which assumptions are not fulfilled and briefly describe some results of alternative methods. A more complete presentation of nonparametric methods is found in Chapter 13. Trimming and other robust methods are not presented in this text (see Koopmans, 1987).

Statistical Significance versus Practical Significance

The use of statistical hypothesis testing provides a powerful tool for decision making. In fact, there really is no other way to determine whether two or more population means differ based solely on the results of one sample or one experiment. However, a statistically significant result cannot be interpreted simply by itself. In fact, we can have a statistically significant result that has no practical implications, or we may not have a statistically significant result, yet useful information may be obtained from the data. For example, a market research survey of potential customers might find that a potential market exists for a particular product. The next question to be answered is whether this market is such that a reasonable expectation exists for making profit if the product is marketed in the area. That is, does the mere existence of a potential market guarantee a profit? Probably not. Further investigation must be done before recommending marketing of the product, especially if the marketing is expensive. The following examples are illustrations of the difference between statistical significance and practical significance.

EXAMPLE 3.7

This is an example of a statistically significant result that is not practically significant.

In the January/February 1992 *International Contact Lens Clinic* publication, there is an article that presented the results of a clinical trial designed to determine the effect of defective disposable contact lenses on ocular integrity

(Efron and Veys, 1992). The study involved 29 subjects, each of whom wore a defective lens in one eye and a nondefective one in the other eye. The design of the study was such that neither the research officer nor the subject was informed of which eye wore the defective lens. In particular, the study indicated that a significantly greater ocular response was observed in eyes wearing defective lenses in the form of corneal epithelial microcysts (among other results). The test had a p value of 0.04. Using a level of significance of 0.05, the conclusion would be that the defective lenses resulted in more microcysts being measured. The study reported a mean number of microcysts for the eyes wearing defective lenses as 3.3 and the mean for eyes wearing the nondefective lenses as 1.6. In an invited commentary following the article, Dr. Michel Guillon makes an interesting observation concerning the presence of microcysts. The commentary points out that the observation of fewer than 50 microcysts per eye requires no clinical action other than regular patient follow-up. The commentary further states that it is logical to conclude that an incidence of microcysts so much lower than the established guideline for action is not clinically significant. Thus, we have an example of the case where statistical significance exists but where there is no practical significance. ■

EXAMPLE 3.8

A major impetus for developing the statistical hypothesis test was to avoid jumping to conclusions simply on the basis of apparent results. Consequently, if some result is not statistically significant the story usually ends. However it is possible to have practical significance but not statistical significance. In a recent study of the effect of a certain diet on weight reduction, a random sample of 10 subjects was weighed, put on a diet for 2 weeks, and weighed again. The results are given in Table 3.2.

Solution A hypothesis test comparing the mean weight before with the mean weight after (see Section 5.4 for the exact procedure for this test) would result in a p value of 0.21. Using a level of significance of 0.05 there would not be sufficient evidence to reject the null hypothesis and the conclusion would be that there is no significant loss in weight due to the diet. However, note that 9 of the 10 subjects lost weight! This means that the diet is probably effective

Table 3.2

Weight Gains (in lbs.)

Subject	Weight Before	Weight After	Difference (Before − After)
1	120	119	+1
2	131	130	+1
3	190	188	+2
4	185	183	+2
5	201	188	+13
6	121	119	+2
7	115	114	+1
8	145	144	+1
9	220	243	−23
10	190	188	+2

in reducing weight, but perhaps does not take a lot of it off. Obviously, the observation that almost all the subjects did in fact lose weight does not take into account the amount of weight lost, which is what the hypothesis test did. So in effect, the fact that 9 of the 10 subjects lost weight (90%) really means that the proportion of subjects losing weight is high rather than that the mean weight loss differs from 0.

We can evaluate this phenomenon by calculating the probability that the results we observed occurred strictly due to chance using the basic principles of probability of Chapter 2. That is, we can calculate the probability that 9 of the 10 differences in before and after weight are in fact positive if the diet does not affect the subjects' weight. If the sign of the difference is really due to chance, then the probability of an individual difference being positive would be 0.5 or 1/2. The probability of 9 of the 10 differences being positive would then be $10(0.5)(0.5)^9$ or 0.009765—a very small value. Thus, it is highly unlikely that we could get 9 of the 10 differences positive due to chance so there is something else causing the differences. That something must be the diet.

Note that although the results appear to be contradictory, we actually tested two different hypotheses. The first one was a test to compare the weight before and after. Thus, if there was a significant increase or decrease in the average weight we would have rejected this hypothesis. On the other hand, the second analysis was really a hypothesis test to determine whether the probability of losing weight is really 0.5 or 1/2. We discuss this type of a hypothesis test in the next chapter. ■

3.6 CHAPTER SUMMARY

The statistical inference principles illustrated in this section, often referred to as the Neyman–Pearson principles, may seem awkward at first. This is especially true of the hypothesis testing procedures, where the null hypothesis is the opposite of what we really want to "prove." These procedures are, however, widely used because of the ease of controlling the type I error, which protects against erroneously announcing a new theory, proposing a large expenditure, or adopting a new policy. Further, it is also useful to be able to specify the degree of trade-off between the precision of the statement and the probability that the statement is incorrect.

At this point it is appropriate to ask "Is it really necessary to go to all this trouble to make inferences?" The answer must obviously be "yes" because, despite all the jokes and sayings about statistics and statisticians, the procedures of statistical inference are designed to avoid lying with statistics. The key to statistical inference is to be able to indicate the reliability of a statistic when it is used to make inferences. In statistical inference the use of random samples allows the use of probability statements to provide a measure of that reliability.

Because statistical significance or confidence is based on probability, it is important to point out the distinction between statistical significance and practical significance. A hypothesis test may, for example, declare that due to a certain plant modification, an average increase of production of 0.04 shirts per day is a statistically significant increase for a large factory. If the modification is very expensive, that relatively small increase is statistically significant, but it is far from being practically significant. This type of result often occurs with very large sample sizes, a situation that sometimes arises from automated data collection. On the other hand, it may happen that some estimated change or difference is of sufficient magnitude to be of practical importance, but is not statistically significant. In such cases the lack of statistical significance provides the necessary protection against that result being taken too seriously.

However, these principles are not wholly suitable for all statistical inference applications. For example, as we noted in Section 3.2, the proper null hypothesis for testing the effectiveness of a drug was that the drug is not effective. It is difficult to state this as a single value of a population parameter. Very few drugs are completely ineffective; hence the hypothesis that p, the proportion of individuals "cured," is zero is not realistic. A more appropriate hypothesis might be H_0: $p > 0.5$, say, but this does not meet the requirement of being a single-valued null hypothesis. Some other inference procedures that are not considered in this text include:

- *The use of penalty or payoff functions.* In the procedures discussed in this book, an incorrect inference is an incorrect inference. There is no "degree" of correctness. In some applications different degrees of being incorrect may incur different magnitudes of penalty. Inference procedures utilizing various expected penalty (or payoff) functions are available. These can be somewhat difficult to use, because the exact nature of the penalty or payoff function is not always known. See, for example, Neter *et al.* (1996).
- *Sequential sampling.* In the standard form of a hypothesis test or estimation problem, the precision is controlled by the selection of the sample size. In some cases where the sample size is not fixed prior to the experiment, a method of inference called "sequential" analysis can be performed. For this procedure sample units are selected in a sequential manner. As each sample unit is selected, the precision of any inference (specifically the actual α and β levels) is checked; if there is sufficient precision (in terms of α and β) the procedure stops and a decision is made; if there is not, the decision is to continue sampling. Sequential analysis has limited uses, however, since the methodology is not easy to implement for all applications, and, in many cases, the very act of sequential sampling is not physically feasible. A bibliography on sequential sampling is found in Wald (1947).

Finally, because the reliability of statistical inferences is expressed in probability terms, it is important to distinguish between **confirmatory** and **exploratory analyses**. Remember that the steps for a hypothesis test are as follows:

- State the hypotheses.
- Collect data and compute statistics.
- Make a decision to confirm or deny hypothesis.

This procedure is a confirmatory statistical analysis since its purpose is to confirm or deny a hypothesis and to provide a probability-based protection against a wrong decision. A very large proportion of statistical analyses does not strictly conform to this scenario. The main reason for this is that most applications do not involve inferences on only one parameter based on a sample from a single population. In multiple-parameter situations inferences are not only concerned with the individual parameters but also with comparisons among parameters. This means that there are many hypotheses and, in order to make inferences more manageable, hypotheses are based on the characteristics of the point estimates of the parameters. This type of situation leads to exploratory analyses, where in effect some hypotheses may be generated by the data. Now, assigning significance probabilities to the results of such tests is like placing a bet on a horse race after a part of the race has already been run.

For example, assume we have samples from t populations having identical population means and we test hypotheses on differences among these population means. (We will do this in Chapter 6.) If we now choose to perform a test involving only the largest and smallest sample means, their difference is likely to be sufficiently large such that they appear to contradict the true null hypothesis that the means are equal. In other words, although the hypothesis test is based on, say, a 0.05 significance level, the probability of rejecting the true null hypothesis greatly exceeds this amount.

There is nothing wrong with exploratory data analysis. Often the complexity and originality of a problem preclude well-formulated specific hypotheses and at least some data-driven analysis procedure must be used. The point to be made here is that results of such analyses should not be embellished with precise statistical significance levels or p values. These statistics, are, however, not useless but should be used in relative context. That is, a p value of 0.0002 most likely means that a result is statistically significant, but the true probability of a type I error is not likely to be as small as 0.0002. Unfortunately, no precise methods exist for obtaining true p values for such situations.

3.7 CHAPTER EXERCISES

CONCEPT QUESTIONS

This section consists of some true/false questions regarding concepts of statistical inference. Indicate whether a statement is true or false and, if false, indicate what is required to make the statement true.

1. _______ In a hypothesis test, the p value is 0.043. This means that the null hypothesis would be rejected at $\alpha = 0.05$.

2. _______ If the null hypothesis is rejected by a one-tailed hypothesis test, then it will also be rejected by a two-tailed test.

3. _______ If a null hypothesis is rejected at the 0.01 level of significance, it will also be rejected at the 0.05 level of significance.

4. _______ If the test statistic falls in the rejection region, the null hypothesis has been proven to be true.

5. _______ The risk of a type II error is directly controlled in a hypothesis test by establishing a specific significance level.

6. _______ If the null hypothesis is true, increasing only the sample size will increase the probability of rejecting the null hypothesis.

7. _______ If the null hypothesis is false, increasing the level of significance (α) for a specified sample size will increase the probability of rejecting the null hypothesis.

8. _______ If we decrease the confidence coefficient for a fixed n, we decrease the width of the confidence interval.

9. _______ If a 95% confidence interval on μ was from 50.5 to 60.6, we would reject the null hypothesis that $\mu = 60$ at the 0.05 level of significance.

10. _______ If the sample size is increased and the level of confidence is decreased, the width of the confidence interval will increase.

PRACTICE EXERCISES

The following exercises are designed to give the reader practice in doing statistical inferences through small examples. The solutions are given in the back of the text.

1. From extensive research it is known that the population of a particular species of fish has a mean length $\mu = 171$ mm and a standard deviation $\sigma = 44$ mm. The lengths are known to have a normal distribution. A sample of 100 fish from such a population yielded a mean length $\bar{y} = 167$ mm. Compute the 0.95 confidence interval for the mean length of the sampled population. Assume the standard deviation of the population is also 44 mm.

2. Using the data in Exercise 1 and using a 0.05 level of significance, test the null hypothesis that the population sampled has a mean of $\mu = 171$. Use a two-tailed alternative.

3. What sample size is required for a maximum error of estimation of 10 for a population whose standard deviation is 40 using a confidence interval of 0.95? How much larger must the sample size be if the maximum error is to be 5?

4. The following sample was taken from a normally distributed population with a known standard deviation $\sigma = 4$. Test the hypothesis that the mean $\mu = 20$ using a level of significance of 0.05 and the alternative that $\mu > 20$:

$$23, 32, 22, 31, 27, 25, 21, 24, 20, 18.$$

MULTIPLE CHOICE QUESTIONS

1. In testing the null hypothesis that $p = 0.3$ against the alternative that $p \neq 0.3$, the probability of a type II error is ______ when the true $p = 0.4$ than when $p = 0.6$.
   (1) the same
   (2) smaller
   (3) larger
   (4) none of the above

2. In a hypothesis test the p value is 0.043. This means that we can find statistical significance at:
   (1) both the 0.05 and 0.01 levels
   (2) the 0.05 but not at the 0.01 level
   (3) the 0.01 but not at the 0.05 level
   (4) neither the 0.05 or 0.01 levels
   (5) none of the above

3. A research report states: The differences between public and private school seventh graders' attitudes toward minority groups was statistically significant at the $\alpha = 0.05$ level. This means that:
   (1) It has been proven that the two groups are different.
   (2) There is a probability of 0.05 that the attitudes of the two groups are different.
   (3) There is a probability of 0.95 that the attitudes of the two groups are different.
   (4) If there is no difference between the groups, the difference observed in the sample would occur by chance with probability of no more than 0.05.
   (5) None of the above is correct.

4. Which of these statements characterizes the outcome if the calculated value of any test statistic falls in the rejection region when a false null hypothesis is being tested?
   (1) The decision is correct.
   (2) A type I error has been committed.
   (3) A type II error has been committed.
   (4) Insufficient information has been given to make a decision.
   (5) None of the above is correct.

5. Which of these statements characterizes the outcome if the calculated value of any test statistic does not fall in the rejection region when a false null hypothesis is being tested?
   (1) The decision is correct.
   (2) A type I error has been committed.
   (3) A type II error has been committed.
   (4) Insufficient information has been given to make a decision.
   (5) None of the above is correct.

6. If the value of any test statistic does not fall in the rejection region, the decision is:
(1) Reject the null hypothesis.
(2) Reject the alternative hypothesis.
(3) Fail to reject the null hypothesis.
(4) Fail to reject the alternative hypothesis.
(5) There is insufficient information to make a decision.

7. For a particular sample, the 0.95 confidence interval for the population mean is from 11 to 17. You are asked to test the hypothesis that the population mean is 18 against a two-sided alternative. Your decision is:
(1) Fail to reject the null hypothesis, $\alpha = 0.05$.
(2) Reject the null hypothesis, $\alpha = 0.05$.
(3) There is insufficient information to decide.

8. Failure to reject the null hypothesis means:
(1) acceptance of the alternative hypothesis
(2) rejection of the null hypothesis
(3) rejection of the alternative hypothesis
(4) absolute acceptance of the null hypothesis
(5) none of the above

9. If we decrease the confidence level, the width of the confidence interval will:
(1) increase
(2) remain unchanged
(3) decrease
(4) double
(5) none of the above

10. If the value of the test statistic falls in the rejection region, then:
(1) We cannot commit a type I error.
(2) We cannot commit a type II error.
(3) We have proven that the null hypothesis is true.
(4) We have proven that the null hypothesis is false.
(5) None of the above is correct.

EXERCISES

1. The following pose conceptual hypothesis test situations. For each situation define H_0 and H_1 so as to provide control of the more serious error. Justify your choice and comment on logical values for α.
(a) You are deciding whether you should take an umbrella to work.
(b) You are planning a proficiency testing procedure to determine whether some employees should be fired.
(c) Same as part (b) except you want to determine whether some employees deserve a special merit raise.

(d) A cigarette manufacturer is conducting a test of nicotine content in order to justify a new advertising claim.
(e) You are considering the procedure to decide guilt or innocence in a court of law.
(f) You are wondering whether you should buy a new battery for your calculator before the next statistics test.
(g) As a university administrator you are considering a policy to restrict student driving in order to improve scholastic achievement.

2. Suppose that in Example 3.3, σ was 0.15 instead of 0.2 and we decided to adjust the machine if a sample of 16 had a mean weight below 7.9 or above 8.1 (same as before).
(a) What is the probability of a type I error now?
(b) Draw the operating characteristic curve using the rejection region obtained in part (a).

3. Assume that a random sample of size 25 is to be taken from a normal population with $\mu = 10$ and $\sigma = 2$. The value of μ, however, is not known by the person taking the sample.
(a) Suppose that the person taking the sample tests H_0: $\mu = 10.4$ against H_1: $\mu \neq 10.4$. Although this null hypothesis is not true, it may not be rejected, and a type II error may therefore be committed. Compute β if $\alpha = 0.05$.
(b) Suppose the same hypothesis is to be tested as that of part (a) but $\alpha = 0.01$. Compute β.
(c) Suppose the person wanted to test H_0: $\mu = 11.2$ against H_1: $\mu \neq 11.2$. Compute β for $\alpha = 0.05$ and $\alpha = 0.01$.
(d) Suppose that the person decided to use H_1: $\mu < 11.2$. Calculate β for $\alpha = 0.05$ and $\alpha = 0.01$.
(e) What principles of hypothesis testing are illustrated by these exercises?

4. Repeat Exercise 3 using $n = 100$. What principles of hypothesis testing do these exercises illustrate?

5. A standardized test for a specific college course is constructed so that the distribution of grades should have $\mu = 100$ and $\sigma = 10$. A class of 30 students has a mean grade of 92.
(a) Test the null hypothesis that the grades from this class are a random sample from the stated distribution. (Use $\alpha = 0.05$.)
(b) What is the p value associated with this test?
(c) Discuss the practical uses of the results of this statistical test.

6. The family incomes in a certain city in 1970 had a mean of \$14,200 with a standard deviation of \$2600. A random sample of 75 families taken in 1975 produced $\bar{y} = \$15,300$ (adjusted for inflation).
(a) Assume σ has remained unchanged and test to see whether mean income has changed using a 0.05 level of significance.

(b) Construct a 99% confidence interval on mean family income in 1975.
(c) Construct the power curve for the test in part (a).

7. Suppose in Example 3.2 we were to reject H_0 if all the jelly beans in a sample of size four were red.
(a) What is α?
(b) What is β?

8. Suppose that for a given population with $\sigma = 7.2$ we want to test H_0: $\mu = 80$ against H_1: $\mu < 80$ based on a sample of $n = 100$.
(a) If the null hypothesis is rejected when $\bar{y} < 76$, what is the probability of a type I error?
(b) What would be the rejection region if we wanted to have a level of significance of exactly 0.05?

9. An experiment designed to estimate the mean reaction time of a certain chemical process has $\bar{y} = 79.6$ s, based on 144 observations. The standard deviation is $\sigma = 8$.
(a) What is the maximum error of estimate at 0.95 confidence?
(b) Construct a 0.95 confidence interval on μ.
(c) How large a sample must be taken so that the 0.95 maximum error of estimate is 1 s or less?

10. A drug company is testing a drug intended to increase heart rate. A sample of 100 yielded a mean increase of 1.4 beats per minute, with a standard deviation known to be 3.6. Since the company wants to avoid marketing an ineffective drug, it proposes a 0.001 significance level. Should it market the drug? (*Hint:* If the drug does not work, the mean increase will be zero.)

11. The manufacturer of auto windows discussed in Exercise 19 of Chapter 2 has developed a new plastic material that can be applied much thinner than the conventional material. To use this material, however, the production machinery must be adjusted. A trial adjustment was made on one of the 10 machines used in production, and a sample of 25 windshields measured. This sample had a mean thickness of 2.9 mm. Using the standard deviation of 0.25 mm, does this adjustment provide for a smaller thickness in the material than the old adjustment (4 mm)? (Use a hypothesis test and level of significance of 0.01. Assume the distribution of thickness is approximately normal.)

12. The manufacturer in Exercise 11 tried another, less expensive adjustment on another machine. A sample of 25 windshields was measured yielding a sample mean thickness of 3.4. Calculate the p value resulting from this mean using the same hypothesis and assumptions as in Exercise 11.

13. An experiment is conducted to determine whether a new computer program will speed up the processing of credit card billing at a large bank. The mean time to process billing using the present program is 12.3 min. with a standard deviation of 3.5 min. The new program is tested with 100 billings and yielded a sample mean of 10.9 min. Assuming the standard deviation of times in the new program is the same as the old, does the new program significantly reduce the time of processing? Use $\alpha = 0.05$.

14. Another bank is experimenting with programs to direct bill companies for commercial loans. They are particularly interested in the number of errors of a billing program. To examine a particular program, a simulation of 1000 typical loans is run through the program. The simulation yielded a mean of 4.6 errors with a standard deviation of 0.5. Construct a 95% confidence interval on the true mean error rate.

15. If the bank wanted to examine a program similar to that of Exercise 14 and wanted a maximum error of estimation of 0.01 with a level of confidence of 95%, how large a sample should be taken? (Assume that the standard deviation of the number of errors remains the same.)

Inferences on a Single Population

EXAMPLE 4.1

How Accurately Are Areas Perceived? The data in Table 4.1 are from an experiment in perceptual psychology. A person asked to judge the relative areas of circles of varying sizes typically judges the areas on a perceptual scale that can be approximated by

$$\text{judged area} = a(\text{true area})^b.$$

For most people the exponent b is between 0.6 and 1. That is, a person with an exponent of 0.8 who sees two circles, one twice the area of the other, would judge the larger one to be only $2^{0.8} = 1.74$ as large. Note that if the exponent is less than 1 a person tends to underestimate the area; if larger than 1, he or she will overestimate the area. The data shown in Table 4.1 are the set of measured exponents for 24 people from one particular experiment (Cleveland *et al.*, 1982). A histogram of this data is given in Figure 4.1.

It may be of interest to estimate the mean value of b for the population from which this sample is drawn; however, because we do not know the value of the population standard deviation we cannot use the methods of Chapter 3. Further, we might be interested in estimating the variance of these measurements as well. This chapter discusses methods for doing inferences on means when the population variance is unknown as well as inferences on the unknown population variance. The inferences for this example are presented in Sections 4.2 and 4.4. ■

4.1 Introduction

The examples used in Chapter 3 to introduce the concepts of statistical inference were neither very interesting nor useful. This was intentional, as we wanted to avoid distractions from issues that were irrelevant to the principles

Table 4.1

Measured Exponents

Note: Reprinted with permission from the American Statistical Association.

0.58	0.63	0.69	0.72	0.74	0.79
0.88	0.88	0.90	0.91	0.93	0.94
0.97	0.97	0.99	0.99	0.99	1.00
1.03	1.04	1.05	1.07	1.18	1.27

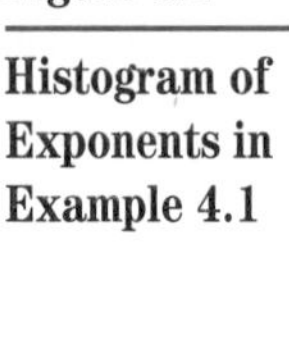

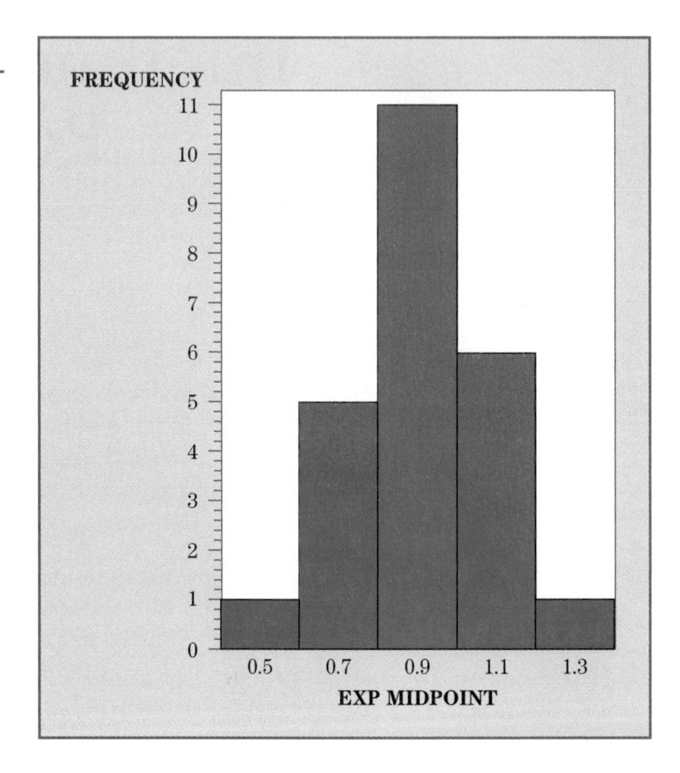

Figure 4.1

Histogram of Exponents in Example 4.1

we were introducing. We will now turn to examples that, although still quite simple, will be have more useful applications. Specifically, we present procedures for

- making inferences on the mean of a normally distributed population where the variance is unknown,
- making inferences on the variance of a normally distributed population, and
- making inferences on the proportion of successes in a binomial population.

Increasing degrees of complexity are added in subsequent chapters. These begin in Chapter 5 with inferences for comparing two populations and in Chapter 6 with inferences on means from any number of populations. In Chapter 7 we present inference procedures for relationships between two variables through what we will refer to as the linear model, which is subsequently used as the common basis for the many other statistical inference procedures. Additional chapters contain brief introductions to other statistical methods that cover different situations as well as methodology that may be used when underlying assumptions cannot be satisfied.

4.2 Inferences on the Population Mean

In Chapter 3 we used the sample mean $\bar{y}$ and its sampling distribution to make inferences on the population mean. For these inferences we used the fact that, for any approximately normally distributed population the statistic[1]

$$z = \frac{(\bar{y} - \mu)}{\sigma/\sqrt{n}}$$

has the standard normal distribution. This statistic has limited practical value because, if the population mean is unknown, it is also likely that the variance of the population is unknown.

In the discussion of the t distribution in Section 2.6 we noted that if, in the above equation, the known standard deviation is replaced by its estimate, s, the resulting statistic has a sampling distribution known as Student's t distribution. This distribution has a single parameter, called **degrees of freedom**, which is $(n - 1)$ for this case. Thus for statistical inferences on a mean from a normally distributed population, we can use the statistic

$$t = \frac{(\bar{y} - \mu)}{\sqrt{s^2/n}},$$

where $s^2 = \sum(y - \bar{y})^2/(n - 1)$.

It is very important to note that the degrees of freedom are based on the denominator of the formula used to calculate s^2, which reflects the general formula for computing s^2,

$$s^2 = \frac{\text{sum of squares}}{\text{degrees of freedom}} = \frac{\text{SS}}{\text{df}},$$

a form that will be used extensively in future chapters.

Inferences on μ follow the same pattern outlined in Chapter 3 with only the test statistic changed, that is, z and σ are replaced by t and s.

Hypothesis Test on μ

To test the hypothesis

$$H_0: \mu = \mu_0 \text{ vs}$$
$$H_1: \mu \neq \mu_0,$$

[1]In Section 2.2 we adopted a convention that used capital letters to designate random variables and lowercase letters to represent realizations of those random variables. At that time we stated that the specificity of this designation would not be necessary after Chapter 3. Therefore, for this and subsequent chapters we will use lower case letters exclusively.

compute the test statistic

$$t = \frac{(\bar{y} - \mu_0)}{\sqrt{s^2/n}} = \frac{\bar{y} - \mu_0}{s/\sqrt{n}}.$$

The decision on the rejection of H_0 follows the rules specified in Chapter 3. That is, H_0 is rejected if the calculated value of t is in the rejection region, as defined by a specified α, found in the table of the t distribution, or if the calculated p value is smaller than a specified value of α. Since most tables of the t distribution have only limited numbers of probability levels available, the calculation of p values is usually provided only when the analysis is being performed on computers, which are not limited to using tables.[2]

Power curves for this test can be constructed; however, they require a rather more complex distribution. Charts do exist for determining the power for selected situations and are available in some texts (see, for example, Neter *et al.*, 1996).

EXAMPLE 4.2

In Example 3.3 we presented a quality control problem in which we tested the hypothesis that the mean weight of peanuts being put in jars was the required 8 oz. We assumed that we knew the population standard deviation, possibly from experience. We now relax that assumption and estimate both mean and variance from the sample. Table 4.2 lists the data from a sample of 16 jars.

Table 4.2

Data for Peanuts Example (oz.)

8.08	7.71	7.89	7.72
8.00	7.90	7.77	7.81
8.33	7.67	7.79	7.79
7.94	7.84	8.17	7.87

Solution We follow the five steps of a hypothesis test (Section 3.2).

1. The hypotheses are

$$H_0\colon \mu = 8,$$
$$H_1\colon \mu \neq 8.$$

2. Specify $\alpha = 0.05$. The table of the t distribution (Appendix Table A.2) provides the t value for the two-tailed rejection region for 15 degrees of freedom as $|t| > 2.1314$.
3. To obtain the appropriate test statistic, first calculate $\bar{y}$ and s^2:

$$\bar{y} = 126.28/16 = 7.8925,$$
$$s^2 = (997.141 - 996.6649)/15 = 0.03174.$$

The test statistic has the value

$$t = (7.8925 - 8)/\sqrt{(0.03174/16)} = (-0.1075)/0.04453 = -2.4136.$$

[2]We noted in Section 2.6 that when the degrees of freedom become large, the t distribution very closely approximates the normal. In such cases, the use of the tables of the normal distribution provides acceptable results even if σ^2 is not known. For this reason many textbooks treat such cases, usually specifying sample sizes in excess of 30, as large sample cases and specify the use of the z statistic for inferences on a mean. Although the results of such methodology are not incorrect, the large sample–small sample dichotomy does not extend to most other statistical methods. In addition, most computer programs correctly use the t distribution regardless of sample size.

4. Since $|t|$ exceeds the critical value of 2.1314, reject the null hypothesis.
5. We will recommend that the machine be adjusted. Note that the chance that this decision is incorrect is at most 0.05, the chosen level of significance.

The actual p value of the test statistic cannot be obtained from Appendix Table A.2. The actual p value, obtained by a computer program, is 0.0290, and we may reject H_0 at any specified α greater than the observed value of 0.0290. ■

EXAMPLE 4.3

An apple buyer is willing to pay a premium price for a load of apples if they have, as claimed, an average diameter of more than 2.5 in. The buyer wants to test the claim of sufficiently large apples, so he takes a random sample of 12 apples from the load and measures their diameters. The results are given in Table 4.3.

Table 4.3

Apple Dimensions (Diameters, in.)

2.9	2.8	2.7	3.2
2.1	3.1	3.0	2.3
2.4	2.8	2.4	3.4

Solution Since getting somewhat smaller apples is not a disaster, the buyer is willing to take a 10% chance of unnecessarily paying the premium price. Therefore the significance level, α, is set at 0.10. Of course, the buyer only pays the premium price if the apples are larger than 2.5 in., which implies a one-tailed test.

1. The hypotheses are

$$H_0\colon \mu = 2.5,$$

$$H_1\colon \mu > 2.5.$$

2. In other words, he will buy the apples only if the null hypothesis is rejected. We have already specified $\alpha = 0.10$. The variance is estimated from the sample of 12; hence the t statistic for the test has 11 degrees of freedom and the one-tailed rejection region is to reject H_0 if the calculated value of t exceeds 1.3634.
3. From the sample, the values of $\bar{y}$ and s^2 are

$$\bar{y} = 2.758,$$

$$s^2 = 0.1554,$$

and the test statistic is

$$t = (2.758 - 2.5)/\sqrt{(0.1554/12)} = 2.267.$$

4. The null hypothesis is rejected.
5. The buyer should be willing to pay the premium price because there is sufficient evidence that the mean diameter of apples from this load is more than 2.5 in.

If this problem had been performed by a computer program, the result of the test would probably be reported in the form of a p value. However, most computer programs automatically give two-tailed probabilities, in which case the correct p value for a one-tailed hypothesis test must be found by dividing the printed p value by 2. For this example, the computer-generated p value is 0.0433; hence the correct one-tailed p value is $0.0443/2 = 0.0222$, which is indeed less than the required 0.10. ■

EXAMPLE 1.2

REVISITED Recall that in Example 1.2, John Mode had been offered a job in a mid-sized east Texas town. Obviously, the cost of housing in this city will be an important consideration in a decision to move. The Modes read an article in the paper from the town in which they presently live that claimed the "average" price of homes was $155,000. The Modes want to know whether the data collected in Example 1.2 indicate a difference between the two cities. They assumed that the "average" price referred to in the article was the mean, and the sample they collected from the new city represents a random sample of all home prices in that city.

For this purpose,

$$H_0\colon \mu = 155, \text{ and}$$

$$H_1\colon \mu \neq 155.$$

They computed the following results from Table 1.2:

$$\sum y = 9755.18, \qquad \sum y^2 = 1{,}876{,}762, \quad \text{and } n = 69.$$

Thus,

$$\bar{y} = 141.4, \qquad \text{SS} = 497{,}580, \quad \text{and } s^2 = 7317.4,$$

and then

$$t = \frac{141.4 - 155.0}{\sqrt{\frac{7317.4}{69}}} = -1.32,$$

which is insufficient evidence (at $\alpha = 0.05$) that the mean price is different. In other words, the mean price of housing appears not to be different from that of the city in which the Modes currently live. ■

Estimation of μ

Confidence intervals on μ are constructed in the same manner as those in Chapter 3 except that σ is replaced with s, and the table value of z for a specified confidence coefficient $(1-\alpha)$ is replaced by the corresponding value from the table of the t distribution for the appropriate degrees of freedom. The general formula of the $(1 - \alpha)$ confidence interval on μ is

$$\bar{y} \pm t_{\alpha/2}\sqrt{\frac{s^2}{n}},$$

where $t_{\alpha/2}$ has $(n - 1)$ degrees of freedom.

A 0.95 confidence interval on the mean weight of peanuts in Example 4.2 (Table 4.2) is

$$7.8925 \pm 2.1314\ (0.04453),$$

$$7.8925 \pm 0.0949,$$

or from 7.798 to 7.987. Remembering the equivalence of hypothesis tests and confidence intervals, we note that this interval does not contain the null hypothesis value of 8 used in Example 4.2, thus agreeing with the results obtained there.

Similarly, the one-sided lower 0.90 confidence interval for the mean apple size is

$$2.758 - 1.3634\sqrt{(0.1554/12)} \text{ or}$$

$$2.758 - 0.155 = 2.603.$$

This is larger than the required value of 2.5, again agreeing with the results of the hypothesis test.

Solution to Example 4.1 We can now solve the problem in Example 4.1 by providing a confidence interval for the mean exponent. We first calculate the sample statistics: $\bar{y} = 0.9225$ and $s = 0.165$. The t statistic is based on $24 - 1 = 23$ degrees of freedom, and since we want a 95% confidence interval we use $t_{0.05/2} = 2.069$ (rounded). The 0.95 confidence interval on μ is given by

$$0.9225 \pm (2.069)(0.165)/\sqrt{24} \text{ or}$$

$$0.9225 \pm 0.070, \quad \text{or from } 0.8525 \text{ to } 0.9925.$$

Thus we are 95% confident that the true mean exponent is between 0.85 and 0.99, rounded to two decimal places. This seems to imply that, on the average, people tend to underestimate the relative areas. ■

Sample Size

Sample size requirements for an estimation problem where σ is not known can be quite complicated. Obviously we cannot estimate a variance before we take the sample; hence the t statistic cannot be used directly to estimate sample size. Iterative methods that will furnish sample sizes for certain situations do exist, but they are beyond the scope of this text. Therefore most sample size calculations simply assume some known variance and proceed as discussed in Section 3.4.

Degrees of Freedom

For the examples in this section the degrees of freedom of the test statistic (the t statistic) have been $(n - 1)$, where n is the size of the sample. It is, however, important to remember that the degrees of freedom of the t statistic are always those used to estimate the variance used in constructing the test statistic. We will see that for many applications this is not $(n - 1)$.

For example, suppose that we need to estimate the average size of stones produced by a gravel crusher. A random sample of 100 stones is to be used. Unfortunately, we do not have time to weigh each stone individually. We can, however, weigh the entire 100 in one weighing, divide the total weight by 100 to obtain an estimate of μ, and call it $\bar{y}_{100}$. We then take a random subsample of 10 stones from the 100, which we weigh individually to compute an estimate of the variance,

$$s^2 = \frac{\sum(y - \bar{y}_{10})^2}{9},$$

where $\bar{y}_{10}$ is calculated from the subsample of 10 observations. The statistic

$$t = \frac{\bar{y}_{100} - \mu}{\sqrt{s^2/100}},$$

will have the t distribution with 9 (not 99) degrees of freedom.

Although situations such as this do not often arise in practice, it illustrates the fact that the degrees of freedom for the t statistic are associated with the calculation of s^2: it is always the denominator in the expression $s^2 = \text{SS/df}$. However, the variance of $\bar{y}_{100}$ is still estimated by $s^2/100$ because the variance of the sampling distribution of the mean is based on the sample size used to calculate that mean.

4.3 Inferences on a Proportion

In a binomial population, the parameter of interest is p, the proportion of "successes." In Section 2.3 we described the nature of a binomial population and provided in Section 2.5 the normal approximation to the distribution of the proportion of successes in a sample of n from a binomial population. This distribution can be used to make statistical inferences about the parameter p, the proportion of successes in a population.

The estimate of p from a sample of size n is the sample proportion, $\hat{p} = y/n$, where y is the number of successes in the sample. Using the normal approximation, the appropriate statistic to perform inferences on p is

$$z = \frac{\hat{p} - p}{\sqrt{p(1 - p)/n}}.$$

Under the conditions for binomial distributions stated in Section 2.3, this statistic has the standard normal distribution, assuming sufficient sample size for the approximation to be valid.

Hypothesis Test on *p*

The hypotheses are

$$H_0: p = p_0,$$

$$H_1: p \neq p_0.$$

The alternative hypothesis may, of course, be one-sided. To perform the test, compute the test statistic

$$z = \frac{\hat{p} - p_0}{\sqrt{p_0(1 - p_0)/n}},$$

which is compared to the appropriate critical values from the normal distribution (Appendix Table A.1), or a p value is calculated from the normal distribution.

Note that we do not use the t distribution here because the variance is not estimated as a sum of squares divided by degrees of freedom. Of course, the use of the normal distribution is an approximation, and it is generally recommended to be used only if $np \geq 5$ and $n(1 - p) \geq 5$.

EXAMPLE 4.4

An advertisement claims that more than 60% of doctors prefer a particular brand of pain killer. An agency established to monitor truth in advertising conducts a survey consisting of a random sample of 120 doctors. Of the 120 questioned, 82 indicated a preference for the particular brand. Is the advertisement justified?

Solution The parameter of interest is p, the proportion of doctors in the population who prefer the particular brand. To answer the question, the following hypothesis test is performed:

$$H_0: p = 0.6,$$

$$H_1: p > 0.6.$$

Note that this is a one-tailed test and that rejection of the hypothesis supports the advertising claim. Is it likely that the manufacturer of the pain killer would use a slightly different set of hypotheses? A significance level of 0.05 is chosen. The test statistic is

$$\begin{aligned} z &= \frac{\frac{82}{120} - 0.6}{\sqrt{0.6(1 - 0.6)/120}} \\ &= \frac{0.083}{0.0447} \\ &= 1.86. \end{aligned}$$

The p value for this statistic (from Appendix Table A.1) is

$$p = P(z > 1.86) = 0.0314.$$

Since this p value is less than the specified 0.05, we reject H_0 and conclude that the proportion is in fact larger than 0.6. That is, the advertisement appears to be justified. ■

Estimation of p

A $(1-\alpha)$ confidence interval on p based on a sample size of n with y successes is

$$\hat{p} \pm z_{\alpha/2}\sqrt{\frac{\hat{p}(1-\hat{p})}{n}}.$$

Note that since there is no hypothesized value of p, the sample proportion $\hat{p}$ is substituted for p in the formula for the variance.

EXAMPLE 4.5

A preelection poll using a random sample of 150 voters indicated that 84 favored candidate Smith, that is, $\hat{p} = 0.56$. We would like to construct a 0.99 confidence interval on the true proportion of voters favoring Smith.

Solution To calculate the confidence interval, we use

$$0.56 \pm (2.576)\sqrt{\frac{(0.56)(1-0.56)}{150}}$$

$$0.56 \pm 0.104,$$

resulting in an interval from 0.456 to 0.664. Note that the interval does contain 50% (0.5) as well as values below 50%. This means that Smith cannot predict with 0.99 confidence that she will win the election. ■

An Alternate Approximation for the Confidence Interval In Agresti and Coull (1998), it is pointed out that the method of obtaining a confidence interval on p presented above tends to result in an interval that does not actually provide the level of confidence specified. This is because the binomial is a discrete random variable and the confidence interval is constructed using the normal approximation to the binomial, which is continuous. Simulation studies reported in Agresti and Coull indicate that even with sample sizes as high as 100 and true proportion of 0.018, the actual number of confidence intervals containing the true p are closer to 84% than the nominal 95% specified.

The solution, as proposed in this article, is to add two successes and two failures and then use the standard formula to calculate the confidence interval. This adjustment results in much better performance of the confidence interval, even with relative small samples. Using this adjustment, the interval is based on a new estimate of p; $\tilde{p} = (y+2)/(n+4)$. For Example 4.5 the interval would be based on $\tilde{p} = (86)/154 = 0.558$. The resulting confidence interval would be

$$0.558 \pm (2.576)\sqrt{\frac{(0.558)(0.442)}{154}}$$

$$0.558 \pm 0.103, \text{ or}$$

the interval would be from 0.455 to 0.661. This interval is not much different from that constructed without the adjustment, mainly because the sample size is large and the estimate of p is close to 0.5. If the sample size were small, this approximation would result in a more reliable confidence interval.

Sample Size

Since estimation on p uses the standard normal sampling distribution, we are able to obtain the required sample sizes for a given degree of precision. In Section 3.4 we noted that for a $(1 - \alpha)$ degree of confidence and a maximum error of estimation E, the required sample size is

$$n = (z_{\alpha/2}\sigma)^2/E^2.$$

This formula is adapted for a binomial population by substituting the quantity $p(1 - p)$ for σ^2.

In most cases we may have an estimate (or guess) for p that can be used to calculate the required sample size. If no estimate is available, then 0.5 may be used for p, since this results in the largest possible value for the variance and, hence, also the largest n for a given E (and, of course, α). In other words, the use of 0.5 for the unknown p provides the most conservative estimate of sample size.

EXAMPLE 4.6

In close elections between two candidates (p approximately 0.5), a preelection poll must give rather precise estimates to be useful. We would like to estimate the proportion of voters favoring the candidate with a maximum error of estimation of 1% (with confidence of 0.95). What sample size would be needed?

Solution To satisfy the criteria specified would require a sample size of

$$n = (1.96)^2(0.5)(0.5)/(0.01)^2 = 9604.$$

This is certainly a rather large sample and is a natural consequence of the high degree of precision and confidence required. ■

4.4 Inferences on the Variance of One Population

Inferences for the variance follow the same pattern as those for the mean in that the inference procedures use the sampling distribution of the point estimate. The point estimate for σ^2 is

$$s^2 = \sum \frac{(y - \bar{y})^2}{n - 1},$$

or more generally SS/df. We also noted in Section 2.6 that the sample quantity

$$\frac{(n - 1)s^2}{o^2} = \frac{\sum(y - \bar{y})^2}{\sigma^2} = \frac{\text{SS}}{\sigma^2}$$

has the χ^2 distribution with $(n-1)$ degrees of freedom, assuming a sample from a normally distributed population. As before, the point estimate and its sampling distribution provide the basis for hypothesis tests and confidence intervals.

Hypothesis Test on σ^2

To test the null hypothesis that the variance of a population is a prescribed value, say σ_0^2, the hypotheses are

$$H_0: \sigma^2 = \sigma_0^2,$$
$$H_1: \sigma^2 \neq \sigma_0^2,$$

with one-sided alternatives allowed. The statistic from Section 2.6 used to test the null hypothesis is

$$X^2 = \mathrm{SS}/\sigma_0^2,$$

where for this case $\mathrm{SS} = \sum(y - \bar{y})^2$. If the null hypothesis is true, this statistic has the χ^2 distribution with $(n-1)$ degrees of freedom.

If the null hypothesis is false, then the value of the quantity SS will tend to reflect the true value of σ^2. That is, if σ^2 is larger (smaller) than the null hypothesis value, then SS will tend to be relatively large (small), and the value of the test statistic will therefore tend to be larger (smaller) than those suggested by the χ^2 distribution. Hence the rejection region for the test will be two-tailed; however, the critical values will both be positive and we must find individual critical values for each tail. In other words, the rejection region is

$$\text{reject } H_0 \text{ if: } \left(\mathrm{SS}/\sigma_0^2\right) > \chi^2_{\alpha/2},$$
$$\text{or if: } \left(\mathrm{SS}/\sigma_0^2\right) < \chi^2_{(1-\alpha/2)}.$$

Like the t distribution, χ^2 is another distribution for which only limited tables are available. Thus it is difficult to calculate p values when performing hypothesis tests on the variance when such tables must be used.

Hypothesis tests on variances are often one-tailed because variability is used as a measure of consistency, and we usually want to maintain consistency, which is indicated by small variance. Thus, an alternative hypothesis of a larger variance implies an unstable or inconsistent process.

EXAMPLE 4.2

REVISITED In filling the jar with peanuts, we not only want the average weight of the contents to be 8 oz., but we also want to maintain a degree of consistency in the amount of peanuts being put in jars. If one jar receives too many peanuts, it will overflow, and waste peanuts. If another jar gets too few peanuts, it will not be full and the consumer of that jar will feel cheated even though *on average* the jars have the specified amount of peanuts. Therefore, a test on the variance of weights of peanuts should also be part of the quality control of the process.

Suppose the weight of peanuts in at least 95% of the jars is required to be within 0.2 oz. of the mean. Assuming an approximately normal distribution we can use the empirical rule to state that the standard deviation should be at most $0.2/2 = 0.10$, or equivalently that the variance be at most 0.01.

Solution We will use the sample data in Table 4.2 to test the hypothesis

$$H_0\colon \sigma^2 = 0.01 \text{ vs}$$
$$H_1\colon \sigma^2 > 0.01,$$

using a significance level of $\alpha = 0.05$. If we reject the null hypothesis in favor of a larger variance we declare that the filling process is not in control. The rejection region is based on the statistic

$$X^2 = \text{SS}/0.01,$$

which is compared to the χ^2 distribution with 15 degrees of freedom. From Appendix Table A.3 the rejection region for rejecting H_0 is for the calculated χ^2 value to exceed 25.00. From the sample, SS = 0.4761, and the test statistic has the value

$$X^2 = 0.4761/0.01 = 47.61.$$

Therefore the null hypothesis is rejected and we recommend the expense of modifying the filling process to ensure more consistency. That is, the machine must be adjusted or modified to reduce the variability. Naturally, after the modification, another series of tests would be conducted to ensure success in reducing variation. ■

EXAMPLE 4.1

REVISITED Suppose in the study in perceptual psychology, the variability of subjects was of concern. In particular, suppose that the researchers wanted to know whether the variance of exponents differed from 0.02.

Solution The hypotheses of interest would then be

$$H_0\colon \sigma^2 = 0.02,$$
$$H_1\colon \sigma^2 \neq 0.02.$$

Using a level of significance of 0.05, the critical region is

reject H_0 if SS/0.02 is larger than 38.08 (rounded)

or smaller than 11.69 (rounded).

The data in Table 4.1 produce SS = 0.628. Hence, the test statistic has a value of $0.628/0.02 = 31.4$, which is not in the critical region; thus, we cannot reject the null hypothesis that $\sigma^2 = 0.02$. ■

Estimation of σ^2

A confidence interval can be constructed for the value of the parameter σ^2 using the χ^2 distribution. Because the distribution is not symmetric, the confidence interval is not symmetric about s^2 and, as in the case of the two-sided hypothesis test, we need two individual values from the χ^2 distribution to calculate the confidence interval.

The lower limit of the confidence interval is

$$L = \text{SS}/\chi^2_{\alpha/2},$$

and the upper limit is

$$U = \text{SS}/\chi^2_{(1-\alpha/2)},$$

where the tail values come from the χ^2 distribution with $(n - 1)$ degrees of freedom. Note that the upper tail value from the χ^2 distribution is used for the lower limit and vice versa.

For Example 4.2 we can calculate a 0.95 confidence interval on σ^2 based on the sample data given in Table 4.2. Since the hypothesis test for this example was one-tailed, we construct a corresponding one-sided confidence interval. In this case we would want the lower 95% limit, which would require the upper 0.05 tail of the χ^2 distribution with 15 degrees of freedom, which we have already seen to be 25.00. The lower confidence limit is $\text{SS}/\chi^2_{\alpha} = 0.4761/25.00 = 0.0190$. The lower 0.95 confidence limit for the standard deviation is simply the square root of the limit for the variance, resulting in the value 0.138. We are therefore 95% confident that the true standard deviation is at least 0.138. This value is larger than that specified by the null hypothesis and again the confidence interval agrees with the result of the hypothesis test.

4.5 Assumptions

Today virtually all statistical analyses are performed by computers. We know that for all practical purposes, computers do not make mistakes, and furthermore the beautifully annotated outputs for such analyses make us believe that the results they produce reveal the ultimate truth. Unfortunately, the results provided by the best computers using the ultimate software only reflect the quality of the submitted data. And if the data are deficient, results of the analysis will be less than useful.

How can data be deficient? There are two major sources:

- sloppy data gathering and recording, and
- failure of the distribution of the variable(s) being studied to conform to the assumptions underlying the statistical inference procedure.

Avoiding sloppy data gathering and recording is mostly a matter of common sense, although the increased use of automatic data gathering and recording increases the chance of undetected errors. For this reason graphical data summarization, including but not limited to stem and leaf, box, and scatter plots should be an integral part of data quality control.

The failure to conform to assumptions is a subtler problem. In this section we briefly summarize the necessary assumptions, suggest a method for detecting violations, and suggest some remedial methods.

Required Assumptions and Sources of Violations

Two major assumptions are needed to assure the correctness for statistical inferences:

- randomness of the sample observations, and
- the distribution of the variable(s) being studied.

We have already noted that randomness is a necessary requirement to define sampling distributions and the consequent use of probabilities associated with these distributions. Another aspect of randomness is that it helps to assure that the observations we obtain have the necessary independence. For example, a failure of the assumption of independence occurs when the sample is selected from the population in some ordered manner. This occurs in some types of economic data obtained on a regular basis at different time periods. These observations then become naturally ordered, and adjacent observations tend to be related, which is a violation of the independence assumption. This does not make the data useless; instead, the user must be aware of the trend and account for it in the analysis (see also Section 7.6).

The distributional assumptions arise from the fact that most of the sampling distributions we use are based on the normal distribution. We know that no "real" data are ever *exactly* normally distributed. However, we also know that the central limit theorem is quite robust so that the normality of the sampling distribution of the mean should not pose major problems except with small sample sizes and/or extremely nonnormal distributions. The χ^2 distribution used for the sampling distribution of the variance and consequently the t distribution are not quite as robust but again, larger sample sizes help.

Outliers or unusual observations are also a major source of nonnormality. If they arise from measurement errors or plain sloppiness, they can often be detected and corrected. However, sometimes they are "real," and no corrections can be made, and they certainly cannot simply be discarded and may therefore pose a problem.

Prevention of Violations

The best method of avoiding violations is to use common sense, diligence, and honesty when collecting, recording, and analyzing data. For example, sloppiness or recording errors may cause extreme values to be included and considered as legitimate data. Improper sampling procedures may result in nonrandom or nonindependent sample observations. Any automated data collection procedure must have close supervision and internal checks. Remember that the very machines that make such data gathering possible also have the ability for error detection and exhaustive data summarization.

Detection of Violations

The exploratory data analysis techniques presented in Chapter 1 should be used as a matter of routine. These techniques not only help to reveal mistakes but can also detect distributional problems. For example, the stem and leaf and

Table 4.4

Exponents from Example 4.1

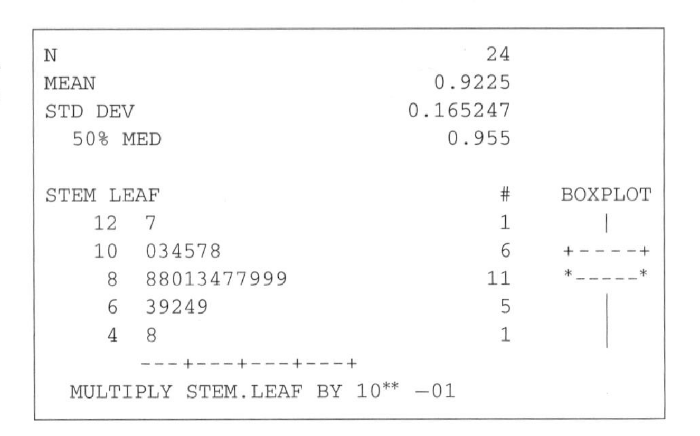

```
N                                        24
MEAN                                 0.9225
STD DEV                            0.165247
  50% MED                             0.955

STEM LEAF                                 #    BOXPLOT
    12  7                                 1       |
    10  034578                            6    +-----+
     8  88013477999                      11    *-----*
     6  39249                             5       |
     4  8                                 1       |
        ---+---+---+---+
  MULTIPLY STEM.LEAF BY 10** -01
```

Figure 4.2

Normal Probability Plot for a Negatively Skewed Distribution

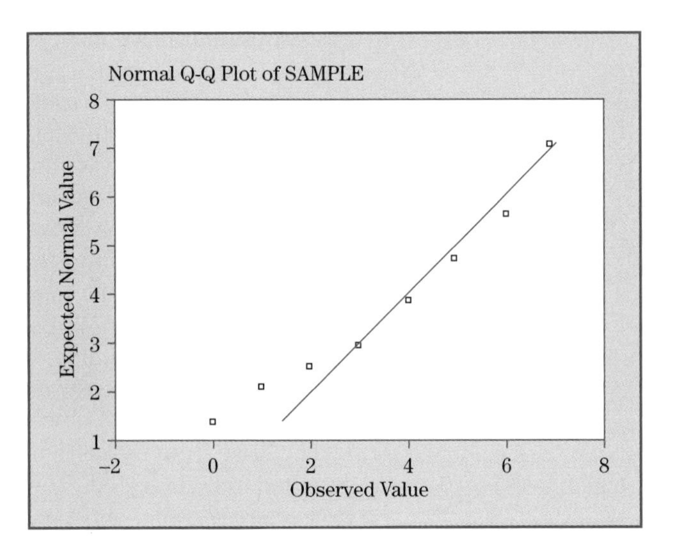

box plots for Example 4.1, shown in Table 4.4, are easily produced and show that there appear to be no obvious problems with the normality assumption.

The use of a **normal probability plot** allows a slightly more rigorous test of the normality assumption. A special plot, called a Q–Q plot (quantile–quantile), shows the observed value on one axis (usually the horizontal axis) and the value that is expected if the data are a sample from the normal distribution on the other axis. The points should cluster around a straight line for a normally distributed variable. If the data are skewed, the normal probability plot will have a very distinctive shape. Figures 4.2, 4.3, and 4.4 were constructed using the Q–Q graphics function in SPSS. Figure 4.2 shows a typical Q–Q plot for a distribution skewed negatively. Note how the points are all above the line for small values. Figure 4.3 shows a typical Q–Q plot for a distribution skewed positively. In this plot the larger points are all below the line. Figure 4.4 shows the Q–Q plot for the data in Example 4.1. Note that the points are reasonably close to the line, and there are no indications of systematic deviations from

Figure 4.3

Normal Probability Plot for a Positively Skewed Distribution

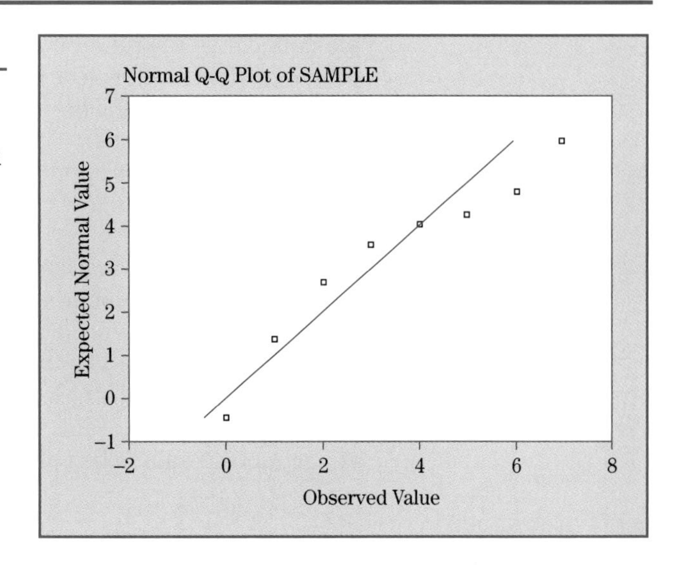

Figure 4.4

Normal Probability Plot for Example 4.1

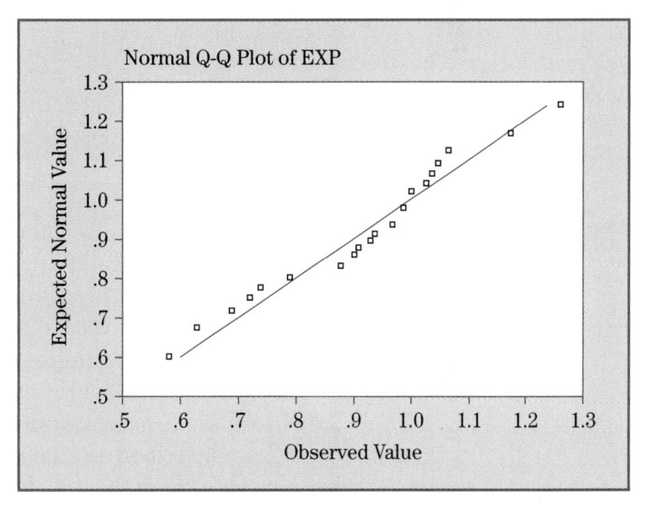

the line, thereby indicating that the distribution of the population is reasonably close to normal.

Tests for Normality

There are formal hypothesis tests that can be used to determine whether a set of observed values fit some specified distribution. Such tests are known as **goodness-of-fit tests**. One such test is the χ^2 test discussed in Section 12.3.

Because tests for distributions are often concerned specifically with the normal distribution and are also not very easy to perform by hand, tests for normality are available in data summarization programs, such as SAS `PROC UNIVARIATE`. One of the most popular tests for normality is the *Kolmogoroff–Smirnoff* test. This test compares the observed cumulative distribution with the cumulative distribution that would occur if the data were normally distributed. The test statistic is based on the maximum difference between these two. For example, using the tree data (Example 1.3) `PROC UNIVARIATE` gives p values for this test as $p > 0.14$ for `HT` and $p < 0.01$ for `HCRN`, which indicates that the height is approximately normally distributed while the height to the crown is not. This test confirms what the histograms in Figs. 1.4 and 1.5 showed.

The sensitivity of such tests is obviously affected by sample size, which means that they may not be sufficiently sensitive for small samples where non-normality may pose a problem, and overly sensitive for large samples where normality may not be very important. Furthermore, most of the procedures used in doing routine statistical inferences are not very sensitive to deviations from normality (Kirk, 1995). A procedure not affected by violations of assumptions is said to be **robust** with respect to these violations.

If Assumptions Fail

Now that we have scared you, we add a few words of comfort. Many statistical methods are reasonably robust and with reasonable care, most statistical analyses can be used as advertised. And even if problems arise, all is not lost. The following example shows the effect of an extreme value on a test for the mean and how an alternate analysis can be used to alleviate the effects of that observation.

EXAMPLE 4.7

A supermarket chain is interested in locating a store in a neighborhood suspected of having families with relatively low incomes, a situation that may cause a store in that neighborhood to be unprofitable. The supermarket chain believes that if the average family income is more than $13,000 the store will be profitable. To determine whether the suspicion is valid, income figures are obtained from a random sample of 20 families in that neighborhood. The data from the sample are given in Table 4.5. Assuming that the conditions for using the t test described in this chapter hold, what can be concluded about the average income in this neighborhood?

Solution The hypotheses

$$H_0\colon \mu = 13.0,$$

$$H_1\colon \mu > 13.0$$

Table 4.5

Data on Household Income (Coded in Units of $1000)

No.	Income	No.	Income	No.	Income	No.	Income
1	17.1	6	12.3	11	15.7	16	16.2
2	12.7	7	13.2	12	93.4	17	13.6
3	16.5	8	13.3	13	14.9	18	12.8
4	14.0	9	17.9	14	13.0	19	13.4
5	14.2	10	12.5	15	13.8	20	16.6

are to be tested using a 0.05 significance level. The estimated mean and variance are

$$\bar{y} = 18.36,$$
$$s^2 = 314.9,$$

resulting in a t statistic of

$$t = (18.36 - 13.0)/\sqrt{314.9/20}$$
$$= 1.351.$$

We compare this with the 0.05 one-tailed t value of 1.729 and the conclusion is to fail to reject the null hypothesis. It appears that the store will not be built.

The developer involved in the proposed venture decides to take another look at the data and immediately notes an obvious anomaly. The observed income values are all less than $20,000 with one exception: One family reported its income as $93,400. Further investigation reveals that the observation is correct. This income belongs to a family of descendants of the original owner of the land on which the neighborhood is located and who are still living in the old family mansion.

The relevant question here is: What effect does this observation have on the conclusion reached by the hypothesis test? One would think that the large value of this observation would inflate the value of the sample mean and therefore tend to increase the probability of finding an adequate mean income in that area. However, the effect of the extreme value is not only on the mean, but also on the variance, and therefore the result is not quite so predictable. To illustrate, assume that the sampling procedure had picked a more typical family with an income of 16.4. This substitution does lower the sample mean from 18.36 to 14.51. However, it also reduces the variance from 314.86 to 3.05! The value of the test statistic now becomes 3.87, and the null hypothesis would be rejected. ■

Alternate Methodology

In the above example we were able to get a different result by replacing an extreme observation with one that seemed more reasonable. Such a procedure is definitely not recommended, because it could easily lead to abuse (data could be changed until the desired result was obtained). There are, however, more legitimate alternative procedures that can be used if the necessary assumptions appear to be unfulfilled. Such methods may be of two types:

1. The data are "adjusted" so that the assumptions fit.
2. Procedures that do not require as many assumptions are used.

Adjusting the data may be accomplished by simply discarding a **prespecified** number of extreme observations (in both tails), and making appropriate (mathematically justified) adjustments in the test statistic. This is referred to as "trimming" the data (see Koopmans, 1987). Trimming is not often used and can be quite difficult to implement in complex situations.

Adjusting the data can also be accomplished by "transforming" the data. For example, the variable measured in an experiment may not have a normal distribution, but the natural logarithm of that variable may. Transformations take many forms, and are discussed in Section 6.4. More complete discussions are given in some texts (see, for example, Neter *et al.*, 1996).

Procedures of the second type are usually referred to as "nonparametric" or "distribution-free" methods since they do not depend on parameters of specified distributions describing the population. For illustration we apply a simple alternative procedure to the data of Example 4.7 that will illustrate the use of a nonparametric procedure for making the decision on the location of the store.

EXAMPLE 4.7

REVISITED In Chapter 1 we observed that for a highly skewed distribution the median may be a more logical measure of central tendency. Remember that the specification for building the store said "average," a term that may be satisfied by the use of the median.

The median (see Section 1.5) is defined as the "middle" value of a set of population values. Therefore, in the population, half of the observations are above and half of the observations are below the median. In a random sample then, observations should be either higher or lower than the median with equal probability. Defining values above the median as successes, we have a sample from a binomial population with $p = 0.5$. We can then simply count how many of the sample values fall above the hypothesized median value and use the binomial distribution to conduct a hypothesis test.

Solution The decision to locate a store in the neighborhood discussed in Example 4.7 is then based on testing the hypotheses

$$H_0\text{: the population median} = 13,$$

$$H_1\text{: the population median} > 13.$$

This is equivalent to testing the hypotheses

$$H_0\text{: } p = 0.5,$$

$$H_1\text{: } p > 0.5,$$

where p is the proportion of the population values exceeding 13.

This is an application of the use of inferences on a binomial parameter. In the sample shown in Table 4.5 we observe that 15 of the 20 values are strictly larger than 13. Thus $\hat{p}$, the sample proportion having incomes greater than 13, is 0.75. Using the normal approximation to the binomial, the value of the test statistic is

$$z = (0.75 - 0.5)/\sqrt{[(0.5)(0.5)/20]} = 2.23.$$

This value is compared with the 0.05 level of the standard normal distribution (1.645), or results in a p value of 0.012. The result is that the null hypothesis is rejected, leading to the conclusion that the store should be built. ■

EXAMPLE 1.2

REVISITED After reviewing the housing data collected in Example 1.1, the Modes realized that the t test they performed might be affected by the small number of very-high-priced homes that appeared in Table 1.2. In fact, they determined that the median price of the data in Table 1.2 was \$119,000, which is quite a bit less than the sample mean of \$141,400 obtained from the data. Further, a re-reading of the article in the paper found that the "average" price of \$155,000 referred to was actually the median price. A quick check showed that 50 of the 69 (or 72.4%) of the housing prices given in Table 1.2 had values below 155. The test for the null hypothesis that the median is \$155,000 gives

$$z = \frac{0.724 - 0.500}{\sqrt{\frac{(0.5)(0.5)}{69}}} = 3.73,$$

which, when compared with the 0.05 level of the standard normal distribution ($z = 1.960$), provides significant evidence that the median price of homes is lower in their prospective new city than that of their current city of residence.

It is necessary to emphasize at this point that, despite its simplicity, this test should not be used if the assumptions necessary for the t test are indeed fulfilled. The reason for this caution is that under the assumption of normality the t test has more power. This is due to the fact that the test on the median does not use all of the information available in the observed values, since the actual values of the observations are not considered when simply counting the number of sample observations larger than the hypothesized median. That is, the ordinal variable describing the median is not as informative as the ratio variable used to compute the mean.

Other nonparametric methods exist for this particular example. Specifically, the Wilcoxon signed rank test (Chapter 13) may be considered appropriate here, but we defer presentation of all nonparametric methods to Chapter 13. ■

4.6 CHAPTER SUMMARY

This chapter provides the methodology for making inferences on the parameters of a single population. The specific inferences presented are

- inferences on the mean, which are based on the Student t distribution,
- inferences on a proportion using the normal approximation to the binomial distribution, and
- inferences on the variance using the χ^2 distribution.

A final section discusses some of the assumptions necessary for ensuring the validity of these inference procedures and provides an example for which a violation has occurred and a possible alternative inference procedure for that situation.

4.7 CHAPTER EXERCISES

CONCEPT QUESTIONS

Indicate true or false for the following statements. If false, specify what change will make the statement true.

1. _______ The t distribution is more dispersed than the normal.
2. _______ The χ^2 distribution is used for inferences on the mean when the variance is unknown.
3. _______ The mean of the t distribution is affected by the degrees of freedom.
4. _______ The quantity

$$\frac{(\bar{y} - \mu)}{\sqrt{\sigma^2/n}}$$

has the t distribution with $(n - 1)$ degrees of freedom.

5. _______ In the t test for a mean, the level of significance increases if the population standard deviation increases, holding the sample size constant.
6. _______ The χ^2 distribution is used for inferences on the variance.
7. _______ The mean of the t distribution is zero.
8. _______ When the test statistic is t and the number of degrees of freedom is >30, the critical value of t is very close to that of z (the standard normal).
9. _______ The χ^2 distribution is skewed and its mean is always 2.
10. _______ The variance of a binomial proportion is npq [or $np(1 - p)$].
11. _______ The sampling distribution of a proportion is approximated by the χ^2 distribution.
12. _______ The t test can be applied with absolutely no assumptions about the distribution of the population.
13. _______ The degrees of freedom for the t test do not necessarily depend on the sample size used in computing the mean.

PRACTICE EXERCISES

The following exercises are designed to give the reader practice in doing statistical inferences on a single population through simple examples with small data sets. The solutions are given in the back of the text.

1. Find the following upper one-tail values:
   (a) $t_{0.05}(13)$
   (b) $t_{0.01}(26)$
   (c) $t_{0.10}(8)$
   (d) $\chi^2_{0.01}(20)$
   (e) $\chi^2_{0.10}(8)$

(f) $\chi^2_{0.975}(40)$
(g) $\chi^2_{0.99}(9)$

2. The following sample was taken from a normally distributed population:

$$3, 4, 5, 5, 6, 6, 6, 7, 7, 9, 10, 11, 12, 12, 13, 13, 13, 14, 15.$$

(a) Compute the 0.95 confidence interval on the population mean μ.
(b) Compute the 0.90 confidence interval on the population standard deviation σ.

3. Using the data in Exercise 2, test the following hypotheses:
(a) H_0: $\mu = 13$,
H_1: $\mu \neq 13$.
(b) H_0: $\sigma^2 = 10$,
H_1: $\sigma^2 \neq 10$.

4. A local congressman indicated that he would support the building of a new dam on the Yahoo River if at least 60% of his constituents supported the dam. His legislative aide sampled 225 registered voters in his district and found 135 favored the dam. At the level of significance of 0.10 should the congressman support the building of the dam?

5. In Exercise 4, how many voters should the aide sample if the congressman wanted to estimate the true level of support to within 1%?

EXERCISES

1. Weight losses of 12 persons in an experimental one-week diet program are given below:

Weight loss in pounds			
3.0	1.4	0.2	−1.2
5.3	1.7	3.7	5.9
0.2	3.6	3.7	2.0

Do these results indicate that a mean weight loss was achieved? (Use $\alpha = 0.05$).

2. In Exercise 1, determine whether a mean weight loss of more than 1 lb. was achieved. (Use $\alpha = 0.01$.)

3. A manufacturer of watches has established that on the average his watches do not gain or lose. He also would like to claim that at least 95% of the watches are accurate to ± 0.2 s per week. A random sample of 15 watches provided the following gains (+) or losses (−) in seconds in one week:

+0.17	−0.07	+0.13	−0.05	+0.23
+0.01	+0.06	+0.08	−0.14	−0.10
+0.08	+0.11	+0.05	−0.87	+0.05

Can the claim be made with a 5% chance of being wrong? (Assume that the inaccurancies of these watches are normally distributed.)

4. A sample of 20 insurance claims for automobile accidents (in \$1000) gives the following values:

1.6	2.0	2.7	1.3	2.0
1.3	0.3	0.9	1.2	1.2
0.2	1.3	5.0	0.8	7.4
3.0	0.6	1.8	2.5	0.3

Construct a 0.95 confidence interval on the mean value of claims. Comment on the usefulness of this estimate (*Hint:* Look at the distribution.)

5. An advertisement for a headache remedy claims that 90% or more of headache sufferers get relief if they use the remedy. A truth in advertising agency is considering a suit for false advertising and obtains a sample of 100 individuals, which shows that 88 indicate that the remedy gave them relief.

(a) Using $\alpha = 0.10$ can the suit be justified?

(b) Comment on the implications of a type I or a type II error in this problem.

(c) Suppose that the company manufacturing the remedy wants to conduct a promotion campaign that claims over 90% of the remedy users get relief from headaches. What would change in the hypotheses statements used in part (a)?

(d) What about the implications discussed in part (b)?

6. Average systolic blood pressure of a normal male is supposed to be about 129. Measurements of systolic blood pressure on a sample of 12 adult males from a community whose dietary habits are suspected of causing high blood pressure are listed below:

115	134	131	143
130	154	119	137
155	130	110	138

Do the data justify the suspicions regarding the blood pressure of this community? (Use $\alpha = 0.01$.)

7. A public opinion poll shows that in a sample of 150 voters, 79 preferred candidate X. If X can be confident of winning, she can save campaign funds by reducing TV commercials. Given the results of the survey should X conclude that she has a majority of the votes? (Use $\alpha = 0.05$.)

8. Construct a 0.95 interval on the true proportion of voters preferring candidate X in Exercise 7.

9. It is said that the average weight of healthy 12-hr-old infants is supposed to be 7.5 lbs. A sample of newborn babies from a low-income neighborhood yielded the following weights (in pounds) at 12 hr after birth:

6.0	8.2	6.4	4.8
8.6	8.0	6.0	
7.5	8.1	7.2	

At the 0.01 significance level, can we conclude that babies from this neighborhood are underweight?

10. Construct a 0.99 confidence interval on the mean weight of 12-hr-old babies in Exercise 9.

11. A truth in labeling regulation states that no more than 1% of units may vary by more than 2% from the weight stated on the label. The label of a product states that units weigh 10 oz. each. A sample of 20 units yielded the following:

10.01	9.92	9.82	10.04
10.04	10.06	9.97	9.94
9.97	9.86	10.02	10.14
9.97	9.97	9.97	10.05
10.19	10.10	9.95	10.00

At $\alpha = 0.05$ can we conclude that these units satisfy the regulation?

12. Construct a 0.95 confidence interval on the variance of weights given in Exercise 11.

13. A production line in a certain factory puts out washers with an average inside diameter of 0.10 in. A quality control procedure that requires the line to be shut down and adjusted when the standard deviation of inside diameters of washers exceeds 0.002 in. has been established. Discuss the quality control procedure relative to the value of the significance level, type I and type II errors, sample size, and cost of the adjustment.

14. Suppose that a sample of size 25 from Exercise 13 yielded $s = 0.0037$. Should the machine be adjusted?

15. Using the data from Exercise 4, construct a stem and leaf plot and a box plot (Section 1.6). Do these graphs indicate that the assumptions discussed in Section 4.5 are valid? Discuss possible alternatives.

16. Using the data from Exercise 11, construct a stem and leaf plot and a box plot. Do these graphs indicate that the assumptions discussed in Section 4.5 are valid? Discuss possible alternatives.

17. In Exercise 13 of Chapter 1 the half-lives of aminoglycosides were listed for a sample of 43 patients, 22 of which were given the drug Amikacin. The data for the drug Amikacin are reproduced in Table 4.6. Use these data to determine a 95% confidence interval on the true mean half-life of this drug.

Table 4.6

Half-Life of Amikacin

2.50	1.20	2.60	1.44	1.87	2.48
2.20	1.60	1.00	1.26	2.31	2.80
1.60	2.20	1.50	1.98	1.40	0.69
1.30	2.20	3.15	1.98		

18. Using the data from Exercise 17, construct a 90% confidence interval on the variance of the half-life of Amikacin.

19. A certain soft drink bottler claims that less than 10% of its customers drink another brand of soft drink on a regular basis. A random sample of 100 customers yielded 18 who did in fact drink another brand of soft drink on a regular basis. Do these sample results support the bottler's claim? (Use a level of significance of 0.05.)

20. Draw a power curve for the test constructed in Exercise 19. (Refer to the discussion on power curves in Section 3.2 and plot $1 - \beta$ versus p = proportion of customers drinking another brand.)

Table 4.7

Data for Exercise 21

Type	Concentration	Differences		
1	1	−0.112	0.163	−0.151
1	2	−0.117	0.072	0.169
1	3	−0.006	−0.092	−0.268
1	4	0.119	0.118	0.051
1	5	−0.272	−0.302	0.343
2	1	−0.094	−0.137	0.308
2	2	−0.238	0.031	0.160
2	3	−0.385	−0.366	−0.173
2	4	−0.259	0.266	−0.303
2	5	−0.125	0.383	0.334
3	1	0.060	0.106	0.084
3	2	−0.016	−0.191	0.097
3	3	−0.024	−0.046	−0.178
3	4	0.040	0.028	0.619
3	5	0.062	0.293	−0.106
4	1	−0.034	0.116	0.055
4	2	−0.023	−0.099	−0.212
4	3	−0.256	−0.110	−0.272
4	4	−0.046	0.009	−0.134
4	5	−0.050	0.009	−0.034

21. This experiment concerns the precision of four types of collecting tubes used for air sampling of hydrofluoric acid. Each type is tested three times at five different concentrations. The data shown in Table 4.7 give the differences between the three observed and true concentrations for each level of true concentration for each of the tubes.

The differences are required to have a standard deviation of no more than 0.1. Do any of the tubes meet this criterion? (*Careful:* What is the most appropriate sum of squares for this test?)

Chapter 5

Inferences for Two Populations

EXAMPLE 5.1

Comparing Costs of an Audit Publicly funded institutions are required to have their financial records periodically audited by independent auditing firms. They are usually free to choose any accredited firm, but there is some inclination to employ a prestigious firm such as one of the "Big Eight." Since there is a suspicion that these firms charge more for their services, the chief accountant of a city conducts a study to investigate this possibility. She obtains information on the cost of their latest audit for a random sample of 25 cities and notes whether the firm was one of the Big Eight. Recognizing that the size of the city also affects the cost of an audit, she also obtains the population of each city. The data are shown in Table 5.1. The population (POP) and audit fee (FEE) are in units of 1000; the columns under BIG8 signify whether the auditing firm is one of the Big Eight by YES or NO.

Figure 5.1 shows the box plots of the fees charged by the two classes of auditing firms. This figure certainly suggests that the BIG8 do charge more; however, the analysis presented in the chapter summary (Section 5.7) provides the surprising result that there is insufficient evidence to conclude that a difference exists. In order to see why this apparent contradiction occurs, we must first explore the method necessary to compare the differences in fees charged by the two classes of auditing firms. This chapter presents methods used to compare two populations. ■

5.1 Introduction

In Chapter 4 we provided methods for inferences on parameters of a single population. A natural extension of these methods occurs when two populations are to be compared. In this chapter we provide the inferential methods

Table 5.1 Audit Fees

POP	FEE	BIG8	POP	FEE	BIG8	POP	FEE	BIG8
25.43	7.50	NO	40.20	20.00	NO	191.00	50.00	YES
25.50	15.00	NO	70.42	30.00	NO	279.27	82.00	YES
26.42	10.00	NO	75.23	44.00	YES	357.87	125.00	YES
27.15	18.00	NO	81.83	32.00	YES	385.46	76.00	YES
29.52	16.00	NO	105.61	48.50	NO	492.37	86.00	YES
30.40	17.62	NO	111.81	65.00	YES	562.99	126.00	YES
32.10	8.45	NO	150.25	90.00	YES	1203.34	177.00	YES
35.81	12.00	NO	164.67	104.50	YES			
36.61	21.50	NO	171.93	95.00	YES			

Figure 5.1

Audit Fees

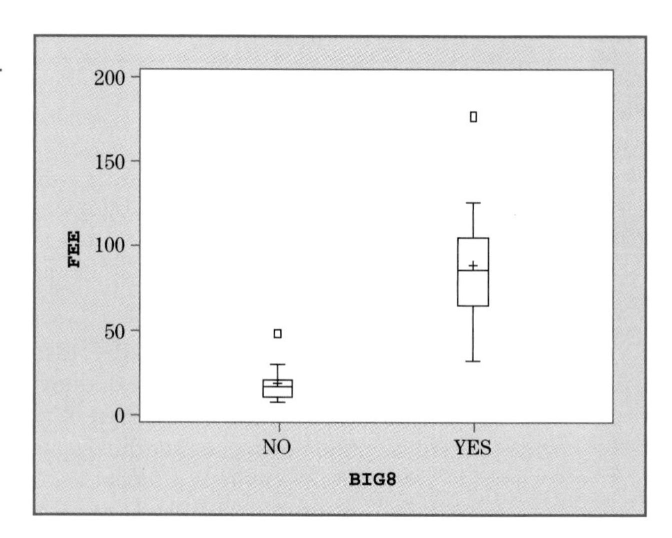

for making comparisons on parameters of two populations. This leads to a natural extension, that of comparing more than two populations, which is presented in Chapter 6. So, why not go directly to comparing parameters of several populations and consider the case of two populations as a special case? There are several good answers to that question:

- Many interesting applications involve only two populations, for example, any comparisons involving differences between the two sexes, comparing a drug with a placebo, comparing old versus new, or before and after some event.
- Some of the concepts underlying comparing several populations are more easily introduced for the two-population case.
- The comparison of two populations results in a single easily understood statistic: the difference between sample means. As we shall see in Chapter 6, such a simple statistic is not available for comparing more than two populations. As a matter of fact, even when we have more than two populations,

we will often want to make comparisons among specific pairs from the set of populations.

Populations that are to be compared arise in two distinct ways:

- The populations are actually different. For example, male and female students, two regions of a state or nation, or two different breeds of cattle. In Section 1.1 we referred to a study involving separate populations as an observational study.
- The populations are a result of an experiment where a single homogeneous population has been divided into two portions where each has been subjected to some sort of modification, for example, a sample of individuals given two different drugs to combat a disease, a field of an agricultural crop where two different fertilizer mixtures are applied to various portions, or a group of school children subjected to different teaching methods. In Section 1.1 this type of study was referred to as a designed experiment.

This latter situation constitutes the more common usage of statistical inference. In such experiments the different populations are usually referred to as "treatments" or "levels of a factor." These terms will be discussed in greater detail in later chapters, especially Chapter 10.

There are also two distinct methods for collecting data on two populations, or equivalently, designing an experiment for comparing two populations. These are called (1) **independent samples** and (2) **dependent** or **paired samples**. We illustrate these two methods with a hypothetical experiment designed to compare the effectiveness of two migraine headache remedies. The response variable is a measure of headache relief reported by the subjects.

Independent Samples A sample of migraine sufferers is randomly divided into two groups. The first group is given remedy A while the other is given remedy B, both to be taken at the onset of a migraine attack. The pills are not identified, so patients do not know which pill they are taking. Note that the individuals sampled for the two remedies are indeed independent of each other.

Dependent or Paired Samples Each person in a group of migraine sufferers is given two pills, one of which is red and the other is green. The group is randomly split into two subgroups and one is told to take the green pill the first time a migraine attack occurs and the red pill for the next one. The other group is told to take the red pill first and the green pill next. Note that both pills are given to each patient so the responses of the two remedies are naturally paired for each patient.

These two methods of comparing the efficacy of the remedies dictate different inferential procedures. The comparison of means, variances, and proportions for independent samples are presented in Sections 5.2, 5.3, and 5.5, respectively, and the comparison of means and proportions for the dependent or paired sample case in Sections 5.4 and 5.5.

5.2 Inferences on the Difference between Means Using Independent Samples

We are interested in comparing two populations whose means are μ_1 and μ_2 and whose variances are σ_1^2 and σ_2^2, respectively. Comparisons may involve the means or the variances (standard deviations). In this section we consider the comparison of means.

For two populations we define the difference between the two means as

$$\delta = \mu_1 - \mu_2.$$

This single parameter δ provides a simple, tractable measure for comparing two population means, not only to see whether they are equal, but also to estimate the difference between the two. For example, testing the null hypothesis

$$H_0: \mu_1 = \mu_2$$

is the same as testing the null hypothesis

$$H_0: \delta = 0.$$

A sample of size n_1 is randomly selected from the first population and a sample of size n_2 is independently drawn from the second. The difference between the two sample means $(\bar{y}_1 - \bar{y}_2)$ provides the unbiased point estimate of the difference $(\mu_1 - \mu_2)$. However, as we have learned, before we can make any inferences about the difference between means, we must know the sampling distribution of $(\bar{y}_1 - \bar{y}_2)$.

Sampling Distribution of a Linear Function of Random Variables

The sampling distribution of the difference between two means from independently drawn samples is a special case of the sampling distribution of a **linear function of random variables**. Consider a set of n random variables $y_1, y_2, \ldots, y_n$, whose distributions have means $\mu_1, \mu_2, \ldots, \mu_n$ and variances $\sigma_1^2, \sigma_2^2, \ldots, \sigma_n^2$. A linear function of these random variables is defined as

$$L = \sum a_i y_i = a_1 y_1 + a_2 y_2 + \cdots + a_n y_n,$$

where the a_i are arbitrary constants. L is also a random variable and has mean

$$\mu_L = \sum a_i \mu_i = a_1 \mu_1 + a_2 \mu_2 + \cdots + a_n \mu_n.$$

If the variables are independent, then L has variance

$$\sigma_L^2 = \sum a_i^2 \sigma_i^2 = a_1^2 \sigma_1^2 + a_2^2 \sigma_2^2 + \cdots + a_n^2 \sigma_n^2.$$

Further, if the y_i are normally distributed, so is L.

The Sampling Distribution of the Difference between Two Means

Since sample means are random variables, the difference between two sample means is a linear function of two random variables. That is,

$$\bar{y}_1 - \bar{y}_2$$

can be written as

$$L = a_1\bar{y}_1 + a_2\bar{y}_2 = (1)\bar{y}_1 + (-1)\bar{y}_2.$$

In terms of the linear function specified above, $n = 2$, and $a_1 = 1$ and $a_2 = -1$. Using these specifications, the sampling distribution of the difference between two means has a mean of $(\mu_1 - \mu_2)$.

Further, since the $\bar{y}_1$ and $\bar{y}_2$ are sample means, the variance of $\bar{y}_1$ is σ_1^2/n_1 and the variance of $\bar{y}_2$ is σ_2^2/n_2. Also, because we have made the assumption that the two samples are independently drawn from the two populations, the two sample means are independent random variables. Therefore, the variance of the difference $(\bar{y}_1 - \bar{y}_2)$ is

$$\sigma_L^2 = (+1)^2\sigma_1^2/n_1 + (-1)^2\sigma_2^2/n_2,$$

or simply

$$= \sigma_1^2/n_1 + \sigma_2^2/n_2.$$

Note that for the special case where $\sigma_1^2 = \sigma_2^2 = \sigma^2$ and $n_1 = n_2 = n$, say, the variance of the difference is $2\sigma^2/n$.

Finally, the central limit theorem states that if the sample sizes are sufficiently large, $\bar{y}_1$ and $\bar{y}_2$ are normally distributed; hence for most applications L is also normally distributed.

Thus, if the variances σ_1^2 and σ_2^2 are known, we can determine the variance of the difference $(\bar{y}_1 - \bar{y}_2)$. As in the one-population case we first present inference procedures that assume that the population variances are known. Procedures using estimated variances are presented later in this section.

Variances Known

We first consider the situation in which both population variances are known. We want to make inferences on the difference

$$\delta = \mu_1 - \mu_2,$$

for which the point estimate is

$$\bar{y}_1 - \bar{y}_2.$$

This statistic has the normal distribution with mean $(\mu_1 - \mu_2)$ and variance $(\sigma_1^2/n_1 + \sigma_2^2/n_2)$. Hence, the statistic

$$z = \frac{\bar{y}_1 - \bar{y}_2 - \delta}{\sqrt{(\sigma_1^2/n_1) + (\sigma_2^2/n_2)}}$$

has the standard normal distribution. Hypothesis tests and confidence intervals are obtained using the distribution of this statistic.

Hypothesis Testing We want to test the hypotheses

$$H_0\colon \mu_1 - \mu_2 = \delta_0,$$

$$H_1\colon \mu_1 - \mu_2 \neq \delta_0,$$

where δ_0 represents the hypothesized difference between the population means. To perform this test, we use the test statistic

$$z = \frac{\bar{y}_1 - \bar{y}_2 - \delta_0}{\sqrt{(\sigma_1^2/n_1) + (\sigma_2^2/n_2)}}.$$

The most common application is to let $\delta_0 = 0$, which is, of course, the test for the equality of the two population means. The resulting value of z is used to calculate a p value (using the standard normal table) or compared with a rejection region constructed for the desired level of significance. One- or two-sided alternative hypotheses may be used.

A confidence interval on the difference $(\mu_1 - \mu_2)$ is constructed using the sampling distribution of the difference presented above. The confidence interval takes the form

$$(\bar{y}_1 - \bar{y}_2) \pm z_{\alpha/2}\sqrt{(\sigma_1^2/n_1) + (\sigma_2^2/n_2)}.$$

EXAMPLE 5.2

A production plant has two fabricating systems: One uses automated equipment, the other is manually operated. Since the automated system costs more to install, we want to know whether it provides increased production in terms of the mean number of finished products fabricated per day. Experience has shown that the daily production of the automated system has a standard deviation of $\sigma_1 = 10$, the manual system, $\sigma_2 = 20$.[1] Independent random samples of 100 days of production are obtained from company records for each system. The sample results are that the automated system had a sample mean production of $\bar{y}_1 = 254$, and the manual system a sample mean of $\bar{y}_2 = 248$. Is the automated system superior to the manual one?

Solution To answer the question, we will test the hypothesis

$$H_0\colon \delta = \mu_1 - \mu_2 = 0 \text{ (or } \mu_1 = \mu_2\text{)},$$

where μ_1 is the average production of the automated system and μ_2 that of the manual system. The alternate hypothesis is

$$H_1\colon \delta = \mu_1 - \mu_2 > 0 \text{ (or } \mu_1 > \mu_2\text{)};$$

that is, the automated system has a higher production rate. Because of the cost of installing the automated system, $\alpha = 0.01$ is chosen to determine whether the manual system should be replaced by an automated system. The test statistic has a value of

$$z = \frac{(254 - 248) - 0}{\sqrt{(10^2/100) + (20^2/100)}} = 2.68.$$

The p value associated with this test statistic is $p = 0.0037$. The null hypothesis is rejected for any significance level exceeding 0.0037; hence we can conclude

[1]The fact that the automated system has a smaller variance is not of interest at this time.

that average daily production will be increased by replacing the manual system with an automated one.

It is also of interest to estimate by what amount the average daily production will be increased. This can be determined by using a one-sided confidence interval similar to that discussed in Section 3.3. In particular, we determine the lower 0.99 confidence limit on the mean as

$$(254 - 248) - 2.326\sqrt{(10)^2/100 + (20)^2/100} = 0.80.$$

This means that the increase may be as low as one unit, which may not be sufficient to justify the expense of installing the new system, illustrating the principle that a statistically significant result does not necessarily imply practical significance as noted in Section 3.6. ■

Variances Unknown but Assumed Equal

The "obvious" methodology for comparing two means when the population variances are not known would seem to be to use the two variance estimates, s_1^2 and s_2^2, in the statistic described in the previous section and determine the significance level from the Student t distribution. This approach will not work because the mathematical formulation of this distribution requires as its single parameter the degrees of freedom for a single variance estimate.

The solution to this problem is to assume that the two-population variances are equal and find an estimate of that variance. The equal variance assumption is actually quite reasonable since in many studies, a focus on means implies that the populations are similar in many respects. Otherwise, it would not make sense to compare just the means (apples with oranges, etc.). If the assumption of equal variances cannot be made, then other methods must be employed, as discussed later in this section.

Assume that we have independent samples of size n_1 and n_2, respectively, from two normally distributed populations with equal variances. We want to make inferences on the difference $\delta = (\mu_1 - \mu_2)$. Again the point estimate of that difference is $(\bar{y}_1 - \bar{y}_2)$.

The Pooled Variance Estimate

The estimate of a common variance from two independent samples is obtained by "pooling," which is simply the weighted mean of the two individual variance estimates with the weights being the degrees of freedom for each variance. Thus the pooled variance, denoted by s_{p}^2, is

$$s_{\mathrm{p}}^2 = \frac{(n_1 - 1)s_1^2 + (n_2 - 1)s_2^2}{(n_1 - 1) + (n_2 - 1)}.$$

We have emphasized that all estimates of a variance have the form

$$s^2 = \mathrm{SS/df},$$

where, for example, df $= (n - 1)$ for a single sample, and consequently SS $= (n-1)s^2$. Using the notation SS_1 and SS_2 for the sums of squares from the two samples, the pooled variance can be defined (and, incidentally, more easily calculated) as

$$s_p^2 = \frac{SS_1 + SS_2}{n_1 + n_2 - 2}.$$

This form of the equation shows that the pooled variance is indeed of the form SS/df, where now df $= (n_1 - 1) + (n_2 - 1) = (n_1 + n_2 - 2)$. The pooled variance is now used in the t statistic, which has the t distribution with $= (n_1 + n_2 - 2)$ degrees of freedom. We will see in Chapter 6 that the principle of pooling can be applied to any number of samples.

The "Pooled" t Test

To test the hypotheses

$$H_0\colon \mu_1 - \mu_2 = \delta_0,$$
$$H_1\colon \mu_1 - \mu_2 \neq \delta_0,$$

we use the test statistic

$$t = \frac{(\bar{y}_1 - \bar{y}_2) - \delta_0}{\sqrt{(s_p^2/n_1) + (s_p^2/n_2)}},$$

or equivalently

$$t = \frac{(\bar{y}_1 - \bar{y}_2) - \delta_0}{\sqrt{s_p^2(1/n_1 + 1/n_2)}}.$$

This statistic will have the t distribution and the degrees of freedom are $(n_1 + n_2 - 2)$ as provided by the denominator of the formula for s_p^2. This test statistic is often called the **pooled *t* statistic** since it uses the pooled variance estimate.

Similarly the confidence interval on $\mu_1 - \mu_2$ is

$$(\bar{y}_1 - \bar{y}_2) \pm t_{\alpha/2}\sqrt{s_p^2(1/n_1 + 1/n_2)},$$

using values from the t distribution with $(n_1 + n_2 - 2)$ degrees of freedom.

EXAMPLE 5.3

Mesquite is a thorny bush whose presence reduces the quality of pastures in the Southwest United States. In a study of growth patterns of this plant, dimensions of samples of mesquite were taken in two similar areas (labeled A and M) of a ranch. In this example, we are interested in determining whether the average heights of the plants are the same in both areas. The data are given in Table 5.2.

Table 5.2

Heights of Mesquite

Location *A* (n_A = 20)		Location *M* (n_M = 26)		
1.70	2.00	1.30	0.90	1.50
3.00	1.30	1.35	1.35	1.50
1.70	1.45	2.16	1.40	1.20
1.60	2.20	1.80	1.00	0.70
1.40	0.70	1.55	1.70	1.20
1.90	1.90	1.20	1.50	0.80
1.10	1.80	1.00	0.65	
1.60	2.00	1.70	1.50	
2.00	2.20	0.80	1.70	
1.25	0.92	1.20	1.70	

Table 5.3

Stem and Leaf Plot for Mesquite Heights

Location *A*	Stem	Location *M*
0	3	
	2	
00022	2	2
6677789	1	5555677778
12344	1	0022223444
79	0	77889

Solution As a first step in the analysis of the data, construction of a stem and leaf plot of the two samples (Table 5.3) is appropriate. The purpose of this exploratory procedure is to provide an overview of the data and look for potential problems, such as outliers or distributional anomalies. The plot appears to indicate somewhat larger mesquite bushes in location *A*. One bush in location *A* appears to be quite large; however, we do not have sufficient evidence that this value represents an outlier or unusual observation that may affect the analysis.

We next perform the test for the hypotheses

$$H_0\colon \mu_A - \mu_M = 0 \text{ (or } \mu_A = \mu_M),$$
$$H_1\colon \mu_A - \mu_M \neq 0 \text{ (or } \mu_A \neq \mu_M).$$

The following preliminary calculations are required to obtain the desired value for the test statistic:

Location *A*	Location *M*
$n = 20$	$n = 26$
$\sum y = 33.72$	$\sum y = 34.36$
$\sum y^2 = 61.9014$	$\sum y^2 = 48.9256$
$\bar{y} = 1.6860$	$\bar{y} = 1.3215$
$SS = 5.0405$	$SS = 3.5175$
$s^2 = 0.2658$	$s^2 = 0.1407$

The computed t statistic is

$$t = \frac{1.6860 - 1.3215}{\sqrt{\frac{5.0495 + 3.5175}{44}\left(\frac{1}{20} + \frac{1}{26}\right)}}$$

$$= \frac{0.3645}{\sqrt{(0.1947)(0.08846)}}$$

$$= \frac{0.3654}{0.1312}$$

$$= 2.778.$$

We have decided that a significance level of 0.01 would be appropriate. For this test we need the t distribution for $20 + 26 - 2 = 44$ degrees of freedom. Because Appendix Table A.2 does not have entries for 44 degrees of freedom, we use the next smaller degrees of freedom, which is 40. This provides for a more conservative test; that is, the true value of α will be somewhat less than the specified 0.01. It is possible to interpolate between 40 and 60 degrees of freedom to provide a more precise rejection region, but such a degree of precision is rarely needed. Using this approximation, we see that the rejection region consists of absolute values exceeding 2.7045.

The value of the test statistic exceeds 2.7045 so the null hypothesis is rejected, and we determine that the average heights of plants differ between the two locations. Using a computer program, the exact p value for the test statistic is 0.008.

The 0.99 confidence interval on the difference in population means, $(\mu_1 - \mu_2)$, is

$$\bar{y}_1 - \bar{y}_2 \pm t_{\alpha/2}\sqrt{s_p^2(1/n_1 + 1/n_2)},$$

which produces the values

$$0.3645 \pm 2.7045\,(0.1312) \quad \text{or} \quad 0.3645 \pm 0.3548,$$

which defines the interval from 0.0097 to 0.7193. The interval does not contain zero, which agrees with the results of the hypothesis test. ■

Variances Unknown but Not Equal

In Example 5.3 we saw that the variance of the heights from location A was almost twice that of location M. The difference between these variances probably is due to the rather large bush measured at location A. Since we cannot discount this observation, we may need to provide a method for comparing means that does not assume equal variances. (A test for equality of variances is presented in Section 5.3 and according to this test these two variances are not significantly different.)

Before continuing, it should be noted that inferences on means may not be useful when variances are not equal. If, for example, the distributions of two populations look like those in Fig. 5.2, the fact that population 2 has a larger

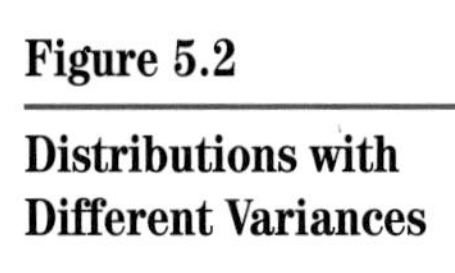

Figure 5.2

Distributions with Different Variances

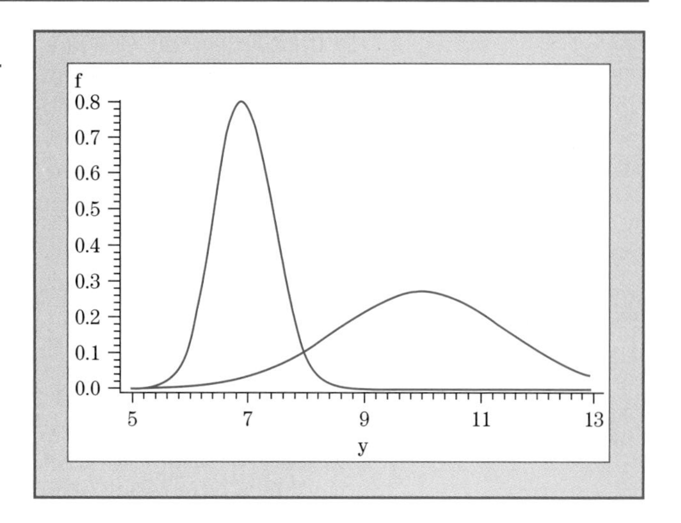

mean is only one factor in the difference between the two populations. In such cases it may be more useful to test other hypotheses about the distributions. Additional comments on this and other assumptions needed for the pooled t test are presented in Section 5.6 and also in Chapter 13.

Sometimes differences in variances are systematic or predictable. For some populations the magnitude of the variance or standard deviation may be proportional to the magnitude of the mean. For example, for many biological organisms, populations with larger means also have larger variances. This type of variance inequality may be handled by making "transformations" on the data, which employ the analysis of some function of the y's, such as log y, rather than the original values. The transformed data may have equal variances and the pooled t test can then be used. The use of transformations is more fully discussed in Section 6.4.

Not all problems with unequal variances are amenable to this type of analysis; hence we need alternate procedures for performing inferences on the means of two populations based on data from independent samples. For this situation we may use one of the following procedures with the choice depending on the sample sizes:

1. If both n_1 and n_2 are large (both over 30) we can assume a normal distribution and compute the test statistic

$$t' = \frac{\bar{y}_1 - \bar{y}_2}{\sqrt{\frac{s_1^2}{n_1} + \frac{s_2^2}{n_2}}}.$$

   Since n_1 and n_2 are large, the central limit theorem will allow us to assume that the difference between the sample means will have approximately the normal distribution. Again, for the large sample case, we can replace σ_1 and σ_2 with s_1 and s_2 without serious loss of accuracy. Therefore, the statistic t' will have approximately the standard normal distribution.

2. If either sample size is not large, compute the statistic t' as in part (1). If the data come from approximately normally distributed populations, this statistic does have an approximate Student t distribution, but the degrees of freedom cannot be precisely determined. A reasonable (and conservative) approximation is to use the degrees of freedom for the smaller sample; however, other approximations may be used (see the example in Section 5.7).

EXAMPLE 5.4

In a study on attitudes among commuters, random samples of commuters were asked to score their feelings toward fellow passengers using a score ranging from 0 for "like" to 10 for "dislike." A sample of 10 city subway commuters (population 1) and an independent sample of 17 suburban rail commuters (population 2) were used for this study. The purpose of the study is to compare the mean attitude scores of the two types of commuters. It can be assumed that the data represent samples from normally distributed populations.

The data from the two samples are given in Table 5.4. Note that the data are presented in the form of frequency distributions; that is, a score of zero was given by three subway commuters and five rail commuters and so forth.

Solution Distributions of scores of this type typically have larger variances when the mean score is near the center (5) and smaller variances when the mean score is near either extreme (0 or 10). Thus, if there is a difference in means, there is also likely to be a difference in variances. We want to test the hypotheses

$$H_0\colon \mu_1 = \mu_2,$$

$$H_1\colon \mu_1 \neq \mu_2.$$

The t' statistic has a value of

$$t' = \frac{3.70 - 1.53}{\sqrt{(13.12/10) + (2.14/17)}} = 1.81.$$

The smaller sample has 10 observations; hence we use the t distribution with 9 degrees of freedom. The 0.05 critical value is ± 2.262. The sample statistic does not lead to rejection at $\alpha = 0.05$; in fact, the p value is somewhat greater than 0.10. Therefore there is insufficient evidence that the attitudes of commuters differ.

Figure 5.3 shows the distributions of the two samples. The plot clearly shows the larger variation for the subway scores, but there does not appear to

Table 5.4

	SCORE										
Commuter Type	**0**	**1**	**2**	**3**	**4**	**5**	**6**	**7**	**8**	**9**	**10**
Subway	3	1		2		1		1		2	
Rail	5	4	5	1	1	1					

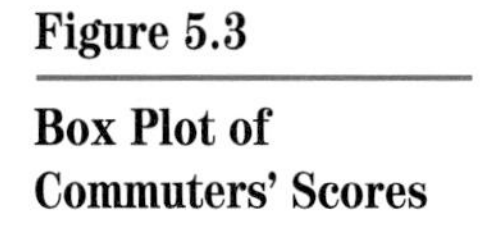

Figure 5.3

Box Plot of Commuters' Scores

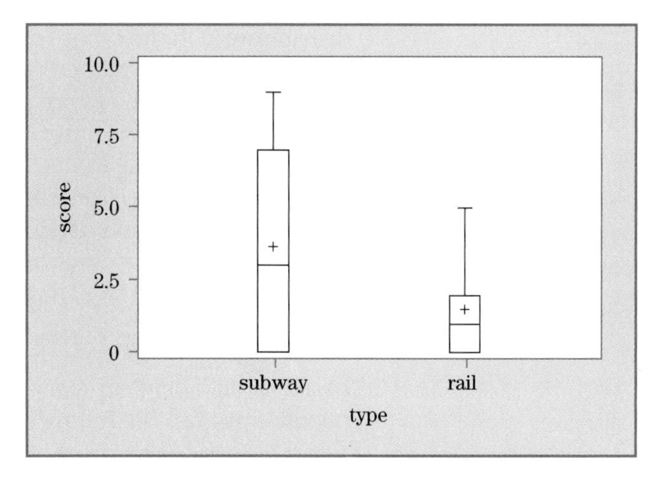

be much difference between the means. Even though the distributions appear to be skewed, Q–Q plots similar to those discussed in Section 4.5 (not shown here) do not indicate any serious deviations from normality.

If this data set had been analyzed using the pooled t test discussed earlier, the t value would be 2.21 with 25 degrees of freedom. The p value associated with this test statistic is about 0.04, which is sufficiently small to result in rejection of the hypothesis at the 0.05 significance level. Thus, if the test had been made under the assumption of equal variances (which in this case is not valid), an incorrect inference may have been made about the attitudes of commuters. ■

Actually the equal variance assumption is only one of several necessary to assure the validity of conclusions obtained by the pooled t test. A brief discussion of these issues and some ideas on remedial or alternate methods is presented in Section 5.6 and also in Chapter 13.

5.3 Inferences on Variances

In some applications it may be important to be able to determine whether the variances of two populations are equal. Such inferences are not only useful to determine whether a pooled variance may be used for inferences on the means, but also to answer more general questions about the variances of two populations. For example, in many quality control experiments, it is important to maintain consistency, and for such experiments inferences on variances are of prime importance, since the variance is a measure of consistency within a population.

In comparing the means of two populations, we are able to use the difference between the two sample means as the relevant point estimate and the sampling distribution of that difference to make inferences. However, the difference between two sample variances does not have a simple, usable distribution. On the other hand, the statistic based on the ratio s_1^2/s_2^2 is, as we saw in Section 2.6, related to the F distribution. Consequently, if we want to state that two variances are equal, we can express this relationship by stating that the ratio σ_1^2/σ_2^2 is unity. The general procedures for performing statistical inference remain the same.

Recall that the F distribution depends on two parameters, the degrees of freedom for the numerator and the denominator variance estimates. Also the F distribution is not symmetric. Therefore the inferential procedures are somewhat different from those for means, but more like those for the variance (Section 4.4).

To test the hypothesis that the variances from two populations are equal, based on independent samples of size n_1 and n_2, from normally distributed populations, use the following procedures:

1. The null hypothesis is

$$H_1: \sigma_1^2 = \sigma_2^2 \quad \text{or} \quad H_0: \sigma_1^2/\sigma_2^2 = 1.$$

2. The alternative hypothesis is

$$H_0: \sigma_1^2 \neq \sigma_2^2 \quad \text{or} \quad H_1: \sigma_1^2/\sigma_2^2 \neq 1.$$

One-tailed alternatives are that the ratio is either greater or less than unity.

3. Independent samples of size n_1 and n_2 are taken from the two populations to provide the sample variances s_1^2 and s_2^2.
4. Compute the ratio $F = s_1^2/s_2^2$.
5. This value is compared with the appropriate value from the table of the F distribution, or a p value is computed from it. Note that since the F distribution is not symmetric, a two-tailed alternative hypothesis requires finding two separate critical values in the table.

As we discussed in Section 2.6 regarding the F distribution, most tables do not have the lower tail values. It was also shown that these values may be found by using the relationship

$$F_{(1-\alpha/2)}(\nu_1, \nu_2) = \frac{1}{F_{\alpha/2}(\nu_2, \nu_1)}.$$

An easier way of obtaining a rejection region for a two-tailed alternative is to always use the larger variance estimate for the numerator, in which case we need only the upper tail of the distribution, remembering to use $\alpha/2$ to find the critical value. In other words, if s_2^2 is larger than s_1^2, use the ratio $F = s_2^2/s_1^2$, and determine the F value for $\alpha/2$ with $(n_2 - 1)$ numerator and $(n_1 - 1)$ denominator degrees of freedom.

For a one-tailed alternative, simply label the populations such that the alternative hypothesis can be stated in terms of "greater than," which then requires the use of the tabled upper tail of the distribution.

Confidence intervals are also expressed in terms of the ratio σ_1^2/σ_2^2. The confidence limits for this ratio are as follows:

Lower limit:

$$\frac{(s_1^2/s_2^2)}{F_{\alpha/2}(n_1-1, n_2-1)}.$$

Upper limit:

$$\frac{(s_1^2/s_2^2)}{F_{(1-\alpha/2)}(n_1-1, n_2-1)}.$$

In this case we must use the reciprocal relationship (Section 2.6) for the two tails of the distribution to compute the upper limit:

$$(s_1^2/s_2^2)F_{\alpha/2}(n_2-1, n_1-1).$$

Alternately, we can compute the lower limit for σ_2^2/σ_1^2, which is the reciprocal of the upper limit for σ_1^2/σ_2^2.

EXAMPLE 5.5

In previous chapters we discussed a quality control example in which we were monitoring the amount of peanuts being put in jars. In situations such as this, consistency of weights is very important and therefore warrants considerable attention in quality control efforts. Suppose that the manufacturer of the machine proposes installation of a new control device that supposedly increases the consistency of the output from the machine. Before purchasing it, the device must be tested to ascertain whether it will indeed reduce variability. To test the device, a sample of 11 jars is examined from a machine without the device (population N), and a sample of 9 jars is examined from the production after the device is installed (population C). The data from the experiment are given in Table 5.5, and Fig. 5.4 shows side-by-side box plots for the weights of the samples. The sample from population C certainly appears to exhibit less variation. The question is, does the control device significantly reduce variation?

Solution We are interested in testing the hypotheses

$$H_0: \sigma_N^2 = \sigma_C^2 \text{ (or } \sigma_N^2/\sigma_C^2 = 1),$$
$$H_1: \sigma_N^2 > \sigma_C^2 \text{ (or } \sigma_N^2/\sigma_C^2 > 1).$$

Table 5.5

Contents of Peanut Jars (oz.)

Population N without Control		Population C with Control	
8.06	8.39	7.99	8.03
8.64	8.46	8.12	8.14
7.97	8.28	8.34	8.14
7.81	8.02	8.17	7.87
7.93	8.39	8.11	
8.57			

Figure 5.4

Box Plots of Weights

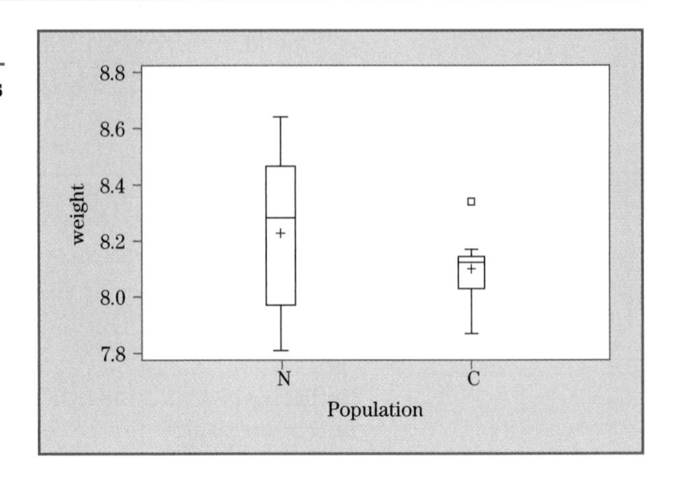

The sample statistics are

$$s_N^2 = 0.07973 \quad \text{and} \quad s_C^2 = 0.01701.$$

Since we have a one-tailed alternative, we place the larger alternate hypothesis variance in the numerator; that is, the test statistic is s_N^2/s_C^2. The calculated test statistic has a value of $F = 0.07973/0.01701 = 4.687$. The rejection region for $\alpha = 0.05$ for the F distribution with 10 and 8 degrees of freedom consists of values exceeding 3.35. Hence the null hypothesis is rejected and the conclusion is that the device does in fact increase the consistency (reduce the variance).

A one-sided interval is appropriate for this example. The desired confidence limit is the lower limit for the ratio σ_N^2/σ_C^2, since we want to be, say, 0.95 confident that the variance of the machine without the control device is larger. The lower 0.95 confidence limit is

$$\frac{(s_N^2/s_C^2)}{F_{0.05}(10, 8)}.$$

The value of $F_{0.05}(10, 8)$ is 3.35; hence the limit is

$$4.687/3.35 = 1.40.$$

In other words we are 0.95 confident that the variance without the control device is at least 1.4 times as large as it is with the control device. As usual, the result agrees with the hypothesis test, which rejected the hypothesis of a unit ratio. ■

5.4 Inferences on Means for Dependent Samples

In Section 5.2 we discussed the methods of inferential statistics as applied to two independent random samples obtained from separate populations. These methods are not appropriate for evaluating data from studies in which each observation in one sample is matched or paired with a particular observation in the other sample. For example, if we are studying the effect of a special diet on weight gains, it is not effective to randomly divide a sample of subjects into two groups and give the special diet to one of these groups and then compare the weights of the individuals from these two groups. Remember that for two independently drawn samples the estimate of the variance is based on the differences in weights among individuals in each sample, and these differences are probably larger than those induced by the special diet. A more logical data collection method is to weigh a random sample of individuals before they go on the diet and then weigh the same individuals after they have been subjected to the diet. The individuals' differences in weight before and after the special diet are then a more precise indicator of the effect of the diet. Of course, these two sets of weights are no longer independent, since the same individuals belong to both. The choice of data collection method (independent or dependent samples in this example) was briefly introduced in Section 5.1 and is an example of the use of a design of an experiment. (Experimental design is discussed briefly in Chapter 6 and more extensively in Chapter 10.)

For two populations, such samples are dependent and are called "paired samples" because our analysis will be based on the differences between pairs of observed values. For example, in evaluating the diet discussed above, the pairs are the weights obtained on individuals before and after the special diet and the analysis is based on the individual weight losses. This procedure can be used in almost any context in which the data can physically be paired.

For example, identical twins provide an excellent source of pairs for studying various medical and psychological hypotheses. Usually each of a pair of twins is given a different treatment, and the difference in response is the basis of the inference. In educational studies, a score on a pretest given to a student is paired with that student's post-test score to provide an evaluation of a new teaching method. Adjacent farm plots may be paired if they are of similar physical characteristics in order to study the effect of radiation on seeds, and so on. In fact, for any experiment where it is suspected that the difference between the two populations may be overshadowed by the variation within the two populations, the paired samples procedure should be appropriate.

Inferences on the difference in means of two populations based on paired samples use as data the simple differences between paired values. For example, in the diet study the observed value for each individual is obtained by subtracting the after weight from the before weight. The result becomes a single sample of differences, which can be analyzed in exactly the same way as any single sample experiment (Chapter 4). Thus the basic statistic is

$$t = \frac{\bar{d} - \delta_0}{\sqrt{s_d^2/n}},$$

where $\bar{d}$ is the mean of the sample differences, d_i; δ_0 is the population mean difference (usually zero); and s_d^2 is the estimated variance of the differences. When used in this way, the t statistic is usually called the "paired t statistic."

Table 5.6

Baseball Attendance (Thousands)

Team	1960	1961	Diff.
1	809	673	−136
2	663	1123	460
3	2253	1813	−440
4	1497	1100	−397
5	862	584	−278
6	1705	1199	−506
7	1096	855	−241
8	1795	1391	−404
9	1187	951	−236
10	1129	850	−279
11	1644	1151	−493
12	950	735	−215
13	1167	1606	439
14	774	683	−91
15	1627	1747	120
16	743	597	−146

Figure 5.5

Baseball Attendance Data

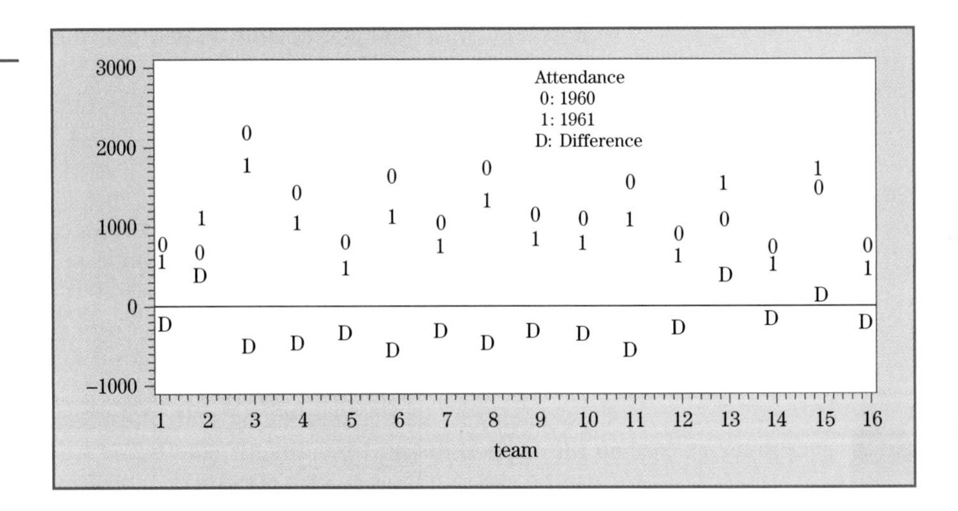

EXAMPLE 5.6

For the first 60 years major league baseball consisted of 16 teams, eight each in the National and the American leagues. In 1961 the Los Angeles Angels and the Washington Senators became the first expansion teams in baseball history. It is conjectured that the main reason that the league allowed expansion teams was the fact that total attendance dropped from 20 million in 1960 to slightly over 17 million in 1961. Table 5.6 shows the total ticket sales for the 16 teams for the two years 1960 and 1961. Examination of the data (helped by Fig. 5.5) shows the reason that a paired t test would be appropriate to determine whether the average attendance did in fact drop significantly from 1960 to 1961. The variation among the attendance figures from team to team is extremely large—going from around 663,000 for team 2 to 2,253,000 for team 3 in 1960, for example. The variation between years by individual teams, on the other hand, is relative small—the largest being 506,000 by team 6.

Solution The attendance data for the 16 major league teams for 1960 and 1961 are given in Table 5.6. The individual differences $d = y_{1961} - y_{1960}$ are used for the analysis. Positive differences indicate increased attendance while negative numbers that predominate here indicate decreased attendance. The hypotheses are

$$H_0: \delta_0 = 0,$$

$$H_1: \delta_0 < 0,$$

where δ_0 is the mean of the population differences. Note that we started out with 32 observations and ended up with only 16 pairs. Thus the mean and variance used to compute the test statistic are based on only 16 observations. This means that the estimate of the variance has 15 degrees of freedom and thus the t distribution for this statistic also has 15 degrees of freedom.

The test statistic is computed from the differences, d_i, using the computations

$$n = 16, \quad \sum d_i = -2843, \quad \sum d_i^2 = 1{,}795{,}451,$$

$$\bar{d} = -177.69, \quad \text{SS}_d = 1{,}290{,}285, \quad s_d^2 = 86{,}019,$$

and the test statistic t has the value

$$t = (-177.69)/\sqrt{(86{,}019/16)} = -2.423.$$

The (one-tailed) 0.05 rejection region for Student t distribution with 15 degrees of freedom is -1.7531; hence we reject the null hypothesis and conclude that average attendance has decreased. The p value for this test statistic (from a computer program) is $p = 0.0150$.

A confidence interval on the mean difference is obtained using the t distribution in the same manner as was done in Chapter 4. We will need the upper confidence limit on the increase (equivalent to lower limit for decrease) from 1960 to 1961. The upper limit is

$$\bar{d} + t_\alpha \sqrt{s_d^2/n},$$

which results in

$$-177.69 + (1.753)\sqrt{(86{,}019/16)} = -49.16;$$

hence, we are 0.95 confident that the true mean decrease is at least 49.16 (thousand).

The benefit of pairing Example 5.6 can be seen by pretending that the data resulted from independent samples. The resulting pooled t statistic would have the value $t = -1.164$ with 30 degrees of freedom. This value would not be significant at the 0.05 level and the test would result in a different conclusion. The reason for this result is seen by examining the variance estimates. The pooled variance estimate is quite large and reflects variation among teams that is irrelevant for studying year-to-year attendance changes. As a result, the paired t statistic will detect smaller differences, thereby providing more power, that is, a greater probability of correctly rejecting the null hypothesis (or equivalently give a narrower confidence interval). ■

It is important to note that while we performed both tests for this example, it was for demonstration purposes only! In a practical application, only procedures appropriate for the design employed in the study may be performed. That is, in this example only the paired t statistic may be used because the data resulted from paired samples.

The question may be asked: "Why not pair all two-population studies?" The answer is that not all experimental situations lend themselves to pairing. In some instances it is impossible to pair the data. In other cases there is not a sufficient physical relationship for the pairing to be effective. In such cases pairing will be detrimental to the outcome because in the act of pairing we "sacrifice" degrees of freedom for the test statistic. That is, assuming equal sample sizes, we go from $2(n-1)$ degrees of freedom in the independent sample case to $(n-1)$ in the paired case. An examination of the t table illustrates the fact that for smaller degrees of freedom the critical value are larger in magnitude, thereby requiring a larger value of the test statistic. Since pairing does not affect the mean difference, it is effective only if the variances of the two populations are definitely larger than the variances among paired differences. Fortunately, the desired condition for pairing often occurs if a physical reason exists for pairing.

EXAMPLE 5.7

Two measures of blood pressure are known as systolic and diastolic. Now everyone knows that high blood pressure is bad news. However, a small difference between the two measures is also of concern. The estimation of this difference is a natural application of paired samples since both measurements are always taken together for any individual. In Table 5.7 are systolic (RSBP) and diastolic (RDBP) pressures of 15 males aged 40 and over participating in a health study. Also given is the difference (DIFF). What we want to do is to construct a confidence interval on the true mean difference between the two pressures.

Table 5.7

Blood Pressures of Males

OBS	RSBP	RDBP	DIFF
1	100	75	25
2	135	85	50
3	110	78	32
4	110	75	35
5	142	96	46
6	120	74	46
7	140	90	50
8	110	76	34
9	122	80	42
10	140	90	50
11	150	110	40
12	120	78	42
13	132	88	44
14	112	72	40
15	120	80	40

Solution Using the differences, we obtain $\bar{d} = 41.0667$ and $s_d^2 = 52.067$, and the standard error of the difference is

$$\sqrt{\frac{52.067}{15}} = 1.863.$$

The 0.95 two-tailed value of the t distribution for 14 degrees of freedom is 2.148. The confidence interval is computed

$$41.0667 \pm (2.1448)(1.863),$$

which produces the interval 37.071 to 45.062.

If we had assumed that these data represented independent samples of 15 systolic and 15 diastolic readings, the standard error of mean difference would be 4.644, resulting in a 0.95 confidence interval from 31.557 to 50.577, which is quite a bit wider. As noted, pairing here is obvious, and it is unlikely that anyone would consider independent samples. ■

5.5 Inferences on Proportions

In Chapter 2 we presented the concept of a binomial distribution, and in Chapter 4 we used this distribution for making inferences on the proportion of "successes" in a binomial population. In this section we present procedures for inferences on differences in the proportions of successes using independent as well as dependent samples from two binomial populations.

Comparing Proportions Using Independent Samples

Assume we have two binomial populations for which the probability of success in population 1 is p_1 and in population 2 is p_2. Based on independent samples of size n_1 and n_2 we want to make inferences on the difference between p_1 and p_2, that is, $(p_1 - p_2)$. The estimate of p_1 is $\hat{p}_1 = y_1/n_1$, where y_1 is the number of successes in sample 1, and likewise the estimate of p_2 is $\hat{p}_2 = y_2/n_2$. Assuming sufficiently large sample sizes (see Section 4.3), the difference $(\hat{p}_1 - \hat{p}_2)$ is normally distributed with mean

$$p_1 - p_2$$

and variance

$$p_1(1 - p_1)/n_1 + p_2(1 - p_2)/n_2.$$

Therefore the appropriate statistic for inferences on $(p_1 - p_2)$ is

$$z = \frac{\hat{p}_1 - \hat{p}_2 - (p_1 - p_2)}{\sqrt{p_1(1 - p_1)/n_1 + p_2(1 - p_2)/n_2}},$$

which has the standard normal distribution.

Note that the expression for the variance of the difference contains the unknown parameters p_1 and p_2. In the single-population case, the null hypothesis value for the population parameter p was used in calculating the variance. In the two-population case the null hypothesis is for equal proportions and we therefore use an estimate of this common proportion for the variance formula. Letting $\hat{p}_1$ and $\hat{p}_2$ be the sample proportions for samples 1 and 2, respectively, the estimate of the common proportion p is a weighted mean of the two-sample proportions,

$$\bar{p} = \frac{n_1\hat{p}_1 + n_2\hat{p}_2}{n_1 + n_2},$$

or, in terms of the observed frequencies,

$$\bar{p} = \frac{y_1 + y_2}{n_1 + n_2}.$$

The test statistic is now computed:

$$z = \frac{\hat{p}_1 - \hat{p}_2 - (p_1 - p_2)}{\sqrt{\bar{p}(1 - \bar{p})(1/n_1 + 1/n_2)}}.$$

In construction of a confidence interval for the difference in proportions, we can not assume a common proportion, hence we use the individual estimates $\hat{p}_1$ and $\hat{p}_2$ in the variance estimate. The $(1 - \alpha)$ confidence interval on the difference $p_1 - p_2$ is

$$(\hat{p}_1 - \hat{p}_2) \pm z_{\alpha/2}\sqrt{(\hat{p}_1(1 - \hat{p}_1)/n_1) + (\hat{p}_2(1 - \hat{p}_2)/n_2)}.$$

As in the one-population case the use of the t distribution is not appropriate since the variance is not calculated as a sum of squares divided by degrees of freedom. However, samples must be reasonably large in order to use the normal approximation.

EXAMPLE 5.8

A candidate for political office wants to determine whether there is a difference in his popularity between men and women. To establish the existence of this difference, he conducts a sample survey of voters. The sample contains 250 men and 250 women, of which 42% of the men and 51% (rounded) of the women favor his candidacy. Do these values indicate a difference in popularity?

Solution Let p_1 denote the proportion of men and p_2 the proportion of women favoring the candidate, then the appropriate hypotheses are

$$H_0\text{: } p_1 = p_2,$$

$$H_1\text{: } p_1 \neq p_2.$$

The estimate of the common proportion is computed using the frequencies of successes:

$$\bar{p} = (105 + 128)/(250 + 250) = 0.466.$$

The test statistic then has the value

$$z = (0.42 - 0.51)/\sqrt{[(0.466)(0.534)(1/250 + 1/250)]}$$
$$= -0.09/0.0446 = -2.02.$$

The two-tailed p value for this test statistic (obtained from the standard normal table) is $p = 0.0434$. Thus the hypothesis is rejected at the 0.05 level, indicating that there is a difference between the sexes in the degree of support for the candidate.

We can construct a 0.95 confidence interval on the difference $(p_1 - p_2)$ as

$$(0.42 - 0.51) \pm (1.96)\sqrt{[(0.42)(0.58)/250] + [(0.51)(0.49)/250]},$$

or

$$-0.09 \pm (1.96)(0.0444).$$

Thus we are 95% confident that the true difference in preference by sex is between 0.003 and 0.177. ■

An Alternate Approximation for the Confidence Interval In Section 4.3 we gave an alternative approximation for the confidence interval on a single proportion. In Agresti and Caffo (2000), it is pointed out that the method of obtaining a confidence interval on the difference between p_1 and p_2 presented previously also tends to result in an interval that does not actually provide the specified level of confidence.

The solution, as proposed by Agresti and Caffo, is to add one success and one failure to each sample, and then use the standard formula to calculate the confidence interval. This adjustment results in much better performance of the confidence interval, even with relative small samples. Using this adjustment, the interval is based on new estimates of p_1, $\tilde{p}_1 = (y_1 + 1)/(n_1 + 2)$ and p_2, $\tilde{p}_2 = (y_2 + 1)/(n_2 + 2)$. For Example 5.8, the interval would be based on $\tilde{p}_1 = 106/252 = 0.417$ and $\tilde{p}_2 = 129/252 = 0.512$. The resulting confidence interval would be

$$0.417 - 0.512 \pm (1.96)\sqrt{\frac{(0.417)(0.583)}{252} + \frac{(0.512)(0.488)}{252}}$$

or

$$-0.095 \pm 0.087, \text{ or}$$

the interval would be from -0.182 to -0.008. As in Chapter 4, this interval is not much different from the one constructed without the adjustment, mainly because the sample sizes are quite large and both sample proportions are close to 0.5. If the sample sizes were small, this approximation would result in a more reliable confidence interval.

Comparing Proportions Using Paired Samples

A binomial response may occur in paired samples and, as is the case for inferences on means, a different analysis procedure that is most easily presented with an example must be used.

EXAMPLE 5.9

In an experiment for evaluating a new headache remedy, 80 chronic headache sufferers are given a standard remedy and a new drug on different days, and the response is whether their headache was relieved. In the experiment 56% or 70% were relieved by the standard remedy and 64% or 80% by the new drug. Do the data indicate a difference in the proportion of headaches relieved?

Solution The usual binomial test is not correct for this situation because it is based on a total of 160 observations, while there are only 80 experimental units (patients). Instead, a different procedure, called McNemar's test, must be used. For this test, the presentation of results is shown in Table 5.8. In this table the 10 individuals helped by neither drug and the 50 who were helped by both are called **concordant pairs**, and do not provide information on the relative merits of the two preparations. Those whose responses differ for the two

Table 5.8

Data on Headache Remedy

	STANDARD REMEDY Headache	No Headache	Totals
New drug			
Headache	10	6	16
No headache	14	50	64
Totals	24	56	80

drugs are called **discordant pairs**. Among these, the 14 who were not helped by the standard but were helped by the new can be called "successes," while the 6 who were helped by the old and not the new can be called "failures." If both drugs are equally effective, the proportion of successes among the discordant pairs should be 0.5, while if the new drug is more effective, the proportion of successes should be greater than 0.5. The test for ascertaining the effectiveness of the new drug, then, is to determine whether the sample proportion of successes, $14/20 = 0.7$, provides evidence to reject the null hypothesis that the true proportion is 0.5. This is a simple application of the one-sample binomial test (Section 4.3) for which the test statistic is

$$z = \frac{0.7 - 0.5}{\sqrt{[(0.5)(0.5)]/20}} = 1.789.$$

Since this is a one-tailed test, the critical value is 1.64485, and we may reject the hypothesis of no effect. ■

5.6 Assumptions and Remedial Methods

This chapter has been largely concerned with the comparison of means and variances of two populations. Yet we noted in Chapter 1 that means and variances are not necessarily good descriptors for populations with highly skewed distributions. This consideration leads to a discussion of assumptions underlying the proper use of the methods presented in this chapter. These assumptions can be summarized as follows.

1. *The pooled t statistic:*
   (a) The two samples are independent.
   (b) The distributions of the two populations are normal or of such a size that the central limit theorem is applicable.
   (c) The variances of the two populations are equal.
2. *The paired t statistic:*
   (a) The observations are paired.
   (b) The distribution of the differences is normal or of such a size that the central limit theorem is applicable.
3. *Inferences on binomial populations:*
   (a) Observations are independent (for McNemar's test *pairs* are independent).
   (b) The probability of success is constant for all observations.

4. *Inferences on variances:*
   (a) The samples are independent.
   (b) The distributions of the two populations are approximately normal.

When assumptions are not fulfilled, the analysis is not appropriate and/or the significance levels (p values) are not as advertised. In other words, conclusions that arise from the inferences may be misleading, which means any recommendations or actions that follow may not have the expected results.

Most of the assumptions are relatively straightforward and violations easily detected by simply examining the data collection procedure. Major problems arise from (1) distributions that are distinctly nonnormal so that the means and variances are not useful measures of location and dispersion and/or the central limit theorem does not work, and, of course, (2) the equal variance assumption does not hold.

Violation of distributional assumptions may be detected by the exploratory data analysis methods described in Chapter 1, which should be routinely applied to all data. The F test for equal variances may be used to detect violation of the equal variance assumption.[2]

What to do when assumptions are not fulfilled is not clear-cut. For the t statistics, minor violations are not particularly serious because these statistics are relatively robust; that is, they do not lose validity for modest departures from the assumptions. The inferences on variances are not quite so robust, because if a distribution is distinctly nonnormal, the variance may not be a good measure of dispersion. Therefore, for cases in which the robustness of the t statistics fails as well as for other cases of violated assumptions, it will be necessary to investigate other analysis strategies. In Section 4.5 we used a test on the median in a situation where the use of the mean was not appropriate. The procedure for comparing two medians is illustrated below.

Comparing medians is, however, not always appropriate. For example, population distributions may have different shapes and then neither means nor variances nor medians may provide the proper comparative measures. A wide variety of analysis procedures, called **nonparametric methods**, are available for such situations and a selection of such methods is presented in Chapter 13, where Section 13.3 is devoted to a two-sample comparison.

EXAMPLE 1.4

REVISITED In Example 4.7 we noted that the existence of extreme observations may compromise the usefulness of inferences on a mean and that an inference on the median may be more useful. The same principle can be applied to inferences for two populations. One purpose of collecting the data for Example 1.4 was to determine whether Cytosol levels are a good indicator of cancer. We noted that the distribution of Cytosol levels (Table 1.11 and Fig. 1.11) is highly skewed and dominated by a few extreme values. For comparing Cytosol levels for patients diagnosed as having or not having cancer, the side-by-side box plots in Fig. 5.6 also show that the variances of the two samples are very different. How can the comparison be made?

[2]Some will argue that one should not test for violation of assumptions. We will not attempt to answer that argument.

Figure 5.6

Box Plot of CYTOSOL

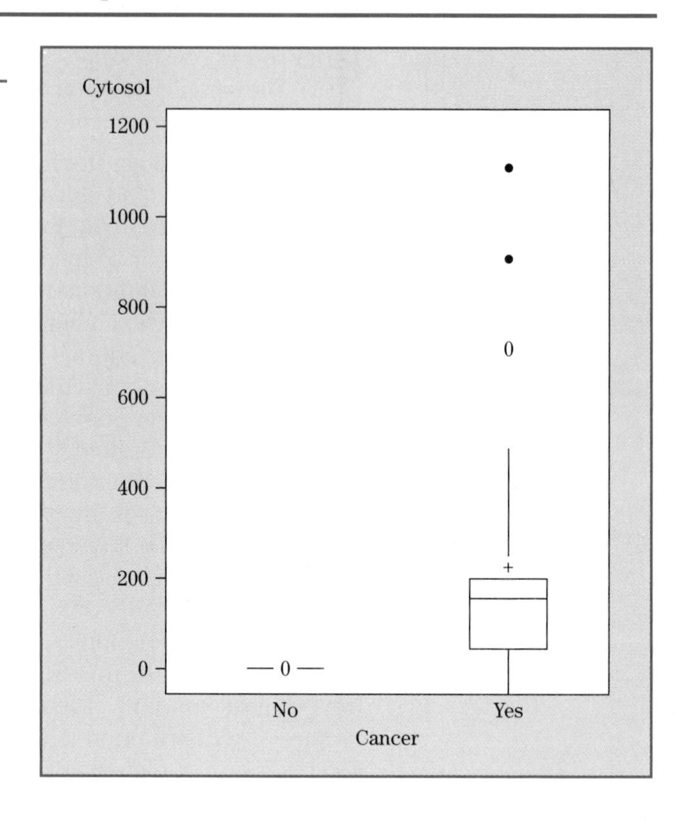

Solution Since we can see that using the t test to compare means is not going to be appropriate, it may be more useful to test the null hypothesis that the two populations have the same median. The test is performed as follows:

1. Find the overall median, which is 25.5.
2. Obtain the proportion of observations above the median for each of the two samples. These are $0/17 = 0.0$ for the no cancer patients and $21/25 = 0.84$ for the cancer patients.
3. Test the hypothesis that the proportion of patients above the median is the same for both populations, using the test for equality of two proportions. The overall proportion is 0.5; hence the test statistic is

$$\begin{aligned} z &= \frac{0.0 - 0.84}{\sqrt{(0.5)(0.5)(1/17 + 1/25)}} \\ &= \frac{-0.84}{0.157} \\ &= -5.35, \end{aligned}$$

   which easily leads to rejection.

In this example the difference between the two samples is so large that any test will declare a significant difference. However, the median test has a useful interpretation in that if the median were to be used as a cancer diagnostic, none of the no-cancer patients and only four of the cancer patients would be misdiagnosed. ■

EXAMPLE 5.4

REVISITED This example had unequal variances and was analyzed using the unequal variance procedure, which resulted in finding inadequate evidence of unequal mean attitude scores for the two populations of commuters. Can we use the procedure above to perform the same analysis? What are the results?

Solution Using the test for equality of medians, we find that the overall median is 2 and the proportions of observations above the median are 0.6 for the subway and 0.38 for the rail commuters. The binomial test, for which sample sizes are barely adequate, results in a z statistic of 1.10, which does not support rejection of the null hypothesis of equal median scores. ■

5.7 CHAPTER SUMMARY

Table 5.9 Audit Fees

TEST PROCEDURE

VARIABLE: FEE

BIG8	N	Mean	Std Dev	Std Error	Minimum	Maximum
NO	12	18.714583	11.2773529	3.25549137	7.50000000	48.5000000
YES	13	88.653846	39.0327771	10.82574457	32.00000000	177.0000000

Variances	T	DF	Prob > }T}
Unequal	−6.1868	14.1	0.0001
Equal	−5.9724	23.0	0.0000

For H0: Variances are equal, F′ = 11.98 DF = (12,11) Prob > F′ = 0.0002

Solution to Example 5.1 In the introduction to this chapter we posed the question of whether the prestigious Big Eight firms charge more for their auditing services. Since the samples of cities are independent, a pooled t test seems in order. Table 5.9 presents the result of this test as provided by `PROC TTEST` of the SAS System. In this output, the first portion provides some standard descriptive statistics for the two samples and the second portion provides information on the t test. Results are provided for both the pooled (variances equal) and unequal variance test[3] and the last line gives the test for equality of variances.

The mean fee for the Big Eight is obviously larger ($88.65) than that charged by the others ($18.71), and the difference appears highly significant (p value <

[3]A different approximation for the degrees of freedom is used by the SAS System, but the hypothesis of equal means will also be rejected using the approximation presented in Section 5.2.

Figure 5.7

Plot of Audit Fees and City Population Sizes

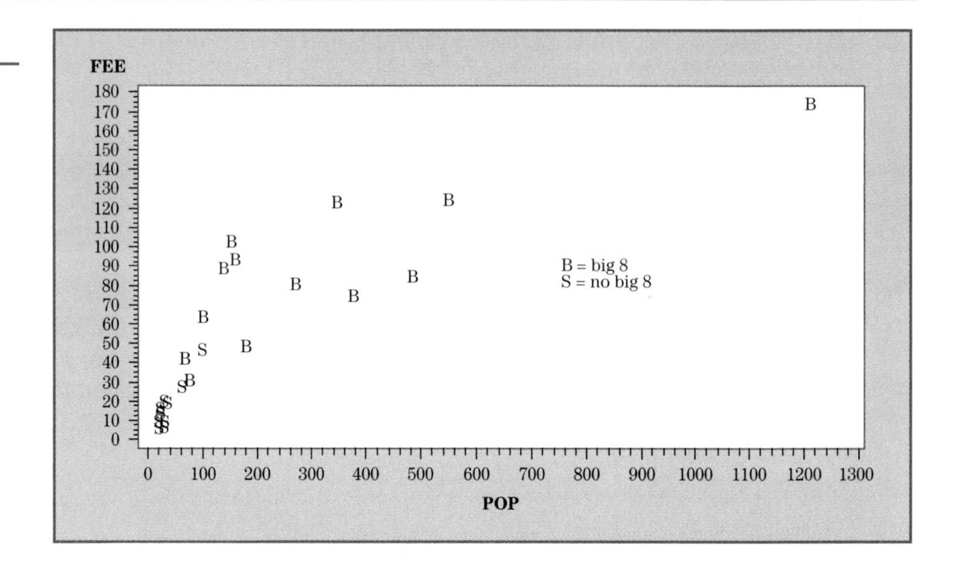

0.0001 assuming variance equal). However, the last line, which gives the F test for equality of variances, shows that variances are not equal. Of course, we can use the unequal variances test, whose results also indicate a difference. However, we have noted that the existence of different variances may imply that the comparisons of means may not be meaningful.

Closer inspection of the data shows that the Big Eight seem to be predominantly used by the larger cities whose audit fees are naturally higher. This is illustrated in Fig. 5.7, which shows the fees and populations of the cities. It is obvious that the larger cities use the Big Eight, the smaller ones do not, and obviously an audit for a larger city will cost more than one for a smaller city. It may therefore be useful to compare the mean per capita audit fees. Using this measure, in cents per person, provides the results given in Table 5.10.

Table 5.10 Audit Fees

TTEST PROCEDURE

VARIABLE: PERCAP

BIG8	N	Mean	Std Dev	Std Error	Minimum	Maximum
NO	12	46.789828	12.9685312	3.74369249	26.32152758	66.30078456
YES	13	38.391125	18.2941248	5.07387733	14.70907201	63.45871237

Variances	T	DF	Prob > }T}
Unequal	1.3320	21.6	0.1967
Equal	1.3137	23.0	0.2019

For H0: Variances are equal, F′ = 1.99 DF = (12, 11) Prob > F′ = 0.2645

In this analysis neither means nor variances appear to differ; hence we cannot infer that using one of the Big Eight firms costs more. Of course, we must add the caution that we have virtually no data on small cities using the Big Eight. ■

This chapter provides the methodology for making inferences on differences between two populations. The focus is on differences in means, variances, and proportions. In performing two-sample inferences it is important to know whether the two samples are independent or dependent (paired). The following specific inference procedures were presented in this chapter:

- Inferences on means based on independent samples where the variances are assumed known use the variance of a linear function of random variables to generate a test statistic having the standard normal distribution. This method has little direct practical application but provides the principles to be used for the methods that follow.
- Inferences on means based on independent samples where the variances can be assumed equal use a single (pooled) estimate of the common variance in a test statistic having Student t distribution.
- Inferences on means based on independent samples where the variances cannot be assumed equal use the estimated variances as if they were the known population variances for large samples. For small samples an approximation must be used.
- Inferences on means based on dependent (paired) samples use differences between the pairs as the variable to be analyzed.
- Inferences on variances use the F distribution, which describes the sampling distribution on the ratio of two estimated variances.
- Inferences on proportions from independent samples use the normal approximation of the binomial to compute a statistic similar to that for inferences on means when variances are assumed known.
- Inferences on proportions from dependent samples use a statistic based on information only on pairs whose responses differ between the two groups.
- Inferences on medians are performed by adapting the method used for inferences on proportions.
- A final section discusses assumptions underlying the various procedures for comparing two populations and a brief discussion of detection of violations and some alternative methods.

5.8 CHAPTER EXERCISES

CONCEPT QUESTIONS

Indicate true or false for the following statements. If false, specify what change will make the statement true.

1. ____________One of the assumptions underlying the use of the (pooled) two-sample test is that the samples are drawn from populations having equal means.

2. ____________In the two-sample t test, the number of degrees of freedom for the test statistic increases as sample sizes increase.

3. ____________ A two-sample test is twice as powerful as a one-sample test.

4. ____________ If every observation is multiplied by 2, then the t statistic is multiplied by 2.

5. ____________ When the means of two independent samples are used to compare two population means, we are dealing with dependent (paired) samples.

6. ____________ The use of paired samples allows for the control of variation because each pair is subject to the same common sources of variability.

7. ____________ The χ^2 distribution is used for making inferences about two population variances.

8. ____________ The F distribution is used for testing differences between means of paired samples.

9. ____________ The standard normal (z) score may be used for inferences concerning population proportions.

10. ____________ The F distribution is symmetric and has a mean of 0.

11. ____________ The F distribution is skewed and its mean is close to 1.

12. ____________ The pooled variance estimate is used when comparing means of two populations using independent samples.

13. ____________ It is not necessary to have equal sample sizes for the paired t test.

14. ____________ If the calculated value of the t statistic is negative, then there is strong evidence that the null hypothesis is false.

PRACTICE EXERCISES

The following exercises are designed to give the reader practice in doing statistical inferences on two populations through the use of sample examples with small data sets. The solutions are given in the back of the text.

1. An engineer was comparing the output from two different processes by independently sampling each one. From process A she took a sample of $n_1 = 64$, which yielded a sample mean of $\bar{y}_1 = 12.5$. Process A has a known standard deviation, $\sigma = 2.1$. From process B she took a sample of $n_2 = 100$, which yielded a sample mean of $\bar{y}_2 = 11.9$. Process B has a known standard deviation of $\sigma = 2.2$. At $\alpha = 0.05$ would the engineer conclude that both processes had the same average output?

2. The results of two independent samples from two populations are listed below:

   Sample 1: 17, 19, 10, 29, 27, 21, 17, 17, 14, 20
   Sample 2: 26, 24, 26, 29, 15, 29, 31, 25, 18, 26.

Use the 0.05 level of significance and test the hypothesis that the two populations have equal means. Assume the two samples come from populations whose standard deviations are equal.

3. Using the data in Exercise 2, compute the 0.90 confidence interval on the difference between the two population means, $\mu_1 - \mu_2$.

4. The following weights in ounces resulted from a sample of laboratory rats on a particular diet. Use $\alpha = 0.05$ and test whether the diet was effective in reducing weight.

Rat	1	2	3	4	5	6	7	8	9	10
Before	14	27	19	17	19	12	15	15	21	19
After	16	18	17	16	16	11	15	12	21	18

5. In a test of a new medication, 65 out of 98 males and 45 out of 85 females responded positively. At the 0.05 level of significance, can we say that the drug is more effective for males?

EXERCISES

1. Two sections of a class in statistics were taught by two different methods. Students' scores on a standardized test are shown in Table 5.11. Do the results present evidence of a difference in the effectiveness of the two methods? (Use $\alpha = 0.05$.)

Table 5.11

Data for Exercise 1

Class A		Class B	
74	76	78	79
97	75	92	76
79	82	94	93
88	86	78	82
78	100	71	69
93	94	85	84
		70	

2. Construct a 95% confidence interval on the mean difference in the scores for the two classes in Exercise 1.

3. Table 5.12 shows the observed pollution indexes of air samples in two areas of a city. Test the hypothesis that the mean pollution indexes are the same for the two areas. (Use $\alpha = 0.05$.)

Table 5.12

Data for Exercise 3

Area A	Area B
2.92	1.84
1.88	0.95
5.35	4.26
3.81	3.18
4.69	3.44
4.86	3.69
5.81	4.95
5.55	4.47

4. A closer examination of the records of the air samples in Exercise 3 reveals that each line of the data actually represents readings on the same day: 2.92 and 1.84 are from day 1, and so forth. Does this affect the validity of the results obtained in Exercise 3? If so, reanalyze.

5. To assess the effectiveness of a new diet formulation, a sample of 8 steers is fed a regular diet and another sample of 10 steers is fed a new diet. The weights of the steers at 1 year are given in Table 5.13. Do these results imply that the new diet results in higher weights? (Use $\alpha = 0.05$.)

Table 5.13

Data for Exercise 5

Regular Diet	New Diet
831	870
858	882
833	896
860	925
922	842
875	908
797	944
788	927
	965
	887

6. Assume that in Exercise 5 the new diet costs more than the old one. The cost is approximately equal to the value of 25 lbs. of additional weight. Does this affect the results obtained in Exercise 5? Redo the problem if necessary.

7. In a test of the reliability of products produced by two machines, machine *A* produced 7 defective parts in a run of 140, while machine *B* produced

10 defective parts in a run of 200. Do these results imply a difference in the reliability of these two machines?

8. In a test of the effectiveness of a device that is supposed to increase gasoline mileage in automobiles, 12 cars were run, in random order, over a prescribed course both with and without the device in random order. The mileages (mpg) are given in Table 5.14. Is there evidence that the device is effective?

9. A new method of teaching children to read promises more consistent improvement in reading ability across students. The new method is implemented in one randomly chosen class, while another class is randomly chosen to represent the standard method. Improvement in reading ability using a standardized test is given for the students in each class in Table 5.15. Use the appropriate test to see whether the claim can be substantiated.

10. The manager of a large office building needs to buy a large shipment of light bulbs. After reviewing specifications and prices from a number of suppliers, the choice is narrowed to two brands whose specifications with respect to price and quality appear identical. He purchases 40 bulbs of each brand and subjects them to an accelerated life test, recording hours to burnout, as shown in Table 5.16.
(a) The manager intends to buy the bulbs with a longer mean life. Do the data provide sufficient evidence to make a choice?

Table 5.14

Data for Exercise 8

Car No.	Without Device	With Device
1	21.0	20.6
2	30.0	29.9
3	29.8	30.7
4	27.3	26.5
5	27.7	26.7
6	33.1	32.8
7	18.8	21.7
8	26.2	28.2
9	28.0	28.9
10	18.9	19.9
11	29.3	32.4
12	21.0	22.0

Table 5.15

Data for Exercise 9

New Method		Standard Method	
13.0	16.7	20.1	27.0
15.1	16.7	16.7	19.2
16.5	18.4	25.6	19.3
19.0	16.6	25.4	26.7
20.2	19.4	22.0	14.7
19.9	23.6	16.8	16.9
23.3	16.5	23.8	23.7
17.3	24.5	23.6	21.7

Table 5.16

Data for Exercise 10

Brand *A* Life (Hours)				Brand *B* Life (Hours)			
915	992	1034	1080	1235	1238	1248	1273
1137	1211	1211	1218	1275	1282	1298	1303
1260	1276	1289	1306	1307	1335	1337	1339
1319	1336	1360	1387	1360	1383	1384	1384
1400	1405	1419	1437	1388	1390	1390	1390
1488	1543	1581	1603	1394	1394	1403	1410
1606	1614	1635	1669	1417	1419	1423	1426
1683	1746	1752	1776	1430	1442	1448	1469
1881	1928	1940	1960	1478	1485	1486	1501
2029	2053	2063	2737	1508	1514	1515	1517

(b) To save labor expense, the owners have decided that all bulbs will be replaced when 10% have burned out. Is the decision in part (a) still valid? Is an alternate test possibly more useful? (Suggest the test only; do not perform.)

11. Chlorinated hydrocarbons (mg/kg) found in samples of two species of fish in a lake are as follows:

Species 1:	34	1	167	20		
Species 2:	45	86	82	70	160	170

Perform a hypothesis test to determine whether there is a difference in the mean level of hydrocarbons between the two species. Check assumptions.

12. Eight samples of effluent from a pulp mill were each divided into 10 batches. From each sample, 5 randomly selected batches were subjected to a treatment process intended to remove toxic substances. Five fish of the same species were placed in each batch, and the mean number surviving in the 5 treated and untreated portions of each effluent sample after 5 days were recorded and are given in Table 5.17. Test to see whether the treatment increased the mean number of surviving fish.

13. In Exercise 13 of Chapter 1, the half-life of aminoglycosides from a sample of 43 patients was recorded. The data are reproduced in Table 5.18. Use

Table 5.17

Data for Exercise 12

	MEAN NUMBER SURVIVING							
Sample No.	**1**	**2**	**3**	**4**	**5**	**6**	**7**	**8**
Untreated	5	1	1.8	1	3.6	5	2.6	1
Treated	5	5	1.2	4.8	5	5	4.4	2

Table 5.18 Half-Life of Aminoglycosides by Drug Type

Pat	Drug	Half-Life	Pat	Drug	Half-Life	Pat	Drug	Half-Life
1	G	1.60	16	A	1.00	31	G	1.80
2	A	2.50	17	G	2.86	32	G	1.70
3	G	1.90	18	A	1.50	33	G	1.60
4	G	2.30	19	A	3.15	34	G	2.20
5	A	2.20	20	A	1.44	35	G	2.20
6	A	1.60	21	A	1.26	36	G	2.40
7	A	1.30	22	A	1.98	37	G	1.70
8	A	1.20	23	A	1.98	38	G	2.00
9	G	1.80	24	A	1.87	39	G	1.40
10	G	2.50	25	G	2.89	40	G	1.90
11	A	1.60	26	A	2.31	41	G	2.00
12	A	2.20	27	A	1.40	42	A	2.80
13	A	2.20	28	A	2.48	43	A	0.69
14	G	1.70	29	G	1.98			
15	A	2.60	30	G	1.93			

these data to see whether there is a significant difference in the mean half-life of Amikacin and Gentamicin. (Use $\alpha = 0.10$.)

14. Draw a stem and leaf plot of half-life for each drug in Exercise 13. Do the assumptions necessary for the test in Exercise 13 seem to be satisfied by the data? Explain.

15. In Exercise 12 of Chapter 1 a study of characteristics of successful salespersons indicated that 44 of 120 sales managers rated reliability as the most important characteristic in salespersons. A study of a different industry showed that 60 of 150 sales managers rated reliability as the most important characteristic of a successful salesperson.

(a) At the 0.05 level of significance, do these opinions differ from one industry to the other?

(b) Construct the power curve for this test. (*Hint:* The horizontal axis will be the difference between the proportions.)

Inferences for Two or More Means

EXAMPLE 6.1

How Do Soils Differ? A study was done to compare soil mapping units on the basis of their lateral variabilities for a single property, silt content. The study area consisted of a sequence of eight contiguous sites extending over the crest and flank of a low rise in a valley plain underlain by marl near Albudeite in the province of Murcia, Spain. The geomorphological sites were the primary mapping units adopted and were small areas of ground surface of uniform shape. Following the delimitation of the sites, soil samples were obtained in each site at 11 random points within a $10 \times 10\text{-m}^2$ area centered on the midpoint of the site. All samples were taken from the same depth. The soil property considered was the silt content, expressed as percentages of the total silt, clay, and sand content. The data are given in Table 6.1. The questions to be answered are as follows:

- Is there a difference in silt content among the soils from different sites?
- If there is a difference, can we identify the sites having the largest and smallest silt content?
- Do the data fit a standard set of assumptions similar to those given in Section 5.6? If not, what is the effect on the analysis?

The solution is given in Section 6.5. ■

6.1 Introduction

Although methods for comparing two populations have many applications, it is obvious that we need procedures for the more general case of comparing several populations. In fact, with the availability of modern technology to acquire, store, and analyze data, there seem to be no limits to the number

Table 6.1

Data on Silt Content of Soils

Source: Adapted from Andrews, D. F., and Herzberg, A. M. (1985), Data: A Collection of Problems from Many Fields for the Student and Research Worker, *pp. 121, 127–130. New York: Springer–Verlag*

Site = 1	Site = 2	Site = 3	Site = 4	Site = 5	Site = 6	Site = 7	Site = 8
46.2	40.0	41.9	41.1	48.6	43.7	47.0	48.0
36.0	48.9	40.7	40.4	50.2	41.0	46.4	47.9
47.3	44.5	44.0	39.9	51.2	44.4	46.3	49.9
40.8	30.3	40.7	41.1	47.0	44.6	47.1	48.2
30.9	40.1	32.3	31.9	42.8	35.7	36.8	40.6
34.9	46.4	37.0	43.0	46.6	50.3	54.6	49.5
39.8	42.3	44.3	42.0	46.7	44.5	43.0	46.4
48.1	34.0	41.8	40.3	48.3	42.5	43.7	47.7
35.6	41.9	41.4	42.2	47.1	48.6	43.7	48.9
48.8	34.1	41.5	50.7	48.8	48.5	45.1	47.0
45.2	48.7	29.7	33.4	38.3	35.8	36.1	37.1

of populations that can be sampled for comparison purposes. This chapter presents statistical methods for comparing means among any number of populations based on samples from these populations.

As we will see, the t test for comparing two means cannot be generalized to the comparing of more than two means. Instead, the analysis most frequently used for this purpose is based on a comparison of *variances*, and is therefore called the *analysis of variance*, often referred to by the acronyms ANOVA or AOV. We will present a motivation for this terminology in Section 6.2. When ANOVA is applied to only two populations, the results are equivalent to those of the t test.

Specifically this chapter covers the following topics:

- the ANOVA method for testing the equality of a set of means,
- the use of the linear model to justify the method,
- the assumptions necessary for the validity of the results of such an analysis and discussion of remedial methods if these assumptions are not met,
- procedures for specific comparisons among selected means, and
- an alternative to the analysis of variance called the analysis of means.

As noted in Section 5.1, comparative studies can arise from either **observational studies** or **designed experiments**, and the methodology in this chapter is applicable to either type of study. Further, in Section 5.1 we indicated that data can be collected in two ways, **independent samples** or **dependent samples**. In this chapter we will consider only the case of independent samples, which in experimental design terminology is called the "completely randomized design" or the CRD. The resulting analysis method is often referred to as a "one-way" or "single-factor" analysis as the single factor consists of the factor levels of the experiment. We will cover the methodology for data having more than one factor, which includes the equivalent of dependent samples, in Chapters 9 and 10.

Using the Computer

Virtually all statistical analyses are now performed with computers. Thus the formulas presented in this chapter (and many others) are rarely implemented by hand on hand-held or desk calculators. The presentation of these formulas is intended as a pedagogical tool, because their use helps to provide an understanding of the methodology.

We assume that anyone using this text has access to a computer and appropriate statistical software for completing assigned exercises as well as duplicating the results of the examples. We do, however, suggest that one or two of the easiest exercises be completed by hand and the results compared to computer outputs.

Most statistical software packages are essentially collections of individual programs or procedures that perform data manipulation and statistical analyses and are typically implemented by a uniform and easy-to-understand instructional format or language. The individual programs or procedures within these packages are usually quite general in scope. For example, most ANOVA programs are designed to do any analysis of variance, regardless of the number of factors. Thus, one program or procedure would probably be capable of doing all the analyses in this chapter as well as those in Chapters 9 and 10 (and many more). However, they may not perform the appropriate analysis for unbalanced data, and may, in fact, provide incorrect answers without comment! Because of this, the user of such programs must be able to implement the program correctly as well as be able to determine what part of the program's output is appropriate for any specific problem. It is important that users of such programs:

- Have data in proper format for the particular package being used. Most, but not all, packages require one observation per line where variables identify both factor levels and the response(s).
- Specify the correct analysis (usually through specification of the model).
- Determine and use only that portion of the output appropriate to the problem at hand.

Despite the generality of most statistical packages, they often do not provide for all aspects of the desired analysis. For example, many programs do not provide a simple way of specifying contrasts to be tested. Yet they do provide for some sort of post hoc multiple-comparison procedure, whether or not it is appropriate (see Section 6.5). Thus, if contrasts are appropriate for a specific problem, the user must either search for a program having that option or implement a separate program or procedure to get the required analyses. The important message here is that "one must not let the computer program dictate the analysis!"

6.2 The Analysis of Variance

We are interested in testing the statistical hypothesis of the equality of a set of population means. At first it might seem logical to extend the two-population procedure of Chapter 5 to the general case by constructing pairwise comparisons on all means; that is, use the two-population t test repeatedly until all possible pairs of population means have been compared. Besides being very awkward (to compare 10 populations would require 45 t tests), fundamental

Table 6.2

Data from Three Populations

	SET 1			SET 2	
Sample 1	**Sample 2**	**Sample 3**	**Sample 1**	**Sample 2**	**Sample 3**
5.7	9.4	14.2	3.0	5.0	11.0
5.9	9.8	14.4	4.0	7.0	13.0
6.0	10.0	15.0	6.0	10.0	16.0
6.1	10.2	15.6	8.0	13.0	17.0
6.3	10.6	15.8	9.0	15.0	18.0
$\bar{y} = 6.0$	$\bar{y} = 10.0$	$\bar{y} = 15.0$	$\bar{y} = 6.0$	$\bar{y} = 10.0$	$\bar{y} = 15.0$

problems arise with such an approach. The main difficulty is that the true level of significance of the analysis as a whole would not be what is specified for each of the individual t tests, but would be considerably distorted; that is, it would not have the value specified by each test. For example, if we were to test the equality of five means, we would have to test 10 pairs. Assuming that α has been specified to be 0.05, then the probability of correctly failing to reject the null hypothesis of equality of each pair is $(1 - \alpha) = 0.95$. The probability of correctly failing to reject the null hypothesis for all 10 tests is then $(0.95)^{10} = 0.60$, assuming the tests are independent. Thus the true value of α for this set of comparisons is at least 0.4 rather than the specified 0.05.

Therefore we will need an alternate approach. We have already noted that the statistical method for comparing means is called the analysis of variance. Now it may seem strange that in order to compare means we study variances. To see why we do this, consider the two sets of contrived data shown in Table 6.2, each having five sample values for each of three populations. Looking *only* at the means we can see that they are identical for the three populations in both sets. Using the means alone, we would state that there is no difference between the two sets.

However, when we look at the box plots of the two sets, as shown in Fig. 6.1, it appears that there is stronger evidence of differences among means in Set 1 than among means in Set 2. That is because the box plots show that the observations *within* the samples are more closely bunched in Set 1 than they are in Set 2, and we know that sample means from populations with smaller variances will also be less variable. Thus, although the variances *among* the means for the two sets are identical, the variance among the observations *within* the individual samples is smaller for Set 1 and is the reason for the apparently stronger evidence of different means. This observation is the basis for using the analysis of variance for making inferences about differences among means: the analysis of variance is based on the comparison of the variance *among* the means of the populations to the variance among sample observations *within* the individual populations.

Notation and Definitions

The purpose of the procedures discussed in this section is to compare sample means of t populations, $t \geq 2$, based on independently drawn random samples from these populations. We assume samples of size n_i are taken from

Figure 6.1

Comparing Populations

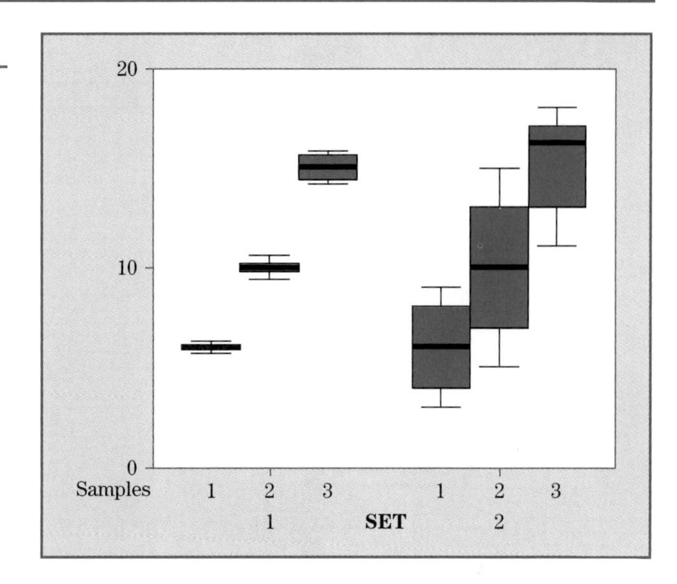

population i, $i = 1, 2, \ldots, t$. An observation from such a set of data is denoted by

$$y_{ij}, \quad i = 1, \ldots, t \quad \text{and} \quad j = 1, \ldots, n_i.$$

There are a total of $\sum n_i$ observations. It is not necessary for all the n_i to be the same. If they are all equal, say, $n_i = n$ for all i, then we say that the data are "balanced."

If we denote by μ_i the mean of the ith population, then the hypotheses of interest are

$$H_0\colon \mu_1 = \mu_2 = \cdots = \mu_t,$$

$$H_1\colon \text{at least one equality is not satisfied.}$$

As we have done in Chapter 5, we assume that the variances are equal for the different populations.

Using the indexing discussed previously, the data set can be listed in tabular form as illustrated by Table 6.3, where the rows identify the populations, which are the treatments or "factor levels." As in previous analyses, the analysis is based on computed sums and means and also sums of squares and variances of observations for each factor level (or sample). Note that we denote totals by capital letters, means by lowercase letters with bars, and that a dot replaces a subscript when that subscript has been summed over. This notation may seem more complicated than is necessary at this time, but we will see later that it is quite useful for more complex situations.

Table 6.3

Notation for One-Way Anova

Factor Levels	Observations				Totals	Means	Sums of Squares
1	y_{11}	y_{12}	$\cdots$	y_{1n_1}	$Y_{1.}$	$\bar{y}_{1.}$	SS_1
2	y_{21}	y_{22}	$\cdots$	y_{2n_2}	$Y_{2.}$	$\bar{y}_{2.}$	SS_2
.	.	.	$\cdots$	.	.	.	.
.	.	.	$\cdots$	.	.	.	.
.	.	.	$\cdots$	.	.	.	.
i	y_{i1}	y_{i2}	$\cdots$	y_{in_i}	$Y_{i.}$	$\bar{y}_{i.}$	SS_i
.	.	.	$\cdots$	.	.	.	.
.	.	.	$\cdots$	.	.	.	.
.	.	.	$\cdots$	.	.	.	.
t	y_{t1}	y_{t2}	$\cdots$	y_{tn_t}	$Y_{t.}$	$\bar{y}_{t.}$	SS_t
Overall					$Y_{..}$	$\bar{y}_{..}$	SS_p

Computing sums and means is straightforward. The formulas are given here to illustrate the use of the notation. The factor level totals are computed as[1]

$$Y_{i.} = \sum_j (y_{ij}),$$

and the factor level means are

$$\bar{y}_{i.} = \frac{Y_{i.}}{n_i}.$$

The overall total is computed as

$$Y_{..} = \sum_i (Y_{i.}) = \sum_i \left[\sum_j (y_{ij}) \right],$$

and the overall mean is

$$\bar{y}_{..} = Y_{..} \Big/ \sum_i (n_i).$$

As for all previously discussed inference procedures, we next need to estimate a variance. We first calculate the corrected sum of squares for each factor level,

$$SS_i = \sum_j (y_{ij} - \bar{y}_{i.})^2, \quad \text{for } i = 1, \ldots, t,$$

or, using the computational form,

$$SS_i = \sum_j y_{ij}^2 - (Y_{i.})^2/n_i.$$

We then calculate a pooled sums of squares,

$$SS_p = \sum_i SS_i,$$

[1] We will use the notation $\sum_i$ to signify the summation is over the "i" index, etc. However, in many cases where the indexing is obvious, we will omit that designation.

which is divided by the pooled degrees of freedom to obtain

$$s_p^2 = \frac{\text{SS}_p}{\sum n_i - t} = \frac{\sum_i \text{SS}_i}{\sum n_i - t}.$$

Note that if the individual variances are available, this can be computed as

$$s_p^2 = \sum_i (n_i - 1)s_i^2 \Big/ \left(\sum n_i - t\right),$$

where the s_i^2 are the variances for each sample.

As in the two-population case, if the t populations can be assumed to have a common variance, say, σ^2, then the pooled sample variance is the proper estimate of that variance. The assumption of equal variances (called **homoscedasticity**) is discussed in Section 6.4.

Heuristic Justification for the Analysis of Variance

In this section, we present a heuristic justification for the analysis of variance procedure for the balanced case (all $n_i = n$). Extension to the unbalanced case involves no additional principles but is algebraically messy. Later in this chapter, we present the "linear model," which provides an alternate (but equivalent) basis for the method and gives a more rigorous justification and readily provides for extensions to many other situations.

For the analysis of variance the null hypothesis is that the means of the populations under study are equal, and the alternative hypothesis is that there are some inequalities among these means. As before, the hypothesis test is based on a test statistic whose distribution can be identified under the null and alternative hypotheses.

In Section 2.5 the sampling distribution of the mean specified that a sample mean computed from a random sample of size n from a population with mean μ and variance σ^2 is a random variable with mean μ and variance σ^2/n. In the present case we have t populations that may have different means μ_i but have the same variance σ^2. If the null hypothesis is true, that is, each of the μ_i has the same value, say, μ, then the distribution of each of the t sample means, $\bar{y}_{i.}$, will have mean μ and variance σ^2/n. It then follows that if we calculate a variance using the sample means as observations

$$s_{\text{means}}^2 = \sum (\bar{y}_{i.} - \bar{y}_{..})^2/(t-1),$$

then this quantity is an estimate of σ^2/n. Hence ns_{means}^2 is an estimate of σ^2. This estimate has $(t-1)$ degrees of freedom, and it can also be shown that this estimate is independent of the pooled estimate of σ^2 presented previously.

In Section 2.6, we introduced a number of sampling distributions. One of these, the F distribution, describes the distribution of a ratio of two independent estimates of a common variance. The parameters of the distribution are the degrees of freedom of the numerator and denominator variances, respectively. Now if the null hypothesis of equal means is true, we use the arguments presented above to compute two estimates of σ^2 as follows:

$$ns^2_{\text{means}} = n\sum(\bar{y}_{i.} - \bar{y}_{..})^2/(t-1) \quad \text{and} \quad s^2_p, \text{ the pooled variance.}$$

Therefore the ratio $(ns^2_{\text{means}}/s^2_p)$ has the F distribution with degrees of freedom $(t-1)$ and $t(n-1)$.

Of course, the numerator is an estimate of σ^2 only if the null hypothesis of equal population means is true. If the null hypothesis is not true, that is, the μ_i are not all the same, we would expect larger differences among sample means, $(\bar{y}_{i.} - \bar{y}_{..})$, which in turn would result in a larger ns^2_{means}, and consequently a larger value of the computed F ratio. In other words, when H_0 is not true, the computed F ratio will tend to have values larger than those associated with the F distribution.

The nature of the sampling distribution of the statistic $(ns^2_{\text{means}}/s^2_p)$ when H_0 is true and when it is not true sets the stage for the hypothesis test. The test statistic is the ratio of the two variance estimates, and values of this ratio that lead to the rejection of the null hypothesis are those that are larger than the values of the F distribution for the desired significance level. (Equivalently p values can be derived for any computed value of the ratio.) That is, the procedure for testing the hypotheses

$$H_0\colon \mu_1 - \mu_2 = \cdots, = \mu_t,$$

$$H_1\colon \text{at least one equality is not satisfied}$$

is to reject H_0 if the calculated value of

$$F = \frac{ns^2_{\text{means}}}{s^2_p}$$

exceeds the α right tail of the F distribution with $(t-1)$ and $t(n-1)$ degrees of freedom.

We can see how this works by returning to the data in Table 6.2. For both sets, the value of ns^2_{means} is 101.67. However, for set 1, $s^2_p = 0.250$, while for set 2, $s^2_p = 10.67$. Thus, for set 1, $F = 406.67$ (p value, 0.0001) and for set 2 it is 9.53 (p value $= 0.0033$), confirming that the *relative* magnitudes of the two variances is the important factor for detecting differences among means (although the means from both sets are significantly different at $\alpha = 0.05$).

EXAMPLE 6.2

An experiment to compare the yield of four varieties of rice was conducted. Each of 16 plots on a test farm where soil fertility was fairly homogeneous was treated alike relative to water and fertilizer. Four plots were randomly assigned each of the four varieties of rice. Note that this is a designed experiment, specifically a completely randomized design. The yield in pounds per acre was

Table 6.4

Rice Yields

Variety	Yields				$Y_{i.}$	$\bar{y}_{i.}$	SS_i
1	934	1041	1028	935	3938	984.50	10085.00
2	880	963	924	946	3713	928.25	3868.75
3	987	951	976	840	3754	938.50	13617.00
4	992	1143	1140	1191	4466	1116.50	22305.00
Overall					15871	991.94	49875.75

Figure 6.2

Box Plots of Rice Yields

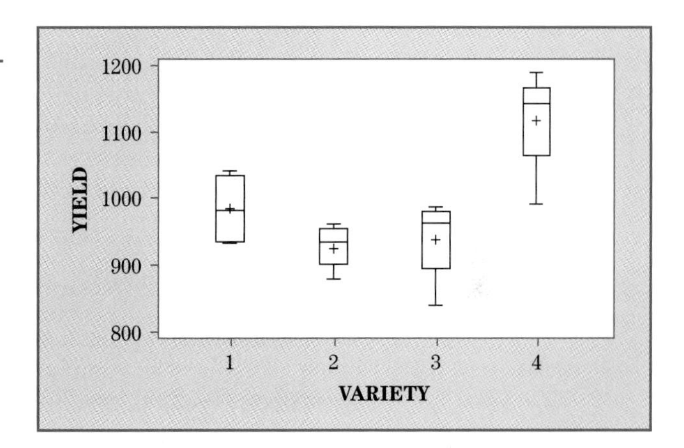

recorded for each plot. Do the data presented in Table 6.4 indicate a difference in the mean yield between the four varieties? The data are shown in Table 6.2 and box plots of the data are shown in Fig. 6.2. Comparing these plots suggests the means may be different. We will use the analysis of variance to confirm or deny this impression.

Solution The various intermediate totals and means and corrected sums of squares (SS_i) are presented in the margin of the table. The hypotheses to be tested are

$$H_0\text{: } \mu_1 = \mu_2 = \mu_3 = \mu_4,$$

$$H_1\text{: not all varieties have the same mean,}$$

where μ_i is the mean yield per acre for variety i.

The value of ns^2_{means} is

$$\begin{aligned} ns^2_{\text{means}} &= n\sum(\bar{y}_{i.} - \bar{y}_{..})^2/(t-1) \\ &= 4[(984.5 - 991.94)^2 + \cdots + (1116.50 - 991.94)^2]/3 \\ &= 29977.06. \end{aligned}$$

The value of s_p^2 is

$$\begin{aligned} s_p^2 &= \sum_i \text{SS}_i/[t(n-1)] \\ &= (10{,}085.00 + \cdots + 22{,}305.00)/12 = 49{,}875.75/12 \\ &= 4156.31. \end{aligned}$$

The calculated F ratio is

$$F = 29{,}977.06/4156.31 = 7.21.$$

The critical region is based on the F distribution with 3 and 12 degrees of freedom. Using an α of 0.01, the critical value is 5.95, and since this value is exceeded by the calculated F ratio we can reject the null hypothesis of equal means, and conclude that a difference exists in the yields of the four varieties. Further analysis will be postponed until Section 6.5 where we will examine these differences for more specific conclusions. ■

Computational Formulas and the Partitioning of Sums of Squares

Calculation of the necessary variance estimates in Example 6.2 is cumbersome. Although the computations for the analysis of variance are almost always done on computers, it is instructive to provide computational formulas that not only make these computations easier to perform but also provide further insight into the structure of the analysis of variance.

Although we have justified the analysis of variance procedure for the balanced case, that is, all n_i are equal, we present the computational formulas for the general case. Note that all the formulas are somewhat simplified for the balanced case.

The Sum of Squares among Means

Remember that the F ratio is computed from two variance estimates, each of which is a sum of squares divided by degrees of freedom. In Chapter 1 we learned a shortcut for computing the sum of squares; that is,

$$\text{SS} = \sum(y - \bar{y})^2$$

is more easily computed by

$$\text{SS} = \sum y^2 - \left(\sum y\right)^2 \Big/ n.$$

In a similar manner, the sum of squares for computing ns^2_{means}, often referred to as the "between groups"[2] or "factor sum of squares," can be obtained by

[2]Students of the English Language recognize that "between" refers to a comparison of two items while "among" refers to comparisons involving more than two items. Statisticians apparently do not recognize this distinction.

using the formula

$$\text{SSB} = \sum \frac{Y_{i.}^2}{n_i} - \frac{Y_{..}^2}{\sum n_i},$$

which is divided by its degrees of freedom, $\text{dfB} = t - 1$, to obtain ns^2_{means}, called the "between groups mean square," denoted by MSB, the quantity to be used for the numerator of the F statistic.

The Sum of Squares within Groups

The sum of squares for computing the pooled variance, often called the "within groups" or the "error sum of squares," is simply the sum of the sums of squares for each of the samples, that is,

$$\text{SSW (or SSE)} = \sum \text{SS}_i = \sum_i \left[\sum_j (y_{ij} - \bar{y}_i)^2 \right] = \sum_{i,j} y_{ij}^2 - \sum_i \frac{Y_{i.}^2}{n_i},$$

where the subscripts under the summation signs indicate the index being summed over. This sum of squares is divided by its degrees of freedom, $\text{df W} = (\sum n_i - t)$, to obtain the pooled variance estimate to be used in the denominator of the F statistic.

The Ratio of Variances

We noted in Chapter 1 that a variance is sometimes called a mean square. In fact, the variances computed for the analysis of variance are always referred to as mean squares. These mean squares are denoted by MSB and MSW, respectively. The F statistic is then computed as MSB/MSW.

Partitioning of the Sums of Squares

If we now consider all the observations to be coming from a single sample, that is, we ignore the existence of the different factor levels, we can measure the overall or total variation by a total sum of squares, denoted by TSS:

$$\text{TSS} = \sum_{\text{all}} (y_{ij} - \bar{y}_{..})^2.$$

This quantity can be calculated by the computational formula

$$\text{TSS} = \sum_{\text{all}} y_{ij}^2 - \frac{Y_{..}^2}{\sum n_i}.$$

This sum of squares has $(\sum n_i - 1)$ degrees of freedom. Using a favorite trick of algebraic manipulation, we subtract and add the quantity $\sum (Y_{i.})^2/n_i$ in this expression. This results in

$$\text{TSS} = \left(\sum_{\text{all}} y_{ij}^2 - \sum \frac{Y_{i.}^2}{n_i} \right) + \left(\sum \frac{Y_{i.}^2}{n_i} - \frac{Y_{..}^2}{\sum n_i} \right).$$

The first term in this expression is SSW and the second is SSB, thus it is seen that

$$\text{TSS} = \text{SSB} + \text{SSW}.$$

This identity illustrates the principle of the partitioning of the sums of squares in the analysis of variance. That is, the total sum of squares, which measures the total variability of the entire set of data, is partitioned into two parts:

1. SSB, which measures the variability among the means, and
2. SSW, which measures the variability within the individual samples.

Note that the degrees of freedom are partitioned similarly. That is, the total degrees of freedom, dfT, can be written

$$\text{dfT} = \text{dfB} + \text{dfW},$$

$$\left(\sum n_i - 1\right) = (t-1) + \left(\sum n_i - t\right).$$

We will see later that this principle of partitioning the sums of squares is a very powerful tool for a large class of statistical analysis techniques.

The **partitioning of the sums of squares** and degrees of freedom and the associated means squares are conveniently summarized in tabular form in the so-called ANOVA (or sometimes AOV) table shown in Table 6.5.

Table 6.5

Tabular Form for the Analysis of Variance

Source	**df**	**SS**	**MS**	***F***
Between groups	$t-1$	SSB	MSB	MSB/MSW
Within groups	$\sum n_i - t$	SSW	MSW	
Total	$\sum n_i - 1$	TSS		

EXAMPLE 6.2 **REVISITED** Using the computational formulas on the data given in Example 6.2, we obtain the following results:

$$\begin{aligned}\text{TSS} &= 934^2 + 1041^2 + \cdots + 1191^2 - (15871)^2/16\\ &= 15{,}882{,}847 - 15{,}743{,}040.06 = 139{,}806.94,\\ \text{SSB} &= 3938^2/4 + \cdots + 4466^2/4 - (15871)^2/16\\ &= 15{,}832{,}971.25 - 15{,}743{,}040.06 = 89{,}931.19.\end{aligned}$$

Because of the partitioning of the sums of squares, we obtain SSW by subtracting SSB from TSS as follows:

$$\text{SSW} = \text{TSS} - \text{SSB} = 139{,}806.94 - 89{,}931.19 = 49{,}875.75.$$

The results are summarized in Table 6.6 and are seen to be identical to the results obtained previously.

The procedures discussed in this section can be applied to any number of populations, including the two-population case. It is not difficult to show that the pooled t test given in Section 5.2 and the analysis of variance F test give identical results. This is based on the fact that the F distribution with 1 and

Table 6.6

Analysis of Variance for Rice Data

Source	df	SS	MS	F
Between varieties	3	89,931.19	29,977.06	7.21
Within varieties	12	49,875.75	4,156.31	
Total	15	139,806.94		

ν degrees of freedom is identically equal to the distribution of the square of t with ν degrees of freedom (Section 2.6). That is,

$$t^2(\nu) = F(1, \nu).$$

Note that in the act of squaring, both tails of the t distribution are placed in the right tail of the F distribution; hence the use of the F distribution automatically provides a two-tailed test. ■

EXAMPLE 6.3

(EXAMPLE 1.2 REVISITED) The Modes were looking at the data on homes given in Table 1.2 and noted that the prices of the homes appeared to differ among the zip areas. They therefore decided to do an analysis of variance to see if their observations were correct. The preliminary calculations are shown in Table 6.7.

Table 6.7

Preliminary Calculations of Prices in Zip Areas

Zip	n	$\sum y$	$\bar{y}$	$\sum y^2$
1	6	521.35	86.892	48912.76
2	13	1923.33	147.948	339136.82
3	16	1543.28	96.455	187484.16
4	34	5767.22	169.624	1301229.07
ALL	69	9755.18	141.379	1876762.82

The column headings are self-explanatory. The sums of squares are calculated as (note that sample sizes are unequal):

$$\text{TSS} = 1{,}876{,}762.82 - (9755.18)^2/69 = 497{,}580.28,$$

$$\text{SSB} = (521.35)^2/6 + \cdots + (5756.22)^2/34 = 77{,}789.84,$$

and by subtraction,

$$\text{SSW} = 497{,}580.28 - 77{,}789.84 = 419{,}790.44.$$

The degrees of freedom for SSB and SSW are 3 and 65, respectively; hence MSB $= 25{,}929.95$ and MSW $= 6458.31$, and then $F = 25{,}929.95/6458.31 = 4.01$. The 0.05 critical value for the F distribution with 3 and 60 degrees of freedom is 2.76; hence we reject the null hypothesis of no price differences among zip areas. The results are summarized in Table 6.8, which shows that the p value is 0.011. ■

Table 6.8

Analysis of Variance for Home Prices

Source	DF	Sum of Squares	Mean Square	F Value	Pr > F
Between zip	3	77789.837369	25929.945790	4.01	0.0110
Within zip	65	419790.437600	6458.3144246		
Total	68	497580.274969			

6.3 The Linear Model

The Linear Model for a Single Population

We introduce the concept of the linear model by considering data from a single population (using notation from Section 1.5) normally distributed with mean μ and variance σ^2. The linear model expresses the observed values of the random variable Y as the following equation or model:

$$y_i = \mu + \varepsilon_i, \quad i = 1, \ldots, n.$$

To see how this model works, consider a population that consists of four values, 1, 2, 3, and 4. The mean of these four values is $\mu = 2.5$. The first observation, whose value is 1, can be represented as the mean of 2.5 plus $\varepsilon_1 = -1.5$. So $1 = 2.5 - 1.5$. The other three observations can be similarly represented as a "function" of the mean and a remainder term that differs for each value. In general, the terms in a statistical model can be described as follows.

The left-hand side of the equation is y_i, which is the ith observed value of the **response variable** Y. The response variable is also referred to as the **dependent** variable.

The right-hand side of the equation is composed of two terms:

- The **functional or deterministic** portion, consisting of functions of parameters. In the single-population case, the deterministic portion is simply μ, the mean of the single population under study.
- The **random** portion, usually consisting of one term, ε_i, measures the difference in the response variable and the functional portion of the model. For example, in the single-population case, the term ε_i can be expressed as $y_i - \mu$. This is simply the difference between the observed value and the population mean. This term accounts for the natural variation existing among the observations. This term is called the **error** term, and is assumed to be a normally distributed random variable with a mean of zero and a variance of σ^2. The variance of this error term is referred to as the **error variance**.

It is important to remember that the nomenclature **error** does not imply any sort of mistake; it simply reflects the fact that variation is an acknowledged factor in any observed data. It is the existence of this variability that makes it necessary to use statistical analyses. If the variation described by this term did not exist, all observations would be the same and a single observation would provide all needed information about the population. Life would certainly be simpler, but unfortunately also very boring.

The Linear Model for Several Populations

We now turn to the linear model that describes samples from $t \geq 2$ populations having means $\mu_1, \mu_2, \ldots, \mu_t$, and common variance σ^2. The linear model describing the response variable is

$$y_{ij} = \mu_i + \varepsilon_{ij}, \quad i = 1, \ldots, t, \quad j = 1, \ldots, n_i,$$

where y_{ij} = jth observed sample value from the ith population, μ_i = mean of the ith population, and ε_{ij} = difference or deviation of the jth observed value from its respective population mean. This error term is specified to be a normally distributed random variable with mean zero and variance σ^2. It is also called the "experimental" error when data arise from experiments.

Note that the deterministic portion of this model consists of the t means, $\mu_1, i = 1, 2, \ldots, t$; hence inferences are made about these parameters. The most common inference is the test that these are all equal, but other inference may be made. The error term is defined as it was for the single population model.

Again, the variance of the ε_{ij} is referred to as the error variance, and the individual ε_{ij} are normally distributed with mean zero and variance σ^2. Note that this specification of the model also implies that there are no other factors affecting the values of the y_{ij} other than the means.

The Analysis of Variance Model

The linear model for samples from several populations can be redefined to correspond to the partitioning of the sum of squares discussed in Section 6.2. This model, called the **analysis of variance model**, is written as

$$y_{ij} = \mu + \tau_i + \varepsilon_{ij},$$

where y_{ij} and ε_{ij} are defined as before, μ = a reference value, usually called the "grand" or overall mean, and τ_i = a parameter that measures the effect of an observation being in the ith population. This effect is, in fact, $(\mu_i - \mu)$, or the difference between the mean of the ith population and the reference value. It is usually assumed that $\sum \tau_i = 0$, in which case μ is the mean of the t populations represented by the factor levels and τ_i is the effect of an observation being in the population defined by factor i. It is therefore called the "treatment effect."

Note that in this model the deterministic component includes μ and the τ_i. When used as the model for the rice yield experiment, μ is the mean yield of the four varieties of rice, and the τ_i indicate by how much the mean yield of each variety differs from this overall mean.

Fixed and Random Effects Model

Any inferences for the parameters of the model for this experiment are restricted to the mean and the effects of these four specific treatment effects, τ_i, $i = 1, 2, 3$, and 4. In other words, the parameters μ and τ_i of this model refer only to the prespecified or fixed set of treatments for this particular experiment. For this reason, the model describing the data from this experiment is called a **fixed effects model**, sometimes called model I, and the parameters (μ and the τ_i) are called **fixed effects**.

In general, a fixed effects linear model describes the data from an experiment whose purpose it is to make inferences only for the specific set of factor levels actually included in that experiment. For example, in our rice yield experiment, all inferences are restricted to yields of the four varieties actually planted for this experiment.

In some applications the τ_i represent the effects of a sample from a population of such effects. In such applications the τ_i are then random variables and the inference from the analysis is on the variance of the τ_i. This application is called the **random effects model**, or model II, and is described in Section 6.6.

The Hypotheses

In terms of the parameters of the fixed effects linear model, the hypotheses of interest can be stated

$$H_0\colon \tau_i = 0 \quad \text{for all } i,$$
$$H_1\colon \tau_i \neq 0 \quad \text{for some } i.$$

These hypotheses are equivalent to those given in Section 6.2 since

$$\tau_1 = \tau_2 = \cdots = \tau_t = 0$$

is the same as

$$(\mu_1 - \mu) = (\mu_2 - \mu) = \cdots = (\mu_t - \mu) = 0,$$

or equivalently

$$\mu_1 = \mu_2 = \cdots = \mu_t = \mu.$$

The point estimates of the parameters in the analysis of variance model are

$$\text{estimate of } \mu = \bar{y}_{..}, \text{ and}$$
$$\text{estimate of } \tau_i = (\bar{y}_{i.} - \bar{y}_{..}),$$

then also

$$\text{estimate of } \mu_i = \mu + \tau_i = \bar{y}_{i.}.$$

Expected Mean Squares

Having defined the point estimates of the fixed parameters, we next need to know what is estimated by the mean squares we calculate for the analysis of variance. In Section 2.2 we defined the expected value of a statistic as the mean of the sampling distribution of that statistic. For example, the expected value of $\bar{y}$ is the population mean, μ. Hence we say that $\bar{y}$ is an unbiased estimate of μ. Using some algebra with special rules about expected values, expressions for the expected values of the mean squares involved in the analysis of variance as functions of the parameters of the analysis of the variance model can be derived. Without proof, these are (for the balanced case)

$$E(\text{MSB}) = \sigma^2 + \frac{n}{t-1}\sum_i \tau_i^2,$$

$$E(\text{MSW}) = \sigma^2.$$

These formulas clearly show that if the null hypothesis is true ($\tau_i = 0$ for all i), then $\sum \tau_i^2 = 0$, and consequently both MSB and MSW are estimates of σ^2. Therefore, if the null hypothesis is true, the ratio MSB/MSW is a ratio of two estimates of σ^2, and is a random variable with the F distribution. If, on the other hand, the null hypothesis is not true, the numerator of that ratio will tend to be larger by the factor $[n/(t-1)]\sum_i \tau_i^2$, which must be a positive quantity that will increase in magnitude with the magnitude of the τ_i. Consequently, large values of τ_i tend to increase the magnitude of the F ratio and will lead to rejection of the null hypothesis. Therefore, the critical value for rejection of the hypothesis of equal means is in the right tail of the F distribution. As this discussion illustrates, the use of the expected mean squares provides a more rigorous justification for the analysis of variance than that of the heuristic argument used in Section 6.2.

The sampling distribution of the ratio of two estimates of a variance is called the "central" F distribution, which is the one for which we have tables. As we have seen, the ratio MSB/MSW has the central F distribution if the null hypothesis of equal population means is true. Violation of this hypothesis causes the sampling distribution of MSB/MSW to be stretched to the right, a distribution that is called a "noncentral" F distribution. The degree to which this distribution is stretched is determined by the factor $[n/(t-1)]\sum_i(\tau_i^2)$, which is therefore called the "noncentrality" parameter. The noncentrality parameter thus shows that the null hypothesis actually tested by the analysis of variance is

$$H_0\text{: } \sum \tau_i^2 = 0;$$

that is, the null hypothesis is that the noncentrality parameter is zero. We can see that this noncentrality parameter increases with increasing magnitudes of the absolute value of τ_i and larger sample sizes, implying greater power of the test as differences among treatments become larger and as sample sizes increase. This is, of course, consistent with the general principles of hypothesis testing presented in Chapter 3. The noncentrality parameter may be used in computing the power of the F test, a procedure not considered in this text (see, for example, Neter *et al.*, 1996).

Notes on Exercises

At this point sufficient background is available to do the basic analysis of variance for Exercises 1 through 8, 11, 14, 16, and 17.

6.4 Assumptions

As in all previously discussed inference procedures, the validity of any inference depends on the fulfillment of certain assumptions about the nature of the data. In most respects, the requirements for the analysis of variance are the same as have been previously discussed for the one- and two-sample procedures.

Assumptions Required

The assumptions in the analysis of variance procedure are usually expressed in terms of the elements of the linear model, and especially the ε_{ij}, the error term. These assumptions can be briefly stated:

1. The specified model and its parameters adequately represent the behavior of the data.
2. The ε_{ij}'s are normally distributed random variables with mean zero and variance σ^2.
3. The ε_{ij}'s are independent in the probability sense; that is, the behavior of one ε_{ij} is not afffected by the behavior value of any other.

The necessity of the first assumption is self-evident. If the model is incorrect, the analysis is meaningless. Of course, we never really know the correct model, but all possible efforts should be made to ensure that the model is relevant to the nature of the data and the procedures used to obtain the data. For example, if the data collection involved a design more complex than the completely randomized design and we attempted to use the one-way analysis of variance procedure to analyze the results, then we would have spurious results and invalid conclusions. As we shall see in later chapters, analysis of more complex data structures requires the specification of more parameters and more complex models. If some parameters have not been included, then the sums of squares associated with them will show up in the error variance, and the error is not strictly random. The use of an incorrect model may also result in biased estimates of those parameters included in the model.

The normality assumption is required so that the distribution of the MSB/MSW ratio will be the required F distribution (Section 2.6). Fortunately, the ability of the F distribution to represent the distribution of a ratio of variances is not severely affected by relatively minor violations of the normality assumption. Because of this, the ANOVA test is known as a relatively robust test. However, extreme nonnormality, especially extremely skewed distributions, or the existence of outliers may result in biased tests. Of course, in such cases, the means may also not be the appropriate set of parameters for description and inferences.

The second assumption also implies that each of the populations has the same variance, which is, of course, the same assumption needed for the pooled t test. As in that case, this assumption is necessary for the pooled variance to be used as an estimate of the variance and, consequently, for the ratio MSB/MSW to be a valid test statistic for the desired hypothesis. Again, minor violations of the equal variance assumptions do not have a significant effect on the analysis, while major violations may cast doubt on the usefulness of inferences on means.

Finally, the assumption of independence is necessary so that the ratio used as the test statistic consists of two independent estimates of the common variance. Usually the requirement that the samples be obtained in a random manner assures that independence. The most frequent violation of this assumption occurs when the observations are collected over some time or space coordinate, in which case adjacent measurements tend to be related. Methodologies for analysis of such data are beyond the scope of this text. See Freund and Wilson (1998, Sections 4 and 5) and Steele and Torrie (1980, Section 11.6) for additional examples.

Detection of Violated Assumptions

Since the assumptions are similar to those discussed previously, the detection methods are also similar. Exploratory data tools, such as stem and leaf and box plots, are useful in identifying outliers and highly skewed distributions. However, in the case of multiple-population data, it is not appropriate to use the observed values because the linear model specifies that these observed values consist of several model components, only one of which is the random error. For example, in the one-way analysis of variance, for the observed values y_{ij}, the model specifies that the observations consist of $(\mu + \tau_i + \varepsilon_{ij})$. Thus any plot of the y_{ij} will exhibit the characteristics of the distribution of $(\tau_i + \varepsilon_{ij})$ and may not reveal anything about the ε_{ij} themselves.

For this reason, the plots that will aid us in detection of violations of the assumptions must be made on estimates of the ε_{ij}. These estimates of the error terms are called "residuals," and are obtained by subtracting from each observation the estimate of $(\mu + \tau_i)$, which, as we have noted, is $\bar{y}_{i.}$. That is, the estimated residuals are $(y_{ij} - \bar{y}_{i.})$ for all observations.

The stem and leaf and box plots for the residuals for the data in Example 6.2 are shown in Table 6.9. Within the limitations imposed by having only 16 observations, these plots do not appear to indicate any serious difficulties. That is, from the shape of the stem and leaf plot we can see no large deviations from normality and the box plot indicates no apparent outliers. The same conclusion is reached for the data in Example 6.3.

Table 6.9

EDA Plots of Residuals for the Rice Data

```
Stem-Leaf                         No. Box Plot
  0 567                            3      |
  0 1223344                        7   +-----+
 -0 0                              1   |  +  |
 -0 555                            3   +-----+
 -1 20                             2      |
    ----+----+----+----+
```

Unequal variances among populations may not be detected by such plots, unless separate plots are made for each sample. Such plots may not be useful for small sample sizes (as in Example 6.2). Occasionally, unequal variances may cause the distribution of the residuals to appear skewed; however, this is not always the case. Therefore, if it is suspected that the variances are not the same for each factor level, it may be advisable to conduct a hypothesis test to verify that suspicion.

The Hartley F-Max Test

A test of the hypothesis of equal variances is afforded by the Hartley F-max test. The test is performed by first calculating the individual variances and computing the ratio of the largest to smallest of these. This ratio is then compared with critical values obtained from Appendix Table A.5. More extensive tables of the F-max distribution can be found in the Pearson and Hartley tables (1972, p. 202).

For the data in Example 6.2 the variances of yields of the four varieties are

$$s_1^2 = 3361.67,$$
$$s_2^2 = 1289.58,$$
$$s_3^2 = 4539.00,$$
$$s_4^2 = 7435.00.$$

The hypotheses of interest are

$$H_0\colon \sigma_1^2 = \sigma_2^2 = \sigma_3^2 = \sigma_4^2,$$
$$H_1\colon \text{at least two variances are not equal.}$$

We specify $\alpha = 0.05$. The parameters for the distribution of the test statistic are t, the number of factor levels, and df, the degrees of freedom of the individual estimated variances. (The test is strictly valid only for balanced data.) For this example, then, $t = 4$ and df $= 3$, and the critical range of the F-max distribution is 39.2 (Appendix Table A.5). The ratio of the largest to the smallest variance, s_4^2/s_2^2, provides the value

$$7435.00/1289.58 = 5.77.$$

Since this is less than the critical value, we have insufficient evidence to reject the hypothesis of equal variances; hence we may conclude that the equal variance assumption is not violated.

While easy to use, the Hartley test strictly requires equal sample sizes and is quite sensitive to departures from the assumption of normal populations. Since the graphic statistics presented in Table 6.9 show no indication of nonnormality, it is appropriate to use the Hartley test. In the case where there is concern about nonnormality, a viable alternative is the Levene test (Levene, 1960). The Levene test is robust against serious departures from normality, and does not require equal sample sizes. To test the same hypothesis of equal variances, the Levene test computes the absolute difference between the value of each observation and its cell mean and performs a one-way analysis of variance on these differences. The ordinary F statistic from this analysis of variance is used as a test for homogeneity of variances. Of course, we would normally not do two tests for the same hypothesis, but for illustration purposes, we present the results of the Levene test using SPSS on the data in Example 6.2. The results are in Table 6.10.

Table 6.10

Test of Homogeneity of Variances

YIELD			
Levene Statistic	df1	df2	Sig.
0.909	3	12	0.465

Note that the p value for the test is 0.465, supporting the conclusion that there is no reason to doubt the assumption of equal variances.

It may come as a surprise that such a wide dispersion of sample variances does not imply heterogeneous population variances. This phenomenon is due to the large dispersion of the sampling distribution of variances especially for small sample sizes.

Violated Assumptions

If it appears that some assumptions may be violated, the first step is, as always, to reexamine closely the data and data collection procedures to determine that the data have been correctly measured and recorded. It is also important to verify the model specification, since defects in the model often show up as violations of assumptions. Since these are subjective procedures and often do not involve any formal statistical analysis, they should be performed by an expert in the subject area in conjunction with the person responsible for the statistical analysis. If none of these efforts succeed in correcting the situation, and a transformation such as that discussed later cannot be used, alternative analyses may be necessary. For example, one of the nonparametric techniques discussed in Chapter 13 may need to be considered.

Variance Stabilizing Transformations

Often when the assumption of unequal variances is not satisfied, the reason is some relationship between the variation among the units and some characteristic of the units themselves. For example, large plants or large animals vary more in size than do small ones. Economic variables such as income or price vary by percentages rather than absolute values. In each of these cases, the standard deviation may be proportional to the magnitude of the response variable. If the response variable consists of frequencies or counts, the underlying distribution may be related to the Poisson distribution (Section 2.3), for which the variance is proportional to the mean. If the response variable consists of percentages or proportions, the underlying distribution may be the binomial (Section 2.3) where the variance is related to the population proportion.

If unequal variation among factor levels is a result of one of these conditions, it may be useful to perform the analysis using transformed values of the observations, which may satisfy the assumption of equal variances. Some transformations that stabilize the variance follow:

1. If σ is proportional to the mean, use the logarithm of the y_{ij} (usually but not necessarily to base e).
2. If σ^2 is proportional to the mean, take the positive square root of the y_{ij}.
3. If the data are proportions or percentages, use arcsin $(\sqrt{y_{ij}})$, where the y_{ij} are the proportions.

Most computer software provides for such transformations.

EXAMPLE 6.4

(EXAMPLE 6.3 REVISITED) We noted in Chapter 1, especially Fig. 1.13, that home prices in the higher priced zip areas seemed to be more variable. Actually, it is quite common that prices behave in this manner: prices of high-cost items vary more than those of items having lower costs. If the variances of home prices are indeed higher for the high-cost zip area, the assumptions underlying the analysis of variance may have been violated. Figure 6.3 is a plot of the standard deviation against the price of homes for the four areas. The association between price and standard deviation is apparent.

We perform the Levene test for homogeneous variances. The analysis of variance of absolute differences gives MSB = 9725.5, MSE = 2619.6, $F = 3.71$, the p value is 0.0158, and we can conclude that variances are different.

Figure 6.3

Plot of Standard Deviations vs Prices

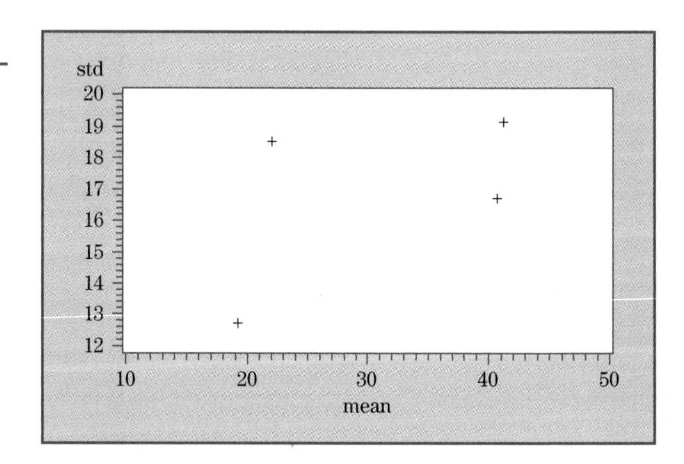

Table 6.11

Means and Standard Deviations

Variable	n	Mean	Standard Deviation
zip = 1			
price	6	86.892	26.877
lprice		4.42	0.324
zip = 2			
price	13	147.948	67.443
lprice		4.912	0.427
zip = 3			
price	16	96.455	50.746
lprice		4.445	0.5231
zip = 4			
price	34	169.624	98.929
lprice		4.988	0.5386

Table 6.12

Analysis of Variance for Logarithm of Prices

Dependent Variable: lprice

Source	df	Sum of Squares	Mean Square	F Value	Pr > F
Model	3	4.23730518	1.41243506	5.60	0.0018
Error	65	16.38771365	0.25211867		
Corrected Total	68	20.62501883			

Because of the obvious relationship between the mean and the standard deviation the logarithmic transformation is likely appropriate. The means and the standard deviations of `price` and of the natural logarithms of the price, labeled `lprice`, are given in Table 6.11. The results of the Levene test for the transformed data are MSB = 0.0905, MSW = 0.0974, $F = 0.93$, which leads to the conclusions that there is no evidence of unequal variances. We now perform the analysis of variance on the logarithm of price (variable `lprice`) with the results shown in Table 6.12. While both analyses indicate a difference in prices among the four zip areas, in this analysis the p value is seen to be considerably smaller than that obtained with the actual prices.

The use of transformations can accomplish more than just stabilizing the variance. Usually unequal variances go hand in hand with nonnormality. That is, unequal variances often cause the underlying distribution to look nonnormal. Thus the transformations listed in this section may often correct both unequal variances and nonnormality at the same time. It should be stressed that just because a transformation appears to have solved some problems, the resulting data should still be examined for other possible violations of assumptions.

The major drawback with using transformed data is that inferences are based on the means of the transformed values. The means of the transformed values are not necessarily the transformed means of the original values. In other words, it is not correct to transform statistics calculated from transformed values back to the original scale. This is easily seen in the data from Example 6.4. The retransformed means of the logarithms are certainly not equal to the means of the original observations (Table 6.11), although the relative magnitudes have been maintained. This will not always be the case. For further information on transformations, see Steel and Torrie (1980, Section 9.16).

Situations occur, of course, in which variable transformations are not helpful. In such cases, inferences on means may not be useful and alternative procedures may be appropriate. For Example 6.4, it may be appropriate to suggest the nonparametric Kruskal–Wallis test, which is detailed in Chapter 13. This method uses the ranks of the values in the data and tests the null hypothesis that the four underlying populations have the same distribution. ■

Notes on Exercises

It is now possible to check assumptions on all exercises previously completed and to perform remedial methods if necessary. In addition, the reader can now do Exercise 9.

6.5 Specific Comparisons

A statistically significant F test in the analysis of variance simply indicates that some differences exist among the means of the responses for the factor levels being considered. That is, the overall procedure tests the null hypothesis

$$H_0\colon \tau_i = 0, \quad i = 1, 2, \ldots, t.$$

However, rejection of that hypothesis does not indicate which of the τ_i are not zero or what specific differences may exist among the μ_i. In many cases we desire more specific information on response differences for different factor levels and, in fact, often have some specific hypotheses in mind. Some examples of specific hypotheses of interest follow:

1. Is the mean response for a specific level superior to that of the others?
2. Is there some trend in the responses to the different factor levels?
3. Is there some natural grouping of factor level responses?

Answers to questions such as these can be obtained by posing specific hypotheses, often called **multiple comparisons**. Multiple-comparison techniques are of two general types:

1. those generated prior to the experiment being conducted, called **preplanned comparisons**, and
2. those that use the result of the analysis (usually the pattern of sample means) to formulate hypotheses. These are called **post hoc comparisons**.

While the term "preplanned" might seem redundant, it is used to reinforce the concept that these contrasts must be specified prior to conducting the experiment or collecting the data. We will adhere to this convention and refer to them as preplanned contrasts throughout the discussion.

By and large, preplanned comparisons should be performed whenever possible. The reasons are as follows:

- **Preplanned comparisons have more power.** Because post hoc comparisons generate hypotheses from the data, rejection regions must be adjusted in order to preserve some semblance of a correct type I error probability. That means that a real difference between means may be found significant using a preplanned comparison but may not be found significant using a post hoc comparison.
- **A post hoc comparison may not provide useful results.** Comparisons of special interest may simply not be tested by a post hoc procedure. For example, if the factor levels are increasing levels of fertilizer on a crop, a post hoc procedure may simply provide the rather uninformative conclusion that the highest fertilizer level produces higher yields than the lowest level. Of course, what we really want to know is by *how much* the yield increases as we add specific amounts of fertilizer.

Most specific comparisons are based on certain types of linear combinations of means called **contrasts**. The presentation of contrasts is organized as follows:

- the definition of a contrast and its use in preplanned comparisons in hypothesis tests using t and F statistics,
- the definition of a special class of contrasts called orthogonal contrasts and how these are used in partitioning sums of squares for the testing of multiple hypotheses, and
- the use of contrasts in a number of different post hoc comparisons that use different statistics based on the "Studentized range."

The various formulas used in this section assume the data are balanced, that is, all $n_i = n$. This is not always a necessary assumption, as we will see in Section 6.7, but is used to simplify computations and interpretation. In fact, most computer software for performing such comparisons do not require this condition and makes appropriate modifications if data are unbalanced.

Contrasts

Consider the rice yield example discussed in Example 6.2 (data given in Table 6.2). The original description simply stated that there are four varieties. This description by itself does not provide a basis for specific preplanned comparisons. Suppose, however, that variety 4 was newly developed and we are interested in determining whether the yield of variety 4 is significantly different from that of the other three. The corresponding statistical hypotheses are stated:

$$H_0\colon \mu_4 = 1/3(\mu_1 + \mu_2 + \mu_3),$$

$$H_1\colon \mu_4 \neq 1/3(\mu_1 + \mu_2 + \mu_3).$$

In other words, the null hypothesis is that the mean yield of the new variety is equal to the mean yield of the other three. Rejection would then mean that the new variety has a different mean yield.[3] We can restate the hypotheses as

$$H_0\colon L = 0,$$
$$H_1\colon L \neq 0,$$

where

$$L = \mu_1 + \mu_2 + \mu_3 - 3\mu_4.$$

This statement of the hypotheses avoids fractions and conforms to the desirable null hypothesis format, which states that a linear function of parameters is equal to 0. This function is estimated by the same function of the sample means:

$$\hat{L} = \bar{y}_{1.} + \bar{y}_{2.} + \bar{y}_{3.} - 3\bar{y}_{4.}.$$

Note that this is a linear function of random variables because each mean is a random variable with mean μ_i and variance σ^2/n. The mean and variance of this function are obtained using the properties of the distribution of a linear function of random variables presented in Section 5.2. The constants (a_i in Section 5.2) of this linear function have the values ($a_1 = 1, a_2 = 1, a_3 = 1, a_4 = -3$). Therefore, the mean of $\hat{L}$ is

$$\mu_1 + \mu_2 + \mu_3 - 3\mu_4,$$

and the variance is

$$[1^2 + 1^2 + 1^2 + (-3)^2]\sigma^2/n = 12\sigma^2/n,$$

where $n = 4$ for this example. Furthermore, the variable $\hat{L}$ has a normal distribution as long as each of the $\bar{y}_{i.}$ are normally distributed. To test the hypotheses

$$H_0\colon L = \mu_1 + \mu_2 + \mu_3 - 3\mu_4 = 0,$$
$$H_1\colon L = \mu_1 + \mu_2 + \mu_3 - 3\mu_4 \neq 0,$$

we use the test statistic

$$t = \frac{\hat{L}}{\sqrt{\text{variance of } \hat{L}}} = \frac{\hat{L}}{\sqrt{\frac{12 \cdot \text{MSW}}{n}}},$$

where the substitution of MSW for σ^2 produces a test statistic that has the Student t distribution with $t(n-1)$ degrees of freedom. As always, the degrees of freedom of the t statistic match those of MSW, the estimate of the variance.

In Example 6.2, $n = 4$ and $t = 4$ so the degrees of freedom are $(3)(4) = 12$. The sample data yield

$$\hat{L} = 984.5 + 928.25 + 938.5 - 3(1116.5) = -498.4$$

[3]A one-sided alternative may be appropriate.

and

$$t = -498.4/\sqrt{(4156.31 \times 12)/4} = -498.4/111.66 = -4.46.$$

To test the hypotheses using $\alpha = 0.01$, we reject the null hypothesis if the t value we calculate exceeds in absolute value 3.0545. Since 4.46 exceeds that value, we reject the null hypothesis and conclude that the mean yield of variety 4 is different from the means of the other three varieties.

DEFINITION 6.1

A **contrast** is a linear function of means whose coefficients add to 0.

That is, a linear function of population means,

$$L = \sum a_i \mu_i,$$

is a contrast if

$$\sum a_i = 0.$$

The linear function of means discussed above satisfies this criterion since

$$\sum a_i = 1 + 1 + 1 - 3 = 0.$$

A contrast is estimated by the same linear function of sample means; hence the estimate of L is

$$\hat{L} = \sum a_i \bar{y}_{i.},$$

and the variance of $\hat{L}$ is

$$\text{var}(\hat{L}) = (\sigma^2/n) \sum a_i^2.$$

To test the hypothesis H_0: $L = 0$ against any alternative, we substitute the estimated variance, in this case MSW, for σ^2 and use the test statistic

$$t = \frac{\sum a_i \bar{y}_{i.}}{\sqrt{(\text{MSW}/n) \sum a_i^2}}.$$

This test statistic has the t distribution if the distributions of the $\bar{y}_{i.}$ are approximately normal, and it has the same degrees of freedom as MSW, which is $t(n-1)$ for the one-way ANOVA.

An equivalent and more informative method for testing hypotheses concerning contrasts uses the fact that $[t(\nu)]^2 = F(1, \nu)$ and performs the test with the F distribution. The appropriate test statistic is

$$t^2 = F = \frac{\left(\sum a_i \bar{y}_{i.}\right)^2}{(\text{MSW}/n) \sum a_i^2}.$$

Remember that the usual expression for an F ratio has the mean square for the hypothesis in the numerator and the error mean square in the denominator. Placing all elements except the error mean square into the numerator produces the mean square due to the hypothesis specified by the contrast as follows:

$$\text{MSL} = \frac{\left(\sum a_i \bar{y}_{i.}\right)^2}{\left(\sum a_i^2/n\right)}.$$

Since this mean square has 1 degree of freedom, it can also be construed as the sum of squares due to the contrast (SSL) with 1 degree of freedom (that is, SSL = MSL).

For the rice yield data, the sum of squares for the contrast for testing the equality of the mean of variety 4 to the others is

$$\text{SSL} = 4(498.25)^2/12 = 82{,}800.8.$$

The resulting F ratio is

$$F = 82{,}800.8/4156.31 = 19.92.$$

The critical value for the F distribution with 1 and 12 degrees of freedom ($\alpha = 0.01$) is 9.33, and the hypothesis is rejected. Note that $\sqrt{19.92} = 4.46$ and $\sqrt{9.33} = 3.055$, which are the values obtained for the test statistic and critical value when using the t statistic for testing the hypothesis.

Orthogonal Contrasts

Additional contrasts may be desired to test other hypotheses of interest. However, conducting a number of simultaneous hypotheses tests may compromise the validity of the stated significance level as indicated in Section 6.2. One method of alleviating this problem is to create a set of **orthogonal contrasts**. (Methods for nonorthogonal contrasts are presented later in this section.)

DEFINITION 6.2
Two contrasts are **orthogonal** if the cross product of their coefficients adds to 0.

Two contrasts,

$$L_1 = \sum a_i \mu_i$$

and

$$L_2 = \sum b_i \mu_i,$$

are orthogonal if

$$\sum (a_i b_i) = 0.$$

Sets of orthogonal contrasts have several interesting and very useful properties. If the data are balanced (all $n_i = n$), then

1. Given t factor levels, it is possible to construct a set of at most $(t - 1)$ mutually orthogonal contrasts. By mutually orthogonal, we mean that every pair of contrasts is orthogonal.
2. The sums of squares for a set of $(t - 1)$ orthogonal contrasts will add to the between sample or factor sum of squares (SSB).

In other words, the $(t-1)$ orthogonal contrasts provide a partitioning of SSB into single degree of freedom sums of squares, SSL_i, each being appropriate for testing one of $(t-1)$ specific hypotheses. Finally, because of this additivity, each of the resulting sums of squares is independent of the other, thus reducing the problem of incorrectly stating the significance level.

The reason for this exact partitioning is that the hypotheses corresponding to orthogonal contrasts are completely independent of each other. This is, the result of a test of any one of a set of orthogonal contrasts is in no way related to the result of the test of any other contrast.

Suppose that in Example 6.2, the problem statement indicated not only that variety 4 was most recently developed, but also that variety 3 was developed in the previous year, variety 2 was developed two years previously, while variety 1 was an old standard. The following hypotheses can be used to test whether each year's new variety provides a change in yield over the mean of those of the previous years:

$$H_{01}\colon \mu_4 = (\mu_1 + \mu_2 + \mu_3)/3,$$

that is, μ_4 is the same as the mean of all other varieties;

$$H_{02}\colon \mu_3 = (\mu_1 + \mu_2)/2,$$

that is, μ_3 is the same as the mean of varieties 1 and 2; and

$$H_{03}\colon \mu_1 = \mu_2.$$

The alternative hypotheses specify "not equal" in each case. The corresponding contrasts are

$$L_1 = \mu_1 + \mu_2 + \mu_3 - 3\mu_4.$$
$$L_2 = \mu_1 + \mu_2 - 2\mu_3,$$
$$L_3 = \mu_1 - \mu_2.$$

The orthogonality of the contrasts can be readily verified. For example, L_1 and L_2 are orthogonal because of the sum of the cross products of the coefficients:

$$(1)(1) + (1)(1) + (1)(-2) + (-3)(0) = 0.$$

The independence of these contrasts is verified by noting that rejecting H_{01} implies nothing about any differences among the means of treatments 1, 2, and 3, which are tested by the other contrasts. Similarly, the test for H_{02} implies nothing for H_{03}.

The sums of squares for the orthogonal contrasts are

$$\begin{aligned}\text{SSL}_1 &= 4[984.50 + 928.25 + 938.50 - 3(1116.50)]^2/(1+1+1+3^2)\\ &= 82{,}751.0\\ \text{SSL}_2 &= 4[984.50 + 928.25 - 2(938.50)]^2/(1+1+2^2) = 852.0\\ \text{SSL}_3 &= 4[984.50 - 928.25]^2/(1+1) = 6328.1.\end{aligned}$$

Table 6.13

Analysis of Variance with Contrasts

Source	df	SS	MS	F
Between varieties	3	89,931.1	29,977.1	7.21
μ_4 versus others	1	82,751.0	82,751.0	19.91
μ_3 versus μ_1 and μ_2	1	852.0	852.0	0.20
μ_2 versus μ_1	1	6,328.1	6,328.1	1.52
Within	12	49,875.75	4,156.3	
Total	15	139,806.9		

Note that $\text{SSL}_1 + \text{SSL}_2 + \text{SSL}_3 = 89{,}931.1$, which is the same as SSB from Table 6.6 (except for round-off).

Because each of the contrast sums of squares has 1 degree of freedom, $\text{SSL}_i = \text{MSL}_i$, and the F tests for testing H_{01}, H_{02}, and H_{03} are obtained by dividing each of the SSL_i by MSW. The results of the entire analysis can be summarized in a single table (Table 6.13). Only the first contrast is significant at the 0.05 level of significance. Therefore we can conclude that the new variety does have a different mean yield, but we cannot detect the specified differences among the others.

Other sets of orthogonal contrasts can be constructed. The choice of contrasts is, of course, dependent on the specific hypotheses suggested by the nature of the treatments. Additional applications of contrasts are presented in the next section and in Chapter 9.

Note, however, that the contrast

$$L_4 = \mu_1 - \mu_3$$

is not orthogonal to all of the above. The reason for the nonorthogonality is that contrasts L_1 and L_2 partially test for the equality of μ_1 and μ_3, which is the hypothesis tested by L_4.

It is important to note that even though we used preplanned orthogonal contrasts, we are still testing more than one hypothesis based on a single set of sample data. That is, the level of significance chosen for evaluating each single degree of freedom test is applicable only for that contrast, and not to the set as a whole. In fact, in the previous example we tested three contrasts, each at the 0.05 level of significance. Therefore, the probability that each test would fail to reject a true null hypothesis is 0.95. Since the tests are independent, the probability that all three would correctly fail to reject true null hypotheses is $(0.95)^3 = 0.857$. Therefore, the probability that at least one of the three tests would falsely reject a true null hypothesis (a type I error) is $1 - 0.857 = 0.143$, not the 0.05 specified for each hypothesis test. This is discussed in more detail in the section on post hoc comparisons.

Sometimes the nature of the experiment does not suggest a full set of $(t-1)$ orthogonal contrasts. Instead, only p orthogonal contrasts may be computed, where $p < (t - 1)$. In such cases it may be of interest of see if that set of contrasts is sufficient to describe the variability among all t factor level means as measured by the factor sum of squares (SSB). Formally the null hypothesis to be tested is that no contrasts exist other than those that have been computed; hence rejection would indicate that other contrasts should be implemented. This **lack of fit** is illustrated in the next section and also in Section 9.4.

Often in designing an experiment, a researcher will have in mind a specific set of hypotheses that the experiment is designed to test. These hypotheses may be expressed as contrasts, and these contrasts may not be orthogonal. In this situation, there are procedures that can be used to control the level of significance to meet the researcher's requirements. For example, we might be interested in comparing a control group with all others, in which case the Dunnett's test would be appropriate. If we have a small group of nonorthogonal preplanned contrasts we might use the Dunn–Sidak test. A detailed discussion of multiple comparison tests can be found in Kirk (1995, Section 4.1).

Fitting Trends

In many problems the levels of the factor represent varying values of a quantitative factor. For example, we may examine the output of a chemical process at different temperatures or different pressures, the effect of varying doses of a drug on patients, or the effect on yield due to increased amounts of fertilizer applied to crops. In such situations, it is logical to determine whether a trend exists in the response variable over the varying levels of the quantitative factor. This type of problem is a special case of multiple regression analysis, which is presented in Section 8.6. However, in cases where the number of factor levels is not large and the magnitudes of the levels are equally spaced, a special set of orthogonal contrasts may be used to establish the nature of such a trend. These contrasts are called "orthogonal polynomial contrasts." The coefficients for these contrasts are available in tables; a short table is given in Appendix Table A.6.

Orthogonal polynomials were originally proposed as a method for fitting polynomial regression curves without having to perform the laborious computations for the corresponding multiple regression (Section 8.6). Although the ready availability of computing power has decreased the usefulness of this application of orthogonal polynomials, it nevertheless provides a method of obtaining information about trends associated with quantitative factor levels with little additional work.

The simplest representation of a trend is a straight line that relates the levels of the factor to the mean response. A straight line is a polynomial of degree 1. This linear trend implies a constant change in the response for a given incremental change in the factor level. The existence of such a linear trend can be tested by using the linear orthogonal polynomial contrast.

If we find that a straight line does not sufficiently describe the relationship between response and factor levels, then we can examine a polynomial of degree 2, called a "quadratic polynomial," which provides a curved line (parabola) to describe the trend. The existence of such a quadratic polynomial can be tested by using the quadratic orthogonal polynomial contrast.

In the same manner higher degree polynomial curves may be included by adding the appropriate contrasts. Since a polynomial of degree $(t-1)$ may be fitted to a set of t data points (or means), the process of increasing the degree of the polynomial curve may be continued until a $(t-1)$ degree curve has been reached. Note that this corresponds to being able to construct at most $(t-1)$ orthogonal contrasts for t factor levels.

However, most practical applications result in responses that can be explained by relatively low-degree polynomials. What we need is a method of determining when to stop adding polynomial terms. Orthogonal polynomial contrasts allow the implementation of such a process by providing the appropriate sums of squares obtained by adding polynomial terms in the fitting of the trend.

The coefficients of these contrasts are given in Appendix Table A.6. A separate set of contrasts is provided for each number of factor levels, ranging from $t = 3$ to $t = 10$. Each column is a set of contrast coefficients, labeled X_i, where the i subscript refers to the degree of the polynomial, whose maximum value in the table is either $(t-1)$ or 4, whichever is smaller (polynomials of degrees higher than 4 are rarely used). The sums of squares for the coefficients, which are required to compute the test statistic, are provided at the bottom of each column.

The question of when to stop adding terms is answered by testing for the statistical significance of each additional contrast as it is added, as well as a lack of fit to see whether additional higher order terms may be needed.

EXAMPLE 6.5

To determine whether the sales of apples can be enhanced by increasing the size of the apple display in supermarkets, 20 large supermarkets are randomly selected from those in a large city. Four stores are randomly assigned to have either 10, 15, 20, 25, or 30 ft^2 of display for apples. Sales of apples per customer for a selected week is the response variable. The data are shown in Table 6.14.

The objective of this experiment is not only to determine whether a difference exists for the five factor levels (display space size), but to determine whether a trend exists and to describe it.

Table 6.14

Sales of Apples per Customer

	DISPLAY SPACE				
	10	**15**	**20**	**25**	**30**
	0.778	0.665	0.973	1.003	1.125
	0.458	0.830	1.029	1.073	1.184
	0.638	0.716	1.106	0.979	0.904
	0.602	0.877	0.964	0.981	0.951
Means	0.619	0.772	1.018	1.009	1.041

Solution We will perform the analysis of variance test for differences among means and, in addition, examine orthogonal contrasts to identify the maximum degree of polynomial that best explains the relationship between sales and display size. Using the method outlined in Section 6.2 we produce the ANOVA table given in Table 6.15. The F ratio for testing the mean sales for the five different display spaces (the line labeled "Space") has a value of 13.72 and a p value of less than 0.0001. We conclude that the amount of display space does affect sales. A cursory inspection of the data (Fig. 6.4, data values indicated by filled circles) indicates that sales appear to increase with space up to 20 ft^2 but sales response to additional space appears to level off.

Table 6.15

Analysis of Apple Sales Data

Source	df	SS	MS	F
Space	4	0.5628	0.1407	13.72
Linear	1	0.4674	0.4674	45.58
Quadratic	1	0.0706	0.0706	6.88
Lack of fit	2	0.0248	0.0124	1.20
Error	15	0.1538	0.0103	
Total	19	0.7166		

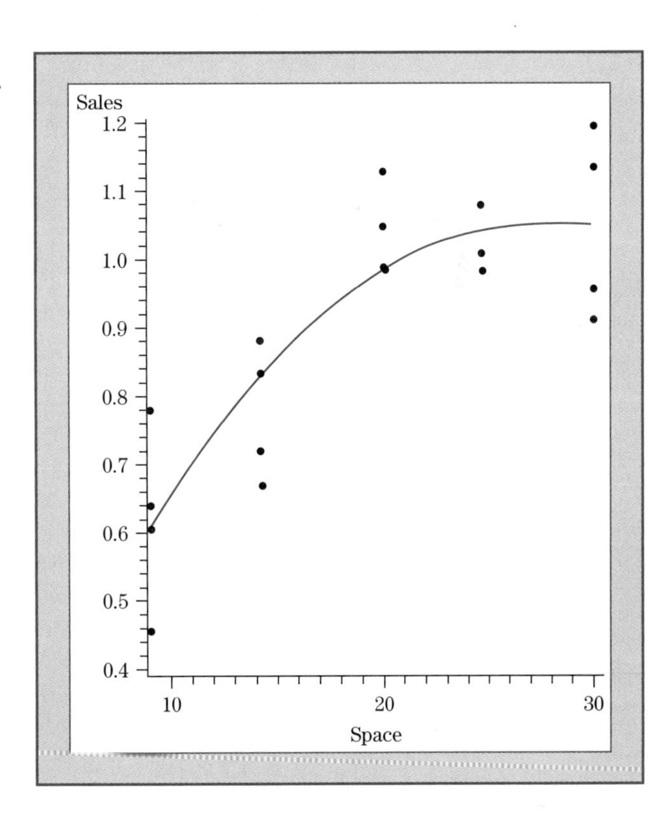

Figure 6.4

Plot of Apple Sales Data

This type of response is typical of a quadratic polynomial. We will use orthogonal polynomials to test for the linear and quadratic effects and also perform a lack of fit test to see if the quadratic polynomial is sufficient.

Contrasts for Trends First, the coefficients of the orthogonal contrasts are obtained from Appendix Table A.6 using five factors ($n = 5$ in the table). The contrasts are

$$L_1 = -2\mu_1 - \mu_2 + \mu_4 + 2\mu_5 \text{ (linear)},$$

$$L_2 = 2\mu_1 - \mu_2 - 2\mu_3 - \mu_4 + 2\mu_5 \text{ (quadratic)}.$$

From the table we also find the sums of squares of the coefficients, which are 10 and 14, respectively. The sums of squares for the contrasts are

$$\text{SSL}_1 = 4[-2(0.619) - 0.772 + 1.009 + 2(1.041)]^2/(10) = 0.4674,$$

$$\text{SSL}_2 = 4[2(0.619) - 0.772 - 2(1.018) - 1.009 + 2(1.041)]^2/(14) = 0.0706.$$

These sums of squares are also listed in Table 6.15 in the lines labeled "Linear" and "Quadratic." Using the MSW as the denominator for the F ratios we obtain the values 45.58 and 6.88 for L_1 and L_2, respectively. Both these values are greater than the critical value of 4.54 ($\alpha = 0.05$); hence we can conclude that a quadratic model may be useful; that is, our first impression is valid.

A graph of the quadratic trend[4] is shown in Fig. 6.4 as the curved line. The results of this analysis confirm the initial impression, which indicated that sales increase with the increased size of the display space up to about 23 or 24 ft^2 and then level off. This should allow supermarket managers to allocate space to apples in such a way as to maximize their sales without using excessive display space. ■

Lack of Fit Test

This test is performed to determine whether a higher degree polynomial is the appropriate next step. We obtain the sums of squares for this test by subtracting $\text{SSL}_1 + \text{SSL}_2$ from SSB. Remember that the sums of squares for a set of orthogonal contrasts add to the treatment sum of squares. Hence this difference is the sum of squares due to all other contrasts that could be proposed. Therefore, the test using this sum of squares is the test of the null hypothesis that other significant contrasts do not exist and, consequently, that the contrasts we have proposed adequately fit the means.

In this example, we have fitted the linear and quadratic polynomials and the other contrasts are those for the third- and fourth-degree polynomials. The subtraction provides a sum of squares of 0.0248 with 2 degrees of freedom, and the mean square for lack of fit is $0.0248/2 = 0.0124$. Again using the MSW value for the denominator we obtain a value for the F ratio of $0.0124/0.0103 = 1.20$,

[4]This plot produced with SAS/GRAPH software.

which is certainly not significant. Thus we can conclude that the quadratic trend adequately describes the relationship of sales to display space.

Notes on Exercises

It is now possible to perform preplanned contrasts or orthogonal polynomials where appropriate in previously worked exercises.

Computer Hint

Many statistical software packages do not have built-in provisions for doing a lack of fit test. Generally you will need to do the analysis of variance first. Then do the contrasts, which may not be available as part of the analysis of variance program and may have to be done by manual calculations. Results of the two analyses must then be combined manually.

Post Hoc Comparisons

In some applications the specifications of the factor levels do not suggest preplanned comparisons. For example, we have noted that the original treatment specification of four unnamed varieties in Example 6.2 did not provide a logical basis for preplanned comparisons. In such cases we employ **post hoc comparisons**, for which specific hypotheses are based on **observed** differences among the estimated factor level means. That is, the hypotheses are based on the sample data.

We noted in Section 3.6 that testing hypotheses based on the data is a form of exploratory analysis for which the use of statistical significance is not entirely appropriate. We also noted at the beginning of this chapter that the testing of multiple hypotheses using a single set of data results in a distortion of the significance level for the experiment as a whole. In other words, the type I error rate for each comparison, called the **comparison-wise** error rate, may be, say, 0.05, but the type I error rate for the analysis of the entire experiment, called the **experiment-wise** error rate, may be much larger. Finally, hypotheses based on the data are usually not independent of each other, which means that rejecting one hypothesis may imply the rejection of another, thereby further distorting the true significance level.

However, tests of this type are often needed; hence a number of methods for at least partially overcoming these distortions have been developed. Unfortunately, test procedures that more closely guarantee the stated experimentwise significance level tend to be less powerful and/or versatile, thus making more difficult the often desired rejection of null hypotheses. In other words, comparison procedures that allow the widest flexibility in the choice of hypotheses may severely compromise the stated significance level, while procedures that guarantee the stated significance level may preclude testing of useful hypotheses. For this reason a number of competing procedures, each of which attempts to provide useful comparisons while making a reasonable compromise between power and protection against the type I error (conservatism), have been developed.

Most post hoc comparison procedures are restricted to testing contrasts that compare pairs of means, that is,

$$H_0: \mu_i = \mu_j, \quad \text{for all values of } i \neq j.$$

Actually, pairwise comparisons are not really that restrictive in that they enable us to "rank" the means, and thus obtain much information about the structure of the means. For example, we can compare all factor levels with a control, determine whether a maximum or minimum value exists among the means, or determine whether a certain group of means are really homogeneous.

Because there is no consensus for a "best" post hoc comparison procedure, most computing packages offer an extensive menu of choices. Presenting such a large number of alternatives is beyond the scope of this book so we will present three of the more popular methods for making paired comparisons:

1. the **Fisher LSD** procedure, which simply does all possible t tests and is therefore least protective in terms of the experiment-wise significance level;
2. the **Tukey** procedure, which indeed assures the stated (usually 5%) experiment-wise significance level but is therefore not very powerful; and
3. the **Duncan multiple range test**, which is one of the many available compromises.

Finally, if the limitation to paired comparisons is too restrictive, the **Scheffé** procedure provides the stated experiment-wise significance level when making any and all possible post hoc contrasts. Of course, this procedure has the least power of all such methods.

The Fisher LSD Procedure The procedure for making all possible pairwise comparisons is attributed to Fisher (1948) and is known as the least significance difference or LSD test.

The LSD method performs a t test for each pair of means using the within mean square (MSW) as the estimate of σ^2. Since all of these tests have the same denominator, it is easier to compute the minimum difference between means that will result in "significance" at some desired level. This difference is known as the least significant difference, and is calculated

$$\text{LSD} = t_{\alpha/2}\sqrt{\frac{2 \cdot \text{MSW}}{n}},$$

where $t_{\alpha/2}$ is the $\alpha/2$ tail probability value from the t distribution, and the degrees of freedom correspond to those of the estimated variance, which for the one-way ANOVA used in this chapter are $t(n-1)$. The LSD procedure then declares as significantly different any pair of means for which the difference between sample means exceeds the computed LSD value.

As we have noted, the major problem with using this procedure is that the experiment-wise error rate tends to be much higher than the comparison-wise error rate. To maintain some control over the experiment-wise error rate, it is strongly recommended that the LSD procedure be implemented only if the hypothesis of equal means has first been rejected by the ANOVA test. This two-step procedure is called the "protected" LSD test. Carmer and Swanson (1973) conducted Monte Carlo simulation studies that indicate that the protected LSD is quite effective in maintaining a reasonable control over false rejection.

For the rice yield data in Example 6.2, the 0.05 level LSD is

$$\text{LSD} = 2.179\sqrt{\frac{2(4156.31)}{4}} = 99.33.$$

Any difference between a pair of sample means exceeding this value is considered to be statistically significant.

Results of paired comparison procedures are usually presented in a manner for reducing the confusion arising from the large number of pairs. First, the sample means are arranged from low to high:

Mean	$\bar{y}_{2.}$	$\bar{y}_{3.}$	$\bar{y}_{1.}$	$\bar{y}_{4.}$
Value	928.25	938.50	984.50	1116.50

A specified sequence of tests is used that employs the fact that no pair of means can be significantly different if the two means fall between two other means that have already been declared not to be significantly different.

We begin by comparing the largest mean to the other means. The first comparison is between the largest and the smallest: $\bar{y}_{4.}$ and $\bar{y}_{2}$. The actual difference, $1116.50 - 928.25 = 188.25$, exceeds the LSD critical value of 99.33; hence we declare that $\mu_4 \neq \mu_2$.

The next comparison is that between the largest and second smallest: $\bar{y}_{4.}$ and $\bar{y}_{3.}$. The actual difference of 178.00 exceeds the critical value 99.33; hence we declare that $\mu_4 \neq \mu_{3.}$. We likewise declare $\mu_4 \neq \mu_1$. This completes all comparisons with $\bar{y}_{4.}$.

We next compare the second largest mean to the others. Again the first comparison is to the smallest: $\bar{y}_1$ and $\bar{y}_2$. The difference is 56.25, which is smaller than the critical value of 99.33; hence we cannot declare $\mu_1 \neq \mu_2$. Since all other comparisons involve means that fall between these two, no other significant differences can be declared to exist.

It is convenient to summarize the results of paired comparisons by listing sample means and connecting with a line those means that are not significantly different. In our example, we found that μ_4 is significantly different from the other three, but that there were no other significant differences. The result can be summarized as:

$$\begin{array}{cccc} \bar{y}_{2.} & \bar{y}_{3.} & \bar{y}_{1.} & \bar{y}_{4.} \\ \underline{928.25 \quad 938.50 \quad 984.50} & & & \underline{1116.5} \end{array}$$

This presentation clearly shows that μ_4 is significantly different from the other three, but there are no other significant differences.

The above result is indeed quite unambiguous and therefore readily interpreted. This is not always true of a set of paired comparisons. For example, it is not unusual to have a pattern of differences result in a summary plot as follows:

Factor Levels

A	B	C	D	E	F

(A–B underlined; B–D underlined; D–F underlined)

This pattern does not really separate groups of means, although it does allow some limited inferences: Level A does have a different mean response from levels C through F, etc. This does not mean that the results are not valid, but does emphasize the fact that we are dealing with statistical rather than numerical differences.

Another convention for presenting the results of a paired comparison procedure is to signify by a specific letter all means that are declared to be not significantly different. An illustration is given in Table 6.16.

Table 6.16

Tukey HSD for Rice Yields

```
Analysis of Variance Procedure
Tukey's Studentized Range (HSD) Test for variable: YIELD
NOTE: This test controls the type I experiment-wise error rate,
but generally has a higher type II error rate than REGWQ.
Alpha = 0.05 df = 12 MSE = 4156.313
Critical Value of Studentized Range = 4.199
Minimum Significant Difference = 135.34
Means with the same letter are not significantly different.
```

Tukey Grouping		Mean	N	VAR
	A	1116.50	4	4
B	A	984.50	4	1
B		938.50	4	3
B		928.25	4	2

Tukey's Procedure As we have seen, the LSD procedure uses the t distribution to declare two means significantly different if the sample means differ by more than

$$\text{LSD} = t_{\alpha/2}\sqrt{2 \cdot \text{MSW}/n},$$

which can be written

$$\text{LSD} = t_{\alpha/2}\sqrt{2}(\text{standard error of } \bar{y}).$$

It is reasonable to expect that using some value greater than $\sqrt{2}t_{\alpha/2}$ as a multiplier of the standard error of the mean will provide more protection in terms of the experiment-wise significance level. The question is: How much larger? One possibility arises through the use of the **Studentized range**.

The Studentized range is the sampling distribution of the sample range divided by the estimated standard deviation. When the range is based on means from samples of size n, the statistic is denoted by

$$q = \frac{(\bar{y}_{\max} - \bar{y}_{\min})}{\sqrt{s^2/n}},$$

where for the one-way ANOVA, $s^2 = \text{MSW}$. Using a critical value from this distribution for a paired comparison provides the appropriate significance level for the worst case; hence it is reasonable to assume that it provides the proper experiment-wise significance level for all paired comparisons.

The distribution of the Studentized range depends on the number of means being compared (t), the degrees of freedom for the error (within) mean square (df), and the significance level (α). Denoting the critical value by $q_\alpha(t, \text{df})$, we can calculate a Tukey W (sometimes called HSD for "honestly significant difference") statistic,

$$W = q_\alpha(t, df)\sqrt{\text{MSW}/n},$$

and declare significantly different any pair of means that differs by an amount greater than W.

For our rice yield data (Example 6.2), with $\alpha = 0.05$, we use the tables of critical values of the Studentized range given in Appendix Table A.7 for two-tailed 0.05 significance level. For this example, $q_{0.05}(4, 12) = 4.20$. Then,

$$W = 4.20\sqrt{\frac{4156.31}{4}} = 135.38.$$

We use this statistic in the same manner as the LSD statistic. The results are shown in Table 6.16 and we can see that this procedure declares μ_4 different only from μ_2 and μ_3. We can no longer declare μ_4 different from μ_1. (Table 6.16 was produced by `PROC ANOVA` of the SAS System.) That is, in guaranteeing a 0.05 experiment-wise type I error rate we have lost some power.

Duncan's Multiple-Range Test It may be argued that the Tukey test guarantee of a stated experiment-wise significance level is too conservative and therefore causes an excessive loss of power. A number of alternative procedures that retain some control over experiment-wise significance levels without excessive power loss have been developed. One of the most popular of these is the Duncan multiple-range test. The justification for the Duncan multiple-range test is based on two considerations (Duncan, 1957):

1. When means are arranged from low to high, the Studentized range statistic is relevant only for the number of means involved in a specific comparison. In other words, when comparing adjacent means, called "comparing means two steps apart," we use the Studentized range for two means (which is identical to the LSD); for comparing means three steps apart we use the Studentized range statistic for three means; and so forth. Since the critical values of the Studentized range distribution are smaller with a lower number of means, this argument allows for smaller differences to be declared significant. However, the procedure maintains the principle that no pair of means is declared significantly different if the pair is within a pair already declared not different.

2. When the sample means have been ranked from lowest to highest, Duncan defines the **protection level** as $(1 - \alpha)^{r-1}$ for two sample means r steps apart. The probability of falsely rejecting the equality of two population means when the sample means are r steps apart can be approximated by $1 - (1 - \alpha)^{r-1}$. So, for adjacent means ($r = 2$) the protection level is $1 - \alpha$, and the approximate experiment-wise significant level is $1 - (1 - \alpha) = \alpha$. Note that the protection level decreases with increasing r. Because of this the Duncan multiple-range test is very powerful—one of the reasons that this test has been extremely popular.

The result is a different set of multipliers for computing an LSD statistic. These multipliers are given in Appendix Table A.8 and are a function of the number of steps apart (r), the degrees of freedom for the variance (df), and the significance level for a single comparison (α).

EXAMPLE 6.4

REVISITED In Example 6.4 we determined that the standard deviation of the home prices were proportionate to the means among the four zip areas. The analysis of variance using the logarithms of prices indicated that the prices do differ among the zip areas (reproduced in Table 6.17 for convenience). Because there is no information to suggest preplanned comparisons, we will perform a Duncan multiple-range test.

Table 6.17

Analysis of Variance for Logarithm of Prices

Dependent Variable: lprice

Source	df	Sum of Squares	Mean Square	F Value	Pr > F
Model	3	4.23730518	1.41243506	5.60	0.0018
Error	65	16.38771365	0.25211867		
Corrected Total	68	20.62501883			

There are four factor levels, so we are comparing four means. The critical values for the statistic can be obtained from Appendix Table A.8, with df = 60. The Duncan multiple-range test is normally applied when sample sizes are equal, in which case the test statistic is obtained by multiplying these values by $\sqrt{\text{MSW}/n}$, when n is the sample size. In this example the sample sizes are not equal, but a procedure that appears to work reasonably well is to define n as the harmonic mean of the sample sizes. This procedure is used by the SAS System with the results in Table 6.18.

The results of the Duncan's test indicate that home prices in zip areas 2 and 4 have prices that are not significantly different but are higher than zip areas 1 and 3.

A number of other procedures based on the Studentized range statistic can be used for testing pairwise comparisons after the data have been examined. One of these is called the **Newman–Keuls test** or sometimes the **Student–Newman–Keuls test**. This test uses the Studentized range that depends on

Table 6.18

Logarithm of Home Prices: Duncan's Multiple Range Test

```
                Duncan's Multiple Range Test for lprice

Note: This test controls the Type I comparisonwise error rate, not
            the experimentwise error rate.
             Alpha                                    0.05
             Error Degrees of Freedom                   65
             Error Mean Square                    0.252119
             Harmonic Mean of Cell Sizes          11.92245

                Note: Cell sizes are not equal.

Number of Means          2         3         4
Critical Range       .4107     .4321     .4462

Means with the same letter are not significantly different.

          Duncan Grouping         Mean       N     zip
-----------------------------------------------------------
                      A        4.9877      34       4
                      A
                      A        4.9119      13       2

                      B        4.4446      16       3
                      B
                      B        4.4223       6       1
```

the number of steps apart, but uses the stated significance level. This test is thus less powerful than Duncan's, but provides more protection against false rejection.

There are also paired comparison tests that have special purposes. For example, Dunnett's multiple-range test is designed to compare only all "factor levels" with a "control"; hence this procedure only makes $(t-1)$ comparisons and therefore has more power, but for a more limited set of hypotheses. All of these procedures, and more, are discussed by Kirk (1995). ■

The Scheffé Procedure So far we have restricted post hoc comparisons to comparing only pairs of means. If we desire to expand a post hoc analysis to include any and all possible contrasts, additional adjustments are required to maintain a satisfactory level of the experiment-wise type I error protection.

Scheffé (1953) has proposed a method for comparing any set of contrasts among factor level means. Scheffé's method is the most conservative of all multiple-comparison tests since it is designed so that the experiment-wise level of significance for all possible contrasts is at most α.

To test the hypotheses

$$H_0: L = 0,$$

$$H_1: L \neq 0,$$

where L is any desired contrast,

$$L = \sum(a_i\mu_i),$$

compute the estimated value of the contrast,

$$\hat{L} = \sum(a_i\bar{y}_i),$$

and compare it with the critical value S, which is computed

$$S = \sqrt{(t-1)F_\alpha \sum a_i^2\left(\frac{\text{MSW}}{n}\right)},$$

where all quantities are as previously defined and F_α is the desired α level critical value of the F distribution with the degrees of freedom for the corresponding ANOVA test, which is $[(t-1), t(n-1)]$ for the one-way ANOVA. If the value of $|\hat{L}|$ is larger than S, we reject H_0.

Consider again the rice yield data given in Example 6.2. Suppose that we decided after examining the data to determine whether the mean of the yields of varieties 1 and 4, which had the highest means in this experiment, differ from the mean of the yields of varieties 2 and 3. In other words, we are interested in testing the hypotheses

$$H_0\colon L = 0,$$

$$H_1\colon L \neq 0,$$

where

$$L = \frac{1}{2}(\mu_1 + \mu_4) - \frac{1}{2}(\mu_2 + \mu_3),$$

which gives the same comparison of means as

$$L = \mu_1 - \mu_2 - \mu_3 + \mu_4.$$

We compute

$$\hat{L} = 984.5 - 928.25 - 938.5 + 1116.5 = 234.25.$$

The 0.05 level critical value of the Scheffé S statistic is

$$S = \sqrt{(3)(3.49)(1+1+1+1)\left(\frac{4156.31}{4}\right)} = 208.61.$$

The calculated value of the contrast is 234.25; hence we reject H_0 and conclude that the mean yield of varieties 4 and 1 is not equal to that of varieties 2 and 3.

Comments

The fact that we have presented four different multiple-comparison procedures makes it obvious that there is no universally best procedure for making post hoc comparisons. In fact, Kirk (1995) points out that there are more than 30 multiple-comparison procedures currently used by researchers. As a result of this, most computer programs offer a wide variety of options. For example, the ANOVA procedure in SAS offers a menu of 16 choices! In general, the different multiple-comparison procedures present various degrees of trade-off between specificity and sensitivity. We trade power for versatility and must be aware of the effect of this on our final conclusions. In any case, the most sensitive (highest power) and most relevant inferences are those based on preplanned orthogonal contrasts, which are tested with single degree of freedom F tests. For this reason, *preplanned contrasts should always be used if possible.* Unfortunately, in most computer packages it is far easier to perform post hoc paired comparisons than to implement contrasts. For this reason, one of the most frequent misuses of statistical methods is the use of post hoc paired comparison techniques when preplanned contrasts should be used. Again it must be emphasized that only one comparison method should be used for a data set. For example, it is normally not recommended to first do preplanned contrasts and then a post-hoc paired comparison, although we do in Example 6.6 to illustrate the procedures.

The most versatile of the post hoc multiple-comparison tests is the Scheffé procedure, which allows any number of post hoc contrasts. For pairwise comparisons after the data have been analyzed, Duncan's multiple-range test seems to be at least as powerful as any other, and is perhaps the most frequently used such test. For a complete discussion, see Montgomery (1984).

As we have noted most statistical computer software offer a variety of post hoc multiple-comparison procedures, often allowing the simultaneous use of several methods, which is inappropriate. For reasons of convenience, we have illustrated several multiple-comparison methods using only two sets of data; however, it is appropriate to perform only one method on one set of data. The method chosen will depend on the requirements of the study and should be decided on prior to starting the statistical analysis.

The use of the analysis of variance as a first step in comparing two or more populations is recommended in almost all situations, even though it is not always necessary. It is, for example, possible to perform Duncan's multiple-range test without first doing the ANOVA. This does not affect the level of significance of the test. However, as we saw in the illustration of Duncan's multiple-range test, it is possible to obtain apparently contradictory results. This occurs because the power of the multiple-range tests is not defined in the same terms as that of the F test. Because of this, we again emphasize that the best results come from the thoroughly planned studies in which specific hypotheses are built into both the design of the experiment and the subsequent statistical analyses.

Solution to Example 6.1 We now return to Example 6.1. To compare the eight sites, a one-way analysis of variance was done. The result of using `PROC ANOVA` of the SAS System is shown in Table 6.19.

A p value of 0.0029 for the test of equal means is certainly small enough to declare that there are some differences in silt content among the locations.

Table 6.19

Example 6.1: Analysis of Variance Procedure

Dependent Variable: SILT

Source	df	Sum of Squares	F Value	Pr > F
Model	7	600.12079545	3.43	0.0029
Error	80	1998.43636364		
Corrected Total	87	2598.55715909		

Source	df	Anova SS	F Value	Pr > F
SITE	7	600.12079545	3.43	0.0029

Table 6.20 Example 6.1: Analysis of Variance Procedure

Duncan's Multiple Range Test for variable: SILT
NOTE: this test controls the type I comparisonwise error rate,
not the experimentwise error rate

Alpha=0.05 df=80 MSE=24.9805

Number of Means	2	3	4	5	6	7	8
Critical Range	4.246	4.464	4.607	4.711	4.799	4.870	4.929

Means with the same letter are not significantly different.

Duncan Grouping			Mean	N	SITE
	A		46.873	11	5
	A		46.473	11	8
B	A		44.527	11	7
B	A	C	43.600	11	6
B		C	41.236	11	1
B		C	41.018	11	2
B		C	40.545	11	4
		C	39.573	11	3

Because the locations are identified only by number, there is no information on which to base specific preplanned contrasts. Therefore, to determine the nature of the differences among the means, Duncan's multiple-range test was done, again using the SAS System. The results of this analysis are shown in Table 6.20. Note that we really do not have a clearly separated set of sites. The results of Duncan's test indicate that sites 5, 8, 7, and 6 are all similar in average silt content, that 7, 6, 1, 2, and 4 are similar, and that 6, 1, 2, 4, and 3 are all similar. This overlapping pattern of means is not uncommon in a multiple-comparison procedure. It simply means that the values of the sample means are such that there is no clear separation. We can, for example, state that sites 5 and 8 do differ from site 3.

It may be argued that since the sites were contiguous, consideration should be given to fitting some sort of trend. However, looking at the means in Table 6.20 indicates that this would not be successful. This is confirmed by the box plots in Fig. 6.5, which show no obvious trend across the sites. ■

Figure 6.5

Plot of Silt at Different Sites

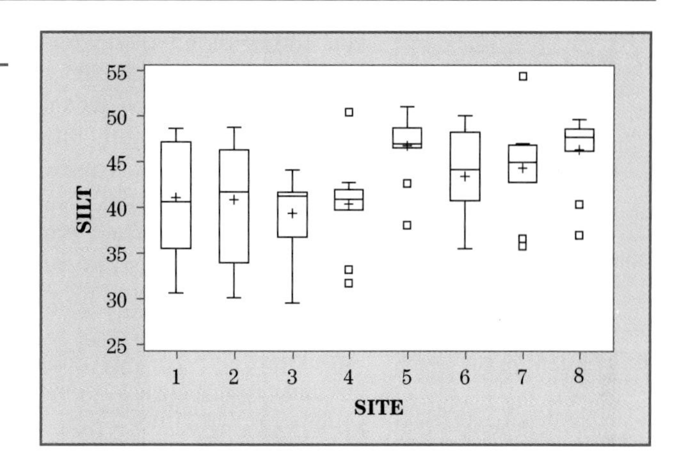

Confidence Intervals

Figure 6.6

Confidence Intervals for Means

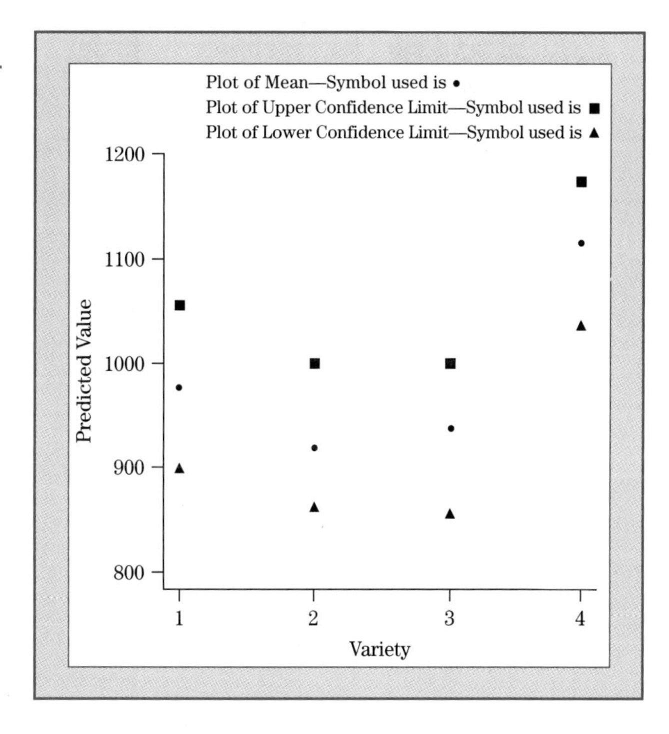

We have repeatedly noted that any hypothesis test has a corresponding confidence interval. It is sometimes useful to compute confidence intervals about factor level means. Using MSW as the estimate of σ^2, such intervals are computed

$$\bar{y}_{i.} \pm t_{\alpha/2}\sqrt{\text{MSW}/n},$$

where $t_{\alpha/2}$ is the $\alpha/2$ critical value for the t distribution with $t(n-1)$ degrees of freedom. An appealing graphical display consists of plotting the factor level means with a superimposed confidence interval indicated. This is presented for the rice data in Fig. 6.6. However, in viewing such a plot we must emphasize that the confidence coefficient is valid only for any one individual mean and not the entire group of means! For this reason it is sometimes recommended that, for example, the Tukey statistic (Studentized range) be used in place of Student t for calculating intervals.

Before leaving the discussion of contrasts, it should be pointed out that contrasts do not always give us the best look at the relationship between a set of means. The following example is an illustration of just such a situation. In addition, we demonstrate a method of using a computer program to calculate the statistics needed to do a Scheffé procedure.

EXAMPLE 6.6

An experiment to determine the effect of various diets on the weight of a certain type of shrimp larvae involved the following seven diets. Five 1-liter containers with 100 shrimp larvae each were fed one of the seven diets in a random assignment.

Experimental diets contained a basal compound diet and

1. corn and fish oil in a 1:1 ratio,
2. corn and linseed oil in a 1:1 ratio,
3. fish and sunflower oil in a 1:1 ratio, and
4. fish and linseed oil in a 1:1 ratio.

Standard diets were

5. basal compound diet (a standard diet),
6. live micro algae (a standard diet), and
7. live micro algae and *Artemia* nauplii.

After a period of time the containers were drained and the dry weight of the 100 larvae determined. The weight of each of the 35 containers is given in Table 6.21.

Table 6.21

Shrimp Weights

Diet	Weights				
1. Corn and fish oil	47.0	50.9	45.2	48.9	48.2
2. Corn and linseed oil	38.1	39.6	39.1	33.1	40.3
3. Fish and sunflower oil	57.4	55.1	54.2	56.8	52.5
4. Fish and linseed oil	54.2	57.7	57.1	47.9	53.4
5. Basal compound	38.5	42.0	38.7	38.9	44.6
6. Live micro algae	48.9	47.0	47.0	44.4	46.9
7. Live micro algae and *Artemia*	87.8	81.7	73.3	82.7	74.8

Solution The analysis attempted to identify the diet(s) that resulted in significantly higher weights in shrimp larvae. Note that the diets are broken up into two groups, the experimental diets and the standard diets. Further,

Table 6.22

Analysis of Variance for Diets

Dependent Variable: Weight

Source	df	Sum of Squares	Mean Square	F Value	Pr > F
Model (diets)	6	5850.774857	975.129143	88.14	0.0001
Error	28	309.792000	11.064000		
Corrected Total	34	6160.566857			

we note that several diets have common ingredients—all of the experimental diets contain the basal compound—hence, it would be useful to extend our analysis to determine how the various diet components affected weight. This is a problem that lends itself to the use of contrasts in the analysis of variance. Even though the questions that we want to ask about the diets can be addressed before the experiment is conducted, these questions will have to be stated in the form of nonorthogonal contrasts. For this reason, our procedure will be to first do the standard ANOVA, and then use the Scheffé procedure to test each of the contrasts.

The analysis of variance results appear in Table 6.22. Note that the p value for the test is 0.0001, certainly a significant result. Our first conclusion is that there is a difference somewhere between the seven diets. To look at the rest of the questions concerning diets, we use the following set of contrasts:

Contrast	**Interpretation**	**COEFFICIENTS OF DIETS** Diet no.: 1	2	3	4	5	6	7
newold	The first four against the three standards	−3	−3	−3	−3	4	4	4
corn	Diets containing corn oil against others	5	5	−2	−2	−2	−2	−2
fish	Diets containing fish oil against others	4	−3	4	4	−3	−3	−3
lin	Diets containing linseed oil against others	−2	5	−2	5	−2	−2	−2
sun	Diets containing sunflower oil against others	−1	−1	6	−1	−1	−1	−1
mic	Diets containing micro algae against others	−2	−2	−2	−2	−2	5	5
art	Diets containing *Artemia* against others	−1	−1	−1	−1	−1	−1	6

As mentioned in the previous "Comments" section, the computation of test statistics for contrasts using a computer program is often not straightforward. The Scheffé procedure is available in the SAS System only for making paired comparisons; however, we can use other procedures to eliminate most of the computational effort and obtain the desired results. Remember that the test for a contrast is

$$F = \frac{\left(\sum a_i \bar{y}_i\right)^2}{\frac{\text{MSW}}{n}\sum a_i^2} = \frac{(\hat{L})^2}{\frac{\text{MSW}}{n}\sum a_i^2}.$$

Now the Scheffé procedure declares a contrast significant if

$$\hat{L}^2 > S^2 = (t-1)F_\alpha \sum a_i^2\left(\frac{\text{MSW}}{n}\right),$$

Table 6.23

Estimates and Tests for Contrasts

Parameter	Estimate	T For H0: Parameter = 0	Pr > |T|	Std Error of Estimate
newold	6.97833333	6.14	0.0001	1.13613379
corn	−12.30000000	−9.88	0.0001	1.24457222
fish	1.06333333	0.94	0.3573	1.13613379
lin	−8.08600000	−6.50	0.0001	1.24457222
sun	3.93666667	2.45	0.0208	1.60673582
mic	16.27400000	13.08	0.0001	1.24457222
art	32.94000000	20.50	0.0001	1.60673582

where F_α is the α level tabulated F value with $t - 1$ and $t(n - 1)$ degrees of freedom. A little manipulation is used to show that this relationship can be restated as

$$F > (t - 1)F_\alpha,$$

where the F on the left-hand side is the calculated F statistic for the contrast.

In this example, $(t - 1) = 6$, and $t(n - 1) = 28$. Therefore $F_{0.05}(6, 28) = 2.49$[5] so $(6)(2.49) = 14.94$. Hence the critical value is 14.94. The contrasts are analyzed using the `ESTIMATE` statement of `PROC GLM` of the SAS System. The results provide the estimates of the contrasts among the groups of means and the corresponding t values used to test the hypothesis that the particular contrast is equal to 0. The results are shown in Table 6.23. Note that the t test for the contrasts is nothing but the square root of the F statistic given above. Therefore, we get the appropriate Scheffé's test by squaring the t value given in the SAS output and comparing it to the critical value of 14.94. The contrasts labeled `newold`, `corn`, `lin`, `mic`, and `art` are significantly different from 0. From examination of the values listed in the "Estimate" column, we observe that (1) the standard diets produce a higher mean weight than those of the experimental group, (2) diets with corn or linseed produce significantly lower mean weight that those without, (3) diets with fish oil or sunflower oil produce weights not significantly different from those of other diets, and (4) diets containing micro algae and *Artemia* produce an average weight higher than those without.

In short, a clear picture of the nature of the relationship between diet and weight cannot be obtained from the use of contrasts. It is, of course, possible to choose other sets of contrasts, but at this point a pairwise comparison may help to clarify the results. Because we have already performed one set of comparison procedures we will use the conservative Tukey procedure to do pairwise comparisons. That way, if any results are significant we can feel confident that it will not be due to chance (recall the discussion in Section 6.2). The results are shown in Table 6.24.

We can use this analysis to interpret the relationship between the diets more readily. For example, diet 7, containing the micro algae and *Artemia*, is

[5]The closest available value in Table A.4A is that for (6, 25) degrees of freedom.

Table 6.24

Tukey Procedure Results

```
 Tukey's studentized Range (HSD Test for variable: WEIGHT
Alpha = 0.05        df = 28        MSE = 11.064
 Critical value of Studentized Range = 4.486
 Minimum Significant Difference = 6.6733
 Means with the same letter are not significantly different.
```

Tukey	Grouping	Mean	N	DIET
	A	80.060	5	7
	B	55.200	5	3
	B			
C	B	54.060	5	4
C				
C	D	48.040	5	1
	D			
E	D	46.840	5	6
E				
E	F	40.540	5	5
	F			
	F	38.040	5	2

by far the best. Interestingly, the diets containing only the micro algae and the basal compound diet do not fare well. Finally, diets with fish oil (diets 1, 3, and 4) do appear to provide some advantages.

Actually, one of the reasons that the results are not easily interpreted is that this is not a very well-planned experiment. An experimental design that would make the results easier to interpret (and might even give more information about the diets) is the factorial experiment discussed in Chapter 9. However, to use the factorial arrangement effectively, more diets would have to be included. This example does illustrate the fact that planning the experiment prior to conducting it pays tremendous dividends when the final analysis is performed. ■

6.6 Random Models

Occasionally we are interested in the effects of a factor that has a large number of levels and our data represent a random selection of these levels. In this case the levels of the factors are a sample from a population of such levels and the proper description requires a **random effects model**, also called model II. For example, if in Example 6.1 the soil samples were a random sample from a population of such samples, the appropriate model for that experiment would be the random effects model.

The objective of the analysis for a random effects model is altered by the fact that the levels of the factor are not fixed. For example, inferences on the effects of individual factor levels are meaningless since the factor levels in a particular set of data are a randomly chosen set. Instead, the objective of the analysis of a random model is to determine the magnitude of the variation among the population of factor levels.

Specifically, the appropriate inferences are on the **variance** of the factor level effects. For example, if we consider Example 6.1 as a random model, the inferences will be on the variance of the means of scores for the population of soil samples.

The random effects model looks like that of the fixed effects model:

$$y_{ij} = \mu + \tau_i + \varepsilon_{ij}, \quad i = 1, \ldots, t, \;\; j = 1, \ldots, n.$$

However, the τ_i now represent a random variable whose distribution is assumed normal with mean zero and variance σ_τ^2. It is this variance, σ_τ^2, that is of interest in a random effects model. Specifically, the hypotheses to be tested are

$$H_0\colon \sigma_\tau^2 = 0,$$
$$H_1\colon \sigma_\tau^2 > 0.$$

The arithmetic for the appropriate analysis of variance is the same as for the fixed model. However, in the random effects model (and balanced data), the expected mean squares are

$$E(\text{MSB}) = \sigma^2 + n\sigma_\tau^2,$$
$$E(\text{MSW}) = \sigma^2.$$

This implies that the F ratio used in the fixed model ANOVA is appropriate for testing $H_0\colon \sigma_\tau^2 = 0$; that is, there is no variation among population means.

If H_0 is rejected, it is of interest to estimate the variances σ^2 and σ_τ^2, which are referred to as variance components. One method of estimating these parameters is to equate the expected mean squares to the mean squares obtained from the data and then solve the resulting equations. This method may occasionally result in a negative estimate for σ_τ^2, in which case the estimate of σ_τ^2 is arbitrarily declared to be zero. An estimate "significantly" less than 0 may indicate a special problem such as correlated errors. A discussion of this matter is found in Ostle (1963).

Table 6.25

Data for Random Model

TEACHER			
A	B	C	D
84	75	72	88
90	85	76	98
76	91	74	70
62	98	85	95
72	82	77	86
81	75	60	80
70	74	62	75

EXAMPLE 6.7

Suppose that a large school district was concerned about the differences in students' grades in one of the required courses taught throughout the district. In particular, the district was concerned about the effect that teachers had on the variation in students' grades. An experiment in which four teachers were randomly selected from the population of teachers in the district was designed. Twenty-eight students who had homogeneous backgrounds and aptitude were then found. Seven of these students were randomly assigned to each of the four teachers, and their final grade was recorded at the end of the year. The grades are given in Table 6.25. Do the data indicate a significant variation in student performance attributable to teacher difference?

Table 6.26

Analysis of Variance, Random Model

Source	df	SS	MS	F
Between sections	3	683.3	227.8	2.57
Within sections	24	2119.7	88.3	
Total	27	2803.0		

Solution The model for this set of data has the form

$$y_{ij} = \mu + \tau_i + \varepsilon_{ij}, \quad i = 1, 2, 3, 4, \quad j = 1, \ldots, 7,$$

where y_{ij} = grade of student j under teacher i, μ = overall mean grade, τ_i = effect of teacher i, a random variable with mean zero and variance σ_τ^2, and ε_{ij} = a random variable with mean zero and variance σ^2.

We are interested in testing the hypotheses

$$H_0\colon \sigma_\tau^2 = 0$$
$$H_1\colon \sigma_\tau^2 > 0.$$

The null hypothesis states that the variability in grades among classes is due entirely to the natural variability among students in these classes, while the alternative hypothesis states that there is additional variability among classes, due presumably to instructor differences.

The calculations are performed as in the fixed effects case and result in the ANOVA table given in Table 6.26. The test statistic is computed in the same manner as for the fixed model,[6] that is, MSB/MSW. The computed F ratio, 2.57, is less than the 0.05 level critical value of 3.01; hence, we cannot conclude that there is variation in mean grades among teachers.

It is of interest to estimate the two variance components: σ_τ^2 and σ^2. Since we have not rejected the null hypothesis that $\sigma_\tau^2 = 0$, we would not normally estimate that parameter, but will do so here to illustrate the method. By equating expected mean squares to sample mean squares we obtain the equations

$$227.8 = \sigma^2 + 7\sigma_\tau^2,$$
$$88.3 = \sigma^2.$$

From these we can solve for $\hat{\sigma}_\tau^2 = 19.9$ and $\hat{\sigma}^2 = 88.3$. The fact that the apparently rather large estimated variance of 19.9 did not lead to rejection of a zero value for that parameter is due to the rather wide dispersion of the sampling distribution of variance estimates, especially for small samples (see Section 2.6). ■

Confidence intervals for variance components may be obtained; see, for example, Neter *et al.* (1996). Methods for obtaining these inferences are beyond the scope of this book.

[6] This is not the case in all ANOVA models. When we have certain experimental designs (Chapter 10), we will see that having one or more random effects may alter the procedure used to construct F ratios.

The validity of an analysis of variance for a random model depends, as it does for the fixed model, on some assumptions about the data. The assumptions for the random model are the same as those for the fixed with the additional assumption that the τ_i are indeed random and independent and have the same variance for the entire population. Also, as in the case of the fixed model, transformations may be used for some cases of nonhomogeneous variances, and the same cautions apply when they are used.

6.7 Unequal Sample Sizes

In most of the previous sections, we have assumed that the number of sample observations for each factor level is the same. This is described as having "balanced" data. We have noted that having balanced data is not a requirement for using the analysis of variance. In fact, the formulas presented for computing the sums of squares (Section 6.2) correspond to the general case using the individual n_i for the sample sizes. However, a few complications do arise when using unbalanced data:

- Contrasts that may be orthogonal with balanced data are usually not orthogonal for unbalanced data. That is, the total of the contrast sums of squares does not add to the factor sum of squares.
- If the sample sizes reflect actual differences in population sizes, which may occur in some situations, the sample sizes may need to be incorporated into the contrasts:

$$\hat{L} = \sum a_i n_i \bar{y}_{i.}.$$

- Post hoc multiple-comparison techniques, such as Duncan's, become computationally more difficult, although computer software will usually perform these calculations.
- Although balanced data are not required for a valid analysis, they do provide more powerful tests for a given total sample size.

6.8 Analysis of Means

The **analysis of means procedure (ANOM)** is a useful alternative to the analysis of variance (ANOVA) for comparing the means of more than two populations. The ANOM method is especially attractive to nonstatisticians because of its ease of interpretation and graphic presentation of results. An ANOM chart, conceptually similar to a control chart (discussed in Chapter 2), portrays decision lines so that magnitude differences and statistical significance may be assessed simultaneously. The ANOM procedure was first proposed by Ott (1967) and has been modified several times since. A complete discussion of the applications of the analysis of means is given in Ramig (1983). The analysis of means uses critical values obtained from a sampling distribution called the **multivariate *t* distribution**. Exact critical values for several common levels of significance are found in Nelson (1983) and reproduced in Appendix Table A.11.

These critical values give the ANOM power comparable to that of the ANOVA under similar conditions (see Nelson, 1985). While ANOM is not an optimal test in any mathematical sense, its ease of application and explanation give it some practical advantage over ANOVA.

This section discusses the application of the ANOM to problems similar to those discussed in Section 6.1. In particular, we will examine an alternative procedure for comparing means that arise from the one-way (or single factor) classification model. The data consist of continuous observations (often called variables data), y_{ij}, $i = 1, \ldots, t$ and $j = 1, \ldots, n$. The factor level means are $\bar{y}_{i.} = \sum y_{ij}/n$. The assumptions on the means are the same as that of the ANOVA; that is, they are assumed to be from normally distributed populations with common variance σ^2. The grand mean is $\bar{y}_{..} = \sum \bar{y}_{i.}/t$, and the pooled estimate of the common but unknown variance is

$$s^2 = \sum s_i^2/t,$$

$$\text{where} \quad s_i^2 = \sum (y_{ij} - \bar{y}_{i.})^2/(n-1).$$

Note that the pooled estimate of the variance is identical to MSW in the ANOVA. Since the ANOVA calculations are not normally done when using the analysis of means procedure, we will refer to the variance estimate as s^2.

We can compare the factor level means with the grand mean using the following steps:

1. Compute the factor level means, $\bar{y}_{i.}$, $i = 1, \ldots, t$.
2. Compute the grand mean, $\bar{y}_{..}$.
3. Compute s, the square root of s^2.
4. Obtain the value h_α from Appendix Table A.11 using $(n-1)t$ as degrees of freedom (df).
5. Compute the upper and lower decision lines, UDL and LDL, where

$$\text{UDL} = \bar{y}_{..} + h_\alpha s\sqrt{(t-1)/(tn)},$$

$$\text{LDL} = \bar{y}_{..} - h_\alpha s\sqrt{(t-1)/(tn)}.$$

6. Plot the means against the decision lines. If any mean falls outside the decision lines, we conclude there is a statistically significant difference among the means.

EXAMPLE 6.8

As an example of the analysis of means, we will again analyze the data from the experiment described in Example 6.2. As always, it is important to say that it is not good practice to do more than one analysis on a given set of data, and we do so only to illustrate the procedure. In this case, the results are the same; however, this is not always the case. Recall that the experiment was a completely randomized design conducted to compare the yield of four varieties of rice. The observations were yields in pounds per acre for each of four different plots of each of the four varieties. The data and summary statistics are given in Table 6.4. Even though the ANOM is a hypothesis test, we rarely state the hypotheses. Instead, we examine the relationship among the four means graphically using the following six steps:

Solution

1. The factor level means are

$$\text{variety 1: } \bar{y}_{1.} = 984.50,$$

$$\text{variety 2: } \bar{y}_{2.} = 928.25,$$

$$\text{variety 3: } \bar{y}_{3.} = 938.50,$$

$$\text{variety 4: } \bar{y}_{4.} = 1116.50.$$

2. The grand mean is

$$\bar{y}_{..} = 991.94.$$

3. The pooled estimate of the variance is

$s_1^2 = (10085.00)/3 = 3361.67,$

$s_2^2 = (3868.75)/3 = 1289.58,$

$s_3^2 = (13617.00)/3 = 4539.00,$

$s_4^2 = (22305.00)/3 = 7435.00,$

$s^2 = (3361.67 + 1289.58 + 4539.00 + 7435.00)/4 = 4156.31$ and $s = 64.47.$

   Again, note that this is the same value that we obtained for MSW in the analysis of variance procedure.

4. Using the standard level of significance of 0.05 and degrees of freedom = 4(3) = 12, we obtain the value $h_{0.05} = 2.85$ from Appendix Table A.11.

5. The upper and lower decision lines are

$$\text{UDL} = 991.94 + (64.47)(2.85)\sqrt{3/16} = 1071.50,$$

$$\text{LDL} = 991.94 - (64.47)(2.85)\sqrt{3/16} = 912.38.$$

6. The plot of the means against the decision lines is given in Fig. 6.7.

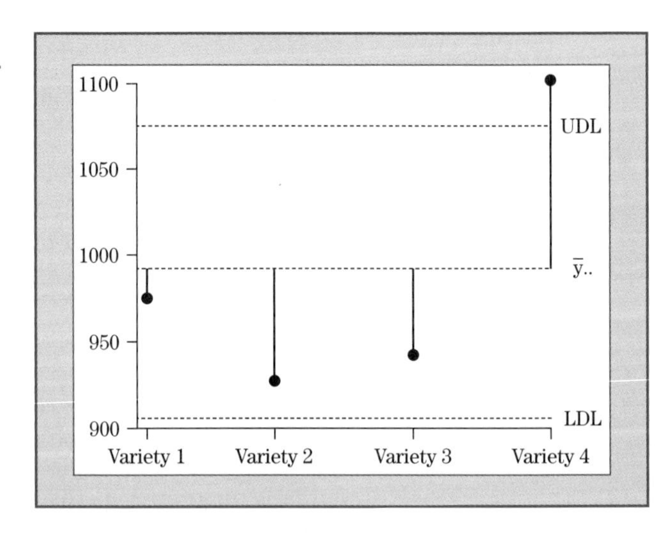

Figure 6.7

Plot of Means against Decision Lines

We observe from Fig. 6.7 that only variety 4 has a value outside the decision limits. Therefore, our conclusion is that the first three varieties do not significantly differ from the grand mean, but that the mean of variety 4 is significantly higher than the grand mean. This is consistent with the results given in Table 6.6 and Section 6.5. Note that we can also make some statements based on this graphic presentation that we could not make without additional analysis using the ANOVA procedure. For example, we might conclude that varieties 1, 2, and 3 all average about the same yield while the fourth variety has a sample average higher than all three. ■

ANOM for Proportions

Many problems arise when the variable of interest turns out to be an attribute, such as a light bulb that will or will not light or a battery whose life is or is not below standard. It would be beneficial to have a simple graphic method, like the ANOM, for comparing the proportion of items with a particular characteristic of this attribute. For example, we might want to compare the proportion of light bulbs that last more than 100 h from four different manufacturers to determine the best one to use in a factory. In Section 6.4 we discussed the problem of comparing several populations when the variable of interest is a proportion or percentage by suggesting a transformation of the data using the arcsin transformation. This approach could be used to do the ANOM procedure presented previously, simply substituting the transformed data for the response variable. There is a simpler method available if the sample size is such that the normal approximation to the binomial can be used.

In Section 2.5 we noted that the sampling distribution of a proportion was the binomial distribution. We also noted that if np and $n(1-p)$ are both greater than 5, then the normal distribution can be used to approximate the sampling distribution of a proportion. If this criterion is met, then we use the following seven-step procedure:

1. Obtain samples of equal size n for each of the t populations. Let the number of individuals having the attribute of interest in each of the t samples be denoted by $x_1, x_2, \ldots, x_t$.
2. Compute the factor level proportions, $p_i = x_i/n$, $i = 1, \ldots, t$.
3. Compute the overall proportion, $p_g = \sum p_i/t$.
4. Compute s, an estimate of the standard deviation of p_i:

$$s = \sqrt{p_g(1 - p_g)/n}.$$

5. Obtain the value h_α from Appendix Table A.11 using infinity as degrees of freedom (because we are using the normal approximation to the binomial, it is appropriate to use df = infinity).
6. Compute the upper and lower decision lines, UDL and LDL, where

$$\text{UDL} = p_g + h_\alpha s\sqrt{(t-1)/(t)},$$

$$\text{LDL} = p_g - h_\alpha s\sqrt{(t-1)/(t)}.$$

7. Plot the proportions against the decision lines. If any proportion falls outside the decision lines, we conclude there is a statistically significant difference in proportions among the t populations.

EXAMPLE 6.9

A problem concerning corrosion in metal containers during storage is discussed in Ott (1975, p. 106). The effect of copper concentration on the failure rate of metal containers after storage is analyzed using an experiment in which three levels of copper concentration, 5, 10, and 15 ppm (parts per million), are used in the construction of containers. Eighty containers ($n = 80$) of each concentration are observed over a period of storage, and the number of failures recorded. The data are given below:

Level of Copper, ppm	Number of Failures, X_i	Proportion of Failures, p_i
5	14	0.175
10	36	0.450
15	47	0.588

Solution We will use the ANOM procedure to determine whether differences in the proportions of failures exist due to the level of copper in the containers. The seven steps are as follows:

1. The three samples of size 80 each yielded

$$x_1 = 14,$$
$$x_2 = 36,$$
$$x_3 = 47.$$

2. The proportions are

$$p_1 = 0.175,$$
$$p_2 = 0.450,$$
$$p_3 = 0.588.$$

3. The overall proportion is

$$p_g = (14 + 36 + 47)/247 = 0.404.$$

4. The estimate of the standard deviation is

$$s = \sqrt{(0.404)(0.596)/80} = 0.055.$$

5. From Appendix Table A.11 using the 0.05 level of significance and df = infinity we get

$$h_{0.05} = 2.34.$$

6. The decision lines are

$$\text{LDL} = 0.404 - (2.34)(0.055)\sqrt{(2)/(3)} = 0.404 - 0.105 = 0.299,$$
$$\text{UDL} = 0.404 + (2.34)(0.055)\sqrt{(2)/(3)} = 0.404 + 0.105 = 0.509.$$

7. The ANOM graph is given in Fig. 6.8.

Figure 6.8

ANOM Graph for Example 6.9

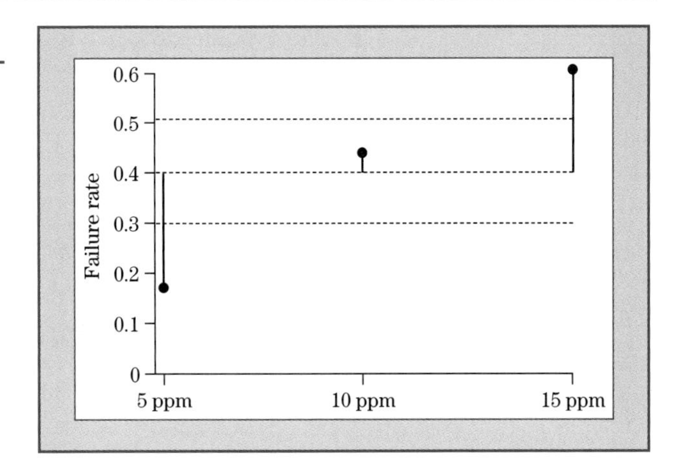

The results are very easy to interpret using the ANOM chart in Fig. 6.8. Even though it was obvious from the data that the more copper in the container, the larger the percent of failure, the ANOM procedure indicates that this difference is indeed statistically significant. Further, we can see from the graph that the increase in failure rate is monotonic with respect to the amount of copper.

That is, containers with 5 ppm copper have a significantly lower failure rate than those with 10 ppm copper, and those with 15 ppm have a significantly higher failure rate than the other two. ■

Analysis of Means for Count Data

Many problems arise in quality monitoring where the variable of interest is the number of nonconformities measured from a sample of items from a production line. If the sample size is such that the normal approximation to the Poisson distribution can be used, an ANOM method for comparing count data can be applied. This procedure is essentially the same as that given for proportions in the previous section, ANOM for proportions, and follows these six steps:

1. For each of the k populations of interest, an "inspection unit" is defined. This inspection unit may be a period of time, a fixed number of items, or a fixed unit of measurement. For example, an inspection unit of "1 h" might be designated as an inspection unit in a quality control monitoring of the number of defective items from a production line. Then a sample of k successive inspection units could be monitored to evaluate the quality of the product. Another example might be to define an inspection unit of "2 ft^2" of material from a weaving loom. Periodically a 2-ft^2 section of material is examined and the number of flaws recorded. The number of items with the attribute of interest (defects) from the ith inspection unit is denoted as c_i, $i = 1, \ldots, k$.

2. The overall average number of items with the attribute is calculated as

$$\bar{c} = \sum c_i/k.$$

3. The estimate of the standard deviation of counts is

$$s = \sqrt{\bar{c}}.$$

4. Obtain the value h_α from Appendix Table A.11 using df = infinity.
5. Compute the upper and lower decision lines, UDL and LDL, where

$$\text{UDL} = \bar{c} + h_\alpha s\sqrt{(k-1)/k}.$$

$$\text{LDL} = \bar{c} - h_\alpha s\sqrt{(k-1)/k},$$

6. Plot the counts, c_i, against the decision lines. If any count falls outside the decision lines we conclude there is a statistically significant difference among the counts.

EXAMPLE 6.10

Ott (1975, p. 107) presents a problem in which a textile mill is investigating an excessive number of breaks in spinning cotton yarn. The spinning is done using frames, each of which contains 176 spindles. A study of eight frames was made to determine whether there were any differences among the frames. When a break occurred, the broken ends were connected and the spinning resumed. The study was conducted over a time period of 2.5 h during the day. The number of breaks for each frame was recorded. The objective was to compare the eight frames relative to the number of breaks using the ANOM procedure.

Solution The results were as follows:

1. The inspection unit was the 150-min. study period. The number of breaks for each frame was recorded:

$$c_1 = 140,$$

$$c_2 = 99,$$

$$c_3 = 96,$$

$$c_4 = 151,$$

$$c_5 = 196,$$

$$c_6 = 124,$$

$$c_7 = 89,$$

$$c_8 = 188.$$

2. $\bar{c} = (140 + 99 + 96 + 151 + 196 + 124 + 89 + 188)/8 = 135.4.$
3. $s = \sqrt{135.4} = 11.64.$
4. From Appendix Table A.11 using $\alpha = 0.05$, $k = 8$, and df = infinity, we get $h_{0.05} = 2.72.$

5. The decision lines are

$$\text{LDL} = 135.4 - (2.72)(11.64)\sqrt{7/8} = 135.4 - 29.62 = 105.78,$$

$$\text{UDL} = 135.4 + (2.72)(11.64)\sqrt{7/8} = 135.4 + 29.62 = 165.02.$$

6. The ANOM chart is

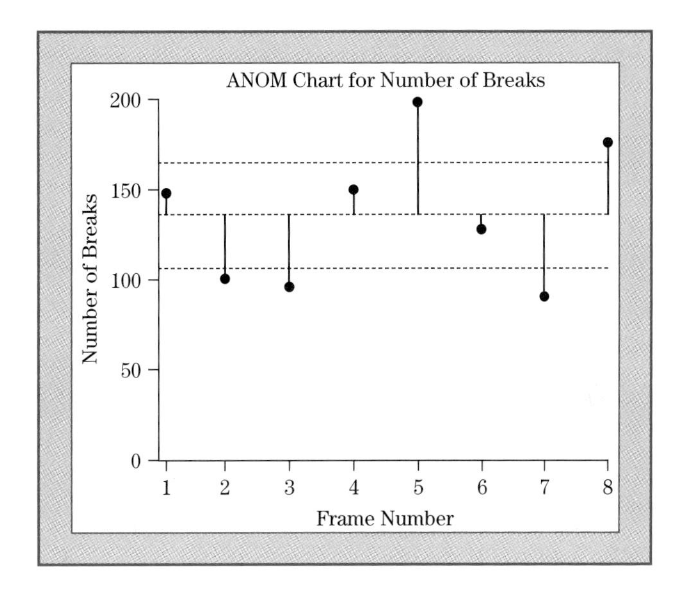

From this plot we can see that there are significant differences among the frames. Frames 2, 3, and 7 are particularly good, and frames 5 and 8 are particularly bad. ■

Most of the time, the ANOVA and the ANOM methods reach the same conclusion. In fact, for only two factor levels the two procedures are identical. However, there is a difference in the two procedures. The ANOM is more sensitive than ANOVA for detecting when *one* mean differs significantly from the others. The ANOVA is more sensitive when groups of means differ. Further, the ANOM can only be applied to fixed effects models, not to random effects models. The ANOM procedure can be extended to many types of experimental designs, including the factorial experiments of Chapter 9. A more detailed discussion of ANOM applied to experimental design problems can be found in Schilling (1973).

6.9 CHAPTER SUMMARY

The analysis of variance provides a methodology for making inferences for means from any number of populations. In this chapter we consider inferences based on data resulting from independently drawn samples from t populations.

This data structure is called a one-way classification or completely randomized design.

The analysis of variance is based on the comparison of the estimated variance among sample means (between mean square or MSB) to the estimated variance of observations within the samples (within mean square or MSW). If the variance among sample means is too large, differences may exist among the population means. The estimated variances or mean squares are derived from a partitioning of sums of squares into two parts corresponding to the variability among means and the variability within samples. The required variances are called mean squares and are obtained by dividing the appropriate sums of squares by their respective degress of freedom. The ratio of these variances is compared to the F distribution to determine whether the null hypothesis of equal means is to be rejected.

The linear model,

$$y_{ij} = \mu + \tau_i + \varepsilon_{ij},$$

is used to describe observations for a one-way classification. In this model the τ_i indicate the differences among the population means. It can be shown that the analysis of variance does indeed test the null hypothesis that all τ_i are zero against the alternative of any violation of equalities. In a fixed model, the τ_i represent a fixed number of populations or factor levels occuring in the sample data and inferences are made only for the parameters for those populations. In a random model, the τ_i represent a sample from a population of τ's and inferences are made on the variance of that population.

As for virtually all statistical analyses, some assumptions must be met in order for the analysis to have validity. The assumptions needed for the analysis of variance are essentially those that have been discussed in previous chapters. Suggestions for detecting violations and some remedial procedures are presented.

The analysis of variance tests only the general hypothesis of the equality of all means. Hence procedures for making more specific inferences are needed. Such inferences are obtained by multiple comparisons of which there are two major types:

- **preplanned comparisons**, which are proposed before the data are collected, and
- **post hoc comparisons**, in which the data are used to propose hypotheses.

Preplanned contrasts, and especially orthogonal contrasts, are preferred because of their greater power and protection against making type I errors. Because post hoc comparisons use data to generate hypotheses, their use tends to increase the so-called experiment-wise error rate, which is the probability of one or more comparisons detecting a difference when none exists. For this reason such methods must embody some means of adjusting stated significance levels. Since no single principle of adjustment has been deemed superior a number of different methods are available, each making some compromise between power and protection against making type I errors. The important message here is that careful considerations must be taken to assure that the most appropriate method is employed and that preplanned comparisons are used whenever possible.

The chapter concludes with short sections covering the random model, unequal sample sizes, analysis of means, and some computing considerations.

6.10 CHAPTER EXERCISES

CONCEPT QUESTIONS

For the following true/false statements regarding concepts and uses of the analysis of variance, indicate whether the statement is true or false and specify what will correct a false statement.

1. ____________ If for two samples the conclusions from an ANOVA and t test disagree, you should trust the t test.

2. ____________ A set of sample means is more likely to result in rejection of the hypothesis of equal population means if the variability within the populations is smaller.

3. ____________ If the treatments in a CRD consist of numeric levels of input to a process, the LSD multiple comparison procedure is the most appropriate test.

4. ____________ If every observation is multiplied by 2, then the value of the F statistic in an ANOVA is multiplied by 4.

5. ____________ To use the F statistic to test for the equality of two variances, the sample sizes must be equal.

6. ____________ The logarithmic transformation is used when the variance is proportional to the mean.

7. ____________ With the usual ANOVA assumptions, the ratio of two mean squares whose expected values are the same has an F distribution.

8. ____________ One purpose of randomization is to remove experimental error from the estimates.

9. ____________ To apply the F test in ANOVA, the sample size for each factor level (population) must be the same.

10. ____________ To apply the F test for ANOVA, the sample standard deviations for all factor levels must be the same.

11. ____________ To apply the F test for ANOVA, the population standard deviations for all factor levels must be the same.

12. ____________ An ANOVA table for a one-way experiment gives the following:

Source	df	SS
Between factors	2	810
Within (error)	8	720

Answer true or false for the following six statements:

______________ The null hypothesis is that all four means are equal.

______________ The calculated value of F is 1.125.

______________ The critical value for F for 5% significance is 6.60.

______________ The null hypothesis can be rejected at 5% significance.

______________ The null hypothesis cannot be rejected at 1% significance.

______________ There are 10 observations in the experiment.

13. ______________ A "statistically significant F" in an ANOVA indicates that you have identified which levels of factors are different from the others.

14. ______________ Two orthogonal comparisons are independent.

15. ______________ A sum of squares is a measure of dispersion.

EXERCISES

1. A study of the effect of different types of anesthesia on the length of post-operative hospital stay yielded the following for cesarean patients:

Group A was given an epidural MS.
Group B was given an epidural.
Group C was given a spinal.
Group D was given general anesthesia.

The data are presented in Table 6.27. In general, the general anesthetic is considered to be the most dangerous, the spinal somewhat less so, and the epidural even less, with the MS addition providing additional safety. Note that the data are in the form of distributions for each group.

(a) Test for the existence of an effect due to anesthesia type.
(b) Does it appear that the assumptions for the analysis of variance are fulfilled? Explain.
(c) Compute the residuals to check the assumptions (Section 6.4). Do these results support your answer in part (b)?

Table 6.27

Data for Exercise 1

	Length of Stay	Number of Patients
Group A	3	6
	4	14
Group B	4	18
	5	2
Group C	4	10
	5	9
	6	1
Group D	4	8
	5	12

(d) What specific recommendations can be made on the basis of these data?

Table 6.28

Data for Exercise 2

Color	Time				
Red	9	11	10	9	15
Green	20	21	23	17	30
Black	6	5	8	14	7

2. Three sets of five mice were randomly selected to be placed in a standard maze but with different color doors. The response is the time required to complete the maze as seen in Table 6.28.
   (a) Perform the appropriate analysis to test whether there is an effect due to door color.
   (b) Assuming that there is no additional information on the purpose of the experiment, should specific hypotheses be tested by a multiple-range test (Duncan's) or orthogonal contrasts? Perform the indicated analysis.
   (c) Suppose now that someone told you that the purpose of the experiment was to see whether the color green had some special effect. Does this revelation affect your answer in part (b)? If so, redo the analysis.

3. A manufacturer of air conditioning ducts is concerned about the variability of the tensile strength of the sheet metal among the many suppliers of this material. Four samples of sheet metal from four randomly chosen suppliers are tested for tensile strength. The data are given in Table 6.29.
   (a) Perform the appropriate analysis to ascertain whether there is excessive variation among suppliers.
   (b) Estimate the appropriate variance components.

Table 6.29

Data for Exercise 3

SUPPLIER			
1	2	3	4
19	80	47	90
21	71	26	49
19	63	25	83
29	56	35	78

4. A manufacturer of concrete bridge supports is interested in determining the effect of varying the sand content of concrete on the strength of the supports. Five supports are made for each of five different amounts of sand in the concrete mix and each support tested for compression resistance. The results are as shown in Table 6.30.
   (a) Perform the analysis to determine whether there is an effect due to changing the sand content.
   (b) Use orthogonal polynomial contrasts to determine the nature of the relationship of sand content and strength. Draw a graph of the response versus sand amount.

Table 6.30

Data for Exercise 4

Percent Sand	Compression Resistance (10,000 psi)				
15	7	7	10	15	9
20	17	12	11	18	19
25	14	18	18	19	19
30	20	24	22	19	23
35	7	10	11	15	11

Table 6.31

Data for Exercise 5

TREATMENT				
1	2	3	4	5
11.6	8.5	14.5	12.3	13.9
10.0	9.7	14.5	12.9	16.1
10.5	6.7	13.3	11.4	14.3
10.6	7.5	14.8	12.4	13.7
10.7	6.7	14.4	11.6	14.9

5. The set of artificial data shown in Table 6.31 is used in several contexts to provide practice in implementing appropriate analyses for different situations. The use of the same numeric values for the different problems will save computational effort.

(a) Assume that the data represent test scores of samples of students in each of five classes taught by five different instructors. We want to reward instructors whose students have higher test scores. Do the sample results provide evidence to reward one or more of these instructors?

(b) Assume that the data represent gas mileage of automobiles resulting from using different gasoline additives. The treatments are:

1. additive type A, made by manufacturer I
2. no additive
3. additive type B, made by manufacturer I
4. additive type A, made by manufacturer II
5. additive type B, made by manufacturer II

Construct three orthogonal contrasts to test meaningful hypotheses about the effects of the additives.

(c) Assume the data represent battery life resulting from different amounts of a critical element used in the manufacturing process.
The treatments are:

1. one unit of the element
2. no units of the element
3. four units of the element
4. two units of the element
5. three units of the element

Analyze for trend using only linear and quadratic terms. Perform a lack of fit test.

6. Do Exercise 3 in Chapter 5 as an analysis of variance problem. You should verify that $t^2 = F$ for the two-sample case.

7. In an experiment to determine the effectiveness of sleep-inducing drugs, 18 insomniacs were randomly assigned to three treatments:

1. placebo (no drug)
2. standard drug
3. new experimental drug

The response as shown in Table 6.32 is average hours of sleep per night for a week. Perform the appropriate analysis and make any specific recommendations for use of these drugs.

Table 6.32

Data for Exercise 7

TREATMENT		
1	2	3
5.6	8.4	10.6
5.7	8.2	6.6
5.1	8.8	8.0
3.8	7.1	8.0
4.6	7.2	6.8
5.1	8.0	6.6

8. The data shown in Table 6.33 are times in months before the paint started to peel for four brands of paint applied to a set of test panels. If all paints cost the same, can you make recommendations on which paint to use? This problem is an example of a relatively rare situation where only the means and variances are provided. For computing the between group sum

Table 6.33

Data for Exercise 8

Paint	Number of Panels	$\bar{y}$	s^2
A	6	48.6	82.7
B	6	51.2	77.9
C	6	60.1	91.0
D	6	55.2	105.2

of squares, simply compute the appropriate totals. For the within sum of squares, remember that $\text{SS}_i = (n_i - 1)s_i^2$, and $\text{SSW} = \sum \text{SS}_i$.

Table 6.34

Data for Exercise 9

INSECTICIDE			
A	B	C	D
85	90	93	98
82	92	94	98
83	90	96	100
88	91	95	97
89	93	96	97
92	81	94	99

9. The data shown in Table 6.34 relate to the effectiveness of several insecticides. One-hundred insects of a particular species were put into a chamber and exposed to an insecticide for 15 s. The procedure was applied in random order six times for each of four insecticides. The response is the number of dead insects. Based on these data, can you make a recommendation? Check assumptions!

10. The data in Table 6.35 are wheat yields for experimental plots having received the indicated amounts of nitrogen. Determine whether a linear or quadratic trend may be used to describe the relationship of yield to amount of nitrogen.

Table 6.35

Data for Exercise 10

NITROGEN					
40	80	120	160	200	240
42	45	46	49	50	46
41	45	48	45	44	45
40	44	46	43	45	45

11. Serious environmental problems arise from absorption into soil of metals that escape into the air from different industrial operations. To ascertain if absorption rates differ among soil types, six soil samples were randomly selected from fields having five different soil types (A, B, C, D, and E) in an area known to have relatively uniform exposure to the metals studied. The 30 soil samples were analyzed for cadmium (Cd) and lead (Pb) content. The results are given in Table 6.36. Perform separate analyses to determine

Table 6.36

Data for Exercise 11

SOIL									
A		B		C		D		E	
Cd	Pb	Cd	Pb	Cd	Pb	Cd	Pb	Cd	Pb
0.54	15	0.56	13	0.39	13	0.26	15	0.32	12
0.63	19	0.56	11	0.28	13	0.13	15	0.33	14
0.73	18	0.52	12	0.29	12	0.19	16	0.34	13
0.58	16	0.41	14	0.32	13	0.28	20	0.34	15
0.66	19	0.50	12	0.30	13	0.10	15	0.36	14
0.70	17	0.60	14	0.27	14	0.20	18	0.32	14

whether there are differences in cadmium and lead content among the soils. Assume that the cadmium and lead content of a soil directly affects the cadmium and lead content of a food crop. Do the results of this study lead to any recommendations?

Check the assumptions for both variables. Does this analysis affect the results in the preceding? If any of the assumptions are violated, suggest an alternative analysis.

Table 6.37

Data for Exercise 12

Medium	Fungus Colony Diameters			
WA	4.5	4.1	4.4	4.0
RDA	7.1	6.8	7.2	6.9
PDA	7.8	7.9	7.6	7.6
CMA	6.5	6.2	6.0	6.4
TWA	5.1	5.0	5.4	5.2
PCA	6.1	6.2	6.2	6.0
NA	7.0	6.8	6.6	6.8

12. For laboratory studies of an organism, it is important to provide a medium in which the organism flourishes. The data for this exercise shown in Table 6.37 are from a completely randomized design with four samples for each of seven media. The response is the diameters of the colonies of fungus.

(a) Perform an analysis of variance to determine whether there are different growth rates among the media.

(b) Is this exercise appropriate for preplanned or post hoc comparisons? Perform the appropriate method and make recommendations.

Table 6.38

Number of Pushups in 60 s by Time with Department

TIME WITH DEPARTMENT (YEARS)			
5	10	15	20
56	64	45	42
55	61	46	39
62	50	45	45
59	57	39	43
60	55	43	41

13. A study of firefighters in a large urban area centered on the physical fitness of the engineers employed by the fire department. To measure the fitness, a physical therapist sampled five engineers each with 5, 10, 15, and 20 years' experience with the department. She then recorded the number of pushups that each person could do in 60 s. The results are listed in Table 6.38. Perform an analysis of variance to determine whether there are differences in the physical fitness of engineers by time with department. Use $\alpha = 0.05$.

14. Using the results of Exercise 13, determine what degree of polynomial curve is required to relate fitness to time with the department. Illustrate the results with a graph.

15. A local bank has three branch offices. The bank has a liberal sick leave policy, and a vice-president was concerned about employees taking advantage of this policy. She thought that the tendency to take advantage depended on the branch at which the employee worked. To see whether there were differences in the time employees took for sick leave, she asked each branch manager to sample employees randomly and record the number of days of sick leave taken during 1990. Ten employees were chosen, and the data are listed in Table 6.39.

Table 6.39

Sick Leave by Branch

Branch 1	Branch 2	Branch 3
15	11	18
20	15	19
19	11	23
14		

(a) Do the data indicate a difference in branches? Use a level of significance of 0.05.

(b) Use Duncan's multiple-range test to determine which branches differ. Explain your results with a summary plot.

16. In Exercise 4 an experiment was conducted to determine the effect of the percent of sand in concrete bridge supports on the strength of these supports. A set of orthogonal polynomial contrasts was used to determine the nature of this relationship. The ANOVA results indicated a cubic polynomial would best describe this relationship. Use the data given and do an analysis of means (Section 6.8). Do the results support the conclusion from the ANOVA? Explain.

17. In Exercise 8 a test of durability of various brands of paint was conducted. The results are given in Table 6.33, which lists the summary statistics only. Perform an analysis of means (Section 6.8) on these data. Do the results agree with those of Exercise 8? Explain.

18. A manufacturing company uses five identical assembly lines to construct one model of an electric toaster. All the toasters produced go to the same retail outlet. A recent complaint from this outlet indicates that there has been an increase in defective toasters in the past month. To determine the location of the problem, complete inspection of the output from each of the five assembly lines was done for a 22-day period.

The number of defective toasters was recorded. The data are given below:

Assembly Line	Number of Defective Toasters
1	123
2	140
3	165
4	224
5	98

Use the ANOM procedure discussed at the end of Section 6.8 to determine whether the assembly lines differ relative to the number of defective toasters produced. Suggest ways in which the manufacturer could prevent complaints in the future.

Linear Regression

EXAMPLE 7.1

Are Suicide Rates Affected by Publicity? Many researchers have proposed that some private plane accidents have a suicidal component. If this conjecture is true, then the number of private airplane crashes should increase significantly after a highly publicized murder–suicide by airplane. The data in Table 7.1 (Phillips, 1978) give the number of multiple-fatality airplane accidents (Crashes) that occurred during the week following a highly publicized murder–suicide by airplane as well as values of a publicity index (Index) measuring the amount, duration, and intensity of the publicity given the murder–suicide. The objective of the study is to determine the nature of the relationship between Crashes and Index.

A scatterplot of these data (see Section 1.7) as shown in Fig. 7.1 appears to indicate an association between newspaper publicity and the number of crashes. The questions to be addressed by a regression analysis are as follows:

- Is this relationship "real"?
- Can we describe this relationship with a model?
- Can we use these data to predict the rate of future crashes?

The regression analysis that provides answers to these questions is presented in Section 7.9. ■

7.1 Introduction

Example 7.1 illustrates a relationship between two quantitative variables. As we saw in Chapter 6, the analysis of variance model allowed us how to make inferences on a population of a quantitative variable identified by levels of a factor, but it does not provide a mechanism for making inferences for a

Table 7.1

Plane Crashes

Adapted from Phillips, D. P. (1979), "Airplane accident fatalities increase just after newspaper stories about murder and suicide." Science ***201***, *748–750*

Index	Crashes	Index	Crashes	Index	Crashes
0	4	44	7	103	6
0	3	63	2	104	4
0	2	82	4	322	8
5	3	85	6	347	5
5	2	96	8	376	8
40	4	98	4		

Figure 7.1

Airplane Crashes

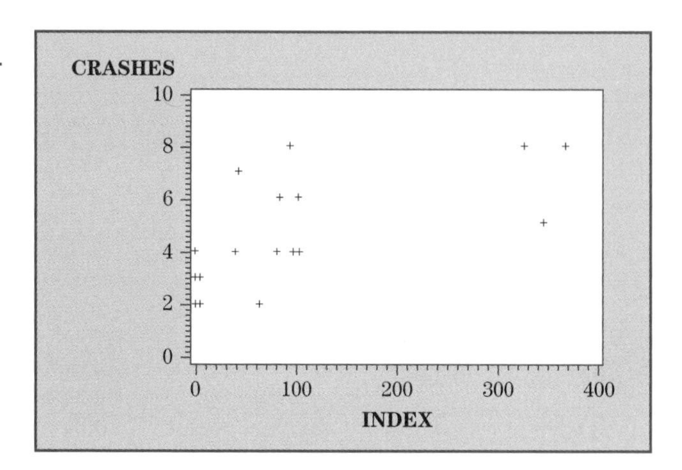

problem like Example 7.1. This chapter introduces the use of the *regression* model, which is used to make inferences on means of populations identified by specified *values* of one of more quantitative factor variables. For example, in an analysis of variance model we may make inferences on the difference in the number of insects killed by different insecticides while in a regression model we want to know what happens to the death rate of insects as we increase the application rate of a specific insecticide.

DEFINITION 7.1
Regression analysis is a statistical method for analyzing a relationship between two or more variables in such a manner that one variable can be predicted or explained by using information on the others.

The term "regression" was first introduced by Sir Francis Galton in the late 1800s to explain the relation between heights of parents and children. He noted that the heights of children of both tall and short parents appeared to "regress" toward the mean of the group. The procedure for actually conducting the regression analysis, called ordinary least squares (see Section 7.3), is generally credited to Carl Friedrich Gauss, who used the procedure in the early part of the nineteenth century. However, there is some controversy concerning this discovery as Adrien Marie Legendre published the first work on its use in 1805. Regression analysis and the method of least squares are generally considered synonymous terms. Note that the definition of regression does not explicitly define the nature of the relation. As we shall see, the relation may take on many different forms and still be analyzed by regression methods.

In the previous chapters, our objective was to sample from one or more populations and to compare certain parameters either with each other or with a specified value. In a regression analysis, the objectives are slightly different. The purpose of a regression analysis is to observe sample measurements taken on different variables, called **factors** or **independent** variables, and to examine the relationship between these variables and a **response** or **dependent** variable. This relationship is then expressed as a statistical model called the regression model. This and several subsequent chapters deal with the regression model.

A regression analysis starts with an estimate of the population mean(s) using a mathematical formula, called a function, which explains the relationship between the factor variable(s) and the response variable. This function is called the **regression model** or **regression function**. This function can be described geometrically by a line if there is only one factor variable or a multidimensional plane if there are several. As in all statistical models, the regression model describes a **statistical relationship**, which we will see, is not a perfect one. That is, if we plot the data (as in Fig. 7.1) and superimpose the line representing the function estimated by a regression analysis, the observed values will certainly not all fall directly on the line described by the regression model.

Some examples of analyses using regression include

- estimating weight gain by the addition to children's diets of different amounts of a dietary supplement,
- predicting scholastic success (grade point ratio) based on students' scores on an aptitude or entrance test,
- estimating changes in sales associated with increased expenditures on advertising,
- estimating fuel consumption for home heating based on daily temperatures, or
- estimating changes in interest rates associated with the amount of deficit spending.

In simple linear regression, which is the topic of this chapter, the relationship is specified to have only one factor variable and the relationship is described by a straight line. This is, as the name implies, the simplest of all regression models. While most relationships between variables are not exactly linear, a straight line often approximates the relationship, especially in a limited or restricted range of values of the variables. For example, the relationship of age and height of children is obviously not linear through the first 15 years of age, but it may be reasonably close to linear from ages 10 to 12.

Symbolically we represent values of the variables involved in regression as follows:

x represents observed values of the factor variable, such as pounds of fertilizer, aptitude test score, or daily temperature. In the context of a regression analysis this variable is called the **independent variable**.

y represents observed values of the response variable, such as yield of corn, grade point averages, or fuel consumption. This variable is called the **dependent variable**.

In a simple linear regression analysis we use a sample of observations on pairs of variables, x and y, to make inferences on the "model." Actually the inferences are made on the parameters that describe the model. These are discussed in Section 7.2 and the remainder of the chapter is devoted to various inferences and further investigations on the appropriateness of the model. Extensions to the use of more factor (independent) variables as well as curvilinear (nonlinear) relationships are presented in Chapter 8.

This chapter starts with the definition and uses of the linear regression model, followed by procedures for estimation of the parameters of that model and the subsequent inferences about those parameters. Also discussed are inferences for the response variable, an introduction to diagnosing possible difficulties in implementing the model, and some hints on computer usage. The related concept of correlation is presented in Section 7.7.

Notes on Exercises

Section 7.3 contains the information and formulas necessary to obtain the regression parameter estimates manually for Exercises 1–4 using a hand-held calculator. Section 7.5 contains the information and formulas necessary to do statistical inferences for these parameters. Using the Computer in Section 7.6 contains the information needed to perform the requested analyses on all other assigned exercises. Section 7.8 provides the tools necessary to review all exercises for possible violations of assumptions.

7.2 The Regression Model

The regression model is similar to the analysis of variance model discussed in Chapter 6 in that it consists of two parts, a **deterministic** or **functional** term and a **random** term. The **simple linear regression model** is of the form

$$y = \beta_0 + \beta_1 x + \epsilon,$$

where x and y represent values[1] of the independent and dependent variables, respectively. This model is often referred to as the **regression of y on x**. The first portion of the model, $\beta_0 + \beta_1 x$, is an equation of the regression line involving the values of the two variables (x and y) and two parameters β_0 and β_1. These two parameters are called the **regression coefficients**. Specifically:

β_1 is the **slope** of the regression line, that is, the change in y corresponding to a unit change in x.

[1]Many textbooks and other references add a subscript i to the symbols representing the variables to indicate that the model applies to individual sample or population observations: $i = 1, 2, \ldots, n$. Since this subscript is always applicable it is not explicitly used here.

β_0, the **intercept**, is the value of the line when $x = 0$. This parameter has no practical meaning if the condition $x = 0$ cannot occur, but is needed to specify the model.

As in the analysis of variance model, the individual values of ε are assumed to come from a population of random variables[2] having the normal distribution with mean zero and variance σ^2.

The interpretation of the model is aided by redefining it as a version of the linear model used for the analysis of variance. Remember that the analysis of variance model can be written

$$y_{ij} = \mu_i + \varepsilon_{ij},$$

where the μ_i refer to the means of the different populations and ε_{ij} are the random errors associated with the individual observations. Equivalently, the regression model can be written

$$y = \mu_{y|x} + \varepsilon,$$

where the symbol $\mu_{y|x}$ represents a mean of y corresponding to a specific value of x. This parameter is known as the **conditional mean** of y and is defined by the relationship

$$\mu_{y|x} = \beta_0 + \beta_1 x.$$

We can now see that this deterministic portion of the model describes a line that is the locus of values of the conditional mean $\mu_{y|x}$ corresponding to all values of x. This is a straight line with an intercept (value of y when $x = 0$) of β_0 and slope of β_1. Combining the two model statements produces the complete regression model:

$$y = \beta_0 + \beta_1 x + \varepsilon.$$

The random error has a mean of zero and variance of σ^2; hence the observed values of the response variable come from a normally distributed population with a mean of $\mu_{y|x}$ and variance of σ^2. This formulation of the regression model is illustrated in Fig. 7.2 with a regression line of $y = x$ ($\beta_0 = 0$ and $\beta_1 = 1$) and showing a normal distribution with unit variance at $x = 2.5$, 5, and 7.5.

In terms of the regression model we can see that the purpose of a regression analysis is to use a set of observed values of x and y to estimate the parameters β_0, β_1, and σ^2, and further to perform hypothesis tests and/or construct confidence intervals on these parameters and also to make inferences on the values of the response variable.

As in previous chapters, the validity of the results of the statistical analysis requires fulfillment of certain assumptions about the data. Those assumptions dealing with the random error are basically the same as they are for the analysis of variance (Section 6.3), with a few additional wrinkles. Specifically we assume the following:

1. The linear model is appropriate.
2. The error terms are independent.

[2]It is the randomness of ε that substitutes for the random sample assumption and allows the use of statistical inferences even when the data are not strictly the result of a random sample.

Figure 7.2

Schematic Representation of Regression Model

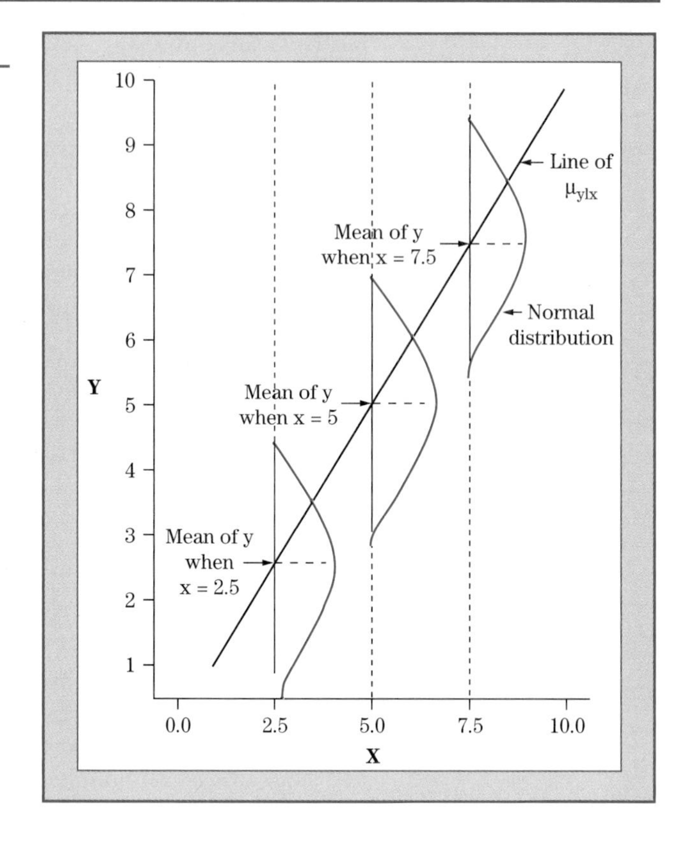

3. The error terms are (approximately) normally distributed.
4. The error terms have a common variance, σ^2.

Aids to the detection of violations of these and other assumptions and some possible remedies are given in Section 7.8. Even if all assumptions are fulfilled, regression analysis has some limitations:

- The fact that a regression relationship has been found to exist does **not**, by itself, imply that x **causes** y. For example, many regression analyses have shown that there is a clear relationship between smoking and lung cancer, but because there are multiple factors affecting the incidence of lung cancer, the results of these regression analyses cannot be used as the sole evidence to prove that smoking causes lung cancer. Basically, to prove cause and effect, it must also be demonstrated that no other factor could cause that result.
- It is not advisable to use an estimated regression relationship for extrapolation. That is, the estimated model should not be used to make inferences on values of the dependent variable beyond the range of observed x values. Such extrapolation is dangerous, because although the model may fit the data quite well, there is no evidence that the model is appropriate outside the range of the existing data.

Table 7.2

Data on Size and Price

obs	size	price	obs	size	price	obs	size	price
1	0.951	30.00	21	1.532	93.500	41	2.336	129.90
2	1.036	39.90	22	1.647	94.900	42	1.980	132.90
3	0.676	46.50	23	1.344	95.800	43	2.483	134.90
4	1.456	48.60	24	1.550	98.500	44	2.809	135.90
5	1.186	51.50	25	1.752	99.500	45	2.036	139.50
6	1.456	56.99	26	1.450	99.900	46	2.298	139.99
7	1.368	59.90	27	1.312	102.000	47	2.038	144.90
8	0.994	62.50	28	1.636	106.000	48	2.370	147.60
9	1.176	65.50	29	1.500	108.900	49	2.921	149.99
10	1.216	69.00	30	1.800	109.900	50	2.262	152.55
11	1.410	76.90	31	1.972	110.000	51	2.456	156.90
12	1.344	79.00	32	1.387	112.290	52	2.436	164.00
13	1.064	79.90	33	2.082	114.900	53	1.920	167.50
14	1.770	79.95	34	.	119.500	54	2.949	169.90
15	1.524	82.90	35	2.463	119.900	55	3.310	175.00
16	1.750	84.90	36	2.572	119.900	56	2.805	179.00
17	1.152	85.00	37	2.113	122.900	57	2.553	179.90
18	1.770	87.90	38	2.016	123.938	58	2.510	189.50
19	1.624	89.90	39	1.852	124.900	59	3.627	199.00
20	1.540	89.90	40	2.670	126.900			

EXAMPLE 7.2

(EXAMPLE 1.2 REVISITED) In previous chapters we have shown some statistical tools the Modes used to investigate the housing market in anticipation of moving to a new city. For example, they used the median test to show that homes in that city appear to cost less than they do in their present location. However, they also know that other factors may have caused that apparent difference. In fact, the well-known association between home size and cost has made the price per square foot a widely used measure of housing costs. An estimate of this cost can be obtained by a regression analysis using `size` as the independent and `price` as the dependent variable.

The scatterplot[3] of home costs and sizes taken from Table 1.2 was shown in Fig. 1.15. This plot shows a reasonably close association between cost and size, except for the higher priced homes. The Modes already know that extreme observations are often a hindrance for good statistical analyses, and besides, those homes were out of their price range. So they decided to perform the regression using only data for homes priced at less than $200,000. We will have more to say about extreme observations later. The data of sizes and prices for the homes, arranged in order of price, are shown in Table 7.2 and the corresponding scatterplot is shown in Fig. 7.3.

Note that one observation does not provide data on size; that observation cannot be used for the regression. The strong association between price and size is evident.

[3]The concept of a scatterplot is presented in Section 1.7.

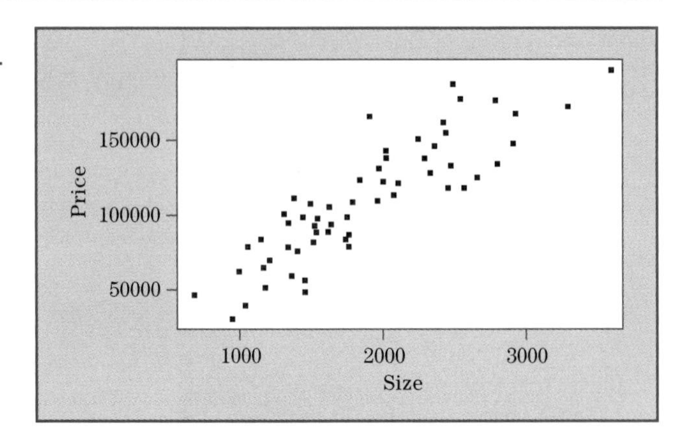

Figure 7.3

Plot of Price and Size

For this example, the model can be written

$$\texttt{price} = \beta_0 + \beta_1 \, \texttt{size} + \varepsilon,$$

or in terms of the generic variable notation

$$y = \beta_0 + \beta_1 x + \varepsilon.$$

In this model β_1 indicates the increase in price associated with a square foot increase in the size of a house.

In the next sections, we will perform the regression analysis in two steps:

1. Estimate the parameters of the model.
2. Perform statistical inferences on these parameters. ■

7.3 Estimation of Parameters β_0 and β_1

The purpose of the estimation step is to find estimates of β_0 and β_1 that produce a set of $\mu_{y|x}$ values that in some sense "best" fit the data. One way to do this would be to lay a ruler on the scatterplot and draw a line that visually appears to provide the best fit. This is certainly not a very objective or scientific method since different individuals would likely define different best fitting lines. Instead we will use a more rigorous method.

Denote the estimated regression line by

$$\hat{\mu}_{y|x} = \hat{\beta}_0 + \hat{\beta}_1 x,$$

where the caret or "hat" over a parameter symbol indicates that it is an estimate. Note that $\hat{\mu}_{y|x}$ is an estimate of the mean[4] of y for any given x. How well the estimate fits the actual observed values of y can be measured by the magnitudes of the differences between the observed y and the corresponding

[4]Many books use $\hat{y}$ for the estimated conditional mean. We will use $\hat{\mu}_{y|x}$ to remind the reader that we are estimating a mean. The symbol $\hat{y}$ will have a special meaning later.

$\hat{\mu}_{y|x}$ values, that is, the individual values of $(y - \hat{\mu}_{y|x})$. These differences are called **residuals**. Since smaller residuals indicate a good fit, the estimated line of best fit should be the line that produces a set of residuals having the smallest magnitudes. There is, however, no universal definition of "smallest" for a collection of values; hence some arbitrary but hopefully useful criterion for this property must first be defined. Some criteria that have been employed are as follows:

1. Minimize the largest absolute residual.
2. Minimize the sum of absolute values of the residuals.

Although both of these (and other) criteria have merit and are occasionally used, we will use the most popular criterion:

3. Minimize the sum of **squared residuals**.

This criterion is called **least squares** and results in an estimated line that minimizes the variance of the residuals. Since we use the variance as our primary measure of dispersion, this estimation procedure minimizes the dispersion of residuals. Estimation using the least squares criterion also has many other desirable characteristics and is easier to implement than other criteria.

The least squares criterion thus requires that we choose estimates of β_0 and β_1 that minimize

$$\sum (y - \hat{\mu}_{y|x})^2 = \sum (y - \hat{\beta}_0 - \hat{\beta}_1 x)^2.$$

It can be shown mathematically, using some elements of calculus, that these estimates are obtained by finding values of β_0 and β_1 that simultaneously satisfy a set of equations, called the **normal equations**:

$$\hat{\beta}_0 n + \hat{\beta}_1 \sum x = \sum y,$$
$$\hat{\beta}_0 \sum x + \hat{\beta}_1 \sum x^2 = \sum xy.$$

By means of a little algebra, the solution to this system of equations produces the least squares estimators[5]:

$$\hat{\beta}_0 = \bar{y} - \hat{\beta}_1 \bar{x},$$
$$\hat{\beta}_1 = \frac{\sum xy - \left(\sum x \sum y / n\right)}{\sum x^2 - \left[\left(\sum x\right)^2 / n\right]}.$$

The estimator of β_1 can also be formulated

$$\hat{\beta}_1 = \frac{\sum (x - \bar{x})(y - \bar{y})}{\sum (x - \bar{x})^2}.$$

This latter formula more clearly shows the structure of the estimate: the sum of products of the deviations of observed values from the means of x and y divided by the sum of squared deviations of the x values. Commonly we call

[5]An estimator is an algebraic expression that provides the actual numeric estimate for a specific set of data.

$\sum(x - \bar{x})^2$ and $\sum(x - \bar{x})(y - \bar{y})$ the corrected or means centered sums of squares and cross products. Since these quantities occur frequently, we will use the notation and computational formulas

$$S_{xx} = \sum(x - \bar{x})^2 = \sum x^2 - \left(\sum x\right)^2 / n,$$

the corrected sum of squares for the independent variable x;

$$S_{xy} = \sum(x - \bar{x})(y - \bar{y}) = \sum xy - \sum x \sum y/n,$$

the corrected sum of products of x and y; and later

$$S_{yy} = \sum(y - \bar{y})^2 = \sum y^2 - \left(\sum y\right)^2 / n,$$

the corrected sum of squares of the dependent variable y. Using this notation, we can write

$$\hat{\beta}_1 = S_{xy}/S_{xx}.$$

The computations are illustrated using the data on homes in Table 7.2. We first perform the preliminary calculations to obtain sums and sums of squares and cross products for both variables:

$$n = 58, \quad \sum x = 109.212, \quad \text{and} \quad \bar{x} = 1.883,$$

$$\sum x^2 = 228.385, \text{ hence}$$

$$S_{xx} = 228.385 - (109.212)^2/58 = 22.743;$$

$$\sum y = 6439.998, \quad \text{and} \quad \bar{y} = 111.034,$$

$$\sum xy = 13,401.788, \text{ hence}$$

$$S_{xy} = 13,401.788 - (109.212)(6439.998)/58 = 1275.494;$$

$$\sum y^2 = 808,293.767, \text{ hence}$$

$$S_{yy} = 808,293.767 - (6439.998)^2/58 = 93,232.142.$$

We can now compute the parameter estimates

$$\hat{\beta}_1 = 1275.494/22.743 = 56.083,$$

$$\hat{\beta}_0 = 111.034 - (56.084)(1.883) = 5.432,$$

and the equation for estimating price is

$$\hat{\mu}_{y|x} = 5.432 + 56.083x.$$

The estimated slope, $\hat{\beta}_1$, is a measure of the change in mean price ($\hat{\mu}_{y|x}$) for a unit change in size. In other words, the estimated price per square foot is \$56.08 (remember both price and space are in units of 1000).

The intercept, $\hat{\beta}_0 = \$5341.57$, is the estimated price of a zero square foot home, which may be interpreted as the estimated price of a lot. However, this value is an extrapolation beyond the reach of the data (there are no lots without houses in this data set) and is of questionable value. We will have more to say about extrapolation later.

A Note on Least Squares

In Chapter 3 we found that for a single sample, the sample mean, $\bar{y}$, was the best estimate of the population mean, μ. Actually we can show that the sample mean is a least squares estimator of the population mean. Consider the regression model without the intercept parameter:

$$y = \beta_1 x + \varepsilon.$$

We will use this model on a set of data for which all values of the independent variable, x, are unity. Now the model is

$$y = \beta_1 + \varepsilon,$$

which is the model for a single population with mean $\mu = \beta_1$. For a model with no intercept the formula for the least squares estimate of β_1 is

$$\hat{\beta}_1 = \frac{\sum xy}{\sum x^2} = \frac{\sum y}{n},$$

which result in the estimate $\hat{\beta}_1 = \bar{y}$. We will extend this principle to show the equivalence of regression and analysis of variance models in Chapter 11.

7.4 Estimation of σ^2 and the Partitioning of Sums of Squares

As we have seen in previous chapters, test statistics for performing inferences require an estimate of the variance of the random error. We have emphasized that any estimated variance is computed as a sum of squared deviations from the estimated population mean(s) divided by the appropriate degrees of freedom. This variance is estimated by a mean square, which is computed as a sum of squared deviations from the estimated population mean(s) divided by degrees of freedom. For example, in one-population inferences (Chapter 4), the sum of squares is $\sum(y - \bar{y})^2$ and the degrees of freedom are $(n - 1)$, since one estimated parameter, $\bar{y}$, is used in the computation of the sum of squares. Using the same principles, in inferences on several populations, the mean square is the sum of squared deviations from the sample means for each of the populations, and the degrees of freedom are the total sample size minus the number of populations, since one parameter (mean) is estimated for each population.

The same principle is used in regression analysis. The estimated means are

$$\hat{\mu}_{y|x} = \hat{\beta}_0 + \hat{\beta}_1 x,$$

for each observed x, and the sum of squares, called the **error** or residual sum of squares, is

$$\text{SSE} = \sum(y - \hat{\mu}_{y|x})^2.$$

This quantity describes the variation in y after estimating the linear relationship of y to x. The degrees of freedom for this sum of squares is $(n - 2)$ since two

Table 7.3

Estimating the Variance (To Save Space, Only a Few of the Observations Are Presented)

Obs	size	price	predict	residual
1	0.951	30.0	58.767	-28.7668
2	1.036	39.9	63.534	-23.6338
3	0.676	46.5	43.344	3.1561
4	1.456	48.6	87.089	-38.4888
5	1.186	51.5	71.946	-20.4463
.	.	.	.	.
.	.	.	.	.
.	.	.	.	.
53	1.920	167.5	113.111	54.3885
54	2.949	169.9	170.821	-0.9212
55	3.310	175.0	191.067	-16.0672
56	2.805	179.0	162.745	16.2548
57	2.553	179.9	148.612	31.2878
58	2.510	189.5	146.201	43.2994
59	3.627	199.0	208.846	-9.8456

estimates, $\hat{\beta}_0$ and $\hat{\beta}_1$, are used to obtain the values of the $\hat{\mu}_{y|x}$. We then define the mean square

$$\begin{aligned} \text{MSE} &= \text{SSE/df} \\ &= \sum (y - \hat{\mu}_{y|x})^2/(n-2). \end{aligned}$$

Table 7.3 provides the various elements needed for computing this estimate of the variance from the house prices data. The first two columns are the observed values of x and y. The third column contains the estimated values ($\hat{\mu}_{y|x}$), which are computed by substituting the individual x values into the model equation.

For example, for the first observation, $\hat{\mu}_{y|x} = 5.432 + 56.083(0.951) = 58.7668$.

The last column contains the residuals $(y - \hat{\mu}_{y|x})$. Again for the first observation,

$$(y - \hat{\mu}_{y|x}) = 30.0 - 58.767 = -28.767.$$

The sum of squares of residuals is

$$\sum (y - \hat{\mu}_{y|x})^2 = (-28.767)^2 + (-23.6338)^2 + \cdots (-9.8456)^2 = 21698.27,$$

hence MSE $= 21698.27/56 = 387.469$. The square root of the variance is the estimated standard deviation, $\sqrt{387.469} = 19.684$. We can now use the empirical rule to state that approximately 95% of all homes will be priced within $2(19.684) = 39.368$ (or \$39,368) of the estimated value ($\hat{\mu}_{y|x}$). Additionally, the sum of residuals $\sum(y - \hat{\mu}_{y|x})$ equals zero, just as $\sum(y - \bar{y})$ equals 0 for the one-sample situation.

This method of computing the variance estimate is certainly tedious, especially for large samples. Fortunately a computational procedure that uses the principle of partitioning sums of squares similar to that found in the analysis of variance exists (Section 6.2). We define the following:

Figure 7.4

Plot of Partitioning of Sums of Squares

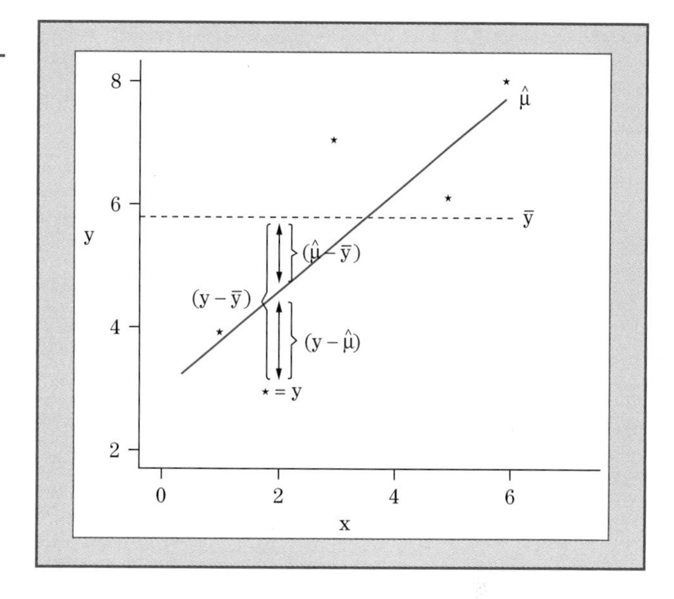

$(y - \bar{y})$ are the deviations of observed values from a model[6] that does not include the regression coefficient β_1.

$(y - \hat{\mu}_{y|x})$ are the deviations of observed values from the estimated values of the regression model.

$(\hat{\mu}_{y|x} - \bar{y})$ are the differences between the estimated population means of the regression and no-regression models.

It is both mathematically and intuitively obvious that

$$(y - \bar{y}) = (y - \hat{\mu}_{y|x}) + (\hat{\mu}_{y|x} - \bar{y}).$$

This relationship is shown for one of the data points in Fig. 7.4 for a typical small data set (the numbers are not reproduced here).

Some algebra and the use of the least squares estimates of the regression parameters provide the not-so-obvious relationship

$$\sum(y - \bar{y})^2 = \sum(y - \hat{\mu}_{y|x})^2 + \sum(\hat{\mu}_{y|x} - \bar{y})^2.$$

The first term is the sum of squared deviations from the mean. This quantity provides the estimate of the total variation if there is only one mean, μ, that does not depend on x; that is, we assume that there is no regression. This is called the TOTAL sum of squares and is denoted by TSS as it was for the analysis of variance. The equation then shows that this total variation is partitioned into two parts:

1. $\sum(y - \hat{\mu}_{y|x})^2$, which we have already defined as the numerator of the estimated variance of the residuals from the means estimated by the regression. This quantity is called the ERROR or RESIDUAL sum of squares and is usually denoted by SSE, and

[6]This model is $y = \beta_0 + \varepsilon$, which is equivalent to $y = \mu + \varepsilon$ and the estimate of μ is $\bar{y}$.

2. $\sum(\hat{\mu}_{y|x} - \bar{y})^2$, which is the difference between the TOTAL and ERROR sum of squares. This difference is the reduction in the variation attributable to the estimated regression and is called the REGRESSION (sometimes called MODEL) sum of squares and is denoted by SSR.

Since these sums of squares are additive, that is, SSR + SSE = TSS, the REGRESSION sum of squares is the indicator of the magnitude of reduction in variance accomplished by fitting a regression. Therefore, large values of SSR (or small values of SSE) relative to TSS indicate that the estimated regression does indeed help to estimate y. Later we will use this principle to develop a formal hypothesis test for the null hypothesis of no relationship.

Partitioning does not by itself assist in the reduction of computations for estimating the variance. However, if we have used least squares, it can be shown that

$$\text{SSR} = (S_{xy})^2/S_{xx} = \hat{\beta}_1^2 S_{xx} = \hat{\beta}_1 S_{xy},$$

all of which use quantities already calculated for the estimation of β_1. It is not difficult to compute TSS $= \sum y^2 - (\sum y)^2/n = S_{yy}$; hence the partitioning allows the computation of SSE by subtracting SSR from TSS.

For our example, we have already computed TSS $= S_{yy} = 93{,}232.142$. The regression sum of squares is

$$\text{SSR} = (S_{xy})^2/S_{xx} = (1275.494)^2/22.743 = 71{,}533.436.$$

Hence,

$$\text{SSE} = \text{TSS} - \text{SSR} = 93{,}232.142 - 71{,}533.436 = 21{,}698.706,$$

which is the same value, except for round-off error, as that obtained directly from the actual residuals (Table 7.3).

The estimated variance, usually called the error mean square, is computed as before:

$$\text{MSE} = \text{SSE/df} = 21{,}698.706/56 = 387.477.$$

The notation of MSE (mean square error) for this quantity parallels the notation for the error sum of squares and is used henceforth.

The formula for the error sum of squares can be represented by a single formula

$$\text{SSE} = \sum y^2 - \left(\sum y\right)^2 / n - S_{xy}^2 / S_{xx},$$

where $\sum y^2$ = total sum of squares of the y values; $(\sum y)^2/n$ = correction factor for the mean, which can also be called the reduction in sum of squares for estimating the mean; and $(S_{xy})^2/S_{xx}$ = additional reduction in the sum of squares due to estimation of a regression relationship.

This **sequential partitioning** of the sums of squares is sometimes used for inferences for regressions involving several independent variables (see Chapter 8).

7.5 Inferences for Regression

The first step in performing inferences in regression is to ascertain if the estimated conditional means, $\hat{\mu}_{y|x}$, provide for a better estimation of the mean of the population of the dependent variable y than does the sample mean $\bar{y}$. This is done by noting that if $\beta_1 = 0$, the estimated conditional mean is the ordinary sample mean, and if $\beta_1 \neq 0$, the estimated conditional mean will provide a better estimate. In this section we first provide procedures for testing hypotheses and subsequently for constructing a confidence interval for β_1.

Other inferences include the estimation of the conditional mean and prediction of the response for individual observations having specific values of the independent variable. Inferences on the intercept are not often performed and are a special case of inference on the conditional mean when $x = 0$ as presented later in this section.

The Analysis of Variance Test for β_1

We have noted that if the regression sum of squares (SSR) is large relative to the total or error sum of squares (TSS or SSE), the hypothesis that $\beta_1 = 0$ is likely to be rejected.[7] In fact, the regression and error sums of squares play the same role in regression as do the factor (SSB) and error (SSW) sums of squares in the analysis of variance for testing hypotheses about the equality of several population means. In each case the sums of squares are divided by the respective degrees of freedom, and the resulting regression or factor mean square is divided by the error mean square to obtain an F statistic. This F statistic is then used to test the hypothesis of no regression or factor effect.

Specifically, for the simple linear regression model, we compute the mean square due to regression,

$$\text{MSR} = \text{SSR}/1,$$

and the error mean square,

$$\text{MSE} = \text{SSE}/(n - 2).$$

As we have noted, MSE is the estimated variance. The test statistic for the null hypothesis $\beta_1 = 0$ against the alternative that $\beta_1 \neq 0$, then, is $F = \text{MSR/MSE}$, which is compared to the tabled F distribution with 1 and $(n - 2)$ degrees of freedom. Because the numerator of this statistic will tend to be large when the null hypothesis is false, the rejection region is in the upper tail.

It is convenient to summarize the statistics resulting in the F statistic in tabular form as was done in Chapter 6. Using the results obtained previously, the analysis of the house prices data are presented in this format in Table 7.4. The 0.01 critical value for the F distribution with df $= (1, 55)$ is 7.12; hence the calculated value of 184.62 clearly leads to rejection of the null hypothesis. This means that we can conclude that home prices are linearly related to size as expressed in square feet. This does not, however, indicate the precision with

[7] For hypothesis tests for nonzero values of β_1, see the next subsection.

Table 7.4

Analysis of Variance of Regression

Source	DF	SS	MS	F
Regression	1	SSE = 71533.436	MSR = 71533.436	184.613
Error	n - 2 = 56	SSE = 21698.706	MSE = 387.477	
Total	n - 1 = 57	TSS = 93232.142		

which selling prices can be estimated by knowing the size of houses. We will do this later.

A more rigorous justification of this procedure is afforded through the use of expected mean squares as was done in Section 6.3 (again without Proof). Using the already defined regression model

$$y = \beta_0 + \beta_1 x + \varepsilon,$$

we can show that

$$E(\text{MSR}) = \sigma^2 + \beta_1^2 S_{xx},$$
$$E(\text{MSE}) = \sigma^2.$$

If the null hypothesis is true, that is, β_1 is zero, the ratio of the two mean squares is the ratio of two estimates of σ^2, and is therefore a random variable with an F distribution with 1 and $(n - 2)$ degrees of freedom. If the null hypothesis is not true, that is, $\beta_1 \neq 0$, the numerator of the ratio will tend to be larger, leading to values of the F statistic in the right tail of the distribution, hence providing for rejection if the calculated value of the statistic is in the right tail rejection region.

The (Equivalent) t Test for β_1

An equivalent test of the hypothesis that $\beta_1 = 0$ is based on the fact that under the assumptions stated earlier, the estimate $\hat{\beta}_1$ is a random variable whose distribution is (approximately) normal with mean $= \beta_1$ and variance $= \sigma^2/S_{xx}$.

The variance of the estimated regression coefficient can also be written

$$\sigma^2/(n - 1)s_x^2,$$

where s_x^2 is the sample variance obtained from the observed set of x values. This expression shows that the variance of $\hat{\beta}_1$ increases with larger values of the population variance, and decreases with larger sample size and/or larger dispersion of the values of the independent variable. This means that the slope of the regression line is estimated with greater precision if

- the population variance is small,
- the sample size is large, and/or
- the independent variable has a large dispersion.

The square root of the variance of an estimated parameter is the standard error of the estimate. Thus the standard error of $\hat{\beta}_1$ is

$$\text{std error of } \hat{\beta}_1 = \sqrt{\sigma^2/S_{xx}}.$$

Hence the ratio

$$z = \frac{\hat{\beta}_1 - \beta_1}{\sqrt{\sigma^2/S_{xx}}}$$

is a standard normal random variable. Substitution of the estimate MSE for σ^2 in the formula for the standard error of $\hat{\beta}_1$ produces a random variable distributed as Student t with $(n-2)$ degrees of freedom. Thus, as in Chapter 4, we have the test statistic necessary for a hypothesis test.

To test the null hypothesis H_0: $\beta_1 = \beta_1^*$ construct the test statistic

$$t = \frac{\hat{\beta}_1 - \beta_1^*}{\sqrt{\text{MSE}/S_{xx}}}.$$

Letting $\beta_1^* = 0$ provides the test for H_0: $\beta_1 = 0$. For the house price data, the test of H_0: $\beta_1 = 0$ produces the values

$$t = \frac{56.083 - 0}{\sqrt{\frac{387.477}{22.743}}} = \frac{56.083}{4.128} = 13.587,$$

which leads to rejection for virtually any value of α. Note that $t^2 = 184.607 = F$ (Table 7.4, except for round-off), confirming that the two tests are equivalent. [Remember, $t^2(v) = F(1, v)$.]

Although the t and F tests are equivalent, the t test has some advantages:

1. It may be used to test a hypothesis for any given value of β_1, not just for $\beta_1 = 0$. For example, in calibration experiments where the reading of a new instrument (y) should be the same as that for the standard (x), the coefficient β_1 should be unity. Hence the test for H_0: $\beta_1 = 1$ is used to determine whether the new instrument is biased.
2. It may be used for a one-tailed test. In many applications a regression coefficient is useful only if the sign of the coefficient agrees with the underlying theory of the model. In this case, the increased power of the resulting one-tailed test makes it appropriate.
3. Remember that the denominator of a t statistic is the standard error of the estimated parameter in the numerator and provides a measure of the precision of the estimated regression coefficient. In other words, the standard error of $\hat{\beta}_1$ is $\sqrt{\text{MSE}/S_{xx}}$.

Confidence Interval for β_1

The sampling distribution of $\hat{\beta}_1$ presented in the previous section is used to construct a confidence interval. Using the appropriate values from the t distribution, the confidence interval for β_1 is computed as

$$\hat{\beta}_1 \pm t_{\alpha/2}\sqrt{\frac{\text{MSE}}{S_{xx}}}.$$

For the home price data, $\hat{\beta}_1 = 56.084$, the standard error is 4.128; hence the 0.95 confidence interval is

$$56.084 \pm (2.004)(4.128),$$

where $t_{0.05}(55) = 2.004$, which is used to approximate $t_{0.05}(56)$ since our table does not have an entry for 56 degrees of freedom. The resulting interval is from 47.811 to 64.357. This means that we can state with 0.95 confidence that the true cost per square foot is between $47.81 and $64.36. Here we can see that although the regression can certainly be called statistically significant, the reliability of the estimate may not be sufficient for practical purposes. That is, the confidence interval is too wide to provide sufficient precision for estimating house prices.

Inferences on the Response Variable

In addition to inferences on the individual parameters, we are also interested in how well the model estimates the response variable. In this context there are two different, but related, inferences:

1. *Inferences on the mean response:* In this case we are concerned with how well the model estimates $\mu_{y|x}$, the conditional mean of the population for any given x value.
2. *Inferences for prediction:* In this case we are interested in how well the model predicts the value of the response variable y for a single randomly chosen future observation having a given value of the independent variable x.

The point estimate for both of these inferences is the value of $\hat{\mu}_{y|x}$ for any specified value of x. However, because the point estimate represents two different inferences, we denote them by different symbols. Specifically, we denote the estimated mean response by $\hat{\mu}_{y|x}$, and the predicted single value by $\hat{y}_{y|x}$. Because these estimates have a different implication, each of these estimates has a different variance. For a specified value of x, say, x^*, the variance for the estimated mean is

$$\text{var}(\hat{\mu}_{y|x}) = \sigma^2\left[\frac{1}{n} + \frac{(x^* - \bar{x})^2}{S_{xx}}\right],$$

and the variance for a single predicted value is

$$\text{var}(\hat{y}_{y|x}) = \sigma^2\left[1 + \frac{1}{n} + \frac{(x^* - \bar{x})^2}{S_{xx}}\right].$$

Both of these variances have their minima when $x^* = \bar{x}$. In other words, when x takes the value $\bar{x}$, the estimated conditional mean is $\bar{y}$ and the variance of the estimated mean is indeed the familiar σ^2/n. The response is estimated with greatest precision when the independent variable is at its mean, with the variance of the estimate increasing as x deviates from its mean. It is also seen that $\text{var}(\hat{y}_{y|x}) > \text{var}(\hat{\mu}_{y|x})$ because a mean is estimated with greater precision than is a single value.

Substituting the error mean square, MSE, for σ^2 provides the estimated variance. The square root is the corresponding standard error used in hypothesis testing or (more commonly) interval estimation with the appropriate value from the t distribution with $(n - 2)$ degrees of freedom.[8] We will obtain the interval estimate for mean and individual predicted values for homes similar to the first home, which had a size of 951 ft^2 for which the estimated price has already been computed to be \$58,767.

All elements of the variance have been obtained previously. The variance of the estimated mean is

$$\begin{aligned}\text{var}(\hat{\mu}_{y|x}) &= 387.469\left[\frac{1}{58} + \frac{(0.951 - 1.883)^2}{22.743}\right]\\ &= 387.469[0.0172 + 0.0382]\\ &= 21.466.\end{aligned}$$

The standard error $\sqrt{21.466} = 4.633$. We now compute the 0.95 confidence interval

$$58.767 \pm (2.004)(4.633),$$

which results in the limits from 49.482 to 68.052. Thus we can state with 0.95 confidence that the mean price of homes with 951 ft^2 of space is between \$49,482 and \$68,052. The width of this interval reinforces the contention that the precision of this regression may be inadequate for practical purposes. The predicted line and confidence interval bands are shown in Fig. 7.5. The tendency for the interval to be narrowest at the center is evident.

The prediction interval for a single observation for the same home is

$$\begin{aligned}\text{var}(\hat{\mu}_{y|x}) &= 387.469\left[1 + \frac{1}{n} + \frac{(0.951 - 1.883)^2}{22.743}\right]\\ &= 387.469[1 + 0.0172 + 0.0382]\\ &= 408.935,\end{aligned}$$

resulting in a standard error of 20.222. The 0.95 prediction interval is

$$58.767 \pm (2.004)(20.222),$$

or from 18.242 to 99.292. Thus we can say with 0.95 confidence that a randomly picked home with 951 ft^2 will be priced between \$18,242 and \$99,292. Again, this interval may be considered too wide to be of practical use.

EXAMPLE 7.3

One aspect of wildlife science is the study of how various habits of wildlife are affected by environmental conditions. This example concerns the effect of air temperature on the time that the "lesser snow geese" leave their overnight roost

[8]Letting $\bar{x} = 0$ in the variance of $\hat{\mu}_{y|x}$ provides the variance for $\hat{\beta}_0$, which can be used for hypothesis tests and confidence intervals for this parameter. As we have noted, in most applications β_0 represents an extrapolation and is thus not a proper candidate for inferences. However, because a computer does not know whether the intercept is a useful statistic for any specific problem, most computer programs do provide that standard error as well as the test for the null hypothesis that $\beta_0 = 0$.

Figure 7.5

Plot of the Predicted Regression Line and Confidence Interval Bands

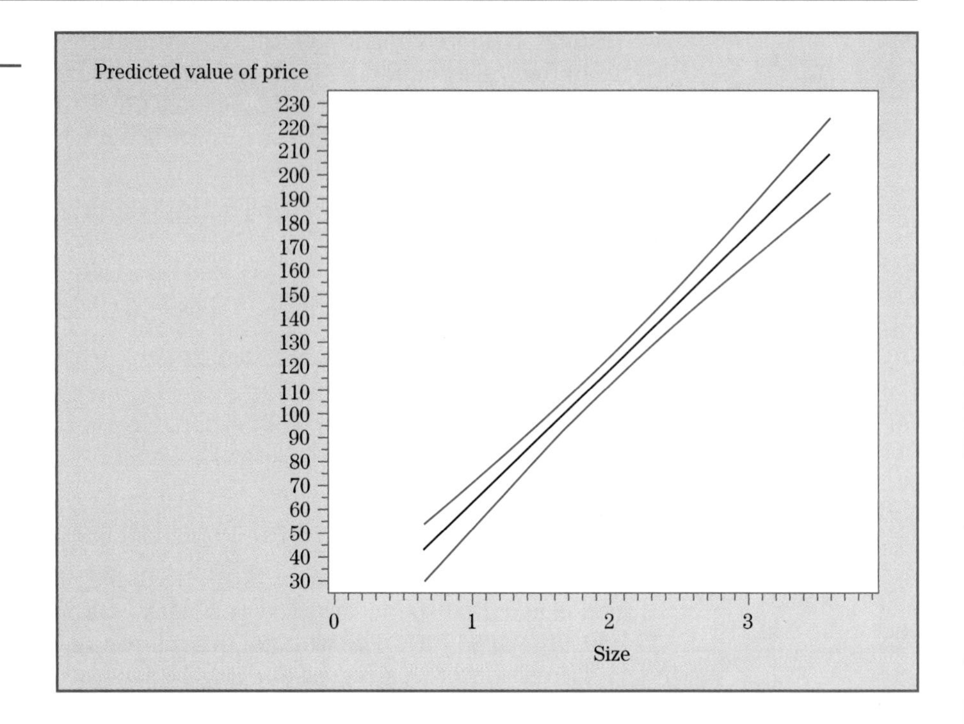

sites to fly to their feeding areas. The data shown in Table 7.5 give departure time (TIME in minutes before (−) and after (+) sunrise) and air temperature (TEMP in degrees Celsius) at a refuge near the Texas Coast for various days of the 1987/88 winter season. A scatterplot of the data, as provided in Fig. 7.6, is useful. The plot does appear to indicate a relationship showing that the geese depart later in warmer weather.

Solution A linear regression relating departure time to temperature should provide useful information on the relationship of departure times. To perform this analysis, the following intermediate results are obtained from the data,

$$\sum x = 334, \quad \bar{x} = 8.79, \quad \sum y = -186, \quad \bar{y} = -4.89,$$
$$S_{xx} = 1834.31, \quad S_{xy} = 3082.84, \quad S_{yy} = 8751.58,$$

resulting in the estimates

$$\hat{\beta}_0 = -19.667 \quad \text{and} \quad \hat{\beta}_1 = 1.681.$$

The resulting regression equation is

$$\widehat{\texttt{TIME}} = -19.667 + 1.681(\texttt{TEMP}).$$

In this case the intercept has a practical interpretation because the condition $\texttt{TEMP} = 0$ (freezing) does indeed occur, and the intercept estimates that the time of departure is approximately 20 min. before sunrise at that temperature.

Table 7.5

Departure Times of Lesser Snow Geese

OBS	DATE	TEMP	TIME
1	11/10/87	11	11
2	11/13/87	11	2
3	11/14/87	11	-2
4	11/15/87	20	-11
5	11/17/87	8	-5
6	11/18/87	12	2
7	11/21/87	6	-6
8	11/22/87	18	22
9	11/23/87	19	22
10	11/25/87	21	21
11	11/30/87	10	8
12	12/05/87	18	25
13	12/14/87	20	9
14	12/18/87	14	7
15	12/24/87	19	8
16	12/26/87	13	18
17	12/27/87	3	-14
18	12/28/87	4	-21
19	12/30/87	3	-26
20	12/31/87	15	-7
21	01/02/88	15	-15
22	01/03/88	6	-6
23	01/04/88	5	-23
24	01/05/88	2	-14
25	01/06/88	10	-6
26	01/07/88	2	-8
27	01/08/88	0	-19
28	01/10/88	-4	-23
29	01/11/88	-2	-11
30	01/12/88	5	5
31	01/14/88	5	-23
32	01/15/88	8	-7
33	01/16/88	15	9
34	01/20/88	5	-27
35	01/21/88	-1	-24
36	01/22/88	-2	-29
37	01/23/88	3	-19
38	01/24/88	6	-9

The regression coefficient indicates that the estimated departure time is 1.681 min. later for each 1° increase in temperature.

The partitioning of the sums of squares and F test for the hypothesis of no regression, that is, H_0: $\beta_1 = 0$, is provided in Table 7.6. This table is adapted from computer output, which also provides the p value. We can immediately see that we reject the null hypothesis $\beta_1 = 0$. The error mean square of 99.18 is the estimate of the variance of the residuals. According to the empirical rule, the resulting standard deviation of 9.96 indicates that 95% of all observed departure times are within approximately 20 min. of the time estimated by the model.

Table 7.6

Analysis of Variance for Goose Data

Source	DF	Sum of Squares	Mean Square	F Value	Prob > F
Regression	1	5181.17736	5181.17736	52.241	0.0001
Error	36	3570.40158	99.17782		
Total	37	8751.57895			

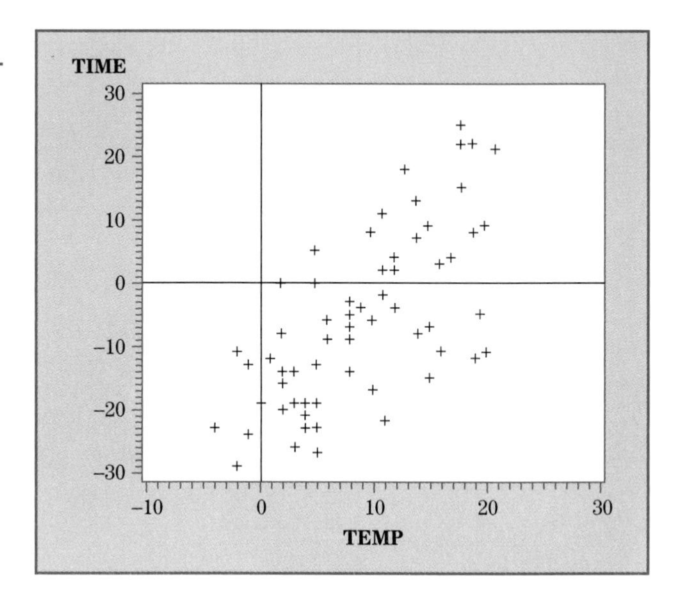

Figure 7.6

Scatterplot of Departure Times and Temperatures

The variance of the estimated regression coefficient, $\hat{\beta}_1$, is $99.178/1834.31 = 0.0541$, resulting in a standard error of 0.2325. We can use this for the t statistic

$$t = 1.681/0.2325 = 7.228,$$

which is the square root of the F value (52.241) and equivalently results in the rejection of the hypothesis that $\beta_1 = 0$. The standard error and 0.05 two-tailed t value of 2.028 for 36 degrees of freedom, obtained from Appendix Table A.2 by interpolation, can be used to compute the 0.95 confidence interval for β_1

$$1.681 \pm (2.028)(0.2325),$$

which results in the interval

$$1.209 \text{ to } 2.153.$$

In other words, we are 95% confident that the true slope of the regression is between 1.209 and 2.153 minutes per degree of temperature increase.

For inferences on the response variable (TIME), We consider the case for which the temperature is 0°C (freezing). The point estimate for the mean response as well as for predicting a single individual is $\hat{\mu}_{y|x=0} = \hat{\beta}_0 = -19.67$

Figure 7.7

Regression Results for Departure Data

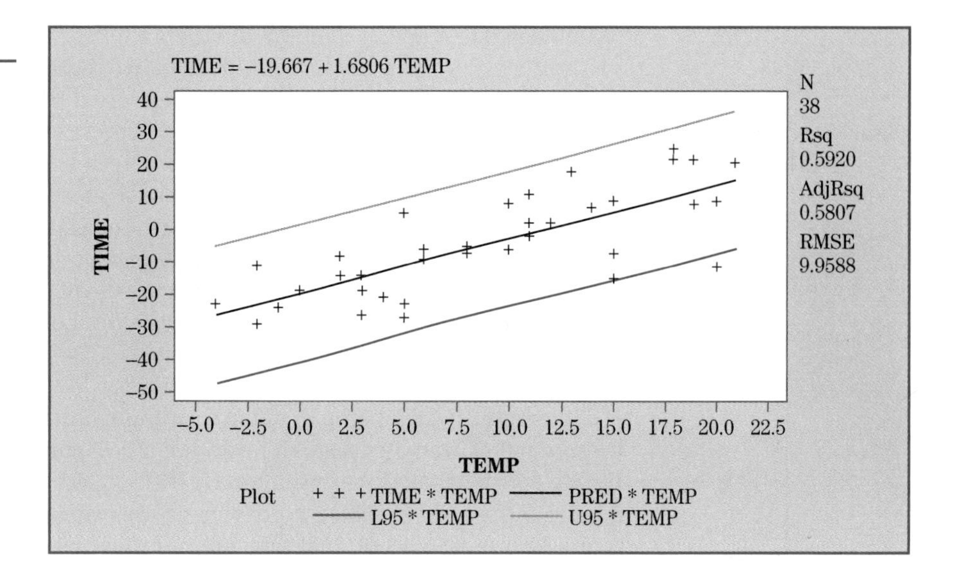

min. after sunrise. The variance of the estimated mean at zero degrees is

$$99.178\left[\frac{1}{38} + \frac{(0 - 8.79)^2}{1834.31}\right] = 6.786,$$

resulting in a standard error of 2.605. The 95% confidence interval, then, is

$$-19.67 \pm (2.028)(2.605),$$

or from -24.95 to -14.38 min. In other words, we are 95% confident that the true mean departure time at 0°C is between 14.38 and 24.95 min. before sunrise.

The plot of the data with the estimated regression line and 95% prediction intervals as produced by SAS PROC REG is shown in Fig. 7.7. In the legend, PRED represents the prediction line and U95 and L95 represent the 0.95 upper and lower prediction intervals, respectively. (When the plot is shown on a computer monitor, the prediction intervals have different colors.)

The 95% prediction interval for 0°C is from -40.54 to $+1.21$ minutes. This means that we are 95% confident that any randomly picked goose will leave within this time frame at 0°C. ■

EXAMPLE 7.4

One interesting application of simple linear regression is to use it to compare two measuring devices or tests relative to their precision and their accuracy. If we define the true value of the characteristic that we are measuring as the independent variable in a regression equation, and the measured value as the response variable, then we can use the procedures previously discussed to evaluate the relative precision and accuracy of the measuring device or test. We define the accuracy of the device or test as its ability to "hit the target." That is, if a test or device is accurate, then we would expect the measured value, on average, to be very close to the actual value. In statistical terms, this is known as unbiasedness, and the amount of bias in a test or device is used as a measure of accuracy. Perfect accuracy would result in a regression equation relating the measured value to the true value that had a zero intercept and a slope of 1. The precision of a measuring device or test is defined as the variation among values recorded by the device or test. In statistical terms, we use the standard deviation as a measure of precision. A very precise measuring device or test would have almost no variation from measurement to measurement. In the regression context, we use the square root of the mean square error from the analysis of variance as a measure of the precision of the device or test. The procedure for comparing two tests or measuring devices would be to compute the accuracy and precision of each, and compare them.

We illustrate these concepts with an example comparing two types of temperature-measuring devices. Suppose that a company is considering two such devices (one labeled A, the other B) to be used to control a temperature-sensitive process. Because no device always records the absolutely correct temperature, we specify that the superior device should be unbiased (i.e., on the average, it records the correct temperature) and also that the device must be precise (i.e., there should be very little variation among readings at any constant temperature). An experiment is conducted by exposing each device randomly three times to each of six known temperatures. The data are shown in Table 7.7. To evaluate these two devices and pick the superior one, we perform a regression analysis using the measured temperature as the response variable and the correct temperature as the independent variable.

Solution The analysis consists of estimating the regression equation for both types of devices, and then performing the hypotheses tests to determine whether $\beta_1 = 1$ and $\beta_0 = 0$. Abbreviated output from `PROC REG` of the SAS System for the two devices is shown in Table 7.8. Note that the analyses assume the straight line (linear) regression models are adequate. The reader is encouraged to perform the lack of fit test, which will support this assumption. Obviously the regressions are significant, but our primary focus is on the regression coefficients. The tests for the hypotheses $\beta_1 = 1$ and $\beta_0 = 0$ are identified as `Test: BIAS_SLO` and `Test: BIAS_INT`, respectively. We see that both hypotheses are rejected for device A but not for device B. Thus it would appear that device B is unbiased and therefore accurate. Our first inclination might be to recommend device B.

Table 7.7

Temperature Readings for Two Devices

Correct Temperature	Readings for Device A			B		
50	50.2	50.4	50.4	49.6	49.9	50.1
70	70.3	70.1	69.9	71.0	70.2	69.2
90	89.6	89.3	89.8	89.1	89.7	90.1
110	109.1	109.2	109.3	110.0	111.1	109.2
130	128.7	129.1	129.1	131.2	131.5	128.9
150	148.5	148.5	148.9	151.2	150.2	149.4

Table 7.8 Comparing Two Temperature Measuring Devices

```
                                DEVICE = A
                           ANALYSIS OF VARIANCE
                       SUM OF          MEAN
SOURCE           DF   SQUARES         SQUARE         F VALUE      PROB > F

MODEL             1  20270.44876   20270.44876    588361.332       0.0001
ERROR            16      0.55124       0.03445
C TOTAL          17  20271.00000

                            PARAMETER ESTIMATES
                     PARAMETER       STANDARD       T FOR H0:
VARIABLE         DF   ESTIMATE         ERROR      PARAMETER = 0  PROB > |T|

INTERCEP          1    1.219048     0.13535109         9.007         0.0001
TEMP              1    0.982476     0.00128086       767.047         0.0001

DEPENDENT VARIABLE: READING
TEST: BIAS_SLO
        NUMERATOR:        6.4488       DF:  1       F VALUE:       187.1790
        DENOMINATOR:    0.034452       DF: 16       PROB > F:        0.0001

DEPENDENT VARIABLE: READING
TEST: BIAS_INT
        NUMERATOR:        2.7947       DF:  1       F VALUE:        81.1181
        DENOMINATOR:    0.034452       DF: 16       PROB > F:        0.0001

                                DEVICE = B
                           ANALYSIS OF VARIANCE
                       SUM OF          MEAN
SOURCE           DF   SQUARES         SQUARE         F VALUE      PROB > F

MODEL             1  21220.57619   21220.57619     31905.881       0.0001
ERROR            16     10.64159       0.66510
C TOTAL          17  21231.21778

                            PARAMETER ESTIMATES
                     PARAMETER       STANDARD       T FOR H0:
VARIABLE         DF   ESTIMATE         ERROR      PARAMETER = 0  PROB > |T|

INTERCEP          1   -0.434921     0.59469645        -0.731         0.4752
TEMP              1    1.005238     0.00562773       178.622         0.0001

DEPENDENT VARIABLE: READING
TEST: BIAS_SLO
        NUMERATOR:        0.5762       DF:  1       F VALUE:         0.8663
        DENOMINATOR:    0.665099       DF: 16       PROB > F:        0.3658

DEPENDENT VARIABLE: READING
TEST: BIAS_INT
        NUMERATOR:        0.3557       DF:  1       F VALUE:         0.5348
        DENOMINATOR:    0.665099       DF: 16       PROB > F:        0.4752
```

Figure 7.8

Scatterplots of Differences

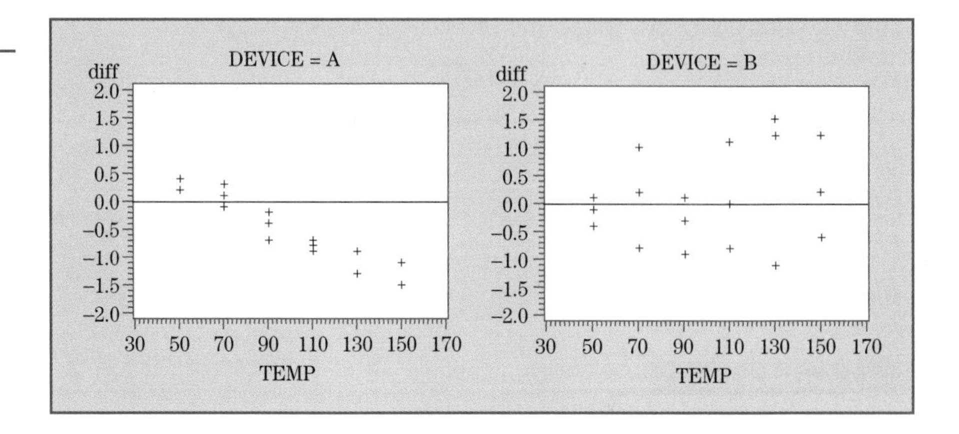

Looking more closely at the parameter estimates, we see that the estimated slope for device A is only 0.0052 units too high, whereas that for device B is 0.0175 units too low. In other words although device A has been shown to be biased, the estimate of the bias ($\hat{\beta}_1 - 1.0$) has a value smaller than that for device B. The reason for the apparent contradiction is that the standard error of the estimated coefficient is much smaller for device A than for device B, resulting in an inflated test statistic. The same applies to the intercept. Note that the square root of the MSE for device A is 0.186, while that for device B is 0.8155, almost five times larger.

What we have here, then, is that device A is biased, but much more precise, while device B is apparently not biased[9] but has much less precision. This is shown in Fig. 7.8, which gives the scatterplots for the *differences* between the reading and the true temperatures for the two devices. Clearly, readings for device A have much less variability but are biased, while those for device B have more variability but are not biased. Now in many cases it is not difficult to recalibrate a device, and if this can be done, device A is a clear winner. However, even if that is not possible, device A may yet be chosen because, as the reader may wish to calculate, the 0.95 prediction interval will always be closer to the true line for device A than that for device B. ■

7.6 Using the Computer

Most statistical calculations, and especially those for regression analyses, are performed on computers. The formulas needed for manual computation of estimates and other inferences are presented in this chapter primarily as a pedagogical device and will not often be used in practice.

As we have noted, most regression analyses are performed by computers using preprogrammed computing software packages. Virtually all such programs for regression analysis are written for a wide variety of analyses of

[9] Remember that we have not accepted the null hypothesis that the device is unbiased.

Table 7.9

Computer Output for Home Price Regression

Dependent Variable: price

Analysis of Variance

Source	DF	Sum of Squares	Mean Square	F Value	Pr > F
Model	1	71534	71534	184.62	<.0001
Error	56	21698	387.46904		
Corrected Total	57	93232			

Root MSE	19.68423	R-Square	0.7673
Dependent Mean	111.03445	Adj R-Sq	0.7631
Coeff Var	17.72804		

Parameter Estimates

Variable	DF	Parameter Estimate	Standard Error	t Value	Pr > |t|
Intercept	1	5.43157	8.19061	0.66	0.5100
Size	1	56.08328	4.12758	13.59	<.0001

which simple linear regression is only a special case. This means that these programs provide options and output statistics that may not be useful for this simple case. The computer output for the regression of selling prices of houses on dwelling square feet, produced by the SAS System `PROC REG`, is given in Table 7.9. There are three sections of this output. The first refers to the partitioning of the sums of squares and the analysis of variance (Table 7.4). The various portions of the output are reasonably well labeled, but we see that the nomenclature is not exactly as we have described in the text.

The second portion contains some miscellaneous statistics:

`Root MSE`, the residual standard deviation,
`R-square`, the coefficient of determination (Section 7.7),
`Dependent mean`, the mean of the dependent variable,
`Coeff var`, the coefficient of variation, which is the residual standard deviation divided by the mean of the dependent variable, and
`Adj R-sq`, a variant of the coefficient of determination, which is useful for multiple-regression models (see Chapter 8).

The last portion of the output contains statistics associated with the regression coefficients, which are called here `Parameter Estimates`. Each line contains statistics for one coefficient, which is identified at the beginning of the line: `Intercept` refers to $\hat{\beta}_0$ and `Size`, the name of the independent variable, refers to $\hat{\beta}_1$. The column headings identify the statistics, which are self-explanatory. Note that the output gives the standard error and test for zero value of the intercept. The reader should compare all of these results with those given in previous sections. Programs such as this one usually have a number of options for additional statistics and further analyses. For example, options specifying the predicted and residual values and the 95% confidence intervals for the conditional mean produce the results shown in Table 7.10. Note that

Table 7.10 Home Prices Regression: Predicted and Residual Values and Confidence Limits

Obs	Dep Var price	Predicted Value	Std Error Mean Predict	95% CL Mean		Residual
1	30.0000	58.7668	4.6344	49.4828	68.0507	-28.7668
2	39.9000	63.5338	4.3476	54.8245	72.2432	-23.6338
3	46.5000	43.3439	5.6124	32.1008	54.5869	3.1561
4	48.6000	87.0888	3.1283	80.8221	93.3556	-38.4888
5	51.5000	71.9463	3.8673	64.1991	79.6936	-20.4463
6	56.9900	87.0888	3.1283	80.8221	93.3556	-30.0988
7	59.9000	82.1535	3.3464	75.4498	88.8572	-22.2535
8	62.5000	61.1783	4.4882	52.1874	70.1693	1.3217
9	65.5000	71.3855	3.8982	63.5766	79.1944	-5.8855
10	69.0000	73.6288	3.7761	66.0643	81.1934	-4.6288
11	76.9000	84.5090	3.2391	78.0203	90.9976	-7.6090
12	79.0000	80.8075	3.4102	73.9760	87.6389	-1.8075
13	79.9000	65.1042	4.2553	56.5799	73.6285	14.7958
14	79.9500	104.6990	2.6264	99.4377	109.9603	-24.7490
15	82.9000	90.9025	2.9792	84.9344	96.8706	-8.0025
16	84.9000	103.5773	2.6423	98.2842	108.8705	-18.6773
17	85.0000	70.0395	3.9728	62.0809	77.9981	14.9605
18	87.9000	104.6990	2.6264	99.4377	109.9603	-16.7990
19	89.9000	96.5108	2.7970	90.9078	102.1138	-6.6108
20	89.9000	91.7998	2.9469	85.8964	97.7033	-1.8998
21	93.5000	91.3512	2.9629	85.4157	97.2866	2.1488
22	94.9000	97.8007	2.7621	92.2676	103.3338	-2.9007
23	95.8000	80.8075	3.4102	73.9760	87.6389	14.9925
24	98.5000	92.3607	2.9273	86.4965	98.2248	6.1393
25	99.5000	103.6895	2.6406	98.3997	108.9792	-4.1895
26	99.9000	86.7523	3.1423	80.4575	93.0472	13.1477
27	102.0000	79.0128	3.4978	72.0059	86.0198	22.9872
28	106.0000	97.1838	2.7784	91.6180	102.7497	8.8162
29	108.9000	89.5565	3.0297	83.4872	95.6257	19.3435
30	109.9000	106.3815	2.6073	101.1585	111.6044	3.5185
31	110.0000	116.0278	2.6107	110.7980	121.2576	-6.0278
32	112.2900	83.2191	3.2972	76.6141	89.8241	29.0709
33	114.9000	122.1970	2.7121	116.7640	127.6299	-7.2970
34	119.5000	.	.	.	.	.
35	119.9000	143.5647	3.5231	136.5070	150.6224	-23.6647
36	119.9000	149.6778	3.8431	141.9792	157.3763	-29.7778
37	122.9000	123.9355	2.7535	118.4195	129.4516	-1.0355
38	123.9380	118.4955	2.6424	113.2022	123.7887	5.4425
39	124.9000	109.2978	2.5878	104.1138	114.4818	15.6022
40	126.9000	155.1739	4.1513	146.8578	163.4900	-28.2739
41	129.9000	136.4421	3.1902	130.0514	142.8328	-6.5421
42	132.9000	116.4765	2.6155	111.2370	121.7160	16.4235
43	134.9000	144.6863	3.5797	137.5153	151.8574	-9.7863
44	135.9000	162.9695	4.6141	153.7262	172.2127	-27.0695
45	139.5000	119.6171	2.6607	114.2870	124.9472	19.8829
46	139.9900	134.3109	3.1008	128.0992	140.5227	5.6791
47	144.9000	119.7293	2.6627	114.3953	125.0633	25.1707
48	147.6000	138.3489	3.2744	131.7895	144.9084	9.2511
49	149.9900	169.2508	5.0038	159.2270	179.2747	-19.2608

(Continued)

Table 7.10 *(continued)*

Obs	Dep Var Price	Predicted Value	Std Error Mean Predict	95% CL Mean		Residual
50	152.5500	132.2919	3.0213	126.2396	138.3443	20.2581
51	156.9000	143.1721	3.5036	136.1536	150.1906	13.7279
52	164.0000	142.0504	3.4484	135.1425	148.9583	21.9496
53	167.5000	113.1115	2.5892	107.9247	118.2982	54.3885
54	169.9000	170.8212	5.1031	160.5984	181.0439	-0.9212
55	175.0000	191.0672	6.4323	178.1817	203.9528	-16.0672
56	179.0000	162.7452	4.6005	153.5293	171.9610	16.2548
57	179.9000	148.6122	3.7854	141.0291	156.1952	31.2878
58	189.5000	146.2006	3.6577	138.8733	153.5279	43.2994
59	199.0000	208.8456	7.6486	193.5236	224.1676	-9.8456
	Sum of residuals				0	
	Sum of squared residuals				21698	
	Predicted residual SS (PRESS)				23201	

Note. The observation with the missing space value is shown. An interesting feature of PROC REG is that if the only dependent variable is missing, the program will provide a predicted value and confidence interval. Also the values of the confidence limits for the first home are somewhat different from those obtained above. The difference is due to round-off, which is more pronounced with manual calculations.

Table 7.11

Minitab Output for Goose Departure Data

The regression equation is
C1 = -19.7" + "1.68 C2

Predictor	Coef	Stdev	t- ratio	p
Constant	−19.667	2.605	−7.55	0.000
C2	1.6806	0.2325	7.23	0.000
s = 9.959	R-sq = 59.2%		R-sq(adj) = 58.1%	

Analysis of Variance

SOURCE	DF	SS	MS	F	p
Regression	1	5181.2	5181.2	52.24	0.000
Error	36	3570.4	99.2		
Total	37	8751.6			

in addition to the requested statistics, a summary, showing that the sum of residuals is indeed zero and that the sum of squared residuals is the same as that computed by the partitioning of sums of squares as seen in Table 7.9, is given. The statistic labeled PRESS is briefly discussed in Chapter 8.

There are, of course, other computer programs for performing statistical analyses. One that is often used as an adjunct to statistics classes is Minitab. Table 7.11 reproduces the output from the REGRESS statement available in this package using the snow geese data presented in Example 7.3. In this output, the variable C1 is time and C2 is temperature, which are default variable names that may be changed by the user with additional programming. It is readily seen that the format of the output is somewhat different from that in Table 7.9, but it does provide essentially the same information. Obviously, the results are identical to those obtained in the original presentation of the example (Table 7.6).

The SAS System requires the submission of a program, which consists of a set of written instruction that requests certain actions by the software. The increasing use of Windows operating systems has given rise to software that use menus and the "point and click" approach to perform statistical analyses. Typically such software provides a menu of analysis options from which a choice is made by a point and click. This produces another menu for specifying the variables to be used and additional menus for specifying other analysis and output options. Final results are, of course, equivalent. This approach is indeed very convenient and avoids the consequences of typographical or other syntax errors. On the other hand, it does not provide the flexibility of the program approach.

Another differences among packages is the default for output options. The SAS System, for example, provides relatively minimal output as the default; additional outputs must be specified as options. Other packages may have a fixed output with no options, while yet others may require specifications of output that are *not* desired.

7.7 Correlation

The purpose of a regression analysis is to estimate the response variable y for a specified value of the independent variable x. Not all relationships between two variables lend themselves to this type of analysis. For example, if we have data on the verbal and quantitative scores on a college entrance exam, we are not usually interested in estimating or predicting one score from another, but are simply interested in ascertaining the strength of the relationship between the two scores.

DEFINITION 7.2
The **correlation coefficient**, measures the strength of the linear relationship between two quantitative (usually ratio or interval) variables.

The correlation coefficient has the following properties:

1. Its value is between $+1$ and -1 inclusive.
2. Values of $+1$ and -1 signify an exact positive and negative relationship, respectively, between the variables. That is, a plot of the values of x and y exactly describes a straight line with a positive or negative slope depending on the sign.
3. A correlation of zero indicates no linear relationship exists between the two variables. This condition does not, however, imply that there is no relationship since correlation does not measure the strength of curvilinear relationships.

4. The correlation coefficient is symmetric with respect to x and y. It is thus a measure of the strength of a linear relationship regardless of whether x or y is the independent variable.

The population correlation coefficient is denoted by ρ. An estimate of ρ may be obtained from a sample of n pairs of observed values of the two variables by Pearson's product moment correlation coefficient, denoted by r. Using the notation of this chapter, this estimate is

$$r = \frac{\sum(x - \bar{x})(y - \bar{y})}{\sqrt{\sum(x - \bar{x})^2 \sum(y - \bar{y})^2}} = \frac{S_{xy}}{\sqrt{S_{xx}S_{yy}}}.$$

The sample correlation coefficient is also a useful statistic in a regression analysis. If we compute the square of r, called "r-square," we get

$$r^2 = \frac{(S_{xy})^2}{S_{xx}S_{yy}}.$$

In Section 7.4 we determined that TSS $= S_{yy}$ and that SSR $= \frac{(S_{xy})^2}{S_{xx}}$. Therefore it can be seen that

$$r^2 = \text{SSR/TSS}.$$

In this context the value of r^2 is known as the coefficient of determination, and is a measure of the relative strength of the corresponding regression. It is therefore widely used to describe the effectiveness of linear regression models. In fact, r^2 is interpreted as the proportional reduction of total variation associated with the regression on x. It can also be shown that

$$F = \frac{\text{MSR}}{\text{MSE}} = \frac{(n-2)r^2}{(1-r^2)},$$

where F is the F statistic from the analysis of variance test for the hypothesis that $\beta_1 = 0$. This relationship shows that large values of the correlation coefficient generate large values of the F statistic, both of which imply a strong linear relationship.

For the home price data, the correlation is computed using quantities previously obtained for the regression analysis

$$\begin{aligned} r &= \frac{1275.494}{\sqrt{(22.743)(93232.142)}} \\ &= \frac{1275.494}{1456.152} \\ &= 0.876. \end{aligned}$$

Equivalently, from Table 7.4 the ratio of SSR to TSS is 0.7673, for which the square root is 0.876, which is the same result. Thus, as noted above, $r^2 = 0.7673$, indicating that approximately 77% of the variation in home prices can be attributed to the linear relationship to space.

The sampling distribution of r cannot be used directly for testing of nonzero values or computing confidence intervals for ρ. Therefore, these tasks are performed by an approximate procedure. The Fisher z transformation states that the random variable

$$z' = 1/2 \log_e \left[\frac{1+r}{1-r} \right]$$

is an approximately normally distributed variable with mean

$$1/2 \log_e \left[\frac{1+\rho}{1-\rho} \right]$$

and variance of $[1/(n-3)]$. The use of this transformation for hypothesis testing is quite straightforward, but the inversion of the transformation required for computing confidence intervals is easiest done using special tables (e.g., Neter *et al.*, 1996, Table B.8).

A confidence interval is obtained by first computing the interval using the z' statistic

$$z' \pm z_{\alpha/2} \sqrt{\frac{1}{n-3}}$$

and using the aforementioned table to obtain the interval for ρ.

EXAMPLE 7.5

The correlation between scores on a traditional aptitude test and scores on a final test is known to be approximately 0.6. A new aptitude test has been developed and is tried on a random sample of 100 students, resulting in a correlation of 0.65. Does this result imply that the new test is better?

Solution The question is answered by testing the hypotheses

$$H_0: \rho = 0.6,$$

$$H_1: \rho > 0.6.$$

Substituting 0.65 for r in the formula for z' gives the value 0.775; substituting the null hypothesis value of 0.6 provides the value 0.693, and the standard error $[1/\sqrt{n-3}] = 0.101$. Substituting these in the standard normal test statistic gives the value 0.81, which does not lead to rejection (one-sided p value is 0.3783).

We can now calculate a 95% confidence interval on ρ. The necessary quantities have already been computed; that is, $z' = 0.775$ and the standard error is 0.101. Assuming a two-sided 0.05 interval, $z_{\alpha/2} = 1.96$ and the interval is from 0.576 to 0.973. The aforementioned table provides the corresponding values of ρ, which are 0.52 and 0.75. Thus we are 0.95 confident that the true correlation between the scores on the new aptitude test and the final test is between 0.52 and 0.75. ■

7.8 Regression Diagnostics

In Section 7.2 we listed the assumptions necessary to assure the validity of the results of a regression analysis and noted that these are essentially the ones that have been used since Chapter 4.[10] As we will see in Chapter 11, this is due to the fact that all of these methods are actually based on linear models.

Violations of these assumptions occur more frequently with regression than with the analysis of variance because regression analyses are often applied to data from operational studies, secondary data, or data that simply "occur." These sources of data may be subject to more unknown phenomena than are found in the results of experiments. In this section we present some diagnostic tools that may assist in detecting such violations, and some suggestions on remedial steps if violations are found. (Additional methodology is presented in Section 8.9.)

In order to carry out these diagnostics, we rearrange assumptions 1, 3, and 4 into four categories that correspond to different diagnostic tools. Violations of assumption 2 (independent errors) occur primarily in studies of time series, which is a topic beyond the scope of this book. See, for example, Freund and Wilson (1998, Section 4.5). The four categories are as follows:

1. The model has been properly specified.
2. The variance of the residuals is σ^2 for all observations.
3. There are no outliers, that is, unusual observations that do not fit in with the rest of the observations.
4. The error terms are at least approximately normally distributed.

If the model is not correctly specified, the analysis is said to be subject to specification error. This error most often occurs when the model should contain additional parameters. It can be shown that a specification error causes estimates of the variance as well as the regression coefficients to be biased, and since the bias is a function of the unknown additional parameters, the magnitude of the bias is not known. A common example of a specification error is for the model to describe a straight line when a curved line should be used.

The assumption of equal variances is, perhaps, the one most frequently violated in practice. The effect of this type of violation is that the estimates of the variances for estimated means and predicted values will be incorrect. The use of transformations for this type of violation was presented in Section 6.4. However, the use of such transformations for regression analysis also changes the nature of the model (an extensive discussion of this topic along with an example is given in Section 8.6). Other remedies include the use of weighted least squares (Section 11.7) and robust estimation, which are beyond the scope of this book (see, for example, Koopmans, 1987).

[10] Not discussed here is the assumption that x is fixed and measured without error. Although this is an important assumption, it is not very frequently violated to the extent that it would greatly influence the results of the analysis. Also diagnostic and remedial methods for violations of this assumption are beyond the scope of this book (Seber, 1977).

Figure 7.9

Residual Plot for House Prices

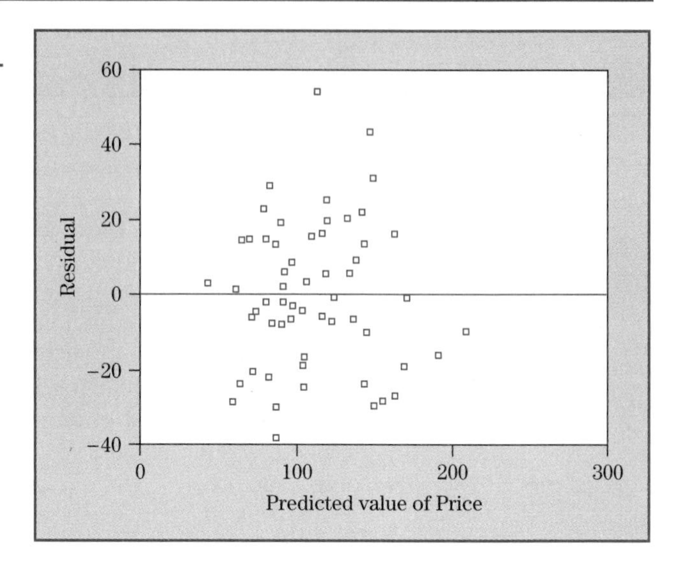

Outliers or unusual observations may be considered a special case of unequal variances, but outliers can cause biased estimates of coefficients as well as incorrect estimates of the variance. It is, however, very important to emphasize that simply discarding observations that appear to be outliers is not good statistical practice. Since any of these violations of assumptions may cast doubt on estimates and inferences, it is important to see whether such violations may have occurred.

A popular tool for detecting violations of assumptions is an analysis of the residuals. Recall that the residuals are the differences between the actual observed y values and the estimated conditional means, $\hat{\mu}_{y|x}$, that is, $(y - \hat{\mu}_{y|x})$. An important part of an analysis of residuals is a residual plot, which is a scatterplot featuring the individual residual values $(y - \hat{\mu}_{y|x})$ on the vertical axis and either the predicted values $(\hat{\mu}_{y|x})$ or x values on the horizontal axis. (See Fig. 7.9.) Occasionally residuals may also be plotted against possible candidates for additional independent variables.

Additional analyses of residuals consist of using descriptive methods, especially the exploratory data analysis techniques such as stem and leaf or box plots described in Chapter 1. Virtually all computer programs for regression provide for the relatively easy implementation of such analyses. Other methods particularly useful for more complicated models are introduced in Section 8.9.

To examine the assumption of normality, we use the Q–Q plot discussed in Section 4.5 and a box plot using the residuals. The Q–Q and box plots for house prices are given in Fig. 7.10.

These three plots do not suggest that any of the assumptions are violated, even though the Q–Q plot does look a little suspicious. It is, however, important to note that the absence of such patterns does not guarantee that there are no violations. For example, outliers may sometimes "pull" the regression line toward themselves, resulting in a biased estimate of that line

Figure 7.10

Q–Q Plot and Boxplot for House Prices

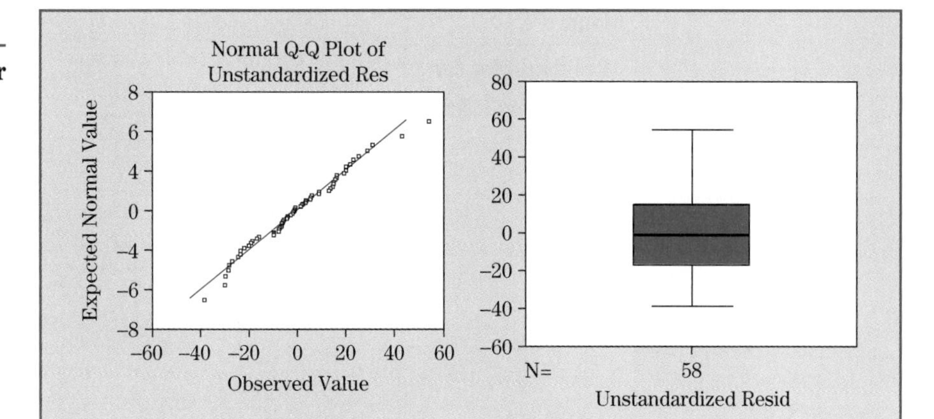

and consequently showing relatively small residuals for those observations. Additional techniques for the detection and treatment of the violations of assumptions are given in Chapter 8, especially Section 8.9.

We illustrate residual plots for some typical violations of assumptions in Figures 7.11, 7.12, and 7.13. For our first example we have generated a set of artificial data using the model

$$y = 4 + x - 0.1x^2 + \varepsilon,$$

where ε is a normally distributed random variable with mean zero and standard deviation of 0.5. (Implementation of such models is presented in Section 8.6.) This model describes a downward curving line. However, assume we have used an incorrect model,

$$y = \beta_0 + \beta_1 x + \varepsilon,$$

which describes a straight line. The plot of residuals against predicted y, shown in Fig. 7.11, shows a curvature pattern typical of this type of misspecification.

Figure 7.11

Residual Plot for Specification Error

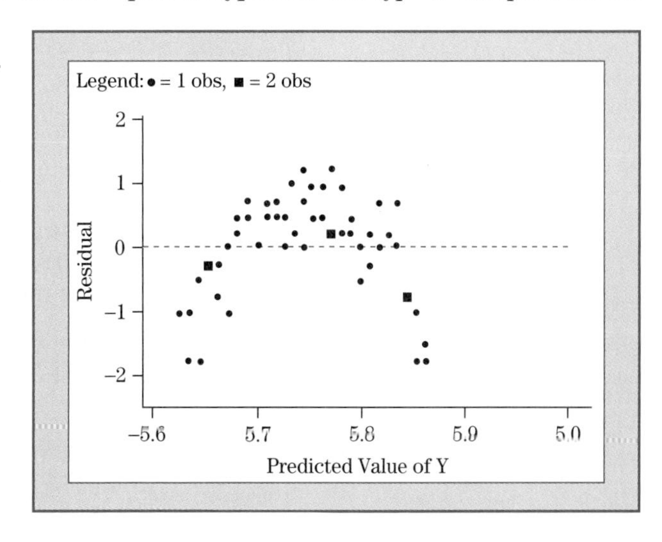

Figure 7.12

Residual Plot for Nonhomogeneous Variance

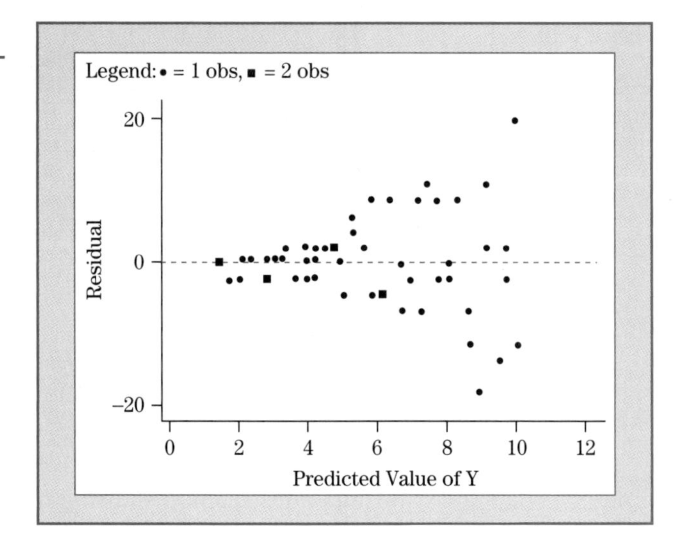

Figure 7.13

Residual Plot for Outliers

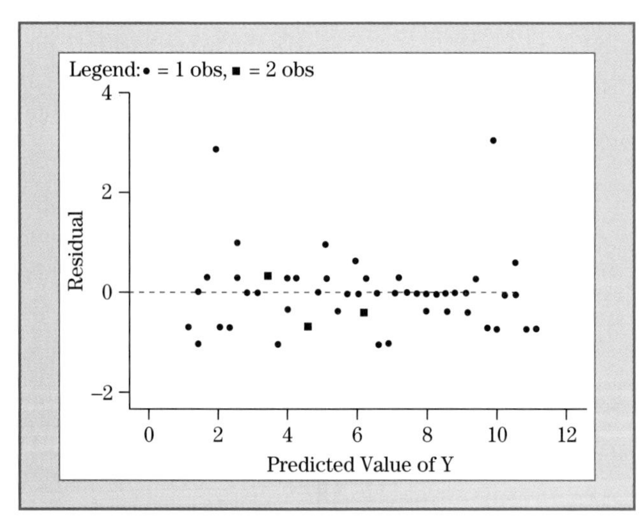

For the second example we have generated data using the model

$$y = x + \varepsilon,$$

where the standard deviation of ε increases linearly with $\mu_{y|x}$. The resulting residuals, shown in Fig. 7.12, show a pattern often described as "fan shaped," which clearly shows larger magnitudes of residuals associated with the larger values of $\hat{\mu}_{y|x}$.

For the last example we have generated data using the model

$$y = x + \varepsilon,$$

where the standard deviation of ε is 0.5, but two values of y are 1.5 units (or 6σ) too large. These two observations are outliers, since they are approximately three standard deviations above the mean (zero) of the residuals. The residual plot is given in Fig. 7.13. The two very large residuals clearly show on this plot.

EXAMPLE 6.5

REVISITED The purpose of the experiment resulting in the data for Example 6.5 was to relate display space to sales of apples in stores. The analysis of variance showed that display space did affect apple sales and the use of orthogonal polynomial contrasts showed that a quadratic trend was appropriate to describe the relationship. Can we use the methods of this chapter to analyze the data?

Solution This data set can also be used for a regression. Using the 20 pairs of observed values of SPACE (the independent variable) and SALES (the dependent variable) we obtain the simple linear regression

$$\widehat{\text{SALES}} = 0.459 + 0.0216 \cdot (\text{SPACE}).$$

The sum of squares due to regression is 0.4674, which is seen to agree with the sum of squares for the linear orthogonal contrast (Table 6.14). The error mean square is 0.0135, and the resulting F statistic is 33.766, easily rejecting the null hypothesis of no linear regression. This F value is not the same as that obtained for the linear contrast because the latter uses the within (or pure error, see Section 6.5 on fitting trends) mean square in the denominator. The regression coefficient indicates an increase of 0.0216 lbs. of apples per square foot of space.

Of course, this regression implies a straight line relationship, while we demonstrated in Chapter 6 that a quadratic model is necessary. In the regression context this misspecification can be verified by the plot of residuals from the linear regression given in Fig. 7.14. The need for a curved line response is

Figure 7.14

Residual Plot for Linear Regression for Apple Sales

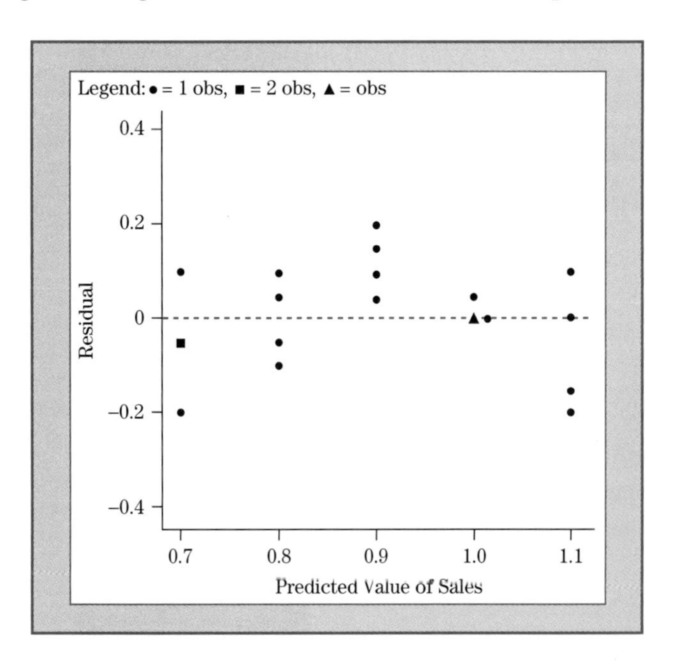

evident, although it is not particularly strong. As we have noted, this agrees with the conclusions of the analysis presented in Chapter 6.

Note that the regression provides the sum of squares for the linear trend obtained by the linear contrast in Section 6.5, reinforcing the statement that these contrasts are indeed a form of regression. In fact, with most computer programs, it is easier to obtain the sums of squares for trends by a regression and the pure error by an analysis of variance and manually combine the results for the lack of fit test. Additional examples are found in Chapter 9. ■

7.9 CHAPTER SUMMARY

Solution to Example 7.1 The effect of newspaper coverage of murder–suicides by airplane crashes on the number of succeeding multiple fatality crashes provides a relatively straightforward application of regression analysis. Using a linear regression model with CRASH as the dependent variable and INDEX as the independent variable produces the computer output using PROC REG from the SAS System shown in Table 7.12.

The F value for testing the model is 10.053 and certainly implies that there is a relationship between these variables and that the index can be used to estimate or predict the number of crashes. The estimated prediction equation is

$$\widehat{\text{CRASHES}} = 3.57 + 0.011 \cdot \text{INDEX}.$$

This equation estimates about 3.6 crashes when there is no publicity, with about one additional crash for every 100 units of the publicity index.

Table 7.12

Regression for Airplane Crash Data

Model: MODEL1
Dependent Variable: CRASH

Analysis of Variance

Source	df	Sum of Squares	Mean Square	F Value	Prob > F
Model	1	28.70256	28.70256	10.053	0.0063
Error	15	42.82685	2.85512		
C Total	16	71.52941			
	Root MSE	1.68971	R-square	0.4013	
	Dep Mean	4.70588	Adj R-sq	0.3614	
	C.V.	35.90636			

Parameter Estimates

Variable	df	Parameter Estimate	Standard Error	T for H0: Parameter = 0	Prob > |T|
INTERCEP	1	3.574149	0.54346601	6.577	0.0001
INDEX	1	0.010870	0.00342825	3.171	0.0063

Figure 7.15

Residual Plot for Airplane Crash Regression

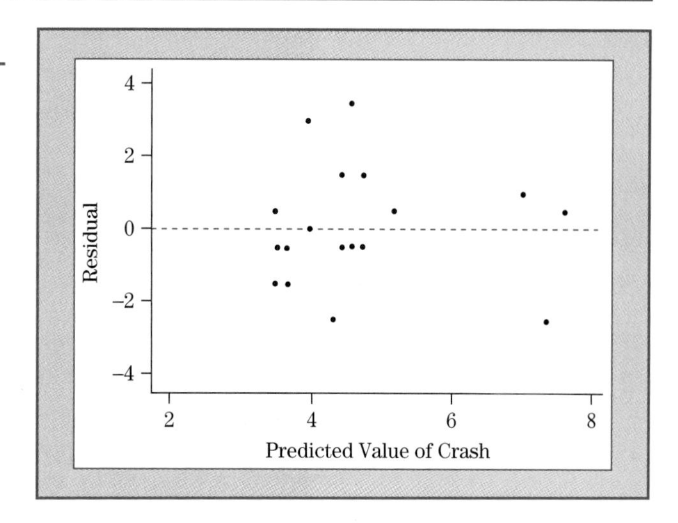

The relatively low value of the coefficient of determination suggests that considerable variation remains in crashes not explained by the model. A plot of prediction intervals (not given here) confirms this result. The residual plot, given in Fig. 7.15, does not indicate any obvious violations of assumptions. ■

The linear regression model,

$$y = \beta_0 + \beta_1 x + \varepsilon,$$

is used as the basis for establishing the nature of a relationship between values of an independent or factor variable, x, and values of a dependent or response variable, y. The model specifies that y is a random variable with a mean that is linearly related to x and has a variance specified by the random variable ε.

The first step in a regression analysis is to use n pairs of observed x and y values to obtain least squares estimates of the model parameters β_0 and β_1.

The next step is to estimate the variance of the random error. This quantity is defined as the variance of the residuals from the regression but is computed from a partitioning of sums of squares. This partitioning is also used for the test of the null hypothesis that the regression relationship does not exist.

An alternate and equivalent test for the hypothesis $\beta_1 = 0$ is provided by a t statistic, which can be used for one-tailed tests and to test for any specified value of β_1 and to construct a confidence interval.

Inferences on the response variable include confidence intervals for the conditional mean as well as prediction intervals for a single observation.

The correlation coefficient is a measure of the strength of a linear relationship between two variables. This measure is also useful when there is no independent/dependent variable relationship. The square of the correlation coefficient is used to describe the effectiveness of a linear regression.

As for most statistical analyses, it is important to verify that the assumptions underlying the model are fulfilled. Of special importance are the assumptions of proper model specification, homogeneous variance, and lack of outliers. In regression, this can be accomplished by examining the residuals. Additional methods are provided in Chapter 8.

7.10 CHAPTER EXERCISES

CONCEPT QUESTIONS

For the following true/false statements regarding concepts and uses of simple linear regression analysis, indicate whether the statement is true or false and specify what will correct a false statement.

1. ______________ The need for a nonlinear regression can only be determined by a lack of fit test.
2. ______________ The correlation coefficient indicates the change in y associated with a unit change in x.
3. ______________ To conduct a valid regression analysis, both x and y must be approximately normally distributed.
4. ______________ Rejecting the null hypothesis of no linear regression implies that changes in x cause changes in y.
5. ______________ In linear regression we may extrapolate without danger.
6. ______________ If x and y are uncorrelated in the population, the expected value of the estimated linear regression coefficient (slope) is zero.
7. ______________ If the true regression of y on x is curvilinear, a linear regression still provides a good approximation to that relationship.
8. ______________ The x values must be randomly selected in order to use a regression analysis.
9. ______________ The error or residual sum of squares is the numerator portion of the formula for the variance of y about the regression line.
10. ______________ The term $\hat{\mu}_{y|x}$ serves as the point estimate for estimating both the mean and individual prediction of y for a given x.
11. ______________ Useful prediction intervals for y can be obtained from a regression analysis.
12. ______________ In a regression analysis, the estimated mean of the distribution of y is the sample mean ($\bar{y}$).
13. ______________ All data points will fit the regression line exactly if the sample correlation is either $+1$ or -1.
14. ______________ The prediction interval for y is widest when x is at its mean.

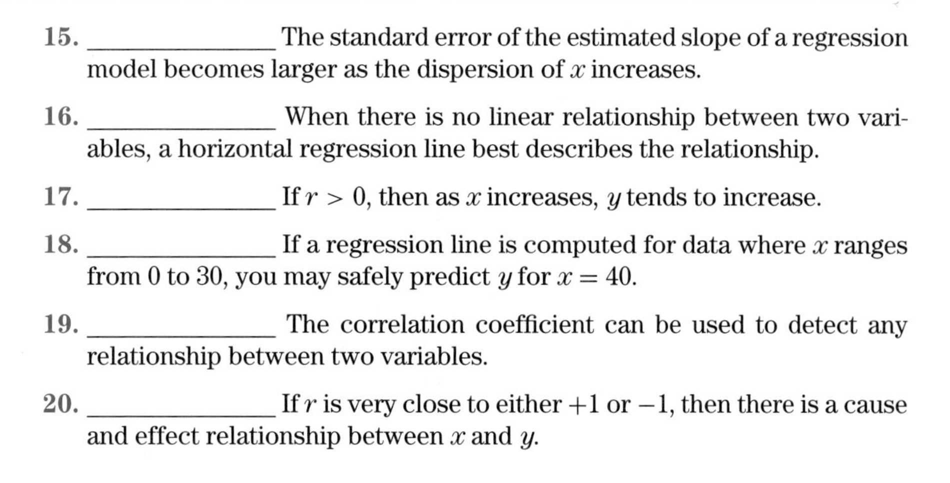

15. ______________ The standard error of the estimated slope of a regression model becomes larger as the dispersion of x increases.

16. ______________ When there is no linear relationship between two variables, a horizontal regression line best describes the relationship.

17. ______________ If $r > 0$, then as x increases, y tends to increase.

18. ______________ If a regression line is computed for data where x ranges from 0 to 30, you may safely predict y for $x = 40$.

19. ______________ The correlation coefficient can be used to detect any relationship between two variables.

20. ______________ If r is very close to either $+1$ or -1, then there is a cause and effect relationship between x and y.

EXERCISES

Note: Exercises 1 through 5 contain very few observations and are suitable for manual computation, which can be checked against computer outputs. The remainder of the problems are best performed by a computer.

Table 7.13

Data for Exercise 1

Oxidation	Temperature
4	−2
3	−1
3	0
2	1
2	2

1. The data of Table 7.13 represent the thickness of oxidation on a metal alloy for different settings of temperature in a curing oven. The values of temperature have been coded so that zero is the "normal" temperature, which makes manual computation easier.
   (a) Calculate the estimated regression line to predict oxidation based on temperature. Explain the meaning of the coefficients and the variance of residuals.
   (b) Calculate the estimated oxidation thickness for each of the temperatures in the experiment.
   (c) Calculate the residuals and make a residual plot. Discuss the distribution of residuals.
   (d) Test the hypothesis that $\beta_1 = 0$, using both the analysis of variance and t tests.

Table 7.14

Data for Exercise 2

Days	Sugar
0	7.9
1	12.0
3	9.5
4	11.3
5	11.8
6	11.3
7	4.2
8	0.4

2. The data of Table 7.14 show the sugar content of a fruit (Sugar) for different numbers of days after picking (Days).
   (a) Obtain the estimated regression line to predict sugar content based on the number of days the fruit is left on the tree.
   (b) Calculate and plot the residuals against days. Do the residuals suggest a fault in the model?

3. The grades for 15 students on midterm and final examinations in an English course are given in Table 7.15.
   (a) Obtain the least-squares regression to predict the score on the final examination from the midterm examination score. Test for significance of the regression and interpret the results.
   (b) It is suggested that if the regression is significant, there is no need to have a final examination. Comment. (*Hint:* Compute one or two 95% prediction intervals.)

Table 7.15

Data for Exercise 3

Midterm	Final
82	76
73	83
95	89
66	76
84	79
89	73
51	62
82	89
75	77
90	85
60	48
81	69
34	51
49	25
87	74

Table 7.16

Data for Exercise 4

x	y
−1	7
−1	3
−1	6
−1	6
−1	7
−1	4
−1	2
1	5
1	8
1	12
1	8
1	6
1	8
1	9

(c) Plot the estimated line and the actual data points. Comment on these results.

(d) Predict the final score for a student who made a score of 82 on the midterm. Check this calculation with the plot made in part (c).

(e) Compute r and r^2 and compare results with the partitioning of sums of squares in part (a).

4. Given the values in Table 7.16 for the independent variable x and dependent variable y:
   (a) Perform the linear regression of y on x. Test H_0: $\beta_1 = 0$.
   (b) Note that half of the observations have $x = -1$ and the rest have $x = +1$. Does this suggest an alternate analysis? If so, perform such an analysis and compare results with those of part (a).

5. It is generally believed that taller persons make better basketball players because they are better able to put the ball into the basket. Table 7.17 lists the heights of a sample of 25 nonbasketball athletes and the number of successful baskets made in a 60-s time period.
   (a) Perform a regression relating Goals to Height to ascertain whether there is such a relationship and, if there is, estimate the nature of that relationship.
   (b) Estimate the number of goals to be made by an athlete who is 60 in. tall. How much confidence can be assigned to that estimate?

6. Table 7.18 gives latitudes (Lat) and the mean monthly range (Range) between mean monthly maximum and minimum temperatures for a selected set of U.S. cities.
   (a) Perform a regression using Range as the dependent and Lat as the independent variable. Does the resulting regression make sense? Explain.
   (b) Compute the residuals; find the largest positive and negative residuals. Do these residuals suggest a pattern? Describe a phenomenon that may explain these residuals.

7. In an effort to determine the cost of air conditioning, a resident in College Station, TX, recorded daily values of the variables

$$\text{Tavg} = \text{mean temperature}$$
$$\text{Kwh} = \text{electricity consumption}$$

   for the period from September 19 through November 4 (Table 7.19).
   (a) Make a scatterplot to show the relationship of power consumption and temperature.
   (b) Using the model

$$\text{Kwh} = \beta_0 + \beta_1(\text{Tavg}) + \varepsilon,$$

   estimate the parameters, test appropriate hypotheses, and write a short paragraph stating your findings.
   (c) If you are doing this with a computer, make a residual plot to see whether the model appears to be appropriately specified.

Table 7.17

Data for Exercise 5: Basket Goals

Obs	Height	Goals
1	71	15
2	74	19
3	70	11
4	71	15
5	69	12
6	73	17
7	72	15
8	75	19
9	72	16
10	74	18
11	71	13
12	72	15
13	73	17
14	72	16
15	71	15
16	75	20
17	71	15
18	75	19
19	78	22
20	79	23
21	72	16
22	75	20
23	76	21
24	74	19
25	70	13

8. In Example 5.1 we posed the question of whether audit fees charged by the Big Eight accounting firms were higher than those charged by others. Perform separate regressions using the audit fee as the dependent variable and population as the independent variable for the cities using the Big Eight and the other cities. Does this analysis shed any light on the question of audit fees for the two groups? (A formal test to answer this question can be obtained by methods presented in Chapter 11.)

9. It has been argued that many cases of infant mortality rates are caused by teenage mothers who, for various reasons, do not receive proper prenatal care. From the *Statistical Abstract of the United States* we have statistics on the teenage birth rate (per 1000) and the infant mortality rate (per 1000 live births) for the 48 contiguous states. The data are given in Table 7.20, where Teen denotes the birthrate for teenage mothers and Mort denotes the infant mortality rate.

(a) Perform a regression to estimate Mort using Teen as the independent variable. Do the results confirm the stated hypothesis? Interpret the results.

(b) Construct a residual plot. Comment on the results.

10. In Exercise 13 of Chapter 1, the half-life of aminoglycosides was measured on 43 patients given either Amikacin or Gentamicin. The data are reproduced in different form in Table 7.21.

(a) Perform a regression to estimate HALF-LIFE using DO_MG_KG for each type of drug separately. Do the drugs seem to have parallel regression lines (a formal test for parallelism is presented in Chapter 11)?

(b) Perform the appropriate inferences on both lines to determine whether the relationship between half-life and dosage is significant. Use $\alpha = 0.05$. Completely explain your results.

(c) Draw a scatter diagram of HALF-LIFE versus DO_MG_KG indexed by type of drug (use A's and G's). Draw the regression lines obtained in part (a) on the same graph.

11. An experimenter is testing a new pressure gauge against a standard (a gauge known to be accurate) by taking three readings each at 50, 100, 150, 200, and 250 lbs./in.^{2}. The purpose of the experiment is to ascertain the precision and accuracy of the new gauge. The data are shown in Table 7.22.

As we saw in Example 7.4 both precision and accuracy are important factors in determining the effectiveness of a measuring instrument. Perform the appropriate analysis to determine the effectiveness of this instrument. However, this device has a shortcoming of a slightly different nature. Perform the appropriate analyses to find the shortcoming.

12. Instructors often suspect that the better students finish tests early. To test this hypothesis an instructor noted both the order (ORDER) and actual time (TIMES) in which students in three sections (SECTN) of a class, numbering 29, 28, and 28 students, respectively, handed in a particular test.

Table 7.18

Data for Exercise 6: Latitudes and Temperature Ranges for U.S. Cities

City	State	Lat	Range	City	State	Lat	Range
Montgome	AL	32.3	18.6	Tuscon	AZ	32.1	19.7
Bishop	CA	37.4	21.9	Eureka	CA	40.8	5.4
San_Dieg	CA	32.7	9.0	San_Fran	CA	37.6	8.7
Denver	CO	39.8	24.0	Washington	DC	39.0	24.0
Miami	FL	25.8	8.7	Talahass	FL	30.4	15.9
Tampa	FL	28.0	12.1	Atlanta	GA	33.6	19.8
Boise	ID	43.6	25.3	Moline	IL	41.4	29.4
Ft_wayne	IN	41.0	26.5	Topeka	KS	39.1	27.9
Louisv	KY	38.2	24.2	New_Orl	LA	30.0	16.1
Caribou	ME	46.9	30.1	Portland	ME	43.6	25.8
Alpena	MI	45.1	26.5	St_cloud	MN	45.6	34.0
Jackson	MS	32.3	19.2	St_Louis	MO	38.8	26.3
Billings	MT	45.8	27.7	N _PLatte	NB	41.1	28.3
L_Vegas	NV	36.1	25.2	Albuquer	NM	35.0	24.1
Buffalo	NY	42.9	25.8	NYC	NY	40.6	24.2
C_Hatter	NC	35.3	18.2	Bismark	ND	46.8	34.8
Eugene	OR	44.1	15.3	Charestn	SC	32.9	17.6
Huron	SD	44.4	34.0	Knoxvlle	TN	35.8	22.9
Memphis	TN	35.0	22.9	Amarillo	TX	35.2	23.7
Brownsvl	TX	25.9	13.4	Dallas	TX	32.8	22.3
SLCity	UT	40.8	27.0	Roanoke	VA	37.3	21.6
Seattle	WA	47.4	14.7	Grn_bay	WI	44.5	29.9
Casper	WY	42.9	26.6				

Table 7.19

Data for Exercise 7: Heating Costs

Mo	Day	Tavg	Kwh	Mo	Day	Tavg	Kwh
9	19	77.5	45	10	13	68.0	50
9	20	80.0	73	10	14	66.5	37
9	21	78.0	43	10	15	69.0	43
9	22	78.5	61	10	16	70.5	42
9	23	77.5	52	10	17	63.0	25
9	24	83.0	56	10	18	64.0	31
9	25	83.5	70	10	19	64.5	31
9	26	81.5	69	10	20	65.0	32
9	27	75.5	53	10	21	66.5	35
9	28	69.5	51	10	22	67.0	32
9	29	70.0	39	10	23	66.5	34
9	30	73.5	55	10	24	67.5	35
10	1	77.5	55	10	25	75.0	41
10	2	79.0	57	10	26	75.5	51
10	3	80.0	68	10	27	71.5	34
10	4	79.0	73	10	28	63.0	19
10	5	76.0	57	10	29	60.0	19
10	6	76.0	51	10	30	64.0	30
10	7	75.5	55	10	31	62.5	23
10	8	79.5	56	11	1	63.5	35
10	9	78.5	72	11	2	73.5	29
10	10	82.0	73	11	3	68.0	55
10	11	71.5	69	11	4	77.5	56
10	12	70.0	38				

Table 7.20

Data for Exercise 9: Birth Rate Statistics

State	Teen	Mort	State	Teen	Mort	State	Teen	Mort
AL	17.4	13.3	MA	8.3	8.5	OH	13.3	10.6
AR	19.0	10.3	MD	11.7	11.7	OK	15.6	10.4
AZ	13.8	9.4	ME	11.6	8.8	OR	10.9	9.4
CA	10.9	8.9	MI	12.3	11.4	PA	11.3	10.2
CO	10.2	8.6	MN	7.3	9.2	RI	10.3	9.4
CT	8.8	9.1	MO	13.4	10.7	SC	16.6	13.2
DE	13.2	11.5	MS	20.5	12.4	SD	9.7	13.3
FL	13.8	11.0	MT	10.1	9.6	TN	17.0	11.0
GA	17.0	12.5	NB	8.9	10.1	TX	15.2	9.5
IA	9.2	8.5	NC	15.9	11.5	UT	9.3	8.6
ID	10.8	11.3	ND	8.0	8.4	VA	12.0	11.1
IL	12.5	12.1	NH	7.7	9.1	VT	9.2	10.0
IN	14.0	11.3	NJ	9.4	9.8	WA	10.4	9.8
KS	11.5	8.9	NM	15.3	9.5	WI	9.9	9.2
KY	17.4	9.8	NV	11.9	9.1	WV	17.1	10.2
LA	16.8	11.9	NY	9.7	10.7	WY	10.7	10.8

Table 7.21

Half-Life of Aminoglycosides: By Dosage and Drug Type

Drug = Amikacin		Drug = Gentamicin	
Half-Life	DO_MG_KG	Half-Life	DO_MG_KG
2.50	7.90	1.60	2.10
2.20	8.00	1.90	2.00
1.60	8.30	2.30	1.60
1.30	8.10	2.50	1.90
1.20	8.60	1.80	2.00
1.60	7.60	1.70	2.86
2.20	6.50	2.86	2.89
2.20	7.60	2.89	2.96
2.60	10.00	1.98	2.86
1.00	9.88	1.93	2.86
1.50	10.00	1.80	2.86
3.15	10.29	1.70	3.00
1.44	9.76	1.60	3.00
1.26	9.69	2.20	2.86
1.98	10.00	2.20	2.86
1.98	10.00	2.40	3.00
1.87	9.87	1.70	2.86
2.31	10.00	2.00	2.86
1.40	10.00	1.40	2.82
2.48	10.50	1.90	2.93
2.80	10.00	2.00	2.95
0.69	10.00		

The dependent variable is the students' average grade (AVERG) at the end of the semester. The data for this exercise are found on the data disk in file FW07P12. For each section as well as for the entire data set perform the regression of final average on both time and order. Do the results confirm instructors' impressions?

Table 7.22

Calibration Data for Exercise 7.11

Standard Gauge	**50**	**100**	**150**	**200**	**250**
New gauge	48	100	154	201	247
	44	100	154	200	245
	46	106	154	205	246

13. Use all of the home data given in Table 1.2 to do a regression of `price` on `space`. Plot the residuals vs the predicted values and comment on the effect the higher priced homes have on the assumptions. Construct a Q–Q plot for the residuals. Does the normality assumption appear to be satisfied with the entire data set? Does the cost per square foot change a lot? What might be the cause of this change?

Multiple Regression

EXAMPLE 8.1

What Factors Win Baseball Games? The game of baseball generates an unbelievable amount of descriptive statistics. Although most of us give these statistics only casual scrutiny, baseball managers may find them quite useful tools for analyzing team performance and consequently implementing policies to improve their team's standing.

Table 8.1 shows some summary statistics about the 10 National League baseball teams for the 1965 through 1968 seasons (Reichler, 1985). The variables collected for this study are

YEAR: the season: 1965–1968,
WIN: the team's winning percentage,
RUNS: the number of runs scored by the team,
BA: the team's overall batting percentage,
DP: the total number of double plays,
WALK: the number of walks given to the other team, and
SO: the number of strikeouts by the team's pitcher.

Obviously the study of the relationships among several variables is much more complicated than that between two variables discussed in Chapter 7. However, it is still useful to examine graphically the relationships among the pairs of variables in this example. Figure 8.1 is a "table" of scatterplots among all pairs of variables in Example 8.1 produced by SAS/INSIGHT. The entries in the diagonal elements (top left to bottom right) identify the variable in the scatterplots on the corresponding rows and columns and the numbers in the corners show the minimum and maximum values of those variables. For example, the first scatterplot in the first row is that between WIN on the vertical axis and

Table 8.1

Winning Baseball Games

OBS	YEAR	WIN	RUNS	BA	DP	WALK	SO
1	1965	0.599	608	0.245	135	425	1079
2	1965	0.586	682	0.252	124	408	1060
3	1965	0.556	675	0.265	189	469	882
4	1965	0.549	825	0.273	142	587	1113
5	1965	0.531	708	0.256	145	541	996
6	1965	0.528	654	0.250	153	466	1071
7	1965	0.497	707	0.254	152	467	916
8	1965	0.444	635	0.238	166	481	855
9	1965	0.401	569	0.237	130	388	931
10	1965	0.309	495	0.221	153	498	776
11	1966	0.586	606	0.256	128	356	1064
12	1966	0.578	675	0.248	131	359	973
13	1966	0.568	759	0.279	215	463	898
14	1966	0.537	696	0.258	147	412	928
15	1966	0.525	782	0.263	139	485	884
16	1966	0.512	571	0.251	166	448	892
17	1966	0.475	692	0.260	133	490	1043
18	1966	0.444	612	0.255	126	391	929
19	1966	0.410	587	0.239	171	521	773
20	1966	0.364	644	0.254	132	479	908
21	1967	0.627	695	0.263	127	431	956
22	1967	0.562	652	0.245	149	453	990
23	1967	0.540	702	0.251	143	463	888
24	1967	0.537	604	0.248	124	498	1065
25	1967	0.506	612	0.242	174	403	967
26	1967	0.500	679	0.277	186	561	820
27	1967	0.475	631	0.240	148	449	862
28	1967	0.451	519	0.236	144	393	967
29	1967	0.426	626	0.249	120	485	1060
30	1967	0.377	498	0.238	147	536	893
31	1968	0.599	583	0.249	135	375	971
32	1968	0.543	599	0.239	125	344	942
33	1968	0.519	612	0.242	149	392	894
34	1968	0.512	690	0.273	144	573	963
35	1968	0.500	514	0.252	139	362	871
36	1968	0.494	583	0.252	162	485	897
37	1968	0.469	470	0.230	144	414	994
38	1968	0.469	543	0.233	163	421	935
39	1968	0.451	473	0.228	142	430	1014
40	1968	0.444	510	0.231	129	479	1021

RUNS on the horizontal axis, and the values of the variable WIN range from 0.309 to 0.627 and RUNS ranges from 470 to 825. Note that each scatterplot is reproduced twice with the axes interchanged.

In this example the focus is on determining the effects of the independent variables (RUNS, BA, DP, WALK, SO) on the winning percentages (WIN). This means that we are interested in the relationships depicted in the first row (or column) of scatterplots. These appear to indicate moderately strong positive

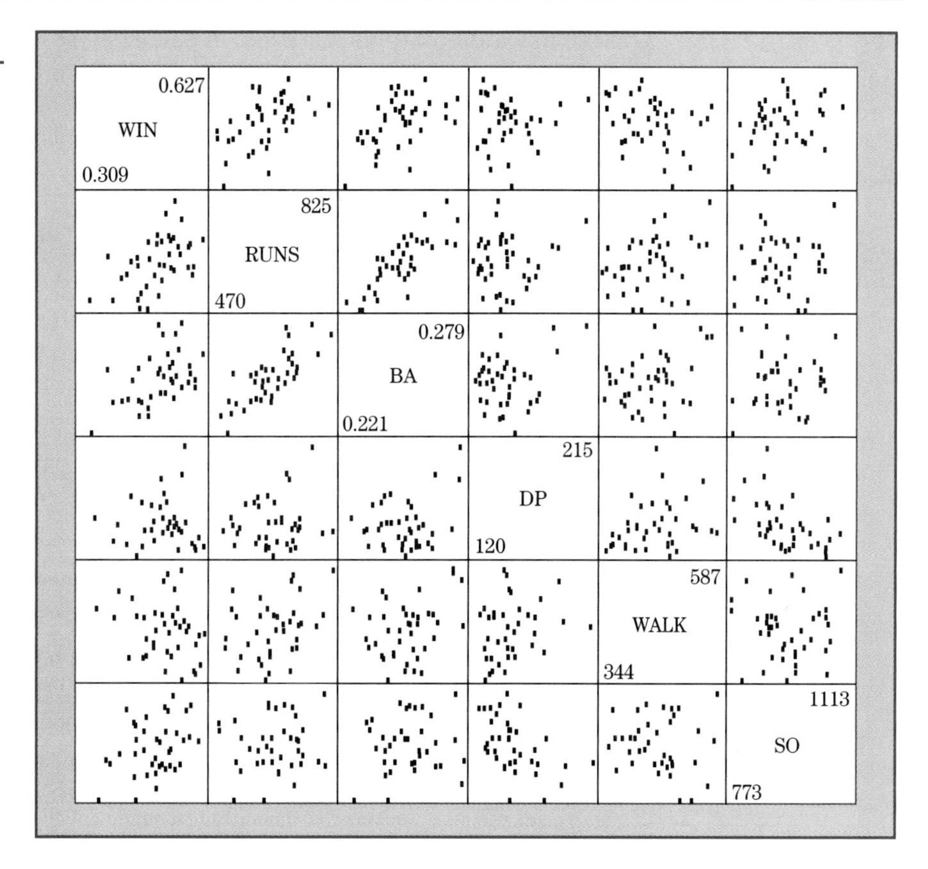

Figure 8.1

Scatterplots of Variables in Example 8.1

relationships of WIN to RUNS, BA, and SO, which appear reasonable. However, looking at the other scatterplots, we see a very strong positive relationship between RUNS and BA. This raises the question whether either or both are responsible for increased winning percentages, since these two variables are closely related. There is also a relatively strong negative relationship between WALK and SO. Could this relationship possibly change the effect of either on the winning percentages?

We will see that multiple regression analysis is designed to help answer these questions. However, because the interplay of so many variables can be very complex, the answers are not always as clear as we would like them to be. The solution to this example is provided in Section 8.10. ■

Notes on Exercises

Computations for all exercises in this chapter require statistical software. In most cases, the same program used for the exercises in Chapter 7 will suffice, the only difference being that more than one independent variable must be specified. After Section 8.2, Exercise 1 can be worked, using software options for the various outputs requested in that exercise. Referring to those outputs will help in understanding the material in Sections 8.1 through 8.4. Section 8.4 is a short review of the interpretation of computer outputs, after which all other assigned exercises except 8.7, 8.9, and 8.10 can be worked. These exercises can be worked after covering Section 8.6.

8.1 The Multiple Regression Model

In Chapter 7 we observed that the simple linear regression model

$$y = \beta_0 + \beta_1 x + \varepsilon,$$

which relates observed values of the dependent or response variable y to values of a single independent variable x, had limited practical application. The extension of this model to allow a number of independent variables is called a **multiple linear regression model**. The multiple regression model is written

$$y = \beta_0 + \beta_1 x_1 + \beta_2 x_2 + \cdots + \beta_m x_m + \varepsilon.$$

As in simple linear regression, y is the dependent or response variable, and the x_i, $i = 1, 2, \ldots, m$, are the m independent variables. The β_i are the (m) parameters or regression coefficients, one for each independent variable, and β_0 is the intercept. Also as in simple linear regression, ε is the random error.

The model is called linear regression because the model is linear in the parameters; that is, the coefficients (β_i) are simple (linear) multipliers of the independent variables and the error term (ε) is added (linearly) to the model. As we will see later, the model need not be linear in the independent variables. Although the model contains $(m + 1)$ parameters, it is often referred to as an m-variable model since the intercept coefficient does not correspond to a variable in the usual sense.

We have already alluded to applications of multiple regression models in Chapter 7. Some other applications include the following:

- A refinement of the fertilizer application example in Section 6.2, which relates yield to amounts applied of the three major fertilizer components: nitrogen, phosphorous, and potash.
- The number of "sick days" of school children is related to various characteristics such as waist circumference, height, weight, and age.
- Students' performances are related to scores on a number of different aptitude or mental ability tests.
- Amount of retail sales by an appliance manufacturer is related to expenditures for radio, television, newspaper, magazine, and direct mail advertising.
- Daily fuel consumption for home heating or cooling is related to temperature, cloud cover, and wind velocity.

In many ways, multiple regression is a relatively straightforward extension of simple linear regression. All assumptions and conditions underlying simple linear regression as presented in Chapter 7 remain essentially the same. The computations are more involved and tedious but computers have made these easier. The use of matrix notation and matrix algebra (Appendix B) makes the computations easier to understand and also illustrates the relationship between simple and multiple linear regression.

The potentially large number of parameters in a multiple linear regression model makes it useful to distinguish three different but related purposes for the use of this model:

1. To estimate the mean of the response variable (y) for a given set of values for the independent variables. This is the conditional mean, $\mu_{y|x}$, presented in Section 7.4, and estimated by $\hat{\mu}_{y|x}$. For example, we may want to estimate the mean fuel consumption for a day having a given set of values for the climatic variables. Associated with this purpose of a regression analysis is the question of whether all of the variables in the model are necessary to adequately estimate this mean.
2. To predict the response of a single unit for a given set of values of the independent variables. The point estimate is $\hat{\mu}_{y|x}$, but, because we are not estimating a mean, we will denote this predicted value by $\hat{y}$.
3. To evaluate the relationships between the response variable and the individual independent variables. That is, to make practical interpretations on the values of the regression coefficients, the β_i. For example, what would it mean if the coefficient for temperature in the above fuel consumption example were negative?

The Partial Regression Coefficient

The interpretation of the individual regression coefficients gives rise to an important difference between simple and multiple regression. In a multiple regression model the regression parameters, β_i, called **partial** regression coefficients, are not the same, either computationally or conceptually, as the so-called **total** regression coefficients obtained by individually regressing y on each x.

DEFINITION 8.1
The **partial regression coefficients** obtained in a multiple regression measure the change in the average value of y associated with a unit change in the corresponding x, **holding constant all other variables**.

This means that normally the individual coefficients of an m-variable multiple regression model will not have the same values nor the same interpretations as the coefficients for the m separate simple linear regressions involving the same variables. Many difficulties in using and interpreting the results of multiple regression arise from the fact that the definition of "holding constant," related to the concept of a partial derivative in calculus, is somewhat difficult to understand.

For example, in the application on estimating sick days of school children, the coefficient associated with the height variable measures the increase in sick days associated with a unit increase in height for a population of children all having identical waist circumference, weight, and age. In this application, the total and partial coefficients for height would differ because the total coefficient for height would measure not only the effect of height, but also indirectly measure the effect of the other related variables.

The application on estimating fuel consumption provides a similar scenario: The total coefficient for temperature would indirectly measure the effect of wind and cloudcover. Again this coefficient will differ from the partial regression coefficient because cloud cover and wind are often associated with lower temperatures.

We will see later that the inferential procedures for the partial coefficients are constructed to reflect this characteristic. We will also see that these inferences and associated interpretations are often made difficult by the existence of strong relationships among the several independent variables, a condition known as multicollinearity (Section 8.7).

Do We Really Need to Study Formulas? All multiple regression analyses are now performed with computers and all examples in this chapter are illustrated with computer outputs. It would therefore seem that the only people who need to know something about the formulas for doing these analyses are professional statisticians and computer programmers who write regression programs. However, we provide these formulas here in order to provide a feel for how multiple regression results are actually obtained and thus make the computer and its software look less like a magic box that somehow digests the data and provides all the necessary answers. We suggest verifying manually some of the computational procedures, although they should not be memorized with the idea that they will be extensively used.

Because the use of multiple regression models entails many different aspects, this chapter is quite long. Section 8.2 presents the procedures for estimating the coefficients, and Section 8.3 presents the procedure for obtaining the error variance and the inferences about model parameter and other estimates. Section 8.4 contains brief descriptions of correlations that describe the strength of linear relationships involving several variables. Section 8.5 provides some ideas on computer usage and presents computer outputs for examples used in previous sections. The last four sections deal with special models and problems that arise in a regression analysis.

8.2 Estimation of Coefficients

In Chapter 7, we showed that the least squares estimates of the parameters of the simple linear regression model are obtained by the solutions to the normal equations:

$$\beta_0 n + \beta_1 \sum x = \sum y,$$

$$\beta_0 \sum x + \beta_1 \sum x^2 = \sum xy.$$

Since there are only two equations in two unknowns, the solutions can be expressed in closed form, that is, as simple algebraic formulas involving the sums, sums of squares, and sums of products of the observed data values of the two variables x and y. These formulas are also used for the partitioning of sums of squares and the resulting inference procedures.

For the multiple regression model with m partial coefficients plus β_0 the least squares estimates are obtained by solving the following set of $(m+1)$ normal equations in $(m+1)$ unknown parameters:

$$\begin{aligned}
\beta_0 n \quad &+ \beta_1 \sum x_1 &&+ \beta_2 \sum x_2 &&+ \cdots + \beta_m \sum x_m &&= \sum y,\\
\beta_0 \sum x_1 &+ \beta_1 \sum x_1^2 &&+ \beta_2 \sum x_1 x_2 &&+ \cdots + \beta_m \sum x_1 x_m &&= \sum x_1 y,\\
\beta_0 \sum x_2 &+ \beta_1 \sum x_2 x_1 &&+ \beta_2 \sum x_2^2 &&+ \cdots + \beta_m \sum x_2 x_m &&= \sum x_2 y,\\
\cdot \quad & \quad \cdot && \quad \cdot && \quad \cdot && \quad \cdot\\
\beta_0 \sum x_m &+ \beta_1 \sum x_m x_1 &&+ \beta_2 \sum x_m x_2 &&+ \cdots + \beta_m \sum x_m^2 &&= \sum x_m y.
\end{aligned}$$

The solution to these normal equations provides the estimated coefficients, which are denoted by $\hat{\beta}_0, \hat{\beta}_1, \ldots, \hat{\beta}_m$. This set of equations is a straightforward extension of the set of two equations for the simple linear regression model. However, because of the large number of equations and variables, it is not possible to obtain simple formulas that directly compute the estimates of the coefficients as we did for the simple linear regression model in Chapter 7. In other words, the system of equations must be specifically solved for each application of this method. Although procedures are available for performing this task with hand-held or desk calculators, the solution is almost always obtained by computers using methods beyond the scope of this book. We do, however, need to represent symbolically the solutions to the set of equations. This is done with matrices and matrix notation.

Appendix B contains a brief introduction to matrix notation and the use of matrices for representing operations involving systems of linear equations. We will not actually be performing many matrix calculations; however, an understanding and appreciation of this material will make more understandable the material in the remainder of this chapter (as well as that of Chapter 11). Therefore, it is recommended Appendix B be reviewed before continuing.

Simple Linear Regression with Matrices

Estimating the coefficients of a simple linear regression produces a system of two equations in two unknowns, which can be solved explicitly and therefore do not require the use of matrix expressions. However, matrices can be used and we will do so here to illustrate this method.

Recall from Chapter 7 that the simple linear regression model for an individual observation is

$$y_i = \beta_0 + \beta_1 x_i + \varepsilon_i, \quad i = 1, 2, \ldots, n.$$

Using matrix notation, the regression model is written

$$\mathbf{Y} = \mathbf{XB} + \mathbf{E},$$

where $\mathbf{Y} = n \times 1$ matrix[1] of observed values of the dependent variable y; $\mathbf{X} = n \times 2$ matrix in which the first column consists of a column of ones[2] and the second column contains the values of the independent variable x; $\mathbf{B} = 2 \times 1$ matrix of the two parameters β_0; and β_1, and $\mathbf{E} = n \times 1$ matrix of the n values of the random error ε_i.

Placing these matrices in the above expression results in the matrix equation

$$\begin{bmatrix} y_1 \\ y_2 \\ \vdots \\ y_n \end{bmatrix} = \begin{bmatrix} 1 & x_1 \\ 1 & x_2 \\ \vdots & \vdots \\ 1 & x_n \end{bmatrix} \cdot \begin{bmatrix} \beta_0 \\ \beta_1 \end{bmatrix} + \begin{bmatrix} \epsilon_1 \\ \epsilon_2 \\ \vdots \\ \epsilon_n \end{bmatrix}.$$

Using the principles of matrix multiplication, we can verify that any row of the resulting matrices reproduces the simple linear regression model for an observation:

$$y_i = \beta_0 + \beta_1 x_i + \varepsilon_i.$$

We want to estimate the parameters of the regression model resulting in the estimating equation

$$\hat{\mathbf{M}}_{y|x} = \mathbf{X}\hat{\mathbf{B}},$$

where $\hat{\mathbf{M}}_{y|x}$ is an $n \times 1$ matrix of the $\hat{\mu}_{y|x}$ values, and $\hat{\mathbf{B}}$ is the 2×1 matrix of the estimated coefficients $\hat{\beta}_0$ and $\hat{\beta}_1$. The set of normal equations that must be solved to obtain the least squares estimates is

$$(\mathbf{X}'\mathbf{X})\hat{\mathbf{B}} = \mathbf{X}'\mathbf{Y},$$

where

$$\mathbf{X}'\mathbf{X} = \begin{bmatrix} 1 & 1 & \dots & 1 \\ x_1 & x_2 & \dots & x_n \end{bmatrix} \cdot \begin{bmatrix} 1 & x_1 \\ 1 & x_2 \\ \vdots & \vdots \\ 1 & x_n \end{bmatrix} = \begin{bmatrix} n & \sum x \\ \sum x & \sum x^2 \end{bmatrix},$$

$$\mathbf{X}'\mathbf{Y} = \begin{bmatrix} 1 & 1 & \dots & 1 \\ x_1 & x_2 & \dots & x_n \end{bmatrix} \cdot \begin{bmatrix} y_1 \\ y_2 \\ \vdots \\ y_n \end{bmatrix} = \begin{bmatrix} \sum y \\ \sum xy \end{bmatrix}.$$

---

[1] We use the convention that a matrix is denoted by the capital letter of the elements of the matrix. Unfortunately, the capital letters corresponding to β and μ are almost indistinguishable from **B** and **M**.

[2] This column may be construed as representing values of an artificial or dummy variable associated with the intercept coefficient, β_0.

The equations can now be written

$$\begin{bmatrix} n & \sum x \\ \sum x & \sum x^2 \end{bmatrix} \cdot \begin{bmatrix} \hat{\beta}_0 \\ \hat{\beta}_1 \end{bmatrix} = \begin{bmatrix} \sum y \\ \sum xy \end{bmatrix}.$$

Again, using the principles of matrix multiplication, we can see that this matrix equation reproduces the normal equations for simple linear regression (Section 7.3). The matrix representation of the solution of the normal equations is

$$\hat{\mathbf{B}} = (\mathbf{X}'\mathbf{X})^{-1}\mathbf{X}'\mathbf{Y}.$$

Since we will have occasion to refer to individual elements of the matrix $(\mathbf{X}'\mathbf{X})^{-1}$, we will refer to it as the matrix **C**, with the subscripts of the elements corresponding to the regression coefficients. Thus

$$\mathbf{C} = \begin{bmatrix} c_{00} & c_{01} \\ c_{10} & c_{11} \end{bmatrix}.$$

The solution can now be represented by the matrix equation

$$\hat{\mathbf{B}} = \mathbf{C}\mathbf{X}'\mathbf{Y}.$$

For the one-variable regression, the $\mathbf{X}'\mathbf{X}$ matrix is a 2×2 matrix and, as we have noted in Appendix B, the inverse of such a matrix is not difficult to compute. Define the matrix

$$\mathbf{A} = \begin{bmatrix} a_{11} & a_{12} \\ a_{21} & a_{22} \end{bmatrix}.$$

Then the inverse is

$$\mathbf{A}^{-1} = \begin{bmatrix} \frac{a_{22}}{k} & \frac{-a_{12}}{k} \\ \frac{-a_{21}}{k} & \frac{a_{11}}{k} \end{bmatrix},$$

where $k = a_{11}a_{22} - a_{12}a_{21}$. Substituting the elements of $\mathbf{X}'\mathbf{X}$, we have

$$(\mathbf{X}'\mathbf{X}^{-1}) = \mathbf{C} = \begin{bmatrix} \frac{\sum x^2}{k} & \frac{-\sum x}{k} \\ \frac{-\sum x}{k} & \frac{n}{k} \end{bmatrix},$$

where $k = n\sum x^2 - (\sum x)^2 = nS_{xx}$. Multiplying the matrices to obtain the estimates,

$$\hat{\mathbf{B}} = (\mathbf{X}'\mathbf{X})^{-1}\mathbf{X}'\mathbf{Y} = \begin{bmatrix} \frac{\sum x^2 \sum y}{nS_{xx}} + \frac{-\sum x \sum xy}{nS_{xx}} \\ \frac{-\sum x \sum y}{nS_{xx}} + \frac{n\sum xy}{nS_{xx}} \end{bmatrix}.$$

The second element of $\hat{\mathbf{B}}$ is

$$\frac{n\sum xy - \sum x \sum y}{nS_{xx}} = \frac{\sum xy - (\sum x \sum y/n)}{S_{xx}} = \frac{S_{xy}}{S_{xx}},$$

which is the formula for $\hat{\beta}_1$ given in Section 7.3. A little more algebra (which is left as an exercise for those who are so inclined) shows that the first element is $(\bar{y} - \hat{\beta}_1\bar{x})$, which is the formula for $\hat{\beta}_0$.

We illustrate the matrix approach with the home price data used to illustrate simple linear regression in Chapter 7 (data in Table 7.2). The data matrices (abbreviated to save space) are

$$\mathbf{X} = \begin{bmatrix} 1 & 0.951 \\ 1 & 1.036 \\ 1 & 0.676 \\ 1 & 1.456 \\ 1 & 1.186 \\ \vdots & \vdots \\ 1 & 1.920 \\ 1 & 2.949 \\ 1 & 3.310 \\ 1 & 2.805 \\ 1 & 2.553 \\ 1 & 2.510 \\ 1 & 3.627 \end{bmatrix} \qquad \mathbf{Y} = \begin{bmatrix} 30.0 \\ 39.9 \\ 46.5 \\ 48.6 \\ 51.5 \\ \vdots \\ 167.5 \\ 169.9 \\ 175.0 \\ 179.0 \\ 179.9 \\ 189.5 \\ 199.0 \end{bmatrix}.$$

Using the transpose and multiplication rules,

$$\mathbf{X'X} = \begin{bmatrix} 58 & 109.212 \\ 109.212 & 228.385 \end{bmatrix}, \quad \text{and} \quad \mathbf{X'Y} = \begin{bmatrix} 6439.998 \\ 13401.788 \end{bmatrix}.$$

The elements of these matrices are the uncorrected or uncentered sums of squares and cross products of the variables x and y and the "variable" represented by the column of ones. For this reason the matrices $\mathbf{X'X}$ and $\mathbf{X'Y}$ are often referred to as the sums of squares and cross-products matrices. Note that $\mathbf{X'X}$ is symmetric. The inverse is

$$(\mathbf{X'X})^{-1} = \mathbf{C} = \begin{bmatrix} 0.17314 & -0.08279 \\ -0.08279 & 0.04397 \end{bmatrix},$$

which can be verified using the special inversion method for a 2×2 matrix, or multiplying $\mathbf{X'X}$ by $(\mathbf{X'X})^{-1}$, which will result in an identity matrix (except for round-off error). Finally,

$$\hat{\mathbf{B}} = (\mathbf{X'X})^{-1}\mathbf{X'Y} = \begin{bmatrix} 5.4316 \\ 56.0833 \end{bmatrix},$$

which reproduces the estimated coefficients obtained using ordinary algebra in Section 7.3.

Estimating the Parameters of a Multiple Regression Model

The use of matrix methods to estimate the parameters of a simple linear regression model may appear to be a rather cumbersome method for getting the same results obtained in Section 7.3. However, if we define the matrices **X** and **B** as

$$\mathbf{X} = \begin{bmatrix} 1 & x_{11} & x_{12} & \cdots & x_{1m} \\ 1 & x_{21} & x_{22} & \cdots & x_{2m} \\ \cdot & \cdot & \cdot & \cdots & \cdot \\ 1 & x_{n1} & x_{n2} & \cdots & x_{nm} \end{bmatrix}, \quad \text{and} \quad \mathbf{B} = \begin{bmatrix} \beta_0 \\ \beta_1 \\ \beta_2 \\ \cdot \\ \beta_m \end{bmatrix},$$

then the multiple regression model,

$$y = \beta_0 + \beta_1 x_1 + \beta_2 x_2 + \cdots + \beta_m x_m + \varepsilon,$$

can be expressed as

$$\mathbf{Y} = \mathbf{XB} + \mathbf{E},$$

and the parameter estimates as

$$\hat{\mathbf{B}} = (\mathbf{X}'\mathbf{X})^{-1}\mathbf{X}'\mathbf{Y}.$$

Note that these expressions are valid for a multiple regression with any number of independent variables. That is, for a regression with m independent variables, the **X** matrix has n rows and $(m + 1)$ columns. Consequently, matrices **B** and $\mathbf{X}'\mathbf{Y}$ are of order $[(m + 1) \times 1]$ and $\mathbf{X}'\mathbf{X}$ and $(\mathbf{X}'\mathbf{X})^{-1}$ are of order $[(m + 1) \times (m + 1)]$.

The procedure for obtaining the estimates of the parameters of a multiple regression model is thus a straightforward application of using matrices to show the solution of a set of linear equations. First compute the $\mathbf{X}'\mathbf{X}$ matrix

$$\mathbf{X}'\mathbf{X} = \begin{bmatrix} n & \sum x_1 & \sum x_2 & \cdots & \sum x_m \\ \sum x_1 & \sum x_1^2 & \sum x_1 x_2 & \cdots & \sum x_1 x_m \\ \sum x_2 & \sum x_2 x_1 & \sum x_2^2 & \cdots & \sum x_2 x_m \\ \cdot & \cdot & \cdot & \cdots & \cdot \\ \sum x_m & \sum x_m x_1 & \sum x_m x_2 & \cdots & \sum x_m^2 \end{bmatrix},$$

that is, the matrix of sums of squares and cross products of all the independent variables. Next compute the $\mathbf{X}'\mathbf{Y}$ matrix

$$\mathbf{X}'\mathbf{Y} = \begin{bmatrix} \sum y \\ \sum x_1 y \\ \sum x_2 y \\ \vdots \\ \sum x_m y \end{bmatrix}.$$

The next step is to compute the inverse of $\mathbf{X'X}$. As we indicated earlier, we do not present here a procedure for this task; instead we assume the inverse has been obtained by a computer, which also provides the estimates by the matrix multiplication

$$\hat{\mathbf{B}} = (\mathbf{X'X})^{-1}\mathbf{X'Y} = \mathbf{CX'Y},$$

where, as previously noted, $\mathbf{C} = (\mathbf{X'X})^{-1}$.

Correcting for the Mean, an Alternative Calculating Method

Recall that the formula for simple linear regression in Chapter 7 used two steps to calculate the estimates of the regression coefficients. That is, the centered or "corrected" sums of squares and cross products was used to calculate $\hat{\beta}_1$, and $\hat{\beta}_0$ was obtained by a separate calculation that involved $\hat{\beta}_1$. This approach can also be used for multiple regression. If we define the elements of $\mathbf{X'X}$, $\mathbf{X'Y}$, and $\mathbf{Y'Y}$ as the corresponding corrected sums of squares and cross products (omitting the elements corresponding to the column of ones in $\mathbf{X}$) and compute the coefficients as shown in the preceding, then $\hat{\mathbf{B}}$ is the $m \times 1$ matrix containing all of the coefficients except $\hat{\beta}_0$. The intercept is calculated separately as

$$\hat{\beta}_0 = \bar{y} - \hat{\beta}_1\bar{x}_1 - \hat{\beta}_2\bar{x}_2 - \cdots - \hat{\beta}_m\bar{x}_m.$$

This method actually corresponds to the usual statistical analyses where the focus is on parameters other than the mean or intercept. In fact using the matrices of corrected sums of squares and cross products was the standard procedure for doing multiple regression calculations before computers. This was because the matrix elements were usually of smaller magnitudes and the order of the matrix to be inverted was one less, which resulted in a moderate saving of calculation time. However, when using computers, it is easier to use the variable represented by the column of ones to incorporate the intercept into the calculation for all coefficients, and is therefore the method we present here.

EXAMPLE 8.2

In Example 7.2 we showed how home prices can be estimated using information on sizes by the use of linear regression. We noted that although the regression was significant, the error of estimation was too large to make the model useful.

It was suggested that the use of other characteristics of houses could make such a model more useful.

Solution In Chapter 7 we used `size` as the single independent variable in a simple linear regression to estimate `price`. To illustrate multiple regression we will estimate `price` using the following five variables:

`age`: age of home, in years,
`bed`: number of bedrooms,
`bath`: number of bathrooms,

`size`: size of home in 1000 ft^2, and
`lot`: size of lot in 1000 ft^2.

In terms of the mnemonic variable names, the model is written

$$\texttt{price} = \beta_0 + \beta_1(\texttt{age}) + \beta_2(\texttt{bed}) + \beta_3(\texttt{bath}) + \beta_4(\texttt{size}) + \beta_5(\texttt{lot}) + \varepsilon.$$

The data for this example are shown in Table 8.2. Note that there is one observation that has no data for `size` as well as several observations with no data on `lot`. Because these observations cannot be used for this regression, the model will be applied to the remaining 51 observations.

Figure 8.2 is a scatterplot matrix of the variables involved in this regression using the same format as in Figure 8.1, except that the dependent variable is in the last row and column. The only strong relationship appears to be between price and size, and there are weaker relationships among `size`, `bed`, `bath`, and `price`.

The first step is to compute the sums of squares and cross products needed for the $\mathbf{X}'\mathbf{X}$ and $\mathbf{X}'\mathbf{Y}$ matrices. Note that for this purpose the $\mathbf{X}$ matrix must contain the column of ones, the dummy variable used for the intercept. Since most computer programs automatically generate this variable, it is not usually listed as part of the data. The results of these computations are shown in the top half of Table 8.3. Normally the intermediate calculations presented in this table are not printed by most software and are available with special options invoked here with `PROC REG` of the SAS System. In this table, each element is the sum of products of the variables listed in the row and column headings. For example, the sum of products of `lot` and `size` is 3558.9235. Note that the first row and column, labeled `intercept`, correspond to the column of ones used to estimate β_0, and the last row and column, labeled `price`, correspond to the dependent variable. Thus the first six rows and columns are $\mathbf{X}'\mathbf{X}$, the first six rows of the last column comprise $\mathbf{X}'\mathbf{Y}$, the first six columns of the last row comprise $\mathbf{Y}'\mathbf{X}$ while the last element is $\mathbf{Y}'\mathbf{Y}$, which is the sum of squares of the dependent variable `price`. Note also that the sum of products of `intercept` and another variable is the sum of values of that variable; the first element is the number of observations used in the analysis, which we have noted is only 51 because of the missing data.

As we have noted, the elements of $\mathbf{X}'\mathbf{X}$ and $\mathbf{X}'\mathbf{Y}$ comprise the coefficients of the normal equations. Specifically, the first equation is

$$51\beta_0 + 1045\beta_1 + 162\beta_2 + 109\beta_3 + 96.385\beta_4 + 1708.838\beta_5 = 5580.958.$$

The other equations follow.

The inverse as well as the solution of the normal equations comprise the second half of Table 8.3. Again the row and column variable names identify the elements. The first six rows and columns are the elements of the inverse, $(\mathbf{X}'\mathbf{X})^{-1}$, which we also denote by $\mathbf{C}$. The first six rows of the last column are the matrix of the estimated coefficients ($\hat{\mathbf{B}}$), the first six columns of the last row are the transpose of the matrix of coefficient estimates ($\hat{\mathbf{B}}'$), and the last

Table 8.2

Data on Home Prices for Multiple Regression

Obs	age	bed	bath	size	lot	price
1	21	3	3.0	0.951	64.904	30.000
2	21	3	2.0	1.036	217.800	39.900
3	7	1	1.0	0.676	54.450	46.500
4	6	3	2.0	1.456	51.836	48.600
5	51	3	1.0	1.186	10.857	51.500
6	19	3	2.0	1.456	40.075	56.990
7	8	3	2.0	1.368	.	59.900
8	27	3	1.0	0.994	11.016	62.500
9	51	2	1.0	1.176	6.256	65.500
10	1	3	2.0	1.216	11.348	69.000
11	32	3	2.0	1.410	25.450	76.900
12	2	3	2.0	1.344	.	79.000
13	25	2	2.0	1.064	218.671	79.900
14	31	3	1.5	1.770	19.602	79.950
15	29	3	2.0	1.524	12.720	82.900
16	16	3	2.0	1.750	130.680	84.900
17	20	3	2.0	1.152	104.544	85.000
18	18	4	2.0	1.770	10.640	87.900
19	28	3	2.0	1.624	12.700	89.900
20	27	3	2.0	1.540	5.679	89.900
21	8	3	2.0	1.532	6.900	93.500
22	19	3	2.0	1.647	6.900	94.900
23	3	3	2.0	1.344	43.560	95.800
24	5	3	2.0	1.550	6.575	98.500
25	5	4	2.0	1.752	8.193	99.500
26	27	3	1.5	1.450	11.300	99.900
27	33	2	2.0	1.312	7.150	102.000
28	4	3	2.0	1.636	6.097	106.000
29	0	3	2.0	1.500	.	108.900
30	36	3	2.5	1.800	83.635	109.900
31	5	4	2.5	1.972	7.667	110.000
32	0	3	2.0	1.387	.	112.290
33	27	4	2.0	2.082	13.500	114.900
34	15	3	2.0	.	269.549	119.500
35	23	4	2.5	2.463	10.747	119.900
36	25	3	2.0	2.572	7.090	119.900
37	24	4	2.0	2.113	7.200	122.900
38	1	3	2.5	2.016	9.000	123.938
39	34	3	2.0	1.852	13.500	124.900
40	26	4	2.0	2.670	9.158	126.900
41	26	3	2.0	2.336	5.408	129.900
42	31	3	2.0	1.980	8.325	132.900
43	24	4	2.5	2.483	10.295	134.900
44	29	5	2.5	2.809	15.927	135.900
45	21	3	2.0	2.036	16.910	139.500
46	10	3	2.0	2.298	10.950	139.990
47	3	3	2.0	2.038	7.000	144.900
48	9	3	2.5	2.370	10.796	147.600
49	29	5	3.5	2.921	11.992	149.990
50	8	3	2.0	2.262	.	152.550
51	7	3	3.0	2.456	.	156.900
52	1	4	2.0	2.436	52.000	164.000

(Continued)

Table 8.2 *(continued)*

Obs	age	bed	bath	size	lot	price
53	27	3	2.0	1.920	226.512	167.500
54	5	3	2.5	2.949	11.950	169.900
55	32	4	3.5	3.310	10.500	175.000
56	29	3	3.0	2.805	16.500	179.000
57	1	3	3.0	2.553	8.610	179.900
58	1	3	2.0	2.510	.	189.500
59	33	3	4.0	3.627	17.760	199.000

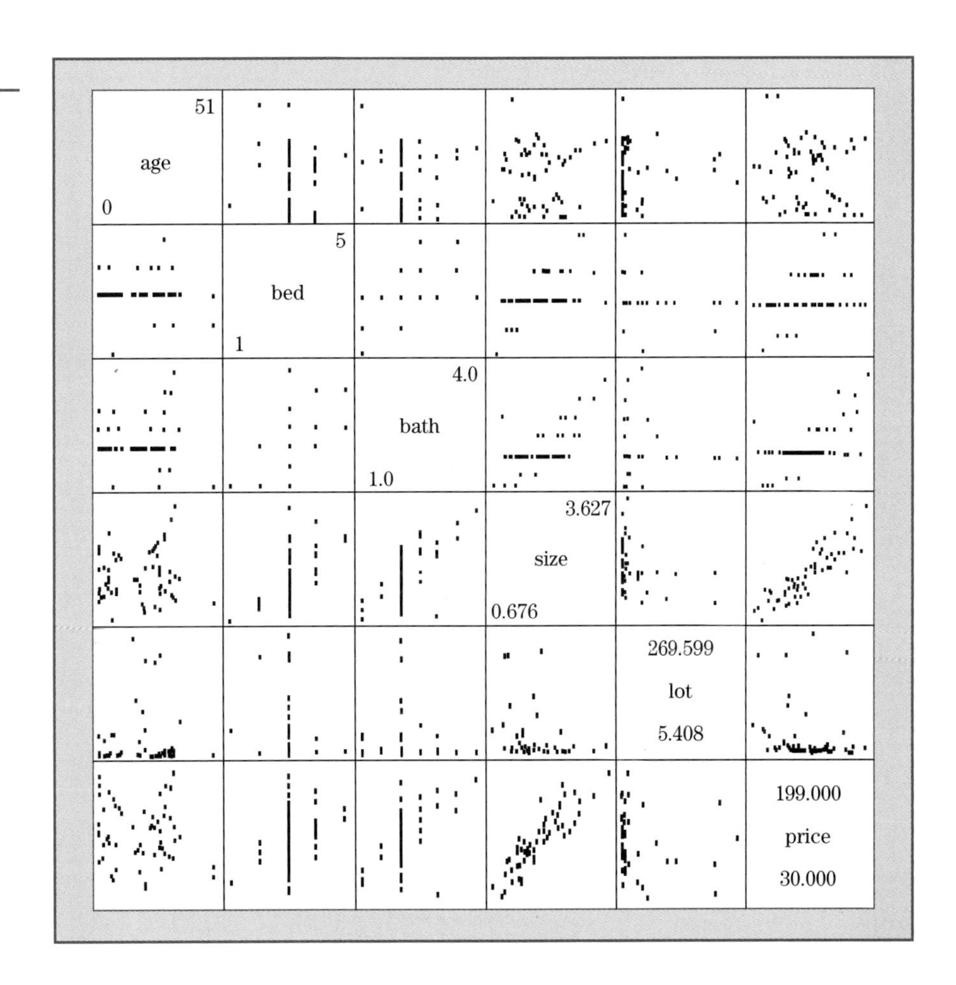

Figure 8.2

Scatterplot Matrix for Home Price Data

element corresponding to the row and column labeled with the dependent variable (`price`) is the residual sum of squares, which is defined in the next section.

A sharp-eyed reader will see the number −2.476418E−6 in the second column of row 6. This is shorthand for saying that the number is to be multiplied by 10^{-6}.

Table 8.3 Matrices for Multiple Regression

The REG Procedure

Model Crossproducts X′X X′Y Y′Y

Variable	Intercept	age	bed	bath	size	lot	price
Intercept	51	1045	162	109	96.385	1708.838	5580.958
age	1045	29371	3313	2199.5	1981.721	36060.245	112308.608
bed	162	3313	538	355	318.762	4981.272	18230.154
bath	109	2199.5	355	250	219.4685	3558.9235	12646.395
size	96.385	1981.721	318.762	219.4685	203.085075	2683.133101	11688.513058
lot	1708.838	36060.245	4981.272	3558.9235	2683.133101	202858.09929	165079.36843
price	5580.958	112308.608	18230.154	12646.395	11688.513058	165079.36843	690197.14064

X′X Inverse, Parameter Estimates, and SSE

Variable	Intercept	age	bed	bath	size	lot	price
Intercept	0.6510931798	−0.003058625	−0.130725187	−0.097462177	0.0383208773	−0.000527955	35.287921644
age	−0.003058625	0.0001293154	0.0000396856	0.0006649237	−0.000558371	−2.476418E−6	−0.349804533
bed	−0.130725187	0.0000396856	0.0640254429	−0.007028134	−0.03218064	0.0000709189	−11.23820158
bath	−0.097462177	0.0006649237	−0.007028134	0.1314351128	−0.087657959	−0.00027108	−4.540152056
size	0.0383208773	−0.000558371	−0.03218064	−0.087657959	0.1328335042	0.0003475797	65.946466578
lot	−0.000527955	−2.476418E−6	0.0000709189	−0.00027108	0.0003475797	8.2341898E−6	0.0620508107
price	35.287921644	−0.349804533	−11.23820158	−4.540152056	65.946466578	0.0620508107	13774.049724

It is instructive to verify the calculation for the estimated coefficients. For example, the estimated coefficient for age is

$$\begin{aligned}\hat{\beta}_1 &= (-0.003058625)(5580.958) + (0.0001293154)(112308.608)\\ &\quad + (0.0000396856)(18230.154) + (0.0006649237)(12646.3950)\\ &\quad + (-0.000558371)(11688.513) + (-2.476418\text{E}{-6})(165079.37)\\ &= -0.349804.\end{aligned}$$

If you try to verify this on a calculator, the result may differ due to round-off. You may also wish to verify some of the other estimates.

We can now write the equations for the estimated regression:

$$\begin{aligned}\widehat{price} &= 35.288 - 0.350(\texttt{age}) - 11.238(\texttt{bed})\\ &\quad - 4.540(\texttt{bath}) + 65.946(\texttt{size}) + 0.062(\texttt{lot}).\end{aligned}$$

This equation may be used to estimate the price for a home having specific values for the independent variables, with the caution that these values are in the range of the values observed in the data set. For example we can estimate the price of the first home shown in Table 8.2 as

$$\begin{aligned}\widehat{price} &= 35.288 - 0.349(21) - 11.238(3) - 4.540(3)\\ &\quad + 65.946(0.951) + 0.062(64.904)\\ &= 47.349,\end{aligned}$$

or $47,349, compared to the actual price of $30,000.

The estimated coefficients are interpreted as follows:

- The intercept ($\hat{\beta}_0 = 35.288$) is the estimated mean price (in $1000) of a home for which the values of all independent variables are zero. As in many applications this coefficient has no practical value, but is necessary in order to specify the equation.
- The coefficient for age ($\hat{\beta}_1 = -0.350$) estimates a decrease of $350 in the average price for each additional year of age, holding constant all other variables.
- The coefficient for bed ($\hat{\beta}_2 = -11.238$) estimates a decrease in price of $11,238 for each additional bedroom, holding constant all other variables.
- The coefficient for bath ($\hat{\beta}_3 = -4.540$) estimates a decrease in price of $4540 for each additional bathroom, holding constant all other variables.
- The coefficient for size ($\hat{\beta}_4 = 65.946$) estimates an increase in price of $65.95 for each additional square foot of the home, holding constant all other variables.
- The coefficient for lot ($\hat{\beta}_5 = 0.062$) estimates an increase in price of 62 cents for each additional square foot of lot, holding constant all other variables.

The coefficients for bed and bath appear to contradict expectations, as one would expect additional bedrooms and bathrooms to increase the price of a home. However, because these are *partial* coefficients, the coefficient for bed estimates the change in price for an additional bedroom *holding constant* size (among others). Now if you increase the number of bedrooms without increasing the size of the home, the bedrooms are smaller and the home seems more crowded and less attractive, hence a lower price. The reason for a negative coefficient for bath is not as obvious.

The values of the partial coefficients are therefore generally different from the corresponding total coefficients obtained with simple linear regression. For example, the coefficient for size in the one variable regression in Chapter 7 was 56.083, which is certainly different from the value of 65.946 in the multiple regression. You may want to verify this for some of the other variables; for example, the coefficient for the regression of price on bed will almost certainly result in a positive coefficient.

Comparison of coefficients across variables can be made by the use of **standardized** coefficients. These are obtained by standardizing all variables to have mean zero and unit variance and using these to compute the regression coefficients. However, they are more easily computed by the formula

$$\hat{\beta}_i^* = \hat{\beta}_i \frac{s_{x_i}}{s_y},$$

where $\hat{\beta}_i$ are the usual coefficient estimates, s_{x_i} is the sample standard deviation of x_i, and s_y is the standard deviation of y. This relationship shows that the standardized coefficient is the usual coefficient multiplied by the ratio of the standard deviations of x_i and y. This coefficient shows the change in standard deviation units of y associated with a standard deviation change in x_i, holding constant all other variables. Standardized coefficients are not often used and are available as special options in most regression programs.

The standardized coefficients for Example 8.2 are shown here as provided by the STB option of SAS System PROC REG:

Variable	Standardized Estimate
Intercept	0
age	−0.11070
bed	−0.19289
bath	−0.06648
size	1.07014
lot	0.08399

The intercept is zero, by definition. We can now see that size has by far the greatest effect, while bath and lot have the least. We will see, however, that this does not necessarily translate into degree of statistical significance (p value). ■

8.3 Inferential Procedures

Having estimated the parameters of the regression model, the next step is to perform the associated inferential procedures. As in simple linear regression, the first step is to obtain an estimate of the variance of the random error ε, which is required for performing these inferences.

Estimation of σ^2 and the Partitioning of the Sums of Squares

As in the case of simple linear regression, the variance of the random error σ^2 is estimated from the residuals

$$s_{y|x}^2 = \frac{\text{SSE}}{\text{df}} = \frac{\sum(y - \hat{\mu}_{y|x})^2}{(n - m - 1)},$$

where the denominator degrees of freedom $(n-m-1) = [n-(m+1)]$ results from the fact that the estimated values, $\hat{\mu}_{y|x}$, are based on $(m+1)$ estimated parameters: $\hat{\beta}_0, \hat{\beta}_1, \ldots, \hat{\beta}_m$.

As in simple linear regression we do not compute the error sum of squares by direct application of the above formula. Instead we use a partitioning of sums of squares:

$$\sum y^2 = \sum \hat{\mu}_{y|x}^2 + \sum (y - \hat{\mu}_{y|x})^2.$$

Note that, unlike the partitioning of sums of squares for simple linear regression, the left-hand side is the uncorrected sum of squares for the dependent variable.[3] Consequently, the term corresponding to the regression sum of squares includes the contribution of the intercept and is therefore not normally used for inferences (see the next subsection).

As with simple linear regression, a shortcut formula is available for the sum of squares due to regression, which is then subtracted from $\sum y^2$ to provide the error sum of squares. Also as in simple linear regression, several equivalent forms are available for computing this quantity, which we will denote by SSR. The most convenient for manual computing is

$$\text{SSR} = \hat{\mathbf{B}}'\mathbf{X}'\mathbf{Y},$$

which results in the algebraic expression

$$\text{SSR} = \hat{\beta}_0 \sum y + \hat{\beta}_1 \sum x_1 y + \cdots + \hat{\beta}_m \sum x_m y.$$

[3]This way of defining these quantities corresponds to the use of matrices consisting of uncorrected sums of squares and cross products with the column of ones for the intercept term. However, using matrices with corrected sums of squares and cross products results in defining TSS and SSR in a manner analogous to those shown in Chapter 7. These different definitions cause minor modifications in computational procedures but the ultimate results are the same.

Note that the individual terms are similar to SSR for the simple linear regression model; other equations for this quantity are

$$\text{SSR} = \mathbf{Y}'\mathbf{X}(\mathbf{X}'\mathbf{X})^{-1}\mathbf{X}'\mathbf{Y} = \hat{\mathbf{B}}'\mathbf{X}'\mathbf{X}\hat{\mathbf{B}}.$$

The quantities needed for the more convenient formula are available in Table 8.3 as

$$\begin{aligned}\sum y^2 &= 690,197.14,\\ \text{SSR} &= (35.288)(5580.958) + (-0.3498)(112308.6) + (-11.2382)(18230.1)\\ &\quad + (-4.5402)(12646.4) + (65.9465)(11688.5) + (0.06205)(165079.4)\\ \text{SSR} &= 676,423.09;\end{aligned}$$

hence by subtraction

$$\text{SSE} = 690,197.14 - 676,423.09 = 13,774.05.$$

This is the same quantity printed as the last element of the inverse matrix portion of the output in Table 8.3. As in simple linear regression, it can also be computed directly from the residuals, which are shown later in Table 8.6. The error degrees of freedom are

$$(n - m - 1) = 51 - 5 - 1 = 45,$$

and the resulting mean square (MSE) provides the estimated variance

$$s^2_{y|x} = 13774.05/45 = 306.09,$$

resulting in an estimated standard deviation of 17.495. This is somewhat smaller than the value of 19.684, which was obtained in Chapter 7 using only `size` as the independent variable. This relatively small decrease suggests that the other variables may contribute only marginally to the fit of the regression equation. The formal test for this is presented in the next subsection.

This estimated standard deviation is interpreted as it was in Section 1.5, and is an often overlooked statistic for assessing the goodness of fit of a regression model. Thus if the distribution of the residuals is reasonably bell shaped, approximately 95% of the residuals will be within two standard deviations of the regression estimates. In the house prices data, the standard deviation is 17.495 ($17,495). Hence using the empirical rule, it follows that approximately 95% of homes are within 2($17,495) or within approximately $35,000 of the values estimated by the regression model.

The Coefficient of Variation

In Section 1.5 we defined the **coefficient of variation** as the ratio of the standard deviation to the mean expressed as a percentage. This measure can also be applied as a measure of residual variation from an estimated regression model. In the house prices example, the mean price of homes is $111,034, and the estimated standard deviation is \$17,495; hence the coefficient of variation is 0.1773, or 17.73%. Again, using the empirical rule, approximately 95% of homes are priced within 35% of the value estimated by the regression model. It should be noted that this statistic is useful primarily when the values of the dependent variable do not span a large range relative to the mean and is useless for variables that can take negative values.

Inferences for Coefficients

We have already noted that we do not get estimates of the partial coefficients by performing m simple linear regressions using the individual independent variables. Likewise we cannot do the appropriate inferences for the partial coefficients by direct application of simple linear regression methods for the individual coefficients.

Instead we will base our inferences on a general principle for testing hypotheses in a linear statistical model for which regression is a special case.

What we do is to define inferences for these parameters in terms of the effect on the model of imposing certain restrictions on the parameters. The following discussion explains this general principle, which is often called the "general linear test."

General Principle for Hypothesis Testing Consider two models: a **full** or **unrestricted model** containing all parameters and a **reduced** or **restricted model**, which places some restrictions on the values of some of these parameters. The effects of these restrictions are measured by the decrease in the effectiveness of the restricted model in describing a set of data. In regression analysis the decrease in effectiveness is measured by the increase in the error sum of squares.

The most common inference is to test the null hypothesis that one or more of the coefficients are restricted to a value of 0. This is equivalent to saying that the corresponding independent variables are not used in the restricted model. The measure of the reduction in effectiveness of the restricted model is the increase in the error sum of squares (or, equivalently, the decrease in the model sum of squares) due to imposing the restriction, that is, due to leaving those variables out of the model.

In more specific terms the testing procedure is implemented as follows:

1. Divide the coefficients in $\mathbf{B}$ into two sets represented by matrices $\mathbf{B}_1$ and $\mathbf{B}_2$. That is,

$$\mathbf{B} = \begin{bmatrix} \mathbf{B}_1 \\ --- \\ \mathbf{B}_2 \end{bmatrix}.$$

   We want to test the hypotheses

$$H_0\colon \mathbf{B}_2 = \mathbf{0},$$

$$H_1\colon \text{ at least one element of } \mathbf{B}_2 \neq \mathbf{0}.$$

Denote the number of coefficients in $\mathbf{B}_1$ by q and the number of coefficients in $\mathbf{B}_2$ by p. Note that $p + q = m + 1$. Since the ordering of elements in the matrix of coefficients is arbitrary, $\mathbf{B}_2$ may contain any desired subset of the entire set of coefficients.[4]

2. Perform the regression using all coefficients, that is, using the full model $\mathbf{Y} = \mathbf{XB} + \mathbf{E}$. The error sum of squares for the full model is SSE(B). As we have noted, this sum of squares has $(n - m - 1)$ degrees of freedom.
3. Perform the regression using only the coefficients in $\mathbf{B}_1$, that is, using the restricted model $\mathbf{Y} = \mathbf{X}_1\mathbf{B}_1 + \mathbf{E}$, which is the model specified by H_0. The error sum of squares for the restricted model is SSE(B_1). This sum of squares has $(n - q)$ degrees of freedom.
4. The difference, SSE(B_1)$-$SSE(B), is the increase in the error sum of squares due to the restriction that the elements in $\mathbf{B}_2$ are zero. This is defined as the **partial** contribution of the coefficients in $\mathbf{B}_2$. Since there are p coefficients in $\mathbf{B}_2$, this sum of squares has p degrees of freedom, which is the difference between the number of parameters in the full and reduced models. For any model TSS = SSR + SSE; hence this difference can also be described as the decrease in the regression (or model) sum of squares due to the deletion of the coefficients in $\mathbf{B}_2$. Dividing the resulting sum of squares by its degrees of freedom provides the corresponding mean square.
5. As before, the ratio of mean squares is the test statistic. In this case the mean square due to the partial contribution of $\mathbf{B}_2$ is divided by the error mean square for the full model. The resulting statistic is compared to the F distribution with $(p, n - m - 1)$ degrees of freedom.

We illustrate with the home prices data. We have already noted that the error mean square for the five variable multiple regression was not much smaller than that using only `size`. It is therefore reasonable to test the hypothesis that the additional four variables do not contribute significantly to the fit of the model. In other words, we want to test the hypothesis that the coefficients for `age`, `bed`, `bath`, and `lot` are all zero.

Formally,

$$H_0\colon \beta_{\text{age}} = 0, \quad \beta_{\text{bed}} = 0, \quad \beta_{\text{bath}} = 0, \quad \beta_{\text{lot}} = 0,$$

$$H_1\colon \text{at least one coefficient is not 0.}$$

Let

$$\mathbf{B}_1 = \begin{bmatrix} \beta_0 \\ \beta_{\text{size}} \end{bmatrix},$$

and

$$\mathbf{B}_2 = \begin{bmatrix} \beta_{\text{age}} \\ \beta_{\text{bed}} \\ \beta_{\text{bath}} \\ \beta_{\text{lot}} \end{bmatrix}.$$

[4]We seldom perform inferences on β_0; hence this coefficient is normally included in $\mathbf{B}_1$.

We have already obtained the full model error sum of squares:

$$\text{SSE}(B) = 13774.05 \text{ with 45 degrees of freedom.}$$

The restricted model is the one obtained for the example in Chapter 7 that used only `size` as the independent variable. However, we cannot use that result directly because that regression was based on 58 observations while the multiple regression was based on the 51 observations that had data on `lot` and `size`. Redoing the simple linear regression with `size` using the 51 observations results in

$$\text{SSE}(B_1) = 17253.47 \text{ with 49 degrees of freedom.}$$

The difference

$$\text{SSE}(B_1) - \text{SS}(B) = 17253.47 - 13774.05 = 3479.42 \text{ with 4 degrees of freedom}$$

is the increase in the error sum of squares due to deleting `age`, `bed`, `bath`, and `lot` from the model and is therefore the partial sum of squares due to those four coefficients. The resulting mean square is 869.855. We use the error mean square for the full model as the denominator for testing the hypothesis that these coefficients are zero, resulting in $F(4, 45) = 869.855/306.09 = 2.842$. The 0.05 critical value for that distribution is 2.58; hence we can reject the hypothesis that all of these coefficients are zero.

Tests Normally Provided by Computer Outputs

Although most computer programs have provisions for requesting almost any kinds of inferences on the regression model, most provide two sets of hypothesis tests as default. These are as follows:

1. H_0: $(\beta_1, \beta_2, \ldots, \beta_m) = 0$, that is, the hypothesis that the entire set of coefficients associated with the m independent variables is zero, with the alternate being that any one or more of these coefficients are not zero. This test is often referred to as the test for the model.
2. H_{0j}: $\beta_j = 0$, $j = 1, 2, \ldots, m$, that is, the m separate tests that each partial coefficient is zero.

The Test for the Model The null hypothesis is

$$H_0: (\beta_1, \beta_2, \ldots, \beta_m) = 0.$$

For this test then, the reduced model contains only β_0. The model is

$$y = \beta_0 + \varepsilon$$

or, equivalently,

$$y = \mu + \varepsilon.$$

The parameter μ is estimated by the sample mean $\bar{y}$, and the error sum of squares of this reduced model is

$$\text{SSE}(B_1) = \sum(y - \bar{y})^2 = \sum y^2 - \left(\sum y\right)^2 \Big/ n,$$

with $(n - 1)$ degrees of freedom.[5] The error sum of squares for the full model is

$$\text{SSE}(B) = \sum y^2 - \hat{\mathbf{B}}'\mathbf{X}'\mathbf{Y}$$

and the difference yields

$$\text{SSR(regression model)} = \hat{\mathbf{B}}\mathbf{X}'\mathbf{Y} - \left(\sum y\right)^2 / n,$$

which has m degrees of freedom. Dividing by the degrees of freedom produces the mean square, which is then divided by the error mean square to provide the F statistic for the hypothesis test.

For the home price data the test for the model is

$$H_0: \begin{bmatrix} \beta_{\text{age}} \\ \beta_{\text{bed}} \\ \beta_{\text{bath}} \\ \beta_{\text{size}} \\ \beta_{\text{lot}} \end{bmatrix} = \begin{bmatrix} 0 \\ 0 \\ 0 \\ 0 \\ 0 \end{bmatrix}.$$

We have already computed the full model error sum of squares: 13,744.05. The error sum of squares for the restricted model using the information from Table 8.3 is

$$690197.14 - (5580.96)^2/51 = 690194.14 - 610727.74 = 79{,}469.40,$$

the difference

$$\text{SS(model)} = 79{,}469.40 - 13{,}774.05 = 65{,}695.36 \text{ with 5 degrees of freedom,}$$

resulting in a mean square of 13,139.07 with 5 degrees of freedom. Using the full model error mean square of 306.09,

$$F(5, 45) = 42.926,$$

which easily leads to rejection of the null hypothesis and we can conclude that at least one of the coefficients in the model is statistically significant.

Although we have presented this test in terms of the difference in error sums of squares, it is normally presented in terms of the partitioning of sums of squares as presented for simple linear regression in Chapter 7. In this presentation the total corrected sum of squares is partitioned into the model sum of squares and error sum of squares. The test is, of course, the same.

For our example then, the total corrected sum of squares is

$$\sum y^2 - \left(\sum y\right)^2 / n = 690197.14 - (5580.96)^2/51 = 690197.14 - 610727.74$$
$$= 79{,}469.40,$$

which is, of course, the error sum of squares for the restricted model with no coefficients (except the intercept). The full model error sum of squares is

[5] We can now see that what we have called the correction factor for the mean (Section 1.5) is really a sum of squares due to the regression for the coefficient μ or, equivalently, β_0.

13,774.05; hence the model sum of squares is the difference, 65,695.34. The results of this procedure are conveniently summarized in the familiar analysis of variance table, which, for this example, is shown in the section dealing with computer outputs (Table 8.6 in Section 8.5).

Tests for Individual Coefficients The testing of hypothesis on the individual partial regression coefficients would seem to require the estimation of m models, each containing $(m-1)$ coefficients. Fortunately a shortcut exists.

It can be shown that the partial sum of squares due to a single partial coefficient, say, β_j, can be computed

$$\text{SSR}(\beta_j) = \hat{\beta}_j^2 / c_{jj}, \quad j = 1, 2, \ldots, m,$$

where c_{jj} is the element on the main diagonal of $\mathbf{C} = (\mathbf{X}'\mathbf{X})^{-1}$ corresponding to the variable x_j. This sum of squares has 1 degree of freedom. This can be used for the test statistic

$$F = \frac{(\hat{\beta}_j^2 / c_{jj})}{\text{MSE}},$$

which has $(1, n-m-1)$ degrees of freedom.[6]

The estimated coefficients and diagonal elements of $\mathbf{C} = (\mathbf{X}'\mathbf{X})^{-1}$ for the home prices data are found in Table 8.3 as

`age`: $\hat{\beta}_1 = -0.3498$, $c_{11} = 0.0001293$,
`bed`: $\hat{\beta}_2 = -11.2383$, $c_{22} = 0.064025$,
`bath`: $\hat{\beta}_3 = -4.5401$, $c_{33} = 0.131435$,
`size`: $\hat{\beta}_4 = 65.9465$, $c_{44} = 0.132834$,
`lot`: $\hat{\beta}_5 = -0.0621$, $c_{55} = 8.2341\text{E}{-6}$.

The partial sums of squares and F statistics are

`age`: $\text{SS} = (-0.3498)^2/0.0001293 = 946.327$, $F = 946.327/306.09 = 3.091$,
`bed`: $\text{SS} = (-11.2383)^2/0.64025 = 1972.657$, $F = 1972.657/306.09 = 6.445$,
`bath`: $\text{SS} = (-4.5401)^2/0.131435 = 156.827$, $F = 156.827/306.09 = 0.512$,
`size`: $\text{SS} = (65.9465)^2/0.132834 = 32739.7$, $F = 32739.7/306.09 = 106.961$,
`lot`: $\text{SS} = (0.06205)^2/8.23418\text{E} - 6 = 467.60$, $F = 467.59/306.09 = 1.528$.

The 0.05 critical value for $F(1, 45)$ is 4.06, and we reject the hypotheses that the coefficients for `bed` and `size` are zero, but cannot reject the corresponding hypotheses for the other variables. This means that the readily explained negative coefficient for `bed` really exists while evidence for the negative coefficient for `bath` is not necessarily confirmed. Note that we can use this same test for H_0: $\beta_0 = 0$, but because the intercept usually has no practical meaning, the test is not often used, although it is normally printed in computer output.

Note that these partial sums of squares do not constitute a partitioning of the model sum of squares. In other words, the sums of squares for the

[6]As labeled in Section 8.2, the first row and column of $\mathbf{C} = (\mathbf{X}'\mathbf{X})^{-1}$ correspond to β_0; hence the row and column corresponding to the jth independent variable will be the $(j+1)$st row and column, respectively. If the computer output uses the names of the independent variable (as in Table 8.3), the desired row and column are easily located.

partial coefficients do not sum to the model sum of squares as was the case with orthogonal contrasts (Section 6.5). This means that, for example, simply because `lot` and `age` cannot individually be deemed significantly different from zero, it does not necessarily follow that the simultaneous addition of these coefficients will not significantly contribute to the model (although they do not in this example).

The Equivalent t Statistic for Individual Coefficients

We noted in Chapter 7 that the F test for the hypothesis that the coefficient is zero can be performed by an equivalent t test. The same relationship holds for the individual partial coefficients in the multiple regression model. The t statistic for testing H_0: $\beta_j = 0$ is

$$t = \frac{\hat{\beta}_j}{\sqrt{c_{jj}\text{MSE}}},$$

where c_{jj} is the jth diagonal element of $\mathbf{C}$, and the degrees of freedom are $(n - m - 1)$. It is easily verified that these statistics are the square roots of the F values obtained earlier and they will not be reproduced here. As in simple linear regression, the denominator of this expression is the standard error (or square root of the variance) of the estimated coefficient, which can be used to construct confidence intervals for the coefficients.

In Chapter 7 we noted that the use of the t statistic allowed us to test for specific (nonzero) values of the parameters, allowed the use of one-tailed tests and the calculation of confidence intervals. For these reasons, most computers provide the standard errors and t tests. A typical computer output for Example 8.2 is shown in Table 8.6. We can use this output to compute the confidence intervals for the coefficients in the regression equation as follows:

`age`: Std. error $= \sqrt{(0.0001293)(306.09)} = 0.199$
0.95 Confidence interval: $-0.3498 \pm (2.0141)(0.199)$: from -0.7506 to 0.051

`bed`: Std. error $= \sqrt{(0.64025)(306.09)} = 4.427$
0.95 Confidence interval: $-11.2382 \pm (2.0141)(4.427)$: from -20.1546 to -2.3218

`bath`: Std. error $= \sqrt{(0.131435)(306.09)} = 6.328$
0.95 Confidence interval: $-4.5401 \pm (2.0141)(6.328)$: from -17.2853 to 8.2051

`size`: Std. error$= \sqrt{(0.132834)(306.09)} = 6.376$
0.95 Confidence interval: $65.9465 \pm (2.0141)(6.376)$: from 53.1045 to 78.7884

`lot`: Std. error $= \sqrt{(8.234189\text{E} - 6)(306.09)} = 0.0502$
0.95 Confidence interval: $0.06205 \pm (2.0141)(0.0502)$: from 0.0391 to 0.1632.

As expected, the confidence intervals of those coefficients deemed statistically significant at the 0.05 level do not include zero.

Finally, note that the tests we have presented are special cases of tests for any linear function of parameters. For example, we may wish to test

$$H_0\colon \beta_4 - 10\beta_5 = 0,$$

which for the home price data tests the hypothesis that the `size` coefficient is ten times larger than the `lot` coefficient. The methodology for this more general hypothesis testing is beyond the scope of this book (see, for example, Freund and Wilson, 1998).

Inferences on the Response Variable

As in the case of simple linear regression, we may be interested in the precision of the estimated conditional mean as well as predicted values of the dependent variable (see Section 7.5). The formulas for obtaining the variances needed for these inferences in multiple regression are quite cumbersome and are not suitable for hand calculation; hence we do not reproduce them here. Most computer programs have provisions for computing confidence and prediction intervals and also for providing the associated standard errors. A computer output showing 95% confidence intervals is presented in Section 8.5. A word of caution: Some computer program documentation may not be clear on which interval (confidence on the conditional mean or prediction) is being produced, so read instructions carefully!

The following example is provided as a review of the various steps for a multiple regression analysis.

EXAMPLE 8.3

Example 7.3 provided a regression model to explain how the departure times (`TIME`) of lesser snow geese were affected by temperature (`TEMP`). Although the results were reasonably satisfactory, it is logical to expect that other environmental factors affect departure times.

Solution Since information on other factors was also collected, we can propose a multiple regression model with the following additional environmental variables:

`HUM`, the relative humidity,
`LIGHT`, light intensity, and
`CLOUD`, percent cloud cover.

The data are given in Table 8.4.

An inspection of the data shows that two observations have missing values (denoted by .) for a variable. This means that these observations cannot be used for the regression analysis. Fortunately, most computer programs recognize missing values and will automatically ignore such observations. Therefore all calculations in this example will be based on the remaining 36 observations.

The first step is to compute $\mathbf{X}'\mathbf{X}$ and $\mathbf{X}'\mathbf{Y}$. We then compute the inverse and the estimated coefficients. As before, we will let the computer do this with the results given in Table 8.5 in the same format as that of Table 8.3.

Table 8.4

Snow Goose Departure Times

DATE	TIME	TEMP	HUM	LIGHT	CLOUD
11/10/87	11	11	78	12.6	100
11/13/87	2	11	88	10.8	80
11/14/87	−2	11	100	9.7	30
11/15/87	−11	20	83	12.2	50
11/17/87	−5	8	100	14.2	0
11/18/87	2	12	90	10.5	90
11/21/87	−6	6	87	12.5	30
11/22/87	22	18	82	12.9	20
11/23/87	22	19	91	12.3	80
11/25/87	21	21	92	9.4	100
11/30/87	8	10	90	11.7	60
12/05/87	25	18	85	11.8	40
12/14/87	9	20	93	11.1	95
12/18/87	7	14	92	8.3	90
12/24/87	8	19	96	12.0	40
12/26/87	18	13	100	11.3	100
12/27/87	−14	3	96	4.8	100
12/28/87	−21	4	86	6.9	100
12/30/87	−26	3	89	7.1	40
12/31/87	−7	15	93	8.1	95
01/02/88	−15	15	43	6.9	100
01/03/88	−6	6	60	7.6	100
01/04/88	−23	5	.	8.8	100
01/05/88	−14	2	92	9.0	60
01/06/88	−6	10	90	.	100
01/07/88	−8	2	96	7.1	100
01/08/88	−19	0	83	3.9	100
01/10/88	−23	−4	88	8.1	20
01/11/88	−11	−2	80	10.3	10
01/12/88	5	5	80	9.0	95
01/14/88	−23	5	61	5.1	95
01/15/88	−7	8	81	7.4	100
01/16/88	9	15	100	7.9	100
01/20/88	−27	5	51	3.8	0
01/21/88	−24	−1	74	6.3	0
01/22/88	−29	−2	69	6.3	0
01/23/88	−19	3	65	7.8	30
01/24/88	−9	6	73	9.5	30

The five elements in the last column, labeled TIME, of the inverse portion contain the estimated coefficients, providing the equation:

$$\widehat{\texttt{TIME}} = -52.994 + 0.9130(\texttt{TEMP}) + 0.1425(\texttt{HUM})$$
$$+ 2.5160(\texttt{LIGHT}) + 0.0922(\texttt{CLOUD}).$$

Unlike the case of the regression involving only TEMP, the intercept now has no real meaning since zero values for HUM and LIGHT cannot exist. The remainder of the coefficients are positive, indicating later departure times for increased values of TEMP, HUM, LIGHT, and CLOUD. Because of the different scales of the independent variables, the relative magnitudes of these

Table 8.5

Snow Goose Departure Times

Model Crossproducts X'X X'Y Y'Y

X'X	INTERCEP	TEMP	HUM
INTERCEP	36	319	3007
TEMP	319	4645	27519
HUM	3007	27519	257927
LIGHT	326.2	3270.3	27822
CLOUD	2280	23175	193085
TIME	−157	1623	−9662

X'X	LIGHT	CLOUD	TIME
INTERCEP	326.2	2280	−157
TEMP	3270.3	23175	1623
HUM	27822	193085	−9662
LIGHT	3211.9	20079.5	−402.8
CLOUD	20079.5	194100	−3730
TIME	−402.8	−3730	9097

X'X Inverse, Parameter Estimates, and SSE

	INTERCEPT	TEMP	HUM
INTERCEP	1.1793413621	0.0085749149	−0.010464297
TEMP	0.0085749149	0.0010691752	0.0000605688
HUM	−0.010464297	0.0000605688	0.0001977643
LIGHT	−0.028115838	−0.00192403	−0.000581237
CLOUD	−0.001558842	−0.000089595	−0.000020914
TIME	−52.99392938	0.9129810924	0.1425316971

	LIGHT	CLOUD	TIME
INTERCEP	−0.028115838	−0.001558842	−52.99392938
TEMP	−0.00192403	−0.000089595	0.9129810924
HUM	−0.000581237	−0.000020914	0.1425316971
LIGHT	0.0086195605	0.0002464973	2.5160019069
CLOUD	0.0002464973	0.0000294652	0.0922051991
TIME	2.5160019069	0.0922051991	2029.6969929

coefficients have little meaning and also are not indicators of relative statistical significance.

Note that the coefficient for TEMP is 0.9130 in the multiple regression model, while it was 1.681 for the simple linear regression involving only the TEMP variable. In this case, the so-called total coefficient for the simple linear regression model includes the indirect effect of other variables, while in the multiple regression model, the coefficient measures only the effect of TEMP by holding constant the effects of other variables.

For the second step we compute the partitioning of the sums of squares. The residual sum of squares

$$\begin{aligned} \text{SSE} &= \sum y^2 - \hat{\mathbf{B}}'\mathbf{X}'\mathbf{Y} \\ &= 9097 - [(-52.994)(-157) + (0.9123)(1623) + (0.1425)(\ 9662) \\ &\quad + (2.5160)(-402.8) + (0.09221)(-3730)], \end{aligned}$$

which is available in the computer output as the last element of the inverse portion and is 2029.70. The estimated variance is MSE $= 2029.70/(36-5) = 65.474$, and the estimated standard deviation is 8.092. This value is somewhat smaller than the 9.96 obtained for the simple linear regression involving only TEMP.

The model sum of squares is

$$\begin{aligned}\text{SSR(regression model)} &= \hat{\mathbf{B}}'\mathbf{X}'\mathbf{Y} - \left(\sum y\right)^2 \Big/ n \\ &= 7067.30 - 684.69 = 6382.61.\end{aligned}$$

The degrees of freedom for this sum of squares is 4; hence the model mean square is $6382.61/4 = 1595.65$. The resulting F statistic is $1595.65/65.474 = 24.371$, which clearly leads to the rejection of the null hypothesis of no regression. These results are summarized in an analysis of variance table shown in Table 8.7 in Section 8.5.

In the final step we use the standard errors and t statistics for inferences on the coefficients. For the TEMP coefficient, the estimated variance of the estimated coefficient is

$$\begin{aligned}\hat{\text{var}}(\hat{\beta}_{\text{TEMP}}) &= c_{\text{TEMP,TEMP}}\text{MSE} \\ &= (0.001069)(65.474) \\ &= 0.0700,\end{aligned}$$

which results in an estimated standard error of 0.2646. The t statistic for the null hypothesis that this coefficient is zero is

$$t = 0.9130/0.2646 = 3.451.$$

Assuming a desired significance level of 0.05, the hypothesis of no temperature effect is clearly rejected. Similarly, the t statistics for HUM, LIGHT, and CLOUD are 1.253, 3.349, and 2.099, respectively. When compared with the tabulated two-tailed 0.05 value for the t distribution with 31 degrees of freedom of 2.040, the coefficient for HUM is not significant, while LIGHT and CLOUD are. The p values are shown later in Table 8.7, which presents computer output for this problem. Basically this means that departure times appear to be affected later with increasing levels of temperature, light, and cloud cover, but there is insufficient evidence to state that humidity affects the departure times. ■

8.4 Correlations

In Section 7.6 we noted that the correlation coefficient provides a convenient index of the strength of the linear relationship between two variables. In multiple regression, two types of correlations describe strengths of linear relationships among the variables in a regression model:

1. multiple correlation, which describes the strength of the linear relationship of the dependent variable with the set of independent variables, and

2. partial correlation, which describes the strength of the linear relationship associated with a partial regression coefficient.

Other types of correlations used in some applications but not presented here are multiple partial and part (or semipartial) correlations (Kleinbaum *et al.*, 1998, Chapter 10).

Multiple Correlation

DEFINITION 8.2
Multiple correlation describes the maximum strength of a linear relationship of one variable with a linear function of a set of variables.

In Section 7.6, the sample correlation between two variables x and y was defined as

$$r_{xy} = \frac{S_{xy}}{\sqrt{S_{xx} \cdot S_{yy}}}.$$

With the help of a little algebra it can be shown that the absolute value of this quantity is equal to the correlation between the observed values of y and $\hat{\mu}_{y|x}$, the values of the variable y estimated by the linear regression of y on x. Thus, for example, the correlation coefficient can also be calculated using the values in the columns labeled `size` and `Predict` in Table 7.3. This definition of the correlation coefficient can be applied to a multiple linear regression and the resulting correlation coefficient is called the **multiple correlation coefficient**, which is usually denoted by R. Also, as in simple linear regression, the square of R, the **coefficient of determination**, is

$$R^2 = \frac{\text{SS due to regression model}}{\text{total SS for } y \text{ corrected for the mean}}.$$

In other words, the coefficient of determination measures the proportional reduction in variability about the mean resulting from the fitting of the multiple regression model. As in simple linear regression there is a correspondence between the coefficient of determination and the F statistic for testing the existence of the model:

$$F = \frac{(n - m - 1)R^2}{m(1 - R^2)}.$$

Also as in simple linear regression, the coefficient of determination must take values between and including 0 and 1 where a value of 0 indicates the linear relationship is nonexistent, and a value of 1 indicates a perfect linear relationship.

How Useful Is the R^2 Statistic?

The apparent simplicity of this statistic, which is often referred to as "R-square," makes it a popular and convenient descriptor of the effectiveness of a multiple regression model. This very simplicity has, however, made the coefficient of determination an often abused statistic. There is no rule or guideline as to what value of this statistic signifies a good regression. For some data, especially that from the social and behavioral sciences, coefficients of determination of 0.3 are often considered quite good, while in fields where random fluctuations are of smaller magnitudes, for example, engineering, coefficients of determination of less than 0.95 may imply an unsatisfactory fit. Incidentally, for the home prices model, the coefficient of determination is 0.9035. This is certainly considered to be high for many applications, yet the residual standard deviation of $4525 leaves much to be desired.

An additional feature of the coefficient of determination is that when a small number of observations are used to estimate an equation, the coefficient of determination may be inflated by having a relatively large number of independent variables. In fact, if n observations are used for an $(n-1)$ variable equation, the coefficient of determination is, by definition, unity! An "adjusted R-square" statistic, which indicates the proportional reduction in the mean square (rather than in the sum of squares), is available to overcome this feature of the coefficient of determination. However, this statistic, although usually available in computer printouts (Section 8.5), has limited usefulness. It also has an interpretive problem due to the fact that it can assume negative values.

As noted in Section 8.3, the residual standard deviation may be a better indicator of the fit of the model.

Partial Correlation

DEFINITION 8.3
A **partial correlation** coefficient describes the strength of a linear relationship between two variables, holding constant a number of other variables.

As noted in Section 7.6, the strength of the linear relationship between x and y was measured by the simple correlation between these variables, and the simple linear regression coefficient described their relationship. Just as a partial regression coefficient shows the relationship of y to one of the independent variables, holding constant the other variables, a **partial correlation** coefficient measures the strength of the relationship between y and one of the independent variables, holding constant all other variables in the model. This means that the partial correlation measures the strength of the linear relationship between two variables after "adjusting" for relationships involving all the other variables.

A partial correlation coefficient has the properties of any correlation coefficient: It takes a value from -1 to $+1$, with a value of 0 indicating no relationship and values of -1 and $+1$ indicating a perfect linear relationship.

In the context of a regression model, the relationship of a partial correlation coefficient to a partial regression coefficient is the same as the relationship between a (simple) correlation coefficient and a regression coefficient in a one-variable regression. Specifically:

- There is an exact relationship to the test statistic of the corresponding regression coefficient. In this case the equivalence is to the t statistic for testing whether a regression coefficient is zero,

$$|t| = \sqrt{\frac{(n-m-1)r^2}{(1-r^2)}},$$

  where r is the partial correlation coefficient corresponding to the coefficient involved in the t statistic.
- The square of the partial correlation of, say, y and x_j, holding constant all other variables in the regression model, is the ratio of the partial sum of squares explained by the estimated coefficient $\hat{\beta}_j$ to the error sum of squares remaining after fitting the model that contains all the *other coefficients* in the model. Thus the partial correlation coefficient has the property of the other correlation coefficients: Its square indicates the portion of the variability explained by a regression. In this case it is the portion of the variability explained by that variable after all the other variables have been included in the model.

For example, suppose that x_1 is the age of a child, x_2 is the number of hours spent watching television, and y is the child's score on an achievement test. The simple correlation between y and x_2 would include the indirect effect of age on the test score and could easily cause that correlation to be positive. However, the partial correlation between y and x_2, holding constant x_1, is the "age-adjusted" correlation between the number of hours spent watching TV and the achievement test score.

The test for the null hypothesis of no partial correlation is the same as that for the corresponding partial regression coefficient. Other inferences are made by an adaptation of the Fisher z transformations (Section 7.6), where the variance of z is $[1/(n-q-3)]$, where q is the number of variables being held constant [usually $(m-2)$].

As an illustration of calculating partial correlation coefficients, we use the data in Example 8.2, for finding the partial correlation between `price` and `size`, holding `age`, `bed`, `bath`, and `lot` fixed. We use the following procedure:

1. Perform the regression of `price` on `age`, `bed`, `bath`, and `lot`. The error sum of squares is 47,747.4.
2. Perform the regression of `price` on `size`, `age`, `bed`, `bath`, and `lot`. The partial sum of squares for `size` is 32,739.7.
3. The square of the partial correlation for `price` and `size`, holding `age`, `bed`, `bath`, and `lot` constant, is $32{,}739.7/47{,}747.4 = 0.686$; the corresponding correlation coefficient is 0.828.

Various more efficient procedures exist for calculating the partial correlation coefficients, but they are not presented here (see, for example, Kleinbaum *et al.*, 1998, Section 10.5). The partial correlation coefficient is not widely used but has application in special situations, such as path analysis (Loehlin, 1987). Finally, partial correlation indicates the strength of a linear relationship between any two variables, holding constant a number of other variables.

8.5 Using the Computer

As noted, almost all statistical analyses are performed on computers using statistical software packages. A comprehensive statistical software package may have several different programs for performing multiple regression. Usually any one of these can be used for performing the analyses presented in previous sections. However, some of these programs may have features designed for special applications or may be quite cumbersome to implement and/or expensive to use in terms of computer resources. It is therefore important to read the documentation of available software systems and pick the program most suited to the desired analysis. For example, one of the available programs may implement a variable selection procedure as described in Section 8.7. If this is not needed for a particular analysis, such a program will provide information that will not be useful and at the same time omit information that may be needed.

Even programs designed for an ordinary regression analysis often have provisions for a number of special options. Using unnecessary options for a particular analysis is a waste of computer resources. Thus it is again important to read the documentation carefully and request only those options germane to the analysis at hand.

EXAMPLE 8.2

REVISITED Table 8.6 contains the output from `PROC REG` of the SAS System for the multiple regression model for the home price data we have been using as an example (we have omitted some of the output to save space). The implementation of this program required the following specifications:

1. The name of the program; in this case it is `PROC REG`.
2. The name of the dependent and independent variables; in this case `price` is the dependent variable and `age`, `bed`, `bath`, `size`, and `lot` are the independent variables. The intercept is not specified since most computer programs automatically assume that an intercept will be included in the model.
3. Options to print, in addition to the standard or default output, the predicted and residual values, the standard errors of the estimated mean, and the 95% confidence intervals for the estimated means.

Although much of the output in Table 8.6 is self-explanatory, a brief summary is presented here. The reader should verify all results that compare with those presented in the previous sections. Also useful are comparisons with output from other computer packages, if available.

Solution The output begins by giving the name of the dependent variable. This identifies the output in case several analyses have been run in one

Table 8.6 Output for Multiple Regression

The REG Procedure
Model: MODEL1
Dependent Variable: price

Analysis of Variance

Source	DF	Sum of Squares	Mean Square	F Value	Pr > F
Model	5	65696	13139	42.93	<.0001
Error	45	13774	306.08999		
Corrected Total	50	79470			

Root MSE	17.49543	R-Square	0.8267
Dependent Mean	109.43055	Adj R-Sq	0.8074
Coeff Var	15.98770		

Parameter Estimates

Variable	DF	Parameter Estimate	Standard Error	t Value	Pr > |t|
Intercept	1	35.28792	14.11712	2.50	0.0161
age	1	−0.34980	0.19895	−1.76	0.0855
bed	1	−11.23820	4.42691	−2.54	0.0147
bath	1	−4.54015	6.34279	−0.72	0.4778
size	1	65.94647	6.37644	10.34	<.0001
lot	1	0.06205	0.05020	1.24	0.2229

Output Statistics

Obs	Dep Var price	Predicted Value	Std Error Mean Predict	95% CL Mean		Residual
1	30.0000	47.3494	10.2500	26.7049	67.9939	−17.3494
2	39.9000	66.9823	9.0854	48.6834	85.2812	−27.0823
3	46.5000	65.0194	8.9813	46.9302	83.1087	−18.5194
4	48.6000	89.6287	4.1333	81.3039	97.9535	−41.0287
5	51.5000	58.0793	8.0053	41.9557	74.2029	−6.5793
6	56.9900	84.3515	3.1288	78.0498	90.6532	−27.3615
7	59.9000	.	.	.	.	.
8	62.5000	53.8228	5.9133	41.9127	65.7329	8.6772
9	65.5000	68.3728	8.5338	51.1849	85.5606	−2.8728
10	69.0000	73.0383	5.7938	61.3689	84.7076	−4.0383
11	76.9000	75.8630	4.3441	67.1135	84.6124	1.0370
.	.	.	.	.	.	.
.	.	.	(Observations Omitted)	.	.	.
.	.	.	.	.	.	.
49	149.9900	146.4358	8.4156	129.4858	163.3858	3.5542
50	152.5500	.	.	.	.	.
51	156.9000	.	.	.	.	.
52	164.0000	144.7772	7.1999	130.2760	159.2785	19.2228
53	167.5000	123.7208	10.2462	103.0839	144.3576	43.7792
54	169.9000	183.6916	7.1420	169.3068	198.0763	−13.7916

(Continued)

Table 8.6 *(continued)*

Obs	Dep Var price	Predicted Value	Std Error Mean Predict	95% CL Mean		Residual
55	175.0000	182.1852	7.0512	167.9834	196.3870	−7.1852
56	179.0000	163.8122	5.9526	151.8230	175.8014	15.1878
57	179.9000	156.4986	6.3606	143.6877	169.3096	23.4014
58	189.5000	.	.	.	.	.
59	199.0000	212.1590	10.5356	190.9392	233.3788	−13.1590

Sum of Residuals	0
Sum of Squared Residuals	13774
Predicted Residual SS (PRESS)	19927

computer job. The first tabular presentation contains the overall partitioning of the sums of squares and the F test for the model. The notation `Corrected Total` is used to denote that this is the total sum of squares corrected for the mean; hence the model sum of squares is presented in the manner we used for simple linear regression. That is, it is the sum of squares due to the regression after the mean has already been estimated.

The next section gives some miscellaneous statistics. `Root MSE` is the residual standard deviation, which is the square root of the error mean square. `Dependent Mean` is $\bar{y}$ and `R-Square` is the coefficient of determination. `Adj R-Sq` is the adjusted coefficient of determination. `Coeff Var` is the coefficient of variation (in %) as defined in Section 8.3.

The third portion contains the parameter (coefficient) estimates and associated statistics: the standard errors and t statistics and their p values, which are labeled `Pr >` $|t|$. The parameter estimates are identified by the names of the corresponding independent variables, and the estimate of β_0 is labeled `Intercept`.

The last portion contains some optional statistics for the individual observations. The values in the columns labeled `Dep Var price` and `Predicted Value` are self-explanatory. The column labeled `Std Error Mean Predict` contains the standard errors of the estimated conditional means. The headings 95% `CL Mean` are the 0.95 confidence limits of the conditional mean.

Finally the sum and sum of squares of the actual residuals are given. The sum should be zero, which it is, and the `Sum of Squared Residuals` should be equal to the error sum of squares obtained in the analysis of variance table.[7] ■

EXAMPLE 8.3

REVISITED Table 8.7 shows the results of implementing the lesser snow geese departure regression on Minitab using the `REGRESS` command. This command required the specification of the name of the dependent variable

[7] If there is more than a minimal difference between the two, severe round-off errors have probably occurred.

Table 8.7 Snow Goose Regression with Minitab

```
 The regression equation is time = −53.0 + 0.913 temp + 0.143 hum + 2.52
 light + 0.0922 cloud 36 cases used 2 cases contain missing values

Predictor        Coef          Stdev      t-ratio      p
-----------------------------------------------------------
Constant       −52.994        8.787      −6.03      0.000
temp             0.9130       0.2646      3.45      0.002
hum              0.1425       0.1138      1.25      0.220
light            2.5160       0.7512      3.35      0.002
cloud            0.09221      0.04392     2.10      0.044

s = 8.092    R-sq = 75.9%    R-sq(adj) = 72.8%

Analysis of Variance
SOURCE          df         SS         MS         F         p
-------------------------------------------------------------
Regression       4      6382.6     1595.7     24.37     0.000
Error           31      2029.7       65.5
Total           35      8412.3

SOURCE     df     SEQ SS
------------------------
temp        1     4996.6
hum         1      633.3
light       1      464.2
cloud       1      288.5

Unusual Observations
Obs.       temp        time      Fit Stdev.     Fit     Residual     St. Resid
-------------------------------------------------------------------------------
 4         20.0       −11.00        12.40       2.84     −23.40       −3.09R
12         18.0        25.00         8.93       2.65      16.07        2.10R

R denotes an obs. with a large st. resid.
```

and the number of independent variables in the model followed by a listing of names of these variables. No additional options were requested.

Solution As we have noted before, the output is somewhat similar to that obtained with the SAS System, and the results are the same as those presented in Example 8.3. This output actually gives the estimated model in equation form as well as a listing of coefficients and their inference statistics. Also the output states that two observations could not be used because of missing values. In the SAS System, this information is given in output we did not present for that example.

In addition, the Minitab output contains two items that were not in the SAS output: a set of sequential sums of squares (`SEQ SS`) and a listing of two unusual observations. The sequential sums of squares are not particularly useful for this example but will be used in polynomial regression, which is presented in Section 8.6. Because these have a special purpose, they must be specifically requested when using the SAS System.

The two unusual observations are identified as having large "Studentized residuals," which are residuals that have been standardized to look like t statistics; hence values exceeding a critical value of t are deemed to be unusual. A discussion of unusual observations is presented in Section 8.9.

Listings of all predicted and residual values, confidence intervals, etc., can be obtained as options for both of these computer programs. In general, we can see that different computer packages generally provide equivalent results, although they may provide different automatic and optional outputs. ■

8.6 Special Models

It is rather well known that straight line relationships of the type described by a multiple linear regression model do not often occur in the real world. Nevertheless, such models enjoy wide use, primarily because they are relatively easy to implement, but also because they provide useful approximations for other functions, especially over a limited range of values of the independent variables. However, strictly linear regression models are not always effective; hence we present in this section some methods for implementing regression models that do not necessarily imply straight line relationships.

As we have noted a linear regression model is constrained to be linear in the **parameters**, that is, the β_i and ε, but not necessarily linear in the independent variables. Thus, for example, the independent variables may be nonlinear functions of observed variables that describe curved responses, such as x^2, $1/x$, $\sqrt{x}$, etc.

The Polynomial Model

The most popular such function is the **polynomial** model, which involves powers of the independent variables. Fitting a polynomial model is usually referred to as "curve fitting" because it is used to fit a curve rather than to explain the relationship between the dependent and independent variable(s). That is, the interest is in the nature of the fitted response curve rather than in the partial regression coefficients. The polynomial model is very useful for this purpose, as it is easy to implement and provides a reasonable approximation to virtually any function within a limited range.

Given observations on a dependent variable y and two independent variables x_1 and x_2, we can estimate the parameters of the polynomial model

$$y = \beta_0 + \beta_1 x_1 + \beta_2 x_1^2 + \beta_3 x_2 + \beta_4 x_2^2 + \beta_5 x_1 x_2 + \varepsilon,$$

by redefining variables

$$w_1 = x_1,$$

$$w_2 = x_1^2,$$

$$w_3 = x_2,$$

$$w_4 = x_2^2,$$

$$w_5 = x_1 x_2,$$

and performing a multiple linear regression using the model

$$y = \beta_0 + \beta_1 w_1 + \beta_2 w_2 + \beta_3 w_3 + \beta_4 w_4 + \beta_5 w_5 + \varepsilon.$$

This is an ordinary multiple linear regression model using the w's as independent variables.

EXAMPLE 8.4

Biologists are interested in the characteristics of growth curves, that is, finding a model for describing how organisms grow with time. Relationships of this type tend to be curvilinear in that the rate of growth decreases with age and eventually stops altogether. A polynomial model is sometimes used for this purpose.

This example concerns the growth of rabbit jawbones. Measurements were made on lengths of jawbones for rabbits of various ages. The data are given in Table 8.8, and the plot of the data is given in Fig. 8.3 where the line is the estimated polynomial regression line described below.

Solution We will use a fourth-degree polynomial model for estimating the relationship of LENGTH to AGE. This model contains as independent variables the first four powers of the variable AGE. Since we will use computer output to show the results, we use the following variable names:

LENGTH, the dependent variable, is the length (in mm) of the jawbone.

AGE is the age (in days) of the rabbits divided by 100. The computations for a polynomial regression model may be subject to considerable round-off error, especially when the independent variable contains both very large and small numbers. Round-off error is reduced if the independent variable can be scaled so that values lie between 0.1 and 10. In this example only one scaled value is outside that recommended range.

A2 = $(\text{AGE})^2$.

A3 = $(\text{AGE})^3$.

A4 = $(\text{AGE})^4$.

Table 8.8

Rabbit Jawbone Length

AGE	LENGTH	AGE	LENGTH	AGE	LENGTH
0.01	15.5	0.41	29.7	2.52	49.0
0.20	26.1	0.83	37.7	2.61	45.9
0.20	26.3	1.09	41.5	2.64	49.8
0.21	26.7	1.17	41.9	2.87	49.4
0.23	27.5	1.39	48.9	3.39	51.4
0.24	27.0	1.53	45.4	3.41	49.7
0.24	27.0	1.74	48.3	3.52	49.8
0.25	26.0	2.01	50.7	3.65	49.9
0.26	28.6	2.12	50.6		
0.34	29.8	2.29	49.2		

Figure 8.3

Polynomial Regression Plot

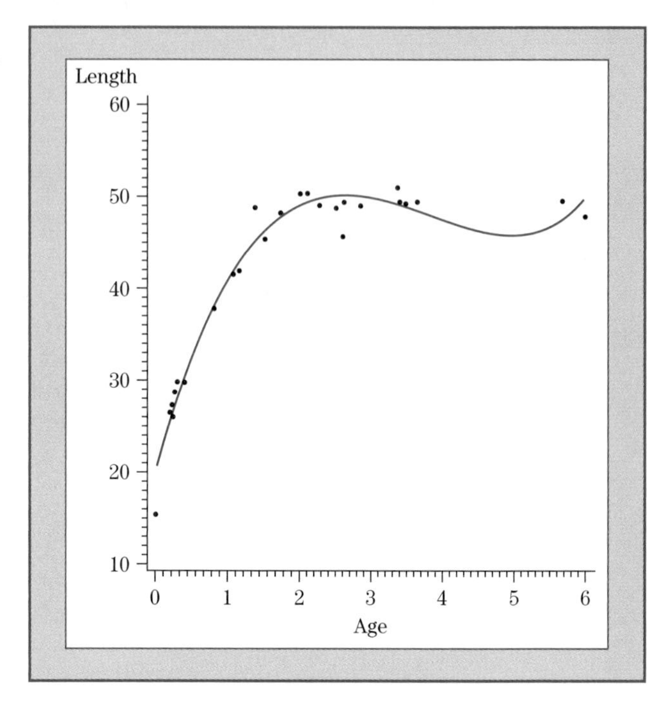

In terms of the computer,[8] the linear regression model now is

$$\text{LENGTH} = \beta_0 + \beta_1(\text{AGE}) + \beta_2(\text{A2}) + \beta_3(\text{A3}) + \beta_4(\text{A4}) + \varepsilon.$$

The results of the regression analysis using this model, again obtained by `PROC REG` of the SAS System, are shown in Table 8.9. The overall statistics for the model in the top portion of the output clearly show that the model is statistically significant, $F(4, 23) = 291.35$, p value < 0.0001. The estimated polynomial equation is

$$\widehat{\text{LENGTH}} = 18.58 + 36.38(\text{AGE}) - 15.69(\text{AGE})^2 + 2.86(\text{AGE})^3 - 0.175(\text{AGE})^4.$$

The individual coefficients in a polynomial equation usually have no practical interpretation; hence the test statistics for these coefficients also have little use. In fact, a pth-degree polynomial should always include all terms with lower powers. It is of interest, however, to ascertain the lowest degree of polynomial required to describe the relationship adequately. To assist in answering this question, many computer programs provide a set of **sequential** sums of squares, which show how the model sum of squares is increased (or error sum of squares is decreased) as higher order polynomial terms are added to the model.[9] In the computer output in Table 8.9, these sequential sums of squares

[8] The powers of AGE are computed in the data input stage. Some computer programs allow the specifications of polynomial terms as part of the regression program.

[9] Sequential sums of squares of this type are automatically provided by orthogonal polynomial contrasts as discussed in Section 6.5. Of course, they cannot be used here because the values of the independent variable are not equally spaced. Furthermore, the ease of direct implementation of polynomial regression on computers make orthogonal polynomials a relatively unattractive alternative except for small experiments such as those presented in Section 6.5 and also Chapter 9.

Table 8.9

Polynomial Regression

Analysis of Variance

Source	DF	Sum of Squares	Mean Square	F Value	Prob > F
Model	4	3325.65171	831.41293	291.346	0.0001
Error	23	65.63507	2.85370		
C Total	27	3391.28679			

Root MSE	1.68929	R-square	0.9806
Dep mean	39.26071	Adj R-sq	0.9773
C.V.	4.30275		

Parameter Estimates

Variable	df	Parameter Estimate	Standard Error	T for H0: Parameter = 0	Prob > |T|
INTERCEP	1	18.583478	1.27503661	14.575	0.0001
AGE	1	36.380515	6.44953987	5.641	0.0001
A2	1	−15.692308	7.54002073	−2.081	0.0487
A3	1	2.860487	3.13335286	0.913	0.3708
A4	1	−0.175485	0.42335354	−0.415	0.6823

Variable	df	Type I SS
INTERCEP	1	43159
AGE	1	2715.447219
A2	1	552.468707
A3	1	57.245461
A4	1	0.490324

are called `Type I SS`.[10] Since these are 1 degree of freedom sums of squares, we can use them to build the most appropriate model by sequentially using an F statistic to test for the significance of each added polynomial term. For this example these tests are as follows:

1. The sequential sum of squares for `INTERCEP` is the correction for the mean of the dependent variable. This quantity can be used to test the hypothesis that the mean of this variable is zero; this is seldom a meaningful test.
2. The sequential sum of squares for `AGE` (2715.4) is divided by the error mean square (2.8537) to get an F ratio of 951.55. We use this to test the hypothesis that a linear regression does not fit the data better than the mean. This hypothesis is rejected.
3. The sequential sums of squares for `A2`, the quadratic term in `AGE`, is divided by the error mean square to test the hypothesis that the quadratic term is not needed. The resulting F ratio of 193.60 rejects this hypothesis.
4. In the same manner, the sequential sums of squares for `A3` and `A4` produce F ratios that indicate that the cubic term is significant but the fourth-degree term is not.

[10]Remember that these were automatically printed with Minitab, while `PROC REG` of the SAS System required a special option. Also in the Minitab output they were called `SEQ SS`. This should serve as a reminder that not all computer programs produce the same default output or use identical terminology!

Sequential sums of squares are additive: They add to the sum of squares for a model containing all coefficients. Therefore they can be used to reconstruct the model and error sums of squares for any lower order model. For example, if we want to compute the mean square error for the third-degree polynomial, we can subtract the sequential sums of squares for the linear, quadratic, and cubic coefficients from the corrected total sum of squares,

$$3391.29 - 2715.44 - 552.47 - 57.241 = 66.12,$$

and divide by the proper degrees of freedom ($n-1-3 = 24$). The result for our example is 2.755.[11] It is of interest to note that this is actually smaller than the error mean square for the full fourth-degree model (2.8537 from Table 8.9). For this reason it is appropriate to reestimate the equation using only the linear, quadratic, and cubic terms. This results in the equation

$$\widehat{\text{LENGTH}} = 18.97 + 33.99(\text{AGE}) - 12.67(\text{AGE})^2 + 1.57(\text{AGE})^3.$$

This equation can be used to estimate the average jawbone length for any age within the range of the data. For example, for AGE $= 0.01$ (one day) the estimated jawbone length is 19.2, compared with the observed value of 15.5. The plot of the estimated jawbone lengths is shown as the solid line in Fig. 8.3. The estimated curve is reasonably close to the observed values with the possible exception of the first observation where the curve overestimates the jawbone length. The nature of the fit can be examined by a residual plot, which is not reproduced here.

We have repeatedly warned that estimated regression equations should not be used for extrapolation. This is especially true of polynomial models, which may exhibit drastic fluctuations in the estimated response beyond the range of the data. For example, using the estimated polynomial regression equation, estimated jawbone lengths for rabbits aged 500 and 700 days are 68.31 and 174.36 mm, respectively!

Although polynomial models are frequently used to estimate responses that cannot be described by straight lines, they are not always useful. For example, the cubic polynomial for the rabbit jawbone lengths shows a "hook" for the older ages, a characteristic not appropriate for growth curves. For this reason, other types of response models are available.

The Multiplicative Model

Another model that describes a curved line relationship is the **multiplicative model**

$$y = e^{\beta_0} x_1^{\beta_1} x_2^{\beta_2} \ldots x_m^{\beta_m} e^{\varepsilon},$$

[11] Equivalently, the sequential sum of squares for the fourth power coefficient may be added to the full model error sum of squares.

where e refers to the Naperian constant used as the basis for natural logarithms. This model is quite popular and has many applications. The coefficients, sometimes called **elasticities**, indicate the *percent* change in the dependent variable associated with a *one-percent* change in the independent variable, holding constant all other variables.

Note that the error term e^{ε} is a multiplicative factor. That is, the value of the deterministic portion is *multiplied* by the error. The expected value of this error, when $\varepsilon = 0$, is one. When the random error is positive the multiplicative factor is greater than 1; when negative it is less than 1. This type of error is quite logical in many applications where variation is proportional to the magnitude of the values of the variable.

The multiplicative model can be made linear by the logarithmic transformation,[12] that is,

$$\log(y) = \beta_0 + \beta_1 \log(x_1) + \beta_2 \log(x_2) + \cdots + \beta_m \log(x_m) + \varepsilon.$$

This model is easily implemented. Most statistical software have provisions for making transformations on the variables in a set of data. ■

EXAMPLE 8.5

We illustrate the multiplicative model with a biological example. It is desired to study the size range of squid eaten by sharks and tuna. The beak (mouth) of squid is indigestible hence it is found in the digestive tracts of harvested fish; hence, it may be possible to predict the total squid weight with a regression that uses various beak dimensions as predictors. The beak measurements and their computer names are

RL = rostral length,
WL = wing length,
RNL = rostral to notch length,
NWL = notch to wing length,
W = width.

The dependent variable WT is the weight of squid.

Data are obtained on a sample of 22 specimens. The data are given in Table 8.10. The specific definitions or meaning of the various dimensions are of little importance for our purposes except that all are related to the total size of the squid.

For simplicity we illustrate the multiplicative model by using only RL and W to estimate WT (the remainder of the variables are used later). First we perform the linear regression with the results in Table 8.11 and the residual plot in Fig. 8.4.

[12]The logarithm base e is used here. The logarithm base 10 (or any other base) may be used; the only difference will be in the intercept.

Table 8.10

Squid Data

Obs	RL	WL	RNL	NWL	W	WT
1	1.31	1.07	0.44	0.75	0.35	1.95
2	1.55	1.49	0.53	0.90	0.47	2.90
3	0.99	0.84	0.34	0.57	0.32	0.72
4	0.99	0.83	0.34	0.54	0.27	0.81
5	1.05	0.90	0.36	0.64	0.30	1.09
6	1.09	0.93	0.42	0.61	0.31	1.22
7	1.08	0.90	0.40	0.51	0.31	1.02
8	1.27	1.08	0.44	0.77	0.34	1.93
9	0.99	0.85	0.36	0.56	0.29	0.64
10	1.34	1.13	0.45	0.77	0.37	2.08
11	1.30	1.10	0.45	0.76	0.38	1.98
12	1.33	1.10	0.48	0.77	0.38	1.90
13	1.86	1.47	0.60	1.01	0.65	8.56
14	1.58	1.34	0.52	0.95	0.50	4.49
15	1.97	1.59	0.67	1.20	0.59	8.49
16	1.80	1.56	0.66	1.02	0.59	6.17
17	1.75	1.58	0.63	1.09	0.59	7.54
18	1.72	1.43	0.64	1.02	0.63	6.36
19	1.68	1.57	0.72	0.96	0.68	7.63
20	1.75	1.59	0.68	1.08	0.62	7.78
21	2.19	1.86	0.75	1.24	0.72	10.15
22	1.73	1.67	0.64	1.14	0.55	6.88

Table 8.11

Linear Regression for Squid Data

Analysis of Variance

Source	DF	Sum of Squares	Mean Square	F value	Pr > F
Model	2	206.74216	103.37108	213.89	<.0001
Error	19	9.18259	0.48329		
Corrected Total	21	215.92475			

Root MSE	0.69519	R-Square	0.9575
Dependent Mean	4.19500	Adj R-Sq	0.9530
Coeff Var	16.57196		

Parameter Estimates

Variable	df	Parameter Estimate	Standard Error	t Value	Pr > |t|
Intercept	1	−6.83495	0.76476	−8.94	<.0001
RL	1	3.27466	1.41606	2.31	0.0321
W	1	13.40078	3.38003	3.96	0.0008

The regression appears to fit well and both coefficients are significant, although the p value for RL is only 0.032. However, the residual plot reveals some problems:

- The residuals have a curved pattern: positive at the extremes and negative in the center. This pattern suggests a curved response.

Figure 8.4

Residual Plot for Linear Regression

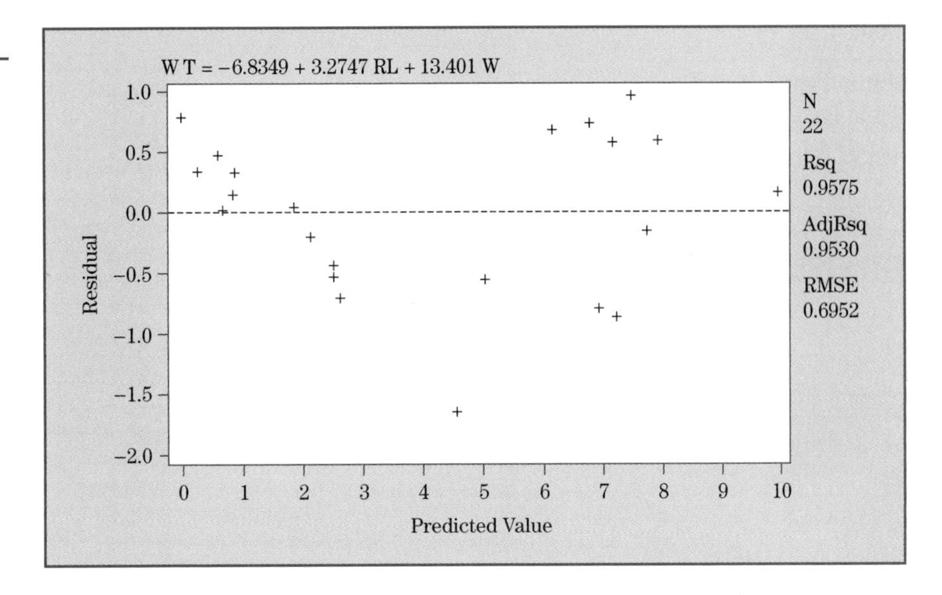

- The residuals are less variable with smaller values of the predicted value and then become increasingly dispersed as values increase. This pattern reveals a heteroscedasticity problem of the type discussed in Section 6.4 where we noted that the logarithmic transformation should be used when the standard deviation is proportional to the mean.

The pattern of residuals for the linear regression would appear to suggest that the variability is proportional to the size of the squid. This type of variability is logical for variables related to sizes of biological specimens, which suggests a multiplicative error. In addition, the multiplicative model itself is appropriate for this example. The dependent variable, the weight of squid, is related to volume, which is a *product* of its dimension. For example, the volume of a cube is d^3, where d is the dimension of a side. The basic shape of a squid is in the form of a cylinder for which the volume is $\pi r^2 l$, where r is the radius and l is the length.

To fit the multiplicative model we first create the variables LWT, LW, and LRL to be the logarithms of WT, W, and RL, respectively, and do a linear regression. The results of fitting the two-variable model using logarithms for the squid data are shown in Table 8.12 and the residual plot is shown in Fig. 8.5.

This model certainly fits better and both coefficients are highly significant. The multiplicative model is

$$\widehat{\mathrm{WT}} = e^{1.169}(\mathrm{RL})^{2.278}(\mathrm{W})^{1.109}.$$

Note that the estimated exponents are close to 2 and unity, which are suggested by the formula for the volume of a cylinder. Finally the residuals appear to have a uniformly random pattern. ■

Table 8.12

Multiplicative Model for Squid Data

Source	DF	Sum of Squares	Mean Square	F Value	Pr > F
Model	2	17.82601	0.91301	400.52	<.0001
Error	19	0.42281	0.02225		
Corrected Total	21	18.24883			

Root MSE	0.14918	R-Square	0.9768
Dependent Mean	1.07156	Adj R-Sq	0.9744
Coeff Var	13.92142		

Parameter Estimates

Variable	DF	Parameter Estimate	Standard Error	t Value	Pr > |t|
Intercept	1	1.16889	0.47827	2.44	0.0245
LRL	1	2.27849	0.49330	4.62	0.0002
LW	1	1.10922	0.37361	2.97	0.0079

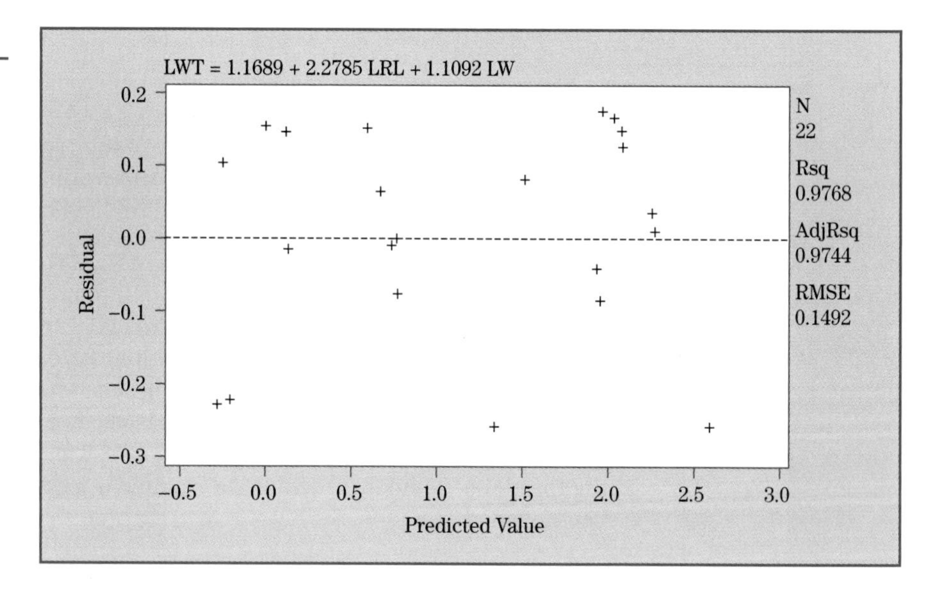

Figure 8.5

Residual Plot for Model Using Logarithms

Nonlinear Models

In some cases no models that are linear in the parameters can be found to provide an adequate description of the data. One such model is the negative exponential model, which is, for example, used to describe the decay of a radioactive substance

$$y = \alpha + \beta e^{\gamma t} + \varepsilon,$$

where y is the remaining weight of the substance at time t. According to the model, $(\alpha + \beta)$ is the initial weight when $t = 0$, α is the ultimate weight of the nondecaying portion of the substance at $t = \infty$, and γ indicates the speed of the decay and is related to the half-life of the substance. Implementation of nonlinear models such as these require methodology beyond the scope of this book (see, for example, Freund and Wilson, 1998).

8.7 Multicollinearity

Often in a multiple regression model, several of the independent variables are measures of similar phenomena. This can result in a high degree of correlation among the set of independent variables. This condition is known as **multicollinearity**. For example, a model used to estimate the total biomass of a plant may include independent variables such as the height, stem diameter, root depth, number of branches, density of canopy, and aerial coverage. Many of these measures are related to the overall size of the plant. All tend to have larger values for larger plants and smaller values for smaller plants and will therefore tend to be highly correlated.

Before the widespread availability of massive computing power made regression analyses easy to perform, much effort was expended in selecting a useful set of independent variables for a regression model. Now, however, it has become customary to put into such a model all possibly relevant variables, as well as polynomial and other curvilinear terms, and then expect the computer to magically reveal the nature of the most appropriate regression model. A natural consequence of having too many variables in a model is the existence of multicollinearity, although this phenomenon is not restricted to this type of situation.

It is certainly true that computers make it easy to perform regressions with large numbers of independent variables. It is also logical to expect that the significance tests for the partial coefficients may be used to determine which of the many variables are actually needed in the model. Unfortunately, the ability of these statistics to perform this task is severely hampered by the existence of multicollinearity.[13]

Remember that a partial coefficient is the change in the dependent variable associated with the change in one of the independent variables, holding constant all other variables. If several variables are closely related it is, by definition, difficult to vary one while holding the others constant. In such cases the partial coefficient is attempting to estimate a phenomenon not exhibited by the data. In a sense such a model is extrapolating beyond the reach of the data.

This extrapolation is reflected by large variances (hence standard errors) of the estimated regression coefficients and a subsequent reduction in the ability to detect statistically significant partial coefficients. A typical result of a regression analysis of data exhibiting multicollinearity is that the overall model is highly significant (has small p value) while few, if any, of the individual partial coefficients are significant (have large p values).

A number of statistics are available for measuring the degree of multicollinearity in a data set. An obvious set of statistics for this purpose is the

[13] In a polynomial regression (Section 8.6), the powers of x are often highly correlated. Technically, this also leads to multicollinearity, which in this case does not have the same implications.

pairwise correlations among all the independent variables. Large magnitudes of these correlations certainly do signify the existence of multicollinearity; however, the lack of large-valued correlations does not guarantee the absence of multicollinearity and for this reason these correlations are not often used to detect multicollinearity.

A very useful set of statistics for detecting multicollinearity is the set of **variance-inflation factors (VIF)**, which indicate, for each independent variable, how much larger the variance of the estimated coefficient is than it would be if the variable were uncorrelated with the other independent variables. Specifically, the VIF for a given independent variable, say, x_j, is $1/(1 - R_j^2)$, where R_j^2 is the coefficient of determination of the regression of x_j on all other independent variables. If R_j^2 is zero, the VIF value is unity and the variable x_j is not involved in any multicollinearity. Any nonzero value of R_j^2 causes the VIF value to exceed unity and indicates the existence of some degree of multicollinearity. For example, if the coefficient of determination for the regression of x_j on all other variables is 0.9, the variance inflation factor will be 10.

There is no universally accepted criterion for establishing the magnitude of a VIF value necessary to identify serious multicollinearity. It has been proposed that VIF values exceeding 10 serve this purpose. However, in cases where the model R^2 is small, smaller VIF values may create problems and vice versa. Finally, if any R_j^2 is 1, indicating an exact linear relationship, VIF $= \infty$, which indicates that $\mathbf{X}'\mathbf{X}$ is singular and thus there is no unique estimate of the regression coefficients.

EXAMPLE 8.5

REVISITED We illustrate multicollinearity with the squid data, using the logarithms of all variables. Because all of these variables are measures of size, they are naturally correlated, suggesting that multicollinearity may be a problem. Figure 8.6 shows the matrix of pairwise scatterplots among the logarithms of the variables. Obviously all variables are highly correlated, and in fact, the correlations with the dependent variable appear no stronger than those among the independent variables. Obviously multicollinearity is a problem with this data set.

We request `PROC REG` of the SAS System to compute the logarithm-based regression using all beak measurements, adding the option for obtaining the variance inflation factors. The results of the regression are shown in Table 8.13. The results are typical of a regression where multicollinearity exists. The test for the model gives a p value of less than 0.0001, while none of the partial coefficients has a p value of less than 0.05. Also, one of the partial coefficient estimates is negative, which is certainly an unexpected result. The variance inflation factors, in the column labeled `VARIANCE INFLATION`, are all in excess of 20 and thus exceed the proposed criterion of 10. The variance inflation factor for the intercept is by definition zero. ■

The course of action to be taken when multicollinearity is found depends on the purpose of the analysis. The presence of multicollinearity is not a violation

Figure 8.6

Scatterplots among Variables in Example 8.5

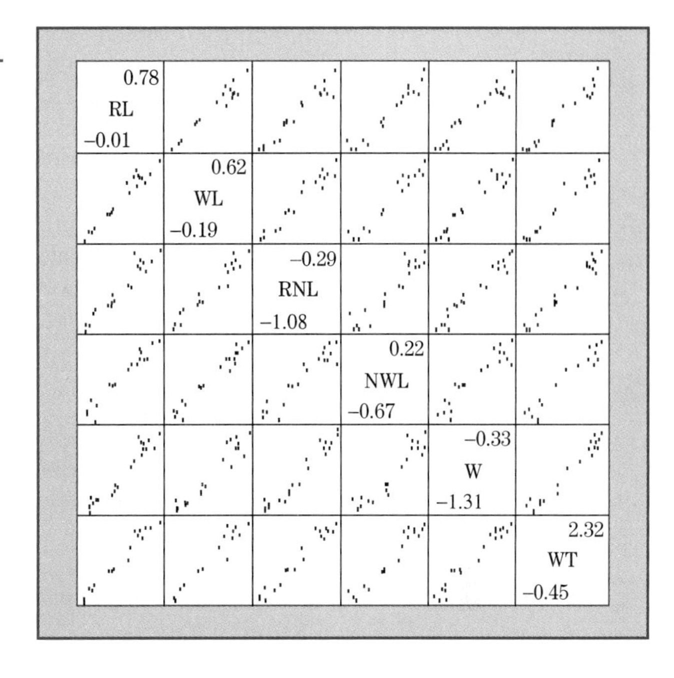

Table 8.13 Regression for Squid Data

DEP VARIABLE: WT

SOURCE	DF	SUM OF SQUARES	MEAN SQUARE	F VALUE	PROB > F
MODEL	5	17.927662	3.585532	178.627	0.0001
ERROR	16	0.321163	0.020073		
C TOTAL	21	18.248825			

ROOT MSE	0.141678	R-SQUARE	0.9824
DEP MEAN	1.071556	ADJR-SQ	0.9769
C.V.	13.22173		

VARIABLE	DF	PARAMETER ESTIMATE	STANDARD ERROR	T FOR H0: PARAMETER = 0	PROB > |T|	VARIANCE INFLATION
INTERCEP	1	2.401917	0.727617	3.301	0.0045	0.000000
RL	1	1.192555	0.818469	1.457	0.1644	43.202506
WL	1	−0.769314	0.790315	−0.973	0.3448	45.184233
RNL	1	1.035553	0.666790	1.553	0.1400	31.309370
NWL	1	1.073729	0.582517	1.843	0.0839	27.486102
W	1	0.843984	0.439783	1.919	0.0730	21.744851

of assumptions and therefore does not, in general, inhibit our ability to obtain a good fit for the model. This can be seen in the above example by the large R-square value and the small residual mean square. Furthermore, the presence of multicollinearity does not affect the inferences about the mean response or prediction of new observations as long as these inferences are made within the range of the observed data. Thus, if the purpose of the analysis is to estimate or predict, then one or more of the independent variables may be dropped from the analysis, using the procedures presented in Section 8.8, to obtain a more efficient model. The purpose of the analysis of the squid data has this objective in mind, and therefore the equation shown in Table 8.10 or the equation resulting from variable selection (Table 8.12) could be effectively used, although care must be taken to avoid any hint of extrapolation.

On the other hand, if the purpose of the analysis is to determine the effect of the various independent variables, then a procedure that simply discards variables is not effective. After all, an important variable may have been discarded because of multicollinearity.

Redefining Variables

One procedure for counteracting the effects of multicollinearity is to redefine some of the independent variables. This procedure is commonly applied in the analysis of national economic statistics collected over time, where variables such as income, employment, savings, etc., are affected by inflation and increases in population and are therefore correlated. Deflating these variables by a price index and converting them to a per capita basis greatly reduces the multicollinearity.

EXAMPLE 8.5

REVISITED In the squid data, all measurements are related to overall size of the beak. It may be useful to retain one measurement of size, say, W, and express the rest as ratios to W. The resulting ratios may then measure shape characteristics and exhibit less multicollinearity. Since the variables used in the regression are logarithms, the logarithms of the ratios are differences. For example, log(RL/W) = log(RL) – log(W). Using these redefinitions and keeping log(W) as is, we obtain the results shown in Table 8.14.

Solution A somewhat unexpected result is that the overall model statistics—the F test for the model, R^2, and the error mean square—have not changed. This is because a linear regression model is not really changed by a linear transformation that retains the same number of variables, as demonstrated by the following simple example. Assume a two-variable regression model:

$$y = \beta_0 + \beta_1 x_1 + \beta_2 x_2 + \varepsilon.$$

Define $x_3 = x_1 - x_2$, and use the model

$$y = \gamma_0 + \gamma_1 x_1 + \gamma_2 x_2 + \varepsilon.$$

In terms of the original variables, this model is

$$y = \gamma_0 + (\gamma_1 + \gamma_2)x_1 - \gamma_2 x_3 + \varepsilon,$$

which is effectively the same model where $\beta_1 = (\gamma_1 + \gamma_2)$ and $\beta_1 = -\gamma_2$.

In the new model for the squid data, we see that the overall width variable (W) clearly stands out as the main contributor to the prediction of weight, and the degree of multicollinearity has been decreased. At the bottom is a test of

Table 8.14 Regression with Redefined Variables

```
Model: MODEL 1
Dependent Variable: WT
                    Analysis of Variance
                 Sum of          Mean
Source      DF   Squares         Square         F Value       Prob > F

Model        5   17.92766        3.58553        178.627        0.0001
Error       16    0.32116        0.02007
C Total     21   18.24883

  Root MSE        0.14168       R-square         0.9824
  Dep Mean        1.07156       Adj R-sq         0.9769
  C.V.           13.22173

                        Parameter Estimates
               Parameter       Standard         T for H0:
Variable   DF   Estimate         Error         Parameter = 0    Prob > |T|

INTERCEP    1    2.401917     0.72761686          3.301           0.0045
RL          1    1.192555     0.81846940          1.457           0.1644
WL          1   -0.769314     0.79031542         -0.973           0.3448
RNL         1    1.035553     0.66679027          1.553           0.1400
NWL         1    1.073729     0.58251746          1.843           0.0839
W           1    3.376507     0.17920582         18.842           0.0001

               Variance
Variable   DF  Inflation

INTERCEP    1  0.00000000
RL          1  8.53690485
WL          1  7.15487734
RNL         1  4.35395220
NWL         1  4.94314166
W           1  3.61063657

Dependent Variable: WT
Test: ALLOTHER   Numerator:     0.1441     df:    4   F value:    7.1790
                 Denominator:   0.020073   df:   16   Prob > F:   0.0016
```

the hypothesis that all other variables contribute nothing to the regression involving W. This test shows that hypothesis to be rejected, indicating the need for at least one of these other variables, although none of the individual coefficients in this set are significant (all p values > 0.05). Variable selection (Section 8.8) may be useful for determining which additional variable(s) may be needed. ■

Other Methods

Another approach is to perform multivariate analyses such as principal components or factor analysis on the set of independent variables to obtain ideas on the nature of the multicollinearity. These methods are beyond the scope of this book (see Freund and Wilson, 1998, Section 5.4).

An entirely different approach is to modify the method of least squares to allow biased estimators of the regression coefficients. Some biased estimators effectively reduce the effect of multicollinearity so that, although the estimates are biased, they have a much smaller variance and therefore have a larger probability of being close to the true parameter value. One such biased regression procedure is called ridge regression (see Freund and Wilson, 1998, Section 5.4).

8.8 Variable Selection

One of the benefits of modern computers is the ability to handle large data sets with many variables. One objective of many experiments is to "filter" these variables to identify those that are most important in explaining a process. In many applications this translates into obtaining a good regression using a minimum number of independent variables. Although the search for this set of variables should use knowledge about the process and its variables, the power of the computer may be useful in implementing a data-driven search for a subset of independent variables that provides adequately precise estimation with a minimum number of variables, which may incidentally provide for less multicollinearity than the full set.

Finding such a model may be accomplished by means of one of a number of **variable selection** techniques. Unfortunately, variable selection is not the panacea it is sometimes ascribed to be. Rather, variable selection is a sort of data dredging that may provide results of spurious validity. Furthermore, if the purpose of the regression analysis is to establish the partial regression relationships, discarding variables may be self-defeating. In other words, variable selection is not always appropriate for the following reasons:

1. It does not help to determine the structure of the relationship among the variables.
2. It uses the power of the computer as a substitute for intelligent study of the problem.
3. The decisions on whether to keep or drop an independent variable from the model are based on the test statistics of the estimated coefficients. Such a procedure is generating hypotheses based on the data, which we have already indicated plays havoc with the specified significance levels. Therefore, just as it is preferable to use preplanned contrasts to automatic post hoc comparisons in the analysis of variance, it is preferable to use knowledge-based selection instead of automatic data-driven selection in regression.

However, despite all these shortcomings, variable selection is widely used, primarily because computers have made it so easy to do. Often there seems to be no reasonable alternative and it actually can produce useful results. For these reasons we present here some variable selection methods together with some aids that may be useful in selecting a useful model.

The purpose of variable selection is to find that subset of the variables in the original model that will in some sense be "optimum." There are two interrelated factors in determining that optimum:

1. For any given subset size (number of variables in the model) we want the subset of independent variables that provides the minimum residual sum of squares. Such a model is considered “optimum” for that subset size.
2. Given a set of such optimum models, select the most appropriate subset size.

One aspect of this problem is that to **guarantee** optimum subsets; all possible subsets must be examined. Hypothetically this method requires that the error sum of squares be computed for 2^m subsets! For example, if $m = 10$, there will be 1024 subsets; for $m = 20$, there will be 1,048,576 subsets!

Modern computers and highly efficient computational algorithms allow some shortcuts, so this problem is not as insurmountable as it may seem. Thus, for example, using the SAS System, the guaranteed optimum subset method can be used for models containing as many as 30 variables. Useful alternatives for models that exceed available computing power are discussed at the end of this selection.

We illustrate the guaranteed optimum subset method with the squid data using the logarithms of the original variables. The program used is `PROC REG` from the SAS System, implementing the `RSQUARE` selection option. The results are given in Table 8.15.

This procedure has examined 31 subsets (not including the null subset), but we have requested that it print results for only the best five for each subset size,

Table 8.15

Variable Selection for Squid Data

Dependent Variable: WT
R-Square Selection Method

Number in Model	R-Square	C(p)	Variables in Model
1	0.9661	12.8361	RL
1	0.9517	25.8810	RNL
1	0.9508	26.7172	W
1	0.9461	30.9861	WL
1	0.9399	36.6412	NWL
2	0.9768	5.0644	RL W
2	0.9763	5.5689	NWL W
2	0.9752	6.5661	RL RNL
2	0.9732	8.3275	RNL NWL
2	0.9682	12.9191	RL NWL
3	0.9797	4.4910	RL NWL W
3	0.9796	4.5603	RNL NWL W
3	0.9786	5.4125	RL RNL W
3	0.9775	6.4971	RL RNL NWL
3	0.9770	6.8654	RL WL W
4	0.9814	4.9478	RL RNL NWL W
4	0.9801	6.1232	WL RNL NWL W
4	0.9797	6.4120	RL WL NWL W
4	0.9787	7.3979	RL WL RNL W
4	0.9783	7.6831	RL WL RNL NWL
5	0.9824	6.0000	RL WL RNL NWL W

which are listed in order from best (optimum) to fifth best. Although we focus on the optimum subsets, the others may be useful, for example, if the second best is almost optimum and contains variables that cost less to measure. For each of these subsets, the procedure prints the R^2 values, the $C(p)$ statistic which is discussed below, and the listing of variables in each selected model.

There are no truly objective criteria for choosing subset size. Statistical significance tests are inappropriate since we generate hypotheses from data. The usual procedure is to plot the behavior of some goodness-of-fit statistic against the number of variables and choose the minimum subset size before the statistic indicates a deterioration of the fit. Virtually any statistic such as MSE or R^2 can be used, but the most popular one currently in use is the $\boldsymbol{C(p)}$ **statistic**.

The $C(p)$ statistic, proposed by Mallows (1973), is a measure of total squared error for a model containing $p(<m)$ independent variables. This total squared error is a measure of the error variance plus a bias due to an underspecified model, that is, a model that excludes variables that should be in the "true" model. Thus, if $C(p)$ is "large" then there is bias due to an underspecified model. The formula for $C(p)$ is of little interest but it is structured so that for a p-variable model:

- if $C(p) > (p + 1)$, the model is underspecified, and
- if $C(p) < (p + 1)$, the model is overspecified; that is, it most likely contains unneeded variables.

By definition, when $p = m$ (the full model), $C(p) = m + 1$. The plot of $C(p)$ values for the variable selections in Table 8.15 is shown in Fig. 8.7; the line plots $C(p)$ against $(p+1)$, which is the boundary between over- and underspecified models.

Figure 8.7

$\boldsymbol{C(p)}$ Plot for Variable Selection

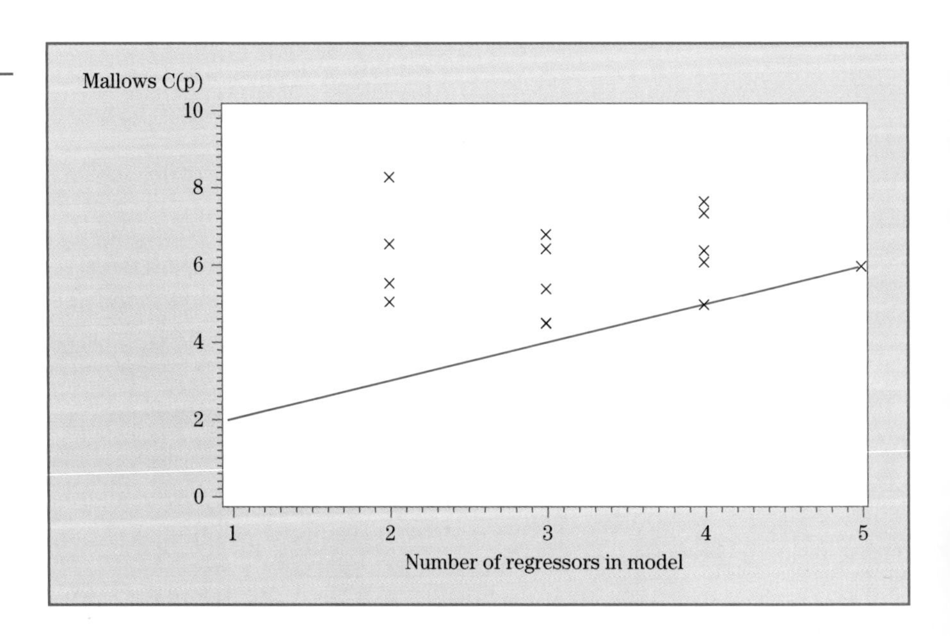

The $C(p)$ plot shows that the four-variable model is slightly overspecified, the three-variable model is slightly underspecified, and the two-variable model is underspecified (the $C(p)$ values for the one-variable model are off the scale). The choice would seem to be the three-variable model. However, note that there are two almost identically fitting "optimum" three-variable models, suggesting that there is still too much multicollinearity. Thus the two-variable model would appear to be a better choice, which is the one used to illustrate the multiplicative model (Table 8.12 and Fig. 8.5). This decision is, of course, somewhat subjective and the researcher can examine the two competing three-variable models and use the one which makes the most sense relative to the problem being addressed.

Other Selection Procedures

We have noted that the guaranteed optimum subset method can be quite expensive to perform. For this reason several alternative procedures that provide nearly optimum models by combining the two aspects of variable selection into a single process exist. Actually these procedures do provide optimum subsets in many cases, but it is not possible to know whether this has actually occurred.

These alternative procedures are also useful as screening devices for models with many independent variables. For example, applying one of these for a 30-variable case may indicate that only about 5 or 6 variables are needed. It is then quite feasible to perform the guaranteed optimum subset method for subsets of size 5 or 6.

The most frequently used alternative methods for variable selection are as follows:

1. *Backward elimination*: Starting with the full model, delete the variable whose coefficient has the smallest partial sum of squares (or smallest magnitude t statistic). Repeat with the resulting $(m-1)$ variable equation, and so forth. Stop deleting variables when all variables contribute some specified minimum partial sum of squares (or have some minimum magnitude t statistic).
2. *Forward selection*: Start by selecting the variable that, by itself, provides the best-fitting equation. Add the second variable whose additional contribution to the regression sum of squares is the largest, and so forth. Continue to add variables, one at a time, until any variable when added to the model contributes less than some specified amount to the regression sum of squares.
3. *Stepwise*: This is an adaptation of forward selection in which, each time a variable has been added, the resulting model is examined to see whether any variable included makes a sufficiently small contribution so that it can be dropped (as in backward elimination).

None of these methods is demonstrably superior for all applications and do not, of course, provide the power of the "all possible" search method.

Although the step methods are usually not recommended for problems with a small number of variables, we illustrate the forward selection method with the transformed squid data, using the forward selection procedure in SPSS Windows. The output is shown in Table 8.16.

The first box in the output summarizes the forward selection procedure. It indicates that two "steps" occurred resulting in two models. The first contained only the variable RL. The second model added W. The box also specifies the method and the criteria used for each step. The next box contains the Model Summary for each model. This box indicates that the R Square for model 1 had a value of .966 and that adding the variable W increased the R Square only to .977.

The third box contains the ANOVA results for both models. Both are significant with a p-value (labeled Sig.) listed as .000 which is certainly less than 0.05.[14] The next box lists the coefficients for the two regression models and the t test for them. Notice that the values of the coefficients for Model 2 are the same as those in Table 8.12.

The final box lists the variables excluded from each model and some additional information about these variables. This table displays information about the variables not in the model at each step. Beta in is the standardized regression coefficient that would result if the variable were entered into the equation at the next step. For example, if we used the model which only contained RL, the variable RNL would result in a regression that had a coefficient for RNL with a value of .382 resulting in a p-value of .016. However, the forward procedure dictated that a better two-variable model would be RL and W. Then when RNL was considered for bringing into the model, it would have a coefficient of .211 but the p-value would be .232.

The last box also includes the partial correlation coefficients (with WT), and something called the "tolerance" which is the reciprocal of the VIF. If the criteria for the VIF is anything larger than 10 then the criteria for the tolerance would be anything less than 0.10.

The forward selection procedure resulted in two "steps" and terminated with a model that contained the variables RL and W. This is, of course, consistent with previous analyses. Normally two different variable selection procedures will result in the same conclusion, but not always, particularly if there is a great deal of multicollinearity present.

In conclusion we emphasize again that variable selection, although very widely used, should be employed with caution. There is no substitute for intelligent, nondata-based variable choices.

8.9 Detection of Outliers, Row Diagnostics

We have repeatedly emphasized that failures of assumptions about the nature of the data may invalidate statistical inferences. For this reason we have encouraged the use of exploratory data analysis of observed or residual values to aid in the detection of failures in assumptions and the use of alternate methods if such failures are found.

[14] Remember that this is not a "true" significance level!

Table 8.16 Results of Forward Selection

Variables Entered/Removed[a]

Model	Variables Entered	Variables Removed	Method
1	RL		Forward (Criterion: Probability-of-F-to-enter <= .050)
2	W		Forward (Criterion: Probability-of-F-to-enter <= .050)

a. Dependent Variable: WT

Model Summary

Model	R	R Square	Adjusted R Square	Std. Error of the Estimate
1	.983[a]	.966	.964	.17592
2	.988[b]	.977	.974	.14918

a. Predictors: (Constant), RL
b. Predictors: (Constant), RL, W

ANOVA[c]

Model		Sum of Squares	df	Mean Square	F	Sig.
1	Regression	17.630	1	17.630	569.664	.000[a]
	Residual	.619	20	.031		
	Total	18.249	21			
2	Regression	17.826	2	8.913	400.524	.000[b]
	Residual	.423	19	.022		
	Total	18.249	21			

a. Predictors: (Constant), RL
b. Predictors: (Constant), RL, W
c. Dependent Variable: WT

Coefficients[a]

Model		Unstandardized Coefficients		Standardized Coefficients	t	Sig.
		B	Std. Error	Beta		
1	(Constant)	−.241	.067		-3.622	.002
	RL	3.690	.155	.983	23.868	.000
2	(Constant)	1.169	.478		2.444	.024
	RL	2.279	.493	.607	4.619	.000
	W	1.109	.374	.390	2.969	.008

a. Dependent Variable: WT

Excluded Variables[c]

Model		Beta In	t	Sig.	Partial Correlation	Collinearity Statistics
						Tolerance
1	WL	.230[a]	1.099	.285	.245	3.835E-02
	RNL	.382[a]	2.639	.016	.518	6.237E-02
	NWL	.212[a]	1.122	.276	.249	4.698E-02
	W	.390[a]	2.969	.008	.563	7.064E-02
2	WL	.079[b]	.414	.684	.097	3.493E-02
	RNL	.211[b]	1.238	.232	.280	4.087E-02
	NWL	.246[b]	1.583	.131	.350	4.675E-02

a. Predictors in the Model: (Constant), RL
b. Predictors in the Model: (Constant), RL, W
c. Dependent Variable: WT

As data and models become more complex, opportunities increase for undetected violations and inappropriate analyses. For example, in regression analysis the misspecification of the model, such as leaving out important independent variables or neglecting the possibility of curvilinear responses, may lead to estimates of parameters exhibiting large variances. The fact that data for regression analysis are usually observed, rather than the result of carefully designed experiments, makes the existence of misspecification, violation of assumptions, and inappropriate analysis more difficult to detect.

For these types of data it is also more difficult to detect outliers. We first discuss the basic reason for this and subsequently present some methodologies that may aid in overcoming the problem.

A Physical Analogue to Least Squares A fundamental law of physics, called Hooke's law, specifies that the tension of a coil spring is proportional to the square of the length that the spring has been stretched (assuming a perfect spring). The least squares estimate of a one-variable regression line is equivalent to hooking a set of springs, perpendicular to the x axis, from the data points to a rigid rod. The equilibrium position of the rod represents the minimum total tension of the springs and thus represents the least squares line (assuming no gravity). This is illustrated in Fig. 8.8.

This analogue is useful for illustrating a number of characteristics of least squares estimation. For example, the amount of force required to pull the rod into a horizontal position ($\beta_1 = 0$) represents the strength or statistical significance of the linear regression of y on x. Remember the estimated variance of β_1 is $(s^2_{y|x}/S_{xx})$, which increases in magnitude as the x values span a narrower range (Section 7.5). Similarly, the force required to pull the rod into the horizontal position is lower if the data values occupy a narrow range in the x direction when the springs are close to the center of the rod.

Figure 8.8

Illustration of Hooke's Law

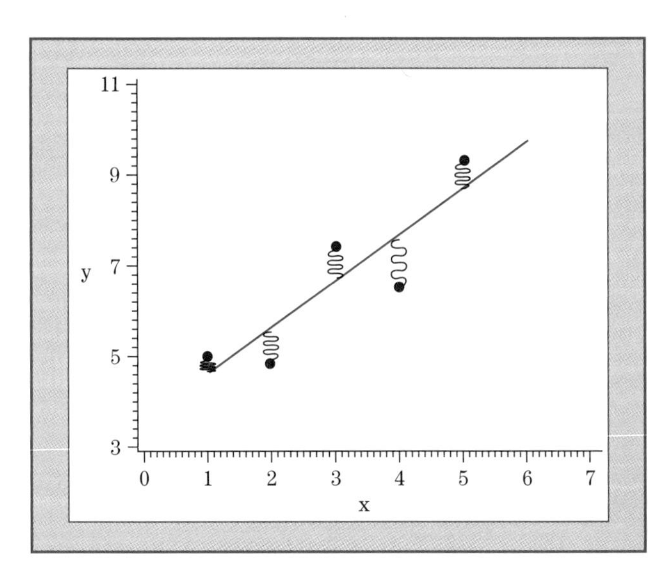

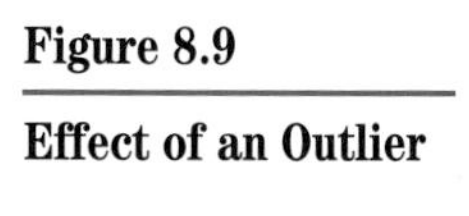

Figure 8.9

Effect of an Outlier

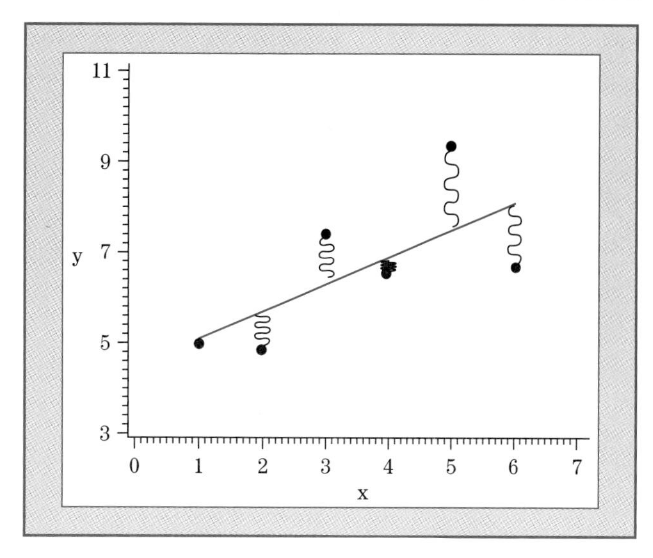

The spring analogue also illustrates the effect of the location of individual observations on the estimated coefficients. For example, an unusual or extreme value for the dependent variable y will tend to exert a relatively large influence or **leverage** on the equilibrium location of the regression line as illustrated in Fig. 8.9, where the data are identical to those of Fig. 8.8 except that the response for $x = 6$ has been decreased by 3 units, making this an outlier. In this case the outlier occurs at the extreme of the range of the x values; hence, the point exerts extreme *leverage* so that the line is forced to pass quite close to that point. Hence the largest observed residual is actually for $x = 5$, which is not an outlier. On the other hand an outlier at the center of the range of x values will not exert such a large leverage on the location of the line. However, the outlier may create a large residual, which when squared, contributes to an overestimate of the variance.

This example shows that the effect of outliers in the response variable depends on where the observation lies in the space of the independent variable(s). This effect is relatively easy to visualize in the case of a simple linear regression but is, obviously, more difficult to "see" when there are many independent variables. While outlier detection statistics tend to focus on outliers in the dependent variable, other statistics focus on outliers in the independent variable, which we have identified as observations having a high degree of leverage. Yet other statistics provide information on both of these aspects. While examining a large number of such statistics can be quite useful, the scope of this book limits our presentation to one of the most frequently used combination statistics. A more complete discussion can be found in Belsley *et al.* (1980).

One important class of statistics that investigate the combined effects of outliers and leverage is known as **influence statistics**. These statistics are based on the question: "What happens if the regression is estimated using the

Table 8.17

Data for Outlier Detection

OBS	Y	X1	X2	X3
1	763	19.8	128	86
2	650	20.9	110	72
3	554	15.1	95	62
4	742	19.8	123	82
5	470	21.4	77	52
6	651	19.5	107	72
7	756	25.2	123	84
8	563	26.2	95	83
9	681	26.8	116	76
10	579	28.8	100	64
11	716	22.0	110	80
12	650	24.2	107	71
13	761	24.9	125	81
14	549	25.6	89	61
15	641	24.7	103	71
16	606	26.2	103	67
17	696	21.0	110	77
18	795	29.4	133	83
19	582	21.6	96	65
20	559	20.0	91	62

data without a particular observation?" We present one such influence statistic and give an example of how it may be useful. The statistic, known as the DFFITS statistic, is the difference between the predicted value for each observation using the model estimated with all data and that using the model estimated with that observation omitted (Belsley *et al.*, 1980). This difference is standardized, using the residual variance as estimated with the observation omitted. Large values of this statistic may indicate suspicious observations. Generally, values exceeding $2\sqrt{(m+1)/n}$ are considered large for this purpose.[15] Actually this criterion is not often needed since outliers having serious effects on model estimates usually have DFFITS values greatly exceeding this criterion.

EXAMPLE 8.6

The production levels of a finished product (produced from sheets of stainless steel) have varied quite a bit, and management is trying to devise a method for predicting the daily amount of finished product. The ability to predict production is useful for scheduling labor, warehouse space, and shipment of raw materials and also to suggest a pricing strategy.

The number of units of the product (Y) that can be produced in a day depends on the width (X1) and the density (X2) of the sheets being processed, and the tensile strength of the steel (X3). The data are taken from 20 days of production. The observations are given in Table 8.17.

[15]Fortunately, it is not necessary to recompute the regression equation omitting each observation in turn. Special algorithms are available that make these computations quite feasible even for rather large problems. We also emphasize that other outlier detection statistics are available and that the DFFITS statistic is not necessarily the best. However, this statistic is quite popular, and to present other statistics at this point may confuse the issue.

Table 8.18 Analysis of Steel Data

SOURCE	DF	SUM OF SQUARES	MEAN SQUARE	F VALUE	PROB > F
MODEL	3	146684.105	48894.702	133.750	0.0001
ERROR	16	5849.095	365.568		
C TOTAL	19	152533.200			

ROOT MSE	19.119844	R-SQUARE	0.9617
DEP MEAN	648.200	ADJ R-SQ	0.9545
C.V.	2.949683		

VARIABLE	DF	PARAMETER ESTIMATE	STANDARD ERROR	T FOR H0: PARAMETER = 0	PROB > |T|	VARIANCE INFLATION
INTERCEP	1	6.383762	40.701546	0.157	0.8773	0.000000
X1	1	−0.916131	1.243010	−0.737	0.4718	1.042464
X2	1	5.409022	0.595196	9.088	0.0001	3.906240
X3	1	1.157731	0.909244	1.273	0.2211	3.896413

OBS	Y	RESIDUALS	DFFITS
1	763	−17.164	−0.596
2	650	−15.586	−0.259
3	554	−24.187	−1.198
4	742	−6.488	−0.175
5	470	6.525	0.263
6	651	0.359	0.007
7	756	10.144	0.218
8	563	−29.330	−12.535
9	681	−16.266	−0.334
10	579	−15.996	−0.592
11	716	42.160	1.138
12	650	4.822	0.064
13	761	7.524	0.167
14	549	14.045	0.380
15	641	17.916	0.261
16	606	−11.078	−0.230
17	696	24.717	0.450
18	795	0.059	0.003
19	582	0.886	0.015
20	559	6.938	0.155

Solution We perform a linear regression of `Y` on `X1`, `X2`, and `X3`, using `PROC REG` of the SAS System. The analysis, including the residuals and `DFFITS` statistics, is shown in Table 8.18. The results appear to be quite reasonable. The regression is certainly significant. Only one coefficient appears to be important and there is little multicollinearity. Thus one would be inclined to suggest a model that includes only `X2` and would probably show increased production with increased values of `X2`. The residuals, given in the column labeled `RESIDUALS`, also show no real surprises. The residual for observation 11 appears quite large, but the residual plot (not reproduced here) does not show it as an extreme value. However, the `DFFITS` statistics show a different story.

Table 8.19 Results When Outlier Is Omitted

DEP VARIABLE: Y

SOURCE	df	SUM OF SQUARES	MEAN SQUARE	F VALUE	PROB > F
MODEL	3	143293.225	47764.408	448.105	0.0001
ERROR	15	1598.880	106.592		
C TOTAL	18	144892.105			

ROOT MSE	10.324340	R-square	0.9890
DEP MEAN	652.684	Adj R-sq	0.9868
C.V.	1.581828		

VARIABLE	df	PARAMETER ESTIMATE	STANDARD ERROR	T FOR H0: PARAMETER = 0	PROB > |T|	VARIANCE INFLATION
INTERCEP	1	−42.267607	23.289383	−1.815	0.0896	0.000000
X1	1	0.982466	0.735468	1.336	0.2015	1.202123
X2	1	1.738216	0.664253	2.617	0.0194	16.053214
X3	1	6.738637	1.011032	6.665	0.0001	15.420233

The value of that statistic for observation 8 is about 10 times that for any other observation. By any criterion this observation is certainly a suspicious candidate.

The finding of a suspicious observation does not, however, suggest what the proper course of action should be. Simply discarding such an observation is usually not recommended. Serious efforts should be made to verify the validity of the data values or to determine whether some unusual event did occur. However, for purposes of illustration here, we do reestimate the regression without that observation. The results of the analysis are given in Table 8.19, where it becomes evident that omitting observation number 8 has greatly changed the results of the regression analysis. The residual variance has decreased from 366 to 106, the F statistic for testing the model has increased from 134 to 448, the estimated coefficients and their p values have changed drastically so that now X3 is the dominant independent variable, and the degree of multicollinearity between X2 and X3 has also increased. In other words, the conclusions about the factors affecting production have changed by eliminating one observation.

The change in the degree of multicollinearity provides a clue to the reasons for the apparent outlier. Figure 8.10 shows the matrix of scatterplots for these variables. The plotting symbol is a period except for observation 8, whose symbol is "8." These plots clearly show that the observed values for X2 and X3 as well as Y and X3 are highly correlated *except* for observation 8. However, that observation appears not to be unusual with respect to the other variables. The conclusion to be reached is that the unusual combination of values x_2 and x_3 that occurred in observation 8 is a combination that does not conform to the normal operating conditions. Or it could be a recording error. ■

Finding and identifying outliers or influential observations does not answer the question of what to do with such observations. Simply discarding or changing such observations is bad statistical practice since it may lead to self-fulfilling

Figure 8.10

Scatterplots of Steel Data

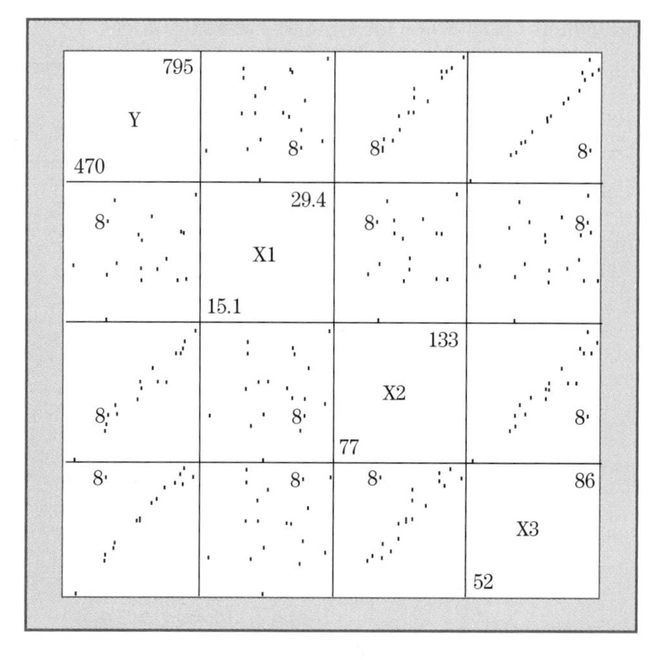

prophesies. Sometimes, when an outlier can be traced to sloppiness or mistakes, deletion or modification may be justified. In the above example, the outlier may have resulted from an unusual product mix that does not often occur. In this case, omission may be justified, but only if the conclusions state that the equation may only be used for the usual product mix and that a close watch must be posted to detect unusual mixes whose costs cannot be predicted by that model. In the previous example, predicting the number of units produced for day 8 without using that day's values provides a predicted value of 702.9, certainly a very bad prediction!

8.10 CHAPTER SUMMARY

Solution to Example 8.1 The effect of performance factors on winning percentages of baseball teams can be studied by a multiple regression using WIN as the dependent variable and the team performance factors as independent variables. Although the data are certainly not random, it is reasonable to assume that the residuals from the model are random and otherwise adhere reasonably to the required assumptions. The output for the regression as produced by `PROC REG` of the SAS System is shown in Table 8.20.

Starting at the top, it is evident that the regression is certainly significant, although the coefficient of determination may not be considered particularly large. The residual standard deviation of 0.045 indicates that about 95% of observed proportion of wins are within 0.09 of the predicted values, which indicates that there are obviously some other factors affecting the winning percentages. The coefficients all have the expected signs, but it appears that

Table 8.20 Regression for Winning Baseball Games

```
Model: MODEL1
Dependent Variable: WIN
                 Analysis of Variance
               Sum of        Mean
Source    DF   Squares       Square     F Value    Prob > F

Model      5   0.12324       0.02465    12.410      0.0001
Error     34   0.06753       0.00199
C Total   39   0.19076

   Root MSE    0.04457     R-square                 0.6460
   Dep Mean    0.50000     Adj R-sq                 0.5940
   C.V.        8.91323

                          Parameter Estimates
                      Parameter       Standard         T for H0:
Variable      DF      Estimate        Error         Parameter = 0      Prob > |T|

INTERCEP       1     -0.277675      0.19131508          -1.451           0.1558
RUNS           1      0.000278      0.00014660           1.895           0.0666
BA             1      1.741999      0.92847059           1.876           0.0692
DP             1      0.000737      0.00045021           1.637           0.1108
WALK           1     -0.000590      0.00012916          -4.566           0.0001
SO             1      0.000346      0.00010441           3.315           0.0022

                      Variance
Variable      DF      Inflation

INTERCEP       1     0.00000000
RUNS           1     2.93207465
BA             1     3.05405561
DP             1     1.61208141
WALK           1     1.21916888
SO             1     1.46815334
```

the only important factors relate to pitching. The variance inflation factors are relatively small, although there appears to be an expected degree of correlation between number of runs and batting average.

It is interesting to investigate the relative importance of the offensive (RUNS, BA) and defensive (DP, WALK, SO) factors. These questions can be answered with this computer program by the so-called TEST commands. The first test, labeled OFFENSE, tests the hypothesis that the coefficients for RUNS and BA are both zero, and the second, labeled DEFENSE, tests the null hypothesis that the coefficients of DP, WALK, and SO are all zero. These commands produce the following results:

```
Test: OFFENSE Numerator:     0.0304    DF:   2  F value:   15.3263
              Denominator:   0.001986  DF:  34  Prob > F:   0.0001
Test: DEFENSE Numerator:     0.0226    DF:   3  F value:   11.3990
              Denominator:   0.001986  DF:  34  Prob > F:   0.0001
```

It appears that both offense and defense contribute to winning, but offense may be more important. This conclusion is not quite consistent with the tests

Table 8.21

Variable Selection for Baseball Regression

The REG Procedure
Model: MODEL1
Dependent Variable: WIN
R-Square Selection Method

Number in Model	R-Square	C(p)	Variables in Model
1	0.2625	34.8352	BA
1	0.2606	35.0174	RUNS
1	0.1793	42.8268	SO
1	0.0691	53.4079	WALK
2	0.4829	15.6621	BA WALK
2	0.4769	16.2464	RUNS WALK
2	0.4069	22.9662	BA SO
2	0.3882	24.7608	RUNS SO
3	0.5856	7.8051	BA WALK SO
3	0.5612	10.1473	RUNS WALK SO
3	0.5313	13.0186	RUNS BA WALK
3	0.4852	17.4423	BA DP WALK
4	0.6181	6.6800	RUNS BA WALK SO
4	0.6094	7.5201	RUNS DP WALK SO
4	0.6086	7.5919	BA DP WALK SO
4	0.5316	14.9882	RUNS BA DP WALK
5	0.6460	6.0000	RUNS BA DP WALK SO

on individual coefficients, a result that may be due to the existence of some correlation among the variables.

Since a number of the individual factors appear to have little effect on the winning percentage, variable selection may be useful. The `RSQUARE` selection of `PROC REG` provides the results shown in Table 8.21. The selection process indicates little loss in the error mean square associated with dropping double plays and runs; hence the remaining three variables may provide a good model. The resulting regression is summarized in Table 8.22.

The model with the three remaining variables fits almost as well as the one with all five variables, and now the effects of the performance factors are more definitive. Additional analysis includes the residual plot, which is shown in Fig. 8.11. Although one team has a rather large negative residual, the overall pattern of residuals shows no major cause for concern about assumptions. ■

The multiple linear regression model

$$y = \beta_0 + \beta_1 x_1 + \cdots + \beta_m x_m + \varepsilon$$

is the extension of the simple linear regression model to more than one independent variable. The basic principles of a multiple regression analysis are the same as for the simple case, but many of the details are different.

The least squares principle for obtaining estimates of the regression coefficients requires the solution of a set of linear equations that can be represented symbolically by matrices and is solved numerically, usually by computers.

Table 8.22

Selected Model for Baseball Regression

```
Model: MODEL1
Dependent Variable: WIN
                         Analysis of Variance
                   Sum of          Mean
Source      DF    Squares         Square           F Value       Prob > F

Model        6    0.11171         0.03724          16.955         0.0001
Error       36    0.07906         0.00220
C Total     39    0.19076
   Root MSE       0.04686       R-Square           0.5856
   Dep Mean       0.50000       Adj R-sq           0.5510
   C.V.           9.37245
                          Parameter Estimates
                 Parameter     Standard       T for H0:
Variable   DF    Estimate       Error      Parameter = 0   Prob > |T|

INTERCEP    1   -0.356943   0.15890423        -2.246          0.0309
BA          1    3.339829   0.60054220         5.561          0.0001
WALK        1   -0.000521   0.00013230        -3.940          0.0004
SO          1    0.000274   0.00009178         2.986          0.0051
```

Figure 8.11

Residual Plot for Baseball Regression

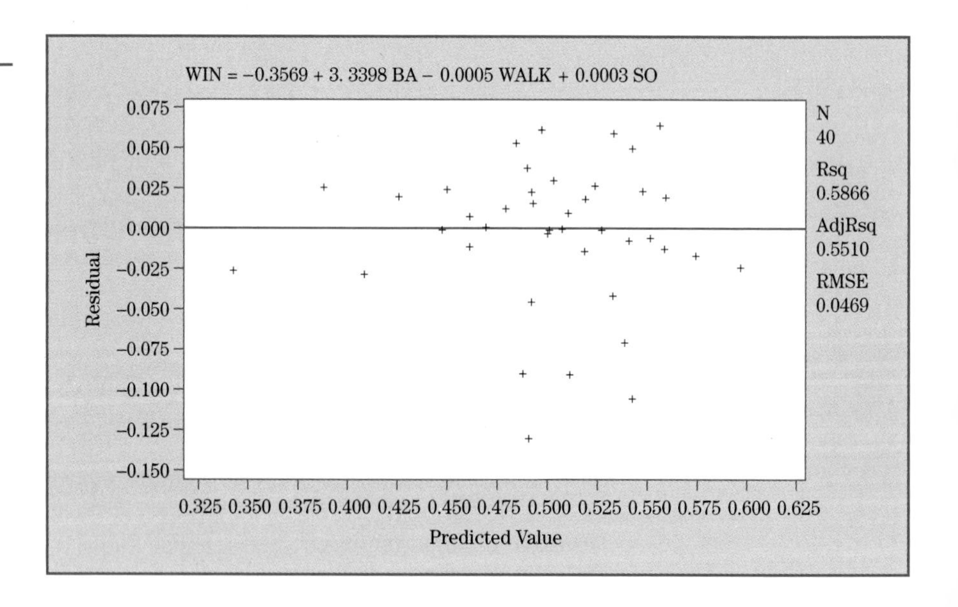

As in simple linear regression, the variance of the random error is based on the sum of squares of residuals and is computed through a partitioning of sums of squares.

Because the partial regression coefficients in a multiple regression model measure the effect of a variable in the presence of all other variables in the model, estimates and inferences for these coefficients are different from the total regression coefficients obtained by the corresponding simple linear regressions. Inference procedures for the partial regression coefficients are therefore based on the comparison of the full model, which includes all coefficients and the restricted model, with the restrictions relating to the inference on specific coefficients.

Inferences for the response have the same connotation as they have for the simple linear regression model.

The multiple correlation coefficient is a measure of the strength of a multiple regression model. The square of the multiple regression coefficient is the ratio of the regression to total sum of squares, as it was for the simple linear regression model. A partial correlation coefficient is a measure of the strength of the relationship associated with a partial regression coefficient.

Although the multiple regression model must be linear in the model parameters, it may be used to describe curvilinear relationships. This is accomplished primarily by polynomial regression, but other forms may be used. A regression linear in the logarithms of the variables has special uses.

Often a proposed regression model has more independent variables than necessary for an adequate description of the data. A side effect of such model specification is that of multicollinearity, which is defined as the existence of large correlations among the independent variables. This phenomenon causes the individual regression coefficients to have large variances, often resulting in an estimated model that has good predictive power but with little statistical significance for the regression coefficients.

One possible solution to an excessive number of independent variables is to select a subset of independent variables for use in the model. Although this is very easy to do, it should be done with caution, because such procedures generate hypotheses with the data.

As in all statistical analyses, it is important to check assumptions. Because of the complexity of multiple regression, simple residual plots may not be adequate. Some additional methods for checking assumptions are presented.

8.11 CHAPTER EXERCISES

CONCEPT QUESTIONS

1. Given that SSR = 50 and SSE = 100, calculate R^2.

2. The multiple correlation coefficient can be calculated as the simple correlation between__________and__________.

3. (a) What value of R^2 is required so that a regression with five independent variables is significant if there are 30 observations? [*Hint:* Use the 0.05 critical value for $F(5, 24)$].
   (b) Answer part (a) if there are 500 observations.
   (c) What do these results tell us about the R^2 statistic?

4. If x is the number of inches and y is the number of pounds, what is the unit of measure of the regression coefficient?

5. What is the common feature of most "influence" statistics?

6. Under what conditions is least squares not the best method for estimating regression coefficients?
7. What is the interpretation of the regression coefficient when using logarithms of all variables?
8. What is the basic principle underlying inferences on partial regression coefficients?
9. Why is multicollinearity a problem?
10. List some reasons why variable selection is not always an appropriate remedial method when multicollinearity exists.

EXERCISES

1. This exercise is designed to provide a review of the mechanics for performing a regression analysis. The data are:

OBS	X1	X2	Y
1	1	5	5.4
2	2	6	8.5
3	4	6	9.4
4	6	5	11.5
5	6	4	9.4
6	8	3	11.8
7	10	3	13.2
8	11	2	12.1

First we compute $\mathbf{X'X}$ and $\mathbf{X'Y}$, the sums of squares and cross products as in Table 8.3. Verify at least two or three of these elements.

MODEL CROSSPRODUCTS X'X X'Y Y'Y

X'X	INTERCEP	X1	X2	Y
INTERCEP	8	48	34	81.3
X1	48	378	171	544.9
X2	34	171	160	328.7
Y	81.3	544.9	328.7	870.27

Next we invert $\mathbf{X'X}$ and compute $\hat{\mathbf{B}} = (\mathbf{X'X})^{-1}\mathbf{X'Y}$, again as in Table 8.3.

X'X INVERSE B, SSE

INVERSE	INTERCEP	X1	X2	Y
INTERCEP	12.73103	−0.762255	−1.89706	−1.44424
X1	−0.762255	0.05065359	0.1078431	1.077859
X2	−1.89706	1.1078431	0.2941176	1.209314
Y	−1.44424	1.077859	1.209314	2.859677

Verify that at least two elements of the matrix product $(\mathbf{X'X})(\mathbf{X'X})^{-1}$ are elements of an identity matrix. We next perform the partitioning of sums of squares and perform the tests for the model and the partial coefficients. Verify these computations.

DEP VARIABLE: Y

SOURCE	DF	SUM OF SQUARES	MEAN SQUARE	F VALUE	PROB > F
MODEL	2	41.199073	20.599536	36.017	0.0011
ERROR	5	2.859677	0.571935		
C TOTAL	7	44.058750			

ROOT MSE	0.756264	R-SQUARE	0.9351
DEP MEAN	10.162500	ADJ R-SQ	0.9091
C.V.	7.441714		

VARIABLE	DF	PARAMETER ESTIMATE	STANDARD ERROR	T FOR H0: PARAMETER = 0	PROB > |T|
INTERCEPT	1	−1.444240	2.701571	−0.535	0.6158
X1	1	1.077859	0.170207	6.333	0.0014
X2	1	1.209314	0.410142	2.949	0.0319

Finally, we compute the predicted and residual values:

OBS	ACTUAL	PREDICT VALUE	RESIDUAL
1	5.400	5.680	−.280188
2	8.500	7.967	0.532639
3	9.400	10.123	−.723080
4	11.500	10.069	0.430515
5	9.400	9.860	−.460172
6	11.800	10.807	0.993423
7	13.200	12.962	0.237704
8	12.100	12.831	−.730842
	SUM OF RESIDUALS		1.e−14
	SUM OF SQUARED RESIDUALS		2.859677

Verify at least two of the predicted and residual values and also that the sum of residuals is zero and that the sum of squares of the residuals is the ERROR sum of squares given in the partitioning of the sums of squares.

2. The complete data set on energy consumption given for Exercise 7 in Chapter 7 contains other factors that may affect power consumption. The following have been selected for this exercise:

TMAX: maximum daily temperature,
TMIN: minimum daily temperature,
WNDSPD: windspeed, coded "0" if less than 6 knots and "1" if 6 or more knots,
CLDCVR: cloud cover coded as follows:
0.0—clear
1.0—less than 0.6 covered
2.0—0.6 to 0.9 covered
3.0—cloudy (increments of 0.5 are used to denote variable cloud cover between indicated codes), and
KWH: electricity consumption.

The data are given in Table 8.23.

Table 8.23

Data for Exercise 2

OBS	MO	DAY	TMAX	TMIN	WNDSPD	CLDCVR	KWH
1	9	19	87	68	1	2.0	45
2	9	20	90	70	1	1.0	73
3	9	21	88	68	1	1.0	43
4	9	22	88	69	1	1.5	61
5	9	23	86	69	1	2.0	52
6	9	24	91	75	1	2.0	56
7	9	25	91	76	1	1.5	70
8	9	26	90	73	1	2.0	69
9	9	27	79	72	0	3.0	53
10	9	28	76	63	0	0.0	51
11	9	29	83	57	0	0.0	39
12	9	30	86	61	1	1.0	55
13	10	1	85	70	1	2.0	55
14	10	2	89	69	0	2.0	57
15	10	3	88	72	1	1.5	68
16	10	4	85	73	0	3.0	73
17	10	5	84	68	1	3.0	57
18	10	6	83	69	0	2.0	51
19	10	7	81	70	0	1.0	55
20	10	8	89	70	1	1.5	56
21	10	9	88	69	1	0.0	72
22	10	10	88	76	1	2.5	73
23	10	11	77	66	1	3.0	69
24	10	12	75	65	1	2.5	38
25	10	13	72	64	1	3.0	50
26	10	14	68	65	1	3.0	37
27	10	15	71	67	0	3.0	43
28	10	16	75	66	1	3.0	42
29	10	17	74	52	1	0.0	25
30	10	18	77	51	0	0.0	31
31	10	19	79	50	0	0.0	31
32	10	20	80	50	0	0.0	32
33	10	21	80	53	0	0.0	35
34	10	22	81	53	1	0.0	32
35	10	22	80	53	0	0.0	34
36	10	24	81	54	1	2.0	35
37	10	25	83	67	0	2.0	41
38	10	26	84	67	1	1.5	51
39	10	27	80	63	1	3.0	34
40	10	28	73	53	1	1.0	19
41	10	29	71	49	0	0.0	19
42	10	30	72	56	1	3.0	30
43	10	31	72	53	1	0.0	23
44	11	1	79	48	1	0.0	35
45	11	2	84	63	1	1.0	29
46	11	3	74	62	0	3.0	55
47	11	4	83	72	1	2.5	56

Table 8.24

Data for Exercise 3: Asphalt Data

Obs	X1	X2	X3	Y1	Y2
1	5.3	0.02	77	42	3.20
2	5.3	0.02	32	481	0.73
3	5.3	0.02	0	543	0.16
4	6.0	2.00	77	609	1.44
5	7.8	0.20	77	444	3.68
6	8.0	2.00	104	194	3.11
7	8.0	2.00	77	593	3.07
8	8.0	2.00	32	977	0.19
9	8.0	2.00	0	872	0.00
10	8.0	0.02	104	35	5.86
11	8.0	0.02	77	96	5.97
12	8.0	0.02	32	663	0.29
13	8.0	0.02	0	702	0.04
14	10.0	2.00	77	518	2.72
15	12.0	0.02	77	40	7.35
16	12.0	0.02	32	627	1.17
17	12.0	0.02	0	683	0.14
18	12.0	0.02	104	22	15.00
19	14.0	0.02	77	35	11.80

Perform a regression analysis to determine how the factors affect fuel consumption (KWH). Include checking for multicollinearity, variable selection (if appropriate), and outlier detection. Finally, interpret the results and assess their usefulness.

3. The data in Table 8.24 represent the results of a test for the strength of an asphalt concrete mix. The test consisted of applying a compressive force on the top of different sample specimens. Two responses occurred: the stress and strain at which a sample specimen failed. The factors relate to mixture proportions, rates of speed at which the force was applied, and ambient temperature. Higher values of the response variables indicate stronger materials.

   The variables are:

   X1: percent binder (the amount of asphalt in the mixture),
   X2: loading rate (the speed at which the force was applied),
   X3: ambient temperature,
   Y1: the stress at which the sample specimen failed, and
   Y2: the strain at which the specimen failed.

   Perform separate regressions to relate stress and strain to the factors of the experiment. Check the residuals for possible specification errors. Interpret all results.

4. The data in Table 8.25 were collected in order to study factors affecting the supply and demand for commercial air travel. Data on various aspects of commercial air travel for an arbitrarily chosen set of 74 pairs of cities were obtained from a 1966 (before deregulation) CAB study. Other data were obtained from a standard atlas. The variables are:

Table 8.25

Data for Exercise 4

CITY1	CITY2	PASS	MILES	INM	INS	POPM	POPS	AIRL
ATL	AGST	3.546	141	3.246	2.606	1270	279	3
ATL	BHM	7.016	139	3.246	2.637	1270	738	4
ATL	CHIC	13.300	588	3.982	3.246	6587	1270	5
ATL	CHST	5.637	226	3.246	3.160	1270	375	5
ATL	CLBS	3.630	193	3.246	2.569	1270	299	4
ATL	CLE	3.891	555	3.559	3.246	2072	1270	3
ATL	DALL	6.776	719	3.201	3.245	1359	1270	2
ATL	DC	9.443	543	3.524	3.246	2637	1270	5
ATL	DETR	5.262	597	3.695	3.246	4063	1270	4
ATL	JAX	8.339	285	3.246	2.774	1270	505	4
ATL	LA	5.657	1932	3.759	3.246	7079	1270	3
ATL	MEM	6.286	336	3.246	2.552	1270	755	3
ATL	NO	7.058	424	3.245	2.876	1270	1050	4
ATL	NVL	5.423	214	3.246	2.807	1270	534	3
ATL	ORL	4.259	401	3.246	2.509	1270	379	3
ATL	PHIL	6.040	666	3.243	3.246	4690	1270	5
ATL	PIT	3.345	521	3.125	3.246	2413	1270	2
ATL	RAL	3.371	350	3.246	2.712	1270	198	3
ATL	SF	4.624	2135	3.977	3.246	3075	1270	3
ATL	SVNH	3.669	223	3.246	2.484	1270	188	1
ATL	TPA	7.463	413	3.246	2.586	1270	881	5
DC	NYC	150.970	205	3.962	2.524	11698	2637	12
LA	BOSTN	16.397	2591	3.759	3.423	7079	3516	4
LA	CHIC	55.681	1742	3.759	3.982	7079	6587	5
LA	DALL	18.222	1238	3.759	3.201	7079	1359	3
LA	DC	20.548	2296	3.759	3.524	7079	2637	5
LA	DENV	22.745	830	3.759	3.233	7079	1088	4
LA	DETR	17.967	1979	3.759	3.965	7079	4063	4
LA	NYC	79.450	2446	3.962	3.759	11698	7079	5
LA	PHIL	14.705	2389	3.759	3.243	7079	4690	5
LA	PHNX	29.002	356	3.759	2.841	7079	837	5
LA	SACR	24.896	361	3.759	3.477	7079	685	3
LA	SEAT	33.257	960	3.759	3.722	7079	1239	2
MIA	ATL	14.242	605	3.246	3.024	1270	1142	4
MIA	BOSTN	21.648	1257	3.423	3.024	3516	1142	5
MIA	CHIC	39.316	1190	3.982	3.124	6587	1142	5
MIA	CLE	13.669	1089	3.559	3.124	2072	1142	4
MIA	DC	14.499	925	3.524	3.024	2637	1142	6
MIA	DETR	18.537	1155	3.695	3.024	4063	1142	5
MIA	NYC	126.134	1094	3.962	3.024	11698	1142	7
MIA	PHIL	21.117	1021	3.243	3.024	4690	1142	7
MIA	TPA	18.674	205	3.024	2.586	1142	881	7
NYC	ATL	26.919	748	3.962	3.246	11698	1270	5
NYC	BOSTN	189.506	188	3.962	3.423	11698	3516	8
NYC	BUF	43.179	291	3.962	3.155	11698	1325	4
NYC	CHIC	140.445	711	3.962	3.982	11698	6587	7
NYC	CLE	53.620	404	3.962	3.559	11698	2072	7
NYC	DETR	66.737	480	3.962	3.695	11698	4063	8
NYC	PIT	53.580	315	3.962	3.125	11698	2413	7
NYC	RCH	31.681	249	3.962	3.532	11698	825	3
NYC	STL	27.380	873	3.962	3.276	11698	2320	5
NYC	SYR	32.502	193	3.962	2.974	11698	515	3

(Continued)

Table 8.25 *(continued)*

CITY1	CITY2	PASS	MILES	INM	INS	POPM	POPS	AIRL
SANDG	CHIC	6.162	1731	3.982	3.149	6587	1173	3
SANDG	DALL	2.592	1181	3.201	3.149	1359	1173	2
SANDG	DC	3.211	2271	3.524	3.149	2637	1173	4
SANDG	LA	21.642	111	3.759	3.149	7079	1173	4
SANDG	LVEG	2.760	265	3.149	3.821	1173	179	5
SANDG	MINP	2.776	1532	3.621	3.149	1649	1173	2
SANDG	NYC	6.304	2429	3.962	3.149	11698	1173	4
SANDG	PHNX	6.027	298	3.149	2.841	1173	837	3
SANDG	SACR	2.603	473	3.149	3.477	1173	685	3
SANDG	SEAT	4.857	1064	3.722	3.149	1239	1173	2
SF	BOSTN	11.933	2693	3.423	3.977	3516	3075	4
SF	CHIC	33.946	1854	3.982	3.977	6587	3075	4
SF	DC	16.743	2435	3.977	3.524	3075	2637	5
SF	DENV	14.742	947	3.977	3.233	3075	1088	3
SF	LA	148.366	347	3.759	3.977	7079	3075	7
SF	LVEG	16.267	416	3.977	3.821	3075	179	6
SF	LVEG	9.410	458	3.977	3.149	3075	1173	5
SF	NYC	57.863	2566	3.962	3.977	11698	3075	5
SF	PORT	23.420	535	3.977	3.305	3075	914	4
SF	RENO	18.400	185	3.977	3.899	3075	109	3
SF	SEAT	41.725	679	3.977	3.722	3075	1239	3
SF	SLC	11.994	598	3.977	2.721	3075	526	3

CITY1 and CITY2: a pair of cities,
PASS: the number of passengers flying between the cities in a sample week,
MILES: air distance between the pair of cities,
INM: per capita income in the larger city,
INS: per capita income in the smaller city,
POPM: population of the larger city,
POPS: population of the smaller city, and
AIRL: the number of airlines serving that route.

(a) Perform a regression relating the number of passengers to the other variables. Check residuals for possible specification errors. Do the results make sense?

(b) Someone suggests using the logarithms of all variables for the regression. Does this recommendation make sense? Perform the regression using logarithms; answer all questions as in part (a).

(c) Another use of the data is to use the number of airlines as the dependent variable. What different aspect of the demand or supply of airline travel is related to this model? Implement that model and relate the results to those of parts (a) and (b).

5. It is beneficial to be able to estimate the yield of useful product of a tree based on measurements of the tree taken before it is harvested. Measurements on four such variables were taken on a sample of trees, which subsequently was harvested and the actual weight of product determined. The variables are:

DBH: diameter at breast height (about 4′ from ground level), in inches,
HEIGHT: height of tree, in feet,
AGE: age of tree, in years,
GRAV: specific gravity of the wood, and
WEIGHT: the harvested weight of the tree (lbs.).

The first two variables (DBH and HEIGHT) are logically the most important and are also the easiest to measure. The data are given in Table 8.26.

(a) Perform a linear regression relating weight to the measured quantities. Plot residuals. Is the equation useful? Is the model adequate?
(b) If the results appear to not be very useful, suggest and implement an alternate model. (*Hint:* Weight is a product of dimensions.)

6. Data were collected to discern environmental factors affecting health standards. For 21 small regions we have data on the following variables:

POP: population (in thousands),
VALUE: value of all residential housing, in millions of dollars; this is the proxy for economic conditions,
DOCT: the number of doctors,
NURSE: the number of nurses,
VN: the number of vocational nurses, and
DEATHS: number of deaths due to health-related causes (i.e., not accidents); this is the proxy for health standards.

The data are given in Table 8.27.

(a) Perform a regression relating DEATHS to the other variables, excluding POP. Compute the variance-inflation factors; interpret all results.
(b) Obviously multicollinearity is a problem for these data. What is the cause of this phenomenon? It has been suggested that all variables should be converted to a per capita basis. Why should this solve the multicollinearity problem?
(c) Perform the regression using per capita variables. Compare results with those of part (a). Is it useful to compare R^2 values? Why or why not?

7. We have data on the distance covered by irrigation water in a furrow of a field. The data are to be used to relate the distance covered to the time since watering began. The data are given in Table 8.28.

(a) Perform a simple linear regression relating distance to time. Plot the residuals against time. What does the plot suggest?
(b) Perform a regression using time and the square of time. Interpret the results. Are they reasonable?
(c) Plot residuals from the quadratic model. What does this plot suggest?

8. Twenty-five volunteer athletes participated in a study of cross-disciplinary athletic abilities. The group was comprised of athletes from football, baseball, water polo, volleyball, and soccer. None had ever played organized basketball, but did acknowledge interest and some social participation in the game.

Table 8.26

Data for Exercise 5: Estimating Tree Weights

OBS	DBH	HEIGHT	AGE	GRAV	WEIGHT
1	5.7	34	10	0.409	174
2	8.1	68	17	0.501	745
3	8.3	70	17	0.445	814
4	7.0	54	17	0.442	408
5	6.2	37	12	0.353	226
6	11.4	79	27	0.429	1675
7	11.6	70	26	0.497	1491
8	4.5	37	12	0.380	121
9	3.5	32	15	0.420	58
10	6.2	45	15	0.449	278
11	5.7	48	20	0.471	220
12	6.0	57	20	0.447	342
13	5.6	40	20	0.439	209
14	4.0	44	27	0.394	84
15	6.7	52	21	0.422	313
16	4.0	38	27	0.496	60
17	12.1	74	27	0.476	1692
18	4.5	37	12	0.382	74
19	8.6	60	23	0.502	515
20	9.3	63	18	0.458	766
21	6.5	57	18	0.474	345
22	5.6	46	12	0.413	210
23	4.3	41	12	0.382	100
24	4.5	42	12	0.457	122
25	7.7	64	19	0.478	539
26	8.8	70	22	0.496	815
27	5.0	53	23	0.485	194
28	5.4	61	23	0.488	280
29	6.0	56	23	0.435	296
30	7.4	52	14	0.474	462
31	5.6	48	19	0.441	200
32	5.5	50	19	0.506	229
33	4.3	50	19	0.410	125
34	4.2	31	10	0.412	84
35	3.7	27	10	0.418	70
36	6.1	39	10	0.470	224
37	3.9	35	19	0.426	99
38	5.2	48	13	0.436	200
39	5.6	47	13	0.472	214
40	7.8	69	13	0.470	712
41	6.1	49	13	0.464	297
42	6.1	44	13	0.450	238
43	4.0	34	13	0.424	89
44	4.0	38	13	0.407	76
45	8.0	61	13	0.508	614
46	5.2	47	13	0.432	194
47	3.7	33	13	0.389	66

Table 8.27

Data for Exercise 6

POP	VALUE	DOCT	NURSE	VN	DEATHS
100	141.83	49	76	221	661
110	246.80	103	250	378	1149
130	238.06	76	140	207	1333
142	265.90	95	150	381	1321
202	397.63	162	324	554	2418
213	464.32	194	282	560	2039
246	409.95	130	211	465	2518
280	556.03	205	383	942	3088
304	711.61	222	461	723	1882
316	820.52	304	469	598	2437
328	709.86	267	525	911	2177
330	829.84	245	639	739	2593
337	465.15	221	343	541	2295
379	839.11	330	714	330	2119
434	792.02	420	865	894	4294
434	883.72	384	601	1158	2836
436	939.71	363	530	1219	4637
447	1141.80	511	180	513	3236
1087	2511.53	1193	1792	1922	7768
2305	6774.16	3450	5357	4125	14590
2637	8318.92	3131	4630	4785	19044

Table 8.28

Distance Covered by Irrigation Water

Obs	Distance	Time
1	85	0.15
2	169	0.48
3	251	0.95
4	315	1.37
5	408	2.08
6	450	2.53
7	511	3.20
8	590	4.08
9	664	4.93
10	703	5.42
11	831	7.17
12	906	8.22
13	1075	10.92
14	1146	11.92
15	1222	13.12
16	1418	15.78
17	1641	18.83
18	1914	21.22
19	1864	21.98

Height, weight, and speed in the 100-yard dash was recorded for each subject. The basketball test consisted of the number of field goals that could be made in a 60-min. period. The data are given in Table 8.29.

(a) Perform the regression relating GOALMADE to the other variables. Comment on the results.
(b) Is there multicollinearity?
(c) Check for outliers.
(d) If appropriate, develop and implement an alternative model.

9. In an effort to estimate the plant biomass in a desert environment, field measurements on the diameter and height and laboratory determination of oven dry weight were obtained for a sample of plants in a sample of transects (area). Collections were made at two times, in the warm and cool seasons. The data are to be used to see how well the weight can be estimated by the more easily determined field observations, and further whether the model for estimation is the same for the two seasons. The data are given in Table 8.30.

(a) Perform separate linear regressions for estimating weight for the two seasons. Plot residuals. Interpret results.
(b) Transform width, height, and weight using the natural logarithm transform discussed in Section 8.6. Perform separate regressions for estimating log–weight for the two seasons. Plot residuals. Interpret results. Compare results with those from part (a). (A formal method for comparing the regressions for the two seasons is presented in Chapter 11 and is applied to this exercise as Exercise 10, Chapter 11.)

Table 8.29

Basket Goals Related to Physique

OBS	WEIGHT	HEIGHT	DASH100	GOALMADE
1	130	71	11.50	15
2	149	74	12.23	19
3	170	70	12.26	11
4	177	71	12.65	15
5	188	69	10.26	12
6	210	73	12.76	17
7	223	72	11.89	15
8	170	75	12.32	19
9	145	72	10.77	16
10	132	74	11.31	18
11	211	71	12.91	13
12	212	72	12.55	15
13	193	73	11.72	17
14	146	72	12.94	16
15	158	71	12.21	15
16	154	75	11.81	20
17	193	71	11.90	15
18	228	75	11.22	19
19	217	78	10.89	22
20	172	79	12.84	23
21	188	72	11.01	16
22	144	75	12.18	20
23	164	76	12.37	21
24	188	74	11.98	19
25	231	70	12.23	13

10. In this problem we are trying to estimate the survival of liver transplant patients using information on the patients collected before the operation. The variables are:

CLOT: a measure of the clotting potential of the patient's blood,
PROG: a subjective index of the patient's prospect of recovery,
ENZ: a measure of a protein present in the body,
LIV: a measure relating to white blood cell count and the response, and
TIME: a measure of the survival time of the patient.

The data are given in Table 8.31.

(a) Perform a linear regression for estimating survival times. Plot residuals. Interpret and critique the model used.

(b) Because the distributions of survival times are often quite skewed, a logarithmic model is often used for such data. Perform the regression using such a model. Compare the results with those of part (a).

11. Considerable variation occurs among individuals in their perception of what specific acts constitute a crime. To obtain an idea of factors that influence this perception, 45 college students were given the following list of acts and asked how many of these they perceived as constituting a crime. The acts were:

Table 8.30

Data for Exercise 9

COOL			WARM		
Width	Height	Weight	Width	Height	Weight
4.9	7.6	0.420	20.5	13.0	6.840
8.6	4.8	0.580	10.0	6.2	0.400
4.5	3.9	0.080	10.1	5.9	0.360
19.6	19.8	8.690	10.5	27.0	1.385
7.7	3.1	0.480	9.2	16.1	1.010
5.3	2.2	0.540	12.1	12.3	1.825
4.5	3.1	0.400	18.6	7.2	6.820
7.1	7.1	0.350	29.5	29.0	9.910
7.5	3.6	0.470	45.0	16.0	4.525
10.2	1.4	0.720	5.0	3.1	0.110
8.6	7.4	2.080	6.0	5.8	0.200
15.2	12.9	5.370	12.4	20.0	1.360
9.2	10.7	4.050	16.4	2.1	1.720
3.8	4.4	0.850	8.1	1.2	1.495
11.4	15.5	2.730	5.0	23.1	1.725
10.6	6.6	1.450	15.6	24.1	1.830
7.6	6.4	0.420	28.2	2.2	4.620
11.2	7.4	7.380	34.6	45.0	15.310
7.4	6.4	0.360	4.2	6.1	0.190
6.3	3.7	0.320	30.0	30.0	7.290
16.4	8.7	5.410	9.0	19.1	0.930
4.1	26.1	1.570	25.4	29.3	8.010
5.4	11.8	1.060	8.1	4.8	0.600
3.8	11.4	0.470	5.4	10.6	0.250
4.6	7.9	0.610	2.0	6.0	0.050
			18.2	16.1	5.450
			13.5	18.0	0.640
			26.6	9.0	2.090
			6.0	10.7	0.210
			7.6	14.0	0.680
			13.1	12.2	1.960
			16.5	10.0	1.610
			23.1	19.5	2.160
			9.0	30.0	0.710

aggravated assault
armed robbery
arson
atheism
auto theft
burglary
civil disobedience
communism
drug addiction
embezzlement
forcible rape
gambling
homosexuality
land fraud
Nazism
payola
price fixing
prostitution
sexual abuse of child
sex discrimination
shoplifting
striking
strip mining
treason
vandalism

The number of activities perceived as crimes is measured by the variable CRIMES. Variables describing personal information that may influence perception are:

Table 8.31

Survival of Liver Transplants

OBS	CLOT	PROG	ENZ	LIV	TIME
1	3.7	51	41	1.55	34
2	8.7	45	23	2.52	58
3	6.7	51	43	1.86	65
4	6.7	26	68	2.10	70
5	3.2	64	65	0.74	71
6	5.2	54	56	2.71	72
7	3.6	28	99	1.30	75
8	5.8	38	72	1.42	80
9	5.7	46	63	1.91	80
10	6.0	85	28	2.98	87
11	5.2	49	72	1.84	95
12	5.1	59	66	1.70	101
13	6.5	73	41	2.01	101
14	5.2	52	76	2.85	109
15	5.4	58	70	2.64	115
16	5.0	59	73	3.50	116
17	2.6	74	86	2.05	118
18	4.3	8	119	2.85	120
19	6.5	40	84	3.00	123
20	6.6	77	46	1.95	124
21	6.4	85	40	1.21	125
22	3.7	68	81	2.57	127
23	3.4	83	53	1.12	136
24	5.8	61	73	3.50	144
25	5.4	52	88	1.81	148
26	4.8	61	76	2.45	151
27	6.5	56	77	2.85	153
28	5.1	67	77	2.86	158
29	7.7	62	67	3.40	168
30	5.6	57	87	3.02	172
31	5.8	76	59	2.58	178
32	5.2	52	86	2.45	181
33	5.3	51	99	2.60	184
34	3.4	77	93	1.48	191
35	6.4	59	85	2.33	198
36	6.7	62	81	2.59	200
37	6.0	67	93	2.50	202
38	3.7	76	94	2.40	203
39	7.4	57	83	2.16	204
40	7.3	68	74	3.56	215
41	7.4	74	68	2.40	217
42	5.8	67	86	3.40	220
43	6.3	59	100	2.95	276
44	5.8	72	93	3.30	295
45	3.9	82	103	4.55	310
46	4.5	73	106	3.05	311
47	8.8	78	72	3.20	313
48	6.3	84	83	4.13	329
49	5.8	83	88	3.95	330
50	4.8	86	101	4.10	398
51	8.8	86	88	6.40	483
52	7.8	65	115	4.30	509
53	11.2	76	90	5.59	574
54	5.8	96	114	3.95	830

AGE: age of interviewee,
SEX: coded 0: female, 1: male,
COLLEGE: year of college, coded 1 through 4, and
INCOME: income of parents ($1000).

Perform a regression to estimate the relationship between the number of activities perceived as crimes and the personal characteristics of the interviewees. Check assumptions and perform any justifiable remedial actions. Interpret the results. The data are given in Table 8.32.

12. Many architects have different tastes compared to the users of their products. This disagreement may result in the design and consequent construction of buildings not appreciated by the public. To provide some idea of the structure of peoples' tastes, a graduate student conducted a survey of 40 individuals (SUBJ) who had no special knowledge of architecture. In the survey the respondents were asked to judge five structures for satisfaction on seven specific characteristics and also give an overall satisfaction index. All responses are on a nine-point scale as shown here. Satisfaction indexes for the seven specific characteristics and overall satisfaction were scored on a scale from 1 to 9, recorded by the variable names as follows:

Characteristic	Var		Scoring 1 2 3 4 5 6 7 8 9	
Beauty	B	ugly	to	beautiful
Function	N	useless	to	useful
Intimacy	I	strange	to	friendly
Dignity	D	humble	to	dignified
Cost	C	cheap	to	expensive
Fashion	F	classic	to	modern
Overall	S	bad	to	good

A condensed version of the data is shown in Table 8.33. (The data set for analysis purposes has 200 records with the scores for each building as rated by each subject.) Perform a regression relating to the specific preferences. Ignore the subjects for now; the variation due to subjects will be discussed in Chapters 10 and 11. Interpret the results.

13. An apartment complex owner is performing a study to see what improvements or changes in her complex may bring in more rental income. From a sample of 34 complexes she obtains the monthly rent on single-bedroom units and the following characteristics:

AGE: the age of the property,
SQFT: square footage of unit,
SD: amount of security deposit,
UNTS: number of units in complex,
GAR: presence of a garage (0–no, 1–yes),
CP: presence of carport (0–no, 1–yes),

Table 8.32

Crimes Perception Data–Exercise 11

OBS	AGE	SEX	COLLEGE	INCOME	CRIMES
1	19	0	2	56	13
2	19	1	2	59	16
3	20	0	2	55	13
4	21	0	2	60	13
5	20	0	2	52	14
6	24	0	3	54	14
7	25	0	3	55	13
8	25	0	3	59	16
9	27	1	4	56	16
10	28	1	4	52	14
11	38	0	4	59	20
12	29	1	4	63	25
13	30	1	4	55	19
14	21	1	3	29	8
15	21	1	2	35	11
16	20	0	2	33	10
17	19	0	2	27	6
18	21	0	3	24	7
19	21	1	2	53	15
20	16	1	2	63	23
21	18	1	2	72	25
22	18	1	2	75	22
23	18	0	2	61	16
24	19	1	2	65	19
25	19	1	2	70	19
26	20	1	2	78	18
27	19	0	2	76	16
28	18	0	2	53	12
29	31	0	4	59	23
30	32	1	4	62	25
31	32	1	4	55	22
32	31	0	4	57	25
33	30	1	4	46	17
34	29	0	4	35	14
35	29	0	4	32	12
36	28	0	4	30	10
37	27	0	4	29	8
38	26	0	4	28	7
39	25	0	4	25	5
40	24	0	3	33	9
41	23	0	3	26	7
42	23	1	3	28	9
43	22	0	3	38	10
44	22	0	3	24	4
45	22	0	3	28	6

SS: security system (0–no, 1–yes),
FIT: fitness facilities (0–no, 1–yes), and
RENT: monthly rental.

The data are presented in Table 8.34, and are available on the data disk file FW08P13.

Table 8.33 Architectural Preferences–Exercise 12

	STRUCTURES																																		
	1							2							3							4							5						
SUBJ	S	B	N	I	D	C	F	S	B	N	I	D	C	F	S	B	N	I	D	C	F	S	B	N	I	D	C	F	S	B	N	I	D	C	F
1	2	4	2	3	4	7	3	5	3	3	4	3	3	7	6	6	7	7	6	6	4	5	7	3	2	6	7	7	7	8	7	7	8	7	6
2	5	3	4	7	4	3	3	6	8	7	4	8	8	3	6	4	8	5	4	6	4	7	8	8	6	8	8	5	4	4	3	7	3	3	3
3	4	4	4	7	3	3	4	7	8	7	7	8	8	4	3	7	7	3	7	7	8	6	8	8	6	8	9	4	7	8	4	4	4	5	8
4	6	6	4	5	5	5	3	4	4	4	5	5	5	6	7	5	8	6	6	6	5	5	3	4	3	4	5	6	7	7	6	8	6	6	5
5	2	1	3	3	2	2	5	6	7	6	6	6	6	5	4	4	3	5	5	3	4	7	5	7	7	7	8	5	5	4	4	5	5	4	6
6	4	3	5	5	4	3	3	6	4	6	6	6	7	4	7	5	7	6	6	6	8	5	6	7	5	7	8	3	3	2	5	4	4	4	5
7	7	7	6	8	8	7	3	5	2	6	2	3	5	6	5	3	7	3	4	5	5	7	7	4	7	8	7	7	4	5	5	7	6	4	5
8	3	3	6	3	3	2	6	7	8	7	8	8	6	4	6	1	5	3	4	5	7	7	7	6	6	6	6	4	6	5	5	6	6	5	7
9	6	3	6	7	5	4	5	7	7	7	7	7	7	6	8	8	6	8	8	6	7	6	6	8	7	7	6	6	3	5	5	4	4	5	5
10	5	7	3	5	5	6	4	3	6	5	2	4	6	7	3	6	7	5	6	4	3	5	8	2	3	3	5	6	6	8	3	6	6	6	5
11	3	3	4	3	2	2	5	6	8	3	6	6	5	5	4	3	7	4	3	3	6	4	4	6	3	5	8	2	3	6	5	2	1	3	8
12	3	6	5	4	2	3	2	6	7	6	7	5	7	4	7	8	6	7	5	5	7	5	6	6	4	7	8	3	3	3	4	3	2	4	6
13	4	8	4	4	6	8	7	1	1	6	2	2	2	2	7	5	7	6	6	6	6	1	2	4	3	3	7	6	6	6	7	7	7	6	6
14	4	4	6	3	3	2	7	6	8	4	6	6	6	5	8	7	8	7	6	6	7	5	6	6	6	5	5	5	1	1	1	1	1	2	6
15	4	5	5	6	3	3	3	4	4	6	6	5	5	4	4	5	3	3	3	3	3	5	6	5	4	7	7	3	5	3	5	4	3	4	5
16	8	9	5	9	7	8	1	5	3	6	2	2	9	9	9	9	9	9	7	9	4	9	7	3	6	5	9	8	5	7	7	6	5	5	4
17	5	3	3	3	3	5	7	5	8	8	7	7	7	2	6	5	5	5	2	8	8	7	7	8	7	8	8	2	5	3	2	5	2	7	9
18	5	3	7	5	4	5	3	4	6	8	6	7	7	1	6	6	7	6	6	8	4	6	7	9	7	8	8	1	6	4	8	5	2	6	8
19	2	7	2	2	7	7	2	5	2	8	2	5	4	8	6	8	8	7	7	3	2	2	7	2	2	2	8	2	8	8	7	7	2	5	2
20	2	2	6	2	2	3	8	8	7	8	7	5	2	2	7	8	8	8	8	7	8	8	8	8	7	8	8	2	2	2	1	2	3	2	8
21	8	7	2	8	6	4	2	8	8	8	8	8	8	2	8	8	8	8	8	3	8	8	9	9	9	9	9	1	3	5	1	2	2	2	8
22	5	7	8	4	7	7	2	4	2	3	2	1	1	9	4	5	7	6	7	6	4	3	3	2	2	3	4	7	5	5	7	7	6	6	6
23	6	3	2	8	2	2	9	6	8	7	9	6	7	1	6	5	4	7	4	4	9	6	7	8	7	8	8	3	4	3	3	2	3	3	8
24	3	4	5	6	3	3	5	5	5	6	6	5	5	6	8	8	8	7	6	7	8	5	6	8	4	6	7	9	6	4	7	9	5	5	8
25	6	7	6	6	8	8	4	2	2	5	2	2	3	7	5	5	7	7	4	4	5	7	5	8	8	3	3	6	6	6	5	4	6	4	5
26	2	3	5	5	2	2	7	7	7	7	7	6	6	5	4	3	5	4	3	5	5	7	5	7	8	6	6	6	2	3	4	4	3	4	7
27	7	4	7	7	6	5	7	7	7	7	7	7	6	5	7	5	6	7	4	5	6	7	6	7	7	7	6	5	7	6	6	7	5	6	7
28	8	8	6	7	6	7	3	5	5	5	5	5	4	8	5	6	5	3	6	4	3	6	7	5	5	5	6	7	3	6	3	3	4	4	3
29	2	2	2	4	3	1	5	6	6	5	6	6	6	3	7	6	5	6	7	5	8	4	7	7	6	8	5	4	5	4	3	3	5	4	7
30	1	4	1	1	1	1	4	5	4	7	5	5	6	3	5	5	4	4	4	3	6	6	6	7	7	6	7	3	6	4	3	3	4	4	6
31	9	8	8	5	5	8	1	7	4	8	2	4	7	9	8	6	8	5	8	3	2	2	3	3	3	2	3	2	8	6	7	9	3	4	8
32	6	3	7	2	4	6	9	9	9	9	8	5	8	1	1	2	9	1	2	7	9	8	9	9	9	9	9	1	4	3	5	7	2	1	9
33	2	1	8	1	2	2	9	6	6	9	9	9	8	1	8	9	8	9	4	4	9	9	9	9	9	9	9	1	8	9	9	1	3	9	9
34	8	7	3	3	6	8	3	4	8	8	2	6	5	8	3	2	7	7	2	3	2	3	3	3	3	7	8	8	9	8	6	8	8	8	8
35	2	2	8	2	2	2	6	7	8	4	8	8	7	8	2	2	8	3	3	2	8	9	9	7	7	8	9	9	2	2	6	2	2	3	8
36	2	4	8	2	2	2	2	9	9	7	8	9	9	8	1	2	8	2	2	2	2	8	7	7	8	8	8	7	3	7	7	2	3	7	6
37	7	8	5	8	2	5	3	2	2	3	3	7	8	8	3	3	3	3	3	3	3	3	2	2	2	6	7	7	6	7	3	6	3	3	3
38	3	3	5	3	4	2	7	7	9	3	9	6	6	3	3	3	3	3	3	3	7	2	6	3	3	7	4	2	2	3	2	2	2	3	7
39	3	5	3	3	3	3	3	3	4	3	6	3	3	3	5	8	4	4	7	7	7	3	8	4	3	8	7	3	3	7	3	3	6	3	8
40	9	9	9	8	9	9	2	5	7	8	2	4	6	7	8	6	9	7	9	8	7	7	7	7	5	6	8	8	8	8	9	8	7	8	8

(a) Perform a regression and make recommendations to the apartment complex owner.

(b) Because there is no way to change some of these characteristics, someone recommends using a model that contains only characteristics that can be modified. Comment on that recommendation.

Table 8.34

Apartment Rent Data

OBS	AGE	SQFT	SD	UNTS	GAR	CP	SS	FIT	RENT
1	7	692	150	408	0	0	1	0	508
2	7	765	100	334	0	0	1	1	553
3	8	764	150	170	0	0	1	1	488
4	13	808	100	533	0	1	1	1	558
5	7	685	100	264	0	0	0	0	471
6	7	710	100	296	0	0	0	0	481
7	5	718	100	240	0	1	1	1	577
8	6	672	100	420	0	1	0	1	556
9	4	746	100	410	1	1	1	1	636
10	4	792	100	404	1	0	1	1	737
11	8	797	150	252	0	0	1	1	546
12	7	708	100	276	0	0	1	0	445
13	8	797	150	252	0	0	0	1	533
14	6	813	100	416	0	1	0	0	617
15	7	708	100	536	0	0	1	1	475
16	16	658	100	188	1	1	1	1	525
17	8	809	150	192	0	0	1	0	461
18	7	663	100	300	0	0	0	1	495
19	1	719	100	300	1	1	1	1	601
20	1	689	100	224	0	1	1	1	567
21	1	737	175	310	1	1	1	1	633
22	1	694	150	476	1	0	1	1	616
23	7	768	150	264	0	0	1	1	507
24	6	699	150	150	0	0	0	0	454
25	6	733	100	260	0	0	1	0	502
26	7	592	100	264	0	0	1	1	431
27	6	589	150	516	0	0	1	1	418
28	8	721	75	216	0	0	1	0	538
29	5	705	75	212	1	0	1	1	506
30	6	772	150	460	0	0	1	1	543
31	7	758	100	260	0	0	1	0	534
32	7	764	100	269	0	0	1	0	536
33	6	722	125	216	0	0	0	1	520
34	1	703	100	248	0	0	1	0	530

14. (a) Use the data set on home prices given in Table 8.2 to do the following:

(i) Use `price` as the dependent variable and the rest of the variables as independent variables and determine the best regression using the stepwise variable selection procedure. Comment on the results.

(ii) The Modes decided to not use the data on homes whose price exceeded $200,000, because the relationship of price to size seemed to be erratic for these homes. Perform the regression using all observations, and compute the outlier detection statistics. Also compare the results of the regression with that obtained using only the under $200,000 homes. Comment on the results. Which regression would you use?

(iii) Compute and study the residuals for the home price regression. Could these be useful for someone who was considering buying one of these homes?

(b) The data originally presented in Chapter 1 (Table 1.2) also included the variables `garage` and `fp`. Perform variable selection that includes these variables as well. Explain the results.

Factorial Experiments

EXAMPLE 9.1

What Makes a Wiring Harness Last Longer? Many electrical wiring harnesses, such as those used in automobiles and airplanes, are subject to considerable stress. Therefore, it is important to design such harnesses to prolong their useful life. The objective of this experiment is to investigate factors affecting the failure of an electrical wiring harness. The factors of the experiment are

STRANDS: the number of strands in the wire, levels are 7 and 9,
SLACK: length of unsoldered, uninsulated wire in 0.01 in., levels are 0, 3, 6, 9, and 12, and
GAGE: a reciprocal measure of the diameter of the wire, levels are 24, 22, and 20.

The response, Cycles, is the number of stress cycles to failure, in 100 s.

The experiment is a completely randomized design with two independent samples for each combination of levels of the three factors, that is, an experiment with a total of $2 \cdot 5 \cdot 3 = 30$ factor levels. The objective of the experiment is to see what combination of these factor levels maximizes the number of cycles to failure. The data are given in Table 9.1, which shows, for example, that 2 and 4 cycles to failure were reported for SLACK $= 0$, STRANDS $= 7$, and GAGE $= 24$ (*source:* Enrick, 1976). ■

9.1 Introduction

In Chapter 6 we presented the methodology for comparing means of populations that represent levels of a single factor. This methodology is based on a one-way or single-factor analysis of variance model. Many data sets, however,

Table 9.1 Cycles to Failure of a Wire Harness

Adapted from Enrick (1976)

	NUMBER OF STRANDS											
Wire Slack			**7**						**9**			
Wire Gage	**24**		**22**		**20**		**24**		**22**		**20**	
0	2	4	14	9	6	8	3	3	10	14	12	11
3	5	2	6	15	5	7	2	5	17	17	16	8
6	6	3	7	7	5	1	5	5	10	10	10	8
9	9	16	12	12	8	12	6	4	16	11	13	7
12	14	12	10	14	12	11	13	15	20	17	12	15

involve two or more factors. This chapter and Chapter 10 present models and procedures for the analysis of multifactor data sets. Such data sets arise from two types of situations:

1. *Factorial experiments:* In many experiments it is desirable to examine the effect of two or more factors on the same type of unit. For example, a crop yield experiment may be conducted to examine the differences in yields of several varieties as well as different levels of fertilizer application. In this experiment, variety is one factor and fertilizer is the other. An experiment that has each combination of all factor levels applied to the experimental units is called a **factorial experiment**. Although data exhibiting a multifactor structure arise most frequently from designed experiments, they may occur in other contexts. For example, data on test scores from a sample survey of students of different ethnic backgrounds from each of several universities may be considered a factorial "experiment," which can be used to ascertain differences on, say, mean test scores among schools and ethnic backgrounds.
2. *Experimental design:* It is often desirable to subdivide the experimental units into groups before assigning them to different factor levels. These groups are defined in such a way as to reduce the estimate of variance used for inferences. This procedure is usually referred to as "blocking," and also results in multifactor data sets. Procedures for the analysis of data arising from experimental designs are presented in Chapter 10.

Actually, a data set may have both a factorial structure and include blocking factors. Such situations are also presented in Chapter 10.

As in the one-way analysis of variance, the analysis of any factorial experiment is the same whether we are considering a designed experiment or an observational study. The interpretation may, however, be different. Also, as in the one-way analysis of variance, the factors in a factorial experiment may have qualitative or quantitative factor levels that may suggest contrasts or trends, or in other cases may be defined in a manner requiring the use of post hoc paired comparisons.

9.2 Concepts and Definitions

In a factorial experiment we apply several factors simultaneously to each experimental unit, which we will again assume to be synonymous with an observational unit.

DEFINITION 9.1
A **factorial experiment** is one in which responses are observed for every combination of factor levels.

We assume (for now) that there are two or more independently sampled experimental units for each combination of factor levels and also that each factor level combination is applied to an equal number of experimental units, resulting in a **balanced** factorial experiment. We relax the assumption of multiple samples per combination in Section 9.5. Lack of balance in a factorial experiment does not alter the basic principles of the analysis of factorial experiments, but does require a different computational approach (see Chapter 11). A factorial experiment may require a large number of experimental units, especially if we have many factors with many levels. Alternatives are briefly noted in Sections 9.6 and 10.5.

A classical illustration of a factorial experiment concerns a study of the crop yield response to fertilizer. The **factors** are the three major fertilizer ingredients: N (nitrogen), P (phosphorus), and K (potassium). The **levels** are the pounds per acre of each of the three ingredients, for example:

N at four levels: 0, 40, 80, and 120 lbs. per acre,
P at three levels: 0, 80, and 160 lbs. per acre, and
K at three levels: 0, 40, and 80 lbs. per acre.

The **response** is yield, which is the variable to be analyzed.

The set of factor levels in the factorial experiment consists of all combinations of these levels, that is, $4 \times 3 \times 3 = 36$ combinations. In other words, there are 36 treatments. This experiment is called a $4 \times 3 \times 3$ factorial experiment, and in this case all three factors have quantitative levels. In this experiment one of these 36 combinations has no fertilizer application, which is referred to as a **control**. However, not all factorial experiments have a control.

The experiment consists of assigning the 36 combinations randomly to experimental units, as was done for the one-way (or CRD) experiment. If five experimental plots are assigned to each factor level combination, 180 such plots would be needed for this experiment.

Consider another experiment intended to evaluate the relationship of the amount of knowledge of statistics to the number of statistics courses to which students have been exposed. The factors are

the number of courses in statistics taken, with levels of 1, 2, 3, or 4, and
the curriculum (major) of the students, with levels of engineering, social science, natural science, and agriculture.

The response variable is the students' scores on a comprehensive statistics test. The resulting data comprise a 4×4 factorial experiment. In this experiment the number of courses is a quantitative factor and the curriculum is a qualitative factor. Note that this data set is not the result of a designed experiment; however, the characteristics of the factorial data set remain.

The statistical analysis of data from a factorial experiment is intended to examine how the behavior of the response variable is affected by the different levels of the factors. This examination takes the form of inferences on two types of phenomena.

DEFINITION 9.2
Main effects are the differences in the mean response across the levels of each factor when viewed individually.

In the fertilizer example, the main effects "nitrogen," "phosphorus," and "potassium" separately compare the mean response across levels of N, P, and K, respectively.

DEFINITION 9.3
Interactions effects are differences or inconsistencies of the main effect responses for one factor across levels of one or more of the other factors.

For example, when applying fertilizer, it is well known that increasing amounts of only one nutrient, say, nitrogen, will have only limited effect on yield. However, in the presence of other nutrients, substantial yield increases may result from the addition of more nitrogen. This result is an example of an interaction among these factors.

In the preceding example of student performance on the test in statistics, interaction may exist because students in disciplines that stress quantitative reasoning will probably show greater improvement with the number of statistics courses taken than will students in curricula having little emphasis on quantitative reasoning.

We will see that the existence of interactions modifies and sometimes even nullifies inferences on main effects. Therefore it is important to conduct experiments that can detect interactions. Experiments that consider only one factor at a time or include only selected combinations of factor levels usually cannot detect interactions. Example 6.6 actually studied seven factors, whose levels were considered in some combinations, but the structure and number of combinations were insufficient to be able to detect interactions among all the factors. Only factorial experiments that simultaneously examine all combinations of factor levels should be used for this purpose.

Table 9.2

Data for Motor Oil Experiment

Oil	Miles Per Gallon					Mean
STANDARD	23.6	21.7	20.3	21.0	22.0	21.72
MULTI	23.5	22.8	24.6	24.6	22.5	23.60
GASMISER	21.4	20.7	20.5	23.2	21.3	21.42

EXAMPLE 9.2

Recently an oil company has been promoting a motor oil that is supposed to increase gas mileage. An independent research company conducts an experiment to test this claim. Fifteen identical cars are used: five are randomly assigned to use a standard single-weight oil (STANDARD), five others a multi-weight oil (MULTI), and the remaining five the new oil (GASMISER). All 15 cars are driven 1000 miles over a controlled course and the gas mileage (miles per gallon) is recorded. This is a one-factor CRD of the type presented in Chapter 6. The data are given in Table 9.2.

Solution We use the analysis of variance to investigate the nature of differences in average gas mileage due to the use of different motor oils. The analysis (not reproduced here) for factor level differences produces an F ratio of 5.75, which has 2 and 12 degrees of freedom. The p value is 0.0177, which provides evidence that the oil types do affect gas mileage. The use of Duncan's multiple range test indicates that at the 5% significance level the only difference is that between MULTI and GASMISER and that between MULTI and STANDARD. Thus, there is insufficient evidence to support the claim of superior gas mileage with the GASMISER oil.

Suppose someone points out that the advertisements for GASMISER also state "specially formulated for the new smaller engines," but it turns out that the experiment was conducted with cars having larger six-cylinder engines. In these circumstances, the decision is made to repeat the experiment using a sample of 15 identical cars having four-cylinder engines. The data from this experiment are given in Table 9.3.

Table 9.3

Data for Motor Oil Experiment on Four-Cylinder Engines

Oil	Miles Per Gallon					Mean
STANDARD	22.6	24.5	23.1	25.3	22.1	23.52
MULTI	23.7	24.6	25.0	24.0	23.1	24.08
GASMISER	26.0	25.0	26.9	26.0	25.4	25.86

The analysis of the data from this experiment produces an F ratio of 7.81 and a p value of 0.0067, and we may conclude that for these engines there is also a difference due to oils. Applications of Duncan's range test shows that for these cars, the GASMISER oil does produce higher mileage, but that there is apparently no difference between STANDARD and MULTI.

The result of these analyses is that the recommendation for using an oil depends on the engine to be used. This is an example of an **interaction** between engine size and type of oil. The existence of this interaction means that we may not be able to make a universal inference of motor oil effect.

That is, any recommendations for oil usage depend on which type of engine is to be used. However, the results of the two separate experiments cannot be used to establish the significance of the interaction because the possible existence of different experimental conditions for the two separate experiments may introduce a confounding effect and thus cloud the validity of inferences. Therefore, such an inference can only be made if a single **factorial experiment** is conducted using both engine types and motor oils as the factors. Such an experiment would be a 2×3 (called "two by three") factorial. ■

9.3 The Two-Factor Factorial Experiment

We present here the principles underlying the analysis of a two-factor factorial experiment and the definitional formulas for performing that analysis. The two factors are arbitrarily labeled A and C. Factor A has levels $1, 2, \ldots, a$, and factor C has levels $1, 2, \ldots, c$, which is referred to as an $a \times c$ factorial experiment. At this point it does not matter if the levels are quantitative or qualitative. There are n independent sample replicates for each of the $a \times c$ factor level combinations; that is, we have a completely randomized design with $a \cdot c$ treatments and $a \cdot c \cdot n$ observed values of the response variable.

The Linear Model

As in the analysis of the completely randomized experiment, the representation of the data by a linear model (Section 6.3) facilitates understanding of the analysis. The linear model for the two-factor factorial experiment specified above is

$$y_{ijk} = \mu + \alpha_i + \gamma_j + (\alpha\gamma)_{ij} + \varepsilon_{ijk},$$

where y_{ijk} = kth observed value, $k = 1, 2, \ldots, n$ of the response variable y for the "cell" defined by the combination of the ith level of factor A and the jth level of factor C; μ = reference value, usually called the "grand" or overall mean; α_i, $i = 1, 2, \ldots, a$ = main effect of factor A, and is the difference between the mean response of the subpopulation comprising the ith level of factor A and the reference value μ; γ_j, $j = 1, 2, \ldots, c$ = main effect of factor C, and is the difference between the mean response of the subpopulation comprising the jth level of factor C and the reference value μ; $(\alpha\gamma)_{ij}$ = interaction between factors A and C, and is the difference between the mean response in the subpopulation defined by the combination of the A_i and C_j factor levels and the main effects α_i and γ_j; and ε_{ijk} = random error representing the variation among observations that have been subjected to the same factor level combinations. This component is a random variable having an approximately normal distribution with mean zero and variance[1] σ^2.

[1] These assumptions about ε were first introduced in Chapter 6. Methods for detection of violations and remedial measures remain the same.

In the linear model for the factorial experiment we consider all factors, including interactions, to be fixed effects (Section 6.3). Occasionally some factors in a factorial experiment may be considered to be random, in which case the inferences are akin to those from certain experimental designs presented in Chapter 10. As in Section 6.4, we add the restrictions[2]

$$\sum_i \alpha_i = \sum_j \gamma_j = \sum_i (\alpha\gamma)_{ij} = \sum_j (\alpha\gamma)_{ij} = 0,$$

which makes μ the overall mean response and α_i, γ_i, and $(\alpha\gamma)_{ij}$, the main and interaction effects, respectively.

We are interested in testing the hypotheses

$$H_0\text{: } \alpha_i = 0,$$

$$H_0\text{: } \gamma_j = 0,$$

$$H_0\text{: } (\alpha\gamma)_{ij} = 0, \quad \text{for all } i \text{ and } j.$$

We have noted that the existence of interaction effects may modify conclusions about the main effects. For this reason it is customary to first perform the test for the existence of interaction and continue with inferences on main effects only if the interaction can be ignored or is too small to hinder the inferences on main effects.

As in the single-factor analysis of variance in Chapter 6, we are also interested in testing specific hypotheses using preplanned contrasts or making post hoc multiple comparisons for responses to the various factor levels (see Sections 9.4 and 9.5).

Notation

The appropriate analysis of data resulting from a factorial experiment is an extension of the analysis of variance presented in Chapter 6. Partitions of the sums of squares are computed using factor level means, and the ratios of corresponding mean squares are used as test statistics, which are compared to the F distribution. The structure of the data from a factorial experiment is more complicated than that presented in Chapter 6; hence the notation presented in Section 6.2 must be expanded.

Consistent with our objective of relying primarily on computers for performing statistical analyses, we present in detail only the definitional formulas for computing sums of squares. These formulas are based on the use of deviations from means and more clearly show the origin of the computed quantities, but are not convenient for manual calculations.

As defined for the linear model, y_{ijk} represents the observed value of the response of the kth unit for the factor level combination represented by the ith level of factor A and jth level of factor C. For example, y_{213} is the third observed value of the response for the treatment consisting of level 2 of factor A and level 1 of factor C. As in the one-way analysis, the computations for the

[2]The notation $\sum_i$ is used to signify summation across the i subscript, etc.

analysis of variance are based on means. In the multifactor case, we calculated a number of means and totals in several different ways. Therefore, we adopt a notation that is a natural extension of the "dot" notation used in Section 6.2:

$\bar{y}_{ij.}$ denotes the mean of the observations occurring in the ith level of factor A and jth level of factor C, and is called the mean of the A_iC_j cell,
$\bar{y}_{i..}$ denotes the mean of all observations for the ith level of factor A, called the A_i main effect mean,
$\bar{y}_{.j.}$ likewise denotes the C_j main effect mean, and
$\bar{y}_{...}$ denotes the mean of all observations, which is called the grand or overall mean.

This notation may appear awkward but is useful for distinguishing the various means, as well as getting a better understanding of the various formulas we will be using. Three important properties underlie this notational system:

1. When a subscript is replaced with a dot, that subscript has been summed over.
2. The number of observations used in calculating a mean is the product of the number of levels (or replications) of the model components represented by the dotted subscripts.
3. It is readily extended to describe data having more than two factors.

Computations for the Analysis of Variance

As in the analysis of variance for the one-way classification, test statistics are based on mean squares computed from factor level means. The computations for performing the analysis of variance for a factorial experiment can be described in two stages:

1. The **between cells analysis**, which is a one-way classification or CRD with factor levels defined by the cells. The cells consist of all combinations of factor levels.
2. The **factorial analysis**, which determines the existence of factor and interaction effects.

This two-stage definition of a factorial experiment provides a useful guide for performing the computations of the sums of squares needed for the analysis of such an experiment. It is also reflected by most computer outputs.

Between Cells Analysis

The first stage considers the variation among the cells for which the model can be written,

$$y_{ijk} = \mu_{ij} + \varepsilon_{ijk},$$

which is the same as it is for the one-way classification, except that μ_{ij} has two subscripts corresponding to the ij cell. The null hypothesis is

$$H_0: \mu_{ij} = \mu_{kl}, \quad \text{all } i, j \neq k, l;$$

that is, all cell means are equal. The test for this hypothesis is obtained using the methodology of Chapter 6 using the cells as treatments. The total sum of squares,

$$\text{TSS} = \sum_{ijk} (y_{ijk} - \bar{y}_{...})^2,$$

represents the variation of observations from the overall mean. The between cell sum of squares,

$$\text{SSCells} = n \sum_{ij} (\bar{y}_{ij.} - \bar{y}_{...})^2,$$

represents the variation among the cell means. The within cell or error sum of squares,

$$\text{SSW} = \sum_{ijk} (y_{ijk} - \bar{y}_{ij.})^2,$$

represents the variation among units within cells. This quantity can be obtained by subtraction:

$$\text{SSW} = \text{TSS} - \text{SSCells}.$$

The corresponding degrees of freedom are

total: the number of observations minus 1, df(total) $= acn - 1$.
between cells: the number of cells minus 1, df(cells) $= ac - 1$.
within cells: $(n - 1)$ degrees of freedom for each cell, df(within) $= ac(n - 1)$.

These quantities provide the mean squares used to test the null hypothesis of no differences among cell means. That is,

$$F = \text{MSCells}/\text{MSW}, \quad \text{with df} = [(ac - 1), ac(n - 1)].$$

This test is sometimes referred to as the test for the model. If the hypothesis of equal cell means is rejected, the next step is to determine whether these differences are due to specific main or interaction effects.[3]

The Factorial Analysis

The linear model for the factorial experiment defines the cell means in terms of the elements of the factorial experiment model as follows:

$$\mu_{ij} = \mu + \alpha_i + \gamma_j + (\alpha\gamma)_{ij}.$$

This model shows that the between cells analysis provides an omnibus test for all the elements of the factorial model, that is,

$$H_0\colon \alpha_i = 0,$$

$$H_0\colon \gamma_j = 0,$$

$$H_0\colon (\alpha\gamma)_{ij} = 0, \quad \text{for all } i \text{ and } j.$$

[3]Failure to reject the hypothesis of equal cell means does not automatically preclude finding significant main effects or interactions, but this is usually the case.

The test for the individual components of the factorial model is accomplished by partitioning the between cells sum of squares into components corresponding to the specific main and interaction effects. This partitioning is accomplished as follows:

1. The sum of squares due to main effect A is computed as if the data came from a completely randomized design with $c \cdot n$ observations for each of the a levels of factor A. Thus,

$$\text{SSA} = cn \sum_i (\bar{y}_{i..} - \bar{y}_{...})^2.$$

2. Likewise, the sum of squares for main effect C is computed as if we had a completely randomized design with $a \cdot n$ observations for each of the c levels of factor C:

$$\text{SSC} = an \sum_j (\bar{y}_{.j.} - \bar{y}_{...})^2.$$

3. The sum of squares due to the interaction of factors A and C is the variation among all cells not accounted for by the main effects. The definitional formula is

$$\text{SSAC} = n \sum_{ij} [(\bar{y}_{ij.} - \bar{y}_{...}) - (\bar{y}_{i..} - \bar{y}_{...}) - (\bar{y}_{.j.} - \bar{y}_{...})]^2.$$

Note that this represents the variation among cells minus the variation due to the main effects. Thus this quantity is most conveniently computed by subtraction:

$$\text{SSAC} = \text{SSCells} - \text{SSA} - \text{SSC}.$$

The degrees of freedom for the main effects are derived as are those for a factor in the one-way case. Specifically,

$$\text{df(A)} = a - 1,$$

$$\text{df(C)} = c - 1.$$

For the interaction, the degrees of freedom are the number of cells minus 1, minus the degrees of freedom for the two corresponding main effects, or equivalently the product of the degrees of freedom for the corresponding main effects:

$$\text{df(AC)} = (ac - 1) - (a - 1) - (c - 1) = (a - 1)(c - 1).$$

As before, all sums of squares are divided by their corresponding degrees of freedom to obtain mean squares, and ratios of mean squares are used as test statistics having the F distribution.

Expected Mean Squares

Since there are now several mean squares that may be used in F ratios, it may not be immediately clear which ratios should be used to test the desired hypotheses. The expected mean squares are useful for determining the appropriate ratios to use for hypothesis testing. Using the already defined model,

$$y_{ijk} = \mu + \alpha_i + \gamma_j + (\alpha\gamma)_{ij} + \varepsilon_{ijk},$$

where μ, α_i, γ_j, and $(\alpha\gamma)_{ij}$ are fixed effects and ε_{ijk} are random with mean zero and variance σ^2, the expected mean squares are[4]

$$E(\text{MSA}) = \sigma^2 + \frac{cn}{a-1}\sum_i \alpha_i^2,$$

$$E(\text{MSC}) = \sigma^2 + \frac{an}{c-1}\sum_j \gamma_j^2,$$

$$E(\text{MSAC}) = \sigma^2 + \frac{n}{(a-1)(c-1)}\sum_{ij} (\alpha\gamma)_{ij}^2,$$

$$E(\text{MSW}) = \sigma^2.$$

As illustrated for the CRD in Section 6.3, the use of expected mean squares to justify the use of the F ratio is based on the following conditions:

- If the null hypothesis is true, both numerator and denominator are estimates of the same variance.
- If the null hypothesis is not true, the numerator contains an additional component, which is a function of the sums of squares of the parameters being tested.

Now if we want to test the hypothesis

$$H_0\colon \alpha_i = 0, \quad \text{for all } i,$$

the expected mean squares show that the ratio MSA / MSW fulfills these criteria. As noted in Section 6.3, we are really testing the hypothesis that $\sum \alpha_i^2 = 0$, which is equivalent to the null hypothesis as originally stated.

Likewise, ratios using MSC and MSAC are used to test for the existence of the other effects of the model. The results of this analysis are conveniently summarized in tabular form in Table 9.4.

Table 9.4

Analysis of Variance Table for Two-Factor Factorial

Source	df	SS	MS	F
Between cells	$ac - 1$	SSCells	MSCells	MSCells / MSW
Factor A	$a - 1$	SSA	MSA	MSA / MSW
Factor C	$c - 1$	SSC	MSC	MSC / MSW
Interaction A*C	$(a-1)(c-1)$	SSAC	MSAC	MSAC / MSW
Within cells (error)	$ac(n-1)$	SSW	MSW	
Total	$acn - 1$	TSS		

[4]Algorithms for obtaining these expressions are available (for example, in Ott, 1988, Section 16.5). They may also be obtained by some computer programs such as `PROC GLM` of the SAS System.

Table 9.5

Data from Factorial Motor Oil Experiment

Note: Variable is MPG.

Engine	STANDARD	MOTOR OIL MULTI	GASMISER	Engine Means $\bar{y}_{i..}$
Six cylinder	23.6	23.5	21.4	22.247
	21.7	22.8	20.7	
	20.3	24.6	20.5	
	21.0	24.6	23.2	
	22.0	22.5	21.3	
Cell means $\bar{y}_{ij.}$	21.72	23.60	21.42	
Four cylinder	22.6	23.7	26.0	24.487
	24.5	24.6	25.0	
	23.1	25.0	26.9	
	25.3	24.0	26.0	
	22.1	23.1	25.4	
Cell means $\bar{y}_{ij.}$	23.52	24.08	25.86	
Oil means $\bar{y}_{.j.}$	22.620	23.840	23.640	$\bar{y}_{...} =$ 23.367

EXAMPLE 9.3

To illustrate the computations for the analysis of a two-factor factorial experiment we assume that the two motor oil experiments were actually performed as a single 2×3 factorial experiment. In other words, treatments correspond to the six combinations of the two engine types and three oils in a single completely randomized design. For the factorial we define

factor A: type of engine with two levels: 4 and 6 cylinders, and
factor C: type of oil with three levels: STANDARD, MULTI, and GASMISER.

The data, together with all relevant means are given in Table 9.5.

Solution The computations for the analysis proceed as follows:

1. The between cells analysis:
   a. The total sum of squares is

$$\begin{aligned}\text{TSS} &= \sum_{ijk}(y_{ijk} - \bar{y}_{...})^2\\ &= (23.6 - 23.367)^2 + (21.7 - 23.367)^2 + \cdots + (25.4 - 23.367)^2\\ &= 92.547.\end{aligned}$$

   b. The between cells sum of squares is

$$\begin{aligned}\text{SSCells} &= n\sum_{ij}(\bar{y}_{ij.} - \bar{y}_{...})^2\\ &= 5[(21.72 - 23.367)^2 + (23.60 - 23.367)^2\\ &\quad + \cdots + (25.86 - 23.367)^2]\\ &= 66.523.\end{aligned}$$

c. The within cells sum of squares is

$$\begin{aligned} \text{SSW} &= \text{TSS} - \text{SSCells} \\ &= 92.547 - 66.523 \\ &= 26.024. \end{aligned}$$

The degrees of freedom for these sums of squares are

$$\begin{aligned} &(a)(c)(n) - 1 = (2)(3)(5) - 1 = 29 \text{ for TSS,} \\ &(a)(c) - 1 = (2)(3) - 1 = 5 \text{ for SSCells, and} \\ &(a)(c)(n - 1) = (2)(3)(5 - 1) = 24 \text{ for SSW.} \end{aligned}$$

2. The factorial analysis:
   a. The sum of squares for factor A (engine types) is

$$\begin{aligned} \text{SSA} &= cn\sum_i (\bar{y}_{i..} - \bar{y}_{...})^2 \\ &= 15[(22.247 - 23.367)^2 + (24.487 - 23.367)^2] \\ &= 37.632. \end{aligned}$$

   b. The sum of squares for factor C (oil types) is

$$\begin{aligned} \text{SSC} &= an\sum_j (\bar{y}_{.j.} - \bar{y}_{...})^2 \\ &= 10[(22.620 - 23.367)^2 + (23.840 - 23.367)^2 + (23.640 - 23.367)^2] \\ &= 8.563. \end{aligned}$$

   c. The sum of squares for interaction, A $\times$ C (engine types by oil types), by subtraction is

$$\begin{aligned} \text{SSAC} &= \text{SSCells} - \text{SSA} - \text{SSC} \\ &= 66.523 - 37.623 - 8.563 \\ &= 20.328. \end{aligned}$$

The sum of these is the same as that for the between sum of squares in part (1).[5] The degrees of freedom are

$$\begin{aligned} (a - 1) &= (2 - 1) = 1 \text{ for SSA,} \\ (c - 1) &= (3 - 1) = 2 \text{ for SSC, and} \\ (a - 1)(c - 1) &= (1)(2) = 2 \text{ for SSAC.} \end{aligned}$$

The mean squares are obtained by dividing sums of squares by their respective degrees of freedom. The F ratios for testing the various hypotheses are

[5] As in Chapter 6, computational formulas are available for computing these sums of squares. These formulas use the cell, factor level, and grand totals, and have the now familiar format of a "raw" sum of squares minus a "correction factor." For details see, for example, Kirk (1995).

Table 9.6 Results of the Analysis of Variance for the Factorial Experiment Analysis of Variance Procedure

Dependent Variable: MPG

Source	df	Sum of Squares	Mean Square	F value	PR > F
Model	5	66.52266667	13.30453333	12.27	0.0001
Error	24	26.02400000	1.08433333		
Corrected Total	29	92.54666667			

R-Square	C.V.	Root MSE	MPG Mean
0.718801	4.4564	1.04131327	23.36666667

Source	df	Anova SS	F Value	PR > F
Cyl	1	37.63200000	34.71	0.0001
Oil	2	8.56266667	3.95	0.0329
Cyl*Oil	2	20.32800000	9.37	0.0010

computed as previously discussed. We confirm the computations for the sums of squares and show the results of all tests by presenting the computer output from the analysis using PROC ANOVA from the SAS System as seen in Table 9.6. In Section 6.1 we presented some suggestions for the use of computers in analyzing data using the analysis of variance. The factorial experiment is simply a logical extension of what was presented in Chapter 6, and the suggestions made in Section 6.1 apply here as well. The similarity of the output to that for regression (Chapter 8) is quite evident and is natural since both the analysis of variance and regression are special cases of linear models (Chapter 11).

The first portion of the output corresponds to what we have referred to as the partitioning of sums of squares due to cells. Here is it referred to as the MODEL, since it is the sum of squares for all parameters in the factorial analysis of variance model. Also, as seen in Chapter 6, ERROR is used for what we have called Within. The resulting F ratio of 12.27 has a p value of less than 0.0001; thus we can conclude that there are some differences among the populations represented by the cell means. Hence it is logical to expect that some of the individual components of the factorial model will be statistically significant.

The next line contains some of the same descriptive statistics we saw in the regression output. They have equivalent implications here.

The final portion is the partitioning of sums of squares for the main effects and interaction. These are annotated by the computer names given the variables that describe the factors: CYL for the number of cylinders in the engine type and OIL for oil type. The interaction is denoted as the product of the two names: CYL*OIL.

We first test for the existence of the interaction. The F ratio of 9.37 with (2,24) degrees of freedom has a p value of 0.0010; hence we may conclude that the interaction exists. The existence of this interaction makes it necessary to be exceedingly careful when making statements about the main effects, even though both may be considered statistically significant (engine types with a

Figure 9.1

Interaction Plots

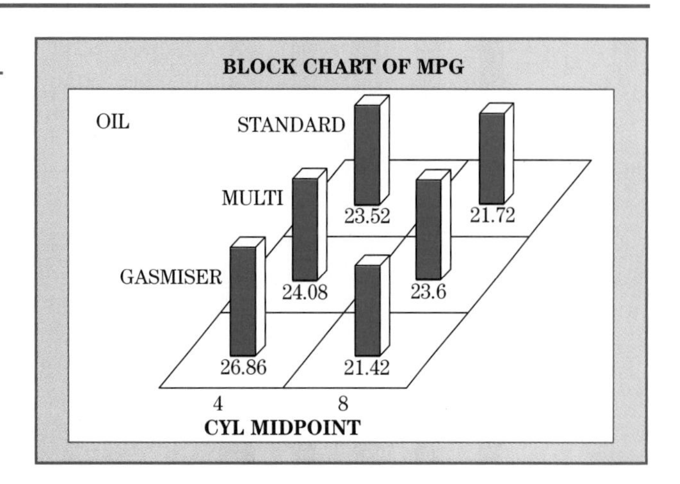

p value of 0.0001 and oil types with $p = 0.0329$). The nature of the conclusions also depends on the relative magnitudes of the interaction and individual main effects.

Graphical representation of the cell means is extremely useful in interpreting the consequences of interaction. A useful plot for illustrating this interaction is provided by a block chart (Section 1.7), where the heights of the blocks represent the means as shown in Fig. 9.1.

The plot shows that four-cylinder engines always get better gas mileage, but the difference is quite small when using the `MULTI` oil. There is, however, no consistent differentiation among the oil types as the relative mileages reverse themselves across the two engine types. More definitive statements about these interactions are provided by the use of contrasts, which are presented in Section 9.4.[6] ■

Notes on Exercises

Exercises 2, 4, 5, 9, and 10 and the basic ANOVA analysis of other exercises can now be worked using the procedures discussed in this section.

9.4 Specific Comparisons

As in Chapter 6 we present techniques for testing two types of hypotheses about differences among means:

1. preplanned hypotheses based on considerations about the structure of the factor levels themselves, and
2. hypotheses generated after examining the data.

[6]These plots were obtained by first creating a data set containing the cell means.

We noted in Chapter 6 that it is generally preferable to use preplanned hypotheses, and this preference is even more pronounced in the case of factorial experiments. Actually, a factorial experiment is a structure imposed on the total set of factor levels. That is, the partitioning of the between cells sum of squares into portions corresponding to main and interaction effects is dictated by the factorial structure. We now want to provide for tests of additional specific hypotheses for both main effects and interactions. As in Chapter 6, the specific structure of main effect factor levels usually suggests certain contrasts or polynomial trends, while the lack of structure may suggest the use of post hoc multiple comparison procedures. And, as before, only one set of comparisons should be performed on any specific problem.

Preplanned Contrasts

We continue the analysis of the motor oil experiment for illustrating the use of contrasts in a factorial experiment. (Refer to the beginning of Section 6.5 for principles of constructing contrasts.) The principles underlying the use of contrasts extend to the factorial experiment, but implementation is somewhat more difficult, especially when performing contrasts for interactions.

Computing Contrast Sums of Squares

In Chapter 6 we presented formulas for manually computing sums of squares for contrasts, and obviously analogous formulas exist for the factorial experiment. However, because we normally perform analyses with computers, we prefer to do the contrasts as part of the computer solution. Unfortunately not all statistical computer software have provisions for directly computing contrast statistics, and programs that do have this capability are sometimes not easy to implement.

Contrast statistics can, however, be computed by the use of regression analysis programs because contrasts are actually regressions using the contrast coefficients as independent variables. Most statistical computer programs have provisions for modifying and creating variables in a data set to be used for an analysis, and these can be used with a multiple regression program to produce contrast sums of squares. However, unless all possible contrasts for an experiment are requested, the residual mean square from the regression analysis will not be appropriate for performing hypothesis tests; therefore we use that from the analysis of variance. We will use this method to illustrate the use of contrasts for a factorial experiment.

If the contrasts are orthogonal, the contrast variables are uncorrelated and the partial sums of squares for the coefficients are the same as if computed by simple linear regression methods. However, since multiple regression does not require uncorrelated independent variables, the regression method may be used with nonorthogonal contrasts, although some caution must be exercised. We present here only the use of orthogonal contrasts.

EXAMPLE 9.3

REVISITED This example concerns the relationship of motor oil type and engine size to gas mileage of cars. The analysis of variance indicated that both main effects and the interaction are statistically significant. Suppose we had decided to use preplanned orthogonal contrasts to provide more specific inferences on these effects.

We consider the following contrasts:

Solution

Engine Types There are only two engine types: four and six cylinders. For two levels of a factor, only one contrast exists, in this case:

$$\mu_6 - \mu_4.$$

For purposes of performing a regression this contrast is represented by the variable

$$L1 = -1 \text{ if engine type is four cylinders, and}$$
$$= +1 \text{ if engine type is six cylinders.}$$

Oil Types There are three oil types, which allows for two orthogonal contrasts. We choose these as follows:

1. Compare STD with the two presumably more expensive oils, MULTI and GASMISER. The contrast is

$$\mu_{\text{STD}} - \frac{1}{2}(\mu_{\text{MULTI}} + \mu_{\text{GASMISER}}).$$

   This contrast is represented by the variable

$$L2 = +1 \text{ if oil type is STANDARD,}$$
$$= -0.5 \text{ if oil type is MULTI, and}$$
$$= -0.5 \text{ if oil type is GASMISER.}$$

2. Compare the two more expensive oils using

$$(\mu_{\text{MULTI}} - \mu_{\text{GASMISER}}).$$

   This contrast is represented by the variable

$$L3 = +1 \text{ if oil type is MULTI,}$$
$$= -1 \text{ if oil type is GASMISER, and}$$
$$= 0 \text{ if oil type is STANDARD.}$$

It is not difficult to verify that the two contrasts for oil types are orthogonal.

Interaction contrasts measure the inconsistencies of the responses to a contrast for one main effect across a contrast for the other main effect. One interaction contrast can be constructed for each degree of freedom for the interaction. For this example, the two degrees of freedom for the interaction between engine and oil types correspond to two contrasts, specifically the interaction between the engine-type contrast ($L1$) and the two oil-type contrasts ($L2$ and $L3$). The variables used in the regression to represent the interaction contrasts are the products of the main effect contrast variables:

$$L1L2 = L1 \cdot L2, \text{ and}$$

$$L1L3 = L1 \cdot L3.$$

The way these interaction contrasts work can be illustrated by placing the contrast coefficients in a two-way table corresponding to the factorial experiment. The $L1L2$ contrast coefficients are as follows:

	OILS		
Cylinders	**STD**	**MULTI**	**GASMISER**
4	−1	0.5	0.5
6	1	−0.5	−0.5

An examination of these coefficients shows that the value of this contrast is zero only if the difference between STD and the mean of the other two is the same for both engine sizes. Nonzero values of this contrast that may lead to rejection are an indication that this difference is inconsistent across engine types. The $L1L3$ contrast does the same for the comparison of MULTI to GASMISER. The resulting set of values of these variables and the response variable, MPG, are listed in Table 9.7.

Interaction contrasts derived from orthogonal main effect contrasts are also orthogonal; hence all five resulting contrasts are mutually orthogonal and provide five 1-df partitions of the between cells (model) sum of squares. These partitions can be obtained by performing a regression using MPG as the dependent variable and all five contrast variables, namely $L1$, $L2$, $L3$, $L1L2$, and $L1L3$, as independent variables. The results, obtained with `PROC REG` of the SAS System, are given in Table 9.8.

Because we have specified all possible contrasts the partitioning of the sums of squares due to the model is identical to that for the analysis of variance (Table 9.6). It can also be verified that the sum of the partial (`TYPE II`) sums of squares is equal to the `MODEL` sum of squares, verifying the orthogonality of the contrasts. Furthermore, the sum of squares for $L1$ (37.632) is the same as for engine types (`CYL`) and the total of the sums of squares for $L2$ and $L3$ (8.363 + 0.200) is the sum of squares for oil types (8.563). Finally, the total of the two interaction contrast sums of squares for $L1L2$ and $L1L3$ (0.726 + 19.602) is the same as the interactions sum of squares (20.382).

Next we examine the t statistics for the individual contrast parameters. Since the interaction is significant, we focus on the interaction contrasts and how they modify main effect conclusions. The interpretation of contrast $L1L2$ is the difference in the mileage between STANDARD and the other two oils across the two engine types. This contrast is not statistically significant; hence we may conclude that the difference between STANDARD and the average of the other two types (contrast $L2$, which is significant) does not differ between

Table 9.7

Listing of Contrast Variables

Obs	Cyl	Oil	MPG	*L1*	*L2*	*L3*	*L1L2*	*L1L3*
1	4	STANDARD	22.6	−1	1.0	0	−1.0	0
2	4	STANDARD	24.5	−1	1.0	0	−1.0	0
3	4	STANDARD	23.1	−1	1.0	0	−1.0	0
4	4	STANDARD	25.3	−1	1.0	0	−1.0	0
5	4	STANDARD	22.1	−1	1.0	0	−1.0	0
6	4	MULTI	23.7	−1	−0.5	1	0.5	−1
7	4	MULTI	24.6	−1	−0.5	1	0.5	−1
8	4	MULTI	25.0	−1	−0.5	1	0.5	−1
9	4	MULTI	24.0	−1	−0.5	1	0.5	−1
10	4	MULTI	23.1	−1	−0.5	1	0.5	−1
11	4	GASMISER	26.0	−1	−0.5	−1	0.5	1
12	4	GASMISER	25.0	−1	−0.5	−1	0.5	1
13	4	GASMISER	26.9	−1	−0.5	−1	0.5	1
14	4	GASMISER	26.0	−1	−0.5	−1	0.5	1
15	4	GASMISER	25.4	−1	−0.5	−1	0.5	1
16	6	STANDARD	23.6	1	1.0	0	1.0	0
17	6	STANDARD	21.7	1	1.0	0	1.0	0
18	6	STANDARD	20.3	1	1.0	0	1.0	0
19	6	STANDARD	21.0	1	1.0	0	1.0	0
20	6	STANDARD	22.0	1	1.0	0	1.0	0
21	6	MULTI	23.5	1	−0.5	1	−0.5	1
22	6	MULTI	22.8	1	−0.5	1	−0.5	1
23	6	MULTI	24.6	1	−0.5	1	−0.5	1
24	6	MULTI	24.6	1	−0.5	1	−0.5	1
25	6	MULTI	22.5	1	−0.5	1	−0.5	1
26	6	GASMISER	21.4	1	−0.5	−1	−0.5	−1
27	6	GASMISER	20.7	1	−0.5	−1	−0.5	−1
28	6	GASMISER	20.5	1	−0.5	−1	−0.5	−1
29	6	GASMISER	23.2	1	−0.5	−1	−0.5	−1
30	6	GASMISER	21.3	1	−0.5	−1	−0.5	−1

the two engine types. In other words, STANDARD is judged inferior regardless of engine type.

The contrast $L1L3$ is the difference in the mileage between MULTI and GASMISER for the two engine types. This contrast is statistically significant ($p = 0.0003$), which means that the difference between GASMISER and MULTI is not the same for the two engine types.

To summarize, STANDARD is always inferior, but the choice among the other two depends on the engine to which it is to be applied. These tests provide a formal test for what we saw in Fig. 9.1. ■

Polynomial Responses

In Chapter 6 we used orthogonal polynomial contrasts to estimate a curve to represent the response to levels of a quantitative factor. We learned in Section 8.6 that multiple regression methods can be used to implement polynomial models by defining a set of independent variables, which are powers and cross products of the numeric values representing the factor levels. In most

Table 9.8 Contrast Regression for Example 9.3

Model: MODEL1
Dependent Variable:MPG

Analysis of Variance

Source	DF	Sum of Squares	Mean Square	F Value	Prob > F
Model	5	66.52267	13.30453	12.270	0.0001
Error	24	26.02400	1.08433		
C Total	29	92.54667			
Root MSE		1.04131	R-square	0.7188	
Dep Mean		23.36667	Adj R-sq	0.6602	
C.V.		4.45640			

Parameter Estimates

Variable	DF	Parameter Estimate	Standard Error	T for H0: Parameter = 0	Prob > |T|
INTERCEP	1	23.366667	0.19011692	122.907	0.0001
L1	1	-1.120000	0.19011692	-5.891	0.0001
L2	1	-0.746667	0.26886593	-2.777	0.0105
L3	1	0.100000	0.23284473	429	0.6714
L1L2	1	0.220000	0.26886593	0.818	0.4213
L1L3	1	0.990000	0.23284473	4.252	0.0003

Variable	DF	Type II SS
INTERCEP	1	16380
L1	1	37.632000
L2	1	8.362667
L3	1	0.200000
L1L2	1	0.726000
L1L3	1	19.602000

computing environments, creating such variables is much easier to implement than is the construction of the orthogonal contrast variables, as shown in the previous section. Furthermore, the direct use of polynomials does not require equal spacing of the numeric factor levels. We will therefore use the straightforward application of polynomial regression to illustrate the estimation of polynomial responses in a factorial experiment.

Note that orthogonal polynomial contrasts do provide for a sequential fitting of polynomial terms in a manner similar to that shown for a one-variable polynomial model in Section 8.6. Thus orthogonal polynomials are useful for hypothesis tests to determine the terms that may be needed to describe the nature of the response surface (Snedecor and Cochran, 1980, Chapter 16).

EXAMPLE 9.4

This experiment concerns the search for some optimum levels of two fertilizer ingredients, nitrogen (N) and phosphorus (P). We know that there is likely to be interaction between these two factors.

Table 9.9

Data and Means for Fertilizer Experiment Response Is Yield

		LEVELS OF P				
		2	**4**	**6**	**8**	**Means**
N = 2		51.85	64.66	68.33	85.63	67.46
		41.30	73.95	75.88	83.32	
	53.18	68.76	67.15	77.71		
	Means	48.78	69.12	70.45	82.22	
N = 4		60.50	75.07	87.49	82.53	77.75
		60.86	75.05	97.21	89.03	
	56.97	82.14	88.95	77.25		
	Means	59.44	77.42	91.22	82.94	
N = 6		56.81	90.91	83.27	79.12	76.50
	52.77	83.44	87.65	77.53		
	51.22	81.54	89.22	84.57		
	Means	53.60	85.30	86.71	80.41	
Means		53.93	77.28	82.80	81.85	73.97

Table 9.10 Analysis of Variance for Fertilizer Experiment

Dependent Variable: Yield

Source	DF	Sum of Squares	Mean Square	F value	PR > F
Model	11	6259.35672222	569.03242929	28.14	0.0001
Error	24	485.37760000	20.22406667		
Corrected Total	35	6744.73432222			

R-Square	C.V.	Root MSE	Yield Mean
0.928036	6.0799	4.49711760	73.96722222

Source	DF	Anova SS	F Value	PR > F
N	2	729.22327222	18.03	0.0001
P	3	4969.73027778	81.91	0.0001
N*P	6	560.40317222	4.62	0.0030

Solution This requires a factorial experiment. The experiment for this study measures crop yield (YIELD) as related to levels of two fertilizer ingredients. The two ingredients are the factors

N, at three levels of 2, 4, and 6 units, and
P, at four levels of 2, 4, 6, and 8 units.

This is a 3×4 factorial experiment with 12 cells. There are three independent replications for each of the 12 cells; that is, there are 36 observations in a completely randomized design. The data are given in Table 9.9 and the computer output for the analysis of variance is given in Table 9.10. The F ratios signify that both main effects and the interaction are highly significant

($p < 0.01$). Since the factor levels comprise quantitative inputs, it is logical to estimate trends, that is, to construct curves showing how the yield responds to increased amounts of either or both fertilizer inputs. We will now build a polynomial regression model to describe this response.

For three levels of the factor N, it is possible to estimate the polynomial function

$$\text{YIELD} = \beta_0 + \beta_1\text{N} + \beta_2\text{N}^2,$$

where N represents the input quantities of ingredient N. For the four levels of factor P, it is possible to add to the model the terms

$$\cdots + \beta_3\text{P} + \beta_4\text{P}^2.$$

Note that with four levels of this factor we could add a cubic term but choose not to do so for this example. The need for this term may be established by a lack of fit test, as is done in the next subsection.

The above polynomial regression reflects only the variation in yield corresponding to the main effects of N and P. Since the interaction is statistically significant in the analysis of variance, it is appropriate to add terms reflecting this interaction. In a polynomial model, these are represented by terms which are the products of the values of the individual factor levels. Thus we can add to the above model the term

$$\cdots + \beta_5\text{NP}.$$

The implication of this term is more readily understood by combining it with the coefficient for P (we omit other terms for simplicity):

$$\cdots + \beta_3\text{P} + \beta_5\text{NP} \cdots = \cdots + (\beta_3 + \beta_5\text{N})\text{P} \cdots.$$

This expression shows that the coefficient for the linear trend for P, $(\beta_3 + \beta_5\text{N})$, changes linearly with N and that β_5 measures by how much that trend changes with N. For example, if β_5 is negative, the linear response to P may look like Fig. 9.2, where the labels (2, 4, and 6) for the lines are the levels for N.

Remember that the definition of interaction is the inconsistency of one main effect across levels of the other main effect(s). The coefficient β_5 represents a very specific interaction: The coefficient of the linear trend for levels of one factor changes linearly with levels of the other factor. This interaction is symmetric in that it also indicates how the linear trend for N changes linearly across levels of P. In this example both interpretations are equivalent, but this is not always the case.

Another term we can add to the model uses the product of N and the square of P,

$$\cdots + \beta_6\text{NP}^2,$$

whose effect can be seen by combining terms,

$$\cdots (\beta_4 + \beta_6\text{N})\text{P}^2,$$

which shows how the quadratic (curvilinear) response to P changes linearly with the levels of N. The response to P may look like Fig. 9.3, where the

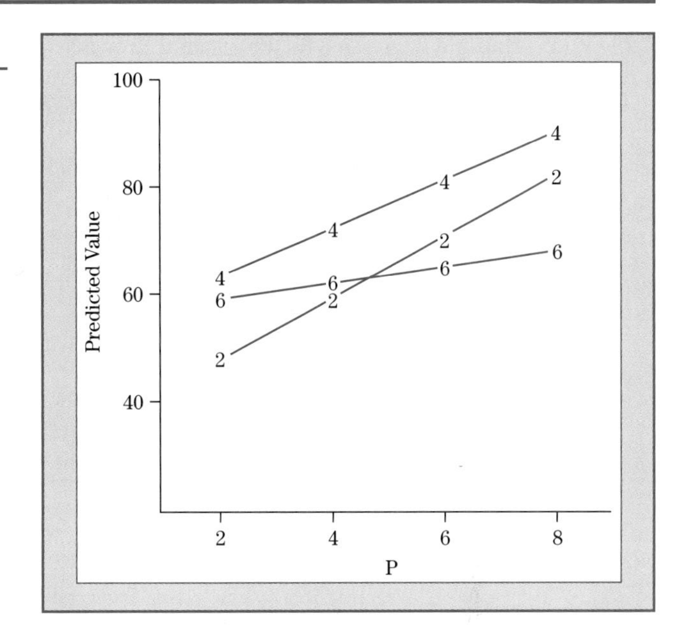

Figure 9.2

Linear by Linear Interaction

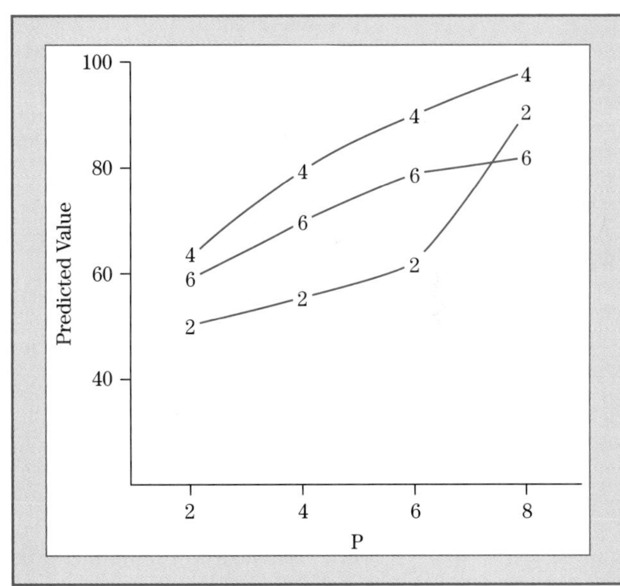

Figure 9.3

Linear by Quadratic Interaction

existence of a negative β_6 causes the response curve to change from convex to concave.

The symmetric definition

$$\cdots (\beta_1 + \beta_6 \mathrm{P}^2)\mathrm{N}$$

is more difficult to interpret in that it indicates that the linear response to N changes with the square of the level of P.

Table 9.11 Multiple Regression for Polynomial Response

Analysis of Variance

Source	DF	Sum of Squares	Mean Square	F Value	Prob > F
Model	7	6000.47220	857.21031	32.249	0.0001
Error	28	744.26212	26.58079009		
C Total	35	6744.73432			
Root MSE		5.155656	R-SQUARE	0.8897	
Dep Mean		73.96722	ADJ R-SQ	0.8621	
C.V.		6.97019			

Parameter Estimates

Variable	DF	Parameter Estimate	Standard Error	T For H0: Parameter = 0	Prob > |T|
INTERCEP	1	19.10416667	19.53790367	0.978	0.3365
N	1	9.35604167	9.39817804	0.996	0.3280
NSQ	1	−1.88708333	1.11623229	−1.691	0.1020
P	1	3.48475000	6.38124547	0.546	0.5893
PSQ	1	0.32145833	0.56835766	0.566	0.5762
NP	1	3.60243750	2.10807397	1.709	0.0985
NSQP	1	0.09339583	0.20379520	0.458	0.6503
NPSQ	1	−0.45973958	0.13154924	−3.495	0.0016

Variable	DF	Type II SS
INTERCEP	1	25.41371494
N	1	26.34297650
NSQ	1	75.96978148
P	1	7.92684265
PSQ	1	8.50303214
NP	1	77.62276451
NSQP	1	5.58258028
NPSQ	1	324.64970

In a similar manner we can add a term involving N^2P. Higher order terms can, of course, be used, but are increasingly difficult to interpret and, therefore, are not frequently used.

We now implement a multiple regression using the variables defined above. Since most computer programs for multiple regression require a specification of independent variables, we must create variables that represent the various squared and product terms. Thus, for this example, we have used mnemonic descriptors for these variables: NPSQ stands for NP^2, and so forth. The results of the regression (using PROC REG from the SAS System) are shown in Table 9.11.

The test for the model is certainly significant. The residual mean square of 26.581 is somewhat higher than that from the analysis of variance (20.224), indicating the possibility that additional polynomial terms may be needed. The lack of fit test for this possibility is given in the next subsection.

As noted in Section 8.6, the individual coefficients in a polynomial regression are not readily interpretable and are used primarily for describing the response curve or surface. Furthermore, the partial (called `Type II` in the output) sums of squares do not sum to the model sums of squares as they did with the use of orthogonal contrasts.

The statistics for the highest order terms in our model, NP, N^2P, and NP^2 can, however, be interpreted. In this case the significant NP^2 term indicates that the quadratic response to P changes negatively with N, while the quadratic response to N is apparently not affected by P.

With a little effort the coefficients can be used to determine some characteristics of the response. Since the term NP^2 was statistically significant, we will examine the nature of the response to P. Using only the terms involving P, and collecting terms, the response to P is

$$(3.485 + 3.602\text{N})\text{P} + (0.321 - 0.460\text{N})\text{P}^2.$$

We can now specify the response curve of P for the values of N = 2, 4, and 6, which were the levels of that factor in the experiment. These equations are

$$\text{N} = 2\text{: } 10.689\text{P} - 0.599\text{P}^2,$$

$$\text{N} = 4\text{: } 17.893\text{P} - 1.519\text{P}^2,$$

$$\text{N} = 6\text{: } 25.092\text{P} - 2.439\text{P}^2.$$

The increasing positive coefficients for P show an increasingly positive linear trend. This is partially offset by an increasingly negative (downward curving) contribution from P^2.

The three-dimensional plot of the polynomial response for Example 9.4 is shown in Fig. 9.4. This plot clearly shows how the response to P becomes

Figure 9.4

Three-Dimensional Plot

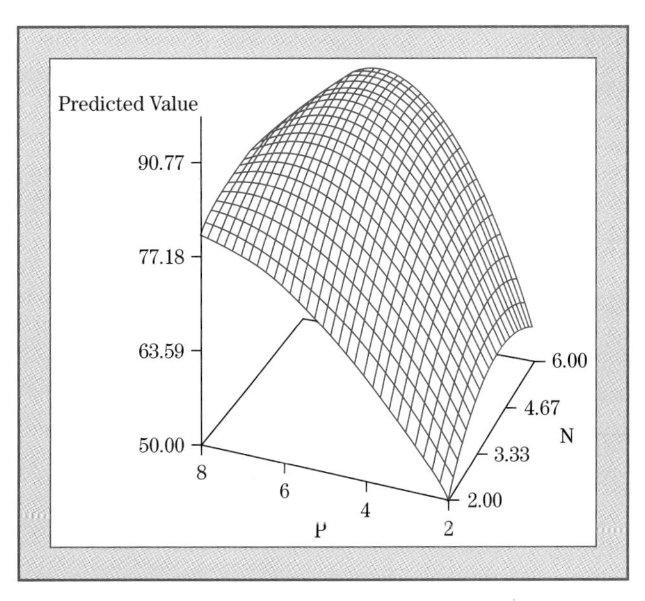

steeper and has more curvature as N increases. On the other hand, the response to N is quite modest but does increase somewhat with P. Graphic representations such as this are increasingly available in today's statistical software packages but may require some programming effort. ■

Lack of Fit Test

Depending on the number of levels of factors in an experiment, polynomial terms of higher order than we have illustrated can be used. In Example 9.4, which is a 3×4 factorial experiment, the following terms may be estimated:

1. linear and quadratic in N,
2. linear, quadratic, and cubic in P, and
3. all six possible products of the terms involving the two terms in N and the three terms in P.

The total number of terms in this polynomial model (11) corresponds exactly to the degrees of freedom for the cells in the analysis of variance for the factorial experiment (Table 9.10). Thus the 11-term polynomial model provides exactly the partitioning for the cell or model sum of squares of the analysis variance.

It is not always desirable to use a model with all possible terms, particularly since some of these may have little useful interpretation. For example, the meaning of interaction of the cubic in P with the quadratic in N would be almost impossible to interpret. If a model with fewer terms has been chosen, it is appropriate to use a lack of fit test to ascertain whether additional terms should be included.

As first presented in Section 6.5 this test is based on the difference between the model sum of squares for the reduced (regression) model and the model sum of squares from the full (analysis of variance) model. As we have noted, the basis for this test is that any factor sum of squares with, say, $(t - 1)$ degrees of freedom, can also be computed as a regression sum of squares for a regression with $(t - 1)$ independent variables (such as contrasts of polynomials) in which the independent variables identify with the factor levels. The difference between the factor level sum of squares and the sum of squares for, say, $k < (t - 1)$ regression coefficients is—by the general principle for hypothesis testing (Section 8.3)—the partial sum of squares for an unspecified set of $(t-k-1)$ additional contrasts or regression coefficients. The difference in sum of squares is divided by the corresponding difference in degrees of freedom to provide mean squares to be used in an F test for ascertaining the possible existence of the additional coefficients.

In the fertilizer experiment we have the following quantities:

Model sum of squares from ANOVA	6259.8, df = 11
Model sum of squares from regression	6000.5, df = 7
Lack of fit (difference)	258.9, df = 4

The F ratio for testing the existence of lack of fit uses the error mean square from the full (ANOVA) model (Table 9.10):

$$F(4, 24) = (258.9/4)/(22.02)$$
$$= 3.20.$$

This value is somewhat larger than the 0.05 value of the F distribution (2.78) but smaller than the 0.01 value (4.22); hence there is some evidence that other terms may be needed.

A logical next step is to examine the residuals. Plots of residuals against values predicted by the polynomial regression and against the individual factor levels may be used. In this example, such plots (not reproduced here) show neither outliers nor evidence of lack of fit. Furthermore, the increase in the residual standard deviation when using the polynomial is quite modest: from 4.50 for the full model ANOVA to 5.16 for the polynomial. These results, coupled with the lack of overwhelming significance of the lack of fit test, lead to the conclusion that the estimated seven-term polynomial model is sufficient.

Multiple Comparisons

When preplanned comparisons are not used, a post hoc comparison procedure, such as the Duncan, Tukey, or Scheffé procedure, may be appropriate.

The mechanics of applying the post hoc comparisons for factorial experiments are adaptations of the procedures applied to one-way analyses presented in Chapter 6. Remember that all multiple comparison methods require the computation of LSD-type statistics, which are of the form

$$\text{LSD} = \text{factor}\sqrt{\frac{\text{estimated variance}}{\text{sample size}}}.$$

The "factor" depends on the specific method and is obtained from tables, and the sample size is the number of observations used to compute the means. Any pair of means differing by more than this statistic is declared to be significantly different.

The same procedure is used for multiple comparisons among the means from a factorial experiment. The comparisons are normally made on main effect means, but may occasionally be performed on the various cell means. Some adaptations of these procedures must be made as follows:

- The estimated variance is the mean square used in the denominator of the F ratio for testing the factor representing the means being compared. As we will see in Chapter 10, this is not always the same for all factors.
- The sample size is the total number of observations for calculating the means involved in the comparison. These are usually not the same for the different main effects and interactions. Remember that this number is the product of the dotted subscript defining the mean, but must be the same for means being compared since we are dealing only with balanced data.

As is true for all inferences from factorial experiments, the use of multiple comparisons in factorial experiments is affected by the possible existence of interactions. As indicated in the preceding, if interaction can be ignored, multiple comparisons may be performed on each main effect separately. However, if an interaction is deemed to exist, multiple comparisons among main effect means may not be useful. In many applications it is instead recommended that multiple comparisons be performed for one set of main effect means for each level of the other factor. For example, we can make separate comparisons for the mean responses among the levels of factor A for each level of factor C.

This may be done by performing separate one-way analyses for each level of factor C. However, this procedure will use the separeted error mean squares for each of the analyses, which have fewer degrees of freedom and may also differ in value among levels of C. The preferred method is to use the error mean square from the factorial analysis for all comparisons. Unfortunately, many computer packages have no provisions for this method, and it must therefore be implemented manually.

In Example 9.3 we found significant interaction between the number of cylinders in the car and the type of oil.[7] To compare the various types of oil (a main effect), we perform a separate set of comparisons for each engine type. For example, to compare STD, MULTI, and GASMISER for four-cylinder cars, we could use pairwise comparisons. If we want to use the Scheffé test, it becomes a test for contrasts where $a_i = 1$ and $a_j = 1$ for all pairs of i and j. The general form of the Scheffé statistic is

$$S = \sqrt{\left[(t-1)F_\alpha \sum a_i^2\right]\left(\frac{\text{MSW}}{n}\right)},$$

where t is the number of factor levels, F_α has the degrees of freedom for the test for the factor, a_i are the contrast coefficients, and n is the number of observations for the means. For this example, the Scheffé significant difference is

$$S = \sqrt{(2)(3.40)(2)\left(\frac{1.084}{5}\right)} = 1.717.$$

The results can be illustrated by the following:

Oil Type	**STD**	**MULTI**	**GASM**
Mean	23.52	24.08	25.86

For four-cylinder cars the GASMISER oil gives significantly higher mileage than the other two.

[7] Since the structure of the main effect factor levels in the example suggests the use of orthogonal contrasts, the use of multiple comparisons is not an appropriate analysis. The use of multiple comparisons would be appropriate if, for example, we simply had comparable oils from three manufacturers. However, we continue with this numeric example to avoid introducing new computations.

The same value of S, 1.717, is used for pairwise comparisons of the three oils for six-cylinder cars. In this case, the results are

Oil Type	GASM	STD	MULTI
Mean	21.42	21.72	23.60

Occasionally the existence of a strong interaction in cases where preplanned comparisons are not suitable suggests that it may be more useful simply to compare all cell means, that is, the means of all factor level combinations, with a multiple comparison test. For example, assume an experiment that compares the effectiveness of three types of insecticides manufactured by four manufacturers. A strong interaction suggests that the types of insecticides may not really be the same for each of the manufacturers, and we may conclude that we simply want to rank the effectiveness of what now appears to be 12 different products.

Of course, when we simply compare the cell means, we are doing what we did for the one-way analysis of Chapter 6 and no longer make use of the factorial structure of the experiment, thereby losing entirely the information on main effects and interactions. Instead we treat the means as if they came from a completely randomized design with $a \cdot c$ levels of a single factor having no particular structure.

EXAMPLE 9.5

A manufacturing plant has had difficulty reproducing good production rates in a catalyst plant. An experiment to investigate the effect of four reagents (A, B, C, and D) and three catalysts (X, Y, and Z) on production rates was initiated.

Because the possibility of interactions exists, a 4×3 factorial experiment was performed. Each of the 12 factor level combinations was run twice in random order (Smith, 1969). The data are given in Table 9.12.

Table 9.12

Production Rates (CODES) in Catalyst Experiment

	CATALYST		
Reagent	X	Y	Z
A	4	11	5
	6	7	9
B	6	13	9
	4	15	7
C	13	15	13
	15	9	13
D	12	12	7
	12	14	9

Solution The analysis of the factorial experiment was performed using `PROC ANOVA` from the SAS System, which produces the output seen in Table 9.13. With these simple numbers the results can be verified manually. The analysis shows that the reagent effect is quite significant while the catalyst and interaction effects are marginally significant. Since no specific information is available, it is logical to perform a paired comparison procedure for the reagent. The result of performing Tukey's test for that factor is given in Table 9.14.

The Tukey test indicates a clear superiority for reagent C over reagents B and A, while we may also state that D is better than A. However, since the interaction was significant with a p value of less than 0.05, a plot of means may be useful. If both factors are qualitative, an informative way to look at the means is through a block chart (Section 1.7). Figure 9.5 is a block chart of the cell means; the height of each block indicates the mean rate, which is also shown at the base. The chart clearly shows the effect of the significant

Table 9.13 Analysis of Variance for Catalyst Data

Analysis of Variance Procedure

Dependent Variable: RATE

Source	DF	Sum of Squares	Mean Square	F value	Pr > F
Model	11	252.0000000	22.9090909	5.73	0.0027
Error	12	48.0000000	4.0000000		
Corrected Total	23	300.0000000			

R Square	C.V.	Root MSE	RATE Mean
0.840000	20.00000	2.000000	10.0000000

Source	DF	Anova SS	Mean Square	F value	Pr > F
REAGENT	3	120.0000000	40.0000000	10.00	0.0014
CATALYST	2	48.0000000	24.0000000	6.00	0.0156
REAGENT*CATALYST	6	84.0000000	14.0000000	3.50	0.0308

Table 9.14

Tukey's Studentized Range (HSD) Test for Variable: Rate

Alpha = 0.05 df = 12 MSE = 4

Critical Value of Studentized Range = 4.199

Minimum Significant Difference = 3.4282

Means with the same letter are not significantly different.

Tukey	Grouping	Mean	N	REAGENT
	A	13.000	6	C
	A			
B	A	11.000	6	D
B				
B	C	9.000	6	B
	C			
	C	7.000	6	A

Figure 9.5

Interaction Plot

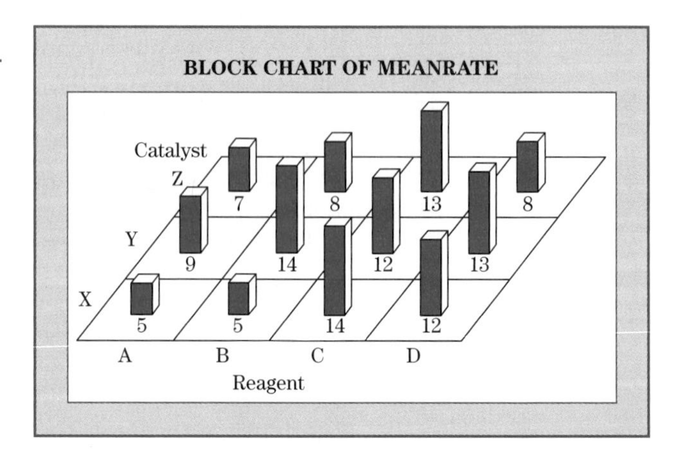

Table 9.15

Tukey's Studentized Range (HSD) Test for All Combinations

Alpha = 0.05 df = 12 MSE = 4
Critical Value of Studentized Range = 5.615
Minimum Significant Difference = 7.9402
Means with the same letter are not significantly different.

Tukey	Grouping	Mean	N	TRT
	A	14.000	2	BY
	A			
	A	14.000	2	CX
	A			
	A	13.000	2	CZ
	A			
	A	13.000	2	DY
	A			
B	A	12.000	2	DX
B	A			
B	A	12.000	2	CY
B	A			
B	A	9.000	2	AY
B	A			
B	A	8.000	2	DZ
B	A			
B	A	8.000	2	BZ
B	A			
B	A	7.000	2	AZ
B				
B		5.000	2	AX
B				
B		5.000	2	BX

interaction. For example, catalyst X has the lowest rates when paired with reagents A and B while it has one of the highest rates when paired with reagent C. On the other hand, reagent C consistently produces high rates. Thus, if we want a reagent that seems to do well in all circumstances, reagent C may be a good choice. The only other reasonably clear conclusion here is to avoid catalyst X with reagents A and B.

As we noted, significant interaction in experiments where contrasts are not indicated may suggest multiple comparison among the cell means, that is, an analysis that ignores the factorial structure. The results of a Tukey comparison among the 12 means for Example 9.5 is shown in Table 9.15.[8]

The 12 factor levels are identified by the two letters representing the reagent and catalyst. This analysis does not appear to provide any definitive conclusions, except maybe to say stay away from catalyst X when using reagents A and B. This example shows how an interaction can seriously affect the ability of an experiment to provide useful inferences about main effects. ■

[8]This table was produced by creating a single variable, called TRT, which provides a unique identification for each cell. This procedure is necessary because computer packages normally do not allow multiple comparisons for cell means.

In some applications the factor levels of one factor may suggest useful preplanned comparisons while those of the second factor do not. If interaction is deemed not to be of importance, the appropriate techniques may be used for the different main effects. If the interaction cannot be ignored, it will probably be more useful to compute separate preplanned comparisons for one factor for each level of the second factor.

9.5 No Replications

So far we have assumed that the factorial experiment is conducted as a completely randomized design providing for an equal number of replicated experimental units of each factor level combination. Since a factorial experiment may be quite large in terms of the total number of cells, it may not be possible to provide for replication. Since the variation among observations within factor level combinations is used as the basis for estimating σ^2, the absence of such replications leaves us without such an estimate.

The usual procedure for such situations is to assume that the interaction does not exist, in which case the interaction mean square provides the estimate of σ^2 to use for the denominator of F ratios for tests for the main effects. Of course, if the interaction does exist, the resulting tests are biased. However, the bias is on the conservative side since the existence of the interaction inflates the denominator of the F ratio for testing main effects.

One possible cause for interaction is that the main effects are multiplicative in a manner suggested by the logarithmic model presented in Section 8.6. The Tukey test for nonadditivity (Kirk, 1995) provides a one degree of freedom sum of squares for an interaction effect resulting from the existence of a multiplicative rather than additive model. Subtracting the sum of squares for the Tukey test from the interaction sum of squares may provide a more acceptable estimate of σ^2 if a multiplicative model exists.

9.6 Three or More Factors

Obviously factorial experiments can have more than two factors. As we have noted, fertilizer experiments are concerned with three major fertilizer ingredients, N, P, and K, whose amounts in a fertilizer are usually printed on the bag. The fundamental principles of the analysis of factorial experiments such as the model describing the data, the partitioning of sums of squares, and the interpretation of results are relatively straightforward extensions of the two-factor case. Since such analyses are invariably performed by computers, computational details are not presented here.

The model for a multifactor factorial experiment is usually characterized by a large number of parameters. Of special concern is the larger number and greater complexity of the interactions. In the three-factor fertilizer experiment, for example, the model contains parameters describing

three main effects: N, P, and K,
three two-factor interactions: N × P, N × K, and P × K, and
one three-factor interaction: N × P × K.

The interpretations of main effects and two-factor interactions remain the same regardless of the number of factors in the experiment. Interactions among more than two factors, which are called higher order interactions, are more difficult to interpret. One interpretation of a three-factor interaction, say, N × P × K, is that it reflects the inconsistency of the N × P interaction across levels of K. Of course, this is equivalent to the inconsistency of the P × K interaction across N, etc.

EXAMPLE 9.6

It is of importance to ascertain how the lengths of steel bars produced by several screw machines are affected by heat treatments and the time of day the bars are produced. A factorial experiment using four machines and two heat treatments was conducted at three different times in one day. This is a three-factor factorial with factors:

Heat treatment, denoted by HEAT, with levels W and L,
Time of experiment, denoted by TIME, with levels 1, 2, and 3 representing 8:00 A.M., 11:00 A.M., and 3:00 P.M., and
Machine, denoted by MACHINE with levels A, B, C, and D.

Each factor level combination was run four times. The response is the (coded) length of the bars. The data are given in Table 9.16.

Solution The analysis of variance for the factorial experiment is performed with PROC ANOVA of the SAS System with the results, which are quite straightforward, shown in Table 9.17. The HEAT and MACHINE effects are clearly significant, with no other factors approaching significance at the 0.05 level. In

Table 9.16

Steel Bar Data for Three-Factor Factorial

	HEAT TREATMENT W MACHINES				HEAT TREATMENT L MACHINES			
Time	**A**	**B**	**C**	**D**	**A**	**B**	**C**	**D**
	6	7	1	6	4	6	−1	4
	9	9	2	6	6	5	0	5
8:00AM	1	5	0	7	0	3	0	5
	3	5	4	3	1	4	1	4
	6	8	3	7	3	6	2	9
	3	7	2	9	1	4	0	4
11:00AM	1	4	1	11	1	1	−1	6
	−1	8	0	6	−2	3	1	3
	5	10	−1	10	6	8	0	4
	4	11	2	5	0	7	−2	3
3:00PM	9	6	6	4	3	10	4	7
	6	4	1	8	7	0	−4	0

Table 9.17 Analysis of Variance for Steel Bar Data

Analysis of Variance Procedure

Dependent Variable: LENGTH

Source	DF	Sum of Squares	Mean Square	F Value	Pr > F
Model	23	590.3333333	25.6666667	4.13	0.0001
Error	72	447.5000000	6.2152778		
Corrected Total	95	1037.8333333			

R Square	C.V.	Root MSE	LENGTH Mean
0.568813	62.98221	2.493046	3.95833333

Source	DF	Anova SS	Mean Square	F value	Pr > F
TIME	2	12.8958333	6.4479167	1.04	0.3596
HEAT	1	100.0416667	100.0416667	16.10	0.0001
TIME*HEAT	2	1.6458333	0.8229167	0.13	0.8762
MACHINE	3	393.4166667	131.1388889	21.10	0.0001
TIME*MACHINE	6	71.0208333	11.8368056	1.90	0.0917
HEAT*MACHINE	3	1.5416667	0.5138889	0.08	0.9693
TIME*HEAT*MACHINE	6	9.7708333	1.6284722	0.26	0.9527

Table 9.18

Analysis of Variance for Steel Bar Data, Duncan's Multiple Range Test for Machine

Alpha = 0.05 cdf = 72 cMSE = 6.215278

Number of Means	2	3	4
Critical Range	1.436	1.510	1.558

Means with the same letter are not significantly different.

Duncan Grouping	Mean	N	MACHINE
A	5.875	24	B
A			
A	5.667	24	D
B	3.417	24	A
C	0.875	24	C

fact, some of the F values are suspiciously small, which may raise doubts about the data collection procedures.

No specifics are given on the structure of the factor levels; hence post hoc paired comparisons are in order. The factor heat has only two levels; hence the only statement to be made is that the sample means of 2.938 and 4.979 for L and W indicate that W produces longer bars. Duncan's multiple range test is applied to the MACHINE factor with results given in Table 9.18.

Figure 9.6 is a block chart illustrating the heat by machine means. In general, for any machine, heat W gives a longer bar and the differences among machines are relatively the same for each heat. This confirms the lack of interaction. ■

Factorial experiments with many factors often produce a large number of factor level combinations. The resulting requirement for a large number of

Figure 9.6

Interaction Plot for Example 9.6

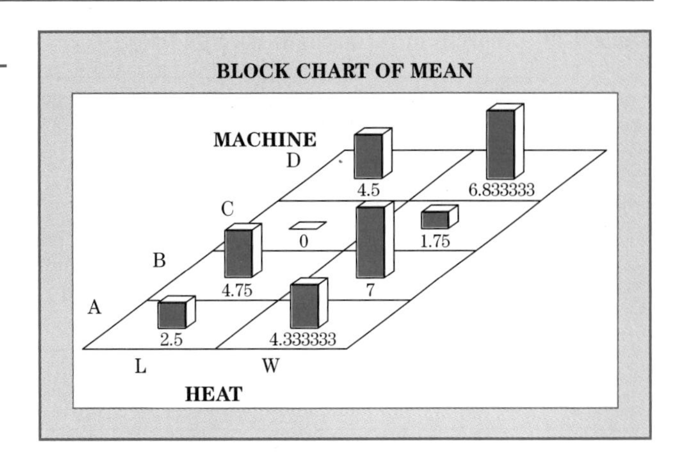

observations may make it impossible to provide for replicated values in the cells. Since higher order interactions are difficult to interpret, their mean squares make good candidates for the estimate of σ^2. Of course, if these interactions do exist, the resulting tests are biased.

Additional Considerations

Special experimental designs are available to overcome partially the often excessive number of experimental units required for factorial experiments. For example, the estimation of a polynomial response regression does not require data from all the factor level combinations provided by the factorial experiment; hence special response surface designs are available for use in such situations. Also, since higher order interactions are often of little or no interest, designs for which a reduction in the number of observations is accomplished in such a manner that sacrifices only the ability to estimate higher order interactions have been developed. For additional information on such topics, refer to a book on experimental design (for example, Kirk, 1995).

9.7 CHAPTER SUMMARY

Solution to Example 9.1 The experiment is a three-factor factorial experiment with factors:

`STRANDS`: number of strands (7 and 9),
`GAGE`: gage of wire (24,22,20), and
`SLACK`: slack in assembly (0,3,6,9,12).

The initial analysis of variance with results obtained from `PROC ANOVA` from the SAS System is shown in Table 9.19. The obvious conclusion is that the major effects are due to `GAGE` and `SLACK`, although `STRANDS` and the

Table 9.19 Analysis of Variance for Wirelife Data

Analysis of Variance Procedure

Dependent Variable: CYCLES

Source	DF	Sum of Squares	Mean Square	F Value	Pr > F
Model	29	979.7333333	33.7839080	4.57	0.0001
Error	30	222.0000000	7.4000000		
Corrected Total	59	1201.7333333			

R-Square	C.V.	Root MSE	CYCLES Mean
0.815267	27.94823	2.720294	9.73333333

Source	DF	Anova SS	Mean Square	F Value	Pr > F
STRANDS	1	45.0666667	45.0666667	6.09	0.0195
GAGE	2	348.6333333	174.3166667	23.56	0.0001
STRANDS*GAGE	2	52.2333333	26.1166667	3.53	0.0420
SLACK	4	307.5666667	76.8916667	10.39	0.0001
STRANDS*SLACK	4	74.7666667	18.6916667	2.53	0.0614
GAGE*SLACK	8	104.0333333	13.0041667	1.76	0.1257
STRANDS*GAGE*SLACK	8	47.4333333	5.9291667	0.80	0.6063

STRANDS*GAGE interaction may need to be examined further. For the time being, we will concentrate on the effect of GAGE and SLACK.

Since all factors are numeric, a polynomial response surface analysis is in order. We will use the response surface model

$$\text{CYCLES} = \beta_0 + \beta_1\text{SLACK} + \beta_2\text{SLSQ} + \beta_3\text{GAGE} + \beta_4\text{GSQ} + \beta_5\text{GSL} + \varepsilon,$$

where SLSQ and GSQ are the squares of SLACK and GAGE, and GSL is the product of GAGE and SLACK. Note that we are including the product term even though the interaction is not significant. The resulting regression is summarized in Table 9.20, produced by PROC REG of the SAS System.

The regression is certainly significant. The statistical significance of the two quadratic terms indicates a curvilinear response and the linear by linear interaction is also significant although apparently not very important.

Since the regression model ignores the STRAND effect as well as other interactions, it is appropriate to perform a lack of fit test. For this test we have

$$\text{FULL MODEL (ANOVA) SS} = 979.73,\ \text{df} = 29$$

$$\text{REDUCED MODEL (regression) SS} = 688.25,\ \text{df} = 5$$

$$\text{LACK OF FIT (difference) SS} = 291.48,\ \text{df} = 24.$$

The mean square for lack of fit is $291.48/24 = 12.145$. This is divided by the error mean square for the full model ($222.00/30 = 7.4$, df $= 30$) to obtain an F ratio of 1.64, which has (24,30) degrees of freedom. The 0.05 upper tail value for the F distribution for those degrees of freedom is 1.89, and we may conclude that

Table 9.20 Response Surface Regression

Model: MODEL1
Dependent Variable:CYCLES

Analysis of Variance

Source	DF	Sum of Squares	Mean Square	F value	Prob > F
Model	5	688.25714	137.65143	14.476	0.0001
Error	54	513.47619	9.50882		
C Total	59	1201.73333			

Root MSE	3.08364	R-Square	0.5727
Dep Mean	9.73333	Adj R-sq	0.5332
C.V.	31.68120		

Parameter Estimates

Variable	DF	Parameter Estimate	Standard Error	T for H0: Parameter = 0	Prob > |T|
INTERCEP	1	-501.140476	102.04729456	-4.911	0.0001
SLACK	1	-3.739683	1.30668919	-2.862	0.0060
SLSQ	1	0.072751	0.02643420	2.752	0.0080
GAGE	1	48.187500	9.29896859	5.182	0.0001
GSQ	1	-1.131250	0.21112221	-5.358	0.0001
GSL	1	0.150000	0.05746019	2.611	0.0117

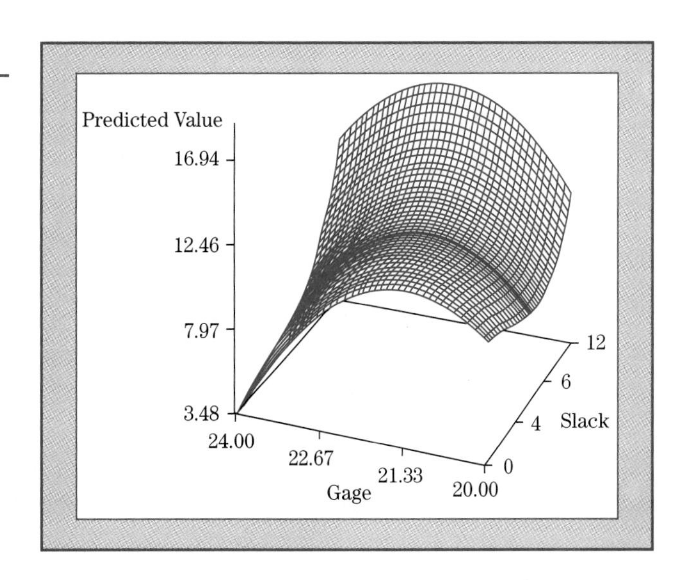

Figure 9.7

Response Surface Plot for Wirelife Data

the response surface regression model is adequate, although some checking or residuals may still be worthwhile.

It is now appropriate to produce a response surface plot, which was produced by PROC G3D of the SAS System and is shown in Fig. 9.7. The minimal degree of interaction is seen by the relatively small changes in the cross section of the response curve for one effect across the other. Since we want maximum cycles to failure, it is evident that 21- to 22-gage wire should be used, and that slack has little effect until about 8 units, at which increased slack definitely helps. In fact, it would appear that increasing slack beyond 12 would be useful, but this conclusion requires extrapolation and, in addition, does not take into account other side effects of too much slack.

Introducing a term for STRANDS would simply produce two identically shaped response curves for the two values of STRANDS as seen in Fig. 9.6, each with slightly different levels of response. However, inclusion of the marginally significant interaction of STRANDS*GAGE (not illustrated) produces a slightly different shape for the two levels of STRANDS. ■

This chapter extends the use of the analysis of variance for comparing means when the populations arise from the use of levels from more than one factor. An important consideration in such analyses is **interaction**, which is defined as the inconsistency of the effects of one factor across levels of another factor. The study of interaction is made possible by the use of factorial experiments in which observations are obtained on all combinations of factor levels.

In most other respects the analysis of data from a factorial experiment is a relatively straightforward generalization of the methods presented in Chapter 6:

- Statistical significance of the factors is determined by the partitioning of sums of squares and computation of F ratios. The major difference is that inferences on main effects must take into account any relevant interactions.
- Contrasts for specific comparisons among means for main effects are constructed as shown in Chapter 6, and interaction contrast coefficients are the products of the main effect contrasts.
- If the factor levels are numeric, polynomial curve called response surfaces can be constructed using regression methods. Lack of fit tests may be used to determine the adequacy of such models.
- If factor levels do not suggest the use of contrasts, post hoc comparison methods may be used but their use may be severely restricted by the existence of interactions. Mixtures of paired comparisons and contrasts or response curves may be used if appropriate.

9.8 CHAPTER EXERCISES

EXERCISES

The most important aspect of the exercises for this chapter is not simply to perform the analysis of variance but to select the appropriate follow-up tests such as contrasts or multiple comparisons and to interpret results. For this reason most exercises consist of a statement and the data and do not provide specific instructions for the analysis.

Exercises 1, 2, and 3 consist of small artificially generated data sets. Most of the computations for these exercises can be performed with the aid of calculators if it is desirable for students to have some practice in applying formulas. Most of the other problems are more efficiently performed with computers.

1. Table 9.21 contains data from a 2 × 4 factorial experiment. The only additional structure is that level C of factor T is a control. This exercise is somewhat unusual in that it will require a contrast to see whether the control is different from the other treatments and also a paired comparison of the treatments.

Table 9.21

Data for Exercise 1

		FACTOR T			
	Levels	**C**	**M**	**N**	**Q**
Factor A	A	3.6	8.9	8.8	8.7
		5.3	8.8	6.8	9.0
	B	3.8	2.5	4.1	3.6
		4.8	3.9	3.4	3.8

2. Table 9.22 contains data from a 3 × 4 factorial experiment in which there is no structure to describe the levels of either factor.

Table 9.22

Data for Exercise 2

		FACTOR C			
	Levels	**1**	**2**	**3**	**4**
Factor A	M	5.6	7.5	7.5	6.2
		6.2	5.8	6.9	4.7
	P	6.4	8.0	11.5	9.2
		8.2	8.5	10.0	7.6
	R	7.2	9.4	11.8	9.1
		6.6	10.1	11.6	7.8

3. Table 9.23 contains data from a 3 × 3 factorial experiment that has equally spaced levels for both factors.

Table 9.23

Data for Exercise 3

		FACTOR C		
	Levels	**1**	**2**	**3**
Factor A	2	3.6	7.1	8.2
		4.5	7.3	6.0
		3.9	5.0	9.8
	4	7.7	9.8	10.3
		7.7	10.0	12.3
		7.7	8.6	8.6
	6	6.5	8.0	6.0
		8.5	6.5	6.9
		6.3	9.6	6.1

Table 9.24

Data for Exercise 4

Oven	COND FRESH	FROZEN	12–THAW	24–THAW
NOT PRE	21.15	25.29	22.38	24.01
	17.82	28.13	21.09	24.42
	23.68	25.87	29.59	23.45
	18.13	26.70	21.84	21.18
	26.72	26.51	26.28	22.53
PRE	19.59	31.05	26.20	22.43
	8.03	19.47	24.21	22.40
	17.78	29.97	23.00	21.71
	9.54	28.56	19.74	26.16
	18.54	24.88	29.10	26.72

4. In an era of increased awareness of the need for energy conservation, it is of interest to find out, when preparing a roast, what the consequences are of (1) thawing a frozen roast before cooking and (2) preheating the oven. The purpose of this 2×4 factorial experiment (data in Table 9.24) is to provide some answers to these questions. The factors and levels are:

   Factor 1 (OVEN): level 1, not preheated (NOT PRE)
   level 2, preheated (PRE)
   Factor 2 (COND): level 1, fresh meat (FRESH)
   level 2, frozen meat (FROZEN)
   level 3, frozen, thawed for 12 h (12-THAW)
   level 4, frozen, thawed for 24 h (24-THAW)

   Starting with 40 roasts of the same size and configuration, five were randomly assigned to each of the eight factor level combinations. Each roast was cooked until the internal temperature reached 160°F. The response is total fuel requirement (GAS).

Table 9.25

Data for Exercise 5

Concentration	FUNGICIDE A	B	C
100	0	0	0
	33	0	0
	0	20	0
	0	0	0
	0	0	0
1000	100	20	0
	40	20	0
	75	0	0
	100	0	50
	60	40	80

5. The data in Table 9.25 are the results of an experiment for studying the effectiveness of two concentrations (100 and 1000 ppm) of three fungicides

for controlling wilt in young watermelon plants. Five pots, having seeds infected with the wilt-causing fungus, were randomly assigned to each of the six factor level combinations. The response is the percent of germinated plants surviving to the 48th day. (*Hint:* Review Section 6.4.)

Table 9.26

Data for Exercise 6

NUTRIENT CONCENTRATION (ppm)								
0.0	**0.5**	**1.0**	**2.5**	**5.0**	**10.0**	**15.0**	**25.0**	**50.0**
3.00	8.00	4.25	3.63	12.33	10.50	16.00	32.75	39.00
7.50	9.00	6.66	8.33	7.00	17.00	24.75	26.50	27.75
3.50	9.50	11.50	12.50	17.50	11.75	31.50	35.00	41.00
15.67	9.75	21.41	15.00	7.67	29.75	31.25	30.66	38.00

6. The data in Table 9.26 do not arise from a factorial experiment but illustrate how a multiple regression model and a lack of fit test can be used. Thirty-six pine seedlings were randomly divided into nine treatment groups. Each group received a different concentration of a complete nutrient solution (in ppm) for a period of several weeks. The response is growth in millimeters during a two-week period. Note that the levels of nutrient are not equally spaced. A curve showing the response to the nutrient solution may be used to determine an economic optimum amount to use.

Table 9.27

Data for Exercise 7

	TEMPR				
Clean	**0.20**	**0.93**	**1.65**	**2.38**	**3.10**
0.0	6.50	6.80	2.55	1.89	1.59
	7.91	4.74	0.29	5.11	5.88
	5.20	7.27	0.39	5.10	1.23
0.5	7.00	8.80	14.60	16.70	10.79
	7.70	3.80	10.23	13.87	9.54
	6.88	10.76	20.68	14.78	12.67
1.0	4.59	31.60	21.70	39.02	26.71
	2.71	28.12	27.00	38.60	34.80
	5.25	27.06	28.83	46.50	31.81
1.5	11.47	39.15	75.41	79.95	59.21
	5.04	47.75	76.81	81.06	63.61
	8.89	41.89	76.15	96.53	60.27
2.0	22.07	77.68	136.79	152.45	93.95
	10.20	71.13	134.30	142.86	104.70
	21.19	82.81	137.74	151.92	112.47

7. The data in Table 9.27 deal with how the quality of steel, measured by ELAST, an index of quality, is affected by two aspects of the processing procedure:

CLEAN: concentration of a cleaning agent, and
TEMPR: an index of temperature and pressure.

The experiment is a 5 × 5 factorial with three independently drawn experimental units for each of the 25 factor level combinations. The factor levels are numeric but equally spaced for only one factor.

Table 9.28

Data for Exercise 8

Nitrogen Variety	60	90	120	150
		Location K		
N	4193	4681	4758	4463
L	5641	5544	6318	6297
B	6129	5697	6853	6457
		Location E		
N	1330	2642	2252	1715
L	4917	5466	4672	5680
B	1561	3088	2869	3957
		Location B		
N	3146	2806	3739	4681
L	2481	3514	3726	4076
B	3910	4015	3894	4870
		Location C		
N	3758	4167	4212	4293
L	4804	4480	4619	4048
B	4340	4024	4306	4479

8. The data in Table 9.28 deal with the effect of location, variety, and nitrogen application on rice yields. There are four locations (K, E, B, and C), three varieties (N, L, and B), and four levels of nitrogen (60, 90, 120, and 150). The response is mean yield of several replicated plots for each factor level combination, so we do not have an estimate of the true error variance.

Table 9.29

Data for Exercise 9

Temp	Day	Volume					
15	1	340	355	370	345	300	310
15	15	318	316	309	324	310	279
25	1	280	255	275	270	250	190
25	15	349	336	342	304	306	379
35	1	335	315	320	315	225	285
35	15	309	309	313	304	292	270

9. After an initial storage of one day at 10°C a sample of eggs was randomly divided into three groups to be stored at temperatures of 15, 25, and 35°C, respectively. Eggs were randomly taken out at 1 and 15 days of storage, the egg whites separated, and six angel food cakes prepared using the pooled egg whites from each storage regime. The volume of each cake is the response variable. The data are given in Table 9.29. Beware of an unexpected result.

Table 9.30

Heat Resistance of Potatoes

Temp	Var	Weights					
15	BUR	0.19	0.00	0.17	0.10	0.21	0.25
20	BUR	0.46	0.42	0.41	0.33	0.27	0.06
25	BUR	0.00	0.14	0.00	0.00	0.00	0.41
30	BUR	0.00	0.00	0.00	0.12	0.00	0.00
15	KEN	0.35	0.36	0.33	0.55	0.38	0.38
20	KEN	0.27	0.39	0.33	0.40	0.44	0.00
25	KEN	0.54	0.28	0.37	0.43	0.19	0.28
30	KEN	0.20	0.00	0.00	0.00	0.17	0.00
15	NOR	0.27	0.33	0.35	0.27	0.40	0.36
20	NOR	0.36	0.40	0.12	0.36	0.26	0.38
25	NOR	0.53	0.51	0.00	0.57	0.28	0.42
30	NOR	0.12	0.00	0.00	0.00	0.15	0.23
15	RLS	0.08	0.29	0.70	0.25	0.19	0.19
20	RLS	0.54	0.23	0.00	0.57	1.25	0.25
25	RLS	0.41	0.39	0.00	0.14	0.16	0.42
30	RLS	0.23	0.00	0.09	0.00	0.09	0.00

10. In a study of heat resistance of potato varieties, six plantlets of four varieties of potatoes were randomly assigned to each of four temperature regimes. Weights of tubers were recorded after 45 days. The resulting experiment is a 4×4 factorial. The data are shown in Table 9.30. Perform the analysis to determine the nature of differences in heat resistance among the varieties. Make recommendations indicated by the results.

11. The nutritive value of a diet for animals is not only a function of the ingredients, but also a function of how the ingredients are prepared. In this experiment three diet ingredients are denoted as factor GRAIN with levels

SORGH: whole sorghum grain,
LYSINE: whole sorghum grain with high lysine content, and
MILLET: whole millet.

Three methods of preparation are denoted as factor PREP with levels

WHOLE: whole grain,
DECORT: decorticated (hull removed), and
BSB: decorticated, boiled, and soaked.

Six rats were randomly assigned to each of 10 diets; the first 9 diets are the nine combinations of the two sets of three factor levels and diet 10 is a control diet. The response variable is biological value (BV). The data are shown in Table 9.31 and are available on the data disk in file FW09P11. Note that for diet 10, the factor levels are shown as blanks.

Perform the appropriate analysis to determine the effects of grain and preparation types. Note that this is a factorial experiment plus a control level. One approach is first to analyze the factorial and then perform a one-way for the 10 treatments with a contrast for control versus all others.

Table 9.31

Data for Exercise 9.11

TRT	GRAIN	PREP	Biological Value					
1	SORGH	WHOLE	40.61	56.78	69.05	39.90	55.06	32.43
2	SORGH	DECORT	74.68	56.33	71.02	53.35	41.43	33.00
3	SORGH	BSB	71.60	62.64	78.95	69.86	60.26	67.05
4	LYSINE	WHOLE	42.46	50.78	48.88	44.12	48.86	43.39
5	LYSINE	DECORT	50.11	57.46	55.36	57.28	51.60	53.96
6	LYSINE	BSB	60.57	62.62	66.20	54.32	47.11	41.56
7	MILLET	WHOLE	45.58	68.51	54.13	45.15	45.03	39.72
8	MILLET	DECORT	46.19	45.54	42.57	30.23	38.83	40.28
9	MILLET	BSB	64.27	56.48	73.24	67.18	51.11	32.97
10			87.77	91.80	81.13	80.88	66.06	73.36

12. In 1937 Raymond Haugh proposed a measurement based on albumen height and egg weight for assessing albumen quality. This measure is known as the Haugh value. In another phase of the experiment described in Exercise 9, 30 eggs were randomly chosen from storage day (DAY, levels 1 and 15) and temperature (TREAT, levels 15, 25, and 35°C) combination. Each egg was weighed (EGGWT) and opened, the albumen height measured (ALBHT), and the Haugh value (HAUGH) determined. The data consisting of 180 observations are on the data disk in file FW09P12.

Perform an analysis to determine how storage time and temperature affect the Haugh measure of the eggs. Also perform the same analysis on the albumen height and egg weight, and relate the results of these analyses to that of the Haugh measure.

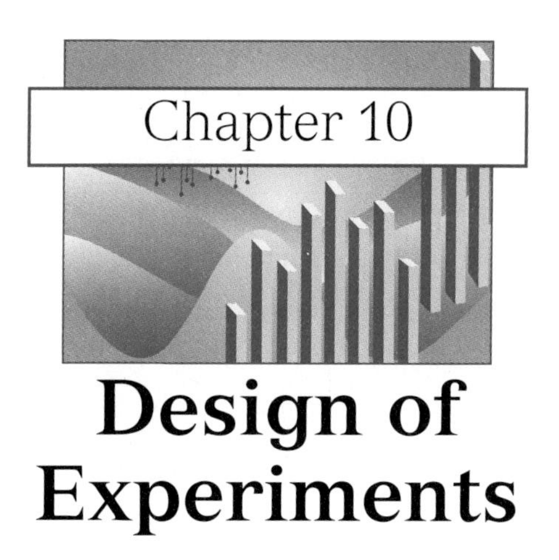

Design of Experiments

EXAMPLE 10.1

A Factorial Experiment with Different Plot Sizes We are interested in the yield response of corn to the following factors:

WTR: levels of irrigation with levels 1 and 2,
NRATE: rate of nitrogen fertilization with levels 1, 2, and 3, and
P: planting rates with levels 5, 10, 20, and 40 plants per experimental plot.

The response variable is total dry matter harvested (TDM).[1] The experiment is a $2 \times 3 \times 4$ factorial experiment. Because of physical limitations the experiment was conducted as follows:

- The experiment used four fields with 24 plots to accommodate all factor level combinations.
- Normally each of the 24 plots would be randomly assigned one factor level combination. However, because it is physically impossible to assign different irrigation levels to the individual plots, each field was divided in half and each half randomly assigned an irrigation level.
- The 12 factor levels of the other factors (NRATE and P) were randomly assigned to each half field.

A possible additional complication arises from the fact that the specified planting rates do not always produce that exact number of plants in each plot. Therefore the actual plants per plot are also recorded. For the time being, we will assume that this complication does not affect the analysis of the data.

[1]Other responses were measured, but are not presented here.

Table 10.1

Example of an Experimental Design

Source: Personal communication from R. M. Jones and M. A. Sanderson, Texas Agricultural Experiment Station, Stephenville, and J. C. Read, Texas Agricultural Experiment Station, Dallas.

		$P = 5$		$P = 10$		$P = 20$		$P = 40$	
WTR	**NRATE**	**NO**	**TDM**	**NO**	**TDM**	**NO**	**TDM**	**NO**	**TDM**
				REP = 1					
1	1	7	3.426	13	2.084	20	2.064	37	2.851
1	2	7	7.070	12	7.323	24	7.321	38	7.865
1	3	6	4.910	10	6.620	22	8.292	43	7.528
2	1	5	2.966	12	3.304	20	4.055	37	2.075
2	2	7	3.484	12	2.894	22	5.662	26	3.485
2	3	5	1.928	10	4.347	20	3.178	33	3.900
				REP = 2					
1	1	6	3.900	11	3.015	27	3.129	38	3.175
1	2	7	5.581	14	7.908	19	6.419	37	7.685
1	3	5	3.350	13	5.986	20	6.515	32	10.515
2	1	5	2.574	12	4.390	20	2.855	42	3.042
2	2	5	3.952	11	4.744	21	5.472	30	5.125
2	3	6	4.494	11	5.480	20	4.871	36	5.294
				REP = 3					
1	1	5	3.829	10	3.173	18	2.741	33	2.166
1	2	5	3.800	13	7.568	19	7.797	34	6.474
1	3	8	6.156	15	7.034	23	7.754	40	8.458
2	1	6	2.872	12	5.759	21	4.512	42	4.864
2	2	5	2.826	14	3.840	21	4.494	30	4.804
2	3	5	3.107	10	3.620	20	4.620	32	5.376
				REP = 4					
1	1	5	3.325	11	4.193	20	3.409	40	4.877
1	2	6	4.984	12	7.627	20	6.562	39	9.093
1	3	6	4.067	12	4.394	20	7.089	28	7.088
2	1	6	2.986	11	5.327	20	5.390	43	5.632
2	2	5	2.417	11	3.592	20	4.311	33	5.975
2	3	9	4.180	12	5.282	19	4.498	35	6.519

We will return to this problem in Chapter 11, Exercise 14 where the effect of the different number of plants in each plot will be examined. The data are shown in Table 10.1: The NRATE and WTR combinations are identified as rows, and the four sets of columns correspond to the four planting rates (P). The two entries in the table are the actual number of plants per plot (NO) and the total dry matter (TDM). The solution is presented in Section 10.6. ■

10.1 Introduction

DEFINITION 10.1
The **design of an experiment** is the process of planning and executing an experiment. While much of the planning of any experiment is technical relative to the discipline (choices of methods and materials), the results

and conclusions depend to a large extent on the manner in which the data are collected. The statistical aspect of experimental design is defined as the set of instructions for assigning treatments to experimental or observational units.

The objective of an experimental design is to provide the maximum amount of reliable information at the minimum cost. In statistical terms, the reliability of information is measured by the standard error of estimates. We know that the standard error of a sample mean is

- directly related to the population variance, and
- inversely related to sample size.

To increase the precision, we want to either reduce the population variance or increase the sample size. We normally take as big a sample as we can afford but it would seem that there is nothing we can do to reduce the population variance. However, it turns out that properly applied experimental designs may be used to effectively reduce that population variance.

For the completely randomized design described in Chapter 6, the within treatment mean square (MSW, also denoted by MSE) is used as the variance for computing standard errors of means. This quantity is the measure of the variation among units treated alike. In this context, this measure is known as the experimental error. Experimental designs structure data collection to reduce the magnitude of the experimental error.

The use of MSW as an estimate of the variance in a CRD assumes a population of units that has a variance of σ^2 everywhere. However, in many populations identifiable subgroups exist that have smaller variances. If we apply all treatments to each subgroup, the variation among units treated alike *within* each subgroup is likely to be smaller, thus reducing the error variance. Such subgroups are referred to as **blocks**, and the act of assigning treatments to blocks is known as **blocking**. Of course, if there is only one replication in a block, we cannot measure that variation directly, but we will see that if we have several blocks, the appropriate error can indeed be estimated. Most experimental designs are concerned with applications of blocking.

Usually data resulting from the implementation of experimental designs are described by linear models and analyzed by the analysis of variance. In fact, the use of blocking results in analyses quite similar to those of the analysis of factorial experiments.

Notes on Exercises

This chapter covers the analysis of a number of different experimental designs; after completing the coverage of a specific design, exercises using this design can be identified and worked. In order to make the exercises more realistic, they are not categorized as to design, although occasional hints are furnished.

10.2 The Randomized Block Design

One of the simplest and probably the most popular experimental design is the randomized block design, usually referred to as the RB design.[2] In this design the sample of experimental units is divided into groups or blocks and then treatments are randomly assigned to units in each block.

Remember that in the completely randomized design (Chapter 6), the variation among observed values was partitioned into two portions:

1. the **assignable** variation due to treatments and
2. the **unassignable** variation among units within treatments.

The unassignable variation among units is deemed to be due to natural or chance variation. It is therefore used as the basis for estimating the underlying population variance and is commonly called the experimental error. This is the statistic used as the denominator in the F ratios used to test for differences in population means and for computing standard errors of estimated population means.

Data resulting from a randomized block design have two sources of assignable variation:

1. the variation due to the treatments and
2. the variation due to blocks.

The remaining unassignable variation is used for estimating experimental error and is the variation among units treated alike within a block. If the blocks have been chosen to contain relative homogeneous units, this variation may be relatively small compared to that of a completely randomized design. In other words, in the RB design the assignable variation due to blocks is removed from the unassignable variation used in the CRD, thereby effectively reducing the magnitude of the estimated experimental error. This results in

- a decrease in the denominator in the F rations used to test for differences in means and
- a smaller estimate of the standard error of the means, thereby resulting in shorter confidence intervals on means.

Note, however, that although randomization of treatments and blocks is required, the randomization occurs after the units have been assigned to blocks. The procedure adds a restriction to the randomization process, which will be accounted for in the analysis and interpretation.

Criteria for the choice of blocks are most frequently different settings or environments for the conduct of the experiment. Examples of blocks may include

[2]Sometimes it is called the randomized complete block (RCB) design, to distinguish it from incomplete block designs (Section 10.5).

- subdivisions of a field,
- litters of animals,
- experiments conducted on different days,
- bricks cured in different kilns, or
- students taught by different instructors.

In any case, blocking criteria should be chosen so that the units within blocks are as homogeneous as possible.

Blocks may also be repetitions or replications of the experiment at another time or place. In such circumstances replications and blocks are synonymous. This is not, however, always the case. In some applications, blocks may be different subpopulations, such as different regions, but in such situations the blocks may more nearly represent a factor in a factorial experiment.

In many applications, an experiment will be conducted with only one application of each treatment per block. In this case, each block acts as one replication of the entire experiment; there are no units treated alike within blocks, and we must estimate the experimental error indirectly. However, even if there are multiple applications of treatments per block, the estimate of variance measuring the variation among units treated alike within blocks is not always the appropriate estimate of experimental error. Such situations are discussed in Section 10.3. Additional uses of blocking are presented in subsequent sections.

EXAMPLE 10.2

Table 10.2 shows hypothetical data that represent yields of three varieties of wheat planted according to a randomized block design with five blocks. The objective is to compare the yields of the varieties. In a field experiment of this type, the blocks are subdivisions of a field and the experimental units, called plots, are indeed small plots of the field in which one variety is planted. Usually the block is composed of a set of contiguous plots, which tend to be more alike or homogeneous than are a set of plots randomly selected from an entire field. The actual field layout would not necessarily look like that implied in Table 10.2 because

- the blocks need not be aligned in a row and
- the varieties are randomly arranged within blocks.

Note that in this experiment there is only one replication of treatments per block.

Table 10.2

Data on Wheat Yields for Randomized Block Design

	BLOCKS					
	1	2	3	4	5	Means
Variety A	31.0	39.5	30.5	35.5	37.0	34.70
B	28.0	34.0	24.5	31.5	31.5	29.90
C	25.5	31.0	25.0	33.0	29.5	28.80
						(grand mean)
Block means	28.17	34.83	26.67	33.33	32.67	31.13

The mean yields for varieties clearly indicate higher yields for variety A. However, the lowest two yields for variety A (30.5 in block 3 and 31.0 in block 1) are met or exceeded by five individual yields of the other varieties. Hence the apparent superiority of variety A does not appear to be clear-cut. However, if we examine the yields of the varieties within each individual block, we see that variety A is a clear winner in every case. Thus the relatively high overall yields in blocks 2, 4, and 5 may be causing varieties B and C in these blocks to have higher yields than variety A in blocks 1 and 3.[3] If the data from this experiment were to be analyzed as a completely randomized design, the variation we see among the blocks would be included in our estimate of the unassigned variation. The resulting variance would tend to be larger, thereby reducing the magnitude of the F statistic for testing the equality among mean yields. However, when we use the randomized blocks analysis, this variation among blocks is now assignable, thereby reducing the unassignable variation. The analysis that accomplishes this is shown later in this section. ■

The Linear Model

The data from a randomized block design can be described by a linear model that suggests the partitioning of the sum of squares and provides a justification for the test statistics. The linear model for the data from a randomized block design with each treatment occurring once in each block is

$$y_{ij} = \mu + \tau_i + \beta_j + \varepsilon_{ij},$$

where y_{ij} = observed response for treatment i in block j; μ = reference value, usually called the "grand" or overall mean; τ_i = effect of treatment i, $i = 1, 2, \ldots, t$; β_j = effect of block j, $j = 1, 2, \ldots, b$; and ε_{ij} = experimental (random) error.

If the block and treatment effects are fixed (see below) we add the restriction

$$\sum \tau_i = \sum \beta_j = 0,$$

in which case μ represents the mean of the population of experimental units.

This model certainly looks like that of the factorial experiment with factors now called treatments and blocks. There is only one replication per cell; hence the interaction mean square is used as the estimate of error (Section 9.5). This analogy is not incorrect. The procedures for the partitioning of the sums of squares and the construction of the analysis of variance table are identical for both cases and are therefore not reproduced here. However, the parameters, especially those involving the blocks, have different implications for the randomized block model.

Generally the blocks in an experiment are considered a random sample from a population of blocks. For that reason, the block parameters β_j represent

[3] We should note at this point that the paired t test (Chapter 5) is actually a randomized block design for two treatments. The use of the randomized block analysis in this chapter provides the equivalent inferences.

Table 10.3

Expected Mean Squares: Randomized Block

Source	df	E (MS)
Treatments	$t - 1$	$\sigma^2 + [b/(t-1)]\sum\tau^2$
Blocks	$b - 1$	$\sigma^2 + t\sigma_\beta^2$
Error	$(t-1)(b-1)$	σ^2

a random effect[4]; that is, they are random variables with mean zero and variance σ_β^2. As noted in Section 6.6, the inferences for a random effect are to the variation among the units of that population. However, the inference on the treatment effects, the τ_i, which are usually fixed, is on the specific set of treatment parameters present in the particular experiment.

The model for the randomized block design contains both fixed and random effects and is an example of a mixed model. In some cases, hypothesis tests and other inferences for a mixed model are different from those of a fixed or random model even though the analysis of variance partitioning is identical.

The importance of the distinction between random and fixed effects is seen in the definition of the experimental error ε_{ij}:

- In the (fixed) model for the factorial experiment, expected mean squares showed that the interaction mean square is an estimate of the experimental error *plus* the interaction effect and is therefore not suitable as an estimate of the experimental error.
- In the mixed model for the randomized block design, the interaction between treatments and blocks measures the inconsistency or variation among treatments effects across the population of blocks.

When blocks are random, this interaction is a random effect and is the measure of the uncertainty of the inferences about the treatment effects based on the sample of blocks. This is why it is called the **experimental error**, and the corresponding mean square is used as the estimate of the variance for hypothesis tests and interval estimates on the treatment effects.

The use of this interaction as the estimate of the error for hypothesis tests is supported by the expected mean squares for this analysis, which are given in Table 10.3. The following features are of interest:

- σ_β^2 and σ^2 are the variances of the (random) block and experimental error effects, respectively.
- The test for H_0: $\sum\tau_i^2 = 0$ is provided by the test statistic

$$F = \frac{\text{treatment mean square}}{\text{error mean square}},$$

which is the same as for the test for a main effect in a factorial experiment with no replications.

[4]It is possible for blocks to be fixed. However, in such cases the blocks are more nearly like the fixed levels of a factor, and the data more appropriately considered as arising from a factorial experiment and analyzed by the methods described in Chapter 9.

- We may also test H_0: $\sigma_\beta^2 = 0$ by the test statistic

$$F = \frac{\text{blocks means square}}{\text{error mean square}}.$$

This test, however, is not overly useful and is considered by some as not strictly valid (Lentner *et al.*, 1989). The value of the F statistic is, however, related to the relative efficiency of the randomized block design discussed below.

We can see that the analysis of the mixed model representing the randomized block design is the same as that for the fixed model representing a factorial experiment. There are some changes in names and a somewhat different interpretation of the inference about the variance of the block effect, but the end product appears identical. It is important to remember that this similarity is deceptive and does not apply to all cases of mixed models. We will see later that it is important to know which effects are random and which are fixed.

Solution Example 10.2 concerns some hypothetical wheat yields for three varieties in a randomized block design with five blocks. The data are shown in Table 10.2.

The computations for the sums of squares and the construction of the analysis of variance table are identical to those of a two-factor factorial experiment with the t treatments corresponding to factor A and the b blocks corresponding to factor C.

The partitioning of the sum of squares is

$$\begin{aligned}
\text{TSS} &= \sum_{ij}(y_{ij} - \bar{y}_{..})^2\\
&= (31.0 - 31.13)^2 + (28.0 - 31.13)^2 + \cdots + (29.5 - 31.13)^2\\
&= 261.733\\
\text{SS(Treatments)} &= b\left[\sum_i(\bar{y}_{i.} - \bar{y}_{..})^2\right]\\
&= 5[(34.70 - 31.13)^2 + (29.90 - 31.13)^2 + (28.80 - 31.13)^2]\\
&= 98.433\\
\text{SS(Blocks)} &= t\left[\sum_j(\bar{y}_{.j} - \bar{y}_{..})^2\right]\\
&= 3[(28.17 - 31.13)^2 + (34.83 - 31.13)^2 + \cdots\\
&\quad + (32.67 - 31.13)^2]\\
&= 148.90\\
\text{SS(Error)} &= \text{TSS} - \text{SS(Treatments)} - \text{SS(Blocks)}\\
&= 261.733 - 98.433 - 148.90\\
&= 14.400.
\end{aligned}$$

Table 10.4

Analysis of Variance

Source	df	SS	MS	F
Varieties	2	98.433	49.217	27.34
Blocks	4	148.900	37.225	20.68
Error	8	14.400	1.800	
Total	14	261.733		

The results are summarized in the tabular analysis of variance format, which is given in Table 10.4. Using the experimental error (1.800) as the denominator, the F ratios from this table lead to rejection ($\alpha < 0.01$) of the null hypothesis of no variety differences as well as the hypothesis of zero block variance.

The lack of treatment structure suggests the use of a post hoc paired comparison procedure to obtain more specific information on treatment differences. The reader may want to verify that the Duncan multiple range test using $s^2 = 1.800$ with 8 degrees of freedom and $\alpha = 0.01$ (Section 6.5, see discussion on post hoc comparisons) requires differences in mean variety yield of 1.95 and 2.04 to detect differences involving two and three means, respectively. Using the means given in Table 10.2, we see that variety A can be said to have a significantly higher yield but we cannot distinguish between the yields of the other two varieties. ■

Relative Efficiency

Having implemented a randomized block design, it is appropriate to ask whether the use of this design did indeed provide for a more powerful test than would have been produced by a completely randomized design. After all, the randomized block design does require somewhat more planning, more careful execution, and somewhat more computing than does the CRD. Further, the RB design has one additional disadvantage in that the error mean square has fewer degrees of freedom; hence a larger p value will result for a given value of the F ratio for the test of treatment effects. Thus, for a given magnitude of treatment difference, the randomized blocks design must provide a smaller variance estimate to maintain a given level of significance.

A formal comparison of the magnitudes of the error mean squares is provided by the **relative efficiency** of the randomized block design, which is obtained as follows:

1. Estimate the error variance that would result from using a completely randomized design for the data. Using the results of the RB analysis this is

$$s_{\text{CR}}^2 = \frac{(b-1)\text{MS}_{\text{blocks}} + [b(t-1)]\text{MS}_{\text{error}}}{bt-1}.$$

2. Compute the relative efficiency

$$\text{RE} = \frac{s_{\text{CR}}^2}{s_{\text{RB}}^2},$$

where s^2_{RB} is the error mean square for the randomized block design. The result indicates how many replications of a CR design are required to obtain the power of the RB design.

3. As we have noted, the advantage accruing to the randomized block design may be compromised by a reduction in the degrees of freedom for estimating the experimental error. Although this reduction causes a loss in efficiency, the loss is usually so small that it may be ignored. A correction factor to be used, especially when the degrees of freedom for the RB error are small (say, <10), is available in Steel and Torrie (1980, Section 9.7).

For our example,

$$s^2_{\text{CR}} = \frac{4(37.225) + 10(1.800)}{14} = 11.921.$$

Then

$$\text{RE} = \frac{11.921}{1.800} = 6.62.$$

Hence over six times as many replications, that is, an experiment using more than 90 experimental units, would be required using the CRD to obtain the same results as the RB design, although the loss of error degrees of freedom from 12 to 8 will minimally decrease the efficiency. It is seen that the use of blocking was quite effective in this case. However, blocking is not always this effective.

Random Treatment Effects in the Randomized Block Design

We noted in Section 6.6 that treatments may represent a random sample for a population of treatments. In such a situation the treatment effects are random, and if they occur in a randomized block design with random block effects, the resulting linear model is a random effects model. The model is

$$y_{ij} = \mu + \tau_i + \beta_j + \varepsilon_{ij},$$

where μ, β_j, and ε_{ij} are as defined in the previous section, and τ_i represents a random variable with mean zero and variance σ^2_τ. The expected mean squares for the analysis of variance are

$$E(\text{MS}_{\text{treatment}}) = \sigma^2 + b\sigma^2_\tau,$$
$$E(\text{MS}_{\text{blocks}}) = \sigma^2 + t\sigma^2_\beta,$$
$$E(\text{MS}_{\text{error}}) = \sigma^2.$$

From these formulas we can see that the analyses for determining the existence of the random treatment and block effects are exactly the same for the mixed or fixed effects models. Of course, the results are interpreted differently. Since the focus is on the variances of the effects the use of multiple comparisons is not logical. The variance components can be estimated by equating the expressions for the expected mean squares to the mean squares obtained by the analysis in the same way as outlined in Section 6.6.

10.3 Randomized Blocks with Sampling

In some experiments blocks may be of sufficient size to allow several units to be assigned to each treatment in a block. Such replication of treatments is referred to as randomized blocks with sampling. Data from such an experiment provide two sources of variation that may be suitable for an estimate of the error variance. The linear model for data from such an experiment is

$$y_{ijk} = \mu + \tau_i + \beta_j + \varepsilon_{ij} + \delta_{ijk},$$

where y_{ijk} = observed value of the response variable in the kth replicate of treatment i in block j; μ = reference value or overall mean; τ_i = fixed effect of treatment i, $i = 1, 2, \ldots, t$; β_j = effect of block j, $j = 1, 2, \ldots, b$, a random variable with mean zero and variance σ_β^2; ε_{ij} = experimental error (as defined in Section 10.2), a random variable with mean zero and variance σ^2; and δ_{ijk} = sampling error, which is the measure of variation among units treated alike within a block, a random variable with mean zero and variance σ_δ^2, $k = 1, 2, \ldots, n$.[5]

The partitioning of the sums of squares and the construction of the analysis of variance table is identical to that for the analysis of the two-factor factorial experiment (Table 9.4), substituting treatments and blocks for factors A and C. As usual, the justification for the appropriate test statistics for the analysis of data from this design is determined by examining the expected mean squares for the analysis of variance. Assuming fixed treatment and random block effects, the analysis of variance and expected mean squares are shown in Table 10.5.

Table 10.5

Analysis of Variance for Randomized Block with Sampling

Source	df	Mean Square	Expected Mean Square
Treatments	$t-1$	MS (Treatments)	$\sigma_\delta^2 + n\sigma^2 + \frac{nb}{t-1}\sum\tau^2$
Blocks	$b-1$	MS (Blocks)	$\sigma_\delta^2 + n\sigma^2 + nt\sigma_\beta^2$
Exp. error	$(t-1)(b-1)$	MS (Exp. error)	$\sigma_\delta^2 + n\sigma^2$
Samp. error	$tb(n-1)$	MS (Samp. error)	σ_δ^2

Up to this point we have become accustomed to use the "bottom line" in the analysis of variance table as the denominator for all hypothesis tests. We will see that this is not correct for this case when we review the basic principle of hypothesis testing.

In Sections 6.3 and 9.3 we noted that the principles of hypothesis testing in the analysis of variance require the following:

[5] As was the case for factorial experiments, here we consider only balanced data.

- If H_0 is true, the numerator and denominator of an F ratio used for a hypothesis test should both be estimates of the same variance (or function of variances).
- If H_0 is not true, the numerator should include, in addition to the estimate of the variance, a positive function involving only those parameters specified in the hypothesis. This function is called the noncentrality parameter of the test, and should have the property that its magnitude increases with larger deviations from the null hypothesis.

We can now see that the ratio we would normally use, that is, MS(Treatments)/MS(Samp. error), provides the test for

$$H_0\colon \left(n\sigma^2 + \frac{nb}{t-1}\sum \tau^2\right) = 0.$$

This is not a particularly useful hypothesis as it provides for a simultaneous test for both treatment effects and the experimental error. However, the test resulting from the ratio MS(Treatments)/MS(Exp. error) provides the test for

$$H_0\colon \left(\frac{nb}{t-1}\sum \tau^2\right) = 0,$$

which is the desired hypothesis for treatment effects. Similarly, the test MS(Exp. error)/(MS Samp. error) provides the test for H_0: $\sigma^2 = 0$ and, if desired, MS(Blocks)/MS(Exp. error) provides the test for block effects.

The distinction between the experimental and sampling errors seen in the model can also be explained by reviewing the sources of the variation and the purpose of the inference:

- The **experimental error** measures the variability among treatment responses across a random sample of blocks. If this had been a factorial experiment, this would in fact be the interaction between blocks and treatments. Since the primary purpose of our inference is to estimate the behavior of the responses for the population of blocks, this source of variation is the correct measure of the uncertainty of this inference.
- The **sampling error** measures the variability of treatment responses within blocks. Since we try to choose blocks that will be relatively homogeneous, this variation may not represent the variability of treatment effects in the population, and is therefore not always the proper error to use for such inferences.
- This is the point at which we make a distinction between **experimental units** and **sampling units**. In Section 1.2 we introduced the concept of experimental units and heretofore we have called any observational unit an experimental unit. For this design the two are not the same. Instead, the experimental units are blocks, and the observational units, called sampling units, are the individual observations within blocks. The distinction occurs because inferences are made on the effects of treatments on the population

of blocks rather than individuals. Sampling units do provide useful information, but inferences are normally not made for these units.

- Just because we do not use the sampling error for tests on treatment effects, it does not mean that having samples is not useful. Note that the magnitude of the noncentrality parameter in the expected mean square for treatment effects increases with n; hence, increasing the number of sample units will tend to magnify the effect of nonzero treatment effects and thereby increase the power of the test.
- Sometimes both the sampling and experimental errors do measure the experimental error. Effectively, then, $\sigma^2 = 0$, and most likely the hypothesis H_0: $\sigma^2 = 0$ will not be rejected. If this has occurred, we may pool the two mean squares and use the resulting pooled mean square as the denominator for F ratios, thus providing more degrees of freedom for the denominator and consequently a more powerful test. However, since failing to reject a null hypothesis does not necessarily imply accepting that hypothesis, pooling is not a universally accepted practice. Pooling may be made more acceptable if the significance level for that test is increased to, say, 0.25 or greater (Bancroft, 1968).
- Other distinctions between experimental and observational units may arise in this type of design (Section 1.2). For example, the replications within blocks may consist of repeated measurements on the same experimental units, or measurements on subunits of the original experimental units. This may occur, for example, in the determination of the radioactivity of a sample of material, where the replications may consist of repeated readings or determinations of the same unit. Such situations do not necessarily invalidate the analysis we outline here, but care must be taken to properly interpret the so-called sampling error.
- If block effects are fixed, the interaction is also fixed and the expected mean squares are those for the two-factor factorial experiment (Section 9.3), and F ratios for all tests use the sampling error in the denominator. If both treatments and blocks are random, the analysis is the same as for the random model with interpretation as outlined in Section 10.2 where random treatment effects are discussed.

EXAMPLE 10.3

We are interested in the stretching ability of different rubber materials as measured by stress at 600% elongation of the materials. Since different testing laboratories often produce different results, four samples of each of seven materials were sent to a sample of 13 laboratories (Mandel, 1976). The data are given in Table 10.6.

Solution In this experiment, the laboratories are the blocks and the materials are the treatments. Manual computations for a data set this large are not feasible, and we simply present the analysis of variance produced by the SAS System as shown in Table 10.7. Note that in this output the sampling error mean square obtained from the analysis for the `Model` is used as the denominator for all F ratios. Virtually all computer programs will do this because, without

Table 10.6

Data on Rubber Stress

	MATERIAL						
Lab	**A**	**B**	**C**	**D**	**E**	**F**	**G**
1	72.0	133.0	37.0	63.0	35.0	31.0	43.0
	79.0	129.0	36.0	49.0	26.0	32.0	40.0
	61.0	123.0	26.0	63.0	24.0	28.0	35.0
	71.0	156.0	24.0	43.0	61.0	26.0	38.0
2	61.0	129.0	20.0	51.0	27.0	22.0	32.0
	49.0	125.0	14.0	52.0	27.0	20.0	29.0
	57.0	136.0	30.0	62.0	26.0	29.0	45.0
	61.0	127.0	27.0	52.0	26.0	28.0	40.0
3	70.0	121.0	33.0	58.0	28.0	27.0	44.0
	62.0	125.0	33.0	64.0	28.0	30.5	44.0
	62.0	109.0	27.0	56.0	27.0	27.0	45.0
	76.0	128.0	29.5	55.0	29.0	27.0	49.0
4	36.0	57.0	27.0	38.0	22.0	22.0	31.0
	39.0	58.0	24.0	38.0	23.0	23.0	31.0
	41.0	59.0	22.0	37.0	20.0	22.0	28.0
	45.0	67.0	25.0	38.0	20.0	22.0	30.0
5	58.0	122.0	34.0	53.0	25.0	26.0	43.0
	57.0	98.0	27.0	47.0	25.0	25.0	35.0
	58.0	107.0	26.0	48.0	21.0	22.0	43.0
	53.0	110.0	26.0	47.0	19.0	18.0	36.0
6	52.0	109.0	30.0	50.0	25.0	24.0	38.0
	56.0	120.0	31.0	50.0	25.0	26.0	41.0
	52.0	112.0	31.0	50.0	26.0	25.0	40.0
	50.0	107.0	28.0	51.0	26.0	26.0	43.0
7	40.7	80.0	26.5	38.8	23.0	22.2	29.4
	45.9	71.9	27.1	39.4	22.9	23.9	31.6
	43.1	75.8	26.6	40.7	22.5	22.6	29.6
	37.3	63.7	25.6	38.0	35.7	25.5	29.3
8	68.1	135.0	38.1	64.5	32.1	32.7	50.2
	69.8	151.0	37.4	65.7	35.2	32.4	50.4
	65.9	143.0	37.9	64.0	33.0	30.3	42.5
	62.1	142.0	37.1	62.5	34.9	35.6	45.0
9	46.0	69.0	26.0	40.0	24.0	23.0	32.0
	47.0	69.0	26.0	38.0	24.0	24.0	31.0
	46.0	73.0	25.0	39.0	24.0	24.0	32.0
	45.0	70.0	25.0	39.0	25.0	23.0	30.0
10	77.0	132.0	45.0	71.0	36.0	38.0	56.0
	74.0	129.0	41.0	69.0	33.0	36.0	48.0
	77.0	141.0	39.0	66.0	35.0	38.0	48.0
	72.0	137.0	38.0	68.0	25.0	38.0	50.0
11	76.0	118.0	27.0	52.0	22.0	23.0	32.0
	55.0	109.0	32.0	45.0	19.0	23.0	37.0
	60.0	115.0	26.0	48.0	18.0	23.0	37.0
	58.0	106.0	26.0	54.0	23.0	24.0	39.0
12	72.5	133.0	32.5	63.0	31.2	30.7	45.8
	76.0	133.0	32.8	64.5	30.2	30.8	45.2
	69.5	128.5	32.9	61.5	29.0	30.0	43.5
	70.5	128.5	34.6	62.7	29.7	29.5	46.5
13	51.0	86.0	24.0	45.0	21.8	24.0	33.0
	50.0	84.0	24.0	43.0	21.8	24.0	33.0
	49.0	96.0	24.0	42.0	24.0	22.0	31.0
	49.0	81.0	26.0	45.0	22.0	24.0	31.0

Table 10.7 Analysis of Variance for Rubber Data

The ANOVA Procedure

Dependent Variable: STRESS

Source	DF	Sum of Squares	Mean Square	F Value	Pr > F
Model	90	322913.2482	3587.9250	177.01	<.0001
Error	273	5533.5800	20.2695		
Corrected Total	363	328446.8282			

R-Square	Coeff Var	Root MSE	STRESS Mean
0.983152	9.253783	4.502169	48.65220

Source	DF	Anova SS	Mean Square	F Value	Pr > F
LAB	12	30328.0547	2527.3379	124.69	<.0001
MATERIAL	6	268778.0771	44796.3462	2210.03	<.0001
LAB*MATERIAL	72	23807.1165	330.6544	16.31	<.0001

Tests of Hypotheses Using the Anova MS for LAB*MATERIAL as an Error Term

Source	DF	Anova SS	Mean Square	F Value	Pr > F
MATERIAL	6	268778.0771	44796.3462	135.48	<.0001

special instructions, they do not know whether the data are from a factorial experiment or a randomized block design. However, special options are normally available for performing the correct tests. Such an option is implemented in this case, producing the test at the bottom of the output. As indicated, using the appropriate error terms, namely the mean square for LAB*MATERIAL, we can reject the hypothesis of no MATERIAL effect with $p < 0.001$.

Since there is no additional information on the materials, a post hoc multiple comparison is indicated. We will use Duncan's multiple range test here with results given in Table 10.8. To produce these results, the computer program was instructed to use the correct error variance (LAB*MATERIAL mean square), which is evidenced by the notation MSE = 330.6544. The conclusion is that material B definitely has the highest mean stress with C, E, and F having the lowest and no distinction among these three.

The relative efficiency is computed as given in Section 10.2. The reconstituted error variance for the completely randomized design is

$$s_{\text{CR}}^2 = \frac{(b-1)\text{MS}_{\text{blocks}} + [b(t-1)]\text{MS}_{\text{error}}}{bt-1}.$$

For this example this quantity is

$$[12(2527.3) + 78(330.65)]/90 = 623.54.$$

The relative efficiency, then, becomes

$$\text{RE} = \frac{s_{\text{CR}}^2}{s_{\text{RB}}^2} = \frac{623.54}{330.7} = 1.89,$$

Table 10.8 Duncan's Multiple Range Test

```
                          The ANOVA Procedure

                Duncan's Multiple Range Test for STRESS

 NOTE: This test controls the Type I comparisonwise error rate, not the
                      experimentwise error rate.

                      Alpha                          0.05
                      Error Degrees of Freedom         72
                      Error Mean Square          330.6544

Number of Means              2        3        4        5        6        7
Critical Range           7.109    7.480    7.725    7.904    8.042    8.153

      Means with the same letter are not significantly different.

      Duncan Grouping         Mean          N    MATERIAL

                    A      108.988         52    B

                    B       58.296         52    A
                    B
                    B       51.621         52    D

                    C       38.692         52    G

                    D       29.435         52    C
                    D
                    D       26.885         52    E
                    D
                    D       26.648         52    F
```

which means that about twice as many observations would be needed for a completely randomized design to obtain the same degree of precision. In this case, however, a completely randomized design would actually be more difficult to implement. ■

10.4 Latin Square Design

Some experimental situations may have more than one factor that may be used for blocking. Consider an experiment that examines the effect of different working conditions, such as types of background music, on productivity in a manufacturing plant. We know that productivity may be affected by time of day as well as by day of the week. We could define every combination of time of day and day of the week as a block, but that would make for a very large experiment, for example, a four-treatment experiment (four types of music), using four daily time periods and four working days as blocks would require 64 experimental units for a single replication.

The Latin square design is specifically constructed for this type of situation. It is based on the so-called Latin square matrix. For example, a 4×4 Latin square is

A B C D

B C D A

C D A B

D A B C

In a Latin square matrix each row and column contains each of the letters A, B, C, and D once and only once. The Latin square design uses the "columns" as one blocking factor, "rows" as the other, and "letters" as the treatment designation. Thus in this design each treatment occurs once and only once in each of the blocking factor level combinations. Of course, the experiment is restricted to having an equal number of row, column, and treatment levels.

A Latin square experiment is conducted as follows:

1. Construct the Latin square for the size of the experiment.
2. Randomly permute the order of the rows.
3. Randomly permute the order of columns.
4. Randomly assign treatments to the letters.

For ease of presentation of the data, the rows and columns are usually presented in the "normal" sequencing, that is, ignoring the randomization. For example, the production experiment could be conducted as

Day of the Week	8 A.M.	10 A.M.	1 P.M.	2 P.M.
Mon	B	C	A	D
Tue	A	B	D	C
Wed	C	D	B	A
Thur	D	A	C	B

where A, B, C, and D are treatments 2, 3, 4, and 1, respectively. Thus treatment 3 will be conducted Monday from 8 to 10 A.M., treatment 4 on Monday from 10 to 12 A.M., and so forth.

The linear model for the Latin square is

$$y_{ijk} = \mu + \rho_i + \gamma_j + \tau_k + \varepsilon_{ijk},$$

where y_{ijk} = observed response of treatment k in row i and column j; μ = reference value or overall mean; ρ_i = effect of row i, $i = 1, 2, \ldots, t$; γ_j = effect of column j, $j = 1, 2, \ldots, t$; τ_k = effect of treatment k, $k = 1, 2, \ldots, t$; and ε_{ijk} = random error.

As before, the blocking effects are usually considered random with means zero and variance σ_ρ^2 and σ_γ^2, respectively, and we add the restriction $\sum \tau_i = 0$ for the fixed treatment effects. Blocking effects may be fixed, but the outline of the analysis is not changed.

There are no interaction terms in the model because interactions between row, column, and treatment effects constitute a violation of assumptions underlying the use of this design. Since violations of this assumption are difficult to detect, care must be taken to use this design only when such an interaction is not expected to exist.

The partitioning of sums of squares for the analysis of variance is relatively straightforward. The row, column, and treatment sums of squares are computed using the respective means and the error sum of squares is obtained by subtraction. The table of expected mean squares (which is not reproduced here) shows that the error mean square obtained in this manner is indeed the proper denominator for F ratios, assuming the assumption of no interaction effects holds. Most computer programs simply require specifying both blocking and experimental factors as sources of variation and the default F statistic uses the residual mean square, which is the correct error term.

EXAMPLE 10.4

This example of a Latin square design concerns a hypothetical experiment on the effect of various types of background music on the productivity of workers in a plant. It is well known that productivity differs among the various days of the week as well as during different times of day. Hence we design the experiment as a Latin square with hours of the day as rows and days of the week as columns. There are five music "treatments":

A: rock and roll
B: country/western
C: easy listening
D: classical
E: no music

The row blocks are five 1-h periods and the column blocks are the five working days. Note that these blocking factors are indeed fixed. The response is the number of parts produced. The data are given in Table 10.9 where the indicator for treatment is given under each response value. Note that the presentation of the data does not show the randomization.

Solution The computations for the analysis of variance are performed as if each factor (rows, columns, and treatments) is a main effect. The error sum

Table 10.9

Data for Latin Square Design

			DAY			
Times	**Mo**	**Tu**	**We**	**Th**	**Fr**	**Means**
9–10	6.3	9.8	14.3	12.3	9.1	10.36
	(A)	(B)	(C)	(D)	(E)	
10–11	7.7	13.5	13.4	12.6	9.9	11.42
	(B)	(C)	(D)	(E)	(A)	
11–12	11.7	10.7	13.8	9.0	10.3	11.10
	(C)	(D)	(E)	(A)	(B)	
1–2	9.0	10.5	9.3	9.8	12.0	10.12
	(D)	(E)	(A)	(B)	(C)	
2–3	4.5	5.3	8.4	9.6	11.0	7.76
	(E)	(A)	(B)	(C)	(D)	
						(overall)
Means	7.84	9.96	11.84	10.66	10.46	10.15
Treatment	A	B	C	D	E	
Means	7.96	9.20	12.22	11.28	10.10	

of squares is obtained by subtraction of all factor sums of squares from the total sum of squares. For this example,

$$\begin{aligned}
\text{TSS} &= (6.3 - 10.15)^2 + (9.8 - 10.15)^2 + \cdots + (11.0 - 10.15)^2\\
&= 154.362\\
\text{SS(Times)} &= 5[(10.36 - 10.15)^2 + (11.42 - 10.15)^2 + \cdots + (7.76 - 10.15)^2]\\
&= 41.362\\
\text{SS(Days)} &= 5[(7.84 - 10.15)^2 + (9.96 - 10.15)^2 + \cdots + (10.46 - 10.15)^2]\\
&= 42.922\\
\text{SS(Music)} &= 5[(7.96 - 10.15)^2 + (9.20 - 10.15)^2 + \cdots + (10.10 - 10.15)^2]\\
&= 56.314.
\end{aligned}$$

The error sum of squares is obtained by subtraction:

$$\begin{aligned}
\text{SSE} &= \text{TSS} - \text{SS(Times)} - \text{SS(Days)} - \text{SS(Music)}\\
&= 13.763.
\end{aligned}$$

The degrees of freedom are 4 for each of the factors, and, by subtraction, 12 for error.

The results are summarized in the usual analysis of variance format in Table 10.10. The F values show that the hypotheses of no treatment, row, or column effect are all rejected ($\alpha < 0.05$). When ranked from high to low, the treatment means are almost equally spaced. In this case post hoc paired comparisons may be appropriate, unless there is some prior information on specific music effects. Therefore, we may use the Duncan multiple range test, with $s^2 = 1.147$, five observations per treatment, and 12 degrees of freedom for s^2. As seen in Table 10.11 no adjacent means may be declared different, but all other pairwise comparisons show differences. We conclude that

Table 10.10

Analysis of Variance for Latin Square

DEPENDENT VARIABLE: PROD

Source	df	Sum of Squares	Mean Square	*F* Value	Pr > *F*
Model	12	140.5992000	11.7166000	10.22	0.0002
Error	12	13.7632000	1.1469333		
Corrected total	24	154.3624000			

R-Square	C.V.	Root MSE	PROD Mean
0.910838	10.54915	1.070950	10.1520000

Source	df	ANOVA SS	Mean Square	*F* Value	Pr > *F*
DAY	4	42.92240000	10.73060000	9.36	0.0011
TIME	4	41.36240000	10.34060000	9.02	0.0013
MUSIC	4	56.31440000	14.07860000	12.27	0.0003

Table 10.11

Duncan's Multiple Range Test for Latin Square

Alpha = 0.05 df = 12 MSE = 1.146933

Number of means	2	3	4	5
Critical range	1.473	1.543	1.590	1.614

Means with the same letter are not significantly different.

Duncan Grouping		Mean	N	MUSIC
	A	12.220	5	C
	A			
B	A	11.280	5	D
B				
B	C	10.100	5	E
	C			
D	C	9.200	5	B
D				
D		7.960	5	A

treatments C and D (easy listening and classical) produce higher productivity while treatments A and B (rock and roll and country/western) produce lower productivity, but we cannot pick within these pairs.

As previously noted, this analysis assumes that there are no interactions among any of the sources of variation. It could, for example, be argued that different types of music may have different effects at different times of day, which constitutes an interaction that would invalidate this analysis.

Unfortunately, there is no test for the hypothesis of no interaction using these data.

The relative efficiency of the Latin square design may be computed using the approach outlined for the randomized block design. These efficiencies can specify the gain in efficiency due to blocking by rows, or columns, or the entire Latin square. For details, see Kuehl (2000). ■

10.5 Other Designs

The randomized block and Latin square designs use the principle of blocking for the purpose of increasing the precision of the analysis. Other, more complex designs, such as the Graeco–Latin square design can be used to eliminate more than two blocking factors (see, for example, Montgomery, 1984, Chapter 5).

Experimental designs can be used to accommodate more complex experimental situations such as factorial experiments. That is, the treatments in a randomized block design may consist of all factor level combinations of a factorial experiment. This application is presented at the beginning of this section.

Another application of experimental design occurs when experimental units contain subunits that are then used as observations. This so-called nested design is outlined later in this section. A factorial experiment requiring experimental units or blocks of different sizes for different factors, called a "split plot" design, is also outlined in this section, along with other considerations of experimental design.

Factorial Experiments in a Randomized Block Design

At this point it may be difficult to differentiate the analysis of a randomized block *design* and a factorial *experiment* because they both result in the same partitioning of the sum of squares. However there are important differences:

- The randomized blocks *design* is concerned with assigning treatments to experimental units in a way that reduces the experimental error. In the analysis, the block effect is a nuisance source of variation that we want to eliminate from the estimate of the experimental error, and the interaction between blocks and treatment is the experimental error.
- The factorial *experiment* is concerned with a factorial structure of the treatments. In the analysis we are interested in determining the effect of each individual factor and the interaction between factors.

We can see the difference when we consider a factorial *experiment* in a randomized block *design*. The conduct of the experiment as well as the analysis of the resulting data is more easily understood if the experiment is considered in two stages. We consider here an A × C factorial experiment with a levels of factor A and c levels of factor C in a randomized block design with b blocks.

Stage One Construct a randomized block design with b blocks and all $a \times c$ factor level combinations as treatments. The analysis of this first stage provides the following partitioning of sums of squares[6]:

Source of Variation	df
Treatments	$ac - 1$
Blocks	$b - 1$
Experimental error	$(ac - 1)(b - 1)$

Stage Two In the second stage, the treatment sum of squares with $(ac - 1)$ degrees of freedom is partitioned according to the factorial structure as presented in Chapter 9, resulting in the following partitioning:

Source of Variation	df
Main effect A	$a - 1$
Main effect C	$c - 1$
Interaction: A × C	$(a - 1)(c - 1)$

[6]It is common practice to give only sources of variation and degrees of freedom when outlining the appropriate analysis of variance.

Final Stage Combine the results of the two stages, which results in the final analysis of variance partitioning:

Source of Variation	**df**
Blocks	$b - 1$
Main effect A	$a - 1$
Main effect C	$c - 1$
Interaction: A × C	$(a - 1)(c - 1)$
Experimental error	$(ac - 1)(b - 1)$

It now becomes clear that the experimental error from the randomized blocks *design* is used for the tests of all effects of the factorial experiment. Note that the $(ac - 1)$ degrees of freedom partition for treatments is not explicitly used in the final partitioning.

If we do not use this two-stage approach the data may appear to arise from a three-factor factorial experiment with factors being blocks, A, and C. In fact, many computer programs will give this as the default analysis. The analysis according to this interpretation is

Source of Variation	**df**
Blocks	$b - 1$
Main effect A	$a - 1$
Main effect C	$c - 1$
Interaction: blocks *A	$(b - 1)(a - 1)$
Interaction: blocks *C	$(b - 1)(c - 1)$
Interaction: A * C	$(a - 1)(c - 1)$
Interaction: blocks * A * C	$(b - 1)(a - 1)(c - 1)$

In this case the three-factor interaction would be used as the error variance for testing all hypotheses, a procedure that produces an incorrect test. Comparing this analysis with the correct one given above, we see that the correct experimental error is obtained by pooling all interactions with blocks. Most computer programs provide options for producing the proper analysis.

EXAMPLE 9.6

REVISITED The three-factor factorial described in Example 9.6 can be considered a randomized block design if the machines are actually a random sample of four machines from a large population of machines at a plant and will be considered blocks. The appropriate analysis can be reconstructed from the analysis of variance in Table 9.17 by pooling all interactions with `MACHINE` to produce the experimental error. The hypothesis tests on the factors (`TIME` and `HEAT`) will use the resulting mean square in the denominator of the F ratios.

Solution For experiments such as this one, it useful to outline the stages of the analysis so we can correctly instruct the computer program to produce the correct analysis.

Stage One We have a randomized blocks design with sampling and six treatments corresponding to the 2 × 3 factorial experiment. The partitioning of sums of squares is

Source of Variation	df
Treatments	5
Blocks (MACHINE)	3
Experimental error	15
Sampling error	24

Stage Two In the second stage, the treatment sum of squares with 5 degrees of freedom is partitioned according to the factorial structure as presented in Chapter 9, resulting in the following partitioning:

Source of Variation	df
TIME	2
HEAT	1
TIME × HEAT	2

Final Stage The results of the two stages are combined, which yields the final analysis of variance table:

Source of Variation	df
Blocks (MACHINE)	3
TIME	2
HEAT	1
TIME × HEAT	2
Experimental error	15
Sampling error	72

The elements in this table will provide the information for a computer program. However, the instructions will need to specify the experimental error because computer programs will not automatically know the construct of that quantity. In this case it is the result of pooling the sums of squares for MACHINE × TIME, MACHINE × HEAT, and MACHINE × TIME × HEAT. Different programs may have different ways of specifying this term, and for some the pooling may need to be done manually. In the SAS System, specifying that the experimental error is MACHINE × TIME × HEAT without specifying the imbedded two-factor interactions, MACHINE × TIME and MACHINE × HEAT, produces the pooled variance estimate. The resulting analysis is shown in Table 10.12. The features of the results are as follows:

- The first portion of the output shows the partitioning for all elements of the experiment.
- The second portion provides the partitioning according to the elements of the experiment. The `MACHINE × TIME × HEAT` (note that the computer program arranges the terms in a different order) is the experimental error. However, the sampling error is used as the default error for the tests in the

Table 10.12 Analysis of Steel Bar Data, Machine Is Block

Source	df	Sum of Squares	Mean Square	F Value	Pr > F
Model	23	590.3333333	25.6666667	4.13	0.0001
Error	72	447.5000000	6.2152778		
Corrected total	95	1037.8333333			

	R-Square	C.V.	Root MSE	LENGTH Mean
	0.568813	62.98221	2.493046	3.95833333

Source	df	Anova SS	Mean Square	F Value	Pr > F
MACHINE	3	393.4166667	131.1388889	21.10	0.0001
TIME	2	12.8958333	6.4479167	1.04	0.3596
HEAT	1	100.0416667	100.0416667	16.10	0.0001
TIME * HEAT	2	1.6458333	0.8229167	0.13	0.8762
TIME * HEAT * MACHINE	15	82.3333333	5.4888889	0.88	0.5851

Tests of Hypotheses using the Anova MS for TIMES * HEAT * MACHINE as an error term

Source	df	Anova SS	Mean Square	F Value	Pr > F
HEAT	1	100.0416667	100.0416667	18.23	0.0007
TIME	2	12.8958333	6.4479167	1.17	0.3358
TIME * HEAT	2	1.6458333	0.8229167	0.15	0.8620

table, which is incorrect except for the test that the experimental error is the same as the sampling error. The test indicates that there is no significant difference between the two. The last portion provides the tests for the factorial effects using the appropriate error mean square. Only the HEAT effect is statistically significant. Of course, since we have found that the sampling and experimental errors may be considered equivalent, a pooled variance estimate may be used. However, the results would not change.

The results are not very different from those obtained when the experiment was considered to be a three-factor factorial (Table 9.17) because the sampling error and experimental error have almost the same value. This does not, however, make the analysis, assuming the factorial experiment correct if machines are really blocks.

The conclusion is that only the heat treatment makes any difference since individual machine differences are of little interest. ■

Nested Designs

In some experimental situations, experimental units may contain sampling units, which may, in turn, contain sample subunits. Such a situation is referred to as a hierarchical or nested design, since the design describes subsamples nested within sample or experimental units.

For example, in a quality control experiment, treatments may be different work environments, which are carried out in different work shifts, the workers are blocks, and randomly sampled units of the product are the experimental or observational units. However, we do not normally have the same workers in different shifts. This type of experimental arrangement is an example of a hierarchical design. Note that if the same workers for each shift could be arranged, we would have a randomized block design with workers as blocks. However, in this case, we have independent samples of workers within the individual shifts and subsamples of units of the product for each of the workers.

EXAMPLE 10.5

In a production plant that operates continuously, quality monitoring requires identification of sources of variation in the production process. For example, it would be important for the quality engineer to know whether there was significant variation between shifts as well as whether there was significant variation between workers during shifts. These questions can be answered by using a nested design experiment. This example discusses one such experiment using a random sample of three shifts taken over a month of production. (Note that a shift is really a combination of time of day and day of the month.) Then a random sample of four workers was taken from each of these three shifts. Five 30-min. production values were randomly selected from the production of each of these workers during that 8-h shift. The number of defective items found in the 30-min. interval was used as a measure of quality. The results of the experiment are shown in Table 10.13. This experiment consists of two factors, Shifts and Workers, both of which are random effects, and five replications of each of the levels of the factors.

Solution This is a nested design for which the linear model is

$$y_{ijk} = \mu + \alpha_i + \beta_{j(i)} + \varepsilon_{k(ij)},$$

where y_{ijk} = kth observation for level i of factor A (shift) and level j of factor B (worker) in shift i; μ = reference value or overall mean; α_i = effect of the ith level of factor A, $i = 1, \ldots, a$; $\beta_{j(i)}$ = effect of level j of factor B nested in the ith level of factor A, $j = 1, \ldots, b$; and $\varepsilon_{k(ij)}$ = variation among sampled units of the product and is the random error, $k = 1, \ldots, n$.

The subscript $j(i)$ is used to denote that different j subscripts occur within each value of i; that is, they are "nested" in i. Likewise, the k subscript is

Table 10.13

Data for Nested Design

Shift	A				B				C			
Worker	**1**	**2**	**3**	**4**	**5**	**6**	**7**	**8**	**9**	**10**	**11**	**12**
Observed values	3	5	0	10	4	5	0	7	14	5	9	9
	5	7	3	7	3	5	1	6	12	2	5	5
	3	4	3	5	4	4	2	5	10	6	2	9
	6	4	4	4	3	7	1	10	9	6	6	4
	4	6	5	7	4	6	1	5	10	3	6	7

"nested" in groups identified by the combined ij subscript. In the example, $a = 3$, $b = 4$, and $n = 5$.

Note that there is no interaction in this model. This is because the levels of B are not the same for each level of A; hence interaction is not definable.

Sums of squares for the analysis of variance are generalizations of the formulas for the one-way (CRD) sums of squares computations and will only be outlined here. The sums of squares for factor A are computed as if there were only the a levels of factor A, that is, disregarding all other factors. The sums of squares for B in A are computed as if there were simply the $a \cdot b$ levels of "factor" B, and subtracted from this quantity is the already computed sum of squares for A. The error sum of squares is obtained by subtraction: SS(Error) = TSS − SSA − SSB(A), where TSS is computed as in all other applications.

The proper test statistics depend on which of the factors are fixed or random. The most frequent application occurs for all random factors, but other combinations are certainly possible. The resulting analysis of variance table and the expected mean squares for the completely random model is as follows:

Source	**df**	**SS**	**E(MS)**
A	$a-1$	SSA	$\sigma^2 + n\sigma_\beta^2 + bn\sigma_\alpha^2$
B(A)	$a(b-1)$	SSB(A)	$\sigma^2 + n\sigma_\beta^2$
Error	$ab(n-1)$	SSE	σ^2

From this table we can see that to test for the A effect we must use the B(A) mean square as the error variance, while to test for B(A) we use the "usual" error variance. Estimates of σ_α^2 and σ_β^2 can be obtained by equating the actual mean squares with the formulas for expected mean squares and solving the equations, as was done for the one-way random effects model outlined in Section 6.6. In many applications these variance components are of considerable importance because they can be used to plan for better designs for future studies. The results of the analysis on the example are provided by an abbreviated output from `PROC NESTED` of the SAS System in Table 10.14.

The first portion of the output gives the coefficients of the mean squares. For this example,

$$MS_{\text{SHIFT}} = 20\sigma^2_{\text{SHIFT}} + 5\sigma^2_{\text{WORKER}} + \sigma^2_{\text{ERROR}}.$$

The next portion gives the analysis of variance with the F ratios obtained by using the denominators indicated in the last column. Here we can see that the shift effect is not significant, while there appears to be significant variation among workers.

The last portion gives the estimates of the variance components and finally the overall mean and standard error of that mean. These statistics are useful when the primary objective is to estimate that mean, as is often the case. ■

Nested designs may be extended to more than two nests or stages and also often do not have an equal number of samples for each factor level. In the case

Table 10.14

Nested Design

Coefficients of Expected Mean Squares

SOURCE	SHIFT	WORKER	ERROR
SHIFT	20	5	1
WORKER	0	5	1
ERROR	0	0	1

Nested Random Effects Analysis of Variance for Variable Y

Variance Source	Degrees of Freedom	Sum of Squares	F Value	Pr > F	Error Term
TOTAL	59	484.183333			
SHIFT	2	86.933333	1.57965	0.258228	WORKER
WORKER	9	247.650000	8.82888	0.00000	ERROR
ERROR	48	149.600000			

Variance Source	Mean Square	Variance Component	Percent of Total
TOTAL	8.206497	8.794167	100.0000
SHIFT	43.466667	0.797500	9.0685
WORKER	27.516667	4.880000	55.4913
ERROR	3.116667	3.116667	35.4402

Mean	5.28333333
Standard error of mean	0.85114302

of unequal sample sizes, the expected mean squares retain the format given above, but the formulas for the coefficients of the individual components are quite complex and are usually derived by computers. Additional information on nested designs can be found in most texts on sampling methodology, for example, Scheaffer *et al.* (1996).

Many applications of nested designs occur with sample surveys. For example, a sample survey for comparing household incomes among several cities may be conducted by randomly sampling areas in each city, then sampling blocks in the sampled areas, and finally sampling households in the sampled blocks. The analysis of variance for the resulting data may be outlined as follows:

Source
Cities
Areas (Cities)
Blocks (Areas, Cities)
Households (Cities, Areas, Blocks)

Analyses such as these are often primarily concerned with estimation of means and variance components. They usually involve large data sets and the use of computers is mandatory. Fortunately, many computer software packages have programs especially designed for such analyses.

Split Plot Designs

Split plot designs occur when different factors in an experiment require different sizes of experimental units. For example, irrigation treatments such as amounts or frequencies of irrigation water require large fields, while different varieties of crops or different fertilizer applications can be applied in smaller plots. In an experiment designed to find the optimum operating conditions for a process, some conditions, such as operating temperature, can only be set at the beginning time of a day, while others, such as pressure or the amount of some ingredient, may be changed at shorter intervals. In such a design, the large units, such as irrigation plots or operating days, are called main plots, while the smaller units are called subplots. Since there are fewer replications of the main plot treatments, the mean responses for the factor levels applied to main plots are estimated with less precision than those for the factor levels applied to subplots. In the analysis of such data, therefore, different error variances are necessary for the main plot and subplot effect tests and estimates.

For example, consider an experiment relating growth of plants to different temperature regimes and amounts of a plant food. Temperature regimes are provided by specially constructed environmental chambers, which accommodate large numbers of plant beds, while different plant foods can be used on individual plant beds. In this experiment environmental chambers are the main plots and plant beds the subplots; hence temperature is the main plot and plant food the subplot factor.

Assume that p temperature regimes are established in each of p environmental chambers and that each of c plant beds in each chamber is given one of c plant foods. The entire experiment is replicated r times, which is analyzed as if there are r blocks. The linear model is

$$y_{ijk} = \mu + \rho_i + \alpha_j + (\rho\alpha)_{ij} + \gamma_k + (\alpha\gamma)_{jk} + (\rho\gamma)_{ik} + (\rho\alpha\gamma)_{ijk},$$

where y_{ijk} = response for plant food k in temperature j of replication i; μ = reference value or overall mean; ρ_i = (usually random) effect of replication i, $i = 1, 2, \ldots, r$; α_j = (fixed) effect of temperature j, $j = 1, 2, \ldots, p$; $(\rho\alpha)_{ij}$ = (random) interaction of replication and temperature; γ_k = (fixed) effect of plant food k, $k = 1, 2, \ldots, c$; $(\alpha\gamma)_{jk}$ = (fixed) interaction between temperature and plant food; $(\rho\gamma)_{ik}$, $(\rho\alpha\gamma)_{ijk}$ = (random) interactions of replication by plant food and replication by plant food by temperature.

The partitioning of sums of squares is accomplished by first pretending that we have a three-factor factorial experiment with factors being replications, temperatures, and plant foods. The analysis of variance is outlined as follows:

Source of Variation	df
Replication	$r-1$
Temperature	$p-1$
Rep * Temperature	$(r-1)(p-1)$
Food	$c-1$
Rep * Food	$(r-1)(c-1)$
Temperature * Food	$(p-1)(c-1)$
Rep * Temp * Food	$(r-1)(p-1)(c-1)$

Table 10.15

Data for Thread Test

	CONE									
	1		2		3		4		5	
	NUMBER OF PINS									
Manuf	**2**	**3**	**2**	**3**	**2**	**3**	**2**	**3**	**2**	**3**
A	270	273	350	326	296	289	277	293	260	269
B	405	392	429	410	450	433	421	431	409	388
C	448	475	439	466	398	401	442	420	432	423
D	298	314	358	363	354	367	339	345	334	331
E	394	417	463	490	419	442	442	477	464	480

As in many other applications, it is useful to consider the analysis in two stages:

1. Stage one considers only the main plot temperature experiment. This is a randomized block design, and replication by temperature is the experimental error used for testing for temperature and replication effects. This experimental error has the same function in this split plot design, where it is commonly referred to as error(a).
2. The second stage concerns the subplot plant food treatment and produces the sum of squares due to plant food and all interactions with plant food. All interactions with replications except temperature by replication are usually pooled[7] to produce the error variance for inferences on the subplot effect (plant food) and its interaction with the main plot effect (plant food by temperature). This error is commonly referred to as error(b).

This analysis may appear complicated but is dictated by the expected mean squares, which we do not reproduce here. Note that many computer programs have the ability to perform this correct analysis if proper instructions are provided by the user. However, the computer has no way of automatically knowing the difference between a three-factor factorial experiment and a split plot design!

As before, multiple comparisons or contrasts may be used for more specific hypothesis tests but these must also use the appropriate variance estimates.

EXAMPLE 10.6

The purpose of this experiment is to test the differences in the strengths of sewing thread made by five different manufacturers. Five cones of thread (`CONES`) were obtained from each of five manufacturers (`MANUF`). The test consists of passing a specimen of thread through some guide pins (`PINS`) and measuring the energy (`ENERGY`) required to rupture the specimen. The lower the required energy, the weaker the thread. In this experiment there are two guide pin treatments using two and three pins, respectively.

Solution Twenty specimens were obtained from each cone with 10 assigned randomly to pass through the two or three guide pins. The data in Table 10.15 give the means of the 10 readings for each cone and number of pins. This is a split plot design with manufacturers as main plots and cones

[7] If the individual mean squares to be pooled are of different magnitudes, pooling may not always be appropriate (Bancroft, 1968).

as subplots. However, in this case the main plot portion is a completely randomized design instead of a randomized block design. In this context it is usually called a nested design, and therefore the main plot analysis is

Source	df
Manufacturers	4
Cones (Manufacturers)	20

The Cones (Manufacturers) is used as the error term for testing manufacturers. The second stage concerns the number of pins and the analysis is

Source	df
Pins	1
Pins * Manufacturers	4
Error	20

The Error is actually Pins by Cones(Manufacturers) and is computed by subtraction. The analysis as provided by `PROC ANOVA` of the SAS System is given in Table 10.16.

As was the case for the randomized block with sampling, the correct test for the null hypothesis of no manufacturer effect must be specifically requested. That is, `CONE(MANUF)` is error(a) for the F ratio for this test, and the F ratio given in the partitioning of sums of squares may not be used. The correct test is provided by the last entry in the table and shows that the strengths do differ among manufacturers ($p < 0.0001$), although the somewhat smaller Manufacturer by Pins interaction ($p < 0.0087$) may modify this conclusion.

The error listed in the analysis of variance table is error(b) and is correctly used for testing all sources associated with the subplot treatment (Pins).

Table 10.16 Analysis of Thread Data

Dependent Variable ENERGY

Source	df	Sum of Squares	Mean Square	F Value	Pr > F
Model	29	212371.2800	7323.1476	74.73	0.0001
Error	20	1960.0000	98.0000		
Corrected Total	49	214331.2800			

	R-Square	C.V.	Root MSE		C2 Mean
	0.990855	2.563839	9.899495		386.120000

Source	df	Anova SS	Mean Square	F Value	Pr > F
MANUF	4	184819.4800	46204.8700	471.48	0.0001
CONE(MANUF)	20	25448.8000	1272.4400	12.98	0.0001
PINS	1	307.5200	307.5200	3.14	0.0917
MANUF*PINS	4	1795.4800	448.8700	4.58	0.0087

Tests of Hypotheses using the Anova MS for CONE(MANUF) as an error term

Source	df	Anova SS	Mean Square	F Value	Pr > F
MANUF	4	184819.4800	46204.8700	36.31	0.0001

Table 10.17

Analysis of Thread Data

```
        Tukey's Studentized Range (HSD) Test for Variable: C
                  Alpha = 0.05 df = 20 MSE = 1272.44
             Critical Value of Studentized Range = 4.232
                Minimum Significant Difference = 47.736

    Means with the same letter are not significantly different.
```

Tukey Grouping	Mean	N	MANUF
A	448.80	10	E
A			
A	434.40	10	C
A			
A	416.80	10	B
B	340.30	10	D
C	290.30	10	A

Figure 10.1

Interaction Chart for Strength of Thread

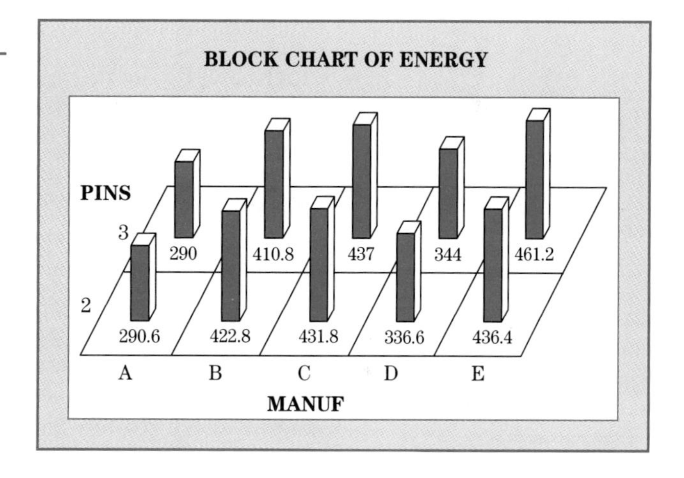

The differences between the number of pins is not statistically significant ($p = 0.0917$), and since an experiment with two pins is easier to do, the recommendation is to use only two pins.

Since the manufacturer differences appear to dominate, we continue with a paired comparison, this time using the Tukey Studentized range test. The results are given in Table 10.17.

It appears that manufacturers A and D produce weaker thread, but that there is no real difference among the others. However, since the `MANUF*PINS` interaction is statistically significant, a block chart is used to examine the effect of this interaction. A block chart of the means shown in Fig. 10.1 shows the interaction. For manufacturers C, D and E, three pins require more energy, for B less energy, and for A there is virtually no difference. However, despite the interaction, we may conclude that D and A are inferior and the others are not much different.

If the observed energy values were available for each specimen, the analysis would have an additional source of variation. This would be the variation of readings among specimens within a cone and pin setting. An additional test would determine whether error(b) is significantly larger than the within cones variance. If not, the two could be pooled, providing an experimental error with larger degrees of freedom. However, even if error(b) is significantly larger, thus precluding pooling, the use of individual observations is likely to lead to results with higher significance because the overall sample size is larger. ■

Additional Topics

As previously noted, the coverage of the field of experimental design in this chapter is quite limited. The purpose here has been to provide some information on the general concepts underlying the construction, use, and analysis of data resulting from such designs. We strongly urge that anyone planning an experiment consult a comprehensive reference or preferably an experienced statistical consultant to assure that the most appropriate design is used for any given situation.

Other considerations in such planning include the following:

- Experimental conditions may dictate block sizes too small to accommodate all treatments. Such situations can be handled by incomplete blocks designs.
- Inadequate block sizes may also occur for factorial experiments. Special types of incomplete blocks designs may be used where the less important effects (such as three-factor interactions) are estimated with less precision.
- Factorial experiments may become so large that it is not even possible to provide for all factor level combinations in the experiment. It is possible to construct a fraction of a factorial that still allows estimation of the more important sources of variation.
- Special designs are available for estimating polynomial response functions. Many of these require far fewer experimental units than the comparable factorial experiment. This is one aspect of experimental design that is not concerned with blocking.
- Split plot designs have many variations. A special application of split plot designs occurs when the subplots consists of measurements on one individual unit taken over a period of time. Data from such an experiment, called a repeated measures design, may be analyzed as if it were a split plot; however, problems may arise due to correlations between measurements made at adjacent time periods. Such correlations violate the assumption of independent random errors and require specialized methodology.

10.6 CHAPTER SUMMARY

Solution to Example 10.1 We can now see that the experiment is an example of a split plot design. The four fields are replicates or blocks, the half fields assigned different irrigation levels are the main plots, and the nitrogen rate and planting rate combinations are applied to subplots.

The first analysis is for the factorial using the model

$$y_{ijkl} = \mu + \rho_i + \alpha_j + \rho\alpha_{ij} + \nu_k + \nu\alpha_{jk} + \gamma_l + \gamma\alpha_{lj} + \gamma\nu_{lk} + \gamma\nu\alpha_{lkj} + \varepsilon_{ijkl},$$

where y_{ijkl} = response for the plot in the ith block (REP), jth WTR level, kth NRATE level, and lth P level; and μ = overall mean.

The main plot elements are

ρ_i = effect of REP i, $i = 1, 2, 3, 4$,

α_j = effect of WTR level j, $j = 1, 2$, and

$\rho\alpha_{ij}$ = interaction of REP and WTR and is the error for testing WTR.

The subplot elements are

ν_k = effect of NRATE level, $k = 1, 2, 3$,

$\nu\alpha_{jk}$ = interaction between WTR and NRATE,

γ_l = effect of P (planting rate), $l = 1, 2, 3, 4$,

$\gamma\alpha_{lj}$ = interaction of P and WTR,

$\gamma\nu_{lk}$ = interaction of P and NRATE,

$\gamma\nu\alpha_{lkj}$ = interaction of P, NRATE, and WTR, and

ε_{ijkl} = error for testing subplot effects. It is actually the interaction of REP with all effects except WTR.

The resulting analysis of variance produced by PROC ANOVA of the SAS System, requesting the appropriate test for the WTR effect, is shown in Table 10.18, where we see that the main effects for NRATE and P are very important. The main effect for WTR is not overwhelming, but its interaction with NRATE is. The interactions of P with NRATE and the three-factor interaction are marginally significant, while the interaction of WTR with P appears to be of no importance.

At this point it is appropriate to obtain the various means and make some preliminary plots, which are not reproduced here. An examination of these plots suggests the following response surface regression to provide more interpretable results:

$$\begin{aligned} y = \beta_0 &+ \beta_1\text{WTR} + \beta_2\text{P} + \beta_3\text{NRATE} + \beta_4(\text{NRATE})^2 + \beta_5(\text{NRATE}) * (\text{P}) + \beta_6\text{P}^2 \\ &+ \beta_7\text{P}^3 + \beta_8(\text{NRATE})^2 * (\text{P}) + \beta_9(\text{NRATE}) * \text{P}^2 + \beta_{10}(\text{WTR}) * (\text{NRATE}) \\ &+ \beta_{11}(\text{WTR}) * (\text{NRATE})^2 + \varepsilon. \end{aligned}$$

The following observations for this model are of interest:

- The model cannot contain quadratics in WTR since there are only two levels of this factor.
- There are no terms for the P by WTR interaction since it is not significant.
- The inclusion of the not frequently used cubic term for P is suggested by a plot of means.

Table 10.18 Analysis of Variance

Analysis of Variance Procedure

Dependent
Variable: TDM

Source	df	Sum of Squares	Mean Square	F Value	Pr > F
Model	29	268.6081192	9.2623439	10.30	0.0001
Error	66	59.3508093	0.8992547		
Corrected Total	95	327.9589285			

R-Square	C.V.	Root MSE	TDM Mean
0.819030	19.42870	0.948290	4.88087500

Source	df		Mean Square	F Value	Pr > F
REP	3	4.59591525	1.53197175	1.70	0.1748
WTR	1	47.90635267	47.90635267	53.27	0.0001
REP*WTR	3	5.34853642	1.78284547	1.98	0.1251
NRATE	2	83.63305575	41.81652788	46.50	0.0001
NRATE*WTR	2	54.96290858	27.48145429	30.56	0.0001
P	3	39.27526683	13.09175561	14.56	0.0001
P*WTR	3	1.35249700	0.45083233	0.50	0.6827
P*NRATE	6	15.08345842	2.51390974	2.80	0.0175
P*NRATE*WTR	6	16.45012825	2.74168804	3.05	0.0108

Test of Hypotheses using the Anova MS for REP*WTR as an error term

Source	df	Anova SS	Mean Squae	F Value	Pr > F
WTR	1	47.90635267	47.90635267	26.87	0.0139

The results of the regression using PROC REG of the SAS System are shown in Table 10.19. Since the computer program does not allow symbols and exponents, in variable names, N2 represents $(\text{NRATE})^2$, NLPL represents (NRATE) * (P), etc.

We first perform a lack of fit test to see whether any additional terms may be needed. In this case it is easier to work with the model rather than the error sums of squares. The full model is the complete factorial portion of the analysis of variance, which is obtained by subtraction of the sums of squares and degrees of freedom for all terms involving REP from the MODEL sum of squares given in the analysis of variance:

$$\text{SS(Full model)} = 258.664, \quad \text{with 23 degrees of freedom.}$$

Note that this is the MODEL sum of squares less the sums of squares for REP and REP*WTR. The reduced model is the regression; hence,

$$\text{SS(Reduced model)} = 239.913, \quad \text{with 11 degrees of freedom.}$$

The difference is

$$\text{SS(Lack of fit)} = 18.751, \quad \text{with 12 degrees of freedom.}$$

Table 10.19 Polynominal Response Regression

```
Model: MODEL1
Dependent Variable: TDM
                         Analysis of Variance

                     Sum of           Mean
Source          df   Squares          Square            F Value     Prob > F

Model           11   239.91297        21.81027          20.808        0.0001
Error           84    88.04596         1.04817
C Total         95   327.95893

    Root MSE          1.02380         R-square           0.7315
    Dep Mean          4.88087         Adj R-sq           0.6964
    C.V.             20.97574
                          Parameter Estimates

                      Parameter    Standard          T for H0:
Variable        df    Estimate       Error         Parameter = 0   Prob > |T|

INTERCEP         1   -13.618687   2.93855382          -4.634        0.0001
WTR              1     8.167250   1.57778158           5.176        0.0001
NRATE            1    17.650473   3.11733375           5.662        0.0001
P                1     0.408540   0.22502853           1.816        0.0730
N2               1    -3.853712   0.76645762          -5.028        0.0001
NLPL             1     0.111415   0.08235249           1.353        0.1797
P2               1    -0.027021   0.01149726          -2.350        0.0211
P3               1     0.000413   0.00017521           2.360        0.0206
N2PL             1    -0.009844   0.01653586          -0.595        0.5532
NLP2             1    -0.000827   0.00103928          -0.796        0.4285
WN               1    -9.470125   1.79164979          -5.286        0.0001
WN2              1     2.005750   0.44331835           4.524        0.0001
```

The lack of fit mean square is 1.563, and using the error mean square from the analysis of variance as the denominator provides an F ratio of 1.738 with (12, 66) degrees of freedom. The closest table entry is 1.92 for $\alpha = 0.05$ for (12, 60) degrees of freedom, and we have insufficient evidence to find the model inadequate.

As previously noted, most of the coefficients in a multiple polynomial regression are not interpretable. It is, however, sometimes useful to ascertain whether a model with fewer terms may be adequate. This is done by examining the test statistics for the highest order terms. In this case, the terms for $(\texttt{NRATE})^2\texttt{P}$ and $(\texttt{NRATE})\texttt{P}^2$ would appear to be unnecessary. The next step would be to refit the reduced model and continue to check for unneeded terms. To save space, we will not do this here, but this is a good exercise for the reader.

The next step is to produce a response surface plot similar to that provided in Fig. 9.4 or Fig. 9.6. Again there are three factors; hence, we must make a response surface plot for two variables for levels of a third factor. Since there are only two levels of WTR, it is most appropriate to produce a response surface for NRATE and P for the two levels of WTR. These are shown in Figs. 10.2 and 10.3. The response surface plots are interpreted as follows:

Figure 10.2

Response Surface Plot for WTR = 1

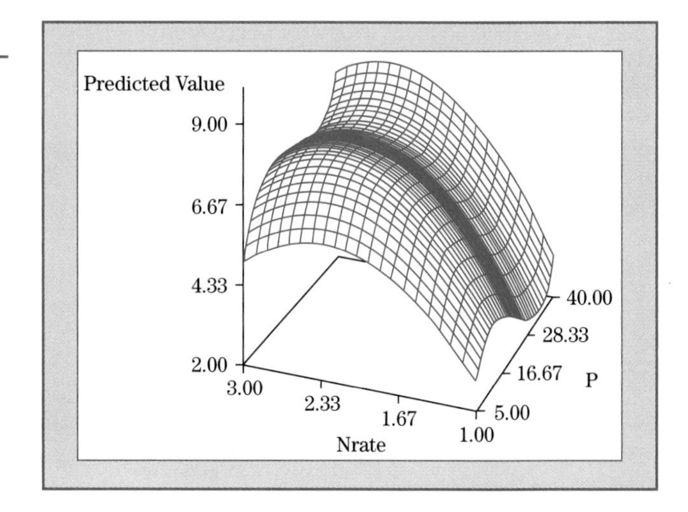

Figure 10.3

Response Surface Plot for WTR = 2

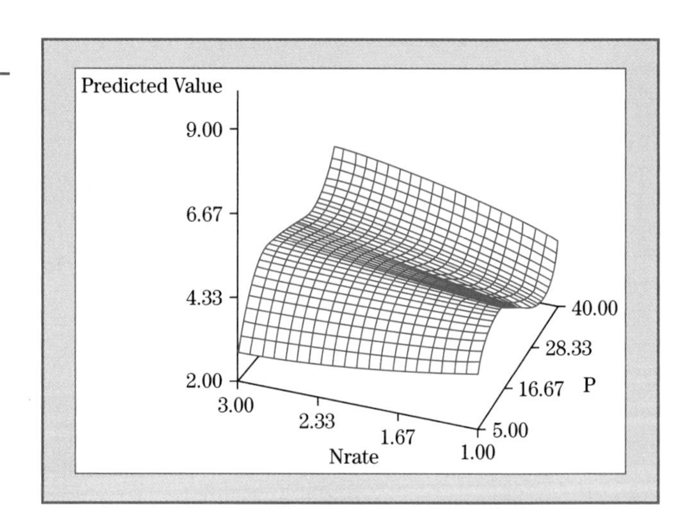

- The responses at WTR = 2 are lower than those at WTR = 1. This contradicts what we would expect.
- The response to nitrogen (NRATE) for WTR = 1 shows the typical pattern of an increase that gradually diminishes with higher rates of NRATE. Also there is virtually no interaction with planting rate (P). The response to NRATE for WTR = 2 is not typical: there is virtually no response for low planting rates and an almost constant increased response for higher planting rates. In other words, nitrogen is only useful for higher planting rates.
- The response to planting rates shows a definite "dip" from about 20 to 30, especially for the higher irrigation rate. This result is quite unusual, but since it is due to the statistically significant cubic term, it cannot be ignored.

Figure 10.4

Residual Plot

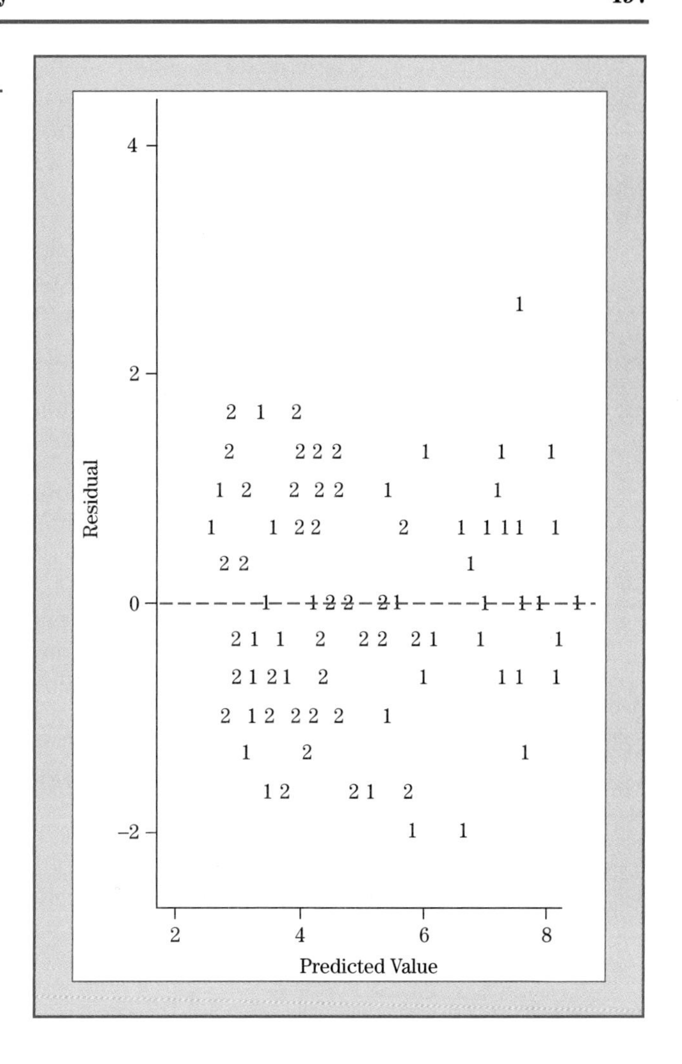

Such unusual responses are often the result of outliers; however, the residual plot shown in Fig. 10.4 and other plots not shown here provide no support for such a phenomenon.

The data in Table 10.1 also gave the actual planting rates. It is not difficult to perform the response surface regression using these rather than the nominal planting rates. This analysis is not shown here but provides essentially the same results.

We should add that the data set for this example represents only a portion of a larger study conducted over several years that also involved other crops and response variables. Actually one reason why this set was used here was that the results were somewhat unusual; most of the other experiments provided more "usual" results. ■

The statistical aspect of experimental design is largely concerned with assigning experimental units to treatments or factor levels. The simplest of all designs, the completely randomized design in which factor levels are randomly assigned to experimental units, is presented in Chapter 6. Designs presented in this chapter include

- the randomized block design, in which the set of experimental units is divided into blocks and factor levels randomly assigned to each block,
- the Latin square design, in which two blocking factors are applied simultaneously,
- the nested design, and
- the split plot design, which accommodates the need for different sizes of blocks for various factors.

Other designs and design principles are briefly discussed.

Because blocks are similar to factor levels in a factorial experiment, the mechanics of the analysis of data from block designs are the same as those from a factorial experiment, resulting in occasional confusion in terminology. There are, however, differences in interpretation and testing procedures that must be observed for the analysis to be correct. Of special importance is the use of the correct "error term" as the denominator in various F ratios.

10.7 CHAPTER EXERCISES

EXERCISES

For all exercises it is important not only to recognize and employ the correct analysis for the design but also to complete the analysis and interpretation with respect to the factors in the experiment.

1. Many organisms show elevated oxygen consumption after periods of anoxia (lack of oxygen). This elevated oxygen consumption is usually interpreted as the result of paying back an oxygen debt. However, preliminary work on a bivalve indicated that the increased oxygen consumption was due to increased ventilatory activity and not a true oxygen debt. The data presented in Table 10.19 are from an experiment designed to determine which of the two possibilities may be correct for oysters. Oysters were exposed to three anoxic regimes:
   (1) nitrogen gas (Gas), where oysters can neither receive oxygen nor ventilate,
   (2) deoxygenated seawater (Water), where they can ventilate but do not receive oxygen, and
   (3) normal seawater, which serves as a control.

   After 24 hours on treatment, the oysters were returned to well-aerated seawater and their oxygen consumption measured. Five oysters were exposed to each treatment, and the entire experiment was replicated twice;

Table 10.20

Oxygen Consumption of Oysters

Control	Treatment Water	Gas
	Replication 1	
1.100	2.403	2.485
0.870	1.649	3.175
1.003	1.678	2.057
1.114	2.546	2.491
1.361	1.180	1.397
	Replication 2	
1.290	1.369	1.715
2.024	0.565	1.285
1.406	0.711	1.693
1.391	0.903	1.202
0.772	1.281	2.026

Table 10.21

Cowpea Data for Exercise 2

		VARIETY			
Rep	**Solution**	**7**	**101**	**289**	**389**
I	5	41	20	24	10
	10	131	69	65	105
	15	175	143	142	112
II	5	52	29	24	5
	10	128	91	92	44
	15	190	178	87	167
III	5	68	27	11	7
	10	94	66	62	103
	15	150	166	199	101

the replicates are the blocks. Which, if either, of the two possibilities do the data of Table 10.20 support?

2. The objective of this experiment was to measure the effect of water stress on nitrogen fixation in four cowpea varieties. Plants were grown in a greenhouse and watered with 5, 10, and 15 ml of the appropriate nutrient solution. Fifty-five days after planting, the nitrogen nodules were removed from the plants and counted and weighed. The entire experiment was replicated three times. The response variable to be analyzed is the weight of the nitrogen nodule. The data are given in Table 10.21. Perform the appropriate analysis to determine the effect of water stress.

3. This problem concerns another rice yield experiment. The response is yield per experimental plot. The experiment was conducted as a randomized block for three years where the blocks are the same for each year. The factors are:

   Var: three varieties, coded A, B, and C, and
   Nit: two levels of nitrogen application, coded 80 and 160.

   This is a split plot with years as main plots and the factorial in the subplots. The data are given in Table 10.22.

Table 10.22

Rice Data for Exercise 3

NIT = 80				NIT = 160			
Yr	Var	Rep	Yield	Yr	Var	Rep	Yield
3	A	1	9.34	3	A	1	9.60
3	A	2	10.41	3	A	2	9.82
3	A	3	10.28	3	A	3	10.43
3	A	4	9.35	3	A	4	10.08
3	B	1	10.23	3	B	1	10.00
3	B	2	9.82	3	B	2	10.05
3	B	3	9.75	3	B	3	9.66
3	B	4	9.81	3	B	4	8.94
3	C	1	9.22	3	C	1	7.92
3	C	2	9.50	3	C	2	8.21
3	C	3	9.31	3	C	3	8.97
3	C	4	8.77	3	C	4	8.17
4	A	1	9.64	4	A	1	10.11
4	A	2	9.12	4	A	2	11.20
4	A	3	8.02	4	A	3	11.21
4	A	4	8.16	4	A	4	11.15
4	B	1	11.21	4	B	1	12.81
4	B	2	10.26	4	B	2	11.71
4	B	3	12.48	4	B	3	11.82
4	B	4	11.40	4	B	4	12.12
4	C	1	10.28	4	C	1	12.16
4	C	2	10.30	4	C	2	11.98
4	C	3	11.53	4	C	3	12.17
4	C	4	10.57	4	C	4	12.67
5	A	1	9.87	5	A	1	9.57
5	A	2	9.01	5	A	2	8.94
5	A	3	9.08	5	A	3	9.89
5	A	4	9.14	5	A	4	9.87
5	B	1	10.77	5	B	1	5.71
5	B	2	11.29	5	B	2	6.52
5	B	3	10.67	5	B	3	9.22
5	B	4	12.80	5	B	4	6.37
5	C	1	10.07	5	C	1	10.77
5	C	2	10.92	5	C	2	11.57
5	C	3	11.04	5	C	3	11.61
5	C	4	11.02	5	C	4	11.30

(a) Perform the analysis using years as a fixed factor. What conclusions can you draw from this analysis? How can these conclusions be used for inferences?

(b) Perform the analysis assuming years as a random factor. [*Hint:* All interactions of fixed factors with blocks and years are error(b).] Draw conclusions. Are they different from those obtained in part (a)? Explain the reasons for any differences.

4. For this experiment on the growth of cotton, the factors are:

Var: six varieties, coded from 1 to 6, and

Trt: a treatment concerning the density of plants: treatment 1 has 32,000 plants per acre and treatment 2 has 64,000 plants per acre.

Table 10.23 Cotton Data for Exercise 4

		REP = 1		REP = 2		REP = 3		REP = 4	
Var	Trt	BRATE	MRATE	BRATE	MRATE	BRATE	MRATE	BRATE	MRATE
1	1	2.98	42	3.11	37	2.25	37	2.22	53
1	2	3.08	42	3.07	42	3.30	37	2.85	37
2	1	2.75	42	3.36	25	2.92	42	2.68	47
2	2	2.75	42	3.56	46	2.90	37	2.92	42
3	1	2.83	42	2.94	58	2.88	42	2.96	32
3	2	3.11	42	2.79	42	3.13	47	2.95	42
4	1	3.11	42	3.76	30	2.70	32	2.87	52
4	2	2.96	42	4.36	30	3.41	42	3.32	47
5	1	3.14	58	3.03	42	3.24	58	3.31	47
5	2	4.12	47	3.49	42	3.86	47	3.94	18
6	1	2.60	37	2.64	42	2.60	32	2.49	37
6	2	2.98	53	2.92	42	2.35	48	2.42	37

There are two responses:

BRATE: the blooming rate, which is an index indicating the rate at which the plants bloom, and
MRATE: an index indicating the rate at which the plants mature.

The experiment is a split plot design with varieties as main plot and planting rate as the subplot treatments. The entire experiment was replicated four times as indicated by the variable REP. The data are given in Table 10.23. Perform the appropriate analysis to ascertain how each of the responses is affected by varieties and treatments. If both high blooming and maturing rates are desired, can you make a recommendation?

5. The purpose of this experiment is to determine the effect of salt in the soil on the emergence of grass seeds. There are three replications of four plots, each treated to contain 0, 8, 16, and 24% salt. The percent of seeds emerging is observed 5, 8, 11, and 14 days after planting. The data are shown in Table 10.24.

The design is a repeated measures design since the subplot treatments have a fixed sequence rather than being randomly assigned. However, as a first approximation, it can be considered a split plot design with salt as main plots and days as subplots.

Note that days is a continuous variable. Analyze the data to determine the effect of salt. Also it is said that the Salt $*$ Day interaction is important; if it is statistically significant, interpret the results.

6. Two each of 40 samples of canned meat were stored at 2, 4.5, 21, and 38°C for periods of 1, 2, 4, 8, and 13 months, respectively. The two samples from each factor combination were randomly given to two taste panels who rated the samples on a continuous scale from 1 (excellent) to 8 (unacceptable). The data are given in Table 10.25; the two numbers in each combination are the ratings of panels 1 and 2, respectively. Analyze the data

Table 10.24

Grass Emergence Data for Exercise 5

Day	Salt	REPLICATION 1	2	3
5	0	68	79	74
8	0	75	89	81
11	0	75	89	82
14	0	75	89	82
5	8	70	55	74
8	8	84	73	87
11	8	87	74	88
14	8	87	75	80
5	16	40	43	36
8	16	78	81	70
11	16	82	85	74
14	16	83	87	74
5	24	11	18	12
8	24	62	75	50
11	24	72	82	62
14	24	72	86	66

Table 10.25

Meat Quality Data for Exercise 6

Time	TEMPERATURE 2	4.5	21	38
1	2.38	2.67	2.93	3.81
	2.19	2.39	2.72	3.07
2	2.74	2.81	2.97	4.14
	2.50	2.64	2.88	3.14
4	2.75	3.00	3.05	4.78
	2.74	2.79	3.21	3.45
8	3.28	3.58	3.68	5.78
	2.83	3.23	3.25	5.28
13	3.81	3.67	4.04	6.05
	3.05	3.61	4.23	7.14

to ascertain the relationship of the quality of meat to time and temperature. Note that both factors have numeric levels (Section 9.4).

7. An experiment was conducted to determine the effect of light and leaf age on the photosynthetic efficiency of a plant. The experiment was conducted as a split plot design with light intensities of 0, 33, and 90 units. The subplots are the sequence number of the first five leaves counted from the top of the plant. The leaf number is proxy for the age of the leaf, with the oldest leaf at the top, etc. There are five replications of the experiment. The data are given in Table 10.26.
   (a) Perform the analysis of variance to test for the existence of the effects of light and age.
   (b) Assuming that the leaf numbers represent equally spaced ages, determine whether a polynomial response can be used for the effects of light and age.
   (c) Perform a lack of fit test.

Table 10.26

Leafweight Data for Exercise 6

Rep	LEAF 1	2	3	4	5
			Light = 0		
1	1.91	2.21	2.01	1.83	2.05
2	1.88	2.12	2.06	1.93	2.23
3	2.03	2.28	2.08	1.81	1.95
4	2.01	2.16	2.08	2.19	1.97
5	2.37	2.13	2.17	2.08	1.94
			Light = 33		
1	2.03	2.59	2.22	2.11	2.15
2	2.27	2.64	2.47	2.41	2.03
3	2.12	2.56	2.49	2.23	2.61
4	2.17	3.06	2.86	2.75	2.86
5	2.32	2.56	2.24	2.30	2.24
			Light = 90		
1	2.40	2.71	2.83	2.80	2.53
2	2.07	2.34	2.10	2.18	1.85
3	2.03	2.37	2.45	2.33	2.19
4	2.27	2.85	2.99	2.55	2.86
5	2.12	2.23	2.23	2.30	2.15

8. This example of a randomized block design with sampling concerns an experiment for testing the effectiveness of three types of gasoline additives for boosting gas mileage on a specific type of car. Three randomly selected cars, which constitute blocks, are purchased for the experiment. Having purchased the cars and made all arrangements for conducting the test, the cost to repeat the individual trials is low. Hence, each additive is tested four (randomly ordered) times on each of the three cars. Thus we have a randomized block design with three treatments, three random effect blocks, and four samples in each treatment–block combination. The data, including cell and marginal totals, are given in Table 10.27. Perform the appropriate test for determining differences due to additives. Also compute the gain in efficiency of the randomized block design compared to the completely randomized design.

9. The productivity of dairy cows is reduced by heat stress. This experiment concerns the effect on body temperature of several types of cooling treatments. Two white and two black cows, identified by the variable Cowid, were randomly assigned to two exposures of four treatments: fan, shade, mist, and sun (control). The experiments were conducted at times for which environmental conditions were essentially identical. Two responses were measured: surface and rectal temperature. The data are show in Table 10.28.

Perform separate analyses for the two response variables for the effect of cow color and shade treatment. Note that this is a split plot with cow color as the main plot effect.

Table 10.27

Data for Randomized Block Design with Sampling (Responses in Miles per Gallon)

	CARS			
	A	B	D	Totals
Additive 1	19.5	21.1	21.1	
	20.3	21.2	22.8	
	19.4	20.4	21.7	
	21.3	20.6	21.6	
	80.5	83.3	87.2	251.0
Additive 2	18.0	21.8	20.5	
	17.8	21.2	20.1	
	17.8	22.7	21.2	
	15.8	23.1	20.9	
	69.4	88.8	82.7	240.9
Additive 3	16.8	21.5	18.5	
	17.0	20.2	19.6	
	16.8	18.6	20.3	
	15.5	20.3	18.8	
	66.1	80.6	77.2	223.9
Totals	216.0	252.7	247.1	715.8

10. One of many activities of birdwatchers is to count the number of birds of various species along a specified route. Table 10.29 shows the total number of birds of all species observed by birdwatchers for routes in three different cities observed at Christmas for each of the 25 years from 1965 through 1989. It is of interest to examine the year-to-year differences and especially to see whether there has been an upward or downward trend over these years. This is a randomized block design with routes (cities) as blocks.

Perform the analysis for studying the effect of years. Check assumptions and perform an alternative analysis if necessary. Interpret results.

11. In Exercise 12 of Chapter 8, 40 respondents (subjects) were asked to judge five structures for satisfaction on seven specific characteristics and also to give an overall satisfaction index. All responses are on a nine-point scale. The data are shown in Table 8.32.

We will now analyze the data to ascertain the nature of the differences among the buildings. For this purpose, the data result from a randomized block design with subjects as blocks and buildings as treatments. Perform separate analyses for each characteristic and the overall score. Summarize results.

12. An experimenter is interested in the effects of electric shock and level of white noise on human galvanic skin response (sweating). Five subjects were each exposed to all combinations of four levels of shock (0.25, 0.50, 0.75, and 1.00 mA) and two levels of noise (40 and 80 dB). The response is a coded indicator of sweat. The data are shown in Table 10.30. Perform the appropriate analysis including means comparison procedures.

Table 10.28

Shading Dairy Cows

Obs	Cowid	Trt	Color	Surface	Rectal
1	2056	fan	white	36.8	39.8
2	2056	fan	white	33.2	39.1
3	2056	mist	white	35.8	40.2
4	2056	mist	white	34.2	39.6
5	2056	shade	white	37.0	39.5
6	2056	shade	white	37.4	39.2
7	2056	sun	white	40.2	40.3
8	2056	sun	white	38.1	39.6
9	4055	fan	black	36.7	40.1
10	4055	fan	black	36.4	39.4
11	4055	mist	black	37.0	40.8
12	4055	mist	black	35.3	39.6
13	4055	shade	black	36.4	40.2
14	4055	shade	black	36.9	39.1
15	4055	sun	black	39.6	41.1
16	4055	sun	black	37.9	39.8
17	5042	fan	black	36.1	39.9
18	5042	fan	black	34.7	39.4
19	5042	mist	black	35.7	40.3
20	5042	mist	black	34.0	39.8
21	5042	shade	black	36.7	39.3
22	5042	shade	black	36.8	39.4
23	5042	sun	black	42.2	40.0
24	5042	sun	black	38.4	39.5
25	5055	fan	white	36.4	39.4
26	5055	fan	white	35.8	39.4
27	5055	mist	white	36.3	40.2
28	5055	mist	white	35.0	39.5
29	5055	shade	white	37.9	40.4
30	5055	shade	white	38.7	40.3
31	5055	sun	white	39.8	40.3
32	5055	sun	white	37.9	39.5

13. An experiment to determine feed efficiency at various feeding levels for the Walleye (fish) was conducted in a controlled environment. Five levels (labeled TREAT with values of 0.25, 0.75, 1.00, 2.00, and 3.00 units) of a major feed component were each randomly assigned to four buckets containing three fish each. After 10 weeks, the following measurements were recorded for each fish:

WEIGHT: weight in grams,
LENGTH: length in millimeters, and
RELWT: relative weight, which is the actual weight divided by a length-specific standard weight.

Perform the appropriate analysis of each response variable independently. Using individual fish measurements makes this a nested design. The data consisting of 60 observations are available on the data disk in file FW10P13. Note that we now can perform multiple regressions (instead

Table 10.29

Bird Counts for Twenty-Five Years

	ROUTE		
Year	A	B	C
65	138	815	259
66	331	1143	202
67	177	607	102
68	446	571	214
69	279	631	211
70	317	495	330
71	279	1210	516
72	443	987	178
73	1391	956	833
74	567	859	265
75	477	1179	348
76	294	772	236
77	292	1224	570
78	201	1146	674
79	267	661	494
80	357	729	454
81	599	845	270
82	563	1166	238
83	481	1854	98
84	1576	835	268
85	1170	968	449
86	1217	907	562
87	377	604	380
88	431	1304	392
89	459	559	425

Table 10.30

Data for Exercise 12

		SUBJECT				
Shock	Noise	1	2	3	4	5
40	0.25	3	7	9	4	1
40	0.50	5	11	13	8	3
40	0.75	9	12	14	11	5
40	1.00	6	11	12	7	4
80	0.25	5	10	10	6	3
80	0.50	6	12	15	9	5
80	0.75	18	18	15	13	9
80	1.00	7	15	14	9	7

of using orthogonal polynomials as an approximation) for estimating the response curves. This means that we do not need equally spaced factor levels.

14. In the experiment described in Exercise 13 the fish were actually measured every other week, identified by the variable WEEK, having values 0, 2, 4, 6, 8, and 10. For this exercise we will analyze the data from the entire experiment. The data consisting of 360 observations are available on the data disk in FW10P14. Analyze each response variable independently.

Table 10.31

Orientation	LOCATION 1	2	3
1	A = 31.0	B = 39.5	C = 30.5
2	B = 34.0	C = 24.5	A = 28.0
3	C = 15.0	A = 25.5	B = 31.0

The complete experiment can be effectively analyzed as a factorial experiment in a split plot design with buckets as main plots and weeks as subplots. (Because the weeks are sequential rather than random, the experiment would probably be better analyzed using a repeated measures design, a design we do not cover. See Winer (1971) for a complete discussion of repeated measures. However, the split plot analysis gives a good approximation.)

15. In Example 10.2 we looked at an analysis of three varieties of wheat in a randomized block design. Suppose that there were three different orientations of the rows in three different locations. Therefore we actually have two "blocking variables" instead of one. An experiment to compare yields of the three varieties would be more appropriately performed using a Latin square design. The results of such a study are given in Table 10.31. The presentation of the data does not show the randomization.

(a) Describe an appropriate randomization scheme for this experiment.

(b) Perform the appropriate analysis and explain the results. Do the two blocking variables appear to be effective in reducing extraneous variation? Explain your answer.

Other Linear Models

EXAMPLE 11.1

Survival of Cancer Patients Medical researchers are interested in determining how the survival time of cancer patients is affected by the grade of the tumor and age of the patient. The data in Table 11.1 show the result of a hospital study, showing the survival time of patients along with their ages and histological grade of their tumor. At first glance a regression of survival on grade and age would appear to be useful. However, histological grade is not strictly a numerical measure and is therefore more like a factor as used in analysis of variance models.

One of the topics of this chapter deals with the analysis of models that include both measured and categorical independent factors. The analysis is given in Section 11.8. ■

11.1 Introduction

The linear model was first introduced in Chapter 6 for justifying the use of the F ratio from the analysis of variance for inferences on a set of means. That model was written

$$y_{ij} = \mu + \tau_i + \varepsilon_{ij},$$

and relates the observed values of the response variable y to parameters representing differences among population means, denoted as $(\mu + \tau_i)$, and a random error, ε_{ij}.

The model is primarily used to make inferences about differences among the τ_i, the so-called treatment or factor level effects. The model was generalized in Chapters 9 and 10 by adding parameters representing different treatment factors and/or experimental conditions. For all such models, called

Table 11.1

Survival of Cancer Patients

OBS	SURVIVAL	AGE	GRADE	OBS	SURVIVAL	AGE	GRADE
1	69	46	1	41	94	31	3
2	80	68	1	42	92	60	3
3	78	58	1	43	1	80	3
4	85	32	1	44	45	84	3
5	30	59	1	45	13	83	3
6	63	61	1	46	8	65	3
7	73	52	1	47	25	80	3
8	100	57	1	48	13	73	3
9	73	65	1	49	72	63	3
10	96	58	1	50	63	62	4
11	72	60	1	51	31	61	4
12	4	61	2	52	16	63	4
13	83	62	2	53	108	58	4
14	38	75	2	54	20	66	4
15	107	64	2	55	1	73	4
16	72	53	2	56	70	82	4
17	82	66	2	57	6	69	4
18	70	62	2	58	13	60	4
19	105	62	2	59	69	67	4
20	8	86	2	60	53	53	4
21	68	67	2	61	13	61	4
22	8	58	2	62	80	56	4
23	67	91	2	63	103	79	4
24	96	58	2	64	63	67	4
25	91	61	2	65	67	63	4
26	90	66	2	66	67	82	4
27	97	68	2	67	75	70	4
28	89	60	2	68	43	72	4
29	89	77	2	69	78	74	4
30	90	44	2	70	41	68	4
31	66	68	2	71	77	61	4
32	100	68	2	72	1	46	4
33	88	42	2	73	64	69	4
34	64	57	2	74	7	62	4
35	30	66	2	75	31	48	4
36	87	59	2	76	18	78	4
37	28	57	2	77	6	60	4
38	73	67	2	78	2	50	4
39	60	70	3	79	1	80	4
40	67	64	3	80	32	62	4

analysis of variance models, the analysis is based on the partitioning of sums of squares and on using ratios of the resulting mean squares as test statistics for making inferences on parameters that represent the various means.

Another linear model, the **regression model**, was introduced in Chapter 7. This model relates the response variable y to an independent variable x using the relationship

$$y = \beta_0 + \beta_1 x + \varepsilon,$$

where the parameters β_0 and β_1 are the regression coefficients that specify the nature of the relationship and ε is the random error. This model was generalized in Chapter 8 to allow several independent variables. Again the inferences are based on the partitioning of sums of squares and computing F ratios using the corresponding mean squares. In the analysis of multiple regression models, it was also necessary to recognize the proper interpretations and inferences for partial coefficients, reflecting the effect of the corresponding independent variable over and above that provided by all the other variables in the model.

Although the analysis of variance and regression models appear to have different applications and somewhat different analysis procedures, we have already noted that we may use aspects of both models in some situations. For example, contrasts and orthogonal polynomials are used in analysis of variance situations but are in reality regressions using specially coded independent variables. In other words we can compute factor sums of squares for an analysis of variance by performing a regression analysis. Furthermore, if, in an analysis of variance problem for a t level factor, we compute the one degree of freedom sums of squares for $(t - 1)$ orthogonal contrasts, the total of the individual contrast sums of squares is equal to the sum of squares for that factor.

The unified treatment of regression and analysis of variance is referred to as the **general linear model**. A comprehensive presentation of statistical analyses based on general linear models is beyond the scope of this book. However, the availability of computer programs for performing complex analyses and the increasing frequency of the use of this methodology makes it desirable to present here an introduction to this topic. In this chapter we

- present the use of dummy variables to construct a linear regression model to represent an analysis of variance model,
- briefly indicate the problem of the singular $\mathbf{X'X}$ matrix resulting from the implementation of the dummy variable model,
- provide a simple example of where this method is needed in place of the standard analysis of variance methodology presented in previous chapters,
- give an example of a computer output using the dummy variable model, and
- extend the method to models containing both dummy and interval independent variables with emphasis on the analysis of covariance.

In addition, this chapter provides an introduction to the methodology for a regression analysis for a binary (binomial) response variable.

The primary purpose of this chapter is to introduce these methods and indicate appropriate uses and potential misuses. More comprehensive presentations can be found in Freund and Wilson (1998) and Littell *et al.* (2002). The underlying theory is described in Graybill (1976).

11.2 The Dummy Variable Model

We have noted that the validity of a regression requires no assumption on the nature of the independent variables except that they be measured without error. For example, we have seen that coefficients of orthogonal contrasts can be used as independent variables to compute sums of squares for such contrasts. In a similar manner, specially constructed variables, called **dummy** or **indicator** variables, can be used to construct a regression model equivalent to an analysis of variance model.

We begin this presentation by showing how such a dummy variable model can be used to obtain the analysis of variance results for a one-way classification or completely randomized design with t treatments or populations. The analysis of variance model as presented in Chapter 6 is written

$$y_{ij} = \mu + \tau_i + \varepsilon_{ij}.$$

The equivalent dummy variable model is

$$y_{ij} = \mu x_0 + \tau_1 x_1 + \tau_2 x_2 + \cdots + \tau_t x_t + \varepsilon_{ij},$$

where the x_i are so-called dummy or indicator variables, indicating the presence or absence of certain conditions for observations,

$$\begin{aligned} x_0 &= 1 \quad \text{for all observations;} \\ x_1 &= 1 \quad \text{for all observations occurring in population 1, and} \\ &= 0 \quad \text{otherwise;} \\ x_2 &= 1 \quad \text{for all observations occurring in population 2, and} \\ &= 0 \quad \text{otherwise;} \end{aligned}$$

and so forth for all t populations. As before, populations may refer to treatments or factor levels. The definitions of μ, τ_i, and ε_{ij} are as before.

This model certainly has the appearance of a regression model. Admittedly, the independent variables are not the usual interval variables that we have become accustomed to using as independent variables, but they do not violate any assumptions. The **X** and **Y** matrices for a set of data described by this model are

$$\mathbf{X} = \begin{bmatrix} 1 & 1 & 0 & \cdots & 0 \\ \cdot & \cdot & \cdot & \cdots & \cdot \\ \cdot & \cdot & \cdot & \cdots & \cdot \\ \cdot & \cdot & \cdot & \cdots & \cdot \\ 1 & 1 & 0 & \cdots & 0 \\ 1 & 0 & 1 & \cdots & 0 \\ \cdot & \cdot & \cdot & \cdots & \cdot \\ \cdot & \cdot & \cdot & \cdots & \cdot \\ \cdot & \cdot & \cdot & \cdots & \cdot \\ 1 & 0 & 1 & \cdots & 0 \\ \cdot & \cdot & \cdot & \cdots & \cdot \\ \cdot & \cdot & \cdot & \cdots & \cdot \\ \cdot & \cdot & \cdot & \cdots & \cdot \\ 1 & 0 & 0 & \cdots & 1 \\ \cdot & \cdot & \cdot & \cdots & \cdot \\ \cdot & \cdot & \cdot & \cdots & \cdot \\ \cdot & \cdot & \cdot & \cdots & \cdot \\ 1 & 0 & 0 & \cdots & 1 \end{bmatrix}, \quad \mathbf{Y} = \begin{bmatrix} y_{11} \\ \cdot \\ \cdot \\ \cdot \\ y_{1n_1} \\ y_{21} \\ \cdot \\ \cdot \\ \cdot \\ y_{2n_2} \\ \cdot \\ \cdot \\ \cdot \\ y_{t1} \\ \cdot \\ \cdot \\ \cdot \\ y_{tn_t} \end{bmatrix}.$$

It is not difficult to compute the $\mathbf{X'X}$ and $\mathbf{X'Y}$ matrices that specify the set of normal equations

$$\mathbf{X'XB} = \mathbf{X'Y}.$$

The resulting matrices are

$$\mathbf{X'X} = \begin{bmatrix} n & n_1 & n_2 & \cdots & n_t \\ n_1 & n_1 & 0 & \cdots & 0 \\ n_2 & 0 & n_2 & \cdots & 0 \\ \cdot & \cdot & \cdot & \cdots & \cdot \\ \cdot & \cdot & \cdot & \cdots & \cdot \\ \cdot & \cdot & \cdot & \cdots & \cdot \\ n_t & 0 & 0 & \cdots & n_t \end{bmatrix}, \quad \mathbf{B} = \begin{bmatrix} \mu \\ \tau_1 \\ \tau_2 \\ \cdot \\ \cdot \\ \cdot \\ \tau_t \end{bmatrix}, \quad \mathbf{X'Y} = \begin{bmatrix} Y_{..} \\ Y_{1.} \\ Y_{2.} \\ \cdot \\ \cdot \\ \cdot \\ Y_{t.} \end{bmatrix}.$$

An inspection of $\mathbf{X'X}$ and $\mathbf{X'Y}$ shows that the sums of elements of rows 2 through $(t+1)$ are equal to the elements of row 1. In other words, the equation represented by the first row contributes no information over and above those provided by the other equations. For this reason, the $\mathbf{X'X}$ matrix is singular (Appendix B); it has no inverse. Hence a unique solution of the set of normal equations to produce a set of parameter estimates is not possible.

The normal equations corresponding to all rows after the first represent equations of the form

$$\mu + \tau_i = \bar{y}_{i.},$$

which reveal the obvious: Each treatment mean $\bar{y}_{i.}$ estimates the mean μ plus the corresponding treatment effect τ_i. We can solve each of these equations for τ_i, producing the estimate

$$\hat{\tau}_i = \bar{y}_{i.} - \hat{\mu}.$$

Note, however, that the solution requires a value for $\hat{\mu}$. It would appear reasonable to use the equation corresponding to the first row to estimate $\hat{\mu}$, but we have already seen that this equation duplicates the rest and is therefore not usable for this task. This is the effect of the singularity of $\mathbf{X'X}$: There are really only t equations for solving for the $t+1$ parameters of the model.

A number of procedures for obtaining useful estimates from this set of normal equations are available. One principle consists of applying restrictions on values of the parameter estimates, a procedure that essentially reduces the number of parameters to be estimated. One popular restriction, which we have indeed used in previous chapters (see especially Section 6.3), is

$$\sum \tau_i = 0,$$

which can be restated

$$\tau_t = -\tau_1 - \tau_2 - \cdots - \tau_{t-1}.$$

This restriction eliminates the need to estimate τ_t from the normal equations; hence the rest of the parameters can be uniquely estimated. The resulting estimates are

$$\hat{\mu} = (1/t)\sum \bar{y}_{i.}$$

$$\hat{\tau}_i = \bar{y}_{i.} - \hat{\mu}, \quad i = 1, 2, \ldots, (t-1),$$

and $\hat{\tau}_t$ is computed by applying the restriction to the estimates, that is,

$$\hat{\tau}_t = -(\hat{\tau}_1 + \hat{\tau}_2 + \cdots + \hat{\tau}_{t-1}).$$

Note that the estimate of μ is not the weighted mean of treatment means we would normally use when sample sizes are unequal.

The inability to estimate directly all parameters and the necessity of applying restrictions are related to the degrees of freedom concept first presented in the estimation of the variance (Section 1.5). There we argued that, having already computed $\bar{y}$, we have lost one degree of freedom when we use that statistic to compute the sum of squared deviations for calculating the variance. The loss of that degree of freedom was supported by noting that $\sum(y - \bar{y}) = 0$. In the dummy variable model we start with t sample statistics, $\bar{y}_1, \bar{y}_2, \ldots, \bar{y}_t$. Having estimated the overall mean (μ) from these statistics, there are only $(t-1)$ degrees of freedom left for computing the estimates of the treatment effect parameters (the t values of the τ_i).

Other sets of restrictions may be used for the solution procedure, which will result in different numerical values of parameter estimates. For this reason, any set of estimates based on implementing a specific restriction is said to be biased. However, the existence of this bias is not in itself a serious detriment to the use of this method since these parameters are by themselves not overly useful. As we have seen, we are usually interested in functions of these parameters, especially contrasts or treatment means, and numerical values of estimates of these functions, called estimable functions, are not affected by the specific restrictions applied.

A simple example illustrates this property. Assume a four-treatment experiment with equal sample sizes for treatments: The means are 4, 6, 7, and 7, respectively. Using the restriction $\sum \tau_i = 0$, that is, the sum of treatment effects is zero, provides treatment effect estimates $\hat{\tau}_i = -2, 0, 1, 1$, respectively, and $\hat{\mu} = 6$. Another popular restriction is to make the last treatment effect zero, that is, $\hat{\tau}_4 = 0$. The resulting estimates of treatment effects are $-3, -1, 0, 0$, respectively, and $\hat{\mu} = 7$.

These two sets of estimates are certainly not the same. However, for example, the estimate of the mean response for treatment one $(\hat{\mu} + \hat{\tau}_1) = 4$, and likewise the estimate of the contrast $(\hat{\tau}_1 - \hat{\tau}_2) = -2$ for both sets of parameter estimates.

Another feature of the implementation of the dummy variable model is that numerical results of the partitioning of sums of squares are not affected by the particular restriction applied. This means that any hypothesis tests based on F ratios using the partitioning of sums of squares is a valid test for the associated hypotheses regardless of the specific restriction applied.

11.3 Unbalanced Data

The dummy variable method of performing an analysis of variance is certainly more cumbersome than the standard methods presented in Chapters 6, 9, and 10. Unfortunately, using those methods for unbalanced data, that is, data with unequal cell frequencies in a factorial or other multiple classification structure, produces incorrect results. However, use of the dummy variable approach does provide correct results for such situations and can also be used for the analysis of covariance (Section 11.5). Therefore, the added complexity required for this method is indeed worthwhile for these applications.

EXAMPLE 11.2

The incorrectness of results obtained by using the standard formulas for partitioning of sums of squares for unbalanced data is illustrated with a small example. Table 11.2 contains data for a 2 × 2 factorial experiment with unequal sample sizes in the cells. The table also gives the marginal means.

For purposes of illustration we want to determine whether there is an effect due to factor A.

Solution Looking only at the data for level 1 of factor C, the difference between the two factor A cell means is

$$\bar{y}_{11.} - \bar{y}_{21.} = \frac{1}{3}(4 + 5 + 6) - 5 = 0.$$

For level 2 of factor C, the difference between the two factor A cell means is

$$\bar{y}_{12.} - \bar{y}_{22.} = 8 - \frac{1}{2}(7 + 9) = 0.$$

Thus we may conclude that there is no difference in response due to factor A.

On the other hand, if we examine the difference between the marginal means for the two levels of factor A,

$$\bar{y}_{1..} - \bar{y}_{2..} = 5.75 - 7 = -1.25,$$

then, based on this result, we may reach the contradictory conclusion that there is a difference in the mean response due to factor A. Furthermore, since the standard formulas for sums of squares (Chapter 9) use these marginal

Table 11.2

Example of Unbalanced Factorial

	FACTOR C		
Factor A	**1**	**2**	**Means**
1	4 5 6	8	5.75
2	5	7 9	7.00
Means	5.00	8.00	6.285

means, the sum of squares for factor A computed in this manner will not be zero, implying that there is a difference due to the levels of factor A.

The reason for this apparent contradiction is found by examining the construct of the marginal means as functions of the model parameters. As presented at the beginning of Section 9.3, the linear model for the factorial experiment (we omit the interaction for simplicity) is

$$y_{ijk} = \mu + \alpha_i + \gamma_j + \varepsilon_{ijk}.$$

Each cell mean is an estimate of

$$\mu + \alpha_i + \gamma_j.$$

The difference between cell means for factor A for level 1 of factor C is

$$\bar{y}_{11.} - \bar{y}_{21.},$$

which is an estimate of

$$(\mu + \alpha_1 + \gamma_1) - (\mu + \alpha_2 + \gamma_1) = (\alpha_1 - \alpha_2),$$

which is the desired difference. Likewise the difference between the cell means for factor A for level 2 of factor C is

$$\bar{y}_{12.} - \bar{y}_{22.},$$

which is also an estimate of $(\alpha_1 - \alpha_2)$.

The marginal means are computed from all observations for each level; hence, they are weighted means of the cell means. In terms of the model parameters, the difference is

$$(\bar{y}_{1..} - \bar{y}_{2..}) = \frac{1}{4}(3\,\bar{y}_{11.} + \bar{y}_{12.}) - \frac{1}{3}(\bar{y}_{21.} + 2\,\bar{y}_{22.}),$$

which is an estimate of

$$\begin{aligned}
&\frac{1}{4}(3\mu + 3\alpha_1 + 3\gamma_1 + \mu + \alpha_1 + \gamma_2) - \frac{1}{3}(\mu + \alpha_2 + \gamma_1 + 2\mu + 2\alpha_2 + 2\gamma_2)\\
&\quad = (\mu + \alpha_1 + 0.75\gamma_1 + 0.25\gamma_2) - (\mu + \alpha_2 + 0.333\gamma_1 + 0.667\gamma_2)\\
&\quad = (\alpha_1 - \alpha_2) + (0.417\gamma_1 - 0.417\gamma_2).
\end{aligned}$$

In other words, the difference between the two marginal factor A means is not an estimate of only the desired difference due to factor A, $(\alpha_1 - \alpha_2)$, but it also contains a function of the difference due to factor C, $(0.417\gamma_1 - 0.417\gamma_2)$. Thus any parameter estimates and sums of squares for a particular factor computed from marginal means of unbalanced data will contain contributions from the parameters of other factors.

In a sense, unbalanced data represent a form of multicollinearity (Section 8.7). If the data are balanced, there is no multicollinearity, and we can estimate the parameters and sums of squares of any factor independent of those for any other factor, just as in regression we can separately estimate each individual regression coefficient if the independent variables are uncorrelated.

We noted in Chapter 8 that for multiple regression we compute partial regression coefficients, which in a sense adjust for the existence of multicollinearity. Therefore, if we use the dummy variable model and implement multiple regression methods, we estimate partial coefficients. This means that the resulting A factor effect estimates hold constant the C factor effects and vice versa.

Extensions of the dummy variable model to more complex models are conceptually straightforward, although the resulting regression models often contain many parameters. For example, interaction dummy variables are created by using all possible pairwise products of the dummy variables for the corresponding main effects. Nested or hierarchical models may also be implemented. ■

11.4 Computer Implementation of the Dummy Variable Model

Because dummy variable models contain a large number of parameters, they are by necessity analyzed by computers using programs specifically designed for such analyses. These programs automatically generate the dummy variables, construct appropriate restrictions, or use other methodology for estimating parameters and computing appropriate sums of squares, and provide, on request, estimates of desired estimable functions. We do not provide here details on the implementation of such programs, but show the results of the implementation of such a program on the 2×2 factorial presented in Table 11.2. A condensed version of the computer output from `PROC GLM` of the SAS System is given in Table 11.3. For this implementation we do specify the inclusion of interaction in the model. From Table 11.3, we obtain the following information:

- The first portion is the partitioning of sums of squares for the whole model. This is the partitioning we would get by performing the analysis of variance for four factor levels representing the four cells. Since this is computed as a regression, the output also gives the coefficient of determination (`R-SQUARE`). The F value, and its p value (`PR > F`), is that for testing for the model, that is, the hypothesis that all four treatment means are equal. The model is not statistically significant at the 0.05 level, a result to be expected with such a small sample.
- The second portion provides the partial sums of squares (which are called `TYPE III` sums of squares in this program) for the individual factors of the model. Note that the sum of squares due to factor A is indeed zero! Note also that the sum of the sums of squares due to the factors does not add to the model sum of squares as it would for the balanced case.
- The final portion provides the estimated treatment means. These are often referred to as adjusted or least squares means (`LSMEAN` in the output). Note that the least squares means for the two levels of factor A are indeed equal.

Table 11.3 Computer Analysis Example of Unbalanced Factorial

SOURCE	DF	SUM OF SQUARES	MEAN SQUARE	F VALUE	PR > F
MODEL	3	15.42857143	5.14285714	3.86	0.1484
ERROR	3	4.00000000	1.33333333		
CORRECTED TOTAL	6	19.42857143			

R-SQUARE = 0.794

SOURCE	DF	TYPE III SS	F VALUE	PR > F
A	1	0.00000000	0.00	1.0000
C	1	12.70588235	9.53	0.0538
A*C	1	0.00000000	0.00	1.0000

LEAST SQUARES MEANS

A	LSMEAN	STD ERR
1	6.50000000	0.66666667
2	6.50000000	0.70710678

C	LSMEAN	STD ERR
1	5.00000000	0.66666667
2	8.00000000	0.70710678

However, the standard errors of these means are not equal to $\sqrt{s^2/n}$, where n is the number of observations in the mean.[1]

11.5 Models with Dummy and Interval Variables

We consider in this section linear models in which some parameters describe effects due to factor levels and others represent regression relationships. Such models include dummy variables representing factor levels as well as interval variables associated with regression analyses. We illustrate with the simplest of these models, which has parameters representing levels of a single factor and a regression coefficient for one independent interval variable. The model is

$$y_{ij} = \beta_0 + \tau_i + \beta_1 x_{ij} + \varepsilon_{ij},$$

where τ_i are the parameters for factor level effects, β_0 and β_1 are the parameters of the regression relationship, and ε_{ij} is the random error.

If in this model we delete the term $\beta_1 x_{ij}$, the model is that for the completely randomized design (replacing β_0 with μ). On the other hand, if we delete the term τ_i, the model is that for a simple linear (one-variable) regression. Thus the

[1]The reader may note that the least squares means are the unweighted means of the cell means. This suggests that an appropriate analysis could be obtained by using these means directly. Although sometimes appropriate, this method is not universally applicable and is therefore not often used.

Figure 11.1

Data and Estimated Model for Analysis of Covariance

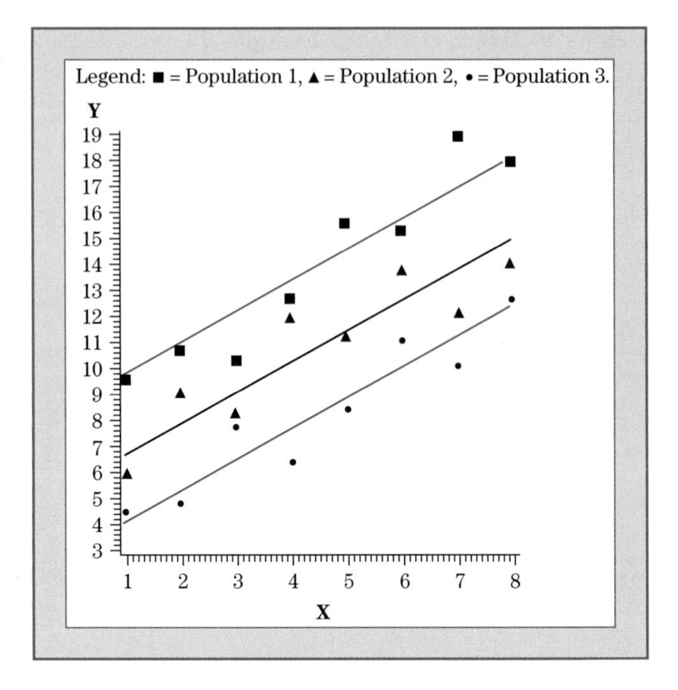

entire model describes a set of data consisting of pairs of values of variables x and y, arranged in a one-way structure or completely randomized design. The interpretation of the model may be aided by redefining parameters

$$\beta_{0i} = \beta_0 + \tau_i, \quad i = 1, 2, \ldots, t,$$

which produces the model

$$y_{ij} = \beta_{0i} + \beta_1 x_{ij} + \varepsilon_{ij}.$$

This model describes a set of t parallel regression lines, one for each treatment. Each has the same slope (β_1), but a different intercept (β_{0i}). A typical plot of such a data set with three treatments is given in Fig. 11.1, where the data points are plotted with different symbols for different populations, and the three lines are the three parallel regression lines.

Analysis of Covariance

The most common application of the model with dummy and interval variables is the analysis of covariance. The simplest case, which has the model described above, is for a completely randomized design (or a single classification survey study), where the values of the response variable are additionally affected by an interval variable. This variable, called a **covariate**, is assumed not to be affected by the factor levels of the experiment and usually reflects prior or environmental conditions of the observational units. Typical examples of covariates include

- aptitude test scores of students being exposed to different teaching strategies,
- blood pressures of patients prior to being treated by different drugs,
- temperatures in a laboratory where an experiment is being conducted, or
- weights of animals prior to a feeding rations experiment.

The purpose of the analysis of covariance is to estimate factor effects over and above the effect of the covariate. In other words, we want to obtain estimates of differences among factor level means that would occur if all observational units had the same value of the covariate. The resulting means are called adjusted treatment means (or least squares means) and are calculated for the mean of the covariate for all observations. Thus in Fig. 11.1, the mean of x is 4.5; hence the adjusted treatment means would be the value of $\hat{\mu}_{y|x}$ for each line at $x = 4.5$.

These inferences are meaningless if the covariate is affected by the factor levels; hence the analysis of covariance requires the assumption that the covariate is not affected by the factor levels. The model may still be useful if this assumption does not hold, but the inferences will be different (Section 11.6).

Another purpose of the analysis of covariance is to reduce the estimated error variance. In a sense, this is similar to the reduction in variance obtained by blocking: In both cases additional model parameters are used to account for known sources of variation that reflect different environments. In the analysis of covariance, an assumed linear relationship exists between the response and the environmental factor, while, for example, in the randomized block design the relationship is reflected by block differences.

EXAMPLE 11.3

We are studying the effect of some knowledge of "computer mathematics" on students' ability to learn trigonometry. The experiment is conducted using students in three classes, which correspond to three treatments (called CLASS) as follows:

CLASS 1: the control class in which students have had no exposure to computer mathematics,
CLASS 2: in which the students were exposed to a course in computer mathematics in the previous semester, and
CLASS 3: in which students have not had a course in computer mathematics, but the first three weeks of the trigonometry class are devoted to an introduction to computer mathematics.

The response variable, called POST, is the students' scores on a standardized test given at the end of the semester.[2] Two variables can be used as a

[2]In this example the experimental unit (Sections 1.2 and 10.3) is a class and the observational unit is a student. The proper variance for testing hypotheses about the teaching method would arise from differences among classes treated alike. However, since we do not have replications of classes for each teaching method, we cannot estimate that variance and must use the variation among students. Thus, the results we will obtain are based on the assumption that variation among classes is reflected by the variation among students, an assumption that is likely but not necessarily valid.

Table 11.4

Data for the Analysis of Covariance

CLASS 1			CLASS 2			CLASS 3		
PRE	POST	IQ	PRE	POST	IQ	PRE	POST	IQ
3	10	122	24	34	129	10	21	114
5	10	121	18	27	114	3	18	114
6	14	101	11	20	116	10	20	110
11	29	131	10	13	126	3	9	94
11	17	129	11	19	110	6	13	102
13	21	115	2	28	138	9	24	128
7	5	122	10	13	119	13	19	111
12	17	112	14	21	123	7	25	119
13	17	123	11	14	115	10	24	120
8	22	119	12	17	116	9	21	112
9	22	122	14	16	125	7	21	105
10	18	111	7	10	122	4	17	120
6	11	117	8	18	120	7	24	120
13	20	112	10	13	111	12	25	118
7	8	122	11	17	127	6	23	110
11	20	124	12	13	122	7	22	127
5	15	118	6	13	127			
9	25	113	3	13	115			
8	25	126	4	13	112			
2	14	132						
11	17	93						

covariate: an aptitude test score (IQ) and a pretest score (PRE) designed to ascertain knowledge of trigonometry prior to the course. The data are shown in Table 11.4. We use the variable PRE as the covariate. The variable IQ is used later in this section.

Solution The analysis of covariance model for these data is

$$y_{ij} = \beta_0 + \tau_i + \beta_1 x_{ij} + \varepsilon_{ij},$$

where y_{ij} = POST score of the jth student in the ith class; β_0 = mean POST score of all students having a PRE score of zero (an estimate having no practical interpretation); τ_i = effect of a student being in the ith section, $i = 1, 2, 3$; β_1 = change in score on the POST test associated with a unit change in the PRE test; x_{ij} = PRE score of the jth student in the ith class; and ε_{ij} = random error associated with each student.

As noted, this model is most efficiently analyzed through the use of computer program; hence we skip computational details and provide typical computer output. The program is the same one used for the analysis of variance for unbalanced data, therefore the output is similar to that of Table 11.3. The results are given in Table 11.5.

The first set of statistics is related to the overall model. The three degrees of freedom for the model are comprised of the two needed for the three treatments (`CLASS`) and one for the covariate (`PRE`). The F ratio has a p value (`PR > F`) of less than 0.0001; hence we conclude that the model can be used to explain variation among the `POST` scores.

Table 11.5 Results of the Analysis of Covariance

SOURCE	df	SUM OF SQUARES	MEAN SQUARE	F VALUE	PR > F
MODEL	3	609.03036550	203.01012183	8.46	0.0001
ERROR	52	1247.09463450	23.98258912		
CORRECTED TOTAL	55	1856.12500000			

R-SQUARE = 0.328119

SOURCE	df	TYPE III SS	F VALUE	PR > F
CLASS	2	228.70912117	4.77	0.0125
PRE	1	493.39220761	20.57	0.0001

PARAMETER	ESTIMATE	T FOR H0: PARAMETER=0	PR > |T|	STD ERROR OF ESTIMATE
PRE	0.77323836	4.54	0.0001	0.17047680

LEAST SQUARES MEANS

CLASS	POST LSMEAN	STD ERR LSMEAN
1	17.2899644	1.0705676
2	16.3334483	1.1512767
3	21.3484519	1.2429693

The second set of statistics relates to the partial contribution of the model factors: CLASS (teaching methods) and PRE, the covariate. Remember again that these partial sums of squares do not total to the model sum of squares.

We first test for the covariate, since if it is not significant we may need only to perform the analysis of variance, whose results are easier to interpret. The F ratio for H_0: $\beta_1 = 0$ is 20.57 with 1 and 52 degrees of freedom. The resulting p value is less than 0.0001; hence the covariate needs to remain in the model, which means that the PRE scores are a factor for estimating the POST scores.

The sum of squares for CLASS provides for the test of no differences among classes, holding constant the effect of the PRE scores. The resulting F ratio of 4.77 with 2 and 52 degrees of freedom results in a p value of 0.0125, and we may conclude that the inclusion of computer mathematics has had some effect on mean POST scores, holding constant individual PRE scores.

The actual coefficient estimates include estimates corresponding to the dummy variables for the CLASS variable and the regression coefficient for PRE. As noted in Section 11.2, the values of the dummy variate coefficients are of little interest since they are a function of the specific restriction employed to obtain a solution; hence they are not reproduced here (although they appear on the complete computer output). The estimate of the coefficient for the covariate is not affected by the nature of the restriction and thus is of interest. For this example the coefficient estimate is 0.773, indicating a 0.773 average increase in the POST score for each unit increase in the PRE score, holding constant the effect of CLASS. The standard error of this estimate (0.170) is used to test H_0: $\beta_1 = 0$. As in one-variable regression, the result is equivalent to that obtained by the F ratio in the preceding. The standard error may also be used for a confidence interval estimate on the regression coefficient.

Finally we have the adjusted treatment means, which are called LSMEAN (for least squares means) in this computer output. These are the estimated mean scores for the three classes at the overall mean PRE score: $\bar{x} = 8.95$. The estimated standard errors of these means may be used for confidence intervals. However, inferences on linear combinations of parameters, such as differences between means, require the use of methods referenced in Section 8.3 in the discussion of inferences for coefficients. Most computer programs can perform such inferences on request, and we illustrate this type of result by testing for all pairwise differences in the least squares means. Adapted from the computer output (which contains other information of no use at this point) the results are

Between Classes	**Estimated Difference**	**Std. Error**	t	**Pr > $\|t\|$**
1 and 2	0.957	1.582	0.60	0.5481
1 and 3	−4.058	1.632	−2.49	0.0161
2 and 3	−5.015	1.726	−2.91	0.0054

These results indicate that CLASS 3 (the one in which some computer mathematics is included at the beginning) appears to have a significantly higher mean score. Of course, these are LSD comparisons; hence p values must be used with caution (Section 6.5). Other multiple comparison techniques, such as Duncan's multiple range test, are usually not performed due to the correlations among the estimated least squares means and different standard errors for the individual comparisons.

The usefulness of the analysis of covariance for this example is seen by performing a simple analysis of variance for the POST scores. The mean square for CLASS is 57.819, and the error mean square is 32.84, which is certainly larger than the value of 23.98 for the analysis of covariance. The F ratio for testing the equality of mean scores is 1.76, which provides insufficient evidence of differences among these means. ■

Multiple Covariates

An obvious generalization of the analysis of covariance model is to have more than one covariate. Conceptually this is a straightforward extension, keeping in mind that the regression coefficients will be partial coefficients and the coefficient estimates may be affected by multicollinearity. Computer implementation is simple since the programs we have been discussing are already adaptations of multiple regression programs.

Table 11.6 Analysis of Covariance; Two Covariates

SOURCE	df	SUM OF SQUARES	MEAN SQUARE	F VALUE	Pr > F
MODEL	4	784.75195702	196.18798926	9.34	0.0001
ERROR	51	1071.37304298	21.00731457		
CORRECTED TOTAL	55	1856.12500000			

R-SQUARE=0.422790

SOURCE	df	TYPE III SS	F VALUE	Pr > F
CLASS	2	333.63171701	7.94	0.0010
PRE	1	502.18880915	23.91	0.0001
IQ	1	175.72159152	8.36	0.0056

PARAMETER	ESTIMATE	T FOR H0: PARAMETER=0	Pr > |T|	STD ERROR OF ESTIMATE
PRE	0.78018937	4.89	0.0001	0.15957020
IQ	0.21286146	2.89	0.0056	0.07359863

LEAST SQUARES MEANS

CLASS	POST LSMEAN	STD ERR LSMEAN
1	17.1760040	1.0027366
2	15.7734395	1.0947585
3	22.1630353	1.1969252

EXAMPLE 11.3

REVISITED We will use the data on the trigonometry classes, using both IQ and the pretest score (PRE) as covariates. The results of the analysis, using the format of Table 11.5, are given in Table 11.6.

Solution The interpretation parallels that of the analysis for the single-covariate model. The overall model remains statistically significant ($F = 9.34$, p value < 0.0001). Addition of the second covariate reduces the error mean square somewhat (from 23.98 to 21.01), indicating that the addition of IQ may be justified. The partial sums of squares (again labeled Type III) show that each of the factors (IQ, PRE, and CLASS) is significant for $\alpha < 0.01$.

The parameter estimates for the covariates have the usual partial regression interpretations: Increases of 0.780 and 0.213 units in the POST score are associated with a unit increase in PRE and IQ, respectively, holding other factors constant. The partial coefficient for PRE has changed little due to the addition of IQ to the model.

The adjusted treatment means have also changed very little from those of the single-covariate model. The standard errors are somewhat smaller, reflecting the decrease in the estimated error variance. The statistics for the pairwise comparisons follows. The implications of these tests are the same as with the single-covariate analysis.

Between Classes	Estimated Difference	Std. Error	t	Pr > $\|t\|$
1 and 2	1.402	1.489	0.94	0.3506
1 and 3	−4.987	1.561	−3.20	0.0024
2 and 3	−6.390	1.684	−3.80	0.0004

■

Unequal Slopes

The analysis of covariance model assumes that the slope of the regression relationship between the covariate and the response is the same for all factor levels. This homogeneity of slopes among factor levels is necessary to provide useful inferences on the adjusted means because, when the regression lines are parallel among groups, differences among means are the same everywhere. On the other hand, if this condition does not hold, differences in factor level means vary according to the value of the covariate. This is readily seen in Fig. 11.2 where, as in Fig. 11.1, the plotting symbols represent observations from three populations, and the lines are the three separate regression lines. We see that the differences in the mean response vary, depending on the value of x. Additional information on this and other problems associated with the analysis of covariance can be found in *Biometrics* **38**(3), 1982, which is entirely devoted to the analysis of covariance.

The existence of different slopes for the covariate among factor levels can be viewed as an interaction between factor levels and the covariate. The model is similar to that used for polynomial responses in a factorial experiment

Figure 11.2

Data and Estimated Model for Different Slopes

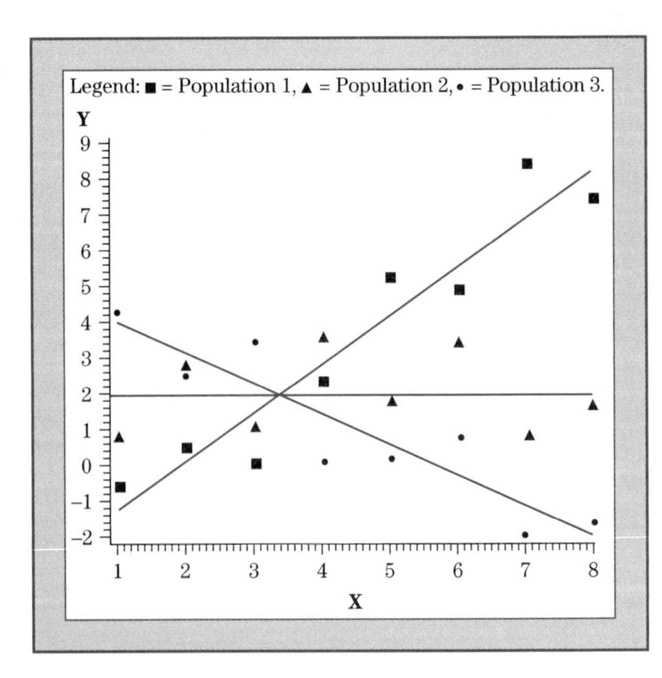

(Section 9.4). For example, consider a model for a single-factor experiment with three levels and a single covariate. The complete dummy variable model is

$$y_{ij} = \mu z_0 + \tau_1 z_1 + \tau_2 z_2 + \tau_3 z_3 + \beta_1 x + \beta_{11} z_1 x + \beta_{12} z_2 x + \beta_{13} z_3 x + \varepsilon,$$

where μ, and τ_i, $i=1,2,3$ are the treatment effects of the factor; z_i, $i = 0, 1, 2, 3$ are the dummy variables as defined in Section 11.2; β_1 is the regression coefficient, which measures the average effect of the covariate; x is the covariate; β_{1i}, $i = 1, 2, 3$ are regression coefficients corresponding to the interactions between the factor levels and the covariate; and $z_1 x$, $z_2 x$, and $z_3 x$ are the products of the dummy variables and the covariate and are measures of the interaction.

The first four terms are those of the analysis of covariance model. The next three terms are the interactions, which allow for different slopes. The slope of the covariate for the first factor level is $(\beta_1 + \beta_{11})$, for the second it is $(\beta_1 + \beta_{12})$, and for the third it is $(\beta_1 + \beta_{13})$. Then the test for the hypothesis of equal slopes is

$$H_0\colon \beta_{11} = \beta_{12} = \beta_{13} = 0.$$

Computer programs for general linear models will normally allow for interactions between factors and covariates and provide for estimating the different factor level slopes.

EXAMPLE 11.3

REVISITED We use PROC GLM of the SAS System to implement the model that includes the interaction terms and allows for the test for different slopes. For simplicity, we only use the variable PRE, the pretest, as a covariate. The results are shown in Table 11.7.

The first portion of the output shows that the model now has five degrees of freedom: two for CLASS, one for PRE, and two for the interaction. The mean square error (24.62) is actually somewhat larger than that for the model without the interaction shown in Table 11.5 (23.98), which suggests that the interaction is not significant. The second portion of the output, which shows that the p value for the interaction is 0.7235, reinforces this conclusion. Finally the last portion gives the estimates of the slopes for the three classes; they are actually somewhat different, but the differences are insufficient to be statistically significant. Recall that the slope of the ith class is really $\beta_1 + \beta_{1i}$.

If a computer program such as PROC GLM is not available, the test for different slopes can be performed, using the unrestricted–restricted model approach. The unrestricted model is that for a different regression for each factor level. The error sum of squares for that model can be obtained by simply running

Table 11.7 Analysis of Covariance: Test for Equal Slopes

The GLM Procedure

Dependent Variable: POST

Source	DF	Sum of Squares	Mean Square	F Value	Pr > F
Model	5	625.070867	125.014173	5.08	0.0008
Error	50	1231.054133	24.621083		
Corrected Total	55	1856.125000			

R-Square	Coeff Var	Root MSE	POST Mean
0.336761	27.37635	4.961964	18.12500

Source	DF	Type III SS	Mean Square	F Value	Pr > F
CLASS	2	37.2900877	18.6450439	0.76	0.4742
PRE	1	421.6803338	421.6803338	17.13	0.0001
PRE*CLASS	2	16.0405020	8.0202510	0.33	0.7235

Parameter	Estimate	Standard Error	t Value	Pr >|t|
pre class1	0.99468792	0.33829074	2.94	0.0050
pre class2	0.66692325	0.22680514	2.94	0.0050
pre class3	0.79790775	0.43280665	2.84	0.0712

regressions for each factor level, and manually combining the sums of squares and degrees of freedom. The restricted model is the analysis of covariance model. The test for the difference is obtained manually.

11.6 Extensions to Other Models

General linear models using dummy and interval variables can be used for virtually any type of data. Obvious extensions are those involving more complex treatment structures and/or experimental designs (see, for example, Littell *et al.*, 2002). Covarience models may be constructed for such situations. Finally, a model containing covariates and treatment factors need not strictly conform to the analysis of covariance model. For example, if the covariate is affected by treatments, the analysis may still be valid except that the interpretations of results will be somewhat more restrictive.

The dummy variable analysis may thus seem to provide a panacea: It seems that one can dump almost any data into such a model and get results. However, this approach must be used with extreme caution:

- Models with dummy variables may easily generate regression models with very many parameters, which may become difficult to implement even on large computers. This is especially true if interactions are included. Therefore, careful model specification is a must.

- Since the dummy variable model also provides correct results for the analysis of variance with balanced data, it is tempting to use this method always. However, computer programs for implementing dummy variable models require considerably greater computer resources than do the conventional analysis of variance programs and also usually produce more confusing output.
- Although dummy variable models provide correct results for unbalanced data, balanced data do provide for greater power of hypothesis tests for a given sample size. Thus proper design to assure balanced data is always a worthwhile effort.
- The existence of missing cells (cells without observations) in factorial experiments is a special case of unbalanced data for which even the dummy variable approach may not yield useful results (Freund, 1980).

11.7 Binary Response Variables

In a variety of applications we may have a response variable that has only two possible outcomes that can be represented by a dummy variable. Such a variable is often called a **quantal** or **binary** response. It is often useful to study the behavior of such a variable as related to one or more independent or factor variables. In other words, we may want to do a regression analysis where the dependent variable is a dummy variable and the independent variable or variables may be interval variables. For example:

- An economist may want to investigate the incidence of failure of savings and loan banks as related to the size of their deposits. The dependent variable can be coded as

  $$\begin{aligned} y &= 1 \quad \text{if the bank succeeded for five years,} \\ &= 0 \quad \text{if it failed within the five-year period.} \end{aligned}$$

  The independent variable is the average size of deposits at the end of the first year of business.
- A biologist investigating the effect of pollution on the survival of a certain species of organism can use the dependent variable

  $$\begin{aligned} y &= 1 \quad \text{if an individual of the species survived to adulthood,} \\ &= 0 \quad \text{if it died prior to adulthood.} \end{aligned}$$

  The independent variable is the level of pollution as measured in the habitat of this particular species.
- A study to determine the effect of an insecticide on insects can use a dependent variable defined as

$$y = 1 \quad \text{if an individual insect exposed to the insecticide dies,}$$
$$= 0 \quad \text{if the individual does not die.}$$

The independent variable is the strength of the insecticide.

In experiments of this type, the independent variable is often called the "dose" and the dependent variable the "response." In fact, this approach to modeling furnishes the foundation for a branch of statistics called **bioassay**. We will briefly discuss some methods used in bioassay later in this section. The reader is referred to Finney (1971) for a complete discussion of this subject.

A number of special procedures for analyzing models with a dichotomous dependent variable have been developed. Two such methods, which we will present in some detail, are as follows:

1. The standard linear model,

$$y = \beta_0 + \beta_1 x + \varepsilon.$$

2. The logistic regression model,

$$y = \frac{\exp(\beta_0 + \beta_1 x)}{1 + \exp(\beta_0 + \beta_1 x)} + \varepsilon.$$

The first model is a straight-line fit of the data, while the second model provides a special curved line. Both have practical applications and have been found appropriate in a wide variety of situations.

Before discussing the procedures for using sample data to estimate the regression coefficients for either model, we will examine the effect of using a dummy response variable.

Linear Model with a Dichotomous Dependent Variable

Dichotomous dependent or response variables occur quite often. For example, medical researchers are interested in determining whether the amount of a certain antibiotic given to mothers after a cesarean delivery affects the incidence of infection. The appropriate regression model for such a study is

$$y = \beta_0 + \beta_1 x + \varepsilon,$$

where

$y = 1$ if infection occurs within 2 weeks,
$= 0$ if not;
$x =$ amount of the antibiotic in ml/hr; and
$\varepsilon =$ random error (see Section 7.2), a random variable with mean zero and variance σ^2.

The researcher is to control x at specified levels for a sample of patients.

In this model the response variable only has values of 0 or 1, but the expected response has a readily interpretable meaning. The expected response is

$$\mu_{y|x} = \beta_0 + \beta_1 x;$$

hence the response variable can be seen to have the discrete probability distribution

y	$p(y)$
0	$1 - p$
1	p

where p is the probability that y takes on the value 1, meaning that the regression model actually predicts the probability that $y = 1$, the probability of a patient suffering a postoperative infection. In other words, the researcher is modeling the probability of postoperative infection for different strengths of the antibiotic.

Unfortunately, special problems arise with the regression process when the response variable is dichotomous. Recall that the error terms in a regression model should have a normal distribution with a constant variance for all observations. In the model that uses a dummy variable for a dependent variable, the error terms are neither normal nor do they have a constant variance.

According to the definition of the dependent variable, the error terms will have the values

$$\varepsilon = 1 - \beta_0 - \beta_1 x, \quad \text{when } y = 1,$$

and

$$\varepsilon = -\beta_0 - \beta_1 x, \quad \text{when } y = 0.$$

Obviously the assumption of normality does not hold for this model. Additionally, since y is a binomial variable, the variance of y is

$$\sigma^2 = p(1 - p).$$

However, $p = \mu_{y|x} = \beta_0 + \beta_1 x$; hence,

$$\sigma^2 = (\beta_0 + \beta_1 x)(1 - \beta_0 - \beta_1 x).$$

Clearly the variance depends on x, which is a violation of the equal variance assumption.

Finally, since $\mu_{y|x}$ is really a probability, its values are bounded by 0 and 1. This imposes a constraint on the regression model that limits the estimation of the regression parameters. In fact, a linear model may predict values for the dependent variable that are negative or larger than 1 even for values of the independent variable that are within the range of the sample data.

While these violations of the assumptions cause a certain amount of difficulty, solutions are available.

- The problem of nonnormality is mitigated by recalling that the central limit theorem indicated that the sampling distribution of the mean will be approximately normal for reasonably large samples. However, even in the case of a small sample, the estimates of the regression coefficients and consequently the estimated responses are unbiased point estimates.

- The problem of unequal variances is solved by the use of a procedure known as **weighted least squares** (discussed in the following subsection).
- If the linear model predicts values for $\mu_{y|x}$ that are outside the interval, we choose a curvilinear model that does not. The logistic regression model is one such choice.

Weighted Least Squares

We have noted that estimates of regression coefficients are those that result from the minimization of the residual sum of squares. This procedure treats each observation alike; that is, each observation is given equal weight in computing the sum of squares. However, when the variances of the observations are not constant, it is appropriate to weight observations differently. In the case of nonconstant variances, the appropriate weight to be assigned to the ith observation is

$$w_i = 1/\sigma_i^2,$$

where σ_i^2 the variance of the ith observation. This weights observations with large variances smaller than those with small variances. In other words, more "reliable" observations provide more information and vice versa.

Weighted least squares estimation is performed by a relatively simple modification of ordinary (unweighted) least squares. Determine the appropriate weights (or obtain reasonable estimates from a sample) and construct the matrix **W**,

$$\mathbf{W} = \begin{bmatrix} w_1 & 0 & \cdots & 0 \\ 0 & w_2 & \cdots & 0 \\ \cdot & \cdot & \cdots & \cdot \\ \cdot & \cdot & \cdots & \cdot \\ \cdot & \cdot & \cdots & \cdot \\ 0 & 0 & \cdots & w_n \end{bmatrix},$$

where w_i is the weight assigned to the ith observation. The weighted least squares estimates of the regression coefficients are then found by

$$\hat{\mathbf{B}} = (\mathbf{X}'\mathbf{W}\mathbf{X})^{-1}\mathbf{X}'\mathbf{W}\mathbf{Y}.$$

The estimated variances of these coefficients are the diagonal elements of

$$s_{\hat{B}}^2 = \text{MSE}(\mathbf{X}'\mathbf{W}\mathbf{X})^{-1}.$$

All other estimation and inference procedures are performed in the usual manner, except that the actual values of sums of squares as well as mean squares reflect the numerical values of the weights; hence they will not have any real interpretation. All computer programs have provisions for performing this analysis.

The problem, of course, is how to find the values of σ_i^2 needed to compute the w_i. In the model with a dichotomous response[3] variable, σ_i^2 is equal to $p_i(1 - p_i)$, where p_i is the probability that the ith observations is 1. We do not know this probability, but according to our model

$$p_i = \beta_0 + \beta_1 x_i.$$

Therefore, a logical procedure for doing weighted least squares to obtain estimates of the regression coefficients is as follows:

1. Use the desired model and perform an ordinary least squares regression to compute the predicted value of y for all x_i. Call these $\hat{\mu}_i$.
2. Estimate the weights by

$$\hat{w}_i = \frac{1}{\hat{\mu}_i(1 - \hat{\mu}_i)}.$$

3. Use these weights in a weighted least squares and obtain estimates of the regression coefficients.
4. This procedure may be iterated until the estimates of the coefficients stabilize. That is, repetition is stopped when estimates change very little from iteration to iteration.

Usually the estimates obtained in this way will stabilize very quickly, making step 4 unnecessary. In fact, in many cases, the estimates obtained from the first weighted least squares will differ very little from those obtained from the ordinary least squares procedure. Thus, ordinary least squares does give satisfactory results in many cases.

Although the estimates of coefficients usually change little due to weighting, the confidence and prediction intervals for the response will reflect the relative degrees of precision based on the appropriate variances. That is, intervals for observations having small variances will be smaller than those observations with large variances. However, even here the differences due to weighting may not be very large.

EXAMPLE 11.4

In a recent study of urban planning in Florida, a survey was taken of 50 cities, 24 of which used tax increment funding (TIF), and 26 did not. One part of the study was to investigate the relationship between the presence or absence of TIF and the median family income of the city. The data are given in Table 11.8.

The linear model chosen to describe these data is

$$y = \beta_0 + \beta_1 x + \varepsilon,$$

[3] Weighted least squares is also used for other applications where the variance of the random error is not constant. See, for example, Freund and Wilson (1998, Section 10.3).

Table 11.8

Data from Urban Planning Study

y	INCOME	y	INCOME
0	9.2	0	12.9
0	9.2	1	9.6
0	9.3	1	10.1
0	9.4	1	10.3
0	9.5	1	10.9
0	9.5	1	10.9
0	9.5	1	11.1
0	9.6	1	11.1
0	9.7	1	11.1
0	9.7	1	11.5
0	9.8	1	11.8
0	9.8	1	11.9
0	9.9	1	12.1
0	10.5	1	12.2
0	10.5	1	12.5
0	10.9	1	12.6
0	11.0	1	12.6
0	11.2	1	12.6
0	11.2	1	12.9
0	11.5	1	12.9
0	11.7	1	12.9
0	11.8	1	12.9
0	12.1	1	13.1
0	12.3	1	13.2
0	12.5	1	13.5

where

$y = 0$ if the city did not use TIF,
$\quad = 1$ if it did;
$x =$ median income of the city; and
$\varepsilon =$ the random error.

We thus have a model with a dichotomous response variable.

Solution The first step in obtaining the desired estimates of the regression coefficients is to perform an ordinary least squares regression. The results are given in Table 11.9. The values of the estimated coefficients are used to obtain the estimated values $\hat{\mu}_i$ for each x, which are then used to calculate weights for performing weighted least squares. *Caution:* The linear model can produce $\hat{\mu}_i$ values less than 0 or greater than 1. If this occurs, the weights will be undefined and an alternative model, such as the logistic model (see next subsection), must be considered. The predicted values and weights are given in Table 11.10. Note that none of the $\hat{\mu}_i$ is less than 0 or more than 1.

The computer output of the weighted least squares regression is given in Table 11.11. Note that these estimates differ very little from the ordinary least squares estimates in Table 11.9. Rounding the parameter estimates in the

Table 11.9

Regression of Income on TIF

```
Dependent Variable: Y
                         Analysis of Variance
                          Sum of          Mean
Source              df    Squares         Square          F Value       Pr > F

Model                1     3.53957       3.53957          19.003        0.0001
Error               48     8.94043       0.18626
C Total             49    12.48000
                         Parameter Estimates
                        Parameter      Standard        T for H0:
Variable            df   Estimate        Error        Parameter=0    Pr > |T|

INTERCEP             1   −1.818872    0.53086972        −3.426        0.0013
INCOME               1    0.205073    0.04704277         4.359        0.0001
```

Table 11.10

Calculated Weights for Weighted Regression

Y	INCOME	PRED	WEIGHT	Y	INCOME	PRED	WEIGHT
0	9.2	0.068	15.821	0	12.9	0.827	6.976
0	9.2	0.068	15.821	1	9.6	0.150	7.850
0	9.3	0.088	12.421	1	10.1	0.252	5.300
0	9.4	0.109	10.312	1	10.3	0.293	4.824
0	9.5	0.129	8.881	1	10.9	0.416	4.115
0	9.5	0.129	8.881	1	10.9	0.416	4.115
0	9.5	0.129	8.881	1	11.1	0.457	4.029
0	9.6	0.150	7.850	1	11.1	0.457	4.029
0	9.7	0.170	7.076	1	11.1	0.457	4.029
0	9.7	0.170	7.076	1	11.5	0.539	4.025
0	9.8	0.191	6.476	1	11.8	0.601	4.170
0	9.8	0.191	6.476	1	11.9	0.622	4.251
0	9.9	0.211	5.999	1	12.1	0.663	4.472
0	10.5	0.334	4.493	1	12.2	0.683	4.619
0	10.5	0.334	4.493	1	12.5	0.745	5.258
0	10.9	0.416	4.115	1	12.6	0.765	5.563
0	11.0	0.437	4.065	1	12.6	0.765	5.563
0	11.2	0.478	4.008	1	12.6	0.765	5.563
0	11.2	0.478	4.008	1	12.9	0.827	6.976
0	11.5	0.539	4.025	1	12.9	0.827	6.976
0	11.7	0.580	4.106	1	12.9	0.827	6.976
0	11.8	0.601	4.170	1	12.9	0.827	6.976
0	12.1	0.663	4.472	1	13.1	0.868	8.705
0	12.3	0.704	4.794	1	13.2	0.888	10.062
0	12.5	0.745	5.258	1	13.5	0.950	20.901

output we get the desired regression equation

$$\hat{\mu}_{y|x} = -1.980 + 0.219(\text{INCOME}).$$

The p value of 0.0001 suggests that median income does have a bearing on the participation in TIF, but the `R-Square` value of 0.45 indicates a rather poor fit. The resulting model can be used, for example, to estimate the probability that

Table 11.11

Weighted Regression of Income on TIF

```
Model: MODEL1
Dependent Variable: Y
                        Analysis of Variance
                     Sum of          Mean
Source       DF      Squares         Square          F Value       Pr > F

Model         1     36.49604        36.49604         38.651        0.0001
Error        48     45.32389         0.94425
C Total      49     81.81993

      Root MSE        0.97172       R-square        0.4461
      Dep Mean        0.45216       Adj R-sq        0.4345
      C.V.          214.90837

                     Parameter Estimates
                   Parameter      Standard        T for H0:
Variable     df    Estimate         Error        Parameter=0    Pr < |T|

INTERCEP      1    -1.979665     0.39479503       -5.014        0.0001
INCOME        1     0.219126     0.03524632        6.217        0.0001
```

a city with median income of \$10,000 uses TIF is $-1.980 + 0.219(10) = 0.213$. That is, there is about a 21% chance that a city with median income of \$10,000 is participating in TIF.

To illustrate the fact that the weighted least squares estimate stabilizes quite rapidly, two more iterations were performed. The results are

$$\text{Iteration 2: } \hat{\mu}_{y|x} = -1.992 + 0.2200(\text{INCOME})$$

and

$$\text{Iteration 3: } \hat{\mu}_{y|x} = -2.015 + 0.2218(\text{INCOME}).$$

The regression estimates change very little, and virtually no benefit in the standard error of the estimates is realized by the additional iterations.

The estimates of parameters using weighted least squares often differ very little from those obtained by ordinary least squares. The major difference in the results occurs in the standard errors of the estimated mean (or predicted individual) values, which are smaller for those observations having larger weights (and vice versa). In other words, observations with smaller variances will be predicted with greater precision. Thus for a binomial response variable this means larger variances for estimates near 0.5 and smaller variances for estimates near 0 or 1.

Figure 11.3 shows the actual, predicted, and 0.95 confidence interval values for the conditional mean for both the unweighted and weighted estimates. The predicted values are indeed very similar, but the confidence intervals are somewhat narrower at the extreme values for the weighted estimates (where the variances are smaller).

Figure 11.3

Plots of Response for Unweighted and Weighted Regressions

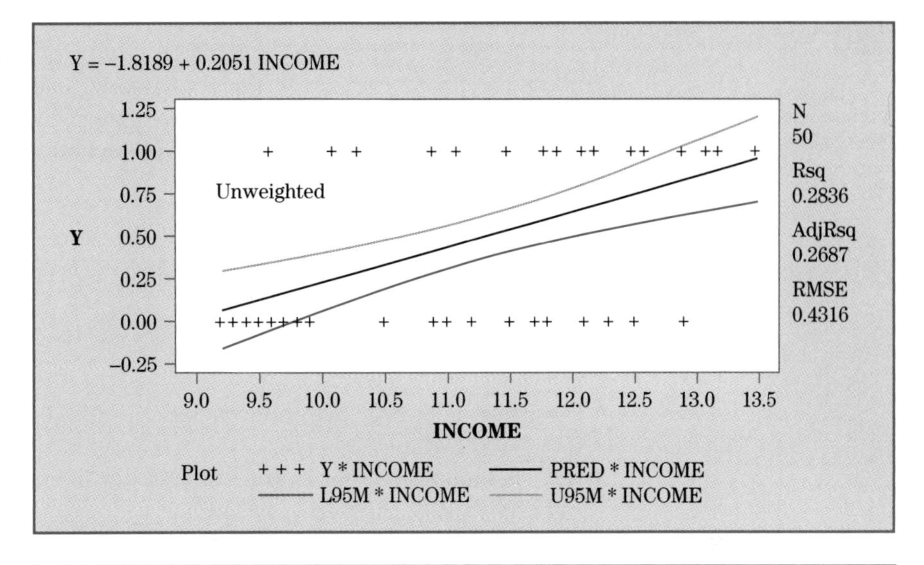

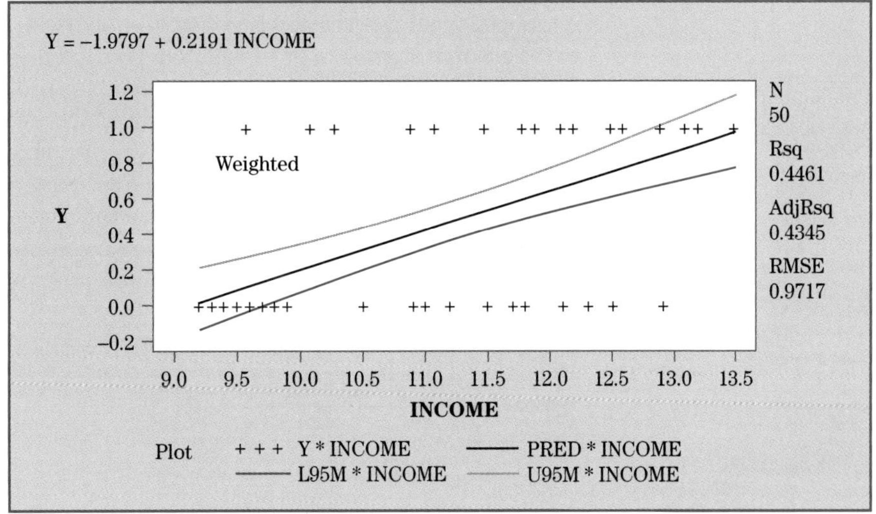

Note that neither the weighted nor unweighted regressions estimate values less than 0 nor greater that 1 in the range of the data (although the confidence intervals do.) However, for the weighted analysis the intervals are slightly narrower at the upper and lower ranges of the response but not sufficiently narrow to avoid including these impossible values.

The fact that the linear model often does not do well when estimating probabilities near 0 and 1 suggests the use of alternate models such as the logistic model presented next. However the linear model can be useful as a first approximation, and is also more easily applied in more complex situations such as multiple regression or analysis of variance models. ■

Logistic Regression

If a simple linear regression equation model using weighted least squares violates the constraints on the model, we may need to use a curvilinear model. One such model with a wide range of applicability is the logistic regression model given early in Section 11.7:

$$\mu_{y|x} = \frac{\exp(\beta_0 + \beta_1 x)}{1 + \exp(\beta_0 + \beta_1 x)}.$$

The curve described by the logistic model has the following properties:

- As x becomes large, $\hat{\mu}_{y|x}$ approaches 1 if $\beta_1 > 0$, or approaches 0 if $\beta_1 < 0$.
- $\mu_{y|x} = 1/2$ when $x = -(\beta_0/\beta_1)$.
- $\mu_{y|x}$ is monotone, that is, the curve either increases (or decreases) everywhere.

A typical logistic regression function is shown in Fig. 11.4. Note that the shape of the graph is sigmoidal or "S" shaped. This feature makes it more useful when there are observations for which the response probability is near 0 or 1 since the curve can never go below 0 or above 1.

While the function itself appears very complex, it is, in fact, relatively easy to use. The model has two unknown parameters, β_0 and β_1. It is not coincidental that these parameters have the same symbols as the simple linear regression model. Estimating the two parameters from sample data is

Figure 11.4

Typical Logistic Curve

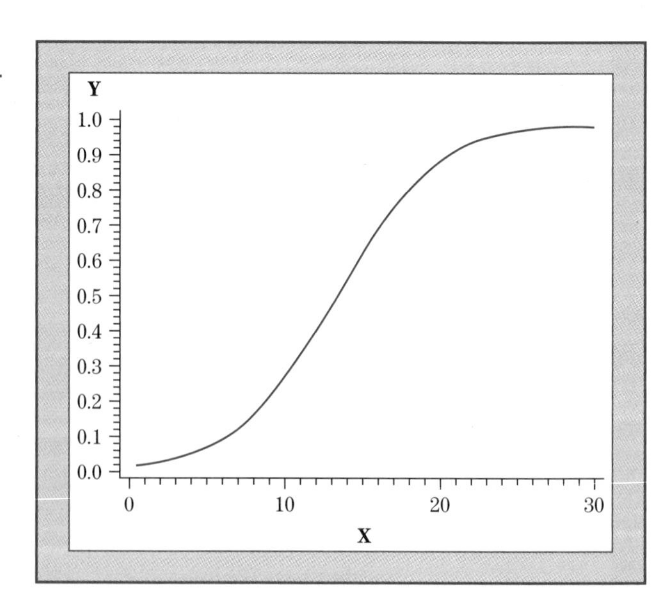

reasonably straightforward. We first make a **logit** transformation of the form

$$\mu_p = \ln\left(\frac{\mu_{y|x}}{1 - \mu_{y|x}}\right),$$

where ln is the natural logarithm. Substituting this transformation for $\mu_{y|x}$ in the logistic model results in a model of the form

$$\mu_p = \beta_0 + \beta_1 x + \varepsilon,$$

which is a simple linear regression model. Of course, the values of the μ_p are usually not known; hence preliminary estimates must be used. If multiple observations exist for each x, preliminary estimates of the $\mu_{y|x}$ are simply the sample proportions. If such multiples are not available, estimates from a preliminary linear model may be used.

The logit transformation linearizes the model, but does not eliminate the problem of nonconstant variance. Therefore, the regression coefficients in this simple linear regression model should be estimated using weighted least squares. We will illustrate the procedure with an example where multiple observations for each value of x are used as preliminary estimates of $\mu_{y|x}$.

EXAMPLE 11.5

A toxicologist is interested in the effect of a toxic substance on tumor incidence in a laboratory animal. A sample of animals is exposed to various concentrations of the substance, and subsequently examined for the presence or absence of tumors. The response for an individual animal is then either 1 if a tumor is present or 0 if not. The independent variable is the concentration of the toxic substance (CONC). The number of animals at each concentration (N) and the number having tumors (NUMBER) comprise the results, which are shown in Table 11.12.

Solution Again we have a model with a dichotomous response variable. However, in this example we have repeated observations for each value of the response variable; hence we can directly use the proportion having tumors as the preliminary estimates of $\mu_{y|x}$.

The first step is to use the logit transformation to "linearize" the model. The second step consists of the use of weighted least squares to obtain estimates of the unknown parameters. Because the experiment was conducted at only six distinct values of the independent variable, concentration of the substance, the task is not difficult.

Table 11.12

Data on Tumors Related to Toxic Substance

OBS	CONC	N	NUMBER
1	0.0	50	2
2	2.1	54	5
3	5.4	46	5
4	8.0	51	10
5	15.0	50	40
6	19.5	52	42

Table 11.13

Data for Linear Regression Estimation

OBS	CONC	N	NUMBER	PHAT	LOG	W
1	0.0	50	2	0.04000	−3.17805	1.92000
2	2.1	54	5	0.09259	−2.28238	4.53704
3	5.4	46	5	0.10870	−2.10413	4.45652
4	8.0	51	10	0.19608	−1.41099	8.03922
5	15.0	50	40	0.80000	1.38629	8.00000
6	19.5	52	42	0.80769	1.43508	8.07692

Table 11.14

Logistic Regression Estimates

```
Dependent Variable: LOG
                        Analysis of Variance
                       Sum of           Mean
Source        df      Squares          Square          F Value         Pr > F

Model          1     97.79495        97.79495           56.063         0.0017
Error          4      6.97750         1.74437
C Total        5    104.77245

          Root MSE           1.32075      R-square         0.9334
          Dep Mean          -0.41382      Adj R-sq         0.9168
          C.V.            -319.15882

                        Parameter Estimates
                      Parameter       Standard         T for H0:
Variable      df      Estimate          Error        Parameter = 0    Pr > |T|

INTERCEP       1     -3.138831     0.42690670           -7.352          0.0018
CONC           1      0.254274     0.03395972            7.488          0.0017
```

We calculate $\hat{p}$, the proportion of ones at each value of CONC. These are given in Table 11.13 under the column PHAT. We then make the logit transformation on the resulting proportions:

$$\text{LOG} = \ln\left[\frac{\hat{p}}{1-\hat{p}}\right] = \ln\left[\frac{\text{PHAT}}{1-\text{PHAT}}\right].$$

These are given in Table 11.13 under the column LOG.

Because the variances are still not constant, it is necessary to perform a weighted least squares regression, using LOG as the dependent variable, concentration as the independent variable, and weights computed as

$$\hat{w}_i = n_i \hat{p}_i(1 - \hat{p}_i),$$

where n_i = total number of animals at concentration x_i, and $\hat{p}_i$ = sample proportion of animals with tumors at concentration x_i. These values are listed in Table 11.13 under the column W.

The results of the weighted least squares estimation are given in Table 11.14. The model is certainly significant with a p value of 0.0017. The coefficient of determination appears to be a respectable 0.93, but this is the fit to the means rather than the original data. The residual variation is somewhat difficult to interpret since we are using the log scale.

Figure 11.5

Plot of Logistic Regression

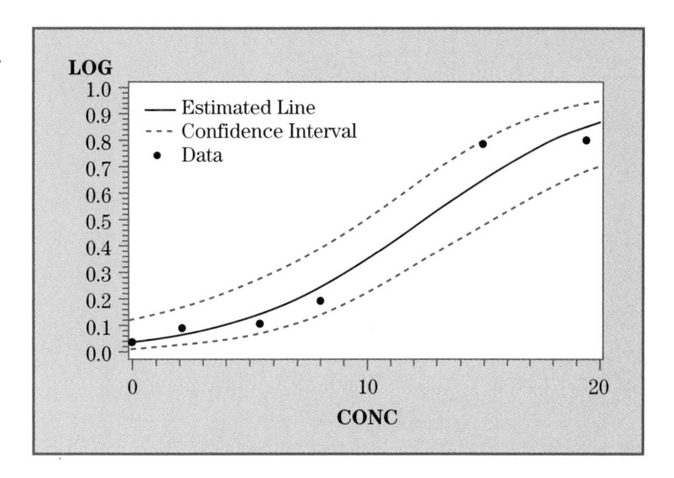

The coefficients of the estimated simple linear regression model are rounded to give

$$\hat{\text{LOG}} = -3.14 + 0.25x.$$

This can be transformed back into the original units using the transformation

$$\text{ESTPROP} = \left[\frac{\text{antilog}(\hat{\text{LOG}})}{1 + \text{antilog}(\hat{\text{LOG}})}\right].$$

The values of the variable ESTPROP provide data to construct the estimated logistic response curve. The data points, estimated response curve, and 0.95 confidence intervals are shown in Fig. 11.5. Note that for this model neither the response curve nor confidence intervals have values less than 0 or greater than 1. From this plot we can see, for example, that the estimated probability of a tumor when the concentration is 10 units is approximately 0.35. For a more precise estimate, we substitute 10 for CONC and obtain

$$\hat{\text{LOG}} = -3.139 + 0.2542(10) = -0.597.$$

This value is then transformed back to the original units by

$$\text{antilog}(-0.596) = 0.551 = \hat{\mu}_{y|x}/(1 - \hat{\mu}_{y|x}).$$

Solving for $\hat{\mu}_{y|x}$ results in the estimate 0.355. This means that, on the average, there is an estimated 35.5% chance that exposure to concentrations of 10 units results in tumors on laboratory animals. ■

Another feature of the logistic regression function is the interpretation of the coefficient β_1. Recall that we defined μ_p as

$$\mu_p = \ln\left(\frac{\mu_{y|x}}{1 - \mu_{y|x}}\right).$$

The quantity $\{\mu_{y|x}/(1 - \mu_{y|x})\}$ is called the odds, that is, the odds in favor of the event, in this case, having a tumor. Then μ_p, the log of the odds at x, is denoted as ln{odds at x}. Suppose we consider the same value at $(x + 1)$. The value

$$\mu_p = \ln\left(\frac{\mu_{y|x+1}}{1 - \mu_{y|x+1}}\right)$$

would be ln {odds at $(x + 1)$}. According to the linear model, ln{odds at x} = $\beta_0 + \beta_1 x$ and ln {odds at $(x + 1)$} = $\beta_0 + \beta_1(x + 1)$. It can be shown that the difference between the odds at $(x + 1)$ and at x is

$$\ln\{\text{odds at } (x + 1)\} - \ln\{\text{odds at } x\} = \beta_1,$$

which is equivalent to

$$\ln\{(\text{odds at } x + 1)/(\text{odds at } x)\} = \beta_1.$$

Taking antilogs of both sides gives the relationship

$$\frac{\text{odds at } (x + 1)}{\text{odds at } x} = e^{\beta_1}.$$

The estimate of this quantity is known as the **odds ratio**, and is interpreted as an increase in the odds for a unit increase in the independent variable. In our example, $\beta_1 = 0.25$; hence the odds ratio is $e^{0.25} = 1.28$. Therefore, the odds of getting a tumor are estimated to increase by 28% with a unit increase in concentration of the toxin.

Other Methods

The procedure presented in this section will not work for data in which one or more of the distinct x values has a $\hat{p}$ of 0 or 1, because the logit is undefined for these values. Modifications in the definition of these extreme values that remedy this problem can be made. One procedure is to define $\hat{p}_i$ to be $1/2n_i$ if the sample proportion is zero and $\hat{p}_i$ to be $(1 - 1/2n_i)$ if the sample proportion is one, where n_i is the number of observations in each factor level.

When we do not have multiple observations at some or all of the x values, and the initial linear model provides estimates of p outside the range 0 to 1, an alternative method is used to obtain estimates of the unknown coefficients in the logistic regression function itself. This method is known as **maximum likelihood**. The method of maximum likelihood uses the logistic function and the distribution of Y to obtain estimates for the coefficients. The procedure is complex and usually requires numerical method algorithms; hence logistic regression of this type is done on a computer. Most computer packages equipped to do logistic regression offer this option.

EXAMPLE 11.4

REVISITED The method we used to estimate the logistic regression for Example 11.5 could not be used for Example 11.4 because we did not have replicated observations for each value of the independent variable. Instead we use the maximum likelihood principle, which is implemented here with PROC

Table 11.15

Maximum Likelihood Estimates

Analysis of Maximum Likelihood Estimates

Parameter	DF	Estimate	Standard Error	Chi-Square	Pr > ChiSq
Intercept	1	−11.3472	3.3511	11.4660	0.0007
INCOME	1	1.0018	0.2954	11.5013	0.0007

Odds Ratio Estimates

Effect	Point Estimate	95% Confidence	Wald Limits
INCOME	2.723	1.526	4.858

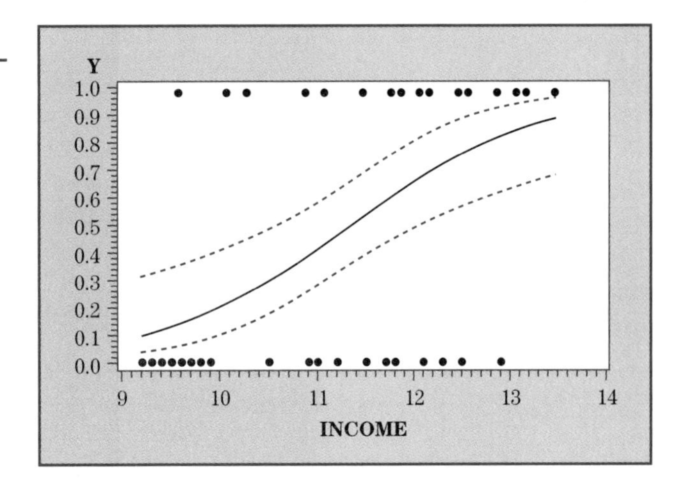

Figure 11.6

Logistic Response Curve for Example 11.4

LOGISTIC from the SAS System. Table 11.15 shows the portion of the output that contains the estimated parameters and hypothesis tests.

The first thing we notice is that the maximum likelihood method does not produce the usual statistics we associate with a regression analysis. Instead the statistical significance tests are provided by χ^2 statistics, which indicate that the coefficient for INCOME is statistically significant. The procedure automatically computes the odds ratio, which indicates that the odds of having TIF funding increases by 272% for each unit ($1000) increase in median income (or more correctly, between 153% and 486% with 95% confidence).

The plot of the response curve and the 0.95 confidence intervals are shown in Fig. 11.6. The results appear to be much more reasonable than they were for the linear regression model.

Another curvilinear model often used with binary responses uses the cumulative normal distribution instead of the logit transformation. This model is known as the **probit model**. The reader is referred to Finney (1971) for a complete discussion of the probit model and its applications. ■

11.8 CHAPTER SUMMARY

Solution to Example 11.1 We can now see that the analysis of covariance, using AGE as the covariate, is appropriate for determining the effect of the grade of tumor on survival. However, it is not at all certain that the effect of age is the same for all tumor grades, so a test for equal slopes is in order. Table 11.16 shows the analysis provided by PROC GLM of the SAS System in which the test for equality of slopes is provided by the test for significant interaction between GRADE and AGE.

We see that the model does not fit particularly well, but this result occurs frequently with this type of data. The interaction is statistically significant at $\alpha = 0.05$; hence we conclude that the slopes are not equal. The estimated slopes, standard errors, and tests for the hypothesis of zero slope for the three tumor grades are shown at the bottom of the output in Table 11.16. The slopes do differ, but only for tumor grade 3 does there appear to be sufficient evidence of a decrease in survival with age. This is illustrated in Fig. 11.7, which plots the estimated survival rates for the four grades. Of course, the standard deviation of the residuals (not shown) is nearly 30, which suggests that there is considerable variation around these lines.

Because survival times tend to have a skewed distribution, stem and leaf and box plots of residuals (reproduced from PROC UNIVARIATE of the SAS System) are shown in Fig. 11.8. These plots do not suggest any major violation of assumptions. ■

Table 11.16 Analysis of Covariance with Unequal Slopes General Linear Models Procedure

Dependent Variable: SURVIVAL

Source	df	Sum of Squares	Mean Square	F value	Pr > F
Model	7	24702.29684	3528.89955	4.07	0.0008
Error	72	62479.65316	867.77296		
Corrected Total	79	87181.95000			

R-Square	C.V.	Root MSE	SURVIVAL Mean
0.283342	52.16111	29.45799	56.4750000

Source	df	Type III SS	Mean Square	F Value	Pr > F
AGE	1	1389.557345	1389.557345	1.60	0.2098
GRADE	3	7937.607739	2645.869246	3.05	0.0340
AGE*GRADE	3	7137.205926	2379.068642	2.74	0.0494

Parameter	Estimate	T for H0: Parameter = 0	Pr > |T|	Std Error of Estimate
AGE, GRADE 1	−0.27049180	−0.29	0.7750	0.94292717
AGE, GRADE 2	−0.61521226	−1.11	0.2705	0.55407143
AGE, GRADE 3	−1.61350036	−2.61	0.0109	0.61710190
AGE, GRADE 4	0.75665954	1.34	0.1848	0.56509924

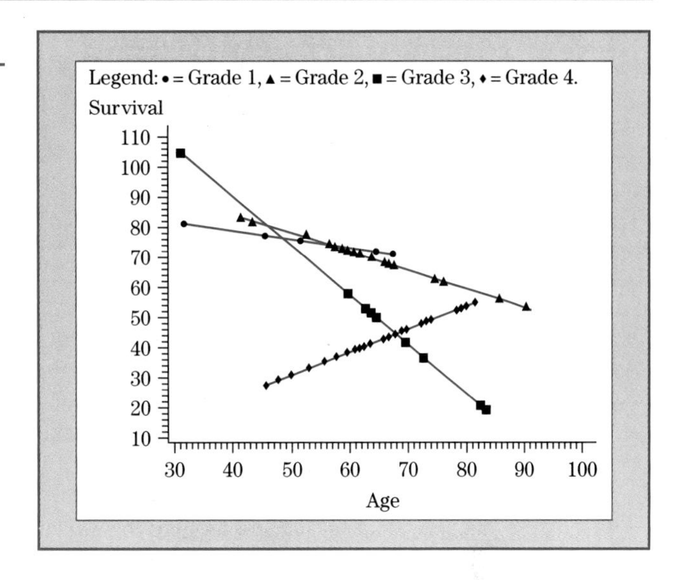

Figure 11.7

Plots of Predicted Survival

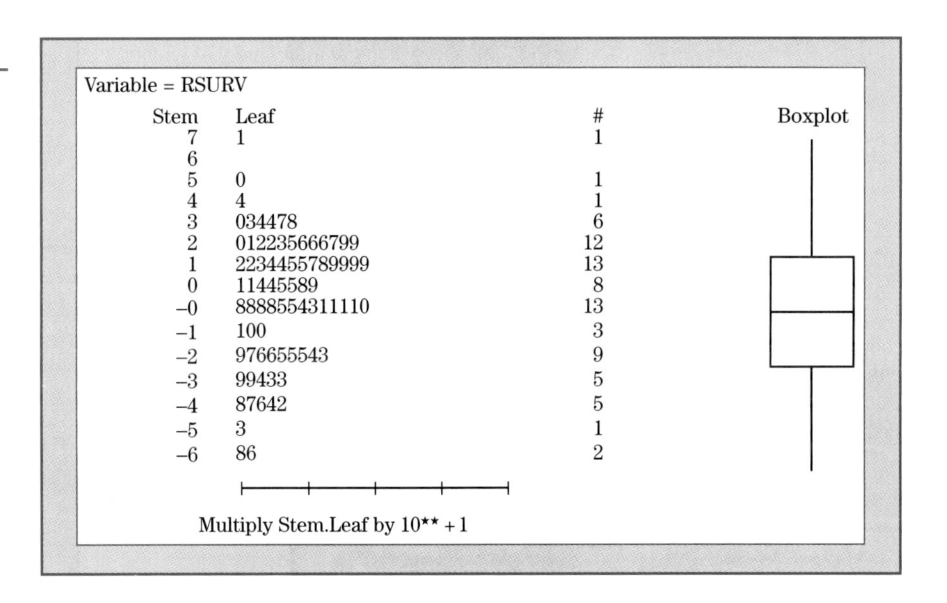

Figure 11.8

Analysis of Residuals

An Example of Extremely Unbalanced Data

EXAMPLE 11.6

Many times an experiment is conducted under conditions in which the physical constraints on the collection of the data result in severely unbalanced data. In fact, often these constraints result in data being completely missing from some combinations of factors. While this problem can be addressed if recognized in advance by using some form of incomplete design, it often is not recognized prior to collection of the data. By using the methods developed in this and previous chapters, we can still obtain extremely useful results. A good example of this is in a study of the relation between longevity (or survival) of patients with a disease and the symptoms presented by the patients. One such experiment is given by C. J. Maloney in the Disease Severity Quantitation, Proceedings of the Seventh Conference on the Design of Experiments in Army Research and Development and Testing (1961). This study presented data on 802 patients who exhibited 71 (out of a possible 128) different combinations of the presence or absence of seven symptoms. The response variable was the logarithm of survival time in months.

Solution The data, available on the data disk in file FW11X05, list the presence (1) or absence (0) of each of the seven symptoms and the mean log survival of all patients with a particular symptom pattern. Excerpts from this data set are presented in Table 11.17. The symptoms are labeled x_1 to x_7. Note that there were two patients who exhibited symptoms 1, 3, 4, and 5, and these two patients' mean log survival time was 1.66.

What we would like to do now is investigate the relationship between symptoms and survival. We can do this using the general linear model with appropriately defined variables. The variables we will use in the model are

x_1 to x_7: indicator variables that indicate presence (1) or absence (0) of each symptom,
LOGSURV: mean log of survival, and
N: the number of patients with that combination of symptoms.

Although the data on the file FW11X05 only contains the mean survival time of patients with the same set of symptoms, the original report gave the analysis of variance results using the individual survival times to test the hypothesis of no differences among symptom patterns. This analysis is reproduced in Table 11.18, which clearly indicates that such differences exist.

We proceed to investigate the relationship between the symptoms and survival. However, what type of analysis would be useful here? The structure of the data represents a 2^7 factorial experiment. That is, we have seven factors

Table 11.17

Excerpts of Cancer Survival Data

Symptom Pattern x_1 to x_7	Mean Log Survival	Patient Count
1011100	1.66	2
1111100	1.48	3
0010010	1.45	1
0000000	1.36	10
.	.	.
.	.	.
.	.	.
1100110	0.00	1
1111101	0.00	1
1101110	0.00	1

Table 11.18

Analysis of Variance of Survival Times

Source	df	SS	F	Pr > F
Between patterns	70	34.06		
Within patterns	731	147.61	2.41	0.005

(symptoms) with two levels of each (presence or absence). As noted in the preceding, there are 128 possible combinations; however, only 71 were observed. Also the number of patients with each of the 71 combinations differs markedly so the data are unbalanced. Now we can use the general linear model to adjust for the unbalance, but that analysis does not adjust for missing combinations. Missing cells restrict the estimation of interaction effects. That is, to estimate any two-factor interaction, the two-way table of main effect combinations must have observations in all cells. Fortunately, in this data set, all such tables are complete; hence, the two factor interactions may be estimated. We cannot, however, do this for any higher order interactions, but since these are difficult to interpret, this inability does not pose a serious problem. A lack of fit test can be used to determine whether the higher order interactions contribute significantly to the model.

Although we can use a general linear models computer program to perform this analysis, it is actually easier to do it by using an ordinary regression program and the original indicator variables to estimate main effects and the 21 products of these variables to estimate the interactions. Because we do not have the original 802 observations, we must use weighted regression with the weights being the number of patients in each of the 71 combinations. The results are shown in Table 11.19.

The model is obviously statistically significant, although one would hope to have a larger coefficient of determination. The lack of fit test is shown in Table 11.20, and the F ratio of 1.005 shows that the model is apparently adequate. In other words, we do not need to look for higher order interactions.

All of the main effect coefficients are significant and, as expected, negative, indicating reduced survival times if a symptom is present. However, several of the interactions are significant and positive, appearing to imply that having some combinations of two symptoms increases survival time. However, we must remember that interactions can only be interpreted in the presence of the involved main effects. To understand this principle better, we show in Table 11.21 the estimated mean of survival times (PSURV) and the 95% confidence intervals of those estimates (LO95 and UP95) for patients having all possible combinations of symptoms 3 and 4, which have the largest positive interaction.

We can see that estimated survival times are decreased by each symptom, and the confidence intervals do not include the estimated survival time for no symptoms. However, patients having both symptoms have estimated survival time not much different from each singly, and the confidence intervals overlap. In other words, it matters little whether a patient has symptom 3, symptom 4, or both.

From here, the modeling effort may be directed toward isolating those symptoms that contribute the most toward survival. The problem with this

Table 11.19 Analysis of the Cancer Data

Analysis of Variance

Source	DF	Sum of Squares	Mean Square	F Value	Prob > F
Model	28	25.53566	0.91199	4.492	0.0001
Error	42	8.52676	0.20302		
C total	70	34.06242			

Root MSE	0.45058	R-square	0.7497
Dep Mean	1.03884	Adj R-sq	0.5828
C.V.	43.37290		

SURVIVAL TIME OF CANCER PATIENTS
Parameter estimates

Variable	DF	Parameter Estimate	Standard Error	T for H0: Parameter = 0	Prob > |T|
INTERCEP	1	1.410840	0.06963577	20.260	0.0001
X1	1	−0.190864	0.06963453	−2.741	0.0090
X2	1	−0.167488	0.07074679	−2.367	0.0226
X3	1	−0.196229	0.07808019	−2.513	0.0159
X4	1	−0.262450	0.07411915	−3.541	0.0010
X5	1	−0.313018	0.07884488	−3.970	0.0003
X6	1	−0.582325	0.09642791	−6.039	0.0001
X7	1	−0.561891	0.15368040	−3.656	0.0007
X1X2	1	0.007425	0.07083303	0.105	0.9170
X1X3	1	0.136356	0.07549957	1.806	0.0781
X1X4	1	−0.056217	0.07842495	−0.717	0.4774
X1X5	1	0.182230	0.08439668	2.159	0.0366
X1X6	1	0.128829	0.10643388	1.210	0.2329
X1X7	1	0.301535	0.18643798	1.617	0.1133
X2X3	1	−0.022920	0.07648145	−0.300	0.7659
X2X4	1	0.053007	0.07878437	0.673	0.5048
X2X5	1	−0.035575	0.08511865	−0.418	0.6781
X2X6	1	0.068764	0.10793314	0.637	0.5275
X2X7	1	−0.003502	0.19043172	−0.018	0.9854
X3X4	1	0.247282	0.08525274	2.901	0.0059
X3X5	1	0.076423	0.09231251	0.828	0.4124
X3X6	1	0.242128	0.13206264	1.833	0.0738
X3X7	1	−0.328161	0.28108994	−1.167	0.2496
X4X5	1	0.166316	0.08628999	1.927	0.0607
X4X6	1	0.153523	0.14087613	1.090	0.2820
X4X7	1	0.199206	0.21549228	0.924	0.3605
X5X6	1	−0.065562	0.14843581	−0.442	0.6610
X5X7	1	−0.020747	0.31871820	−0.065	0.9484
X6X7	1	0.678453	0.29574035	2.294	0.0269

Table 11.20

Lack of Fit Test

Source	df	SS	MS	F
Full model (ANOVA)	70	34.06		
Restricted Model (Regression)	28	25.54		
Lack of Fit	42	8.52	0.2029	1.005
Error, Full Model	731	147.61	0.2019	

Table 11.21

Interpreting Interaction: Survival Time of Cancer Patients

X3	X4	X3X4	PSURV	LO95	UP95
0	0	0	1.41084	1.27031	1.55137
1	0	0	1.21461	1.06471	1.36452
0	1	0	1.14839	1.03115	1.26563
1	1	1	1.19944	1.01397	1.38491

approach is that even with all symptoms in the model, the overall fit to the data may be somewhat suspect as indicated by the low coefficient of determination. ■

The primary purpose of this chapter is to present the **general linear model** as a unifying principle underlying most of the statistical methods presented so far. Although this very general approach is not needed for all applications, it is very useful in the following situations:

- By considering the analysis of variance as an application of regression, the concept of inferences for partial coefficients allows the analysis of unbalanced factorial and other multifactor analyses of variance.
- Allowing the use of factor levels and quantitative independent variables in a model opens up a wide variety of applications, including the analysis of covariance.
- Also presented in this chapter are some methods used when the dependent variable is dichotomous, including applications of linear and logistic models.

11.9 CHAPTER EXERCISES

EXERCISES

1. In a study for determining the effect of weaning conditions on the weight of 9-week-old pigs, data on weaning (WWT) and 9-week (FWT) weights were recorded for pigs from three litters. One of these litters was weaned at approximately 21 days (EARLY), the second at about 28 days (MEDIUM), and the third at about 35 days (LATE). The data are given in Table 11.22. Perform an analysis of covariance using FWT as the response, weaning time as the factor, and WWT as the covariate. Comment on the results. Is there a problem with assumptions?

2. In a study of livestock pricing, data were collected on cattle being sold at two auction markets. The response is PRICE (dollars per hundred pounds), the factors are the MARKET, the CLASS of livestock (calf or heifer), the GRADE (good, choice, and prime), and the weight (WGT) of the animal (100 pounds). These types of data are invariably unbalanced (Table 11.23).

Table 11.22

Data for Exercise 1

EARLY		MEDIUM		LATE	
WWT	**FWT**	**WWT**	**FWT**	**WWT**	**FWT**
9	37	16	48	18	45
9	28	16	45	17	38
12	40	15	47	16	35
11	45	14	46	15	38
15	44	14	40	14	34
15	50	12	36	14	37
14	45	10	33	13	37

(a) Perform an analysis of variance using MARKET, CLASS, and GRADE as factors. Depending on the computer program you use, you may not wish to estimate all interactions.
(b) Add weight as a covariate. Interpret results.
(c) Perform an analysis to determine whether the effect of weight is the same for both calves and heifers. Why could this be a useful analysis?

3. Perform a logistic regression for the data in Table 11.24 using the regression method presented in the text. If a computer program for the maximum likelihood method is available, use it and compare the results.

4. The data in Table 11.25 concern the growth of pines in Colorado. The variables for a set of randomly sampled trees are:

RINGS: the number of rings, which is the age of the tree,
CIRCUM: the circumference of the tree at four feet, and
SIDE: the side of the mountain on which the tree is found:
NORTH: the north slope,
SOUTH: the south slope.

We want to establish the relationship of tree size to age and determine how that relationship may be affected by the side on which the tree is located.
(a) Perform an analysis of covariance for CIRCUM using SIDE as the factor and RINGS as the covariate. Interpret results as they apply to the questions asked.
(b) Perform a test to see whether the relationship of RINGS to CIRCUM is different for the two sides of the mountain. How do these results change the interpretations in part (a)?
(c) Define GROWTH as the ratio (CIRCUM)/(RINGS). What does this variable mean and how can it be used to answer the questions posed in the problem statement?

5. It has been proposed that the size of the ventricle, a physiological feature of the brain as measured by X-rays, may be associated with abnormal EEG (brain wave) readings. Table 11.26 shows ventricle sizes and results of EEG readings (coded 0 for normal and 1 for abnormal) for a set of 71 elderly patients.

Table 11.23

Data for Exercise 2

MARKET 14				MARKET 21			
CLASS	GRADE	PRICE	WGT	CLASS	GRADE	PRICE	WGT
CALF	CHOICE	47.82	1.60	CALF	CHOICE	46.65	2.06
CALF	CHOICE	51.82	2.20	CALF	CHOICE	48.00	2.90
CALF	CHOICE	56.43	2.10	CALF	CHOICE	46.88	2.26
CALF	CHOICE	56.34	2.05	CALF	CHOICE	48.75	2.05
CALF	CHOICE	57.63	1.90	CALF	CHOICE	32.64	1.98
CALF	CHOICE	55.00	2.10	CALF	CHOICE	45.75	2.12
CALF	CHOICE	49.09	2.20	CALF	CHOICE	54.00	2.00
CALF	GOOD	46.24	2.30	CALF	CHOICE	53.25	2.05
CALF	GOOD	38.47	2.30	CALF	CHOICE	52.50	1.85
CALF	GOOD	41.08	2.30	CALF	CHOICE	50.25	2.45
CALF	GOOD	40.90	1.65	CALF	CHOICE	51.00	1.95
CALF	GOOD	45.91	2.45	CALF	GOOD	18.40	2.45
CALF	GOOD	45.00	1.90	CALF	GOOD	43.50	2.20
CALF	GOOD	42.96	2.20	CALF	GOOD	40.88	3.45
CALF	PRIME	58.13	2.40	CALF	GOOD	29.25	2.00
CALF	PRIME	63.75	2.35	CALF	GOOD	43.50	2.25
CALF	PRIME	60.00	2.20	CALF	GOOD	45.75	2.40
HEIFER	CHOICE	48.24	2.55	CALF	GOOD	36.75	1.65
HEIFER	CHOICE	38.35	3.05	CALF	GOOD	22.50	1.10
HEIFER	GOOD	39.00	2.50	CALF	PRIME	62.25	2.35
HEIFER	GOOD	44.11	2.55	CALF	PRIME	60.00	2.10
HEIFER	GOOD	45.00	2.60	CALF	PRIME	57.75	1.85
HEIFER	GOOD	44.11	2.55	CALF	PRIME	60.60	2.10
HEIFER	GOOD	41.38	2.90	CALF	PRIME	60.75	1.95
HEIFER	GOOD	34.41	3.40	CALF	PRIME	56.25	2.35
HEIFER	PRIME	58.23	2.55	CALF	PRIME	63.00	2.00
				CALF	PRIME	59.63	2.25
				CALF	PRIME	59.25	2.00
				CALF	PRIME	56.63	2.35
				CALF	PRIME	52.88	2.15
				CALF	PRIME	58.50	2.05
				CALF	PRIME	55.88	1.80
				CALF	PRIME	55.88	2.25
				CALF	PRIME	46.50	2.20
				HEIFER	CHOICE	40.50	2.60
				HEIFER	CHOICE	40.88	3.35
				HEIFER	CHOICE	32.47	4.23
				HEIFER	CHOICE	37.88	3.10
				HEIFER	CHOICE	36.75	3.75
				HEIFER	CHOICE	37.13	3.60
				HEIFER	CHOICE	44.25	2.70
				HEIFER	CHOICE	40.50	2.70
				HEIFER	CHOICE	39.75	3.05
				HEIFER	CHOICE	34.50	3.65
				HEIFER	GOOD	33.38	2.02
				HEIFER	GOOD	33.00	3.95
				HEIFER	PRIME	57.75	2.55
				HEIFER	PRIME	42.00	2.70
				HEIFER	PRIME	42.38	2.90
				HEIFER	PRIME	60.00	2.65

Table 11.24

Data for Exercise 3

Concentration	Sample Size	Response
2.6	50	6
3.8	48	16
5.1	46	24
7.7	49	42
10.2	50	44

Table 11.25 Data on Tree Rings and Circumference

SIDE	CIRCUM	RINGS	SIDE	CIRCUM	RINGS	SIDE	CIRCUM	RINGS
NORTH	93	33	NORTH	70	25	SOUTH	155	62
NORTH	164	52	NORTH	44	8	SOUTH	34	27
NORTH	138	43	NORTH	44	10	SOUTH	58	24
NORTH	125	23	NORTH	63	14	SOUTH	55	13
NORTH	129	25	NORTH	133	32	SOUTH	105	39
NORTH	65	19	NORTH	239	42	SOUTH	66	24
NORTH	193	44	NORTH	133	25	SOUTH	70	29
NORTH	68	12	SOUTH	35	20	SOUTH	56	26
NORTH	139	32	SOUTH	30	25	SOUTH	38	11
NORTH	81	20	SOUTH	42	35	SOUTH	43	23
NORTH	73	16	SOUTH	30	18	SOUTH	47	33
NORTH	130	26	SOUTH	21	18	SOUTH	157	65
NORTH	147	44	SOUTH	79	30	SOUTH	100	52
NORTH	51	9	SOUTH	60	29	SOUTH	22	16
NORTH	56	15	SOUTH	63	20	SOUTH	105	52
NORTH	61	7	SOUTH	53	28			
NORTH	115	11	SOUTH	131	52			

(a) Perform an unweighted linear regression to predict the probability of an abnormal EEG. Plot the predicted values.

(b) Because the predicted values from the linear regression are not bounded by 0 and 1, it is not possible to use the predicted values for weighted regression or the linearized version of the logistic regression. An ad hoc procedure is to assign arbitrarily minimum and maximum predicted values of 0.05 and 0.95, respectively, for this purpose. Perform both a weighted linear and linearized logistic regression using this procedure.

(c) If a program is available, perform a logistic regression. Plot residuals and compare results with those from part (b).

6. We desire to test the effectiveness of a sales promotion on frozen peas. In two cities (CITY), eight stores (STORE) (numbered 1 to 8 for each city) were randomly divided into two groups of 4. In one group (stores labeled 1–4) a 4-week promotion campaign was conducted, while nothing was done in the other stores. The variable TRT is used to distinguish the

Table 11.26 Data for Exercise 5

VENT	EEG	VENT	EEG	VENT	EEG	VENT	EEG	VENT	EEG	VENT	EEG
53	0	37	0	63	0	25	0	60	0	58	0
56	0	59	0	50	0	58	1	70	0	68	1
50	0	59	0	51	0	76	0	74	1	62	1
41	0	65	0	50	0	94	1	73	1	72	0
45	1	56	0	56	0	75	0	76	0	78	1
50	0	68	0	47	0	66	0	42	1	76	1
57	0	65	0	51	0	83	1	51	0	80	1
70	0	68	1	49	0	56	1	58	1	58	1
64	1	60	1	57	0	54	0	58	0	63	1
61	0	70	0	40	0	51	1	58	1	70	1
57	1	84	0	58	0	51	1	57	0	85	1
50	0	48	0	67	1	62	0	65	0		

Table 11.27 Sales of Peas

			WEEK							
CITY	STORE	TRT	SALES	CUST	SALES	CUST	SALES	CUST	SALES	CUST
A	1	PROM	463	409	809	557	531	605	563	415
A	2	PROM	958	796	1219	880	890	901	1287	870
A	3	PROM	2051	2067	1947	1502	1863	1984	2597	1770
A	4	PROM	786	601	837	597	733	805	1965	673
A	5	CONT	1000	1305	1295	1597	1193	1201	1145	1059
A	6	CONT	635	775	608	807	858	957	1293	1021
A	7	CONT	112	143	223	257	288	307	152	146
A	8	CONT	826	958	1314	1276	531	757	1400	1159
B	1	PROM	1294	706	1395	897	1014	509	1131	651
B	2	PROM	1570	942	1039	719	1188	567	1506	801
B	3	PROM	3042	1506	2626	1795	2894	1474	2650	1345
B	4	PROM	1738	1005	1223	897	2467	1304	2103	1249
B	5	CONT	1139	480	741	497	1045	658	1028	805
B	6	CONT	1228	887	1588	1047	1237	936	1402	1003
B	7	CONT	2642	1706	2476	1972	2509	1679	2510	2056
B	8	CONT	4319	2807	3654	2476	3743	2911	3139	2517

promotion stores (PROM), while the others are labeled CONT. No price specials or other marketing efforts were conducted during the time period. The response variable is the number of 10-oz. packages (equivalent) sold weekly. Recognizing that sales are affected by store size and general sales activity, the weekly customer count (CUST) was also recorded. The data are given in Table 11.27.

There are two aspects to the analysis of this data set:

(1) the design of the experiment and
(2) the use of the customer count variable.

The design is a split plot, with stores as the main plots and weeks as subplots.[4] Note that there are four independently chosen stores for each city and sales cell.

(a) Perform an analysis of covariance of sales, using customer count as the covariate. Are the results useful? (*Hint:* Compare estimated control versus promotion sales for a small and large customer count.)

(b) Check for parallel lines for control versus treatment.[5] What do these results imply?

(c) Perform an analysis of variance using sales per customer as the response variable. Compare results with those of parts (a) and (b). Which of these analyses appears to be the most useful?

7. Skidding is a major contributor to highway accidents. The following experiment was conducted to estimate the effect of pavement and tire tread depth on spinout speed, which is the speed (in mph) at which the rear wheels lose friction when negotiating a specific curve. There are two asphalt (ASPHALT1 and ASPHALT2) pavements and one concrete pavement and three tire tread depths (1-, 2-, and 6-sixteenths of an inch.). This is a factorial experiment, but the number of observations per cell is not the same. The data are given in Table 11.28.

(a) Perform the analysis of variance using both the dummy variable and "standard" approaches. Note that the results are not the same although the differences are not very large.

(b) The tread depth is really a measured variable. Perform any additional or alternative analysis to account for this situation.

(c) It is also known that the pavement types can be characterized by their coefficient of friction at 40 mph as follows:

ASPHALT1: 0.35,
ASPHALT2: 0.24,
CONCRETE: 0.48.

Again, perform an alternative analysis suggested by this information. Which of the three analyses is most useful?

8. In Exercise 13 of Chapter 1, a study to examine the difference in half-life of the aminoglycosides Amikacin (A) and Gentamicin (G) was done. DO_MG_KG is the dosage of the drugs. The data are reproduced in Table 11.29.

(a) Perform an analysis of covariance using DRUG as the treatment and DO_MG_KG as covariate with HALF-LIFE as the response variable.

(b) Test for parallel slopes. (See discussion of unequal slopes in Section 11.5.)

[4]Actually this may be considered a repeated measurement design, but the split plot analogy works well here.

[5]This may be done by a separate analysis of convariance for each treatment or by a program that allows the specification of an interaction between customer count and treatment.

Table 11.28

Spinout Speeds

OBS	PAVE	TREAD	SPEED
1	ASPHALT1	1	36.5
2	ASPHALT1	1	34.9
3	ASPHALT1	2	40.2
4	ASPHALT1	2	38.2
5	ASPHALT1	2	38.2
6	ASPHALT1	6	43.7
7	ASPHALT1	6	43.0
8	CONCRETE	1	40.2
9	CONCRETE	1	41.6
10	CONCRETE	1	42.6
11	CONCRETE	1	41.6
12	CONCRETE	2	40.9
13	CONCRETE	2	42.3
14	CONCRETE	2	45.0
15	CONCRETE	6	47.1
16	CONCRETE	6	48.4
17	CONCRETE	6	51.2
18	ASPHALT2	1	33.4
19	ASPHALT2	1	38.2
20	ASPHALT2	1	34.9
21	ASPHALT2	2	36.8
22	ASPHALT2	2	35.4
23	ASPHALT2	2	35.4
24	ASPHALT2	6	40.2
25	ASPHALT2	6	40.9
26	ASPHALT2	6	43.0

Table 11.29 Half-Life of Aminoglycosides and Dosage by Drug Type

PAT	DRUG	HALF-LIFE	DO_MG_KG	PAT	DRUG	HALF-LIFE	DO_MG_KG
1	G	1.60	2.10	23	A	1.98	10.00
2	A	2.50	7.90	24	A	1.87	9.87
3	G	1.90	2.00	25	G	2.89	2.96
4	G	2.30	1.60	26	A	2.31	10.00
5	A	2.20	8.00	27	A	1.40	10.00
6	A	1.60	8.30	28	A	2.48	10.50
7	A	1.30	8.10	29	G	1.98	2.86
8	A	1.20	2.60	30	G	1.93	2.86
9	G	1.80	2.00	31	G	1.80	2.86
10	G	2.50	1.90	32	G	1.70	3.00
11	A	1.60	7.60	33	G	1.60	3.00
12	A	2.20	6.50	34	G	2.20	2.86
13	A	2.20	7.60	35	G	2.20	2.86
14	G	1.70	2.86	36	G	2.40	3.00
15	A	2.60	10.00	37	G	1.70	2.86
16	A	1.00	9.88	38	G	2.00	2.86
17	G	2.86	2.89	39	G	1.40	2.82
18	A	1.50	10.00	40	G	1.90	2.93
19	A	3.15	10.29	41	G	2.00	2.95
20	A	1.44	9.76	42	A	2.80	10.00
21	A	1.26	9.69	43	A	0.69	10.00
22	A	1.98	10.00				

Table 11.30

Recognition Value for Preschool Children

Medium Used	TIME OF EXPOSURE 5 min	10 min	15 min	20 min
TV:	49	50	43	53
	39	55	38	48
Audio tape:	55	67	53	85
	41	58		
Written material:	66	85	69	85
	68	92	62	

9. In many studies using preschool children as subjects,"missing" data are a problem. For example, a study that measured the effect of length of exposure to material on learning was hampered by the fact that the small children fell asleep during the period of exposure, thereby resulting in unbalanced data. The results of one such experiment are shown in Table 11.30. The measurement was based on a "recognition" value, which consists of the number of objects that can be associated with words. The factors were (1) the length of time of exposure and (2) the medium used to educate the children:

(a) Using the dummy variable model, test for differences in time of exposure, medium used, and interaction. Explain your results.

(b) Do you think that the pattern of missing data is related to the factors? Explain. How does this effect the analysis?

10. In Exercise 9 of Chapter 8, field measurements on the diameter and height and laboratory determination of oven dry weight were obtained for a sample of plants in the warm and cool seasons. The data are given in Table 11.31.

In Chapter 8 the data were used to see how well linear and log–linear models estimated the weight using the more easily determined field observations for the two seasons. Using the methods presented in this chapter, determine for both models whether the equations are different for the two seasons. Comment on the results.

11. In Exercise 12 of Chapter 8 and Exercise 11 of Chapter 10, 40 respondents (subjects) were asked to judge five structures for satisfaction on seven specific characteristics and also give an overall satisfaction index. All responses are on a nine-point scale. The data are shown in Table 8.32.

In Chapter 8 we used regression to determine how well the overall score is predicted by the various characteristics scores, and in Chapter 10 we analyzed the data to determine the differences in total scores among buildings and subjects.

(a) Using the general linear model, analyze the total scores using both the subjects and individual characteristics scores as independent factors. Compare results with those from the regression in Chapter 8.

(b) Using the general linear model, analyze the total scores using subjects, buildings, and individual characteristics scores as independent factors. Compare the results of this analysis, and the analyses in Exercise 12 of Chapter 8, Exercise 11 of Chapter 10, and (a) above, and suggest the

Table 11.31

Data for Exercise 10

COOL			WARM		
WIDTH	HEIGHT	WEIGHT	WIDTH	HEIGHT	WEIGHT
4.9	7.6	0.420	20.5	13.0	6.840
8.6	4.8	0.580	10.0	6.2	0.400
4.5	3.9	0.080	10.1	5.9	0.360
19.6	19.8	8.690	10.5	27.0	1.385
7.7	3.1	0.480	9.2	16.1	1.010
5.3	2.2	0.540	12.1	12.3	1.825
4.5	3.1	0.400	18.6	7.2	6.820
7.1	7.1	0.350	29.5	29.0	9.910
7.5	3.6	0.470	45.0	16.0	4.525
10.2	1.4	0.720	5.0	3.1	0.110
8.6	7.4	2.080	6.0	5.8	0.200
15.2	12.9	5.370	12.4	20.0	1.360
9.2	10.7	4.050	16.4	2.1	1.720
3.8	4.4	0.850	8.1	1.2	1.495
11.4	15.5	2.730	5.0	23.1	1.725
10.6	6.6	1.450	15.6	24.1	1.830
7.6	6.4	0.420	28.2	2.2	4.620
11.2	7.4	7.380	34.6	45.0	15.310
7.4	6.4	0.360	4.2	6.1	0.190
6.3	3.7	0.320	30.0	30.0	7.290
16.4	8.7	5.410	9.0	19.1	0.930
4.1	26.1	1.570	25.4	29.3	8.010
5.4	11.8	1.060	8.1	4.8	0.600
3.8	11.4	0.470	5.4	10.6	0.250
4.6	7.9	0.610	2.0	6.0	0.050
			18.2	16.1	5.450
			13.5	18.0	0.640
			26.6	9.0	2.090
			6.0	10.7	0.210
			7.6	14.0	0.680
			13.1	12.2	1.960
			16.5	10.0	1.610
			23.1	19.5	2.160
			9.0	30.0	0.710

most appropriate one. Are there any surprises? (*Hint:* The last analysis combines all the other analyses.)

12. Exercise 12 of Chapter 9 concerned an experiment to determine how storage time and temperature affect the Haugh measure of egg quality. Because the Haugh measure is based on, but not directly related to, albumen height and egg weight, it was proposed that all three responses should be analyzed and the results compared. Another way to assess possible differences between the Haugh measure and the two individual response variables is to perform an analysis of covariance, using the Haugh measure as the response and egg weight and albumen height as covariates. The data are available on the data disk in file FW11P12. Perform such an analysis to see whether the two components provide additional information on the effect of storage times and temperatures.

13. It is of importance to the fishing industry to determine the effectiveness of various types of nets. Effectiveness includes not only quantities of fish caught, but also the net selectivity for different sizes and species. In this experiment gill–net characteristics compose a two-factor factorial experiment with factors:

SIZE: two mesh sizes, 1 and 2 in. and
TYPE: material, monofilament or multifilament thread.

Four nets, composed of four panels randomly assigned to a factor level combination, are placed in four locations in a lake. After a specific time, the nets were retrieved and fish harvested. Data were recorded for each fish caught as follows:

Species
- bb: black bullhead
- bg: bluegill
- cc: channel catfish
- fwd: freshwater drum
- gs: gizzard shad
- lmb: largemouth bass

Size
- length in millimeters.

The data, comprising measurements of the 261 fish caught, are available on the data disk in file FW11P13. Of that total, 226 (85.8%) were gizzard shad.

(a) Using data for gizzard shad only, perform the appropriate analysis to determine the effects of net characteristics and location on the length of fish caught.

(b) Combine data for all other species and perform the same analysis.

14. In Example 10.1 an experiment was conducted to determine the effect of irrigation, nitrogen fertilizer, and planting rates on the yield of corn. One possible complication of the study was the fact that the specified planting rates did not always produce the exact number of plants in each plot. Analyze the data using the actual number of plants per plot as a covariate. Compare your results with those given in Section 10.6.

15. In Chapter 6 the analysis of variance was used to show that home prices differed among the zip areas, while in Chapter 8 multiple regression was used to show than home size was the most important factor affecting home prices.

(a) Using the data in Table 1.2 analyze home prices $200,000 or less with the general linear model using `zip` as a categorical variable and `size`, `bed`, and `bath` as continuous variables. Does this analysis change any previously stated conclusions?

(b) Using the data in Table 1.2 analyze the data using a model with all variables in the data set. Write a short report stating all relevant conclusions and how the Modes may use the analyses for making decisions about their impending move and subsequent home buying.

Categorical Data

EXAMPLE 12.1 **Developmental Research** A study by Aylward *et al.* (1984), reported in Green (1988), examines the relationship between neurological status and gestational age. The researchers were interested in determining whether knowing an infant's gestational age can provide additional information regarding the infant's neurological status. For this study, 505 newborn infants were cross-classified on two variables: overall neurological status, as measured by the Prechtl examination, and gestational age. The data are shown in Table 12.1.

Note that the response variable, Prechtl status, is a categorical variable; hence a linear model of the type we have been using is not appropriate. Additionally, in this example, the independent variable, the age of the infant, is recorded by intervals and can therefore also be considered a categorical variable. We will return to this example in Section 12.5. ■

12.1 Introduction

Up to this point we have been primarily concerned with analyses in which the response variable is ratio or interval and usually continuous in nature. The only exceptions occurred in Sections 4.3 and 5.5, where we presented methods for inferences on the binomial parameter p, which is based on a nominal response variable, with only two classes that we arbitrarily called "success" and "failure," and in Section 11.7, where we presented the problem of modeling a binary response using a continuous dependent variable.

Nominal variables are certainly not restricted to having only two categories: Variables such as flower petal color, geographic region, and plant or animal species, for example, are described by many categories. When we deal with

Table 12.1

Number of Infants

Prechtl Status	GESTATIONAL AGE (IN WEEKS) 31 or Less	32–33	34–36	37 or More	All Infants
Normal	46	111	169	103	409
Dubious	11	15	19	11	56
Abnormal	8	5	4	3	20
All infants	65	131	192	117	505

variables of this nature we are usually interested in the frequencies or counts of the number of observations occurring in each of the categories; hence, these types of data are often referred to as categorical data.

This chapter covers the following topics:

- Hypothesis tests for a multinomial population.
- The use of the χ^2 distribution as a goodness-of-fit test.
- The analysis of contingency tables.
- An introduction to the loglinear model to analyze categorical data.

12.2 Hypothesis Tests for a Multinomial Population

When the response variable has only two categories, we have used the binomial distribution to describe the sampling distribution of the number of "successes" in n trials. If the number of trials is sufficiently large, the normal approximation to the binomial is used to make inferences about the single parameter p, the proportion of successes in the population.

When we have more than two categories, the underlying distribution is called the **multinomial distribution**. For a multinomial population with k categories, the distribution has k parameters, p_i, which are the probabilities of an observation occurring in category i. Since an observation must fall in one category, $\sum p_i = 1$. The actual function that describes the multinomial distribution is of little practical use for making inferences. Instead we will use large sample approximations, which use the χ^2 distribution presented in Section 2.6.

When making inferences about a multinomial population, we are usually interested in determining whether the probabilities p_i have some prespecified values or behave according to some specified pattern. The hypotheses of interest are

$$H_0\colon p_i = p_{i0}, \quad i = 1, 2, \ldots, k,$$

$$H_1\colon p_i \neq p_{i0}, \quad \text{for at least two } i,$$

where p_{i0} are the specified values for the parameters.

The values of the p_{i0} may arise either from experience or from theoretical considerations. For example, a teacher may suspect that the performance of a particular class is below normal. Past experience suggests that the percentages of letter grades A, B, C, D, and F are 10, 20, 40, 20, and 10%, respectively.

The hypothesis test is used to determine whether the grade distribution for the class in question comes from a population with that set of proportions. In genetics, the "classic phenotypic ratio" states that inherited characteristics, say, A, B, C, or D, should occur with a 9:3:3:1 ratio if there are no crossovers. In other words, on the average, $9/16$ of the offspring should have characteristic A, $3/16$ should have B, $3/16$ should have C, and $1/16$ should have D. Based on sample data on actual frequencies, we use this hypothesis test to determine whether crossovers have occurred.

The test statistic used to test whether the parameters of a multinomial distribution match a set of specified probabilities is based on a comparison between the actually observed frequencies and those that would be expected if the null hypothesis were true. Assume we have n observations classified according to k categories with observed frequencies $n_1, n_2, \ldots, n_k$. The null hypothesis is

$$H_0\colon p_i = p_{i0}, \quad i = 1, 2, \ldots, k.$$

The alternate hypothesis is that at least two of the probabilities are different. The expected frequencies, denoted by E_i, are computed by

$$E_i = n\, p_{i0}, \quad i = 1, 2, \ldots k.$$

Then the quantities $(n_i - E_i)$ represent the magnitudes of the differences and are indicators of the disagreement between the observed values and the expected values if the null hypothesis were true. The formula for the test statistic is

$$X^2 = \sum \frac{(n_i - E_i)^2}{E_i},$$

where the summation is over all k categories. We see that the squares of these differences are used to eliminate the sign of the differences, and the squares are "standardized" by dividing by the E_i. The resulting quantities are then summed over all the categories.

If the null hypothesis is true, then this statistic is approximately distributed as χ^2 with $(k - 1)$ degrees of freedom, with the approximation being sufficiently close if sample sizes are sufficiently large. This condition is generally satisfied if the smallest expected frequency is five or larger. The rationale for having $(k - 1)$ degrees of freedom is that if we know the sample size and any $(k - 1)$ frequencies, the other frequency is uniquely determined. As you can see, the argument is similar to that underlying the degrees of freedom for an estimated variance.

If the null hypothesis is not true, the differences between the observed and expected frequencies would tend to be larger and the χ^2 statistic will tend to become larger in magnitude. Hence the test has a one-tailed rejection region, even though the alternative hypothesis is one of "not equal." In other words, p values are found from the upper tail of the χ^2 distribution. This test is known as the χ^2 test.

EXAMPLE 12.2

Suppose we had a genetic experiment where we hypothesize the 9:3:3:1 ratio of characteristics A, B, C, D. The hypotheses to be tested are

$$H_0\text{: } p_1 = 9/16, \quad p_2 = 3/16, \quad p_3 = 3/16, \quad p_4 = 1/16,$$

$$H_1\text{: at least two proportions differ from those specified.}$$

A sample of 160 offspring are observed and the actual frequencies are 82, 35, 29, and 14, respectively.

Solution Using the formula $E_i = np_{i0}$, the expected values are 90, 30, 30, and 10, respectively. The calculated test statistic is

$$\begin{aligned} X^2 &= \frac{(82-90)^2}{90} + \frac{(35-30)^2}{30} + \frac{(29-30)^2}{30} + \frac{(14-10)^2}{10} \\ &= 0.711 + 0.833 + 0.0033 + 1.600 \\ &= 3.177. \end{aligned}$$

Since there are four categories, the test statistic has 3 degrees of freedom. At a level of significance of 0.05, Appendix Table A.3 shows that we will reject the null hypothesis if the calculated value exceeds 7.81. Hence, we cannot reject the hypothesis of the 9:3:3:1 ratio at the 0.05 significance level. In other words, there is insufficient evidence that crossover has occurred. ■

EXAMPLE 12.3

Recall that in Example 4.4 we tested the hypothesis that 60% of doctors preferred a particular brand of painkiller. The response was whether a doctor preferred a particular brand of painkiller. The null hypothesis was that p, the proportion of doctors preferring the brand, was 0.6. The hypotheses can be written as

$$H_0\text{: } p_1 = 0.6 \quad (\text{hence } p_2 = 0.4),$$

$$H_1\text{: } p_1 \neq 0.6 \quad (\text{hence } p_2 \neq 0.4),$$

where p_1 is the probability that a particular doctor will express preference for this particular brand and p_2 is the probability that the doctor will not. A random sample of 120 doctors indicated that 82 preferred this particular brand of painkiller, which means that 38 did not. Thus $n_1 = 82$ and $n_2 = 38$.

Solution We can test this hypothesis by using the χ^2 test. The expected frequencies are $120(0.6) = 72$ and $120(0.4) = 48$. The test statistic is

$$X^2 = \frac{(82-72)^2}{72} + \frac{(38-48)^2}{48} = 3.47.$$

The test statistic for this example is χ^2 with one degree of freedom. Using Appendix Table A.3, we find that the null hypothesis would be rejected if the test statistic is greater than 3.841. Since the test statistic does not fall in the rejection region, we fail to reject the null hypothesis and conclude that we have insufficient evidence to back the claimed preference for the painkiller.

Note that this is automatically a two-tailed test since deviations in any direction will tend to make the test statistic fall in the rejection region. The test presented in Example 4.4 was specified as a one-tailed alternative hypothesis, but if we had used a two-tailed test, these two tests would have given identical results. This is due to the fact (Section 2.6) that the distribution of z^2 is χ^2 with one degree of freedom. We can see this by noting that the two-tailed rejection region based on the normal distribution (Appendix Table A.1) is $z > 1.96$ or $z < -1.96$ while the rejection region above was $\chi^2 > 3.84$. Since $(1.96)^2 = 3.84$, the two regions are really the same.[1] ■

12.3 Goodness of Fit Using the χ^2 Test

Suppose we have a sample of observations from an unknown probability distribution. We can construct a frequency distribution of the observed variable and perform a χ^2 test to determine whether the data "fit" a similar frequency distribution from a specified probability distribution.

This approach is especially useful for assessing the distribution of a discrete variable where the categories are the individual values of the variable, such as we did for the multinomial distribution. Occasionally some categories may need to be combined to obtain minimum cell frequencies required by the test. The χ^2 test is not quite as useful for continuous distributions since information is lost by having to construct class intervals to construct a frequency distribution. As the discussion in Section 4.5 and at the end of Example 12.5 indicates, an alternative test for continuous distributions usually offers better choices (see Daniel, 1990).

Test for a Discrete Distribution

EXAMPLE 12.4

In Section 2.3 we introduced the Poisson distribution to describe data that are "counts," that is, it describes the probability of observing frequencies of occurrences of some event. The Poisson distribution is often derived as the limit of the binomial distribution as the number of trials (n) approaches infinity and the probability of an event (p) goes to 0. For this reason, and because Poisson probabilities are easier to calculate, the Poisson distribution is often used to approximate binomial probabilities for large n and small p. We will illustrate this approach by generating 200 observations from a binomial distribution with $p = 0.05$ and $n = 200$. The observations theoretically will range from 0 to 200 with the expected number of successes equal to 5. We will use the goodness-of-fit test to see how well this distribution is approximated by the Poisson distribution with $\mu = 5$.

[1]The χ^2 test can be used for a one-tailed alternative by comparing the statistic with a 2α value from the table and rejecting only if the deviation from the null hypothesis is in the correct direction.

Table 12.2

Using the Poisson to Approximate the Binomial

y	COUNT	FREQ
0	1	1.35
1	5	6.74
2	18	16.84
3	32	28.07
4	36	35.09
5	28	35.09
6	32	29.24
7	23	20.89
8	15	13.06
9	5	7.25
10	4	3.63
11	1	1.65
12	.	0.69
13	.	0.26
14	.	0.09
15	.	0.03

The first column of Table 12.2 gives the values of y, the number of successes, and the second column the distribution of successes (COUNT) for the 200 samples. Note that since p is quite small, the number of successes actually only ranges from 0 to 11 in the sample. The second column gives the expected values (FREQ), which are obtained by multiplying the Poisson probabilities by 200. Since there is a measurable probability that y takes on the values 12 through 15, these values are included. The agreement between the two appears reasonable. We will use the χ^2 test for a multinomial population to test the hypotheses:

H_0: The observed distribution is Poisson with $\mu = 5$, versus

H_1: The observed distribution is not Poisson with $\mu = 5$.

We can immediately see that the minimum expected cell frequency of 5 is not met for a number of cells. We therefore combine the first two rows, giving 6 observed and 8.09 expected frequencies of 0 and 1 successes, and combine the last five rows, giving 1 observed and 2.72 expected values for 11 or more successes. We can now compute the test statistic:

$$X^2 = \frac{(6 - 8.09)^2}{8.09} + \frac{(18 - 16.84)^2}{16.84} + \cdots + \frac{(1 - 2.72)^2}{2.72} = 5.21.$$

The statistic has 10 degrees of freedom and obviously does not lead to rejection of the hypothesis that the Poisson approximation fits. ■

Test for a Continuous Distribution

If the observed values from a continuous distribution are available in the form of a frequency distribution, the χ^2 goodness-of-fit test can be used to determine whether the data come from some specified theoretical distribution (such as the normal distribution).

EXAMPLE 12.5

A certain population is believed to exhibit a normal distribution with $\mu = 100$ and $\sigma = 10$. A sample of 100 from this population yielded the frequency distribution shown in Table 12.3. The hypotheses of interest are then

H_0 = the distribution is normal with $\mu = 100$, $\sigma = 10$, and

H_1 = the distribution is different.

Solution The first step is to obtain the expected frequencies, which are those that would be expected with a normal distribution. These are obtained as follows:

1. Standardize the class limits. For example, the value of $y = 130$ becomes $z = (130 - 100)/10 = 3$.
2. Appendix Table A.1, the table of the normal distribution, is then used to find probabilities of a normally distributed population falling within the limits. For example, the probability of $Z > 3$ is 0.0013, which is the probability

Table 12.3

Observed Distribution

Class Interval[a]	Frequency
Less than 70	1
70–79	4
80–89	15
90–99	32
100–109	33
110–119	12
120–129	3
Greater than 130	0

[a]Assuming integer values of the observations.

Table 12.4

Expected Probabilities and Frequencies

Y	Z	Probability	Expected Frequency	Actual Frequency
<70	<−3	0.0013	0.1	1
70–79	−2 to −3	0.0215	2.2	4
80–89	−1 to −2	0.1359	13.6	15
90–99	−1 to 0	0.3413	34.1	32
100–109	0 to 1	0.3413	34.1	33
110–119	1 to 2	0.1359	13.6	12
120–129	2 to 3	0.0215	2.2	3
>130	>3	0.0013	0.1	0

that an observation exceeds 130. Similarly, the probability of an observation between 120 and 130 is the probability of Z being between 2 and 3, which is 0.0215 and so on.

3. The expected frequencies are the probabilities multiplied by 100.

Results of the above procedure are listed in Table 12.4.

We have noted that the use of the χ^2 distribution requires that cell frequencies generally exceed five, and we can see that this requirement is not met for two cells and only marginally met for two others. We must therefore combine cells, which we do here by redefining the first class to have an interval of less than or equal to 79, with the last being greater than or equal to 120. The resulting distribution still has two cells with expected values of 2.3, which are less than the suggested minimum of 5. These cells are in the "tail" of the normal distribution. That is, they represent the two ends of the data values. Recalling the shape of the normal distribution, we would expect these to have a smaller number of observations. Therefore, we will use the distribution as is. Using the data from Table 12.4, we obtain a value of $X^2 = 3.88$. These class intervals provide for six groups; hence, we will compare the test statistic with the χ^2 with five degrees of freedom. At a 0.05 level of significance, we will reject the null hypothesis if the value of χ^2 exceeds 11.07. Therefore, there is insufficient reason to believe that the data do not come from a normal population with mean 100 and standard deviation 10.

This goodness of fit test is very easy to perform and can be used to test for just about any distribution. However, it does have limitations. For example, the number and values of the classes used are subjective choices. By decreasing the number, we lose degrees of freedom, but by increasing the number we may end up with classes that have too small expected frequencies for the χ^2 approximation to be valid. For example, if we had not combined classes, the test statistic would have a value of 10.46, which, although not significant at the 0.05 level, is certainly larger.

Further, if we want to test for a "generic" distribution, such as the normal distribution with unspecified mean and standard deviation, we must first estimate these parameters from the data. In doing so, we lose an additional two degrees of freedom as a penalty for having to estimate the two unknown parameters. Because of this, it is probably better to use an alternative method when testing for a distribution with unspecified parameters. One such alternative is the Kolmogorov–Smirnov test discussed in Section 4.5. ■

12.4 Contingency Tables

Suppose that a set of observations is classified according to two categorical variables and the resulting data are represented by a two-way frequency table as illustrated in Section 1.7 (Table 1.13). For such a data set we may not be as interested in the marginal frequencies of the two individual variables as we are in the combinations of responses for the two categories. Such a table is referred to as a **contingency table**. In general, if one variable has r categories and the other c categories, then the table of frequencies can be formed to have r rows and c columns and is called an $r \times c$ contingency table. The general form of a contingency table is given in Table 12.5. In the body of the table, the n_{ij} represent the number of observations having the characteristics described by the ith category of the row variable and the jth category of the column variable. This frequency is referred to as the ijth "cell" frequency. The R_i and C_j are the total or marginal frequencies of occurrences of the row and column categories, respectively. The total number of observations n is the last entry.

Table 12.5

Representation of a Contingency Table

	COLUMNS					
Rows	**1**	**2**	**3**	$\cdots$	c	**Totals**
1	n_{11}	n_{12}	n_{13}	$\cdots$	n_{1c}	R_1
2	n_{21}	n_{22}	n_{23}	$\cdots$	n_{2c}	R_2
.	.	.	.	$\cdots$	.	.
.	.	.	.	$\cdots$	.	.
.	.	.	.	$\cdots$	.	.
r	n_{r1}	n_{r2}	n_{r3}	$\cdots$	n_{rc}	R_r
Totals	C_1	C_2	C_3	$\cdots$	C_c	n

In this section we examine two types of hypotheses concerning the contingency table:

1. the test for **homogeneity** and
2. the test for **independence**.

The test for homogeneity is a generalization of the procedure for comparing two binomial populations discussed in Section 5.5. Specifically, this test assumes independent samples from r multinomial populations with c classes. The null hypothesis is that all rows come from the same multinomial population (identified by the r rows) or, equivalently, that all come from the same distribution. In terms of the contingency table, the hypothesis is that the proportions in each row are equal. Note that the data represent samples from r potentially different populations.

The test for independence determines whether the frequencies in the column variable in a contingency table are independent of the row variable in which they occur or vice versa. This procedure assumes that one sample is taken from a population, and that all the elements of the sample are then put into exactly one level of the row category and the column category. The null hypothesis is that the two variables are independent, and the alternative hypothesis is that they are dependent.

The difference between these two tests may appear a bit fuzzy and there are situations where it may not be obvious which hypothesis is appropriate. Fortunately both hypotheses tests are performed in exactly the same way, but it is important that the conclusions be appropriately stated. As in the test for a specified multinomial distribution, the test statistic is based on differences between observed and expected frequencies.

Computing the Test Statistic

If the null hypothesis of **homogeneity** is true, then the relative frequencies in any row, that is, the E_{ij}/R_i, should be the same for each row. In this case, they would be equal to the marginal column frequencies, that is,

$$E_{ij}/R_i = C_j/n, \text{ hence}$$

$$E_{ij} = R_i C_j/n.$$

If the null hypothesis of **independence** is true, then each cell probability is a product (Section 2.2) of its marginal probabilities. That is,

$$E_{ij}/n = (R_i/n)(C_j/n), \text{ hence}$$

$$E_{ij} = R_i C_j/n.$$

Thus the expected frequencies for both the homogeneity and independence tests are computed by

$$E_{ij} = R_i C_j/n.$$

That is, the expected frequency for the ijth cell is a product of its row total and its column total divided by the sample size.

To test either of these hypotheses we use the test statistic

$$X^2 = \sum_{ij} \frac{(n_{ij} - E_{ij})^2}{E_{ij}},$$

where $i = 1, \ldots, r$, $j = 1, \ldots, c$; n_{ij} = observed frequency for cell ij; and E_{ij} = expected frequency for cell ij. If either null hypothesis of (homogeneity or independence) is true, this statistic X^2 has the χ^2 distribution with $(r - 1)(c - 1)$ degrees of freedom. For example, a 4×5 contingency table results in a χ^2 distribution with $(4 - 1)(5 - 1) = 12$ degrees of freedom.

As in the test for multinomial proportions, the distribution of the test statistic is only approximately χ^2, but the approximation is adequate for sufficiently large sample sizes. Minimum expected cell frequencies exceeding five are considered adequate but it has been shown that up to 20% of the expected frequencies can be smaller than 5 and cause little difficulty in cases where there are a large number of cells. As in the case of testing for a multinomial population, the rejection region is in the upper tail of the distribution.

Test for Homogeneity

As noted, for this test we assume a sample from each of several multinomial populations having the same classification categories and perform the test to ascertain whether the multinomial probabilities are the same for all populations.

EXAMPLE 12.6

A study was performed to determine whether the type of cancer differed between blue collar, white collar, and unemployed workers. A sample of 100 of each type of worker diagnosed as having cancer was categorized into one of three types of cancer. The results are shown in Table 12.6. The hypothesis to be tested is that the proportions of the three cancer types are the same for all three occupation groups. That is,

$$H_0\text{: } p_{1j} = p_{2j} = p_{3j} \text{ for all } j \text{ (types of cancer)}$$

$$H_1\text{: } p_{ij} \neq p_{kj} \text{ for some } j \text{ and some pair } i \text{ and } k,$$

where p_{ij} = the probability of occupation i having cancer type j.

Solution The expected frequencies are obtained as described above, that is,

$$E_{ij} = R_i C_j / n.$$

Table 12.6

Cancer Occurrence for Different Populations

	TYPE OF CANCER			
Occupation	**Lung**	**Stomach**	**Other**	**Total**
Blue collar	53	17	30	100
White collar	10	67	23	100
Unemployed	30	30	40	100
Total	93	114	93	300

Table 12.7

Expected Frequencies

	TYPE OF CANCER		
Occupation	**Lung**	**Stomach**	**Other**
Blue collar	31	38	31
White collar	31	38	31
Unemployed	31	38	31

Table 12.7 gives the expected frequencies. The test statistic is

$$X^2 = (53 - 31)^2/31 + (17 - 38)^2/38 + \cdots + (40 - 31)^2/31 = 70.0.$$

The rejection region for this test is $X^2 > 9.488$ for $\alpha = 0.05$ (χ^2 with degrees of freedom $(3 - 1)(3 - 1) = 4$). We reject the null hypothesis and conclude that the distribution of cancer is not homogeneous among types of workers. In fact, the data indicate that more blue collar workers have lung cancer, while more white collar workers have stomach cancer. ■

Table 12.8

Example 5.7 as a Contingency Table

Sex	**Favor**	**Do Not Favor**	**Total**
Men	105	145	250
Women	128	122	250
Total	233	267	500

EXAMPLE 12.7

To illustrate that the test for homogeneity is an extension of the two-sample test for proportions of Section 5.5, we reanalyze Example 5.6 using the χ^2 test of homogeneity. Table 12.8 gives the data from Example 5.6 written as a contingency table. The hypotheses statements are the same as in Chapter 5; that is, the null hypothesis is that the proportion of men favoring the candidate is the same as the proportion of women.

Solution The test statistic is

$$X^2 = (105 - 116.5)^2/116.5 + (128 - 116.5)^2/116.5 + (145 - 133.5)^2/133.5 + (122 - 133.5)^2/133.5 = 4.252.$$

The rejection region for $\alpha = 0.05$ for χ^2 for one degree of freedom is 3.84, and as in Example 5.6 we reject the null hypothesis and assume that the proportion of men favoring the candidate differs from that of women. Note that, except for round-off differences, the test statistic X^2 is the square of the test statistic from Example 5.6. That is, $z = -2.02$ from Example 5.6 squared is almost equal to $X^2 = 4.2$. Recall from Section 2.6 that the χ^2 with one degree of freedom is the same as the distribution of z^2. ■

Test for Independence

As noted, the test for independence can be used to determine whether two categorical variables are related. For example, we may want to know whether the sex of a person is related to opinion about abortion or whether the performance of a company is related to its organizational structure. In Chapter 7 we discussed the correlation coefficient, which measured the strength of association between two variables measured in the interval or ratio scale. The association or relationship between two categorical variables is not as easy to quantify. That is, we must be careful when we talk about the strength of association between two variables that are only qualitative in nature. To say that one increases as the other increases (or decreases) may not mean anything if one variable is hair color and the other is eye color! We can, however, determine whether the two are related by using the test for independence.

The test for independence is conducted by taking a sample of size n and assigning each individual to one and only one level of each of two categorical variables. The hypotheses to be tested are

H_0: the two variables are independent, and

H_1: the two variables are related.

EXAMPLE 12.8

Opinion polls often provide information on how different groups' opinions vary on controversial issues. A random sample of 102 registered voters was taken from the Supervisor of Election's roll. Each of the registered voters was asked the following two questions:

1. What is your political party affiliation?
2. Are you in favor of increased arms spending?

The results are given in Table 12.9.

The null hypothesis we want to test is that the opinions of individuals concerning increased military spending are independent of party affiliation. That is, the null hypothesis states that the opinions of people concerning increased military spending do not depend on their party affiliation. The alternative hypothesis is that opinion and party affiliation are dependent.

Solution The expected frequencies are obtained as before, that is, by multiplying row total by column total and then dividing by n. The results are given

Table 12.9

Table of Opinion by Party

		PARTY		
OPINION	DEM	REP	NONE	TOTAL
FAVOR	16	21	11	48
NOFAVOR	24	17	13	54
TOTAL	40	38	24	102

Table 12.10

Results of χ^2 Test

TABLE OF OPINION BY PARTY

OPINION Frequency Expected Cell Chi-Square	PARTY DEM	 REP	 NONE	 Total
FAVOR	16 18.824 0.4235	21 17.882 0.5435	11 11.294 0.0077	48
NOFAVOR	24 21.176 0.3765	17 20.118 0.4831	13 12.706 0.0068	54
Total	40	38	24	102

STATISTICS FOR TABLE OF OPINION BY PARTY

Statistic	df	Value	Prob
Chi-Square	2	1.841	0.398
Likelihood Ratio Chi-Square	2	1.846	0.397
Mantel-Haenszel Chi-Square	1	0.414	0.520
Phi Coefficient		0.134	
Contingency Coefficient		0.133	
Cramer's V		0.134	

Sample Size = 102

as the second entry in each cell of Table 12.10, which is obtained from PROC FREQ of the SAS System.

The third entry in each cell is its contribution to the test statistic, that is,

$$\frac{(n_{ij} - E_{ij})^2}{E_{ij}},$$

(rounded to four decimal places). The test statistic (computed from the non-rounded values) has the value 1.841, which is shown as the first entry at the bottom of the computer output.[2] This value is compared with the χ^2 statistic with two degrees of freedom. We will reject the null hypothesis if the value of our test statistic is larger than 5.99 for a level of significance of 0.05. We fail to reject the null hypothesis, which is confirmed in the computer output with a p value of 0.398. There is insufficient evidence to suggest that party affiliation affects opinions on this issue. ■

The major difference between the test for homogeneity and the test for independence is the method of sampling. In the test for homogeneity, the number of observations from each sample is "fixed" and each observation is assigned to the appropriate level of the other variable. Thus, we say that the row totals (or column totals) are fixed. This is not the case in the test for independence where only the total sample size, n, is fixed and observations are classified in two "directions," one corresponding to rows, the other to columns of the table. Therefore, only the total sample size is fixed prior to the experiment.

[2]Some of the other test statistics shown in the output are discussed later.

Measures of Dependence

In many cases, we are interested in finding a measure of the degree of dependence between two categorical variables. As noted, the precise meaning of dependence may be hard to interpret; however, a number of statistics can be used to quantify the degree of dependence between two categorical variables. For example, in Example 12.8, we may be interested in the degree of association or dependence between the political affiliation and feelings about increased military spending. A large degree of dependence may indicate a potential "split" along party lines.

Several statistics are used to quantify this dependence between two categorical variables. One such statistic is called **Pearson's contingency coefficient**, or simply the contingency coefficient. This coefficient is calculated as

$$t = \sqrt{\frac{X^2}{n + X^2}},$$

where X^2 is the value of the computed χ^2 statistic and n is the total sample size. The coefficient is similar to the coefficient of determination where the value 0 implies independence and 1 means complete dependence. For Example 12.8 the contingency coefficient, given as the third entry at the bottom of Table 12.9, has a value of 0.133. Since we failed to reject the hypothesis of independence, we expected the value of the coefficient to be quite low and indeed it is.

Because a number of different interpretations are available for defining the association between two categorical variables, other measures of that degree of dependence exist. Some of these are

1. Cramer's contingency coefficient (Cramer's V in Table 12.10),
2. the mean square contingency coefficient (given in Table 12.10),
3. Tschuprow's coefficient (not given), and
4. the phi coefficient (given in Table 12.10).

Note that for this example they are all almost exactly the same, but this is not always the case. A complete discussion of these coefficients is given in Conover (1999).

Other Methods

As was previously noted, the χ^2 statistic is a large sample approximation to the true distribution of X^2. The obvious question to be considered is "What happens if the sample is small?" In most cases, the problem cannot be readily addressed. One suggestion, which we have already used, is to collapse rows or columns, thus getting larger cell frequencies. There is, however, a special method for calculating the test statistic for a 2×2 contingency table with small samples. This method, called the "Fisher exact test" (Kendall and Stuart, 1979), computes the exact probability of obtaining any particular table, and computes a p value by enumerating all tables more contradictory to H_0 than the observed table. Since occasions for the use of this test do not arise with great frequency, we do not provide details for its use. Fisher's exact test is an option available in many computer programs.

Another test statistic that can be used to test for homogeneity or independence is called the likelihood ratio test statistic. This test statistic has the form

$$X_2^2 = 2\sum_{ij} n_{ij} \ln\left(\frac{n_{ij}}{E_{ij}}\right).$$

The likelihood ratio test statistic is also compared to the χ^2 distribution with $(r-1)(c-1)$ degrees of freedom. This statistic is also given at the bottom of Table 12.9, and is seen to be almost exactly equal to the "usual" χ^2 statistic. This is the case, unless there are one or more very small expected frequencies in the table. Investigations of both test statistics in tables with small sample sizes by Feinberg (1980), Koehler and Larntz (1980), and Larntz (1978) indicate that the X^2 statistic is usually more appropriate for tables with very small expected frequencies.

The likelihood ratio statistic does, however, have a particular additive property that makes it more desirable for performing the analyses presented in the next section.

Most computer programs used to analyze contingency tables automatically compute a wide variety of different test statistics. Sometimes the array of available options can be overwhelming; therefore it is important to use only those that best suit the problem at hand.

12.5 Loglinear Model

The majority of the statistical analysis discussed up to this chapter involved the use of models, and most inferences were made on the parameters of these models. The analysis of data from a contingency table presented in this chapter thus far did not involve specifying a model nor any parameters, and therefore was concerned with less specific hypotheses. Further, we have been concerned with contingency tables for only two variables. A more general strategy for the analysis of any size contingency table involves specifying a series of models, and testing these models to determine which one is most appropriate. This series includes not only the model of independence but also models that represent various associations or interactions among the variables. Each model generates expected cell frequencies that are compared with the observed frequencies. The model that best fits the observed data is chosen. This allows for the analysis of problems with more than two variables and the identification of simple and complex associations among these variables.

One such way of analyzing contingency tables is known as **loglinear modeling**. The difference in this approach and that discussed in Section 12.4 is in the manner in which the expected frequencies are obtained. In the loglinear modeling approach, the expected frequencies are computed under the assumption that a certain specified model is appropriate for explaining the relationship among variables. The complexity of this model usually results in computational problems in obtaining the expected frequencies that can be resolved only through the use of iterative methods. As a consequence of this, most analyses are done with computers.

As an example of a loglinear model, consider the problem in Example 12.8. The variables are "party affiliation" and "opinion." We will designate the probability of an individual belonging to the ijth cell as p_{ij}, the marginal probability of belonging to the ith row (opinion) as p_i, and the marginal probability of belonging to the jth column (party) as p_j. From Chapter 2 the condition of independence allows us to write

$$p_{ij} = p_i p_j.$$

Under this condition the expected frequencies are

$$E_{ij} = np_{ij} = np_i p_j.$$

Taking natural logs of both sides results in the relationship

$$\ln(E_{ij}) = \ln(n) + \ln(p_i) + \ln(p_j).$$

Therefore, if the two variables are independent, the log of the expected frequencies is a linear function of the marginal probabilities. We turn this around and see that the test for independence is really a test to see whether the log of the expected frequencies is a linear function of the logs of the marginal probabilities. Define

$$\mu_{ij} = \ln(E_{ij}), \quad \ln(n) = \mu, \quad \ln(p_i) = \lambda_i^{\mathrm{A}}, \quad \text{and} \quad \ln(p_j) = \lambda_j^{\mathrm{B}}.$$

Then the model can be written as

$$\mu_{ij} = \mu + \lambda_i^{\mathrm{A}} + \lambda_j^{\mathrm{B}}.$$

Note that A and B are superscripts, not exponents.

This model closely resembles the ANOVA model of Chapter 9, and in fact the analysis very closely resembles that of a two-way ANOVA model. The terms λ_i^{A} represent the effects of the variable designated as "rows" (opinion), and the terms λ_j^{B} represent the effects of the variable "columns" (party affiliation).

Notice that the model is constructed under the assumption that rows and columns of the contingency table are independent. If they are not independent, this model requires an additional term, which can be called an "association" factor. Using consistent notation, this term may be designated $\lambda_{ij}^{\mathrm{AB}}$. This term is analogous to the interaction term in the ANOVA model and has a similar interpretation. The test for independence then becomes one of determining whether the association factor should be in the model. This is done by what is called a "lack of fit" test, usually using the likelihood ratio statistic.

This test follows the same pattern as the test for interaction in the factorial ANOVA model, and the results are usually displayed in a table very similar to the ANOVA table. Instead of using sums of squares and the F distribution to test hypotheses about the parameters in the model, we use the likelihood ratio statistic and the χ^2 distribution. The likelihood ratio test statistic is used because it can be subdivided, corresponding to the various terms in the model, whereas the χ^2 statistic, X^2, cannot.

EXAMPLE 12.9

In Example 12.8 we examined the relationship between party affiliation and opinion. To determine whether the two were independent, we did the "usual" χ^2 test and failed to reject the hypothesis of independence. We can do the same test using a loglinear model. If we specify the model as outlined, the hypothesis of independence becomes

$$H_0\colon \lambda_{ij}^{AB} = 0, \quad \text{for all } i \text{ and } j, \text{ and}$$

$$H_1\colon \lambda_{ij}^{AB} \neq 0, \quad \text{for some } i \text{ and } j.$$

Table 12.11

Loglinear Model Analysis for Example 12.9

Source	df	χ^2	Prob
Party	2	4.74	0.1117
Opinion	1	0.35	0.5527
Likelihood ratio	2	1.85	0.3972

Solution The analysis is performed by `PROC CATMOD` from the SAS System with results shown in Table 12.11. The last item is the likelihood ratio test for goodness of fit, which has a value of 1.85 and a p value of 0.3972. Thus, we cannot reject H_0, and we conclude the independence model fits. Note that this is the same value as the likelihood ratio statistic for the test of independence given in Table 12.9 (as it should be).

The other items in the table are the tests on the "main effects," which are a feature of the use of this type of analysis. It is interesting to note that both the opinion and the party likelihood ratio statistics are not significant. While the exact hypotheses tested by these statistics are expressed in terms of means of logarithms of expected frequencies, the general interpretation is that there is no difference in the marginal values for opinion nor for party. The interpretation here is that the 54 to 48 majority for `NOFAVOR` (Table 12.10) is insufficient evidence for declaring a majority on that side of the issue and, likewise, the party affiliation proportions are insufficient evidence of one party having a plurality. In conclusion, there is nothing about this table that differs significantly![3] ■

Solution to Example 12.1 The example presented at the beginning of this chapter is now analyzed using the loglinear modeling approach. That is, we develop a set of hierarchial models, starting with the simplest, which may be of little interest, and going to the most complex, testing each model for goodness of fit. The model that best fits the data will be adopted. Some of the

[3] In some applications, these main effects may not be of interest.

Table 12.12

Expected Frequencies

Prechtl Status	AGE GROUP 1	2	3	4	Total
Normal	42	42	42	42	168
Dubious	42	42	42	42	168
Abnormal	42	42	42	42	168

Table 12.13

Expected Frequencies

Prechtl Status	AGE GROUP 1	2	3	4	Total
Normal	107	107	107	107	429
Dubious	14	14	14	14	56
Abnormal	5	5	5	5	20

computations will be done by hand for illustration purposes only. All loglinear modeling is normally done with a computer.

We start with the simplest model, one that contains only the overall mean. This model has the form

$$\ln(E_{ij}) = \mu_{ij} = \mu.$$

The expected frequencies under this model are given in Table 12.12. Note that all the expected frequencies are the same, 42. This is because the model assumes all the cells have the same value, μ. The expected frequencies are then the total divided by the number of cells, or $505/12 = 42$ (rounded to integers). The likelihood ratio statistic for testing the lack of fit of this model, obtained by `PROC CATMOD` from the SAS System, has a huge value of 252.7. This value obviously exceeds the 0.05 table value of 12.59 for the χ^2 distribution with 6 degrees of freedom; hence we readily reject the model and go to the next model.

The next model has only one term in addition to the mean. That is, we could choose a model that had only the grand mean and a row effect, or we could choose a model with only the grand mean and a column effect. For the purposes of this example, we choose the model with a grand mean and a row effect. This model is

$$\ln(E_{ij}) = \mu_{ij} = \mu + \lambda_i^{\text{A}}.$$

The term λ_i^{A} represents the effect due to Prechtl scores. Note that there is no effect due to age groups in the model.

The expected frequencies are listed in Table 12.13. They are obtained by dividing each row total by 4, the number of columns. For example, the first row is obtained by dividing 429 by 4 (rounded to integers). The likelihood ratio test has a value of 80.85. Again, the model obviously does not fit, so we must go to the next model. The next model has both age and Prechtl as factors. That is, the model is

$$\ln(E_{ij}) = \mu_{ij} = \mu + \lambda_i^{\text{P}} + \lambda_j^{\text{A}}.$$

Table 12.14

Expected Frequencies

Prechtl	AGE GROUP				
Status	**1**	**2**	**3**	**4**	**Total**
Normal	55	111	163	99	429
Dubious	7	15	21	13	56
Abnormal	3	5	8	5	20

Note that this is the same model we used to test for independence in Example 12.9. Therefore, we will be testing the goodness of fit of the model, but really we will be testing for independence. This is because this is the lack of fit test for the model that contains all possible terms except the "interaction" term, λ_{ij}^{AB}.

The expected frequencies are given in Table 12.14. The values are calculated by multiplying row totals by column totals and dividing by the total in exactly the same way they were calculated for the χ^2 tests. The likelihood ratio test statistic for testing the goodness of fit of this model has a value of 14.30. This exceeds the critical value of 12.59 that we obtain from the χ^2 table, so this model does not fit either. That is, there is a significant relationship between the gestational age of newborn infants and their neurological status. Examination of Table 12.1 indicates that 40% of abnormal infants were less than 31 weeks of age, and that the percentage of abnormal infants decreases across age. ■

The extension of the loglinear model to more than two categorical variables is relatively straightforward, and most computer packages offer this option. There are also many variations of the modeling approach to the analysis of categorical data. These topics are discussed in various texts including Bishop *et al.* (1975) and Upton (1978). A discussion of categorical data with ordered categories is given in Agresti (1984).

A methodology that clearly distinguishes between independent and dependent variables is given in Grizzle *et al.* (1969). This methodology is often called the "linear model" approach and emphasizes estimation and hypothesis testing of the model parameters. Therefore, it is easily used to test for differences among probabilities, but is awkward to use for tests of independence. Conversely, the loglinear model is relatively easy to use to test independence but not so easy to test for differences among probabilities. Most computer packages offer the user a choice of approaches. As in all methodology that relies heavily on computer calculations, the user should be sure that the analysis is really what is expected by carefully reading documentation on the particular program used.

12.6 CHAPTER SUMMARY

This chapter deals with problems concerned with categorical or count data. That is, the variables of interest are usually nominal in scale, and the measurement of interest is the frequency of occurrence. For a single category, we saw that questions of goodness of fit could be answered by use of the χ^2

distribution. This test is also used to determine whether sample frequencies associated with categories of the variable agree with what could be expected according to the null hypothesis. We also saw that this test could be used to determine whether the sample values fit a prescribed probability distribution, such as the normal distribution.

When observations are made on two variables, we were concerned with frequencies associated with the cells of the contingency table formed by cross-tabulating observations. Again we used a χ^2 test that measured the deviation from what was expected under the null hypothesis by the observed samples. If the data represented independent samples from more than one population, the test was a test of homogeneity. If the data represented one sample cross-classified into two categories, the test was a test of independence. Both these tests were conducted in an identical manner, with only the interpretation differing.

The loglinear modeling approach to contingency table analysis was briefly discussed. This procedure allows for more flexibility in the analysis, and allows for analyses with more than two categorical variables.

12.7 CHAPTER EXERCISES

EXERCISES

1. To reduce the use of drugs and other harmful substances, some public schools have started to use dogs to locate undesirable substances. Many arguments have been directed against this practice, including the allegations that (1) the dogs too often point at suspects (or their lockers or cars) where there are no contraband substances and (2) that there is too much difference in the abilities of different dogs.

   In this experiment, four different dogs were randomly assigned to different schools such that each dog visited each school the same number of times. The dogs pointed to cars in which they smelled a contraband substance. Permission was then obtained from the owners of these cars, and they were then searched. A "success" was deemed to consist of a car that contained, or was admitted by the owner to have recently contained, a contraband substance.

   Cars that for some reason could not be searched have been deleted from the study. The resulting data are given in Table 12.15.

Table 12.15

Data for Exercise 1

Dog	RESULT		
Frequency	Fail	Success	Total
A	51	103	154
G	43	103	146
K	79	192	271
M	40	126	166
Total	213	524	737

(a) Give a 0.99 confidence interval for the proportion of success for the set of dogs (see Chapter 4).
(b) Test the hypothesis that the dogs all have the same proportion of success.

2. A newspaper story gave the frequencies of armed robbery and auto theft for two neighboring communities. Do the data of Table 12.16 suggest different crime patterns of the communities?

Table 12.16

Data for Exercise 2 (Table of City by Type)

City Frequency	TYPE Auto	Robbery	Total
B	175	54	229
C	97	11	108
Total	272	65	337

3. An apartment owner believes that more of her poolside apartments are leased by single occupants than by those who share an apartment. The data in Table 12.17 were collected from current occupants. Do the data support her hypothesis?

Table 12.17

Data for Exercise 3

Pool Frequency	TYPE Single	Multiple	Total
YES	22	23	45
NO	24	31	55
Total	46	54	100

4. A serious problem that occurs when conducting a sample survey by mail is that of delayed or no response. Late respondents delay the processing of data, while nonrespondents may bias results, unless a costly personal follow-up is conducted.

A firm that specializes in mail surveys usually experiences the following schedule of replies:

25% return in week 1,
20% return in week 2,
10% return in week 3,

and the remainder fail to return (or return too late). The firm tries to improve this return schedule by placing a dollar bill and a message of thanks in each questionnaire. In a sample of 500 questionnaires, there were

156 returns in week 1,
149 in week 2,
100 in week 3,

and the remainder were not returned or arrived too late to be processed.

Test the hypotheses that (1) the overall return schedule has been improved and (2) the rate of nonrespondents has been decreased. (*Note:* These are not independent hypotheses.)

5. Use the data on tree diameters given in Table 1.7 to test whether the underlying distribution is normal. Estimate the mean and variance from the data. Combine intervals to avoid small cell frequencies if necessary.

6. Out of a class of 40 students, 32 passed the course. Of those that passed, 24 had taken the prerequisite course, while of those that failed, only 1 had taken the prerequisite. Test the hypothesis that taking the prerequisite course did not help to pass the course.

7. A machine has a record of producing 80% excellent, 18% good, and 2% unacceptable parts. After extensive repairs, a sample of 200 produced 157 excellent, 42 good, and 1 unacceptable part. Have the repairs changed the nature of the output of the machine?

8. To determine the gender balance of various job positions the personnel department of a large firm took a sample of employees from three job positions. The three job positions and the gender of employees from the sample are shown in Table 12.18. Use the hierarchical approach to log-linear modeling to determine which model best fits the data. Explain the results.

Table 12.18

Gender and Job Positions

Job Position	Males	Females
Accountant	60	20
Secretarial	10	90
Executive	20	20

9. The market research department for a large department store conducted a survey of credit card customers to determine whether they thought that buying with a credit card was quicker than paying cash. The customers were from three different metropolitan areas. The results are given in Table 12.19. Test the hypothesis that there is no difference in proportions of ratings among the three cities.

Table 12.19

Survey Results

Rating	City 1	City 2	City 3
Easier	62	51	45
Same	28	30	35
Harder	10	19	20

10. In Exercise 12 of Chapter 1, the traits of salespersons considered most important by sales managers were listed in Table 1.23. These data are condensed in Table 12.20. Test the hypothesis that there is no difference in

Table 12.20

Traits of Salespersons

Trait	Number of Responses
Reliability	44
Enthusiasm	30
Other	46

the proportions of sales managers that rated the three categories as most important.

11. A sample of 100 public school tenth graders, 80 private school tenth graders, and 50 technical school tenth graders was taken. Each student was asked to identify the category of person that most affected their life. The results are listed in Table 12.21.

Table 12.21

Sample of Tenth Graders

Person	Public School	Private School	Tech School
Clergy	50	44	10
Parent	30	25	33
Politician	19	10	5
Other	1	1	2

(a) Do the data indicate that there is a difference in the way the students answered the question? (Use $\alpha = 0.05$.)
(b) Does there seem to be a problem with using the χ^2 test to answer part (a)? What is the problem and how would you solve it? Reanalyze the data after applying your solution. Do the results change?

12. In the study discussed in Exercise 10, the sales managers were also asked what traits they considered most important in a sales manager. The results are given in Table 12.22.

Table 12.22

Traits of Salespersons

	SALESPERSON		
Sales Manager	Reliability	Enthusiasm	Other
Reliability	12	18	20
Enthusiasm	23	7	11
Other	9	5	15

(a) Are the two independent? Explain.
(b) Calculate Pearson's contingency coefficient. Is there a strong relationship between the traits the sales managers think are important for salespersons and those for sales managers?

13. Exercise 13 in Chapter 11 looked at a gill–net experiment designed to determine the effect of net characteristics on the size of fish caught. The data are on the data disk in FW11P13. Using the methods of this chapter, we can see how the relative frequencies of species caught are related to

net characteristics. The data showed that out of a total of 261 fish caught, 224 were gizzard shad. To satisfy the minimum cell size requirements necessary to use the χ^2 statistic, we will have to combine species. Use two species categories, Shad and Other, to do the following analyses:

(a) Perform separate χ^2 tests for independence to relate species to mesh size and net type.

(b) If an appropriate computer package is available, construct the loglinear model of Section 12.5 using both mesh size and net type. Interpret the results.

Nonparametric Methods

EXAMPLE 13.1

Quality Control A large company manufacturing rubber windshield wipers for use on automobiles was involved in a research project for improving the quality of their standard wiper. An engineer developed four types of chemical treatments that were thought to increase the lifetime of the wiper. An experiment was performed in which samples of 15 blades were treated with each of these chemical treatments and measured for the amount of wear (in mm) over a period of 2 h on a test machine. The results are shown in Table 13.1.

An analysis of variance was performed (see Chapter 6) to test for difference in average wear over the four treatments. The results are shown at the bottom of Table 13.1. The engineer, however, did not believe that the assumption of normality was valid (see Section 6.4). That is, she suspected that the error terms were probably distributed more like a uniform distribution. The histogram of the residuals given in Fig. 13.1 appears to justify the concern of the engineer. To further check the assumption of normality, she performed a goodness of fit test and rejected the null hypothesis of normality (see Section 12.2). An approach for solving this problem is presented in the material covered in this chapter, and we will return to this example in Section 13.7. ■

13.1 Introduction

As Chapter 11 demonstrated, most of the statistical analysis procedures presented in Chapters 4 through 11 are based on the assumption that some form of linear model describes the behavior of a ratio or interval response variable. That is, the behavior of the response variable is approximated by a linear model and inferences are made on the parameters of that model. Because the primary focus is on the parameters, including those describing the distribution of the

Table 13.1

Wear Data for Window Wipers for Four Teatments (in mm)

TREAT = 1	TREAT = 2	TREAT = 3	TREAT = 4
11.5	14.3	13.7	17.0
11.5	12.7	14.8	14.7
10.1	14.3	13.5	16.5
11.6	13.1	14.2	15.5
11.2	14.3	14.7	14.2
10.6	14.7	14.4	16.6
11.2	12.5	14.2	14.5
11.5	14.0	14.8	16.6
10.3	15.0	14.0	14.9
11.8	13.2	14.8	16.5
11.3	13.9	15.0	16.5
10.1	14.9	13.2	14.2
10.9	12.6	14.2	16.4
11.2	14.2	13.3	14.6
10.4	12.8	13.5	15.3

ANOVA for Wear Data

Source	df	SS	F	Pr > F
Treat	3	165.3	88.65	0.0001
Error	56	34.8		
Total	59	200.11		

Figure 13.1

Histogram of Residuals

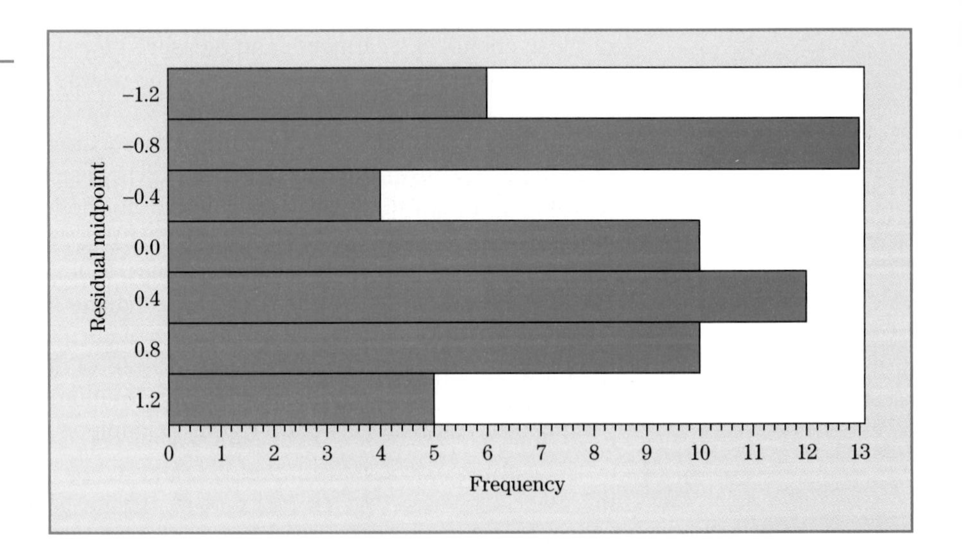

random error, statistical methods based on linear models are often referred to as "parametric" methods.

We have repeatedly noted that the correctness of an analysis depends to some degree on certain assumptions about the model. One major assumption is that the errors have a nearly normal distribution with a common variance, so that the normal, t, χ^2, and F distributions properly describe the distribution of the test statistics. This assures that the probabilities associated with the inferences are as stated by significance levels or p values. Fortunately, these methods are reasonably "robust" so that most estimates and test statistics give sufficiently valid results even if the assumptions on the model are not exactly satisfied (as they rarely are).

Obviously, there are situations in which the assumptions underlying an analysis are not satisfied and remedial methods such as transformations do not work, in which case there may be doubt as to whether the significance levels or confidence coefficients are really correct. Therefore, an alternative approach for which the correctness of the stated significance levels and confidence coefficients is not heavily dependent on rigorous distributional assumptions is needed. Such methods should not depend on the distribution of the random error nor necessarily make inferences on any particular parameter. Such procedures are indeed available and are generally called "nonparametric" or "distribution-free" methods. These procedures generally use simple, tractable techniques for obtaining exact error probabilities while not assuming any particular form of the model.

Many of the tests discussed in Chapter 12 may be considered nonparametric methods. For example, the contingency table analysis makes no assumptions about the underlying probability distribution. The p values for this test statistic were obtained by using a large sample χ^2 approximation. Obviously this type of methodology does not fit the "normal" theory of parametric statistics because the scale of measure used was at best nominal (categorical). As a matter of fact, one of the desirable characteristics of most nonparametric methods is that they do not require response variables to have an interval or ratio scale of measurement.

A brief introduction to the concept of nonparametric statistics was presented in Section 3.5 where we noted that a wide spectrum of methods is available when the assumptions on the model are not fulfilled. Examples of nonparametric tests were presented in Sections 4.5 and 5.6 where the tests were done on medians. In both of these examples, we noted that the distributions of the observations were skewed, therefore making the "parametric" t tests suspect. In this chapter, we present some additional examples of nonparametric methods.

The methods presented in Chapter 12 were used to analyze response variables that are categorical in nature; that is, they are measured in a nominal scale. Of course, data of the higher order scales can be artificially converted to nominal scale, simply by grouping observations. That is, ordinal data and interval or ratio scale measurements can be "categorized" into nominal-looking data. Interval or ratio measurements can also be changed into ordinal scale measurements by simply ranking the observations.[1] A number of nonparametric statistical methods are, in fact, based on ranks. The methods presented in this chapter are mostly of this type. These methods work equally well on

[1]This was illustrated in Chapter 1.

variables originally measured in the ordinal scale as well as on variables measured on ratio or interval scales and subsequently converted to ranks.

Ranks may actually be preferable to the actual data in many cases. For example, if numerical measurements assigned to the observations have no meaning by themselves, but only have meaning in a comparison with other observations, then the ranks convey all the available information. An example of this type of variable is the "scores" given to individual performers in athletic events such as gymnastics or diving. In such situations the measurements are essentially ordinal in scale from the beginning. Even when measurements are actually in the interval scale, the underlying probability distribution may be intractable. That is, we are not able to use the additional information in a statistical inference because we cannot evaluate the resulting sampling distribution. In this case, switching to ranks allows us to use the relatively simple distributions associated with ranks.

To convert interval or ratio data into ranks, we must have a consistent procedure for ordering data. This ordering is called ranking and the ranking procedure normally used in statistics orders data from "smallest" to "largest" with a "1" being the smallest and an "n" being the largest (where n is the size of the data set being ranked). This ranking does not necessarily imply a numerical relationship, but may represent another ordinality such as "good," "better," and "best," or "sweet" to "sour," "preferred" to "disliked," or some other relative ranking. In other words, any ratio, interval, or ordinal variable can usually be converted to ranks.

As indicated in Section 1.3, a special problem in ranking occurs when there are "ties," that is, when a variable contains several identically recorded values. As a practical solution, ties are handled by assigning mean ranks to tied values. While the methodology of rank-based nonparametric statistics usually assumes no ties, a reasonably small number of ties have minimal effect on the usefulness of the resulting statistics.

As we will see, most tables used in nonparametric statistics are for use with small samples and are exact only when there are no ties in the data. Approximate distributions used for larger samples also assume no ties. When the number of ties is excessive, adjustments must be made either in the tables or in the test statistic. These adjustments are usually based on large sample approximations. Such adjustments are provided in this chapter for some of the procedures; others are available in most texts on nonparametric statistics (Conover, 1999).

Before presenting specific rank-based nonparametric test procedures, we must understand how parametric and nonparametric tests compare.

- *Power:* Conversion of a set of interval or ratio values to ranks involves a loss of information. This loss of information usually results in a loss of power in a hypothesis test. As a result, most rank-based nonparametric tests will be less powerful when used on interval or ratio data, especially if the distributional assumptions of parametric tests are not severely violated. Of course, we cannot compare parametric and nonparametric tests if the observed variables are already nominal or ordinal in scale.

- *Interpretation:* By definition, nonparametric tests do not state hypotheses in terms of parameters; hence inferences for such tests must be stated in other terms. For some of these tests the inference is on the median, but for others the tests may specify location or difference in distribution. Fortunately, this is not usually a problem, because the research hypothesis can be adjusted if a nonparametric test is to be used.
- *Application:* As will be seen, comprehensive rank-based methodology analogous to that for the linear model is not available.[2] Instead nonparametric tests are usually designed for a specific experimental situation. In fact, nonparametric methods are most frequently applied to small samples and simple experimental situations, where violations of assumptions are likely to have more serious consequences.
- *Computing:* While many nonparametric test statistics are simple in nature, those based on ranks require that interval or ratio scale variables be ranked. This can be time consuming if done by hand for large data sets. Fortunately many statistical computer programs include rank-based and other nonparametric options.
- *Robustness:* While few assumptions are made for nonparametric methods, these methods are not uniformly insensitive to all types of violations of assumptions nor are they uniformly powerful against all alternative hypotheses.
- *Experimental design:* The use of nonparametric methods does not eliminate the need for careful planning and execution of the experiment or other data accumulation. Good design principles are important regardless of the method of analysis.

The various restrictions and disadvantages of nonparametric methods would appear to severely limit their usefulness. This is not the case, however, and they should be used (and usually are) when the nature of the population so warrants. Additionally, many of the nonparametric methods are extremely easy to perform, especially on small data sets. They are therefore an attractive alternative for "on the spot" analyses. In fact, one of the earlier books on nonparametric methods is called *Some Rapid Approximate Statistical Procedures* (Wilcoxon and Wilcox, 1964).

The following sections discuss some of the more widely used rank-based nonparametric hypothesis testing procedures. In the illustration of these procedures, we use some examples previously analyzed with parametric techniques. The purpose of this is to compare the two procedures. Of course, in actual practice, only one method should be applied to any one set of data. Because it is so important to use only one procedure for analyzing a set of data, it is imperative to know the requisites of each procedure as well as the limitations. That way, we can correctly perform either a parametric or a nonparametric test. That is, it must be done right the first time!

[2] It has been suggested that converting the response variable to ranks and then performing standard linear model analyses will produce acceptable results. Of course, such analyses do not produce the usual estimates of means and other parameters. This is briefly discussed in Section 13.7; a more comprehensive discussion is found in Conover and Iman (1981).

We emphasize that the methods presented in this chapter represent only a few of the available nonparametric techniques. If none of these are suitable, additional nonparametric or other robust methods may be found in the literature, such as in Huber (1981), or in texts on the subject, such as Conover (1999).

13.2 One Sample

In Section 4.5 we considered an alternative approach to analyzing some income data that had a single extreme observation. This approach was based on the fact that the median is not affected by extreme observations. Recall that in this example we converted the income values into either a "success" (if above the specified median) or a "failure" (if below), and the so-called sign test was based on the proportion of successes. In other words, the test was performed on a set of data that were converted from the ratio to the nominal scale.

Of course the conversion of the variable from a ratio to a nominal scale with only two values implies a loss of information; hence, the resulting test is likely to have less power. However, converting a nominal variable to ranks preserves more of the information and thus a test based on ranks should provide more power. One such test is known as the Wilcoxon signed rank test.

The Wilcoxon signed rank test is used to test that a distribution is symmetric about some hypothesized value, which is equivalent to the test for location. We illustrate with a test of a hypothesized median, which is performed as follows:

1. Rank the magnitudes (absolute values) of the deviations of the observed values from the hypothesized median, adjusting for ties if they exist.
2. Assign to each rank the sign (+ or −) of the deviation (thus, the name "signed rank").
3. Compute the sum of positive ranks, $T(+)$, or negative ranks, $T(-)$, the choice depending on which is easier to calculate. The sum of $T(+)$ and $T(-)$ is $n(n+1)/2$, so either can be calculated from the other.
4. Choose the smaller of $T(+)$ and $T(-)$, and call this T.
5. For small samples ($n \leq 50$), critical values are found in Appendix Table A.9. If n is large, the sampling distribution of T is approximately normal with

$$\mu = n\,(n+1)/4, \quad \text{and}$$
$$\sigma^2 = n\,(n+1)(2n+1)/24,$$

which can be used to compute a z statistic for the hypothesis test.

EXAMPLE 13.2

Example 4.7, particularly the data in Table 4.5, concerned a test for the mean family income of a neighborhood whose results were unduly influenced by an extreme outlier. A test for the median was used to overcome the influence of that observation. We now use that example to illustrate the Wilcoxon signed

Table 13.2

Deviations from the Median and Signed Ranks

Obs	Diff	Signed Rank	Obs	Diff	Signed Rank
1	4.1	18	11	2.7	14
2	−0.3	−4.5	12	80.4	20
3	3.5	16	13	1.9	13
4	1.0	11	14	0.0	1
5	1.2	12	15	0.8	10
6	−0.7	−9	16	3.2	15
7	0.2	2.5	17	0.6	8
8	0.3	4.5	18	−0.2	−2.5
9	4.9	19	19	0.4	6
10	−0.5	−7	20	3.6	17

rank test. The hypothesis of interest is

H_0: the distribution on incomes is symmetric about 13.0,

with a two-tailed alternative,

H_1: the distribution is symmetric about some other value.

Solution The deviations of the observed values from 13.0 (the specified H_0 value) are given in Table 13.2 in the column labeled "Diff," followed by the signed ranks corresponding to the differences. Note that several ties are given average ranks, and that zero is arbitrarily given a positive sign. A quick inspection shows that there are fewer negative signed ranks so we first compute $T(-)$:

$$T(-) = 4.5 + 9 + 7 + 2.5 = 23.$$

The total sum of n ranks is $(n)(n+1)/2$; hence, it follows that $T(+) + T(-) = (20)(21)/2 = 210$. Thus $T(+) = 210 - 23 = 187$. The test statistic is the smaller, $T = T(-) = 23$. From Appendix Table A.9, using $n = 20$ and $\alpha = 0.01$, we see that the critical value is 37. We reject H_0 if the calculated value is less than 37; hence, we reject the hypothesis and conclude that the population is not symmetric about 13.0.

Alternately, we can use the large sample normal approximation. Under the null hypothesis, T is approximately normally distributed with

$$\mu = (20)(21)/4 = 105, \quad \text{and}$$
$$\sigma^2 = (20)(21)(41)/24 = 717.5;$$

hence $\sigma = 26.79$. These values are used to compute the test statistic

$$z = \frac{23 - 105}{26.79} = -3.06.$$

Using Appendix Table A.1, we find a (two-tailed) p value of approximately 0.002; hence, the null hypothesis is readily rejected. However, the sample is rather small; hence, the p value calculated from the large sample approximation should not be taken too literally.

Table 13.3

Effect of Diet on Activity

Child	Before Rating	After Rating	|Difference|	Signed Rank
1	19	11	8	−10
2	14	15	1	+1
3	20	17	3	−3.5
4	6	12	6	+8
5	12	8	4	−5
6	4	9	5	+6.5
7	10	7	3	−3.5
8	13	6	7	−9
9	15	10	5	−6.5
10	9	11	2	+2

The p value obtained for the sign test in Section 4.5 was 0.012. Thus, for $\alpha = 0.01$ the Wilcoxon signed rank test rejected the null hypothesis while the sign test did not.[3] ■

A popular application of the signed rank test is for comparing means from paired samples. In this application the differences between the pairs are computed as is done for the paired t test (Section 5.4). The hypothesis to be tested is that the distribution of differences is symmetric about 0.

EXAMPLE 13.3

To determine the effect of a special diet on activity in small children, 10 children were rated on a scale of 1 to 20 for degree of activity during lunch hour by a school psychologist. After 6 weeks on the special diet, the children were rated again. The results are give in Table 13.3. We test the hypothesis that the distribution of differences is symmetric about 0 against the alternative that it is not.

Solution The sum of the positive ranks is $T(+) = 17.5$; hence $T(-) = 55 - 17.5 = 37.5$. Using $\alpha = 0.05$, the rejection region is for the smaller of $T(+)$ and $T(-)$ to be less than 8 (from Appendix Table A.9). Using $T(+)$ as our test statistic, we cannot reject the null hypothesis, so we conclude that there is insufficient evidence to conclude that the diet affected the level of activity. ■

13.3 Two Independent Samples

The Mann–Whitney test (also called the Wilcoxon two-sample test) is a rank-based nonparametric test for comparing the location of two populations using independent samples. Note that this test does not specify an inference to any particular parameter of location. Using independent samples of n_1

[3]The problem as stated in Example 4.7 had a one-sided alternative while the procedure for the two-sided alternative is presented here since it is more general.

and n_2, respectively, the test is conducted as follows:

1. Rank all $(n_1 + n_2)$ observations as if they came from one sample, adjusting for ties.
2. Compute T, the sum of ranks for the smaller sample.
3. Compute $T' = n_1(n_1 + n_2 + 1) - T$, the sum of ranks for the larger sample. This is necessary to assure a two-tailed test.
4. For small samples $(n_1 + n_2 \leq 30)$, compare the smaller of T and T' with the rejection region consisting of values less than the critical values given in Appendix Table A.10. If either T or T' falls in the rejection region, we reject the null hypothesis. Note that even though this is a two-tailed test, we only use the lower quantiles of the tabled distribution.
5. For large samples, the statistic T or T' (whichever is smaller) has an approximately normal distribution with

$$\mu = n_1(n_1 + n_2 + 1)/2 \quad \text{and}$$

$$\sigma^2 = n_1 n_2(n_1 + n_2 + 1)/12.$$

These parameter values are used to compute a test statistic having a standard normal distribution. We then reject the null hypothesis if the value of the test statistic is smaller than $-z_{\alpha/2}$. Modifications are available when there are a large number of ties (for example, Conover, 1999).

The procedure for a one-sided alternative hypothesis depends on the direction of the hypothesis. For example, if the alternative hypothesis is that the location of population 1 has a smaller value than that of population 2 (a one-sided hypothesis), then we would sum the ranks from sample 1 and use that sum as the test statistic. We would reject the null hypothesis of equal distributions if this sum is less than the $\alpha/2$ quantile of the table. If the one-sided alternative hypothesis is the other direction, we would use the sum of ranks from sample 2 with the same rejection criteria.

Table 13.4

Hamburger Taste Test

Type of Burger	Score
A	1
A	2
A	3
B	4
A	5
A	6
B	7
B	8
B	9
B	10

EXAMPLE 13.4

Because the taste of food is impossible to quantify, results of tasting experiments are often given in ordinal form, usually expressed as ranks or scores. In this experiment two types of hamburger substitutes were tested for quality of taste. Five sample hamburgers of type A and five of type B were scored from best (1) to worst (10). Although these responses may appear to be ratio variables (and are often analyzed using this definition), they are more appropriately classified as being in the ordinal scale. The results of the taste test are given in Table 13.4. The hypotheses of interest are

H_0: the types of hamburgers have the same quality of taste, and

H_1: they have different quality of taste.

Solution Because the responses are ordinal, we use the Mann–Whitney test. Using these data we compute

$$T = 1 + 2 + 3 + 5 + 6 = 17 \quad \text{and}$$

$$T' = 5(11) - 17 = 38.$$

Choosing $\alpha = 0.05$ and using Appendix Table A.10, we reject H_0 if the smaller of T or T' is less than or equal to 17. The computed value of the test statistic is 17; hence we reject the null hypothesis at $\alpha = 0.05$, and conclude that the two types differ in quality of taste. If we had to choose one or the other, we would choose burger type A based on the fact that it has the smaller rank sum. ■

13.4 More Than Two Samples

The extension to more than two independent samples provides a nonparametric analog for the one-way analysis of variance, which can be used with a completely randomized design experiment or a t sample observational study. That is, we test the null hypothesis that t independent samples come from t populations with identical distributions against the alternative that they do not, with the primary differences being in location. A test for this hypothesis is provided by a rank-based nonparametric test called the Kruskal–Wallis t sample test. The procedure for this test follows the same general pattern as that for two samples. The Kruskal–Wallis test is conducted in the following manner:

1. Rank all observations. Denote the ijth rank by R_{ij}.
2. Sum the ranks for each sample (treatment), denote these totals by T_i.
3. The test statistic is

$$H = \frac{1}{S^2}\left[\sum \frac{T_i^2}{n_i} - \frac{n(n+1)^2}{4}\right],$$

where

$$S^2 = \frac{1}{n-1}\left[\sum R_{ij}^2 - \frac{n(n+1)^2}{4}\right],$$

and where the R_{ij} are the actual ranks,[4] and n_i are the sizes of the ith sample, and $n = \sum n_i$. If no ties are present in the ranks, then the test statistic takes on the simpler form

$$H = \frac{12}{n(n+1)}\sum \frac{T_i^2}{n_i} - 3(n+1).$$

For a select group of small sample sizes, there exist specialized tables of rejection regions for H. For example, some exact tables are given in Iman *et al.* (1975). Usually, however, approximate values based on the χ^2 distribution with $(t-1)$ degrees of freedom are used. This test is similar to the Wilcoxon in that it uses only the upper tail of the distribution of the test statistic. Therefore, we would reject H_0 if the value of H exceeded the α level of the χ^2 distribution with $(t-1)$ degrees of freedom. If this hypothesis is rejected, we would naturally like to be able to determine where the

[4]If there are no ties, $\sum R_{ij}^2$ is more easily computed by $[n(n+1)(2n+1)]/6$. This is also a rather good approximation if there are few ties.

differences are. Since no parameters such as means are estimated in this procedure, we cannot construct contrasts or use differences in means to isolate those populations that differ. Therefore, we will use a pairwise comparison method based on the average ranks. This is done in the following manner.

We infer at the α level of significance that the locations of the response variable for factor levels i and j differ if

$$\left|\frac{T_i}{n_i} - \frac{T_j}{n_j}\right| > t_{\alpha/2}\sqrt{S^2\left(\frac{n-1-H}{n-t}\right)\left(\frac{1}{n_i}+\frac{1}{n_j}\right)},$$

where $t_{\alpha/2}$ is the $\alpha/2$ critical value from the t distribution with $(n-t)$ degrees of freedom.

EXAMPLE 13.5

A psychologist is trying to determine whether there is a difference in three methods of training six-year-old children to learn a foreign language. A random selection of 10 six-year-old children with similar backgrounds is assigned to each of three different methods. Method 1 uses the traditional teaching format. Method 2 uses repeated listening to tapes of the language along with classroom instruction. Method 3 uses videotapes exclusively. At the end of a 6-week period, the children were given identical, standardized exams. The exams were scored, with high scores indicating a better grasp of the language. Because of attrition, method 1 had 7 students finishing, method 2 had 8, and method 3 only 6. It is, however, important to note that we must assume that attrition was unrelated to performance. The data and associated ranks are given in Table 13.5.

Solution Although the test scores may be considered ratio variables, concerns about the form of the distribution suggest the use of the Kruskal–Wallis nonparametric method. Since there are few ties, we will use the simpler form of the test statistic, resulting in

$$H = \left[\frac{12}{(21)(22)}\right]\left(\frac{116.5^2}{7}+\frac{82.0^2}{8}+\frac{32.5^2}{6}\right) - 3(22)$$
$$= 10.76.$$

Table 13.5

Data and Ranks for Example 13.5

TEACHING METHOD					
1		2		3	
y	Rank	y	Rank	y	Rank
78	12.5	70	2.5	60	1
80	14	72	5.5	70	2.5
83	16	73	7	71	4
86	17	74	8.5	72	5.5
87	18	75	10	74	8.5
88	19	78	12.5	76	11
90	20	82	15		
		95	21		
$n_1 = 7$		$n_2 = 8$		$n_3 = 6$	
$T_1 = 116.5$		$T_2 = 82.0$		$T_3 = 32.5$	

From Appendix Table A.3, we see that $\chi^2(2)$ for $\alpha = 0.05$ is 5.99; hence we reject the null hypothesis of equal location and conclude that there is a difference in the distributions of test scores for the different teaching methods.

To determine where the differences lie, we perform the multiple comparison procedure based on the average ranks discussed in the preceding. Using the ranks in Table 13.5 we obtain $\sum R_{ij}^2 = 3309$, so[5] that

$$S^2 = (1/20)[3309 - 21(22)^2/4] = 38.4.$$

The mean ranks are

Method 1: $116.5/7 = 16.64$,
Method 2: $82.0/8 = 10.25$, and
Method 3: $32.5/6 = 5.42$.

From Appendix Table A.2, the appropriate t value for a 5% significance level is 2.101. We will compare the difference between method 1 and method 2 with

$$(2.101)\sqrt{38.4\left(\frac{20 - 10.76}{18}\right)\left(\frac{1}{8} + \frac{1}{7}\right)} = 4.83.$$

The mean rank difference between methods 1 and 2 has a value of 6.39, which exceeds this quantity; hence we conclude the distributions of test scores for methods 1 and 2 may be declared different. Similarly, for comparing methods 1 and 3 the mean difference of 11.22 exceeds the required value of 5.18; hence we conclude that the distributions of scores differ. Finally, the mean difference between methods 2 and 3 is 4.83, which is less than the required difference of 5.03; hence there is insufficient evidence to declare different distributions between methods 2 and 3. The psychologist can conclude that the results of using method 1 differ from those of both the other methods, but that the effect of the other two may not. ■

We have noted that the Kruskal–Wallis test is primarily designed to detect differences in "location" among the populations. In fact, theoretically, the Kruskal–Wallis test requires that the underlying distribution of each of the populations be identical in shape, differing only by their location. Fortunately, the test is rather insensitive to moderate differences in the shape of the underlying distributions, and this assumption can be relaxed in all but the most extreme applications. However, it is not useful for detecting differences in variability among populations having similar locations.

There are many nonparametric tests available for the analysis of t independent samples designed for a wide variety of alternative hypotheses. For example, there are tests to detect differences in scale (or shape) of the distributions, tests to detect differences in the skewness (symmetry) of the distributions, and tests to detect differences in the kurtosis (convexity) of the distributions. There are also so-called omnibus tests that detect any differences

[5] Using the shortcut formula for $\sum R_{ij}^2$ gives 3311.

in the distributions, no matter what that difference may be. A good discussion of many of these tests can be found in Boos (1986).

13.5 Randomized Block Design

Data from a randomized block design may be analyzed by a nonparametric rank-based method known as the Friedman test. The Friedman test for the equality of treatment locations in a randomized block design is implemented as follows:

1. Rank treatment responses within each block, adjusting in the usual manner for ties. These ranks will go from 1 to t, the number of treatments, in each block. These are denoted R_{ij}.
2. Obtain the sum of ranks for each treatment. This means that we add one rank value from each block, for a total of b (the number of blocks) ranks. Call this sum R_i for the ith treatment.
3. The test statistic is

$$T^* = (b-1)\frac{\left[B - \frac{bt(t+1)^2}{4}\right]}{A - B},$$

where $A = \sum\sum R_{ij}^2$, which, if there are no ties, simplifies to

$$A = bt(t+1)(2t+1)/6$$

and $B = \frac{1}{b}\sum R_i^2$.

The test statistic, T^*, is compared to the F distribution with $[t-1, (b-1)(t-1)]$ degrees of freedom.

Some references give the Friedman test statistic as

$$T_1 = \frac{12}{bt(t+1)}\sum R_i^2 - 3b(t+1),$$

where t and b represent the number of treatments and blocks, respectively. This test statistic is compared with the χ^2 distribution with $(t-1)$ degrees of freedom. However, the T^* test statistic using the F distribution has been shown to be superior to the χ^2 approximation (Iman and Davenport 1980), and we therefore recommend the use of that statistic.

Pairwise comparisons can be performed using the R_i in the following manner. For a significance level of α, we can declare that the distributions of treatments i and j differ in location if

$$|R_i \quad R_j| > t_{\alpha/2}\sqrt{\frac{2b(A-B)}{(b-1)(t-1)}},$$

where $t_{\alpha/2}$ has $(b-1)(t-1)$ degrees of freedom.

Table 13.6

Percent Weeds Killed

Note: Ranks are in parentheses.

Treatment	BLOCKS 1	2	3	R_i
1	16 (4.5)	51 (5)	11 (4.5)	14.0
2	1 (1)	29 (4)	2 (2)	7.0
3	16 (4.5)	24 (3)	11 (4.5)	12.0
4	4 (2.5)	11 (2)	5 (3)	7.5
5	4 (2.5)	1 (1)	1 (1)	4.5

EXAMPLE 13.6

Responses given in terms of proportions will follow the binomial distribution, which can be quite nonnormal and also exhibit heterogeneous variances. This experiment is concerned with the effectiveness of five weed killers. The experiment was conducted in a randomized block design with five treatments and three blocks, which corresponded to plots in the test area. The response is the percentage of weeds killed. The hypothesis that the killers (treatments) have equal effects on weeds is tested against an alternative that there are some differences. The data are given in Table 13.6, along with the ranks in parentheses.

Solution The Friedman test is appropriate for this example. Using the ranks from Table 13.6 we obtain the values

$$\text{A} = 163.5 \quad \text{and} \quad \text{B} = 155.17.$$

The test statistic is

$$T^* = 2\left[\frac{155.17 - \frac{(3)(5)(6)^2}{4}}{163.5 - 155.17}\right] = 4.84.$$

The null hypothesis is rejected if the test statistic is in the rejection region of the F distribution with 4 and 8 degrees of freedom. Using the 0.05 level of significance in Appendix Table A.4 we find the critical value of 3.84. Therefore, we reject the null hypothesis and conclude there is a difference among the killers tested.

To identify the nature of the differences we perform a multiple comparison test. We compare the pairwise differences among the R_i with

$$(2.306)\sqrt{\frac{(2)(3)(8.33)}{(2)(4)}} = 5.76.$$

The differences and conclusions of the multiple comparisons among the R_i are given in Table 13.7, where it is seen that treatment 1 differs from treatments 2, 4, and 5, and that treatment 3 differs from treatment 5. No other differences are significant. Using the traditional schematic (see discussion of post hoc comparisons in Section 6.5), the results can be presented as

Treatments	1	3	4	2	5
	———	———			
		———	———	———	
			———	———	———

■

Table 13.7

Differences among Treatments

Treatments	Differences	Significant or Not
1 vs 5	$14 - 4.5 = 9.5$	yes
3 vs 5	$12 - 4.5 = 7.5$	yes
4 vs 5	$7.5 - 4.5 = 3$	no
2 vs 5	$7 - 4.5 = 2.5$	no
1 vs 2	$14 - 7 = 7$	yes
3 vs 2	$12 - 7 = 5$	no
4 vs 2	$7.5 - 7 = 0.5$	no
1 vs 4	$14 - 7.5 = 6.5$	yes
3 vs 4	$12 - 7.5 = 4.5$	no
1 vs 3	$14 - 12 = 2$	no

13.6 Rank Correlation

The concept of correlation as a measure of association between two variables was presented in Section 7.6 where correlation was estimated by the Pearson product moment correlation coefficient. The value of this statistic is greatly influenced by extreme observations, and the test for significance is sensitive to deviations from normality. A correlation coefficient based on the ranked, rather than the originally observed, values would not be as severely affected by extreme or influential observations. One such rank-based correlation coefficient is obtained by simply using the formula given for the correlation coefficient in Section 7.6 on the ranks rather than the individual values of the observations. This rank-based correlation coefficient is known as Spearman's coefficient of rank correlation, which can, of course, also be used with ordinal variables. For reasonably large samples, the test statistic for determining the existence of significant correlation is the same as that for linear correlation given in Chapter 7,

$$F = (n - 2)r^2/(1 - r^2),$$

where r^2 is the square of the rank-based correlation coefficient.

Because the data consist of ranks, a shortcut formula exists for computing the Spearman rank correlation. This shortcut is useful for small data sets that have few ties. First, separately rank the observations in each variable (from 1 to n). Then for each observation compute the difference between the ranks of the two variables, ignoring the sign. Denote these differences as d_i. The correlation coefficient is then computed:

$$r = 1 - \frac{6 \sum d_i^2}{n(n^2 - 1)}.$$

EXAMPLE 13.7

The data from Exercise 2 of Chapter 1 described the abundance of waterfowl at different lakes. It was noted that the distributions of both waterfowl abundance and lake size were dominated by one very large lake. We want to determine the correlation between the water area (WATER) and the number

Table 13.8

Waterfowl Data for Spearman Rank Correlation

OBS	RWATER	RFOWL	DIFF	OBS	RWATER	RFOWL	DIFF
1	20.5	8.5	12.0	27	6.5	8.5	2.0
2	6.5	24.0	17.5	28	28.0	50.0	22.0
3	20.5	42.5	22.0	29	33.0	19.0	14.0
4	46.0	36.0	10.0	30	50.0	8.5	41.5
5	20.5	8.5	12.0	31	52.0	52.0	0.0
6	51.0	37.0	14.0	32	20.5	8.5	12.0
7	15.5	31.0	15.5	33	12.0	29.0	17.0
8	15.5	8.5	7.0	34	28.0	31.0	3.0
9	33.0	28.0	5.0	35	6.5	8.5	2.0
10	28.0	33.0	5.0	36	6.5	8.5	2.0
11	20.5	8.5	12.0	37	15.5	42.5	27.0
12	47.5	48.0	0.5	38	6.5	19.0	12.5
13	6.5	25.5	19.0	39	24.5	8.5	16.0
14	39.0	49.0	10.0	40	41.0	46.0	5.0
15	45.0	22.0	23.0	41	33.0	41.0	8.0
16	24.5	35.0	10.5	42	39.0	44.0	5.0
17	12.0	21.0	9.0	43	33.0	8.5	24.5
18	47.5	40.0	7.5	44	6.5	25.5	19.0
19	33.0	8.5	24.5	45	39.0	51.0	12.0
20	28.0	38.0	10.0	46	42.5	39.0	3.5
21	12.0	27.0	15.0	47	44.0	47.0	3.0
22	15.5	34.0	18.5	48	1.5	8.5	7.0
23	6.5	17.0	10.5	49	1.5	8.5	7.0
24	49.0	19.0	30.0	50	42.5	45.0	2.5
25	36.0	31.0	5.0	51	37.0	8.5	28.5
26	28.0	23.0	5.0	52	20.5	8.5	12.0

of waterfowl (FOWL). The magnitude of the Pearson correlation is easily seen to be dominated by the values of the variables for the one large pond (observation 31) and may therefore not reflect the true magnitude of the relationship between these two variables.

Solution The Spearman correlation may be a better measure of association for these variables. Table 13.8 gives the ranks of the two variables, labeled RWATER and RFOWL, and the absolute values of differences in the ranks, DIFF.

The correlation coefficient computed directly from the ranks is 0.490. Using the F statistic, we are able to test this correlation for significance. The p value for this test is 0.006, so we conclude that the correlation is in fact significant. The shortcut formula using the differences among ranks results in a correlation coefficient of 0.4996. The difference is due to a small number of ties in the data. Of course, for this large data set the special formula represents no savings in computational effort.

The Pearson correlation coefficient computed from the observed values results in a value of 0.885. The fact that this value is much larger than the Spearman correlation is the result of the highly skewed nature of the distributions of the variables in this data set. ■

13.7 CHAPTER SUMMARY

Solution to Example 13.1 The distribution of the residuals from the ANOVA model for Example 13.1 did not have the assumed normal probability distribution. This leads us to suspect the results of the F test, particularly the p value. This problem, however, does fit the criteria for the use of a Kruskal–Wallis test. The data, the ranks and the result of using `PROC NONPAR1WAY` in SAS are given in Table 13.9. Note that the printout gives the Kruskal–Wallis test statistic along with the p value calculated from the χ^2 approximation. In this example, the p value is quite small so we reject the null hypothesis of equal treatment distributions.

Note that the output also gives the sums and means of the ranks (called scores). The sums are the $\sum R_i$ in the formula for the test statistic. Also provided are the expected sum and the standard deviations if the null hypothesis is true. These are identical because the sample sizes are equal (each is 15), and the null hypothesis is that of equality. That is, we expect all four of the treatments to have equal sums of ranks if the populations are identical.

The mean scores given in Table 13.9 can be used to make pairwise comparisons (Section 13.4). The least significant difference between average ranks for $\alpha = 0.05$ is 6.69. From Table 13.9 we can see that treatment 1 is significantly smaller than the other three, and that treatment 4 is significantly larger than the other three. Treatments 2 and 3 are not significantly different. Since we wanted to minimize the amount of wear, chemical treatment number 1 seems to be the best.

It is interesting to note that these results are quite similar to those obtained by the analysis of variance. This is because, unlike highly skewed or fat-tailed distributions, the uniform distribution of the random error does not pose a very serious violation of asumptions. ■

Table 13.9

Windshield Wipers

N P A R 1 W A Y P R O C E D U R E
Wilcoxon Scores (Rank Sums) for Variable WEAR
Classified by Variable TRT

TRT	N	Sum of Scores	Expected Under H0	Std Dev Under H0	Mean Score
1	15	120.000000	457.500000	58.5231375	8.0000000
2	15	452.500000	457.500000	58.5231375	30.1666667
3	15	516.000000	457.500000	58.5231375	34.4000000
4	15	741.500000	457.500000	58.5231375	49.4333333

Average Scores were used for Ties

Kruskal-Wallis Test (Chi - Square Approximation)
CHISQ = 43.360 DF = 3 Pr > CHISQ = 0.0001

Nonparametric methods provide alternative statistical methodology when assumptions necessary for the use of linear model-based methods fail as well as provide procedures for making inferences when the scale of mesurement is ordinal or nominal. Generally, nonparametric methods use functions of observations, such as ranks, and therefore make no assumptions about underlying probability distributions. Previous chapters have presented various nonparametric procedures (for example, Chapter 12) usually used for handling nominal scale data. This chapter discusses rank-based nonparametric methods for one, two, and more than two independent samples, paired samples, randomized block designs, and correlation.

While it may not be apparent from the form of most of the statistics presented in this chapter, most rank-based nonparametric statistics use formulas identical to the parametric statistics with ranks substituted for the original data. For example, a one-way analysis of the variance applied to ranks is equivalent to the Kruskal–Wallis test. The difference in the two procedures is the probability distribution of the test statistic. Nonparametric statistics usually require a separate table of critical values or use a large sample approximation involving the χ^2 or F distribution.

To illustrate this, we will work Example 13.6 using SAS by first ranking the observations within blocks using `PROC RANK`, and then analyzing the ranks using `PROC ANOVA` with the randomized block design model. The results are given in Table 13.10. Note from Table 13.10 that the test statistic for testing `TREAT` is 4.84, exactly the value obtained in Example 13.6. Further, since we used the F approximation with 4 and 8 degrees of freedom, the p value given on the printout is the appropriate one for the test statistic. Note, also, that the sums of squares for `BLOCK` is exactly 0. This is because the Friedman test requires that we rank within each block. Therefore, each block will have exactly the same values (in differing order); therefore the sums of squares will always be zero.

This procedure will work on the Kruskal–Wallis test, the Spearman correlation, and even the Wilcoxon test. For this reason, some computer packages provide a procedure for ranking data for use in such analyses rather than providing separate procedures for nonparametric methods.

Table 13.10

Example 13.6

ANALYSIS OF VARIANCE PROCEDURE
DEPENDENT VARIABLE: RANKPCT

SOURCE	DF	SUM OF SQUARES	MEAN SQUARE
MODEL	6	20.16666667	3.36111111
ERROR	8	8.33333333	1.04166667
CORRECTED TOTAL	14	28.50000000	

SOURCE	DF	ANOVA SS	F VALUE	PR > F
TREAT	4	20.16666667	4.84	0.0280
BLOCK	2	0.00000000	0.00	1.0000

13.8 CHAPTER EXERCISES

EXERCISES

1. In 11 test runs a brand of harvesting machine operated for 10.1, 12.2, 12.4, 12.4, 9.4, 11.2, 14.8, 12.6, 10.1, 9.2, and 11.0 h on a tank of gasoline.
   (a) Use the Wilcoxon signed rank test to determine whether the machine lives up to the manufacturer's claim of an average of 12.5 h on a tank of gasoline. (Use $\alpha = 0.05$.)
   (b) For the sake of comparison, use the one-sample t test and compare results. Comment on which method is more appropriate.

2. Twelve adult males were put on a liquid diet in a weight-reducing plan. Weights were recorded before and after the diet. The data are shown in Table 13.11. Use the Wilcoxon signed rank test to ascertain whether the plan was successful. Do you think the use of this test is appropriate for this set of data? Comment.

Table 13.11 Data for Exercise 2

	SUBJECT											
	1	2	3	4	5	6	7	8	9	10	11	12
Before	186	171	177	168	191	172	177	191	170	171	188	187
After	188	177	176	169	196	172	165	190	165	180	181	172

Table 13.12

Data for Exercise 3

PROFESSOR	
A	B
74	75
78	80
68	87
72	81
76	72
69	73
71	80
74	76
77	68
71	78

3. The test scores shown in Table 13.12 were recorded by two different professors for two sections of the same course. Using the Mann–Whitney test and $\alpha = 0.05$, determine whether the locations of the two distributions are equal. Why might the median be a better measure of location than the mean for these data?

4. Inspection of the data for Exercise 11 in Chapter 5 suggests that the data may not be normally distributed. Redo the problem using the Mann–Whitney test. Compare the results with those obtained by the pooled t test.

5. Eight human molar teeth were sliced in half. For each tooth, one randomly chosen half was treated with a compound designed to slow loss of minerals; the other half served as a control. All tooth halves were then exposed to a demineralizing solution. The response is percent of mineral content remaining in the tooth enamel. The data are given in Table 13.13.
   (a) Perform the Wilcoxon signed rank test to determine whether the treatment maintained a higher mineral content in the enamel.
   (b) Compute the paired t statistic and compare the results. Comment on the differences in the results.

Table 13.13

Data for Exercise 5

	Mineral Content							
Control	66.1	79.3	55.3	68.8	57.8	71.8	81.3	54.0
Treated	59.1	58.9	55.0	65.9	54.1	69.0	60.2	55.5

Table 13.14

Data for Exercise 6

METHOD 1	2	3
94	82	89
87	85	68
90	79	72
74	84	76
86	61	69
97	72	
	80	

6. Three teaching methods were tested on a group of 18 students with homogeneous backgrounds in statistics and comparable aptitudes. Each student was randomly assigned to a method and at the end of a 6-week program was given a standardized exam. Because of classroom space, the students were not equally allocated to each method. The results are shown in Table 13.14.

(a) Test for a difference in distributions of test scores for the different teaching methods using the Kruskal–Wallis test.

(b) If there are differences, explain the differences using a multiple comparison test.

Table 13.15

Data for Exercise 7

County	YEAR 1	2	3
P	49	141	82
B	13	64	8
C	175	30	7
R	179	9	7

7. Hail damage to cotton, in pounds per planted acre, was recorded for four counties for three years. The data are shown in Table 13.15. Using years as blocks use the Friedman test to determine whether there was a difference in hail damage among the four counties. If a difference exists, determine the nature of this difference with a multiple comparison test. Also discuss why this test was recommended.

8. To be as fair as possible, most county fairs employ more than one judge for each type of event. For example, a pie-tasting competition may have two judges testing each entered pie and ranking it according to preference. The Spearman rank correlation coefficient may be used to determine the consistency between the judges (the interjudge reliability). In one such competition there were 10 pies to be judged. The results are given in Table 13.16.

(a) Calculate the Spearman correlation coefficient between the two judges' rankings.

(b) Test the correlation for significance at the 0.05 level.

Table 13.16

Ranking of Pies by Judges

Pie	Judge A	Judge B
1	4	5
2	7	6
3	5	4
4	8	9
5	10	8
6	1	1
7	2	3
8	9	10
9	3	2
10	6	7

9. An agriculture experiment was conducted to compare four varieties of sweet potatoes. The experiment was conducted in a completely randomized design with varieties as the treatment. The response variable was yield in tons per acre. The data are given in Table 13.17. Test for a difference in distributions of yields using the Kruskal–Wallis test. (Use $\alpha = 0.01$.)

10. In a study of student behavior, a school psychologist randomly sampled four students from each of five classes. He then gave each student one of four different tasks to perform and recorded the time, in seconds, necessary to complete the assigned task. The data from the study are listed in Table 13.18. Using classes as blocks use the Friedman test to determine whether there is a difference in tasks. Use a level of significance of 0.10. Explain your results.

Table 13.17

Yield of Sweet Potatoes

Variety A	Variety B	Variety C	Variety D
8.3	9.1	10.1	7.8
9.4	9.0	10.0	8.2
9.1	8.1	9.6	8.1
9.1	8.2	9.3	7.9
9.0	8.8	9.8	7.7
8.9	8.4	9.5	8.0
8.9	8.3	9.4	8.1

Table 13.18

Time of Perform Assigned Task

	TASK			
Class	1	2	3	4
1	43.2	45.8	45.4	44.7
2	48.3	48.7	46.9	48.8
3	56.6	56.1	55.3	54.6
4	72.0	74.1	89.5	82.7
5	88.0	88.6	91.5	88.2

Table 13.19

Bird Counts for Twenty-Five Years

	ROUTE		
Year	A	B	C
65	138	815	259
66	331	1143	202
67	177	607	102
68	446	571	214
69	279	631	211
70	317	495	330
71	279	1210	516
72	443	987	178
73	1391	956	833
74	567	859	265
75	477	1179	348
76	294	772	236
77	292	1224	570
78	201	1146	674
79	267	661	494
80	357	729	454
81	599	845	270
82	563	1166	238
83	481	1854	98
84	1576	835	268
85	1170	968	449
86	1217	907	562
87	377	604	380
88	431	1304	392
89	459	559	425

11. Table 13.19 shows the total number of birds of all species observed by birdwatchers for routes in three different cities observed at Christmas for each of the 25 years from 1965 through 1989.

    An inspection of the data indicates that the counts are not normally distributed. Since the responses are frequencies, a possible alternative is to use the square root transformation, but another alternative is to use a nonparametric method. Perform the analysis using the Friedman test. Compare results with those obtained in Exercise 10, Chapter 10. Which method appears to provide the most useful results?

Sampling and Sample Surveys

14.1 Introduction

An almost daily occurrence in our news media is a report of some opinion poll giving percentages of individuals favoring some issue or political candidate. Usually these reports are given with a great deal of authority and may even be accompanied by a rather small "margin of error." Rarely will the report contain any explanation as to the method of selection of the sample used in this poll or just exactly what the margin of error really means.

Actually, the statistics quoted in reports such as these are based on some type of sample from a target population (registered voters, for example) and are nothing more than point estimates of parameters. The percentage of individuals favoring some issue is nothing more than a point estimate of the proportion of individuals in the population favoring this issue. The margin of error is usually based on a 0.95 confidence interval on this proportion and is simply one-half the width of this interval. Therefore, if a report shows 45% of the people polled favored an issue with a margin of error of 2%, then we know that the true proportion of people in the population favoring this issue will be somewhere between 0.43 and 0.47 with 0.95 confidence. In other words, opinion polls are essentially exercises in statistical methodology based on the principles we have been using throughout this book. The major difference is that the sample is likely not taken in a completely random manner. Recall that we have made the assumption that our samples were what we called completely random samples (except when the sample came from a designed experiment). This chapter gives a brief introduction to some of the possible sampling schemes used in such polls.

Opinion polls are an application of a subfield of the discipline of statistics usually called **sampling** or **sample survey methodology**. In many ways this sampling methodology is similar to methods we have presented in previous chapters. There are, however, sufficient differences and special considerations such that the theory and methods of sampling justify a wide variety of books, journal articles, and formal courses. A classic reference work in this area is *Sampling Techniques* by Cochran (1977).

We present in this chapter a very brief overview of the methodology of this subfield. The presentation in this chapter is intended only to acquaint the reader with this subject area and is not expected to provide sufficient information to carry out and analyze a sample survey.[1]

Sampling methodology is not confined to opinion polls. Other applications of sampling are numerous and include

- samples of trees from a forest to estimate timber yield of the entire forest,
- sample surveys of individuals and/or families to assess social and economic characteristics such as unemployment, income, purchasing habits, etc.,
- sample surveys of farms and businesses to assess economic conditions such as production, costs, profits, etc., and
- inspection of samples of output from a manufacturing process to determine the quality of products produced.

Applications of the second and third types listed are employed by government agencies to produce many of the series of indexes of economic activity. Many important applications of sampling methodology involve samples of persons from whom information is solicited; hence the material in this chapter is geared to that type of application. Most of the basic principles do, however, apply to all types of sampling.

The methodology of sampling might, at first, appear to be similar to that of experimental design. There is a difference, however, both in the objectives of and in the procedure used to obtain the sample.

Some of the principal features that distinguish sampling methodology from experimental design methodology follow:

- Normally the population is finite, such as, for example, the population of voters in a state. This is in contrast to most of what we have been doing where the population is conceptually infinite, such as, for example, all possible repetitions of a particular experiment.
- The units of the population cannot be manipulated or controlled as is done in experimental design. Rather, the sample is selected from a defined, existing population. The difference between sampling and experimental design can be restated as the following:

Sample: examination of what is present in the population.
Experiments: examination of what may happen in the population if conditions change.

[1]Therefore no exercises are given at the end of this chapter.

- Often some known information exists about the population. For example, a city can be divided into sections representing households of different income classes, or states can be divided into regions representing different types of agricultural practices.
- The primary purpose of sample surveys is to estimate population parameters, usually the means or totals of several variables. Hypothesis testing or comparative studies are relatively rare applications of sampling.

In summary, then, sampling methodology is distinguished by

- the existence of a defined finite population,
- an inability to manipulate or change the population, and
- primarily being interested in describing the population in terms of estimates of some of its parameters.

On the other hand, most of the basic principles of statistical inference do apply to sampling, including

- the importance of randomness in the selection of units,
- the use of the central limit theorem, and, in general,
- close scrutiny of data collection procedures and the nature of the data.

We begin the chapter with a discussion of some practical considerations in planning and executing a sample survey. This is followed by a brief introduction to the use of two of the more popular sampling designs: simple random sampling and stratified random sampling. A final section mentions some additional topics related to sample survey design.

14.2 Some Practical Considerations

We have already noted that the population for a sample survey cannot be controlled or manipulated. Also, sample surveys often involve large sample sizes: Samples in excess of 1000 observations are quite frequent. For these reasons the data collection phase of a sample survey is more complicated than it is for most experiments and must therefore be more carefully planned and executed. Many aspects of this phase are not strictly "statistical" in that they do not involve principles of statistical inference. Rather they consist of principles and methodology that assure that the sample provides useful and reliable data, which are compatible with the purposes of the data collection and can be analyzed by the appropriate statistical methodology. Many books and articles have been written on this subject (for example, Sudman, 1976). We present in this section highlights of the fundamental aspects of this phase of sampling.

- *Definition of the population:* In previously discussed applications of statistical methods, the definition of the population of units and the values of the variables describing these units have not caused much difficulty: yields of plots of rice, test scores of students, weights of peanuts in jars, etc. The focus of sample surveys is often on populations and variables that are more difficult to define. For example, what is a family? Is a group of students in a sorority a family? Does the income of a family include earnings of a student away at college who works part time? How do you record the production of a farm whose owner cultivates some crops on nonowned land on a share basis? Is a five-acre vegetable garden a farm?

- *The frame:* The existence of a defined population would seem to imply a list of population units from which the sample can be drawn. Unfortunately, this is not always the case. For example, there is usually no complete list of individuals or households in a city. Instead a frame is defined as an available list (or equivalent) used as the vehicle to provide the basis for a sample that will represent the desired population. For example, a "list" of city blocks can be obtained from an aerial photograph, and sampling families from a random sample of blocks may provide an acceptable sample of families. The definition of a useful frame is not always straightforward. For example, a list of farm addresses is not necessarily a good frame for a sample of farmers.
- *Method of data collection:* Data from experiments are usually obtained by measurement or counting, while data from sample surveys are often obtained as answers to questions posed to sampled individuals. One method of collecting such data is for an interviewer to contact the sample of individuals and record answers to questions listed in a questionnaire. This method is quite expensive, largely due to the time and effort required to locate and contact the individuals selected for the sample. Using the mails or telephone incurs lower costs; however, these data collection methods have serious drawbacks because lists of mailing addresses or telephone numbers often comprise an incomplete frame. Furthermore, mail samples, and to a lesser degree telephone samples, suffer from nonresponse, the failure of all contacted individuals to provide answers. Since the failure to respond is not a random act, a sample that consists only of volunteer responders is not a random sample from the entire population and should not be used to make inferences to the population.
- *Questionnaire construction:* The instrument used to obtain information about respondents in a sample survey is the questionnaire. A poorly conceived or executed questionnaire can ruin an otherwise well-planned and executed sample survey. Principles for good questionnaire construction involve the use of clearly worded unambiguous questions that elicit the desired information, avoidance of "loaded" questions that suggest a "correct" answer, construction of questions whose answers can be translated into variables suitable for statistical analysis, and finally assurance that the questionnaire obtains all the needed information without being unduly long. An important vehicle in questionnaire construction is a pretest where a questionnaire is used on some individuals who are typical (but not a part) of the sample respondents. Also important is careful selection and training of the interviewers who will administer the questionnaire.

- *Data analysis:* Because sample surveys usually involve rather large amounts of data, careful planning of the analysis phase is very important. Aspects that require special care include coding of answers to questions to allow statistical analyses and careful data editing to assure high data quality.

We cannot emphasize too strongly that these nonstatistical aspects of planning and executing a sample survey are not trivial and must be carefully considered and implemented. For the remainder of this chapter we assume that these aspects have been addressed, and concentrate on the **sample survey design**, which specifies the procedures for selecting the sample units from the population and the subsequent calculation of statistics and making the desired inferences.

14.3 Simple Random Sampling

The simplest and most obvious sampling design is called simple random sampling, which consists of randomly sampling units from the population at large.

Notation

Consider a population of N units. One or more variables are associated with each unit. For simplicity we focus on the variable we denote by y. The N values of y,

$$y_i, \quad i = 1, \ldots, N,$$

constitute the values of the variable for the population. The distribution of the variable in the population is characterized by the usual parameters,[2] the mean μ and the variance σ^2.

DEFINITION 14.1
A **simple random sample** of n units is defined such that each possible sample of size n is equally likely to be drawn. This sampling principle assures that each unit in the population has probability (n/N) of being selected for the sample.

DEFINITION 14.2
The quantity (n/N) is referred to as the **sampling fraction**. The reciprocal (N/n) is known as the **expansion factor**.

[2]Many references (for example, Cochran, 1977) define the population variance $\sum(y_i - \mu)^2/(N-1)$. This formulation of the population variance is used as the basis for the other formulas presented in this chapter. Some of these references will, in fact, use $\bar{y}$ (or $\bar{Y}$) and S^2 to denote the population mean and variance, Since the population size (N) is usually quite large, the use of (N) rather than $(N-1)$ will make no significant difference.

Sampling Procedure

The idealized procedure for selecting a simple random sample of n units is to sequentially identify the population units with integers from 1 to N, and select units corresponding to n random integers, chosen **without duplication** from either a table of random numbers or generated by a computer.

As we noted earlier, the idealized procedure may need to be adapted according to the manner in which individual population units can be identified.

Estimation

The point estimates for the population parameters μ and σ^2 are the corresponding sample statistics $\bar{y}$ and s^2, computed in the usual manner. In some applications, for example in estimating crop yields, we may want to estimate the population **total**, $N\mu$, for which the estimate is $N\bar{y}$.

The sampling distribution of the mean tends to be, as before, normal, with

$$\text{mean}\,(\bar{y}) = \mu \quad \text{and}$$

$$\text{var}(\bar{y}) = \left(\frac{N-n}{N}\right)\left(\frac{\sigma^2}{n}\right).$$

The factor $(N - n)/N$, which can also be written $(1 - n/N)$, reduces the magnitude of the variance of the mean by the sampling fraction from what it would be if the population were infinite. This reduction factor is called the **finite population correction factor** (fpc). The fpc reflects the additional information we obtain by sampling a portion of a finite population. The larger n is relative to N, the more information we have about the whole population. In fact, if we were to sample the entire population, the fpc would be zero, because then we would know exactly the population mean and the variance of that "estimate" is zero. The fpc has minimal effect and is usually ignored if $n/N < 0.05$.

The estimated population total is $N\bar{y}$. Since N is a known constant, the variance of the estimated total is N^2 times the variance of the mean:

$$\text{var}(N\bar{y}) = N(N-n)\frac{\sigma^2}{n}.$$

Of course, the population variance is usually not known and the estimated variance s^2 is substituted to obtain the estimated variance of the mean. The use of the estimated variance requires the use of the t distribution for confidence interval estimates of the mean:

$$\bar{y} \pm t_{\alpha/2}\sqrt{\left(\frac{N-n}{N}\right)\left(\frac{s^2}{n}\right)}.$$

Usually sample sizes are sufficiently large to allow use of the appropriate percentage points of the normal (rather than the t) distribution. In fact, it is customary to use the factor 2 as a close approximation to $z_{\alpha/2}$ for the 0.95 confidence interval.

In some applications, such as opinion polls, the response variable is binomial. The normal approximation to the binomial is used as outlined in Section 4.3 with the addition of the fpc. Thus, for example, the $(1 - \alpha)$ confidence interval for the proportion of successes is

$$\hat{p} \pm z_{\alpha/2}\sqrt{\left(\frac{N - n}{N}\right)\left[\frac{\hat{p}(1 - \hat{p})}{n}\right]},$$

where $\hat{p}$ is the sample proportion of successes.

Systematic Sampling

The actual process of drawing a random sample as described in the previous section can be quite tedious, especially for large populations and large samples. An alternative procedure, which is often easier to implement and works quite well in most cases, is **systematic sampling**. The systematic sampling procedure is implemented as follows:

1. Identify all units of the population: $i = 1, 2, \ldots, N$.
2. Calculate the expansion factor (N/n) and round to the nearest integer. Call this r.
3. Choose a random integer, r_0, $1 \leq r_0 \leq (r - 1)$.
4. Identify units to be included in the sample by $i = r_0, r_0 + r, r_0 + 2r, \ldots$.

The sampling method will result in a sample of n units (± 1 due to rounding of r), and each unit of the population has an equal chance of being selected. A major reason for the popularity of systematic sampling is not having to select n different random numbers. It is also often more convenient to locate every rth unit rather than to select units randomly spaced.

The systematic sampling procedure can, however, produce a biased sample if the list of population units contains cycles or trends. For example, if $r = 16$, and we are sampling a set of 16-unit apartment buildings, all sampled units will be in the same location in each apartment building. Thus, if $r_0 = 1$, then all sampled units would be the first apartment unit in each building, which will most likely be on the first floor and may, for example, contain a larger proportion of aged and handicapped persons. Such extreme biases rarely occur but less obvious cycles may occur when least expected; hence caution is advised.

If the population is indeed in random order, sample estimates and variances are computed as if simple random sampling has been used.

Sample Size

In simple random sampling, calculation of the sample sizes required for specified precision uses the same principles presented in Sections 3.4 and 4.3. However, because of the fpc, the formula for sample size determination is somewhat more complicated. Since the fpc is needed only if $n/N > 0.05$, it is customary to first calculate sample size ignoring the fpc; then if it appears that

n/N may exceed 0.05 an adjustment is made for this factor. The initial sample size estimate, n_0, ignoring fpc, is the one presented in Section 3.4,

$$n_0 = \frac{z_{\alpha/2}^2 \sigma^2}{E^2},$$

where E is the maximum error of estimate (half the width of the $(1 - \alpha)$ confidence interval). If $n_0/N > 0.05$, the final sample size estimate, n, is

$$n = \frac{n_0}{1 + n_0/N}.$$

Sample surveys are often used to estimate parameters from distributions of several variables. Calculation of minimum required sample sizes for each of the variables is likely to produce different sample size requirements. The safest recommendation is, of course, to take the largest of these required sizes; however, this may not be feasible. Compromises on the ultimate sample size must reflect the relative importance of the required precision for the different variables.

All of these formulas involve the population variance σ^2, which is normally not known. In practice, an estimate of the variance, usually based on prior samples of similar populations, is used.

14.4 Stratified Sampling

In experimental design (Chapter 10) we use the principle of blocking to reduce the error variance, thereby increasing the precision of estimates and/or power of hypothesis tests.

DEFINITION 14.3
Stratified sampling is a sampling method in which the population is divided into portions, called **strata**, which are expected to contain relatively homogeneous units, and samples (either random or systematic) are taken independently in each stratum.

The use of stratified sampling is based on the premise that in many populations definable portions that consist of relatively homogeneous units exist. For example, crop yields in a specific area of a state will be relatively more uniform than those of the entire state, or voters in a specific precinct will tend to have a more consistent voting pattern than do the voters of an entire city. As is the case for blocking designs, the variances of estimates are based on variances within strata, and are therefore likely to be smaller than those resulting from a random sample of the entire population.

Estimation

Consider a population with s strata. There are N_i, $i = 1, 2, \ldots, s$ units in stratum i, and $\sum N_i, = N$ is the total population size. The stratum means and variances of the response variable are μ_i and σ_i^2, respectively. The overall mean can be computed as the weighted mean of stratum means: $\mu = \sum N_i \mu_i / N$.

A sample of n_i units is drawn from each stratum; hence $\sum n_i = n$ is the total sample size. For each stratum we compute the sample mean $\bar{y}_i$ and variance s_i^2. The estimate of the overall population mean μ is

$$\bar{y}_{\text{strat}} = \frac{1}{N} \sum N_i \bar{y}_i,$$

which is an unbiased estimate of the population mean. The variance of the sampling distribution of this estimate[3] is

$$\text{var}(\bar{y}_{\text{strat}}) = \frac{1}{N^2} \sum N_i (N_i - n_i) \left(\frac{\sigma_i^2}{n_i} \right).$$

In contrast to analysis of variance methods, stratum variances are not required to be equal. Hence there is no pooled variance estimate, and the t distribution cannot be used to compute confidence intervals. However, sample sizes are almost always sufficiently large so that the normal distribution may be safely used for the construction of confidence intervals. As in simple random sampling, the factor 2 is usually used in place of the z value for constructing a 0.95 confidence interval.

Sample Sizes

The formula for the variance of the estimated mean from a stratified sample involves not only the overall sample size but also the individual strata sample sizes. Therefore the determination of sample size needed for a given precision must be done in two stages:

1. determining the size of the entire sample n and
2. the **allocation** of the sample to the different strata, that is, determining n_i/n.

Since the variance of the estimate of the overall mean is influenced by the nature of the allocation, it is necessary to first consider the allocation problem.

A very logical allocation principle is that of **proportional allocation**, for which the sampling fraction is the same for all strata, that is,

$$\frac{n_i}{N_i} = \frac{n}{N}, \quad i = 1, 2, \ldots, s.$$

Proportional allocation is easy to implement since the selection procedure follows the same rules for each stratum. Furthermore, estimation is simplified because the sample is self-weighting; that is, the estimate of the population mean is the simple (unweighted) mean of all observed values,

$$\bar{y} = \frac{1}{n} \sum y,$$

[3]It is customary to present the formulas for variance of estimates in terms of the population variances. Sample estimates are substituted for actual calculations.

where the summation is over all sample values. The formula for the variance of the estimated mean is also simplified:

$$\operatorname{var}(\bar{y}_{\text{prop}}) = \frac{N-n}{N}\frac{1}{n}\sum \frac{N_i}{N}\sigma_i^2.$$

Another allocation principle is based on the idea that more precise estimates can be obtained if relatively larger samples are taken from strata having large within-stratum variances. The fact that this principle does indeed provide smaller variance estimates is demonstrated with the aid of calculus, where it can be shown that the smallest possible variance of the mean is obtained if the sampling fractions of the strata are specified according to

$$\frac{n_i}{n} = \frac{N_i \sigma_i}{\sum N_i \sigma_i}.$$

This so-called **optimum allocation** is seen to cause sample sizes larger than those of proportional allocation for strata exhibiting larger within-stratum standard deviations. Optimum allocation does not, however, provide a self-weighting sample; hence the estimated mean and variance are obtained by the general formula given earlier in this section. If within-stratum variances are equal, the optimum allocation is equivalent to proportional allocation.

Here again, the individual stratum variances are normally not known and estimates based on previous studies are used in practice. Note that poor estimates of these variances may undo the increased precision implied for optimum allocation.

Substituting the optimum allocation fractions into the formula for the variance of the estimated mean, we obtain

$$\operatorname{var}(\bar{y}_{\text{opt}}) = \frac{1}{n}\left[\frac{1}{N^2}\left(\sum N_i \sigma_i\right)^2\right] - \frac{1}{N^2}\sum N_i \sigma_i^2,$$

where the second portion is the finite population correction factor.

Note that once the allocations are determined, the variance of the estimated mean is a function of the overall sample size. Also the contribution of the fpc is a separate part of the function. This means that the overall sample size for a given degree of precision can be obtained in two stages as was done for the simple random sampling procedure (see the following example).

Refinements exist for modifying the allocations to reflect differential costs for collecting data among strata (Cochran, 1977, Section 5.5). Occasionally optimum allocation may specify $n_i/N_i \geq 1$ for some strata, in which case those strata are sampled at 100% and the remainder of the sample is allocated to the other strata as indicated above. In this case, the strata sampled at a 100% rate contribute nothing to the variance of the estimated mean.

For a binomial response variable, the values $p_i(1 - p_i)$, where p_i are the stratum population proportions, are substituted for the σ_i^2 in the above formulas. For this case it is readily seen that strata with p_i near 0 or unity will have low allocations relative to strata with p_i near 0.5. Here again, values of the p_i based on previous studies will need to be used.

Efficiency

It can be shown that for a given total sample size n,

$$\text{var}(\bar{y}_{\text{SRS}}) \geq \text{var}(\bar{y}_{\text{prop}}) \geq \text{var}(\bar{y}_{\text{opt}}),$$

where $\bar{y}_{\text{SRS}}$ represents simple random sampling, that is, a sampling procedure that ignores the existence of the strata. The advantages that accrue to stratified sampling are similar to those for blocking: They are due to relatively large variation among strata; that is, a large value for $\sum(\mu_i - \mu)^2$. Additional advantages accrue to optimum allocation due to large differences in variances within the different strata.

Of course, these advantages are functions of the population variances, which are unknown. Educated guesses using information from previous studies are normally used to determine the stratum parameters and the relative allocations. Often these work quite well, but bad guesses can nullify the expected advantages.

An Example

Most sample surveys are too large to be used as textbook examples. We therefore illustrate the principles of stratified sampling with a small artificial population of 110 units with three strata. The data and population parameters are presented in Table 14.1. In this example, we know the population variances; hence, we can show the relative efficiencies of the different sampling designs.

We first compare the variance of the estimated mean using simple random and proportional and optimum allocated stratified samples with a total sample size of 30. For simple random sampling, the variance is

$$\text{var}(\bar{y}_{\text{SRS}}) = \left(\frac{110-30}{110}\right)\left(\frac{23283.6}{30}\right) = 564.45.$$

For the proportionally allocated stratified sample, the sampling ratio of $n/N = 30/110 = 0.273$ is applied to each of the strata:

$$n_1 = 5.46 \sim 6,$$
$$n_2 = 8.19 \sim 8,$$
$$n_3 = 16.38 \sim 16.$$

The variance of the estimated mean is

$$\text{var}(\bar{y}_{\text{prop}}) = \left(\frac{110-30}{110}\right)\left(\frac{1}{30}\right)\left[\frac{20}{110}(8721.2) + \frac{30}{110}(2536.0) + \frac{60}{110}(303.5)\right]$$
$$= 59.22.$$

The variance of the estimated mean is seen to be reduced considerably by using stratification. Examination of the data clearly shows the reason: There are large differences among the μ_i.

Table 14.1

Population Data for Stratified Sample

STRATUM			
1	**2**	**3**	
385	241	73	69
473	204	109	121
371	190	92	85
321	303	108	95
413	312	83	130
495	320	102	124
701	368	79	84
474	257	119	102
494	352	85	117
501	299	100	119
618	237	117	76
480	310	98	136
457	280	85	56
379	287	120	126
461	267	92	109
328	307	83	94
540	216	92	96
519	279	120	120
469	277	113	132
607	243	92	115
	379	114	104
	319	77	102
	286	88	91
	311	123	116
	340	98	87
	401	111	107
	265	104	116
	306	93	116
	317	93	95
	371	97	135

	STRATUM			
Parameters	**1**	**2**	**3**	**All**
N_i	20	30	60	110
μ	474.3	294.8	102.2	222.41
σ^2	8721.2	2536.0	303.5	23283.6
N_i/N	0.182	0.273	0.545	

The computations for optimum allocation are summarized as follows:

Stratum	$N_i\sigma_i$	$(N_i\sigma_i)/\sum N_i\sigma_i$	n_i
1	1867.7	0.422	$12.66 \sim 13$
2	1510.8	0.342	$10.26 \sim 10$
3	1045.3	0.236	$7.08 \sim 7$
Total	4423.8	1.000	30

We can see that although stratum 1 is the smallest, it gets the largest sample size due to the large within-stratum variance. The variance of the estimated mean is

$$\text{var}(\bar{y}_{\text{opt}}) = \left(\frac{1}{30}\right)\left[\frac{1}{110^2}(4423.8)^2\right] - \left(\frac{1}{110^2}\right)$$

$$[20(8721.2) + 30(2536.0) + 60(303.5)] = 31.70.$$

We can see that there is an additional reduction in the variance due to optimum allocation, although the decrease is not as dramatic as that accomplished by stratification.

Another way to see the advantage of stratification is to compare required sample sizes for a desired precision of the estimated mean. As we have noted, sample size requirements are determined by adapting the principles presented in Section 3.4.

We continue the example by determining the required sample size for a maximum error of estimate of 20, which is equivalent to a maximum width of 40 for a 0.95 confidence interval (Section 3.3). Remember that a maximum error of estimate of E implies a standard error of the estimated mean of $E/2$ (using the value 2 as an approximation for the 0.025 value of the normal distribution), which implies that the variance of the mean is $E^2/4$. For our required E of 20, the variance of the estimated mean must be less than or equal to $20^2/4 = 100$. We will obtain the required sample size by equating the expression for the variance of the mean to 100 and solving the resulting equation for n.

All quantities for these computations are available in Table 14.1. For the simple random sample,

$$\text{var}(\bar{y}_{\text{SRS}}) = \sigma^2/n.$$

Substituting the value of 100 for the variance, we have

$$100 = 23283.6/n_0; \text{ hence}$$

$$n_0 = 233.$$

This is certainly more than 5% of the population; hence

$$n = \frac{n_0}{1 + n_0/N} = \frac{233}{1 + 233/110} = 75.$$

For stratified sampling with proportional allocation,

$$\text{var}(\bar{y}_{\text{prop}}) = \frac{1}{n_0}\sum\left(\frac{N_i}{N}\right)\sigma_i^2;$$

hence

$$100 = \frac{2442.8}{n_0}$$

$$n_0 = 25.$$

Again $n_0/N > 0.05$; hence

$$n = \frac{25}{1 - (25/110)} = 21.$$

Finally, for stratified sampling with optimum allocation,

$$\text{var}(\bar{y}_{\text{opt}}) = \left(\frac{1}{n_0}\right)\left(\frac{1}{N^2}\right)\left(\sum N_i\sigma_i\right)^2;$$

hence

$$100 = \frac{1617.4}{n_0}$$

$$n_0 = 17.$$

As before, $n_0/N > 0.05$; hence

$$n = \frac{17}{1 + (17/110)} = 15.$$

These results confirm the fact that stratified sampling and optimum allocation are effective for sampling from this population.

Again, keep in mind that values of unknown population parameters were used to obtain the optimum allocation. If the informed guesses on these values used for allocation are not close to the population values, the optimum allocation may not be helpful.

Additional Topics in Stratified Sampling

An apparent necessity for stratified sampling is to have readily identifiable strata such as political or geographic regions or subdivisions. For some situations, strata are not easily identified: Income or occupational groups are recognized as good criteria for stratification, but there are no universally accepted definitions of the exact delineation of such strata. Some general methods are available (for example, Cochran, 1977, Section 5A.8), but they are not always practically useful.

Another difficulty with stratified sampling occurs when the stratum identification of a unit cannot be determined prior to taking the sample. For example, political party identification of an individual cannot be determined until **after** the individual has been interviewed. The most popular solution for this situation is called **post stratification**, and consists of taking a simple random sample, stratifying the sample after it has been taken, and using the formulas for stratified sampling to obtain estimates of the mean and variance. For example, we could stratify an individual by his or her political party affiliation after interviewing the individual. Of course, stratum sample sizes cannot be determined in advance and are, in fact, themselves random variables, which complicates estimation procedures.

An alternative to post stratification is **quota sampling**, which requires a quota of units to be sampled for each stratum. A simple random sample is taken, and sampling continues until the quota has been met for all strata. This method may become expensive, since a number of excess units (from whom information is not used) may need to be interviewed before all quotas are met. Many public opinion polls use adaptations of quota sampling.

14.5 Other Topics

A large variety of additional sampling methodologies is available for obtaining greater precision or to cope with special situations. We briefly mention a few of these here, emphasizing that the coverage is entirely too brief and incomplete to guide actual implementation of these methods.

Multistage sampling is used when units can be divided into subunits. An example of this occurs where city blocks are units, families are subunits, and individuals in families are sub-subunits. The first stage of the sample selects a sample of city blocks, in the second stage a sample of families is selected in the sampled blocks, and in a third stage samples of individuals are selected in the sampled families. In a sense, stratified sampling can be considered a special case of two-stage sampling where units (strata) are sampled at a 100% rate. Multistage sampling can be quite efficient since it allows great flexibility in allocating samples among units as well as subunits, etc. It can, however, also become quite complicated, especially if units vary greatly in size.

Prior information can often be used to increase efficiency of estimates. For example, in preelection polls the actual (population) vote in the previous election is known, and the sample for the current election is used to estimate the change in voting behavior. Such an estimate of change usually has much smaller variance than an estimate of voting behavior itself. Such estimates are called ratio or regression estimates.

Multiple frame sampling may be used when an easily identified frame provides a sample for a large but incomplete portion of the population, while the frame that covers the entire population is more difficult to define and/or sample. For example, a list of farmers receiving price support payments provides an up-to-date list of a major fraction of all farmers. A list of all farmers is, however, not as easily obtained and may need to be drawn by physically locating farmers in sampled areas. In a multiple frame survey, the major portion of the sample is drawn from the easily identified frame while a small sample is drawn from the other. Special instruments must be used to identify the overlap of the two frames.

Special populations often require specialized techniques. For example, in sampling wildlife populations, a technique called line transect sampling consists of having an observer walk a predetermined (sample) path and record the number of animals sighted. This method is greatly affected by the visibility of animals and density of competing structures (plants); hence estimation can become quite complicated. In **capture–recapture** sampling, an initial sample of animals is tagged, and in a second sample the proportion of tagged animals found is used to obtain an estimate of total population size.

Cluster sampling, a variation of stratified sampling, can be used when the cost of sampling is greatly reduced if a whole stratum can be sampled. In this scheme, the population is first divided into strata. A random selection of strata is then obtained and completely sampled. For example, in a study to determine substandard housing stock in a city, the city might be stratified into blocks.

A random selection of blocks is taken, and each block totally sampled. This greatly reduces the time and cost of sampling.

14.6 CHAPTER SUMMARY

We again emphasize that the purpose of this chapter is to acquaint the reader with the field of sampling and sample surveys. There is obviously insufficient information here to provide anyone with the knowledge needed to plan and implement even a small sample survey. Instead the information in this chapter should be used to aid in the initial decisions of whether a sample survey is appropriate for a specific study, and also to aid in understanding the recommendations of a qualified sample survey statistician.

Appendix A

Table A.1

The Normal Distribution—Probabilities Exceeding Z

Z	Prob > Z	Z	Prob > Z	Z	Prob > Z	Z	Prob > Z
−3.99	1.0000	−3.49	0.9998	−2.99	0.9986	−2.49	0.9936
−3.98	1.0000	−3.48	0.9997	−2.98	0.9986	−2.48	0.9934
−3.97	1.0000	−3.47	0.9997	−2.97	0.9985	−2.47	0.9932
−3.96	1.0000	−3.46	0.9997	−2.96	0.9985	−2.46	0.9931
−3.95	1.0000	−3.45	0.9997	−2.95	0.9984	−2.45	0.9929
−3.94	1.0000	−3.44	0.9997	−2.94	0.9984	−2.44	0.9927
−3.93	1.0000	−3.43	0.9997	−2.93	0.9983	−2.43	0.9925
−3.92	1.0000	−3.42	0.9997	−2.92	0.9982	−2.42	0.9922
−3.91	1.0000	−3.41	0.9997	−2.91	0.9982	−2.41	0.9920
−3.90	1.0000	−3.40	0.9997	−2.90	0.9981	−2.40	0.9918
−3.89	0.9999	−3.39	0.9997	−2.89	0.9981	−2.39	0.9916
−3.88	0.9999	−3.38	0.9996	−2.88	0.9980	−2.38	0.9913
−3.87	0.9999	−3.37	0.9996	−2.87	0.9979	−2.37	0.9911
−3.86	0.9999	−3.36	0.9996	−2.86	0.9979	−2.36	0.9909
−3.85	0.9999	−3.35	0.9996	−2.85	0.9978	−2.35	0.9906
−3.84	0.9999	−3.34	0.9996	−2.84	0.9977	−2.34	0.9904
−3.83	0.9999	−3.33	0.9996	−2.83	0.9977	−2.33	0.9901
−3.82	0.9999	−3.32	0.9995	−2.82	0.9976	−2.32	0.9898
−3.81	0.9999	−3.31	0.9995	−2.81	0.9975	−2.31	0.9896
−3.80	0.9999	−3.30	0.9995	−2.80	0.9974	−2.30	0.9893
−3.79	0.9999	−3.29	0.9995	−2.79	0.9974	−2.29	0.9890
−3.78	0.9999	−3.28	0.9995	−2.78	0.9973	−2.28	0.9887
−3.77	0.9999	−3.27	0.9995	−2.77	0.9972	−2.27	0.9884
−3.76	0.9999	−3.26	0.9994	−2.76	0.9971	−2.26	0.9881
−3.75	0.9999	−3.25	0.9994	−2.75	0.9970	−2.25	0.9878
−3.74	0.9999	−3.24	0.9994	−2.74	0.9969	−2.24	0.9875
−3.73	0.9999	−3.23	0.9994	−2.73	0.9968	−2.23	0.9871
−3.72	0.9999	−3.22	0.9994	−2.72	0.9967	−2.22	0.9868
−3.71	0.9999	−3.21	0.9993	−2.71	0.9966	−2.21	0.9864
−3.70	0.9999	−3.20	0.9993	−2.70	0.9965	−2.20	0.9861
−3.69	0.9999	−3.19	0.9993	−2.69	0.9964	−2.19	0.9857
−3.68	0.9999	−3.18	0.9993	−2.68	0.9963	−2.18	0.9854
−3.67	0.9999	−3.17	0.9992	−2.67	0.9962	−2.17	0.9850
−3.66	0.9999	−3.16	0.9992	−2.66	0.9961	−2.16	0.9846
−3.65	0.9999	−3.15	0.9992	−2.65	0.9960	−2.15	0.9842
−3.64	0.9999	−3.14	0.9992	−2.64	0.9959	−2.14	0.9838
−3.63	0.9999	−3.13	0.9991	−2.63	0.9957	−2.13	0.9834
−3.62	0.9999	−3.12	0.9991	−2.62	0.9956	−2.12	0.9830
−3.61	0.9998	−3.11	0.9991	−2.61	0.9955	−2.11	0.9826
−3.60	0.9998	−3.10	0.9990	−2.60	0.9953	−2.10	0.9821
−3.59	0.9998	−3.09	0.9990	−2.59	0.9952	−2.09	0.9817
−3.58	0.9998	−3.08	0.9990	−2.58	0.9951	−2.08	0.9812
−3.57	0.9998	−3.07	0.9989	−2.57	0.9949	−2.07	0.9808
−3.56	0.9998	−3.06	0.9989	−2.56	0.9948	−2.06	0.9803
−3.55	0.9998	−3.05	0.9989	−2.55	0.9946	−2.05	0.9798
−3.54	0.9998	−3.04	0.9988	−2.54	0.9945	−2.04	0.9793
−3.53	0.9998	−3.03	0.9988	−2.53	0.9943	−2.03	0.9788
−3.52	0.9998	−3.02	0.9987	−2.52	0.9941	−2.02	0.9783
−3.51	0.9998	−3.01	0.9987	−2.51	0.9940	−2.01	0.9778
−3.50	0.9998	−3.00	0.9987	−2.50	0.9938	−2.00	0.9772

Table A.1 *(continued)*

Z	Prob > Z	Z	Prob > Z	Z	Prob > Z	Z	Prob > Z
−1.99	0.9767	−1.49	0.9319	−0.99	0.8389	−0.49	0.6879
−1.98	0.9761	−1.48	0.9306	−0.98	0.8365	−0.48	0.6844
−1.97	0.9756	−1.47	0.9292	−0.97	0.8340	−0.47	0.6808
−1.96	0.9750	−1.46	0.9279	−0.96	0.8315	−0.46	0.6772
−1.95	0.9744	−1.45	0.9265	−0.95	0.8289	−0.45	0.6736
−1.94	0.9738	−1.44	0.9251	−0.94	0.8264	−0.44	0.6700
−1.93	0.9732	−1.43	0.9236	−0.93	0.8238	−0.43	0.6664
−1.92	0.9726	−1.42	0.9222	−0.92	0.8212	−0.42	0.6628
−1.91	0.9719	−1.41	0.9207	−0.91	0.8186	−0.41	0.6591
−1.90	0.9713	−1.40	0.9192	−0.90	0.8159	−0.40	0.6554
−1.89	0.9706	−1.39	0.9177	−0.89	0.8133	−0.39	0.6517
−1.88	0.9699	−1.38	0.9162	−0.88	0.8106	−0.38	0.6480
−1.87	0.9693	−1.37	0.9147	−0.87	0.8078	−0.37	0.6443
−1.86	0.9686	−1.36	0.9131	−0.86	0.8051	−0.36	0.6406
−1.85	0.9678	−1.35	0.9115	−0.85	0.8023	−0.35	0.6368
−1.84	0.9671	−1.34	0.9099	−0.84	0.7995	−0.34	0.6331
−1.83	0.9664	−1.33	0.9082	−0.83	0.7967	−0.33	0.6293
−1.82	0.9656	−1.32	0.9066	−0.82	0.7939	−0.32	0.6255
−1.81	0.9649	−1.31	0.9049	−0.81	0.7910	−0.31	0.6217
−1.80	0.9641	−1.30	0.9032	−0.80	0.7881	−0.30	0.6179
−1.79	0.9633	−1.29	0.9015	−0.79	0.7852	−0.29	0.6141
−1.78	0.9625	−1.28	0.8997	−0.78	0.7823	−0.28	0.6103
−1.77	0.9616	−1.27	0.8980	−0.77	0.7794	−0.27	0.6064
−1.76	0.9608	−1.26	0.8962	−0.76	0.7764	−0.26	0.6026
−1.75	0.9599	−1.25	0.8944	−0.75	0.7734	−0.25	0.5987
−1.74	0.9591	−1.24	0.8925	−0.74	0.7704	−0.24	0.5948
−1.73	0.9582	−1.23	0.8907	−0.73	0.7673	−0.23	0.5910
−1.72	0.9573	−1.22	0.8888	−0.72	0.7642	−0.22	0.5871
−1.71	0.9564	−1.21	0.8869	−0.71	0.7611	−0.21	0.5832
−1.70	0.9554	−1.20	0.8849	−0.70	0.7580	−0.20	0.5793
−1.69	0.9545	−1.19	0.8830	−0.69	0.7549	−0.19	0.5753
−1.68	0.9535	−1.18	0.8810	−0.68	0.7517	−0.18	0.5714
−1.67	0.9525	−1.17	0.8790	−0.67	0.7486	−0.17	0.5675
−1.66	0.9515	−1.16	0.8770	−0.66	0.7454	−0.16	0.5636
−1.65	0.9505	−1.15	0.8749	−0.65	0.7422	−0.15	0.5596
−1.64	0.9495	−1.14	0.8729	−0.64	0.7389	−0.14	0.5557
−1.63	0.9484	−1.13	0.8708	−0.63	0.7357	−0.13	0.5517
−1.62	0.9474	−1.12	0.8686	−0.62	0.7324	−0.12	0.5478
−1.61	0.9463	−1.11	0.8665	−0.61	0.7291	−0.11	0.5438
−1.60	0.9452	−1.10	0.8643	−0.60	0.7257	−0.10	0.5398
−1.59	0.9441	−1.09	0.8621	−0.59	0.7224	−0.09	0.5359
−1.58	0.9429	−1.08	0.8599	−0.58	0.7190	−0.08	0.5319
−1.57	0.9418	−1.07	0.8577	−0.57	0.7157	−0.07	0.5279
−1.56	0.9406	−1.06	0.8554	−0.56	0.7123	−0.06	0.5239
−1.55	0.9394	−1.05	0.8531	−0.55	0.7088	−0.05	0.5199
−1.54	0.9382	−1.04	0.8508	−0.54	0.7054	−0.04	0.5160
−1.53	0.9370	−1.03	0.8485	−0.53	0.7019	−0.03	0.5120
−1.52	0.9357	−1.02	0.8461	−0.52	0.6985	−0.02	0.5080
−1.51	0.9345	−1.01	0.8438	−0.51	0.6950	−0.01	0.5040
−1.50	0.9332	−1.00	0.8413	−0.50	0.6915	0.00	0.5000

(Continued)

Table A.1 *(continued)*

Z	Prob > *Z*	*Z*	Prob > *Z*	*Z*	Prob > *Z*	*Z*	Prob > *Z*
0.01	0.4960	0.51	0.3050	1.01	0.1562	1.51	0.0655
0.02	0.4920	0.52	0.3015	1.02	0.1539	1.52	0.0643
0.03	0.4880	0.53	0.2981	1.03	0.1515	1.53	0.0630
0.04	0.4840	0.54	0.2946	1.04	0.1492	1.54	0.0618
0.05	0.4801	0.55	0.2912	1.05	0.1469	1.55	0.0606
0.06	0.4761	0.56	0.2877	1.06	0.1446	1.56	0.0594
0.07	0.4721	0.57	0.2843	1.07	0.1423	1.57	0.0582
0.08	0.4681	0.58	0.2810	1.08	0.1401	1.58	0.057
0.09	0.4641	0.59	0.2776	1.09	0.1379	1.59	0.0559
0.10	0.4602	0.60	0.2743	1.10	0.1357	1.60	0.0548
0.11	0.4562	0.61	0.2709	1.11	0.1335	1.61	0.0537
0.12	0.4522	0.62	0.2676	1.12	0.1314	1.62	0.0526
0.13	0.4483	0.63	0.2643	1.13	0.1292	1.63	0.0516
0.14	0.4443	0.64	0.2611	1.14	0.1271	1.64	0.0505
0.15	0.4404	0.65	0.2578	1.15	0.1251	1.65	0.0495
0.16	0.4364	0.66	0.2546	1.16	0.1230	1.66	0.0485
0.17	0.4325	0.67	0.2514	1.17	0.1210	1.67	0.0475
0.18	0.4286	0.68	0.2483	1.18	0.1190	1.68	0.0465
0.19	0.4247	0.69	0.2451	1.19	0.1170	1.69	0.0455
0.20	0.4207	0.70	0.2420	1.20	0.1151	1.70	0.0446
0.21	0.4168	0.71	0.2389	1.21	0.1131	1.71	0.0436
0.22	0.4129	0.72	0.2358	1.22	0.1112	1.72	0.0427
0.23	0.4090	0.73	0.2327	1.23	0.1093	1.73	0.0418
0.24	0.4052	0.74	0.2296	1.24	0.1075	1.74	0.0409
0.25	0.4013	0.75	0.2266	1.25	0.1056	1.75	0.0401
0.26	0.3974	0.76	0.2236	1.26	0.1038	1.76	0.0392
0.27	0.3936	0.77	0.2206	1.27	0.1020	1.77	0.0384
0.28	0.3897	0.78	0.2177	1.28	0.1003	1.78	0.0375
0.29	0.3859	0.79	0.2148	1.29	0.0985	1.79	0.0367
0.30	0.3821	0.80	0.2119	1.30	0.0968	1.80	0.0359
0.31	0.3783	0.81	0.2090	1.31	0.0951	1.81	0.0351
0.32	0.3745	0.82	0.2061	1.32	0.0934	1.82	0.0344
0.33	0.3707	0.83	0.2033	1.33	0.0918	1.83	0.0336
0.34	0.3669	0.84	0.2005	1.34	0.0901	1.84	0.0329
0.35	0.3632	0.85	0.1977	1.35	0.0885	1.85	0.0322
0.36	0.3594	0.86	0.1949	1.36	0.0869	1.86	0.0314
0.37	0.3557	0.87	0.1922	1.37	0.0853	1.87	0.0307
0.38	0.3520	0.88	0.1894	1.38	0.0838	1.88	0.0301
0.39	0.3483	0.89	0.1867	1.39	0.0823	1.89	0.0294
0.40	0.3446	0.90	0.1841	1.40	0.0808	1.90	0.0287
0.41	0.3409	0.91	0.1814	1.41	0.0793	1.91	0.0281
0.42	0.3372	0.92	0.1788	1.42	0.0778	1.92	0.0274
0.43	0.3336	0.93	0.1762	1.43	0.0764	1.93	0.0268
0.44	0.3300	0.94	0.1736	1.44	0.0749	1.94	0.0262
0.45	0.3264	0.95	0.1711	1.45	0.0735	1.95	0.0256
0.46	0.3228	0.96	0.1685	1.46	0.0721	1.96	0.0250
0.47	0.3192	0.97	0.1660	1.47	0.0708	1.97	0.0244
0.48	0.3156	0.98	0.1635	1.48	0.0694	1.98	0.0239
0.49	0.3121	0.99	0.1611	1.49	0.0681	1.99	0.0233
0.50	0.3085	1.00	0.1587	1.50	0.0668	2.00	0.0228

Table A.1 *(continued)*

Z	Prob > Z	Z	Prob > Z	Z	Prob > Z	Z	Prob > Z
2.01	0.0222	2.51	0.0060	3.01	0.0013	3.51	0.0002
2.02	0.0217	2.52	0.0059	3.02	0.0013	3.52	0.0002
2.03	0.0212	2.53	0.0057	3.03	0.0012	3.53	0.0002
2.04	0.0207	2.54	0.0055	3.04	0.0012	3.54	0.0002
2.05	0.0202	2.55	0.0054	3.05	0.0011	3.55	0.0002
2.06	0.0197	2.56	0.0052	3.06	0.0011	3.56	0.0002
2.07	0.0192	2.57	0.0051	3.07	0.0011	3.57	0.0002
2.08	0.0188	2.58	0.0049	3.08	0.0010	3.58	0.0002
2.09	0.0183	2.59	0.0048	3.09	0.0010	3.59	0.0002
2.10	0.0179	2.60	0.0047	3.10	0.0010	3.60	0.0002
2.11	0.0174	2.61	0.0045	3.11	0.0009	3.61	0.0002
2.12	0.0170	2.62	0.0044	3.12	0.0009	3.62	0.0001
2.13	0.0166	2.63	0.0043	3.13	0.0009	3.63	0.0001
2.14	0.0162	2.64	0.0041	3.14	0.0008	3.64	0.0001
2.15	0.0158	2.65	0.0040	3.15	0.0008	3.65	0.0001
2.16	0.0154	2.66	0.0039	3.16	0.0008	3.66	0.0001
2.17	0.0150	2.67	0.0038	3.17	0.0008	3.67	0.0001
2.18	0.0146	2.68	0.0037	3.18	0.0007	3.68	0.0001
2.19	0.0143	2.69	0.0036	3.19	0.0007	3.69	0.0001
2.20	0.0139	2.70	0.0035	3.20	0.0007	3.70	0.0001
2.21	0.0136	2.71	0.0034	3.21	0.0007	3.71	0.0001
2.22	0.0132	2.72	0.0033	3.22	0.0006	3.72	0.0001
2.23	0.0129	2.73	0.0032	3.23	0.0006	3.73	0.0001
2.24	0.0125	2.74	0.0031	3.24	0.0006	3.74	0.0001
2.25	0.0122	2.75	0.0030	3.25	0.0006	3.75	0.0001
2.26	0.0119	2.76	0.0029	3.26	0.0006	3.76	0.0001
2.27	0.0116	2.77	0.0028	3.27	0.0005	3.77	0.0001
2.28	0.0113	2.78	0.0027	3.28	0.0005	3.78	0.0001
2.29	0.0110	2.79	0.0026	3.29	0.0005	3.79	0.0001
2.30	0.0107	2.80	0.0026	3.30	0.0005	3.80	0.0001
2.31	0.0104	2.81	0.0025	3.31	0.0005	3.81	0.0001
2.32	0.0102	2.82	0.0024	3.32	0.0005	3.82	0.0001
2.33	0.0099	2.83	0.0023	3.33	0.0004	3.83	0.0001
2.34	0.0096	2.84	0.0023	3.34	0.0004	3.84	0.0001
2.35	0.0094	2.85	0.0022	3.35	0.0004	3.85	0.0001
2.36	0.0091	2.86	0.0021	3.36	0.0004	3.86	0.0001
2.37	0.0089	2.87	0.0021	3.37	0.0004	3.87	0.0001
2.38	0.0087	2.88	0.0020	3.38	0.0004	3.88	0.0001
2.39	0.0084	2.89	0.0019	3.39	0.0003	3.89	0.0001
2.40	0.0082	2.90	0.0019	3.40	0.0003	3.90	0.0000
2.41	0.0080	2.91	0.0018	3.41	0.0003	3.91	0.0000
2.42	0.0078	2.92	0.0018	3.42	0.0003	3.92	0.0000
2.43	0.0075	2.93	0.0017	3.43	0.0003	3.93	0.0000
2.44	0.0073	2.94	0.0016	3.44	0.0003	3.94	0.0000
2.45	0.0071	2.95	0.0016	3.45	0.0003	3.95	0.0000
2.46	0.0069	2.96	0.0015	3.46	0.0003	3.96	0.0000
2.47	0.0068	2.97	0.0015	3.47	0.0003	3.97	0.0000
2.48	0.0066	2.98	0.0014	3.48	0.0003	3.98	0.0000
2.49	0.0064	2.99	0.0014	3.49	0.0002	3.99	0.0000
2.50	0.0062	3.00	0.0013	3.50	0.0002	4.00	0.0000

Table A.1A

Selected Probability Values for the Normal Distribution—Values of Z Exceeded with Given Probability

Prob	Z
0.5000	0.00000
0.4000	0.25335
0.3000	0.52440
0.2000	0.84162
0.1000	1.28155
0.0500	1.64485
0.0250	1.95996
0.0100	2.32635
0.0050	2.57583
0.0020	2.87816
0.0010	3.09023
0.0005	3.29053
0.0001	3.71902

Table A.2 The t Distribution—Values of t Exceeded with Given Probability

df	$P = 0.25$	$P = 0.10$	$P = 0.05$	$P = 0.025$	$P = 0.01$	$P = 0.005$	$P = 0.001$	$P = 0.0005$	df
1	1.0000	3.0777	6.3138	12.706	31.821	63.657	318.31	636.62	1
2	0.8165	1.8856	2.9200	4.3027	6.9646	9.9248	22.327	31.599	2
3	0.7649	1.6377	2.3534	3.1824	4.5407	5.8409	10.215	12.924	3
4	0.7407	1.5332	2.1318	2.7764	3.7469	4.6041	7.1732	8.6103	4
5	0.7267	1.4759	2.0150	2.5706	3.3649	4.0321	5.8934	6.8688	5
6	0.7176	1.4398	1.9432	2.4469	3.1427	3.7074	5.2076	5.9588	6
7	0.7111	1.4149	1.8946	2.3646	2.9980	3.4995	4.7853	5.4079	7
8	0.7064	1.3968	1.8595	2.3060	2.8965	3.3554	4.5008	5.0413	8
9	0.7027	1.3830	1.8331	2.2622	2.8214	3.2498	4.2968	4.7809	9
10	0.6998	1.3722	1.8125	2.2281	2.7638	3.1693	4.1437	4.5869	10
11	0.6974	1.3634	1.7959	2.2010	2.7181	3.1058	4.0247	4.4370	11
12	0.6955	1.3562	1.7823	2.1788	2.6810	3.0545	3.9296	4.3178	12
13	0.6938	1.3502	1.7709	2.1604	2.6503	3.0123	3.8520	4.2208	13
14	0.6924	1.3450	1.7613	2.1448	2.6245	2.9768	3.7874	4.1405	14
15	0.6912	1.3406	1.7531	2.1314	2.6025	2.9467	3.7329	4.0728	15
16	0.6901	1.3368	1.7459	2.1199	2.5835	2.9208	3.6862	4.0150	16
17	0.6892	1.3334	1.7396	2.1098	2.5669	2.8982	3.6458	3.9652	17
18	0.6884	1.3304	1.7341	2.1009	2.5524	2.8784	3.6105	3.9217	18
19	0.6876	1.3277	1.7291	2.0930	2.5395	2.8609	3.5794	3.8834	19
20	0.6870	1.3253	1.7247	2.0860	2.5280	2.8453	3.5518	3.8495	20
21	0.6864	1.3232	1.7207	2.0796	2.5176	2.8314	3.5272	3.8193	21
22	0.6858	1.3212	1.7171	2.0739	2.5083	2.8188	3.5050	3.7922	22
23	0.6853	1.3195	1.7139	2.0687	2.4999	2.8073	3.4850	3.7677	23
24	0.6848	1.3178	1.7109	2.0639	2.4922	2.7969	3.4668	3.7454	24
25	0.6844	1.3163	1.7081	2.0595	2.4851	2.7874	3.4502	3.7252	25
26	0.6840	1.3150	1.7056	2.0555	2.4786	2.7787	3.4350	3.7066	26
27	0.6837	1.3137	1.7033	2.0518	2.4727	2.7707	3.4210	3.6896	27
28	0.6834	1.3125	1.7011	2.0484	2.4671	2.7633	3.4082	3.6739	28
29	0.6830	1.3114	1.6991	2.0452	2.4620	2.7564	3.3963	3.6594	29
30	0.6828	1.3104	1.6973	2.0423	2.4573	2.7500	3.3852	3.6460	30
35	0.6816	1.3062	1.6896	2.0301	2.4377	2.7238	3.3401	3.5912	35
40	0.6807	1.3031	1.6839	2.0211	2.4233	2.7045	3.3069	3.5510	40
45	0.6800	1.3006	1.6794	2.0141	2.4121	2.6896	3.2815	3.5203	45
50	0.6794	1.2987	1.6759	2.0086	2.4033	2.6778	3.2614	3.4960	50
55	0.6790	1.2971	1.6730	2.0040	2.3961	2.6682	3.2452	3.4764	55
60	0.6786	1.2958	1.6706	2.0003	2.3901	2.6603	3.2317	3.4602	60
65	0.6783	1.2947	1.6686	1.9971	2.3851	2.6536	3.2204	3.4466	65
70	0.6780	1.2938	1.6669	1.9944	2.3808	2.6479	3.2108	3.4350	70
75	0.6778	1.2929	1.6654	1.9921	2.3771	2.6430	3.2025	3.4250	75
90	0.6772	1.2910	1.6620	1.9867	2.3685	2.6316	3.1833	3.4019	90
105	0.6768	1.2897	1.6595	1.9828	2.3624	2.6235	3.1697	3.3856	105
120	0.6765	1.2886	1.6577	1.9799	2.3578	2.6174	3.1595	3.3735	120
∞	0.6745	1.2816	1.6449	1.9600	2.3263	2.5758	3.0902	3.2905	∞

Table A.3 χ^2 **Distribution—**χ^2 **Values Exceeded with Given Probability**

df	0.995	0.99	0.975	0.95	0.90	0.75	0.50	0.25	0.10	0.05	0.025	0.01	0.005
1	0.000	0.000	0.001	0.004	0.016	0.102	0.455	1.323	2.706	3.841	5.024	6.635	7.879
2	0.010	0.020	0.051	0.103	0.211	0.575	1.386	2.773	4.605	5.991	7.378	9.210	10.579
3	0.072	0.115	0.216	0.352	0.584	1.213	2.366	4.108	6.251	7.815	9.348	11.345	12.838
4	0.207	0.297	0.484	0.711	1.064	1.923	3.357	5.385	7.779	9.488	11.143	13.277	14.860
5	0.412	0.554	0.831	1.145	1.610	2.675	4.351	6.626	9.236	11.070	12.833	15.086	16.750
6	0.676	0.872	1.237	1.635	2.204	3.455	5.348	7.841	10.645	12.592	14.449	16.812	18.548
7	0.989	1.239	1.690	2.167	2.833	4.255	6.346	9.037	12.017	14.067	16.013	18.475	20.278
8	1.344	1.646	2.180	2.733	3.490	5.071	7.344	10.219	13.362	15.507	17.535	20.090	21.955
9	1.735	2.088	2.700	3.325	4.168	5.899	8.343	11.389	14.684	16.919	19.023	21.666	23.589
10	2.156	2.558	3.247	3.940	4.865	6.737	9.342	12.549	15.987	18.307	20.483	23.209	25.188
11	2.603	3.053	3.816	4.575	5.578	7.584	10.341	13.701	17.275	19.675	21.920	24.725	26.757
12	3.074	3.571	4.404	5.226	6.304	8.438	11.340	14.845	18.549	21.026	23.337	26.217	28.300
13	3.565	4.107	5.009	5.892	7.042	9.299	12.340	15.984	19.812	22.362	24.736	27.688	29.819
14	4.075	4.660	5.629	6.571	7.790	10.165	13.339	17.117	21.064	23.685	26.119	29.141	31.319
15	4.601	5.229	6.262	7.261	8.547	11.037	14.339	18.245	22.307	24.996	27.488	30.578	32.801
16	5.142	5.812	6.908	7.962	9.312	11.912	15.338	19.369	23.542	26.296	28.845	32.000	34.267
17	5.697	6.408	7.564	8.672	10.085	12.792	16.338	20.489	24.769	27.587	30.191	33.409	35.718
18	6.265	7.015	8.231	9.390	10.865	13.675	17.338	21.605	25.989	28.869	31.526	34.805	37.156
19	6.844	7.633	8.907	10.117	11.651	14.562	18.338	22.718	27.204	30.144	32.852	36.191	38.582
20	7.434	8.260	9.591	10.851	12.443	15.452	19.337	23.828	28.412	31.410	34.170	37.566	39.997
21	8.034	8.897	10.283	11.591	13.240	16.344	20.337	24.935	29.615	32.671	35.479	38.932	41.401
22	8.643	9.542	10.982	12.338	14.041	17.240	21.337	26.039	30.813	33.924	36.781	40.289	42.796
23	9.260	10.196	11.689	13.091	14.848	18.137	22.337	27.141	32.007	35.172	38.076	41.638	44.181
24	9.886	10.856	12.401	13.848	15.659	19.037	23.337	28.241	33.196	36.415	39.364	42.980	45.559
25	10.520	11.524	13.120	14.611	16.473	19.939	24.337	29.339	34.382	37.652	40.646	44.314	46.928
26	11.160	12.198	13.844	15.379	17.292	20.843	25.336	30.435	35.563	38.885	41.923	45.642	48.290
27	11.808	12.879	14.573	16.151	18.114	21.749	26.336	31.528	36.741	40.113	43.195	46.963	49.645
28	12.461	13.565	15.308	16.928	18.939	22.657	27.336	32.620	37.916	41.337	44.461	48.278	50.993
29	13.121	14.256	16.047	17.708	19.768	23.567	28.336	33.711	39.087	42.557	45.722	49.588	52.336
30	13.787	14.953	16.791	18.493	20.599	24.478	29.336	34.800	40.256	43.773	46.979	50.892	53.672
35	17.192	18.509	20.569	22.465	24.797	29.054	34.336	40.223	46.059	49.802	53.203	57.342	60.275
40	20.707	22.164	24.433	26.509	29.051	33.660	39.335	45.616	51.805	55.758	59.342	63.691	66.766
45	24.311	25.901	28.366	30.612	33.350	38.291	44.335	50.985	57.505	61.656	65.410	69.957	73.166
50	27.991	29.707	32.357	34.764	37.689	42.942	49.335	56.334	63.167	67.505	71.420	76.154	79.490
55	31.735	33.570	36.398	38.958	42.060	47.610	54.335	61.665	68.796	73.311	77.380	82.292	85.749
60	35.534	37.485	40.482	43.188	46.459	52.294	59.335	66.981	74.397	79.082	83.298	88.379	91.952
65	39.383	41.444	44.603	47.450	50.883	56.990	64.335	72.285	79.973	84.821	89.177	94.422	98.105
70	43.275	45.442	48.758	51.739	55.329	61.698	69.334	77.577	85.527	90.531	95.023	100.425	104.215
75	47.206	49.475	52.942	56.054	59.795	66.417	74.334	82.858	91.061	96.217	100.839	106.393	110.286
80	51.172	53.540	57.153	60.391	64.278	71.145	79.334	88.130	96.578	101.879	106.629	112.329	116.321
85	55.170	57.634	61.389	64.749	68.777	75.881	84.334	93.394	102.079	107.522	112.393	118.236	122.325
90	59.196	61.754	65.647	69.126	73.291	80.625	89.334	98.650	107.565	113.145	118.136	124.116	128.299
95	63.250	65.898	69.925	73.520	77.818	85.376	94.334	103.899	113.038	118.752	123.858	129.973	134.247
100	67.328	70.065	74.222	77.929	82.358	90.133	99.334	109.141	118.498	124.342	129.561	135.807	140.169

Table A.4 The F Distribution, $p = 0.1$

Denominator df	NUMERATOR df 1	2	3	4	5	6	7	8	9	10	11
1	39.9	49.5	53.6	55.8	57.2	58.2	58.9	59.4	59.9	60.2	60.5
2	8.53	9.00	9.16	9.24	9.29	9.33	9.35	9.37	9.38	9.39	9.40
3	5.54	5.46	5.39	5.34	5.31	5.28	5.27	5.25	5.24	5.23	5.22
4	4.54	4.32	4.19	4.11	4.05	4.01	3.98	3.95	3.94	3.92	3.91
5	4.06	3.78	3.62	3.52	3.45	3.40	3.37	3.34	3.32	3.30	3.28
6	3.78	3.46	3.29	3.18	3.11	3.05	3.01	2.98	2.96	2.94	2.92
7	3.59	3.26	3.07	2.96	2.88	2.83	2.78	2.75	2.72	2.70	2.68
8	3.46	3.11	2.92	2.81	2.73	2.67	2.62	2.59	2.56	2.54	2.52
9	3.36	3.01	2.81	2.69	2.61	2.55	2.51	2.47	2.44	2.42	2.40
10	3.29	2.92	2.73	2.61	2.52	2.46	2.41	2.38	2.35	2.32	2.30
11	3.23	2.86	2.66	2.54	2.45	2.39	2.34	2.30	2.27	2.25	2.23
12	3.18	2.81	2.61	2.48	2.39	2.33	2.28	2.24	2.21	2.19	2.17
13	3.14	2.76	2.56	2.43	2.35	2.28	2.23	2.20	2.16	2.14	2.12
14	3.10	2.73	2.52	2.39	2.31	2.24	2.19	2.15	2.12	2.10	2.07
15	3.07	2.70	2.49	2.36	2.27	2.21	2.16	2.12	2.09	2.06	2.04
16	3.05	2.67	2.46	2.33	2.24	2.18	2.13	2.09	2.06	2.03	2.01
17	3.03	2.64	2.44	2.31	2.22	2.15	2.10	2.06	2.03	2.00	1.98
18	3.01	2.62	2.42	2.29	2.20	2.13	2.08	2.04	2.00	1.98	1.95
19	2.99	2.61	2.40	2.27	2.18	2.11	2.06	2.02	1.98	1.96	1.93
20	2.97	2.59	2.38	2.25	2.16	2.09	2.04	2.00	1.96	1.94	1.91
21	2.96	2.57	2.36	2.23	2.14	2.08	2.02	1.98	1.95	1.92	1.90
22	2.95	2.56	2.35	2.22	2.13	2.06	2.01	1.97	1.93	1.90	1.88
23	2.94	2.55	2.34	2.21	2.11	2.05	1.99	1.95	1.92	1.89	1.87
24	2.93	2.54	2.33	2.19	2.10	2.04	1.98	1.94	1.91	1.88	1.85
25	2.92	2.53	2.32	2.18	2.09	2.02	1.97	1.93	1.89	1.87	1.84
30	2.88	2.49	2.28	2.14	2.05	1.98	1.93	1.88	1.85	1.82	1.79
35	2.85	2.46	2.25	2.11	2.02	1.95	1.90	1.85	1.82	1.79	1.76
40	2.84	2.44	2.23	2.09	2.00	1.93	1.87	1.83	1.79	1.76	1.74
45	2.82	2.42	2.21	2.07	1.98	1.91	1.85	1.81	1.77	1.74	1.72
50	2.81	2.41	2.20	2.06	1.97	1.90	1.84	1.80	1.76	1.73	1.70
55	2.80	2.40	2.19	2.05	1.95	1.88	1.83	1.78	1.75	1.72	1.69
60	2.79	2.39	2.18	2.04	1.95	1.87	1.82	1.77	1.74	1.71	1.68
75	2.77	2.37	2.16	2.02	1.93	1.85	1.80	1.75	1.72	1.69	1.66
100	2.76	2.36	2.14	2.00	1.91	1.83	1.78	1.73	1.69	1.66	1.64
∞	2.71	2.30	2.08	1.94	1.85	1.77	1.72	1.67	1.63	1.60	1.57

(Continued)

Table A.4 *(continued)*

Denominator df	NUMERATOR df 12	13	14	15	16	20	24	30	45	60	120
1	60.7	60.9	61.1	61.2	61.3	61.7	62	62.3	62.6	62.8	63.1
2	9.41	9.41	9.42	9.42	9.43	9.44	9.45	9.46	9.47	9.47	9.48
3	5.22	5.21	5.20	5.20	5.20	5.18	5.18	5.17	5.16	5.15	5.14
4	3.90	3.89	3.88	3.87	3.86	3.84	3.83	3.82	3.80	3.79	3.78
5	3.27	3.26	3.25	3.24	3.23	3.21	3.19	3.17	3.15	3.14	3.12
6	2.90	2.89	2.88	2.87	2.86	2.84	2.82	2.80	2.77	2.76	2.74
7	2.67	2.65	2.64	2.63	2.62	2.59	2.58	2.56	2.53	2.51	2.49
8	2.50	2.49	2.48	2.46	2.45	2.42	2.40	2.38	2.35	2.34	2.32
9	2.38	2.36	2.35	2.34	2.33	2.30	2.28	2.25	2.22	2.21	2.18
10	2.28	2.27	2.26	2.24	2.23	2.20	2.18	2.16	2.12	2.11	2.08
11	2.21	2.19	2.18	2.17	2.16	2.12	2.10	2.08	2.04	2.03	2.00
12	2.15	2.13	2.12	2.10	2.09	2.06	2.04	2.01	1.98	1.96	1.93
13	2.10	2.08	2.07	2.05	2.04	2.01	1.98	1.96	1.92	1.90	1.88
14	2.05	2.04	2.02	2.01	2.00	1.96	1.94	1.91	1.88	1.86	1.83
15	2.02	2.00	1.99	1.97	1.96	1.92	1.90	1.87	1.84	1.82	1.79
16	1.99	1.97	1.95	1.94	1.93	1.89	1.87	1.84	1.80	1.78	1.75
17	1.96	1.94	1.93	1.91	1.90	1.86	1.84	1.81	1.77	1.75	1.72
18	1.93	1.92	1.90	1.89	1.87	1.84	1.81	1.78	1.74	1.72	1.69
19	1.91	1.89	1.88	1.86	1.85	1.81	1.79	1.76	1.72	1.70	1.67
20	1.89	1.87	1.86	1.84	1.83	1.79	1.77	1.74	1.70	1.68	1.64
21	1.87	1.86	1.84	1.83	1.81	1.78	1.75	1.72	1.68	1.66	1.62
22	1.86	1.84	1.83	1.81	1.80	1.76	1.73	1.70	1.66	1.64	1.60
23	1.84	1.83	1.81	1.80	1.78	1.74	1.72	1.69	1.64	1.62	1.59
24	1.83	1.81	1.80	1.78	1.77	1.73	1.70	1.67	1.63	1.61	1.57
25	1.82	1.80	1.79	1.77	1.76	1.72	1.69	1.66	1.62	1.59	1.56
30	1.77	1.75	1.74	1.72	1.71	1.67	1.64	1.61	1.56	1.54	1.50
35	1.74	1.72	1.70	1.69	1.67	1.63	1.60	1.57	1.52	1.50	1.46
40	1.71	1.70	1.68	1.66	1.65	1.61	1.57	1.54	1.49	1.47	1.42
45	1.70	1.68	1.66	1.64	1.63	1.58	1.55	1.52	1.47	1.44	1.40
50	1.68	1.66	1.64	1.63	1.61	1.57	1.54	1.50	1.45	1.42	1.38
55	1.67	1.65	1.63	1.61	1.60	1.55	1.52	1.49	1.44	1.41	1.36
60	1.66	1.64	1.62	1.60	1.59	1.54	1.51	1.48	1.42	1.40	1.35
75	1.63	1.61	1.60	1.58	1.57	1.52	1.49	1.45	1.40	1.37	1.32
100	1.61	1.59	1.57	1.56	1.54	1.49	1.46	1.42	1.37	1.34	1.28
∞	1.55	1.52	1.50	1.49	1.47	1.42	1.38	1.34	1.28	1.24	1.17

Table A.4A The F Distribution, $p = 0.05$

Denominator df	NUMERATOR df 1	2	3	4	5	6	7	8	9	10	11
1	161	199	216	225	230	234	237	239	241	242	243
2	18.5	19	19.2	19.2	19.3	19.3	19.4	19.4	19.4	19.4	19.4
3	10.1	9.55	9.28	9.12	9.01	8.94	8.89	8.85	8.81	8.79	8.76
4	7.71	6.94	6.59	6.39	6.26	6.16	6.09	6.04	6.00	5.96	5.94
5	6.61	5.79	5.41	5.19	5.05	4.95	4.88	4.82	4.77	4.74	4.70
6	5.99	5.14	4.76	4.53	4.39	4.28	4.21	4.15	4.10	4.06	4.03
7	5.59	4.74	4.35	4.12	3.97	3.87	3.79	3.73	3.68	3.64	3.60
8	5.32	4.46	4.07	3.84	3.69	3.58	3.50	3.44	3.39	3.35	3.31
9	5.12	4.26	3.86	3.63	3.48	3.37	3.29	3.23	3.18	3.14	3.10
10	4.96	4.10	3.71	3.48	3.33	3.22	3.14	3.07	3.02	2.98	2.94
11	4.84	3.98	3.59	3.36	3.20	3.09	3.01	2.95	2.90	2.85	2.82
12	4.75	3.89	3.49	3.26	3.11	3.00	2.91	2.85	2.80	2.75	2.72
13	4.67	3.81	3.41	3.18	3.03	2.92	2.83	2.77	2.71	2.67	2.63
14	4.60	3.74	3.34	3.11	2.96	2.85	2.76	2.70	2.65	2.60	2.57
15	4.54	3.68	3.29	3.06	2.90	2.79	2.71	2.64	2.59	2.54	2.51
16	4.49	3.63	3.24	3.01	2.85	2.74	2.66	2.59	2.54	2.49	2.46
17	4.45	3.59	3.20	2.96	2.81	2.70	2.61	2.55	2.49	2.45	2.41
18	4.41	3.55	3.16	2.93	2.77	2.66	2.58	2.51	2.46	2.41	2.37
19	4.38	3.52	3.13	2.90	2.74	2.63	2.54	2.48	2.42	2.38	2.34
20	4.35	3.49	3.10	2.87	2.71	2.60	2.51	2.45	2.39	2.35	2.31
21	4.32	3.47	3.07	2.84	2.68	2.57	2.49	2.42	2.37	2.32	2.28
22	4.30	3.44	3.05	2.82	2.66	2.55	2.46	2.40	2.34	2.30	2.26
23	4.28	3.42	3.03	2.80	2.64	2.53	2.44	2.37	2.32	2.27	2.24
24	4.26	3.40	3.01	2.78	2.62	2.51	2.42	2.36	2.30	2.25	2.22
25	4.24	3.39	2.99	2.76	2.60	2.49	2.40	2.34	2.28	2.24	2.20
30	4.17	3.32	2.92	2.69	2.53	2.42	2.33	2.27	2.21	2.16	2.13
35	4.12	3.27	2.87	2.64	2.49	2.37	2.29	2.22	2.16	2.11	2.07
40	4.08	3.23	2.84	2.61	2.45	2.34	2.25	2.18	2.12	2.08	2.04
45	4.06	3.20	2.81	2.58	2.42	2.31	2.22	2.15	2.10	2.05	2.01
50	4.03	3.18	2.79	2.56	2.40	2.29	2.20	2.13	2.07	2.03	1.99
55	4.02	3.16	2.77	2.54	2.38	2.27	2.18	2.11	2.06	2.01	1.97
60	4.00	3.15	2.76	2.53	2.37	2.25	2.17	2.10	2.04	1.99	1.95
75	3.97	3.12	2.73	2.49	2.34	2.22	2.13	2.06	2.01	1.96	1.92
100	3.94	3.09	2.70	2.46	2.31	2.19	2.10	2.03	1.97	1.93	1.89
∞	3.84	3.00	2.60	2.37	2.21	2.10	2.01	1.94	1.88	1.83	1.79

(Continued)

Table A.4A *(continued)*

Denominator df	NUMERATOR df 12	13	14	15	16	20	24	30	45	60	120
1	244	245	245	246	246	248	249	250	251	252	253
2	19.4	19.4	19.4	19.4	19.4	19.4	19.5	19.5	19.5	19.5	19.5
3	8.74	8.73	8.71	8.70	8.69	8.66	8.64	8.62	8.59	8.57	8.55
4	5.91	5.89	5.87	5.86	5.84	5.80	5.77	5.75	5.71	5.69	5.66
5	4.68	4.66	4.64	4.62	4.60	4.56	4.53	4.50	4.45	4.43	4.40
6	4.00	3.98	3.96	3.94	3.92	3.87	3.84	3.81	3.76	3.74	3.70
7	3.57	3.55	3.53	3.51	3.49	3.44	3.41	3.38	3.33	3.30	3.27
8	3.28	3.26	3.24	3.22	3.20	3.15	3.12	3.08	3.03	3.01	2.97
9	3.07	3.05	3.03	3.01	2.99	2.94	2.90	2.86	2.81	2.79	2.75
10	2.91	2.89	2.86	2.85	2.83	2.77	2.74	2.70	2.65	2.62	2.58
11	2.79	2.76	2.74	2.72	2.70	2.65	2.61	2.57	2.52	2.49	2.45
12	2.69	2.66	2.64	2.62	2.60	2.54	2.51	2.47	2.41	2.38	2.34
13	2.60	2.58	2.55	2.53	2.51	2.46	2.42	2.38	2.33	2.30	2.25
14	2.53	2.51	2.48	2.46	2.44	2.39	2.35	2.31	2.25	2.22	2.18
15	2.48	2.45	2.42	2.40	2.38	2.33	2.29	2.25	2.19	2.16	2.11
16	2.42	2.40	2.37	2.35	2.33	2.28	2.24	2.19	2.14	2.11	2.06
17	2.38	2.35	2.33	2.31	2.29	2.23	2.19	2.15	2.09	2.06	2.01
18	2.34	2.31	2.29	2.27	2.25	2.19	2.15	2.11	2.05	2.02	1.97
19	2.31	2.28	2.26	2.23	2.21	2.16	2.11	2.07	2.01	1.98	1.93
20	2.28	2.25	2.22	2.20	2.18	2.12	2.08	2.04	1.98	1.95	1.90
21	2.25	2.22	2.20	2.18	2.16	2.10	2.05	2.01	1.95	1.92	1.87
22	2.23	2.20	2.17	2.15	2.13	2.07	2.03	1.98	1.92	1.89	1.84
23	2.20	2.18	2.15	2.13	2.11	2.05	2.01	1.96	1.90	1.86	1.81
24	2.18	2.15	2.13	2.11	2.09	2.03	1.98	1.94	1.88	1.84	1.79
25	2.16	2.14	2.11	2.09	2.07	2.01	1.96	1.92	1.86	1.82	1.77
30	2.09	2.06	2.04	2.01	1.99	1.93	1.89	1.84	1.77	1.74	1.68
35	2.04	2.01	1.99	1.96	1.94	1.88	1.83	1.79	1.72	1.68	1.62
40	2.00	1.97	1.95	1.92	1.90	1.84	1.79	1.74	1.67	1.64	1.58
45	1.97	1.94	1.92	1.89	1.87	1.81	1.76	1.71	1.64	1.60	1.54
50	1.95	1.92	1.89	1.87	1.85	1.78	1.74	1.69	1.61	1.58	1.51
55	1.93	1.90	1.88	1.85	1.83	1.76	1.72	1.67	1.59	1.55	1.49
60	1.92	1.89	1.86	1.84	1.82	1.75	1.70	1.65	1.57	1.53	1.47
75	1.88	1.85	1.83	1.80	1.78	1.71	1.66	1.61	1.53	1.49	1.42
100	1.85	1.82	1.79	1.77	1.75	1.68	1.63	1.57	1.49	1.45	1.38
∞	1.75	1.72	1.69	1.67	1.64	1.57	1.52	1.46	1.37	1.32	1.22

Table A.4B The F Distribution, $p = 0.025$

Denominator df	NUMERATOR df 1	2	3	4	5	6	7	8	9	10	11
1	648	800	864	900	922	937	948	957	963	969	973
2	38.5	39	39.2	39.2	39.3	39.3	39.4	39.4	39.4	39.4	39.4
3	17.4	16	15.4	15.1	14.9	14.7	14.6	14.5	14.5	14.4	14.4
4	12.2	10.6	9.98	9.60	9.36	9.20	9.07	8.98	8.90	8.84	8.79
5	10	8.43	7.76	7.39	7.15	6.98	6.85	6.76	6.68	6.62	6.57
6	8.81	7.26	6.60	6.23	5.99	5.82	5.70	5.60	5.52	5.46	5.41
7	8.07	6.54	5.89	5.52	5.29	5.12	4.99	4.90	4.82	4.76	4.71
8	7.57	6.06	5.42	5.05	4.82	4.65	4.53	4.43	4.36	4.30	4.24
9	7.21	5.71	5.08	4.72	4.48	4.32	4.20	4.10	4.03	3.96	3.91
10	6.94	5.46	4.83	4.47	4.24	4.07	3.95	3.85	3.78	3.72	3.66
11	6.72	5.26	4.63	4.28	4.04	3.88	3.76	3.66	3.59	3.53	3.47
12	6.55	5.10	4.47	4.12	3.89	3.73	3.61	3.51	3.44	3.37	3.32
13	6.41	4.97	4.35	4.00	3.77	3.60	3.48	3.39	3.31	3.25	3.20
14	6.30	4.86	4.24	3.89	3.66	3.50	3.38	3.29	3.21	3.15	3.09
15	6.20	4.77	4.15	3.80	3.58	3.41	3.29	3.20	3.12	3.06	3.01
16	6.12	4.69	4.08	3.73	3.50	3.34	3.22	3.12	3.05	2.99	2.93
17	6.04	4.62	4.01	3.66	3.44	3.28	3.16	3.06	2.98	2.92	2.87
18	5.98	4.56	3.95	3.61	3.38	3.22	3.10	3.01	2.93	2.87	2.81
19	5.92	4.51	3.90	3.56	3.33	3.17	3.05	2.96	2.88	2.82	2.76
20	5.87	4.46	3.86	3.51	3.29	3.13	3.01	2.91	2.84	2.77	2.72
21	5.83	4.42	3.82	3.48	3.25	3.09	2.97	2.87	2.80	2.73	2.68
22	5.79	4.38	3.78	3.44	3.22	3.05	2.93	2.84	2.76	2.70	2.65
23	5.75	4.35	3.75	3.41	3.18	3.02	2.90	2.81	2.73	2.67	2.62
24	5.72	4.32	3.72	3.38	3.15	2.99	2.87	2.78	2.70	2.64	2.59
25	5.69	4.29	3.69	3.35	3.13	2.97	2.85	2.75	2.68	2.61	2.56
30	5.57	4.18	3.59	3.25	3.03	2.87	2.75	2.65	2.57	2.51	2.46
35	5.48	4.11	3.52	3.18	2.96	2.80	2.68	2.58	2.50	2.44	2.39
40	5.42	4.05	3.46	3.13	2.90	2.74	2.62	2.53	2.45	2.39	2.33
45	5.38	4.01	3.42	3.09	2.86	2.70	2.58	2.49	2.41	2.35	2.29
50	5.34	3.97	3.39	3.05	2.83	2.67	2.55	2.46	2.38	2.32	2.26
55	5.31	3.95	3.36	3.03	2.81	2.65	2.53	2.43	2.36	2.29	2.24
60	5.29	3.93	3.34	3.01	2.79	2.63	2.51	2.41	2.33	2.27	2.22
75	5.23	3.88	3.30	2.96	2.74	2.58	2.46	2.37	2.29	2.22	2.17
100	5.18	3.83	3.25	2.92	2.70	2.54	2.42	2.32	2.24	2.18	2.12
∞	5.02	3.69	3.12	2.79	2.57	2.41	2.29	2.19	2.11	2.05	1.99

(Continued)

Table A.4B *(continued)*

Denominator df	NUMERATOR df 12	13	14	15	16	20	24	30	45	60	120
1	977	980	983	985	987	993	997	1001	1007	1010	1014
2	39.4	39.4	39.4	39.4	39.4	39.4	39.5	39.5	39.5	39.5	39.5
3	14.3	14.3	14.3	14.3	14.2	14.2	14.1	14.1	14	14	13.9
4	8.75	8.71	8.68	8.66	8.63	8.56	8.51	8.46	8.39	8.36	8.31
5	6.52	6.49	6.46	6.43	6.40	6.33	6.28	6.23	6.16	6.12	6.07
6	5.37	5.33	5.30	5.27	5.24	5.17	5.12	5.07	4.99	4.96	4.90
7	4.67	4.63	4.60	4.57	4.54	4.47	4.41	4.36	4.29	4.25	4.20
8	4.20	4.16	4.13	4.10	4.08	4.00	3.95	3.89	3.82	3.78	3.73
9	3.87	3.83	3.80	3.77	3.74	3.67	3.61	3.56	3.49	3.45	3.39
10	3.62	3.58	3.55	3.52	3.50	3.42	3.37	3.31	3.24	3.20	3.14
11	3.43	3.39	3.36	3.33	3.30	3.23	3.17	3.12	3.04	3.00	2.94
12	3.28	3.24	3.21	3.18	3.15	3.07	3.02	2.96	2.89	2.85	2.79
13	3.15	3.12	3.08	3.05	3.03	2.95	2.89	2.84	2.76	2.72	2.66
14	3.05	3.01	2.98	2.95	2.92	2.84	2.79	2.73	2.65	2.61	2.55
15	2.96	2.92	2.89	2.86	2.84	2.76	2.70	2.64	2.56	2.52	2.46
16	2.89	2.85	2.82	2.79	2.76	2.68	2.63	2.57	2.49	2.45	2.38
17	2.82	2.79	2.75	2.72	2.70	2.62	2.56	2.50	2.42	2.38	2.32
18	2.77	2.73	2.70	2.67	2.64	2.56	2.50	2.44	2.36	2.32	2.26
19	2.72	2.68	2.65	2.62	2.59	2.51	2.45	2.39	2.31	2.27	2.20
20	2.68	2.64	2.60	2.57	2.55	2.46	2.41	2.35	2.27	2.22	2.16
21	2.64	2.60	2.56	2.53	2.51	2.42	2.37	2.31	2.23	2.18	2.11
22	2.60	2.56	2.53	2.50	2.47	2.39	2.33	2.27	2.19	2.14	2.08
23	2.57	2.53	2.50	2.47	2.44	2.36	2.30	2.24	2.15	2.11	2.04
24	2.54	2.50	2.47	2.44	2.41	2.33	2.27	2.21	2.12	2.08	2.01
25	2.51	2.48	2.44	2.41	2.38	2.30	2.24	2.18	2.10	2.05	1.98
30	2.41	2.37	2.34	2.31	2.28	2.20	2.14	2.07	1.99	1.94	1.87
35	2.34	2.30	2.27	2.23	2.21	2.12	2.06	2.00	1.91	1.86	1.79
40	2.29	2.25	2.21	2.18	2.15	2.07	2.01	1.94	1.85	1.80	1.72
45	2.25	2.21	2.17	2.14	2.11	2.03	1.96	1.90	1.81	1.76	1.68
50	2.22	2.18	2.14	2.11	2.08	1.99	1.93	1.87	1.77	1.72	1.64
55	2.19	2.15	2.11	2.08	2.05	1.97	1.90	1.84	1.74	1.69	1.61
60	2.17	2.13	2.09	2.06	2.03	1.94	1.88	1.82	1.72	1.67	1.58
75	2.12	2.08	2.05	2.01	1.99	1.90	1.83	1.76	1.67	1.61	1.52
100	2.08	2.04	2.00	1.97	1.94	1.85	1.78	1.71	1.61	1.56	1.46
∞	1.94	1.90	1.87	1.83	1.80	1.71	1.64	1.57	1.45	1.39	1.27

Table A.4C The F Distribution, $p = 0.01$

Denominator df	NUMERATOR df 1	2	3	4	5	6	7	8	9	10	11
1	4052	5000	5403	5625	5764	5859	5928	5981	6022	6056	6083
2	98.5	99	99.2	99.2	99.3	99.3	99.4	99.4	99.4	99.4	99.4
3	34.1	30.8	29.5	28.7	28.2	27.9	27.7	27.5	27.3	27.2	27.1
4	21.2	18	16.7	16	15.5	15.2	15	14.8	14.7	14.5	14.5
5	16.3	13.3	12.1	11.4	11	10.7	10.5	10.3	10.2	10.1	9.96
6	13.7	10.9	9.78	9.15	8.75	8.47	8.26	8.10	7.98	7.87	7.79
7	12.2	9.55	8.45	7.85	7.46	7.19	6.99	6.84	6.72	6.62	6.54
8	11.3	8.65	7.59	7.01	6.63	6.37	6.18	6.03	5.91	5.81	5.73
9	10.6	8.02	6.99	6.42	6.06	5.80	5.61	5.47	5.35	5.26	5.18
10	10	7.56	6.55	5.99	5.64	5.39	5.20	5.06	4.94	4.85	4.77
11	9.65	7.21	6.22	5.67	5.32	5.07	4.89	4.74	4.63	4.54	4.46
12	9.33	6.93	5.95	5.41	5.06	4.82	4.64	4.50	4.39	4.30	4.22
13	9.07	6.70	5.74	5.21	4.86	4.62	4.44	4.30	4.19	4.10	4.02
14	8.86	6.51	5.56	5.04	4.69	4.46	4.28	4.14	4.03	3.94	3.86
15	8.68	6.36	5.42	4.89	4.56	4.32	4.14	4.00	3.89	3.80	3.73
16	8.53	6.23	5.29	4.77	4.44	4.20	4.03	3.89	3.78	3.69	3.62
17	8.40	6.11	5.19	4.67	4.34	4.10	3.93	3.79	3.68	3.59	3.52
18	8.29	6.01	5.09	4.58	4.25	4.01	3.84	3.71	3.60	3.51	3.43
19	8.18	5.93	5.01	4.50	4.17	3.94	3.77	3.63	3.52	3.43	3.36
20	8.10	5.85	4.94	4.43	4.10	3.87	3.70	3.56	3.46	3.37	3.29
21	8.02	5.78	4.87	4.37	4.04	3.81	3.64	3.51	3.40	3.31	3.24
22	7.95	5.72	4.82	4.31	3.99	3.76	3.59	3.45	3.35	3.26	3.18
23	7.88	5.66	4.76	4.26	3.94	3.71	3.54	3.41	3.30	3.21	3.14
24	7.82	5.61	4.72	4.22	3.90	3.67	3.50	3.36	3.26	3.17	3.09
25	7.77	5.57	4.68	4.18	3.85	3.63	3.46	3.32	3.22	3.13	3.06
30	7.56	5.39	4.51	4.02	3.70	3.47	3.30	3.17	3.07	2.98	2.91
35	7.42	5.27	4.40	3.91	3.59	3.37	3.20	3.07	2.96	2.88	2.80
40	7.31	5.18	4.31	3.83	3.51	3.29	3.12	2.99	2.89	2.80	2.73
45	7.23	5.11	4.25	3.77	3.45	3.23	3.07	2.94	2.83	2.74	2.67
50	7.17	5.06	4.20	3.72	3.41	3.19	3.02	2.89	2.78	2.70	2.63
55	7.12	5.01	4.16	3.68	3.37	3.15	2.98	2.85	2.75	2.66	2.59
60	7.08	4.98	4.13	3.65	3.34	3.12	2.95	2.82	2.72	2.63	2.56
75	6.99	4.90	4.05	3.58	3.27	3.05	2.89	2.76	2.65	2.57	2.49
100	6.90	4.82	3.98	3.51	3.21	2.99	2.82	2.69	2.59	2.50	2.43
∞	6.63	4.61	3.78	3.32	3.02	2.80	2.64	2.51	2.41	2.32	2.25

(Continued)

Table A.4C *(continued)*

Denominator df	NUMERATOR df 12	13	14	15	16	20	24	30	45	60	120
1	6106	6126	6143	6157	6170	6209	6235	6261	6296	6313	6339
2	99.4	99.4	99.4	99.4	99.4	99.4	99.5	99.5	99.5	99.5	99.5
3	27.1	27	26.9	26.9	26.8	26.7	26.6	26.5	26.4	26.3	26.2
4	14.4	14.3	14.2	14.2	14.2	14	13.9	13.8	13.7	13.7	13.6
5	9.89	9.82	9.77	9.72	9.68	9.55	9.47	9.38	9.26	9.20	9.11
6	7.72	7.66	7.60	7.56	7.52	7.40	7.31	7.23	7.11	7.06	6.97
7	6.47	6.41	6.36	6.31	6.28	6.16	6.07	5.99	5.88	5.82	5.74
8	5.67	5.61	5.56	5.52	5.48	5.36	5.28	5.20	5.09	5.03	4.95
9	5.11	5.05	5.01	4.96	4.92	4.81	4.73	4.65	4.54	4.48	4.40
10	4.71	4.65	4.60	4.56	4.52	4.41	4.33	4.25	4.14	4.08	4.00
11	4.40	4.34	4.29	4.25	4.21	4.10	4.02	3.94	3.83	3.78	3.69
12	4.16	4.10	4.05	4.01	3.97	3.86	3.78	3.70	3.59	3.54	3.45
13	3.96	3.91	3.86	3.82	3.78	3.66	3.59	3.51	3.40	3.34	3.25
14	3.80	3.75	3.70	3.66	3.62	3.51	3.43	3.35	3.24	3.18	3.09
15	3.67	3.61	3.56	3.52	3.49	3.37	3.29	3.21	3.10	3.05	3.96
16	3.55	3.50	3.45	3.41	3.37	3.26	3.18	3.10	2.99	2.93	2.84
17	3.46	3.40	3.35	3.31	3.27	3.16	3.08	3.00	2.89	2.83	2.75
18	3.37	3.32	3.27	3.23	3.19	3.08	3.00	2.92	2.81	2.75	2.66
19	3.30	3.24	3.19	3.15	3.12	3.00	2.92	2.84	2.73	2.67	2.58
20	3.23	3.18	3.13	3.09	3.05	2.94	2.86	2.78	2.67	2.61	2.52
21	3.17	3.12	3.07	3.03	2.99	2.88	2.80	2.72	2.61	2.55	2.46
22	3.12	3.07	3.02	2.98	2.94	2.83	2.75	2.67	2.55	2.50	2.40
23	3.07	3.02	2.97	2.93	2.89	2.78	2.70	2.62	2.51	2.45	2.35
24	3.03	2.98	2.93	2.89	2.85	2.74	2.66	2.58	2.46	2.40	2.31
25	2.99	2.94	2.89	2.85	2.81	2.70	2.62	2.54	2.42	2.36	2.27
30	2.84	2.79	2.74	2.70	2.66	2.55	2.47	2.39	2.27	2.21	2.11
35	2.74	2.69	2.64	2.60	2.56	2.44	2.36	2.28	2.16	2.10	2.00
40	2.66	2.61	2.56	2.52	2.48	2.37	2.29	2.20	2.08	2.02	1.92
45	2.61	2.55	2.51	2.46	2.43	2.31	2.23	2.14	2.02	1.96	1.85
50	2.56	2.51	2.46	2.42	2.38	2.27	2.18	2.10	1.97	1.91	1.80
55	2.53	2.47	2.42	2.38	2.34	2.23	2.15	2.06	1.94	1.87	1.76
60	2.50	2.44	2.39	2.35	2.31	2.20	2.12	2.03	1.90	1.84	1.73
75	2.43	2.38	2.33	2.29	2.25	2.13	2.05	1.96	1.83	1.76	1.65
100	2.37	2.31	2.27	2.22	2.19	2.07	1.98	1.89	1.76	1.69	1.57
∞	2.18	2.13	2.08	2.04	2.00	1.88	1.79	1.70	1.55	1.47	1.32

Table A.4D The F Distribution, $p = 0.005$

Denominator df	NUMERATOR df 1	2	3	4	5	6	7	8	9	10	11
1	6,000	20,000	22,000	22,000	23,000	23,000	24,000	24,000	24,000	24,000	24,000
2	199	199	199	199	199	199	199	199	199	199	199
3	55.6	49.8	47.5	46.2	45.4	44.8	44.4	44.1	43.9	43.7	43.5
4	31.3	26.3	24.3	23.2	22.5	22	21.6	21.4	21.1	21	20.8
5	22.8	18.3	16.5	15.6	14.9	14.5	14.2	14	13.8	13.6	13.5
6	18.6	14.5	12.9	12	11.5	11.1	10.8	10.6	10.4	10.3	10.1
7	16.2	12.4	10.9	10.1	9.52	9.16	8.89	8.68	8.51	8.38	8.27
8	14.7	11	9.60	8.81	8.30	7.95	7.69	7.50	7.34	7.21	7.10
9	13.6	10.1	8.72	7.96	7.47	7.13	6.88	6.69	6.54	6.42	6.31
10	12.8	9.43	8.08	7.34	6.87	6.54	6.30	6.12	5.97	5.85	5.75
11	12.2	8.91	7.60	6.88	6.42	6.10	5.86	5.68	5.54	5.42	5.32
12	11.8	8.51	7.23	6.52	6.07	5.76	5.52	5.35	5.20	5.09	4.99
13	11.4	8.19	6.93	6.23	5.79	5.48	5.25	5.08	4.94	4.82	4.72
14	11.1	7.92	6.68	6.00	5.56	5.26	5.03	4.86	4.72	4.60	4.51
15	10.8	7.70	6.48	5.80	5.37	5.07	4.85	4.67	4.54	4.42	4.33
16	10.6	7.51	6.30	5.64	5.21	4.91	4.69	4.52	4.38	4.27	4.18
17	10.4	7.35	6.16	5.50	5.07	4.78	4.56	4.39	4.25	4.14	4.05
18	10.2	7.21	6.03	5.37	4.96	4.66	4.44	4.28	4.14	4.03	3.94
19	10.1	7.09	5.92	5.27	4.85	4.56	4.34	4.18	4.04	3.93	3.84
20	9.94	6.99	5.82	5.17	4.76	4.47	4.26	4.09	3.96	3.85	3.76
21	9.83	6.89	5.73	5.09	4.68	4.39	4.18	4.01	3.88	3.77	3.68
22	9.73	6.81	5.65	5.02	4.61	4.32	4.11	3.94	3.81	3.70	3.61
23	9.63	6.73	5.58	4.95	4.54	4.26	4.05	3.88	3.75	3.64	3.55
24	9.55	6.66	5.52	4.89	4.49	4.20	3.99	3.83	3.69	3.59	3.50
25	9.48	6.60	5.46	4.84	4.43	4.15	3.94	3.78	3.64	3.54	3.45
30	9.18	6.35	5.24	4.62	4.23	3.95	3.74	3.58	3.45	3.34	3.25
35	8.98	6.19	5.09	4.48	4.09	3.81	3.61	3.45	3.32	3.21	3.12
40	8.83	6.07	4.98	4.37	3.99	3.71	3.51	3.35	3.22	3.12	3.03
45	8.71	5.97	4.89	4.29	3.91	3.64	3.43	3.28	3.15	3.04	2.96
50	8.63	5.90	4.83	4.23	3.85	3.58	3.38	3.22	3.09	2.99	2.90
55	8.55	5.84	4.77	4.18	3.80	3.53	3.33	3.17	3.05	2.94	2.85
60	8.49	5.79	4.73	4.14	3.76	3.49	3.29	3.13	3.01	2.90	2.82
75	8.37	5.69	4.63	4.05	3.67	3.41	3.21	3.05	2.93	2.82	2.74
100	8.24	5.59	4.54	3.96	3.59	3.33	3.13	2.97	2.85	2.74	2.66
∞	7.88	5.30	4.28	3.72	3.35	3.09	2.90	2.74	2.62	2.52	2.43

(Continued)

Table A.4D *(continued)*

Denominator df	NUMERATOR df 12	13	14	15	16	20	24	30	45	60	120
1	24,000	25,000	25,000	25,000	25,000	25,000	25,000	25,000	25,000	25,000	25,000
2	199	199	199	199	199	199	199	199	199	199	199
3	43.4	43.3	43.2	43.1	43	42.8	42.6	42.5	42.3	42.1	42
4	20.7	20.6	20.5	20.4	20.4	20.2	20	19.9	19.7	19.6	19.5
5	13.4	13.3	13.2	13.1	13.1	12.9	12.8	12.7	12.5	12.4	12.3
6	10	9.95	9.88	9.81	9.76	9.59	9.47	9.36	9.20	9.12	9.00
7	8.18	8.10	8.03	7.97	7.91	7.75	7.64	7.53	7.38	7.31	7.19
8	7.01	6.94	6.87	6.81	6.76	6.61	6.50	6.40	6.25	6.18	6.06
9	6.23	6.15	6.09	6.03	5.98	5.83	5.73	5.62	5.48	5.41	5.30
10	5.66	5.59	5.53	5.47	5.42	5.27	5.17	5.07	4.93	4.86	4.75
11	5.24	5.16	5.10	5.05	5.00	4.86	4.76	4.65	4.52	4.45	4.34
12	4.91	4.84	4.77	4.72	4.67	4.53	4.43	4.33	4.19	4.12	4.01
13	4.64	4.57	4.51	4.46	4.41	4.27	4.17	4.07	3.94	3.87	3.76
14	4.43	4.36	4.30	4.25	4.20	4.06	3.96	3.86	3.73	3.66	3.55
15	4.25	4.18	4.12	4.07	4.02	3.88	3.79	3.69	3.55	3.48	3.37
16	4.10	4.03	3.97	3.92	3.87	3.73	3.64	3.54	3.40	3.33	3.22
17	3.97	3.90	3.84	3.79	3.75	3.61	3.51	3.41	3.28	3.21	3.10
18	3.86	3.79	3.73	3.68	3.64	3.50	3.40	3.30	3.17	3.10	2.99
19	3.76	3.70	3.64	3.59	3.54	3.40	3.31	3.21	3.07	3.00	2.89
20	3.68	3.61	3.55	3.50	3.46	3.32	3.22	3.12	2.99	2.92	2.81
21	3.60	3.54	3.48	3.43	3.38	3.24	3.15	3.05	2.91	2.84	2.73
22	3.54	3.47	3.41	3.36	3.31	3.18	3.08	2.98	2.84	2.77	2.66
23	3.47	3.41	3.35	3.30	3.25	3.12	3.02	2.92	2.78	2.71	2.60
24	3.42	3.35	3.30	3.25	3.20	3.06	2.97	2.87	2.73	2.66	2.55
25	3.37	3.30	3.25	3.20	3.15	3.01	2.92	2.82	2.68	2.61	2.50
30	3.18	3.11	3.06	3.01	2.96	2.82	2.73	2.63	2.49	2.42	2.30
35	3.05	2.98	2.93	2.88	2.83	2.69	2.60	2.50	2.36	2.28	2.16
40	2.95	2.89	2.83	2.78	2.74	2.60	2.50	2.40	2.26	2.18	2.06
45	2.88	2.82	2.76	2.71	2.66	2.53	2.43	2.33	2.19	2.11	1.99
50	2.82	2.76	2.70	2.65	2.61	2.47	2.37	2.27	2.13	2.05	1.93
55	2.78	2.71	2.66	2.61	2.56	2.42	2.33	2.23	2.08	2.00	1.88
60	2.74	2.68	2.62	2.57	2.53	2.39	2.29	2.19	2.04	1.96	1.83
75	2.66	2.60	2.54	2.49	2.45	2.31	2.21	2.10	1.96	1.88	1.74
100	2.58	2.52	2.46	2.41	2.37	2.23	2.13	2.02	1.87	1.79	1.65
∞	2.36	2.29	2.24	2.19	2.14	2.00	1.90	1.79	1.63	1.53	1.36

Table A.5 The F_{max} Distribution—Percentage Points of $F_{max} = s^2_{max}/s^2_{max}$

					t						
df$_2$	**2**	**3**	**4**	**5**	**6**	**7**	**8**	**9**	**10**	**11**	**12**
					Upper 5% points						
2	39.0	87.5	142	202	266	333	403	475	550	626	704
3	15.4	27.8	39.2	60.7	62.0	72.9	83.5	93.9	104	114	124
4	9.60	15.5	20.6	26.2	29.5	33.6	37.5	41.1	44.6	48.0	51.4
5	7.15	10.3	13.7	16.3	18.7	20.8	22.9	24.7	26.5	28.2	29.9
6	5.82	8.38	10.4	12.1	13.7	15.0	16.3	17.5	18.6	19.7	20.7
7	4.99	6.94	8.44	9.70	10.8	11.8	12.7	13.5	14.3	15.1	15.8
8	4.43	6.00	7.18	8.12	9.03	9.78	10.5	11.1	11.7	12.2	12.7
9	4.03	5.34	6.31	7.11	7.80	8.41	8.95	9.45	9.91	10.3	10.7
10	3.72	4.85	5.67	6.34	6.92	7.42	7.87	8.28	8.66	9.01	9.34
12	3.28	4.16	4.79	5.30	5.72	6.09	6.42	6.72	7.00	7.25	7.48
15	2.86	3.54	4.01	4.37	4.68	4.95	5.19	5.40	5.59	5.77	5.93
20	2.46	2.95	3.29	3.54	3.76	3.94	4.10	4.24	4.37	4.49	4.59
30	2.07	2.40	2.61	2.78	2.91	3.02	3.12	3.21	3.29	3.36	3.39
60	1.67	1.85	1.96	2.04	2.11	2.17	2.22	2.26	2.30	2.33	2.36
∞	1.00	1.00	1.00	1.00	1.00	1.00	1.00	1.00	1.00	1.00	1.00
					Upper 1% points						
2	199	448	729	1036	1362	1705	2063	2432	2813	3204	3605
3	47.5	85	120	151	184	21(6)	24(9)	28(1)	31(0)	33(7)	36(1)
4	23.2	37	49	59	69	79	89	97	106	113	120
5	14.9	22	28	33	38	42	46	50	54	57	60
6	11.1	15.5	19.1	22	25	27	30	32	34	36	37
7	8.89	12.1	14.5	16.5	18.4	20	22	23	24	26	27
8	7.50	9.9	11.7	13.2	14.5	15.8	16.9	17.9	18.9	19.8	21
9	6.54	8.5	9.9	11.1	12.1	13.1	13.9	14.7	15.3	16.0	16.6
10	5.85	7.4	8.6	9.6	10.4	11.1	11.8	12.4	12.9	13.4	13.9
12	4.91	6.1	6.9	7.6	8.2	8.7	9.1	9.5	9.9	10.2	10.6
15	4.07	4.9	5.5	6.0	6.4	6.7	7.1	7.3	7.5	7.8	8.0
20	3.32	3.8	4.3	4.6	4.9	5.1	5.3	5.5	5.6	5.8	5.9
30	2.63	3.0	3.3	3.4	3.6	3.7	3.8	3.9	4.0	4.1	4.2
60	1.96	2.2	2.3	2.4	2.4	2.5	2.5	2.6	2.6	2.7	2.7
∞	1.00	1.0	1.0	1.0	1.0	1.0	1.0	1.0	1.0	1.0	1.0

Note: s^2_{max} is the largest and s^2_{min} the smallest in a set of t independent mean squares, each based on $df_2 = n - 1$ degrees of freedom. Values in the column $t = 2$ and in the rows $df_2 = 2$ and ∞ are exact. Elsewhere the third digit may be in error by a few units for the 5% points and several units for the 1% points. The third digit figures in brackets for $df_2 = 3$ are the most uncertain.

Source: From Pearson and Hartley (1966). Reproduced by permission of the *Biometrika* Trustees.

Table A.6 Orthogonal Polynomials (Tables of Coefficients for Polynomial Trends)

$t = 3$	x_1	x_2
	−1	1
	0	−2
	1	1
$\sum x_i^2$	2	6

$t = 4$	x_1	x_2	x_3
	−3	1	−1
	−1	−1	3
	1	−1	−3
	3	1	1
$\sum x_i^2$	20	4	20

$t = 5$	x_1	x_2	x_3	x_4
	−2	2	−1	1
	−1	−1	2	−4
	0	−2	0	6
	1	−1	−2	−4
	2	2	1	1
$\sum x_i^2$	10	14	10	70

$t = 6$	x_1	x_2	x_3	x_4
	−5	5	−5	1
	−3	−1	7	−3
	−1	−4	4	2
	1	−4	−4	2
	3	−1	−7	−3
	m5	5	5	1
$\sum x_i^2$	70	84	180	28

$t = 7$	x_1	x_2	x_3	x_4
	−3	5	−1	3
	−2	0	1	−7
	−1	−3	1	1
	0	−4	0	6
	1	−3	−1	1
	2	0	−1	−7
	3	5	1	3
$\sum x_i^2$	28	84	6	154

$t = 8$	x_1	x_2	x_3	x_4
	−7	7	−7	7
	−5	1	5	−13
	−3	−3	7	−3
	−1	−5	3	9
	1	−5	−3	9
	3	−3	−7	−3
	5	1	−5	−13
	7	7	7	7
$\sum x_i^2$	168	168	264	616

$t = 9$	x_1	x_2	x_3	x_4
	−4	28	−14	14
	−3	7	7	−21
	−2	−8	13	−11
	−1	−17	9	9
	0	−20	0	18
	−1	−17	−9	9
	2	−8	−13	−11
	3	7	−7	−21
	4	28	14	14
$\sum x_i^2$	60	2,772	990	2,002

$t = 10$	x_1	x_2	x_3	x_4
	−9	6	−42	18
	−7	2	−14	−22
	−5	−1	35	−17
	−3	−3	31	3
	−1	−4	12	18
	1	−4	−12	18
	3	−3	−31	3
	5	−1	−35	−17
	7	2	−14	−22
	9	6	42	18
$\sum x_i^2$	330	132	8,580	2,860

Source: Abridged from Pearson and Hartley (1966), Table 47, p. 236. Reproduced by permission of the *Biometrika* Trustees.

Table A.7 Percentage Points of the Studentized Range

		t = NUMBER OF TREATMENT MEANS									
Error df	α	**2**	**3**	**4**	**5**	**6**	**7**	**8**	**9**	**10**	**11**
5	0.05	3.64	4.60	5.22	5.67	6.03	6.33	6.58	6.80	6.99	7.17
	0.01	5.70	6.98	7.80	8.42	8.91	9.32	9.67	9.97	10.24	10.48
6	0.05	3.46	4.34	4.90	5.30	5.63	5.90	6.12	6.32	6.49	6.65
	0.01	5.24	6.33	7.03	7.56	7.97	8.32	8.61	8.87	9.10	9.30
7	0.05	3.34	4.16	4.68	5.06	5.36	5.61	5.82	6.00	6.16	6.30
	0.01	4.95	5.92	6.54	7.01	7.37	7.68	7.94	8.17	8.37	8.55
8	0.05	3.26	4.04	4.53	4.89	5.17	5.40	5.60	5.77	5.92	6.05
	0.01	4.75	5.64	6.20	6.62	6.96	7.24	7.47	7.68	7.86	8.03
9	0.05	3.20	3.95	4.41	4.76	5.02	5.24	5.43	5.59	5.74	5.87
	0.01	4.60	5.43	5.96	6.35	6.66	6.91	7.13	7.33	7.49	7.65
10	0.05	3.15	3.88	4.33	4.65	4.91	5.12	5.30	5.46	5.60	5.72
	0.01	4.48	5.27	5.77	6.14	6.43	6.67	6.87	7.05	7.21	7.36
11	0.05	3.11	3.82	4.26	4.57	4.82	5.03	5.30	5.35	5.49	5.61
	0.01	4.39	5.15	5.62	5.97	6.25	6.48	6.67	6.84	6.99	7.13
12	0.05	3.08	3.77	4.20	4.52	4.75	4.95	5.12	5.27	5.39	5.51
	0.01	4.32	5.05	5.50	5.84	6.10	6.32	6.51	6.67	6.81	6.94
13	0.05	3.06	3.73	4.15	4.45	4.69	4.88	5.05	5.19	5.32	5.43
	0.01	4.26	4.96	5.40	5.73	5.98	6.19	6.37	6.53	6.67	6.79
14	0.05	3.03	3.70	4.11	4.41	4.64	4.83	4.99	5.13	5.25	5.36
	0.01	4.21	4.89	5.32	5.63	5.88	6.08	6.26	6.41	6.54	6.66
15	0.05	3.01	3.67	4.08	4.37	4.59	4.78	4.94	5.08	5.20	5.31
	0.01	4.17	4.84	5.25	5.56	5.80	5.99	6.16	6.31	6.44	6.55
16	0.05	3.00	3.65	4.05	4.33	4.56	4.74	4.90	5.03	5.15	5.26
	0.01	4.13	4.79	5.19	5.49	5.72	5.92	6.08	6.22	6.35	6.46
17	0.05	2.98	3.63	4.02	4.30	4.52	4.70	4.86	4.99	5.11	5.21
	0.01	4.10	4.74	5.14	5.43	5.66	5.85	6.01	6.15	6.27	6.38
18	0.05	2.97	3.61	4.00	4.28	4.49	4.67	4.82	4.96	5.07	5.17
	0.01	4.07	4.70	5.09	5.38	5.60	5.79	5.94	6.08	6.20	6.31
19	0.05	2.96	3.59	3.98	4.25	4.47	4.65	4.79	4.92	5.04	5.14
	0.01	4.05	4.67	5.05	5.33	5.55	5.73	5.89	6.02	6.14	6.25
20	0.05	2.95	3.58	3.96	4.23	4.45	4.62	4.77	4.90	5.01	5.11
	0.01	4.02	4.64	5.02	5.29	5.51	5.69	5.84	5.97	6.09	6.19
24	0.05	2.92	3.53	3.90	4.17	4.37	4.54	4.68	4.81	3.92	5.01
	0.01	3.96	4.55	4.91	5.17	5.37	5.54	5.69	5.81	5.92	6.02
30	0.05	2.89	3.49	3.85	4.10	4.30	4.46	4.60	4.72	4.82	4.92
	0.01	3.89	4.45	4.80	5.05	5.24	5.40	5.54	5.65	5.76	5.85
40	0.05	2.86	3.44	3.79	4.04	4.23	4.39	4.52	4.63	4.73	4.82
	0.01	3.82	4.37	4.70	4.93	5.11	5.26	5.39	5.50	5.60	5.69
60	0.05	2.83	3.40	3.74	3.98	4.16	4.31	4.44	4.55	4.65	4.73
	0 .01	3.76	4.28	4.59	4.82	4.99	5.13	5.25	5.36	5.45	5.53
120	0.05	2.80	3.36	3.68	3.92	4.10	4.24	4.36	4.47	4.56	4.64
	0.10	3.70	4.20	4.50	4.71	4.87	5.01	5.12	5.21	5.30	5.37
∞	0.05	2.77	3.31	3.63	3.86	4.03	4.17	4.29	4.39	4.47	4.55
	0.01	3.64	4.12	4.40	4.60	4.76	4.88	4.99	5.08	5.16	5.23

(Continued)

Table A.7 *(continued)*

t = NUMBER OF TREATMENT MEANS										
12	13	14	15	16	17	18	19	20	α	Error df
7.32	7.47	7.60	7.72	7.83	7.93	8.03	8.12	8.21	0.05	5
10.70	10.89	11.08	11.24	11.40	11.55	11.68	11.81	11.93	0.01	
6.79	6.92	7.03	7.14	7.24	7.34	7.43	7.51	7.59	0.05	6
9.48	9.65	9.81	9.95	10.08	10.21	10.32	10.43	10.54	0.01	
6.43	6.55	6.66	6.76	6.85	6.94	7.02	7.10	7.17	0.05	7
8.71	8.86	9.00	9.12	9.24	9.35	9.46	9.55	9.65	0.01	
6.18	6.29	6.39	6.48	6.57	6.65	6.73	6.80	6.87	0.05	8
8.18	8.31	8.44	8.55	8.66	8.76	8.85	8.94	9.03	0.01	
5.98	6.09	6.19	6.28	6.36	6.44	6.51	6.58	6.64	0.05	9
7.78	7.91	8.03	8.13	8.23	8.33	8.41	8.49	8.57	0.01	
5.83	5.93	6.03	6.11	6.19	6.27	6.34	6.40	6.47	0.05	10
7.49	7.60	7.71	7.81	7.91	7.99	8.08	8.15	8.23	0.01	
5.71	5.81	5.90	5.98	6.06	6.13	6.20	6.27	6.33	0.05	11
7.25	7.36	7.46	7.56	7.65	7.73	7.81	7.88	7.95	0.01	
5.61	5.71	5.80	5.88	5.95	6.02	6.09	6.15	6.21	0.05	12
7.06	7.17	7.26	7.36	7.44	7.52	7.59	7.66	7.73	0.01	
5.53	5.63	5.71	5.79	5.86	5.93	5.99	6.05	6.11	0.05	13
6.90	7.01	7.10	7.19	7.27	7.35	7.42	7.48	7.55	0.01	
5.46	5.55	5.64	5.71	5.79	5.85	5.91	5.97	6.03	0.05	14
6.77	6.87	6.96	7.05	7.13	7.20	7.27	7.33	7.39	0.01	
5.40	5.49	5.57	5.65	5.72	5.78	5.85	5.90	5.96	0.05	15
6.66	6.76	6.84	6.93	7.00	7.07	7.14	7.20	7.26	0.01	
5.35	5.44	5.52	5.59	5.66	5.73	5.79	5.84	5.90	0.05	16
6.56	6.66	6.74	6.82	6.90	6.97	7.03	7.09	7.15	0.01	
5.31	5.39	5.47	5.54	5.61	5.67	5.73	5.79	5.84	0.05	17
6.48	6.57	6.66	6.73	6.81	6.87	6.94	7.00	7.05	0.01	
5.27	5.35	5.43	5.50	5.57	5.63	5.69	5.74	5.79	0.05	18
6.41	6.50	6.58	6.65	6.73	6.79	6.85	6.91	6.97	0.01	
5.23	5.31	5.39	5.46	5.53	5.59	5.65	5.70	5.75	0.05	19
6.34	6.43	6.51	6.58	6.65	6.72	6.78	6.84	6.89	0.01	
5.20	5.28	5.36	5.43	5.49	5.55	5.61	5.66	5.71	0.05	20
6.28	6.37	6.45	6.52	6.59	6.65	6.71	6.77	6.82	0.01	
5.10	5.18	5.25	5.32	5.38	5.44	5.49	5.55	5.59	0.05	24
6.11	6.19	6.26	6.33	6.39	6.45	6.51	6.56	6.61	0.01	
5.00	5.08	5.15	5.21	5.27	5.33	5.38	5.43	5.47	0.05	30
5.93	6.01	6.08	6.14	6.20	6.26	6.31	6.36	6.41	0.01	
4.90	4.98	5.04	5.11	5.16	5.22	5.27	5.31	5.36	0.05	40
5.76	5.83	5.90	5.96	6.02	6.07	6.12	6.16	6.21	0.01	
4.81	4.88	4.94	5.00	5.06	5.11	5.15	5.20	5.24	0.05	60
5.60	5.67	5.73	5.78	5.84	5.89	5.93	5.97	6.01	0.01	
4.71	4.78	4.84	4.90	4.95	5.00	5.04	5.09	5.13	0.05	120
5.44	5.50	5.56	5.61	5.66	5.71	5.75	5.79	5.83	0.01	
4.62	4.68	4.74	4.80	4.85	4.89	4.93	4.97	5.01	0.05	∞
5.29	5.35	5.40	5.45	5.49	5.54	5.57	5.61	5.65	0.01	

Source: Abridged from Pearson and Hartley (1958), Table 29. Reproduced with the permission of the editors and the trustees of *Biometrika*.

Table A.8 Percentage Points of the Duncan Multiple Range Test

		r = NUMBER OF ORDERED STEPS BETWEEN MEANS													
Error df	α	2	3	4	5	6	7	8	9	10	12	14	16	18	20
1	0.05	18.0	18.0	18.0	18.0	18.0	18.0	18.0	18.0	18.0	18.0	18.0	18.0	18.0	18.0
	0.01	90.0	90.0	90.0	90.0	90.0	90.0	90.0	90.0	90.0	90.0	90.0	90.0	90.0	90.0
2	0.05	6.09	6.09	6.09	6.09	6.09	6.09	6.09	6.09	6.09	6.09	6.09	6.09	6.09	6.09
	0.01	14.0	14.0	14.0	14.0	14.0	14.0	14.0	14.0	14.0	14.0	14.0	14.0	14.0	14.0
3	0.05	4.50	4.50	4.50	4.50	4.50	4.50	4.50	4.50	4.50	4.50	4.50	4.50	4.50	4.50
	0.01	8.26	8.5	8.6	8.7	8.8	8.9	8.9	9.0	9.0	9.0	9.1	9.2	9.3	9.3
4	0.05	3.93	4.01	4.02	4.02	4.02	4.02	4.02	4.02	4.02	4.02	4.02	4.02	4.02	4.02
	0.01	6.51	6.8	6.9	7.0	7.1	7.1	7.2	7.2	7.3	7.3	7.4	7.4	7.5	7.5
5	0.05	3.64	3.74	3.79	3.83	3.83	3.83	3.83	3.83	3.83	3.83	3.83	3.83	3.83	3.83
	0.01	5.70	5.96	6.11	6.18	6.26	6.33	6.40	6.44	6.5	6.6	6.6	6.7	6.7	6.8
6	0.05	3.46	3.58	3.64	3.68	3.68	3.68	3.68	3.68	3.68	3.68	3.68	3.68	3.68	3.68
	0.01	5.24	5.51	5.65	5.73	5.81	5.88	5.95	6.00	6.0	6.1	6.2	6.2	6.3	6.3
7	0.05	3.35	3.47	3.54	3.58	3.60	3.61	3.61	3.61	3.61	3.61	3.61	3.61	3.61	3.61
	0.01	4.95	5.22	5.37	5.45	5.53	5.61	5.69	5.73	5.8	5.8	5.9	5.9	6.0	6.0
8	0.05	3.26	3.39	3.47	3.52	3.55	3.56	3.56	3.56	3.56	3.56	3.56	3.56	3.56	3.56
	0.01	4.74	5.00	5.14	5.23	5.32	5.40	5.47	5.51	5.5	5.6	5.7	5.7	5.8	5.8
9	0.05	3.20	3.34	3.41	3.47	3.50	3.52	3.52	3.52	3.52	3.52	3.52	3.52	3.52	3.52
	0.01	4.60	4.86	4.99	5.08	5.17	5.25	5.32	5.36	5.4	5.5	5.5	5.6	5.7	5.7
10	0.05	3.15	3.30	3.37	3.43	3.46	3.47	3.47	3.47	3.47	3.47	3.47	3.47	3.47	3.48
	0.01	4.48	4.73	4.88	4.96	5.06	5.13	5.20	5.24	5.28	5.36	5.42	5.48	5.54	5.55
11	0.05	3.11	3.27	3.35	3.39	3.43	3.44	3.45	3.46	3.46	3.46	3.46	3.46	3.47	3.48
	0.01	4.39	4.63	4.77	4.86	4.94	5.01	5.06	5.12	5.15	5.24	5.28	5.34	5.38	5.39
12	0.05	3.08	3.23	3.33	3.36	3.40	3.42	3.44	3.44	3.46	3.46	3.46	3.46	3.47	3.48
	0.01	4.32	4.55	4.68	4.76	4.84	4.92	4.96	5.02	5.07	5.13	5.17	5.22	5.23	5.26
13	0.05	3.06	3.21	3.30	3.35	3.38	3.41	3.42	3.44	3.45	3.45	3.46	3.46	3.47	3.47
	0.01	4.26	4.48	4.62	4.69	4.74	4.84	4.88	4.94	4.98	5.04	5.08	5.13	5.14	5.15
14	0.05	3.03	3.18	3.27	3.33	3.37	3.39	3.41	3.42	3.44	3.45	3.46	3.46	3.47	3.47
	0.01	4.21	4.42	4.55	4.63	4.70	4.78	4.83	4.87	4.91	4.96	5.00	5.04	5.06	5.07
15	0.05	3.01	3.16	3.25	3.31	3.36	3.38	3.40	3.42	3.43	3.44	3.45	3.46	3.47	3.47
	0.01	4.17	4.37	4.50	4.58	4.64	4.72	4.77	4.81	4.84	4.90	4.94	4.97	4.99	5.00
16	0.05	3.00	3.15	3.23	3.30	3.34	3.37	3.39	3.41	3.43	3.44	3.45	3.46	3.47	3.47
	0.01	4.13	4.34	4.45	4.54	4.60	4.67	4.72	4.76	4.79	4.84	4.88	4.91	4.93	4.94
17	0.05	2.98	3.13	3.22	3.28	3.33	3.36	3.38	3.40	3.42	3.44	3.45	3.46	3.47	3.47
	0.01	4.10	4.30	4.41	4.50	4.56	4.63	4.68	4.72	4.75	4.80	4.83	4.86	4.88	4.89
18	0.05	2.97	3.12	3.21	3.27	3.32	3.35	3.37	3.39	3.41	3.43	3.45	3.46	3.47	3.47
	0.01	4.07	4.27	4.38	4.46	4.53	4.59	4.64	4.68	4.71	4.76	4.79	4.82	4.84	4.85
19	0.05	2.96	3.11	3.19	3.26	3.31	3.35	3.37	3.39	3.41	3.43	3.44	3.46	3.47	3.47
	0.01	4.05	4.24	4.35	4.43	4.50	4.56	4.61	4.64	4.67	4.72	4.76	4.79	4.81	4.82
20	0.05	2.95	3.10	3.18	3.25	3.30	3.34	3.36	3.38	3.40	3.43	3.44	3.46	3.46	3.47
	0.01	4.02	4.22	4.33	4.40	4.47	4.53	4.58	4.61	4.65	4.69	4.73	4.76	4.78	4.79
22	0.05	2.93	3.08	3.17	3.24	3.29	3.32	3.35	3.37	3.39	3.42	3.44	3.45	3.46	3.47
	0.01	3.99	4.17	4.28	4.36	4.42	4.48	4.53	4.57	4.60	4.65	4.68	4.71	4.74	4.75
24	0.05	2.92	3.07	3.15	3.22	3.28	3.31	3.34	3.37	3.38	3.41	3.44	3.45	3.46	3.47
	0.01	3.96	4.14	4.24	4.33	4.39	4.44	4.49	4.53	4.57	4.62	4.64	4.67	4.70	4.72
26	0.05	2.91	3.06	3.14	3.21	3.27	3.30	3.34	3.36	3.38	3.41	3.43	3.45	3.46	3.47
	0.01	3.93	4.11	4.21	4.30	4.36	4.41	4.46	4.50	4.53	4.58	4.62	4.65	4.67	4.69
28	0.05	2.90	3.04	3.13	3.20	3.26	3.30	3.33	3.35	3.37	3.40	3.43	3.45	3.46	3.47
	0.01	3.01	3.08	4.18	4.28	4.34	4.39	4.43	4.47	4.51	4.56	4.60	4.62	4.65	4.67

(Continued)

Table A.8 *(continued)*

		r = NUMBER OF ORDERED STEPS BETWEEN MEANS													
Error df	α	**2**	**3**	**4**	**5**	**6**	**7**	**8**	**9**	**10**	**12**	**14**	**16**	**18**	**20**
30	0.05	2.89	3.04	3.12	3.20	3.25	3.29	3.32	3.35	3.37	3.40	3.43	3.44	3.46	3.47
	0.01	3.89	4.06	4.16	4.22	4.32	4.36	4.41	4.45	4.48	4.54	4.58	4.61	4.63	4.65
40	0.05	2.86	3.01	3.10	3.17	3.22	3.27	3.30	3.33	3.35	3.39	3.42	3.44	3.46	3.47
	0.01	3.82	3.99	4.10	4.17	4.24	4.30	4.34	4.37	4.41	4.46	4.51	4.54	4.57	4.59
60	0.05	2.83	2.98	3.08	3.14	3.20	3.24	3.28	3.31	3.33	3.37	3.40	3.43	3.45	3.47
	0.01	3.76	3.92	4.03	4.12	4.17	4.23	4.27	4.31	4.34	4.39	4.44	4.47	4.50	4.53
100	0.05	2.80	2.95	3.05	3.12	3.18	3.22	3.26	3.29	3.32	3.36	3.40	3.42	3.45	3.47
	0.01	3.71	3.86	3.93	4.06	4.11	4.17	4.21	4.25	4.29	4.35	4.38	4.42	4.45	4.48
∞	0.05	2.77	2.92	3.02	3.09	3.15	3.19	3.23	3.26	3.29	3.34	3.38	3.41	3.44	3.47
	0.01	3.64	3.80	3.90	3.98	4.04	4.09	4.14	4.17	4.20	4.26	4.31	4.34	4.38	4.41

Source: Reproduced from Duncan (1955) with permission from the Biometric Society and the author.

Table A.9 Critical Values for the Wilcoxon Signed Rank Test $N = 5(1)50$

One-sided	Two-sided	N = 5	N = 6	N = 7	N = 8	N = 9	N = 10	N = 11	N = 12
$P = 0.05$	$P = 0.10$	1	2	4	6	8	11	14	17
$P = 0.025$	$P = 0.05$		1	2	4	6	8	11	14
$P = 0.01$	$P = 0.02$			0	2	3	5	7	10
$P = 0.005$	$P = 0.01$				0	2	3	5	7
One-sided	**Two-sided**	**N = 13**	**N = 14**	**N = 15**	**N = 16**	**N = 17**	**N = 18**	**N = 19**	**N = 20**
$P = 0.05$	$P = 0.10$	21	26	30	36	41	47	54	60
$P = 0.025$	$P = 0.05$	17	21	25	30	35	40	46	52
$P = 0.01$	$P = 0.02$	13	16	20	24	28	33	38	43
$P = 0.005$	$P = 0.01$	10	13	16	19	23	28	32	37
One-sided	**Two-sided**	**N = 21**	**N = 22**	**N = 23**	**N = 24**	**N = 25**	**N = 26**	**N = 27**	**N = 28**
$P = 0.05$	$P = 0.10$	68	75	83	92	101	110	120	130
$P = 0.025$	$P = 0.05$	59	66	73	81	90	98	107	117
$P = 0.01$	$P = 0.02$	49	56	62	69	77	85	93	102
$P = 0.005$	$P = 0.01$	43	49	55	61	68	76	84	92
One-sided	**Two-sided**	**N = 29**	**N = 30**	**N = 31**	**N = 32**	**N = 33**	**N = 34**	**N = 35**	**N = 36**
$P = 0.05$	$P = 0.10$	141	152	163	175	188	201	214	228
$P = 0.025$	$P = 0.05$	127	137	148	159	171	183	195	208
$P = 0.01$	$P = 0.02$	111	120	130	141	151	162	174	186
$P = 0.005$	$P = 0.01$	100	109	118	128	138	149	160	171
One-sided	**Two-sided**	**N = 37**	**N = 38**	**N = 39**	**N = 40**	**N = 41**	**N = 42**	**N = 43**	**N = 44**
$P = 0.05$	$P = 0.10$	242	256	271	287	303	319	336	353
$P = 0.025$	$P = 0.05$	222	235	250	264	279	295	311	327
$P = 0.01$	$P = 0.02$	198	211	224	238	252	267	281	297
$P = 0.005$	$P = 0.01$	183	195	208	221	234	248	262	277
One-sided	**Two-sided**	**N = 45**	**N = 46**	**N = 47**	**N = 48**	**N = 49**	**N = 50**		
$P = 0.05$	$P = 0.10$	371	389	408	427	446	466		
$P = 0.025$	$P = 0.05$	344	361	379	397	415	434		
$P = 0.01$	$P = 0.02$	313	329	345	362	380	398		
$P = 0.005$	$P = 0.01$	292	307	323	339	356	373		

Source: Reproduced from Wilcoxon and Wilcox (1964), with permission of the American Cyanamid Company.

Table A.10 The Mann–Whitney Two-Sample Test

n_2	n_1: 2	3	4	5	6	7	8	9	10	11	12	13	14	15
							5% Critical points of rank sums							
4			10											
5		6	11	17										
6		7	12	18	26									
7		7	13	20	27	36								
8	3	8	14	21	29	38	49							
9	3	8	15	22	31	40	51	63						
10	3	9	15	23	32	42	53	65	78					
11	4	9	16	24	34	44	55	68	81	96				
12	4	10	17	26	35	46	58	71	85	99	115			
13	4	10	18	27	37	48	60	73	88	103	119	137		
14	4	11	19	28	38	60	63	76	91	106	123	141	160	
15	4	11	20	29	40	52	65	79	94	110	127	145	164	185
16	4	12	21	31	42	54	67	82	97	114	131	150	169	
17	5	12	21	32	43	56	70	84	100	117	135	154		
18	5	13	22	33	45	58	72	87	103	121	139			
19	5	13	23	34	46	60	74	90	107	124				
20	5	14	24	35	48	62	77	93	110					
21	6	14	25	37	50	64	79	95						
22	6	15	26	38	51	66	82							
23	6	15	27	39	53	68								
24	6	16	28	40	55									
25	6	16	28	42										
26	7	17	29											
27	7	17												
28	7													
							1% Critical points of rank sums							
5				15										
6			10	16	23									
7			10	17	24	32								
8			11	17	25	34	43							
9		6	11	18	26	35	45	56						
10		6	12	19	27	37	47	58	71					
11		6	12	20	28	38	49	61	74	87				
12		7	13	21	30	40	51	63	76	90	106			
13		7	14	22	31	41	53	65	79	93	109	125		
14		7	14	22	32	43	54	67	81	96	112	129	147	
15		8	15	23	33	44	56	70	84	99	115	133	151	171
16		8	15	24	34	46	58	72	86	102	119	137	155	
17		8	16	25	36	47	60	74	89	105	122	140		
18		8	16	26	37	49	62	76	92	108	125			
19	3	9	17	27	38	50	64	78	94	111				
20	3	9	18	28	39	52	66	81	97					
21	3	9	18	29	40	53	68	83						
22	3	10	19	29	42	55	70							
23	3	10	19	30	43	57								
24	3	10	20	31	44									
25	3	11	20	32										
26	3	11	21											
	4	11												
	4													

Note: n_1 and n_2 are the numbers of cases in the two groups. If the groups are unequal in size, n_1 refers to the smaller.
Source: Reproduced from White, C. (1956). The use of ranks in a test of significance for comparing two treatments. *Biometrics* **8**, 33–41, with permission of the Biometrics Society.

Table A.11 Exact Critical Values for Use with the Analysis of Means

NUMBER OF MEANS, t

A. $h_{0.10}$, Significance level = 0.10

df	3	4	5	6	7	8	9	10	11	12	13	14	15	16	17	18	19	20	df
3	3.16																		3
4	2.81	3.10																	4
5	2.63	2.88	3.05																5
6	2.52	2.74	2.91	3.03															6
7	2.44	2.65	2.81	2.92	3.02														7
8	2.39	2.59	2.73	2.85	2.94	3.02													8
9	2.34	2.54	2.68	2.79	2.88	2.95	3.01												9
10	2.31	2.50	2.64	2.74	2.83	2.90	2.96	3.02											10
11	2.29	2.47	2.60	2.70	2.79	2.86	2.92	2.97	3.02										11
12	2.27	2.45	2.57	2.67	2.75	2.82	2.88	2.93	2.98	3.02									12
13	2.25	2.43	2.55	2.65	2.73	2.79	2.85	2.90	2.95	2.99	3.03								13
14	2.23	2.41	2.53	2.63	2.70	2.77	2.83	2.88	2.92	2.96	3.00	3.03							14
15	2.22	2.39	2.51	2.61	2.68	2.75	2.80	2.85	2.90	2.94	2.97	3.01	3.04						15
16	2.21	2.38	2.50	2.59	2.67	2.73	2.79	2.83	2.88	2.92	2.95	2.99	3.02	3.05					16
17	2.20	2.37	2.49	2.58	2.65	2.72	2.77	2.82	2.86	2.90	2.93	2.97	3.00	3.03	3.05				17
18	2.19	2.36	2.47	2.56	2.64	2.70	2.75	2.80	2.84	2.88	2.92	2.95	2.98	3.01	3.03	3.06			18
19	2.18	2.35	2.46	2.55	2.63	2.69	2.74	2.79	2.83	2.87	2.90	2.94	2.96	2.99	3.02	3.04	3.06		19
20	2.18	2.34	2.45	2.54	2.62	2.68	2.73	2.78	2.82	2.86	2.89	2.92	2.95	2.98	3.00	3.03	3.05	3.07	20
24	2.15	2.32	2.43	2.51	2.58	2.64	2.69	2.74	2.78	2.82	2.85	2.88	2.91	2.93	2.96	2.98	3.00	3.02	24
30	2.13	2.29	2.40	2.48	2.55	2.61	2.66	2.70	2.74	2.77	2.81	2.84	2.86	2.89	2.91	2.93	2.96	2.98	30
40	2.11	2.27	2.37	2.45	2.52	2.57	2.62	2.66	2.70	2.73	2.77	2.79	2.82	2.85	2.87	2.89	2.91	2.93	40
60	2.09	2.24	2.34	2.42	2.49	2.54	2.59	2.63	2.66	2.70	2.73	2.75	2.78	2.80	2.82	2.84	2.86	2.88	60
120	2.07	2.22	2.32	2.39	2.45	2.51	2.55	2.59	2.62	2.66	2.69	2.71	2.74	2.76	2.78	2.80	2.82	2.84	120
∞	2.05	2.19	2.29	2.36	2.42	2.47	2.52	2.55	2.59	2.62	2.65	2.67	2.69	2.72	2.74	2.76	2.77	2.79	∞

B. $h_{0.05}$, Significance level = 0.05

df	3	4	5	6	7	8	9	10	11	12	13	14	15	16	17	18	19	20	df
3	4.18																		3
4	3.56	3.89																	4
5	3.25	3.53	3.72																5
6	3.07	3.31	3.49	3.62															6
7	2.94	3.17	3.33	3.45	3.56														7
8	2.86	3.07	3.21	3.33	3.43	3.51													8
9	2.79	2.99	3.13	3.24	3.33	3.41	3.48												9
10	2.74	2.93	3.07	3.17	3.26	3.33	3.40	3.45											10
11	2.70	2.88	3.01	3.12	3.20	3.27	3.33	3.39	3.44										11
12	2.67	2.85	2.97	3.07	3.15	3.22	3.28	3.33	3.38	3.42									12
13	2.64	2.81	2.94	3.03	3.11	3.18	3.24	3.29	3.34	3.38	3.42								13
14	2.62	2.79	2.91	3.00	3.08	3.14	3.20	3.25	3.30	3.34	3.37	3.41							14
15	2.60	2.76	2.88	2.97	3.05	3.11	3.17	3.22	3.26	3.30	3.34	3.37	3.40						15
16	2.58	2.74	2.86	2.95	3.02	3.09	3.14	3.19	3.23	3.27	3.31	3.34	3.37	3.40					16
17	2.57	2.73	2.84	2.93	3.00	3.06	3.12	3.16	3.21	3.25	3.28	3.31	3.34	3.37	3.40				17
18	2.55	2.71	2.82	2.91	2.98	3.04	3.10	3.14	3.18	3.22	3.26	3.29	3.32	2.35	3.37	3.40			18
19	2.45	2.70	2.81	2.89	2.96	3.02	3.08	3.12	3.16	3.20	3.24	3.27	3.30	3.32	3.35	3.37	3.40		19
20	2.53	2.68	2.79	2.88	2.95	3.01	3.06	3.11	3.15	3.18	3.22	3.25	3.28	3.30	3.33	3.35	3.37	3.40	20
24	2.50	2.65	2.75	2.83	2.90	2.96	3.01	3.05	3.09	3.13	3.16	3.19	3.22	3.24	3.27	3.29	3.31	3.33	24
30	2.47	2.61	2.71	2.79	2.85	2.91	2.96	3.00	3.04	3.07	3.10	3.13	3.16	3.18	3.20	3.22	3.25	3.27	30
40	2.43	2.57	2.67	2.75	2.81	2.86	2.91	2.95	2.98	3.01	3.04	3.07	3.10	3.12	3.14	3.16	3.18	3.20	40
60	2.40	2.54	2.63	2.70	2.76	2.81	2.86	2.90	2.93	2.96	2.99	3.02	3.04	3.06	3.08	3.10	3.12	3.14	60
120	2.37	2.50	2.59	2.66	2.72	2.77	2.81	2.84	2.88	2.91	2.93	2.96	2.98	3.00	3.02	3.04	3.06	3.08	120
∞	2.34	2.47	2.56	2.62	2.68	2.72	2.76	2.80	2.83	2.86	2.88	2.90	2.93	2.95	2.97	2.98	3.00	3.02	∞

(Continued)

Table A.11 *(continued)*

	NUMBER OF MEANS, t																	
df	**3**	**4**	**5**	**6**	**7**	**8**	**9**	**10**	**11**	**12**	**13**	**14**	**15**	**16**	**17**	**18**	**19**	**20**	**df**
	C. $h_{0.01}$, Significance level = 0.05																		
3	7.51																		3
4	5.74	6.21																	4
5	4.93	5.29	5.55																5
6	4.48	4.77	4.98	5.16															6
7	4.18	4.44	4.63	4.78	4.90														7
8	3.98	4.21	4.38	4.52	4.63	4.72													8
9	3.84	4.05	4.20	4.33	4.43	4.51	4.59												9
10	3.73	3.92	4.07	4.18	4.28	4.36	4.43	4.49											10
11	3.64	3.82	3.96	4.07	4.16	4.23	4.30	4.36	4.41										11
12	3.57	3.74	3.87	3.98	4.06	4.13	4.20	4.25	4.31	4.35									12
13	3.51	3.68	3.80	3.90	3.98	4.05	4.11	4.17	4.22	4.26	4.30								13
14	3.46	3.63	3.74	3.84	3.92	3.98	4.04	4.09	4.14	4.18	4.22	4.26							14
15	3.42	3.58	3.69	3.79	3.86	3.92	3.98	4.03	4.08	4.12	4.16	4.19	4.22						15
16	3.38	3.54	3.65	3.74	3.81	3.87	3.93	3.98	4.02	4.06	4.10	4.14	4.17	4.20					16
17	3.35	3.50	3.61	3.70	3.77	3.83	3.89	3.93	3.98	4.02	4.05	4.09	4.12	4.14	4.17				17
18	3.33	3.47	3.58	3.66	3.73	3.79	3.85	3.89	3.94	3.97	4.01	4.04	4.07	4.10	4.12	4.15			18
19	3.30	3.45	3.55	3.63	3.70	3.76	3.81	3.86	3.90	3.94	3.97	4.00	4.03	4.06	4.08	4.11	4.13		19
20	3.28	3.42	3.51	3.61	3.67	3.73	3.78	3.83	3.87	3.90	3.94	3.97	4.00	4.02	4.05	4.07	4.09	4.12	20
24	3.21	3.35	3.45	3.52	3.58	3.64	3.69	3.73	3.77	3.80	3.83	3.86	3.89	3.91	3.94	3.96	3.98	4.00	24
30	3.15	3.28	3.37	3.44	3.50	3.55	3.59	3.63	3.67	3.70	3.73	3.76	3.78	3.81	3.83	3.85	3.87	3.89	30
40	3.09	3.21	3.29	3.36	3.42	3.46	3.50	3.54	3.58	3.60	3.63	3.66	3.68	3.70	3.72	3.74	3.76	3.78	40
60	3.03	3.14	3.22	3.29	3.34	3.38	3.42	3.46	3.49	3.51	3.54	3.56	3.59	3.61	3.63	3.64	3.66	3.68	60
120	2.97	3.07	3.15	3.21	3.26	3.30	3.34	3.37	3.40	3.42	3.45	3.47	3.49	3.51	3.53	3.55	3.56	3.58	120
∞	2.91	3.01	3.08	3.14	3.18	3.22	3.26	3.29	3.32	3.34	3.36	3.38	3.40	3.42	3.44	3.45	3.47	3.48	∞
	D. $h_{0.001}$, Significance level = 0.001																		
3	16.4																		3
4	10.6	11.4																	4
5	8.25	8.79	9.19																5
6	7.04	7.45	7.76	8.00															6
7	6.31	6.65	6.89	7.09	7.25														7
8	5.83	6.12	6.32	6.49	6.63	6.75													8
9	5.49	5.74	5.92	6.07	6.20	6.30	6.40												9
10	5.24	5.46	5.63	5.76	5.87	5.97	6.05	6.13											10
11	5.05	5.25	5.40	5.52	5.63	5.71	5.79	5.86	5.92										11
12	4.89	5.08	5.22	5.33	5.43	5.51	5.58	5.65	5.71	5.76									12
13	4.77	4.95	5.08	5.18	5.27	5.35	5.42	5.48	5.53	5.58	5.63								13
14	4.66	4.83	4.96	5.06	5.14	5.21	5.28	5.33	5.38	5.43	5.48	5.51							14
15	4.57	4.74	4.86	4.95	5.03	5.10	5.16	5.21	5.26	5.31	5.35	5.39	5.42						15
16	4.50	4.66	4.77	4.86	4.94	5.00	5.06	5.11	5.16	5.20	5.24	5.28	5.31	5.34					16
17	4.44	4.59	4.70	4.78	4.86	4.92	4.98	5.03	5.07	5.11	5.15	5.18	5.22	5.25	5.28				17
18	4.38	4.53	4.63	4.72	4.79	4.85	4.90	4.95	4.99	5.03	5.07	5.10	5.14	5.16	5.19	5.22			18
19	4.33	4.47	4.58	4.66	4.73	4.79	4.84	4.88	4.93	4.96	5.00	5.03	5.06	5.09	5.12	5.14	5.17		19
20	4.29	4.42	4.53	4.61	4.67	4.73	4.78	4.83	4.87	4.90	4.94	4.97	5.00	5.03	5.05	5.08	5.10	5.12	20
24	4.16	4.28	4.37	4.45	4.51	4.56	4.61	4.65	4.69	4.72	4.75	4.78	4.81	4.83	4.86	4.88	4.90	4.92	24
30	4.03	4.14	4.23	4.30	4.35	4.40	4.44	4.48	4.51	4.54	4.57	4.60	4.62	4.64	4.67	4.69	4.71	4.72	30
40	3.91	4.01	4.09	4.15	4.20	4.25	4.29	4.32	4.35	4.38	4.40	4.43	4.45	4.47	4.49	4.50	4.52	4.54	40
60	3.80	3.89	3.96	4.02	4.06	4.10	4.14	4.17	4.19	4.22	4.24	4.27	4.29	4.30	4.32	4.33	4.35	4.37	60
120	3.69	3.77	3.84	3.89	3.93	3.96	4.00	4.03	4.05	4.07	4.09	4.11	4.13	4.15	4.16	4.17	4.19	4.21	120
∞	3.58	3.66	3.72	3.76	3.80	3.84	3.87	3.89	3.91	3.93	3.95	3.97	3.99	4.00	4.02	4.03	4.04	4.06	∞

Source: Reproduced from Nelson (1983), with permission from the American Society for Quality Control.

Appendix B

A Brief Introduction to Matrices

This section provides a brief introduction to matrix notation and the use of matrices for representing operations involving systems of linear equations. The purpose here is not to provide a manual for performing matrix calculations but rather to provide for an understanding and appreciation of the various matrix operations as they apply to regression analysis.

DEFINITION
A **matrix** is a rectangular array of elements arranged in rows and columns.

A matrix is much like a table and can be thought of as a many-dimensional number. Matrix algebra consists of a set of operations or algebraic rules that allow the manipulation of matrices. In this section we present those operations that will enable the reader to understand the fundamental building blocks of a multiple regression analysis. Additional information is available in a number of texts (such as Graybill, 1983).

The elements of a matrix usually consist of numbers or symbols representing numbers. Each element is indexed by its location within the matrix, which is identified by its row and column (in that order). For example, matrix **A** has 3 rows and 4 columns:

$$\mathbf{A} = \begin{bmatrix} a_{11} & a_{12} & a_{13} & a_{14} \\ a_{21} & a_{22} & a_{23} & a_{24} \\ a_{31} & a_{32} & a_{33} & a_{34} \end{bmatrix}.$$

The element a_{ij} identifies the element in the ith row and jth column. Thus the element a_{21} identifies the element in the second row and first column. The notation for this matrix follows the usual convention of denoting a matrix by a boldface capital letter and its elements by the same lowercase letter with the appropriate row and column subscripts.

An example of a matrix with three rows and columns is

$$\mathbf{B} - \begin{bmatrix} 3 & 7 & 9 \\ 1 & 4 & 2 \\ 9 & 15 & 3 \end{bmatrix}.$$

In this matrix, $b_{22} = 4$ and $b_{23} = -2$.

A matrix is characterized by its **order**, which is the number of rows and columns it contains. The matrix **B** (in the preceding) is a 3×3 matrix since it contains three rows and three columns. A matrix with equal numbers of rows and columns, such as **B**, is called a **square matrix**. A 1×1 matrix is known as a **scalar**. A scalar is, in fact, an ordinary number, and a matrix operation performed on a scalar is the same arithmetic operation done on ordinary numbers.

In a matrix, the elements whose row and column indicators are equal, say, a_{ii}, are known as **diagonal elements** and lie on the **main diagonal** of the matrix. For example, in matrix **B**, the main diagonal consists of the elements $b_{11} = 3$, $b_{22} = 4$, and $b_{33} = 3$.

A square matrix that contains nonzero elements only on the main diagonal is a **diagonal matrix**. A diagonal matrix whose nonzero elements are all unity is an **identity matrix**. It has the same function as the scalar "one" in that if a matrix is multiplied by an identity matrix it is unchanged.

Matrix Algebra

Two matrices **A** and **B** are **equal** only if all corresponding elements of **A** are the same as those of **B**. Thus $\mathbf{A} = \mathbf{B}$ implies $a_{ij} = b_{ij}$ for i and j. It follows that two equal matrices must be of the same order.

The **transpose** of a matrix **A** of order $(r \times c)$ is defined as a matrix $\mathbf{A}'$ of order $(c \times r)$ such that

$$a'_{ij} = a_{ji}.$$

For example, if

$$\mathbf{A} = \begin{bmatrix} 1 & -5 \\ 2 & 2 \\ 4 & 1 \end{bmatrix}, \quad \text{then} \quad \mathbf{A}' = \begin{bmatrix} 1 & 2 & 4 \\ -5 & 2 & 1 \end{bmatrix}.$$

In other words, the rows of **A** are the columns of $\mathbf{A}'$ and vice versa. This is one matrix operation that is not relevant to scalars.

A matrix **A** for which $\mathbf{A} = \mathbf{A}'$ is said to be **symmetric**. A symmetric matrix must obviously be square, and each row has the same elements as the corresponding column. For example, the following matrix is symmetric:

$$\mathbf{C} = \begin{bmatrix} 5 & 4 & 2 \\ 4 & 6 & 1 \\ 2 & 1 & 8 \end{bmatrix}.$$

The operation of **matrix addition** is defined as

$$\mathbf{A} + \mathbf{B} = \mathbf{C}$$

if $a_{ij} + b_{ij} = c_{ij}$, for all i and j. Thus, the addition of matrices is accomplished by the addition of corresponding elements. For example, let

$$\mathbf{A} = \begin{bmatrix} 1 & 2 \\ 4 & 9 \\ -5 & 4 \end{bmatrix} \quad \text{and} \quad \mathbf{B} = \begin{bmatrix} 4 & -2 \\ 1 & 2 \\ 5 & -6 \end{bmatrix},$$

then

$$\mathbf{C} = \mathbf{A} + \mathbf{B} = \begin{bmatrix} 5 & 0 \\ 5 & 11 \\ 0 & -2 \end{bmatrix}.$$

If two matrices are to be added, they must, be **conformable** for addition, that is, they must have the same order. Subtraction of matrices follows the same rules.

The process of **matrix multiplication** is more complicated. The definition of matrix multiplication is

$$\mathbf{C} = \mathbf{A} \cdot \mathbf{B}$$

if

$$c_{ij} = \sum_k a_{ik} b_{kj}.$$

The operation may be better understood when expressed in words: The element of the ith row and jth column of the product matrix **C**, c_{ij}, is the pairwise sum of products of the corresponding elements of the ith row of **A** and the jth column of **B**.

For **A** and **B** to be conformable for multiplication, then the number of columns of **A** must be equal to the number of rows of **B**. The order of the product matrix **C** will be equal to the number of rows of **A** by the number of columns of **B**.

As an example, let

$$\mathbf{A} = \begin{bmatrix} 2 & 1 & 6 \\ 4 & 2 & 1 \end{bmatrix} \quad \text{and} \quad \mathbf{B} = \begin{bmatrix} 4 & 1 & -2 \\ 1 & 5 & 4 \\ 1 & 2 & 6 \end{bmatrix}.$$

Note that matrix **A** has three columns and that **B** has three rows; hence these matrices are conformable for multiplication. Also since **A** has two rows and **B** has three columns, the product matrix **C** will have two rows and three columns. The elements of $\mathbf{C} = \mathbf{AB}$ are obtained as follows:

$$\begin{aligned} c_{11} &= a_{11}b_{11} + a_{12}b_{21} + a_{13}b_{31} \\ &= (2)(4) + (1)(1) + (6)(1) = 15, \\ c_{12} &= a_{11}b_{12} + a_{12}b_{22} + a_{13}b_{32} \\ &= (2)(1) + (1)(5) + (6)(2) = 19, \\ &\quad .. \quad .. \quad .. \quad .. \quad .. \\ c_{23} &= a_{21}b_{13} + a_{22}b_{23} + a_{23}b_{33} \\ &= (4)(-2) + (2)(4) + (1)(6) = 6. \end{aligned}$$

The entire matrix **C** is

$$\mathbf{C} = \begin{bmatrix} 15 & 19 & 36 \\ 19 & 16 & 6 \end{bmatrix}.$$

Note that even if **A** and **B** are conformable for the multiplication **AB**, it may not be possible to perform the operation **BA**. However, even if the matrices are conformable for both operations, usually

$$\mathbf{AB} \neq \mathbf{BA},$$

although exceptions occur for special cases.

An interesting corollary of the rules for matrix multiplication is that

$$(\mathbf{AB})' = \mathbf{B}'\mathbf{A}';$$

that is, the transpose of a product is the product of the individual transposed matrices in reverse order.

There is no matrix division as such. If we require matrix **A** to be divided by matrix **B**, we first obtain the **inverse** of **B**. Denoting that matrix by **C**, we then multiply **A** by **C** to obtain the desired result.

The inverse of a matrix **A**, denoted $\mathbf{A}^{-1}$, is defined by the property:

$$\mathbf{AA}^{-1} = \mathbf{I},$$

where **I** is the identity matrix which, as defined above, has the role of the number 1. Inverses are defined only for square matrices. However, not all square matrices are invertible, as discussed later.

Unfortunately, the definition of the inverse of a matrix does not suggest a procedure for computing it. In fact, the computations required to obtain the inverse of a matrix are quite tedious. Procedures for inverting matrices using hand or desk calculators are available but are not presented here. Instead we always present inverses that have been obtained by a computer.

The following serves as an illustration of the inverse of a matrix. Consider two matrices **A** and **B**, where $\mathbf{A}^{-1} = \mathbf{B}$:

$$\mathbf{A} = \begin{bmatrix} 9 & 27 & 45 \\ 27 & 93 & 143 \\ 45 & 143 & 245 \end{bmatrix}, \quad \mathbf{B} = \begin{bmatrix} 1.47475 & -0.113636 & -0.204545 \\ -0.113636 & 0.113636 & -0.045455 \\ -0.204545 & -0.0454545 & 0.068182 \end{bmatrix}.$$

The fact that **B** is the inverse of **A** is verified by multiplying the two matrices. The first element of the product **AB** is the sum of products of the elements of the first row of **A** with the elements of the first column of **B**:

$$(9)(1.47475) + (27)(-0.113636) + (45)(-0.2054545) = 1.000053.$$

This element should be unity; the difference is due to round-off error, which is a persistent feature of matrix calculations. Most modern computers carry sufficient precision to make round-off error insignificant, but this is not always guaranteed (see Section 8.7). The reader is encouraged to verify the correctness of the above inverse for at least a few other elements.

Other properties of matrix inverses are as follows:

1. $\mathbf{AA}^{-1} = \mathbf{A}^{-1}\mathbf{A}$.
2. If $\mathbf{C} = \mathbf{AB}$ (all square), then $\mathbf{C}^{-1} = \mathbf{B}^{-1}\mathbf{A}^{-1}$. Note the reversal of the ordering, just as for transposes.
3. If $\mathbf{B} = \mathbf{A}^{-1}$, then $\mathbf{B}' = (\mathbf{A}')^{-1}$.

4. If $\mathbf{A}$ is symmetric, then $\mathbf{A}^{-1}$ is also symmetric.
5. If an inverse exists, it is unique.

Certain matrices do not have inverses; such matrices are called **singular**. For example, the matrix

$$\mathbf{A} = \begin{bmatrix} 2 & 1 \\ 4 & 2 \end{bmatrix}$$

cannot be inverted, because the elements in row two are simply twice the elements in row one. We can better see why a matrix such as this cannot be inverted in the context of linear equations presented in the next section.

Solving Linear Equations

Matrix algebra is of interest in performing regression analyses because it provides a shorthand description for the solution to a set of linear equations. For example, assume we want to solve the set of equations

$$\begin{aligned} 5x_1 + 10x_2 + 20x_3 &= 40, \\ 14x_1 + 24x_2 + 2x_3 &= 12, \\ 5x_1 - 10x_2 &= 4. \end{aligned}$$

This set of equations can be represented by the matrix equation

$$\mathbf{A} \cdot \mathbf{X} = \mathbf{B},$$

where

$$\mathbf{A} = \begin{bmatrix} 5 & 10 & 20 \\ 14 & 24 & 2 \\ 5 & -10 & 0 \end{bmatrix}, \quad \mathbf{X} = \begin{bmatrix} x_1 \\ x_2 \\ x_3 \end{bmatrix}, \quad \text{and} \quad \mathbf{B} = \begin{bmatrix} 40 \\ 12 \\ 4 \end{bmatrix}.$$

The solution to this set of equations can now be represented by some matrix operations. Premultiply both sides of the matrix equation by $\mathbf{A}^{-1}$ as follows:

$$\mathbf{A}^{-1} \cdot \mathbf{A} \cdot \mathbf{X} = \mathbf{A}^{-1} \cdot \mathbf{B}.$$

Now $\mathbf{A}^{-1} \cdot \mathbf{A} = \mathbf{I}$, the identity matrix; hence, the equation can be written

$$\mathbf{X} = \mathbf{A}^{-1} \cdot \mathbf{B},$$

which is a matrix equation representing the solution.

We can now see the implications of the singular matrix shown above. Using that matrix for the coefficients and adding a right-hand side produces the equations

$$\begin{aligned} 2x_1 + x_2 &= 3, \\ 4x_1 + 2x_2 &= 6. \end{aligned}$$

Note that these two equations are really equivalent; therefore, any of an infinite number of combinations of x_1 and x_2 that satisfy the first equation are also a solution to the second equation. On the other hand, changing the right-hand side produces the equations:

$$2x_1 + x_2 = 3,$$

$$4x_1 + 2x_2 = 10,$$

which are inconsistent and have no solution. In regression applications it is not possible to have inconsistent sets of equations.

References

Agresti, A. (1984). *Analysis of Ordinal Categorical Data.* New York: Wiley.

Agresti, A., and Caffo, B. (2000). Simple and effective confidence intervals for proportions and differences of proportion result from adding two successes and two failures. *Am. Statist.* **54**, 280–288.

Agresti, A., and Coull, B. A. (1998). Approximation is better than exact for interval estimation of binomial proportions. *Am. Statist.* **52**, 119–126.

American Council of Life Insurance (1988). *1988 Life Insurance Fact Book.*

Andrews, D. F., and Herzberg, A. M. (1985). *Data: A Collection of Problems from Many Fields for the Student and Research Worker.* New York: Springer-Verlag.

Aylward, G. Pl., Harcher, R. P., Leavitt, L. A., Rao, V., Bauer, C. R., Brennan, M. J., and Gustafson, N. F. (1984). Factors affecting neobehavioral responses of preterm infants at term conceptual age. *Child Develop.* **55**, 1155–1165.

Bancroft, T. A. (1968). *Topics in Intermediate Statistical Methods.* Ames, IA: Iowa State Univ. Press.

Barnett, V., and Lewis, T. (1994). *Outliers in Statistical Data*, 3rd ed. New York; Wiley.

Belsley, D. A., Kuh, E., and Welsch, R. E. (1980). *Regression Diagnostics.* New York: Wiley.

Bishop, Y. M. M., Fienberg, S. E., and Holland, P. W. (1975). *Discrete Multivariate Analysis.* Cambridge, MA: MIT Press.

Boos, D. D. (1986). Comparing k populations with linear rank statistics. *J. Am. Statist. Assoc.* **81**, 1018–1025.

Carmer, S. G., and Swanson, M. R. (1973). An evaluation of ten pairwise multiple comparison procedures by Monte Carlo methods. *J. Am. Statist. Assoc.* **68**, 66–74.

Chambers, J. M., Cleveland, W. S., Kliener, B., and Tukey, P. A. (1983). *Graphical Methods for Data Analysis.* Pacific Grove, CA: Wadsworth.

Cleveland, W. S., Harris, C. S., and McGill, R. (1982). Judgments of circle sizes on statistical maps. *J. Am. Statist. Assoc.* **77**, 541–547.

Cochran, W. G. (1977). *Sampling Techniques.* New York: Wiley.

Cochran, W. G., and Cox, G. M. (1957). *Experimental Design.* New York: Wiley.

Collmann, R. D., and Stoller, A. (1962). A survey of mongoloid births in Victoria, Australia, 1942–1957. *Am. J. Pub. Health.* **57**, 813–829.

Conover, W. J. (1999). *Practical Nonparametric Statistics*, 3rd ed. New York: Wiley.

Conover, W. J., and Iman, R. L. (1981). Rank transformations as a bridge between parametric and nonparametric statistics. *Am. Statist.* **35**, 124–133.

Daniel, W. W. (1990). *Applied Nonparametric Statistics*, 2nd ed. Boston: PWS-Kent.

Draper, N. R., and Smith, H. (1981). *Applied Regression Analysis*, 2nd ed. New York: Wiley.

Duncan, D. B. (1955). Multiple range and multiple A tests. *Biometrics* **11**, 1–42.

Duncan, D. B. (1957). Multiple range tests for correlated and heteroscedastic means. *Biometrics* **13**, 164–176.

Efron, N., and Veys, J. (1992). Defects in disposable contact lenses can compromise ocular integrity. *Internat. Contact Lens Clinic* **19**, 8–18.

Enrick, N. L. (1976). An analysis of means in a three way factorial. *J. Quality Technol.* **8**, 189.

Feinberg, S. (1980). *The Analysis of Cross-Classified Categorical Data*, 2nd ed. Cambridge, MA: MIT Press.

Finney, D. J. (1971). *Probit Analysis*, 3rd ed. Cambridge, UK: Cambridge Univ. Press.

Fisher, R. A. (1948). *The Design of Experiments*. Edinburgh: Oliver and Boyd.

Freund, J. E., and Williams, F. J. (1982). *Elementary Business Statistics: The Modern Approach*, 4th ed. Englewood Cliffs, NJ: Prentice Hall.

Freund, R. J. (1980). The case of the missing cell. *Am. Statist.* **24**, 94–98.

Freund, R. J., and Minton, P. D. (1979). *Regression Methods*. New York: Marcel Dekker.

Freund, R., and Wilson, W. (1998). *Regression Analysis: Statistical Modeling of a Response Variable.* San Diego: Academic Press.

Glucksberg, H., Cheever, M. A., Farewell, V. T., Fefer, A., Sale, G. E., and Thomas, E. D. (1981). High dose combination chemotherapy for acute nonlymphoblastic leukemia in adults. *Cancer* **48**, 1073–1081.

Graybill, F. A. (1976). *Theory and Application of the Linear Model*. Boston: Duxbury Press.

Graybill, F. A. (1983). *Matrices with Applications in Statistics*, 2nd ed. Pacific Grove, CA: Wadsworth.

Green, J. A. (1988). Loglinear analysis of cross-classified ordinal data: Applications in developmental research. *Child Develop.* **59**, 1–25.

Grizzle, J. E., Starmer, C. F., and Koch, G. G. (1969). Analysis of categorical data by linear models. *Biometrics* **25**, 489–504.

Huber, P. J. (1981). *Robust Statistics*. New York: Wiley.

Huff, D. (1982). *How to Lie with Statistics*. New York: Norton.

Huntsberger, D. V., and Billingsley, P. (1977). *Elements of Statistical Inference*, 4th ed. Boston: Allyn and Bacon.

Iman, R. L., and Davenport, J. M. (1980). Approximations of the critical region of the Friedman statistic. *Comm. Statistics—Theory and Methods* **9**, 571–595.

Iman, R. L., Quade, D., and Alexander (1975). Exact probability levels for the Kruskal–Wallis test statistic. *Selected Tables in Math. Statist.* **3**, 329–384.

Kendall, M., and Stuart, A. (1979). *The Advanced Theory of Statistics*, Vol. 2. New York: Macmillan.

Kirk, R. (1995). Experimental Design, 3rd ed. Pacific Grove, CA: Brooks/Cole.

Kleinbaum, D. G., Kupper, L. L., and Muller, K. E. (1988). *Applied Regression Analysis and Other Multivariable Methods*, 2nd ed. Boston: PWS-Kent.

Kleinbaum, D., Kupper, L., Muller, K., and Nizam, A. (1998). *Applied Regression Analysis and Other Multivariate Methods*, 3rd ed. Pacific Grove, CA: Duxbury Press.

Koehler, K. J., and Larntz, K. (1980). An empirical investigation of goodness-of-fit statistics for sparse multinomials. *J. Am. Statist. Assoc.* **75**, 336–344.

Koopmans, L. H. (1981). *An Introduction to Contemporary Statistics*. Boston: Duxbury Press.

Koopmans, L. H. (1987). *An Introduction to Contemporary Statistics*, 2nd ed. Boston: Duxbury Press.

Kuel, R. O. (2000). *Design of Experiments Statistica Principles of Research Design and Analysis*, 2nd ed. Pacific Grove, CA: Duxbury Press.

Larntz, K. (1978). Small-sample comparisons of exact levels for chi-squared goodness-of-fit statistics. *J. Am. Statist. Assoc.* **73**, 253–263.

Lentner, J., Arnold, J., and Hinkelmann, K. (1989). The efficiency of blocking: How to use MS(blocks)/ MS(error) correctly. *Am. Statist.* **43**, 106–108.

Levene, H. A. (1960). Robust tests for quality of variances. In *Contributions to Probability and Statistics* (I. Olkin, Ed.). Palo Alto, CA: Stanford Univ. Press.

Littell, R. C., Stoup, W. W., and Freund, R. J. (2002). *SAS for Linear Models*, Cary, NC: SAS Institute.

Loehlin, J. C. (1987). *Latent Variable Models*. Hillsdale, NJ: Lawrence Erlbaum.

Mallows, C. L. (1973). Some comments on C(p). *Technometrics* **15**, 661–675.

Mandel, J. (1976). Models, transformations of scale and weighting. *J. Qual. Technol.* **8**.

Masood, S. (1989). Use of monoclonal antibody for assessment of estrogen receptor content in fine-needle aspiration biopsy specimen from patients with breast cancer. *Arch. Pathol. Lab. Med.* **113**, 26–30.

Masood, S., and Johnson, H. (1987). The value of imprint cytology in cytochemical detection of steroid hormone receptors in breast cancer. *Am. J. Clin. Pathol.* **87**, 30–36.

Mattson, G. A., Twogood, R. P., and Wilson, W. J. (1991). Professionalism and capacity building: impediments to economic innovation in small Florida cities. In *Proceedings: Rural Planning and Development: Visions of the 21st Century*. Gainesville, FL: Univ. of Florida Press.

Maxwell, S. E., and Delaney, H. D. (2000). *Designing Experiments and Analyzing Data: A Model Comparison Perspective.* Mahwah, NJ: Lawrence Erlbaum.

Montgomery, D. C. (1984). *Design and Analysis of Experiments.* New York: Wiley.

Nelson, L. S. (1983). Exact critical values for use with the analysis of means. *J. Quality Technol.* **15**, 40–44.
Nelson, P. R. (1985). Power curves for the analysis of means. *Technometrics* **27**, 65–73.
Neter, J., Wasserman, W., and Kutner, M. H. (1985). *Applied Linear Statistical Models*, 2nd ed. Homewood, IL: Irwin.
Neter, J., Wasserman, W., and Kutner, M. H. (1990). *Applied Linear Statistical Models*, 3rd ed. Homewood, IL: Irwin.
Neter, J., Kutner, M., Nachtsheim, C., and Wasserman, W. (1996). *Applied Linear Statistical Models*, 4th ed. Chicago: Irwin.
Ostle, B. (1963). *Statistics in Research*, 2nd ed. Ames, IA: Iowa State Univ. Press.
Ott, E. R. (1967). Analysis of means—A graphical procedure. *Indust. Qual. Control* **24**, 101–109.
Ott, E. R. (1975). *Process Quality Control.* New York: McGraw-Hill.
Ott, L. (1977). *An Introduction to Statistical Methods and Data Analysis.* Boston: Duxbury Press.
Ott, L. (1988). *An Introduction to Statistical Methods and Data Analysis*, 3rd ed. Boston: PWS-Kent.
Ott, L. (1993). *An Introduction to Statistical Methods and Data Analysis*, 4th ed. Belmont, CA: Duxbury Press.
Owen, D. B. (1962). *Handbook of Statistical Tables.* Reading, MA: Addison-Wesley.
Pearson, E. S., and Hartley, H. O. (Eds.) (1958). *Biometrika Tables for Statisticians*, 2nd ed., Vol. I. New York: Cambridge Univ. Press.
Pearson, E. S., and Hartley, H. O. (Eds.) (1966). *Biometrika Tables for Statisticians*, 3rd ed., Vol. I. New York: Cambridge Univ. Press.
Pearson, E. S., and Hartley, H. O. (Eds.) (1972). *Biometrika Tables for Statisticians*, 4th ed., Vol. II. Reprinted with corrections 1976. Cambridge, UK: Cambridge Univ. Press.
Phillips, D. P. (1978). Airplane accident fatalities increase just after newspaper stories about murder and suicide. *Science* **201**, 748–750.
Ramig, P. F. (1983). Applications of the analysis of means. *J. Quality Technol.* **15**, 19–25.
Rawlings, J. (1988). *Applied Regression Analysis: A Research Tool.* Pacific Grove, CA: Wadsworth.
Reichler, J. L. (Ed.) (1985). *The Baseball Encyclopedia*, 6th ed. New York: Macmillan.
Ross, S. (1984). *A First Course in Probability*, 2nd ed. New York: MacMillan.
Ross, S. (1994). *A First Course in Probability*, 4th ed. New York: Macmillan College.
Ryan, T. A., Joiner, B. L., and Ryan, B. F. (1985). *Minitab Handbook.* Boston: Duxbury Press.
SAS Institute (1985). *SAS User's Guide: Statistics.* Cary, NC: SAS Institute.
Scheaffer, R. L., Mendenhall, W., and Ott, L. (1979). *Elementary Survey Sampling*, 3rd ed. Boston: Duxbury Press.

Scheaffer, R. L., Mendenhall, W., and Ott, L. (1990). *Elementary Survey Sampling*, 4th ed. Boston: Duxbury Press.

Scheaffer, R. L., Mendenhall, W., and Ott, L. (1996). *Elementary Survey Sampling*, 5th ed. Belmont, CA: Duxbury Press.

Scheffé, H. (1953). A method for judging all contrasts in an analysis of variance. *Biometrika* **40**, 87–104.

Schilling, E. G. (1973). A systematic approach to the analysis of means. *J. Quality Technol.* **5**, 93–108, 147–159.

Seber, G. A. F. (1977). *Linear Regression Analysis*. New York: Wiley.

Smith, H. (1969). The analysis of data from a designed experiment. *J. Quality Technol.* **1**, 4.

Snedecor, G. W., and Cochran, W. G. (1980). *Statistical Methods*, 7th ed. Ames, IA: Iowa State Univ. Press.

Steel, R. G. D., and Torrie, J. H. (1980). *Principles and Procedures of Statistics*, 2nd ed. New York: McGraw-Hill.

Sudman, S. (1976). *Applied Sampling*. New York: Academic Press.

Tukey, J. W. (1977). *Exploratory Data Analysis*. Reading, MA: Addison–Wesley.

Upton, G. J. G. (1978). *The Analysis of Cross-Tabulated Data*. New York: Wiley.

Wackerly, D. D., Mendenhall, W., and Scheaffer, R. (1996). *Mathematical Statistics with Applications*, 5th ed. Belmont, CA: Duxbury Press.

Wald, A. (1947). *Sequential Analysis*. New York: Wiley.

White, C. (1956). The use of ranks in a test of significance for comparing two treatments. *Biometrics* **8**, 33–41.

Wilcoxon, F., and Wilcox, R. A. (1964). *Some Rapid Approximate Statistical Procedures*. Pearl River, NY: Lederle Laboratories.

Winer, B. J. (1971). *Statistical Principles in Experimental Design*. New York: McGraw-Hill.

Wright, S. P., and O'Brien, R. G. (1988). Power analysis in an enhanced GLM procedure: What it might look like. *Proceedings SUGI* **13**, 1097–1102.

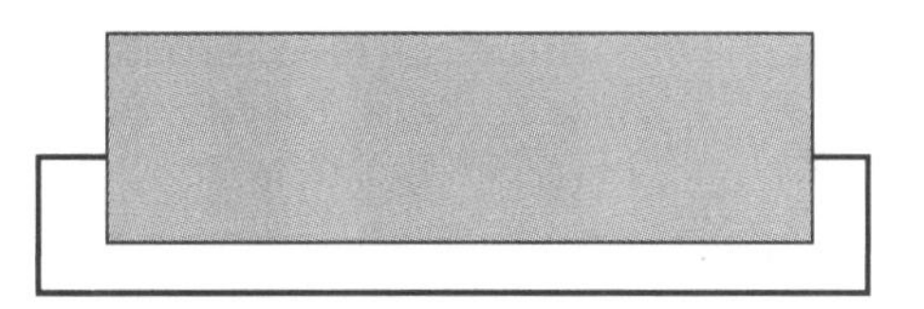

Solutions to Selected Exercises

Chapter 1

PRACTICE EXERCISES

1.

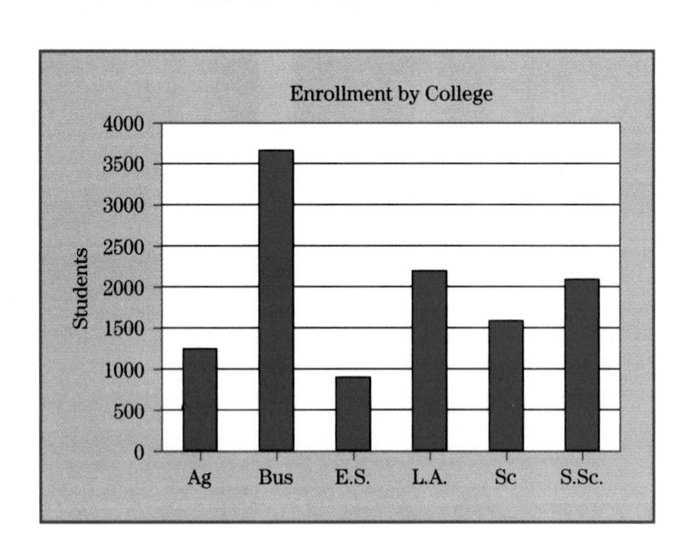

2. Mean = 10.9, median = 12.5, std dev = 5.3427, variance = 28.54444.

3. Mean = 2, std dev = 2.94.

4. (a)

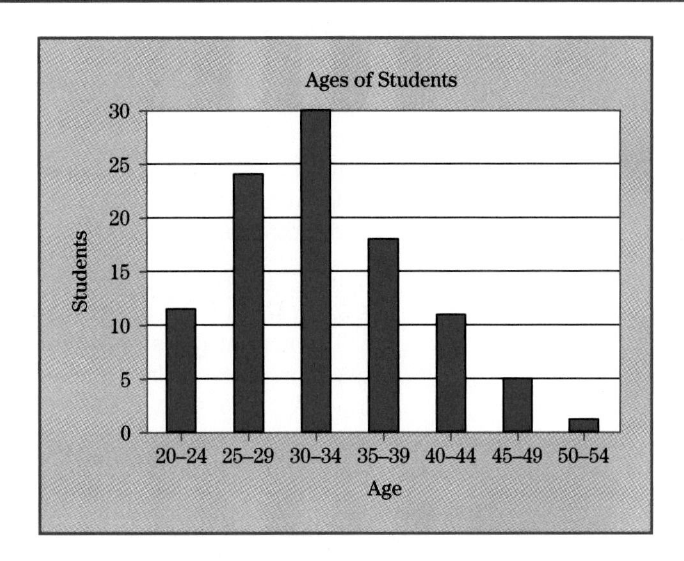

(b) Mean = 32.65, std dev = 6.91.

5. (a)

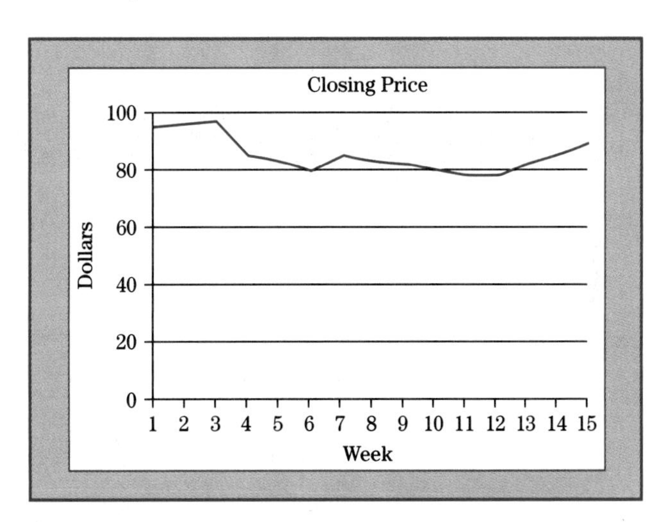

(b) Stem and leaf

9	5
9	430
8	97755
8	4420
7	88

(c) Mean = 86.06667, median = 85.

(d) Std dev = 5.417784, variance = 29.35238.

EXERCISES

1. (a) Mean = 17, median = 17, variance = 30.57, range = 22, interquartile range = 7.
   (c) The stem and leaf reveals a heavier concentration of data in the range 10–19, piling up around the mean and median, indicative of a symmetrical distribution.

3. (a) FUTURE: Mean = −0.20848, median = −0.3, variance = 0.601018.
   INDEX: Mean = −0.14935, median = −0.155, variance = 1.770753.
   (b) Yes. The scatterplot shows that, in general, as the NYSE Composite Index increases so does the price movements of the general stock market.

5. (a) DAYS: Mean = 15.85366, median = 17, variance = 24.324.
   TEMP: Mean = 39.34756, median = 40, variance = 11.15191.
   (b) From the scatterplot, there appears to be no definitive relationship between the average temperature and the number of rainy January days.

7. The strongest relationship exists between DFOOT, the diameter of the tree at one foot above ground level and HT, the total height of the tree. One would expect that as the base of a tree increases in diameter the tree would increase in height as well.

9. (a) The mean is larger than (to the right of) the median, indicating a distribution skewed to the right. Yes, both the stem and leaf plot and the box plot reveal the skewness of the distribution.
   (b) The outliers 955 and 1160 may have resulted from younger patients.
   (c) Approximately 75% or 38 of the 51 patients were in remission for less than one year.

11. (a) Mean = 7.2695, std dev = 1.633599.
   (b) Median = 7.1 occurred near the end of 1977 or the beginning of 1978.

13. (b) Initial doses for drug G are lower than drug A, which also tends to have a shorter half-life.
   (c) Drug A: Mean = 1.88, std dev = 0.63.
   Drug B: Mean = 2.02, std dev = 0.40.
   This supports the conclusion in part (b).

Chapter 2

PRACTICE EXERCISES

1. (a) $P(\text{Both}) = (0.4)(0.3) = 0.12$
   (b) $P(\text{Neither}) = (0.6)(0.7) = 0.42$
   (c) $P(\text{At Least One}) = (0.4) + (0.3) - (0.12) = 0.58$

2. (a) $P(A) = 0.2$
$P(B) = 0.3$
(b) $P(A \text{ and } B) = 0$
(c) $P(A \text{ or } B) = (0.2) + (0.3) = 0.50$

3. (a) $\mu = 1.0$
$\sigma^2 = 1.5$
(b) 0.03125

4. \$1,450

5. (a) $(0.1587)(0.5793) = 0.0919$
(b) 0.6461
(c) $(0.5)(0.5) = 0.25$

EXERCISES

1. (a)

Y	\$0	\$10	\$1,000	\$10,000	\$50,000
$p(Y)$	$\frac{148,969}{150,000}$	$\frac{1000}{150,000}$	$\frac{25}{150,000}$	$\frac{5}{150,000}$	$\frac{1}{150,000}$

(b) $\mu = \$0.90$
(c) Expected net winnings are −\$0.10; therefore a purchase is not worthwhile from a strictly monetary point of view.
(d) $\sigma^2 = \$142.00$

3. (a) $\mu = 1$
$\sigma^2 = 0.8$
(b) Yes

5. Arrangement I: P(system fail) = 0.0001999
Arrangement II: P(system fail) = 0.0394

7. (a) 0.1586
(b) 0.3108
(c) 16.4

9. 0.0571

11. 39

13. 0.5762

15. $\mu = 75$
$\sigma^2 = (11.72)^2 = 137.36$

17. (a) 7.15×10^{-8}
(c) 0.0186

Chapter 3

PRACTICE EXERCISES

1. (158.4, 175.6)

2. $Z = -0.91$

3. $n \approx 62$, $n \approx 246$

4. $z = 3.40$

EXERCISES

3. (a) $\beta = 0.8300$
(b) $\beta = 0.9428$
(c) $\beta = 0.3372$
(d) $\beta = 0.0877$

5. (a) $z = -4.38$
(b) p value $\simeq 0$

7. (a) $\alpha = 0.0256$
(b) $\beta = 0.8704$

9. (a) $E = 1.31$
(b) (78.3, 80.9)
(c) $n = 246$

11. $z = -22$
p value $\simeq 0$

13. $z = -4.0$

15. $n = 9604$

Chapter 4

PRACTICE EXERCISES

1(a) 1.7709
(b) 2.4786
(c) 1.3968
(d) 37.566
(e) 13.362
(f) 24.433
(g) 2.088

2. (a) (7.2, 10.8)
(b) (3.0, 5.3)

3. (a) $t = -4.59$
(b) $X^2 = 26$

4. $z = 0.0$

5. $n = 9220$

EXERCISES

1. $t = 3.99$

3. $\chi^2 = 92.17$

5. (a) $z = -0.67$

7. $z = 0.65$

9. $t = -1.08$

11. $\chi^2 = 23.96$

15. Normality assumption violated

17. (1.596, 2.155)

19. $z = 2.67$

21. Type 1: $\chi^2 = 50.32$, Type 2: $\chi^2 = 102.13$
Type 3: $\chi^2 = 58.88$, Type 4: $\chi^2 = 24.84$

Chapter 5

PRACTICE EXERCISES

1. $z = 1.24$

2. $t = -2.44$

3. $(-9.92, -1.68)$

4. $t = 1.94$

5. $z = 1.84$

EXERCISES

1. $t = 1.223$

3. $t = 1.479$

5. $t = 3.136$

7. $z = 0$

9. $F = 1.513$

11. $t = -1.1798$

13. $t = -0.8862$

15. (a) $z = -0.56$

Chapter 6

EXERCISES

1. (a) $F = 15.32$

3. (a) $F = 19.04$
(b) $\text{Est}(\sigma_s^2) = 629.94$
$\text{Est}(\sigma^2) = 139.65$

5. (a) $F = 53.55$
(b) Control: $F = 12.8$
MFG: $F = 36.09$
ADD: $F = 137.5$
(c) LIN: $F = 195.61$
QUAD: $F = 13.60$
LOF: $F = 2.489$

7. $F = 15.04$

9. $F = 17.99$

11. CD: $F = 58.0$
PB: $F = 13.82$

13. $F = 30.40$

15. $F = 6.21$

Chapter 7

EXERCISES

1. (a) $\hat{\mu}_{y|x} = 2.8 - 0.5X$
(b) 3.8, 3.3, 2.8, 2.3, 1.8
(c) 0.2, −0.3, 0.2, −0.3, 0.2
(d) $t = -5.0$

3. (a) $\hat{\mu}_{y|x} = 14.396 + 0.765X$
(d) 77.1328

5. (a) $\hat{\mu}_{y|x} = -71.451 + 1.209X$
(b) 1.17

7.(b) $\hat{\mu}_{y|x} = -97.924 + 2.001X$

9. (a) $\hat{\mu}_{y|x} = 7.526 + 0.225X$

11. Relationship is not linear.

Chapter 8

EXERCISES

3. $\hat{\mu}_{y1|x} = 700.62 - 1.526X1 + 175.984X2 - 6.697X3.$
$\hat{\mu}_{y2|x} = -5.611 + 0.668X1 - 1.235X2 + 0.073X3.$

5. (a) $\hat{\mu}_{y|x} = -379.248 + 170.220\text{DBH} + 1.900\text{HEIGHT} + 8.146\text{AGE} - 1192.868\text{GRAV}.$

7. (a) $\hat{\mu}_{y|x} = 219.275 + 77.725X.$
(b) $\hat{\mu}_{y|x} = 178.078 + 93.106X - 0.729X^2.$

9. (a) COOL: $\hat{\mu}_{y|x} = -2.638 + 0.439\text{WIDTH} + 0.110\text{HEIGHT}.$
WARM: $\hat{\mu}_{y|x} - 2.117 + 0.207\text{WIDTH} + 0.118\text{HEIGHT}.$
(b) COOL: $\hat{\mu}_{y|x} = -4.597 + 1.571\text{LWIDTH} + 0.747\text{LHEIGHT}.$
WARM: $\hat{\mu}_{y|x} = -4.421 + 1.669\text{LWIDTH} + 0.209\text{LHEIGHT}.$

11. $\hat{\mu}_{y|x} = -10.305 + 0.378\text{AGE} + 2.294\text{SEX} + 0.179\text{COLLEGE} + 0.293\text{INCOME}.$

13. $\hat{\mu}_{y|x} = 104.906 - 6.682\text{AGE} + 0.636\text{SQFT} - 0.403\text{SD} + 0.098\text{UNTS} + 65.000\text{GAR} + 33.051\text{CP} - 9.977\text{SS} + 14.327\text{FIT}.$

Chapter 9

EXERCISES

1.

Source	df	*F* Value
A	1	85.88
T	3	4.41
A*T	3	9.48

3.

Source	df	*F* Value
A	2	13.93
C	2	6.75
A*C	4	3.12

$\hat{\mu}_{y|x} = -11.107 + 6.486\text{A} - 0.631\text{A}^2 + 6.322\text{C} - 0.739\text{C}^2 - 0.596\text{AC}.$
F for LOF $= 0.458.$

5.

Source	df	*F* Value
FUNGICID	2	6.57
CONCENTR	1	21.50
FUNGICID*CONCENTR	2	4.91

7.

Source	df	*F* Value
TEMPR	4	323.05
CLEAN	4	1233.17
TEMPR*CLEAN	16	86.09

$$\hat{\mu}_{y|x} = -8.035 + 36.275\text{T} - 12.460\text{T}^2 - 30.952\text{C} + 23.787\text{C}^2 + 17.654\text{TC}.$$

9.

Source	df	*F* Value
TEMPL	1	4.28
TEMPSQ	1	2.78
DAY	1	3.95
TEMPL*DAY	1	1.46
TEMPSQ*DAY	1	23.58

11.

Source	df	*F* Value
GRAIN	2	2.93
PREP	2	7.33
GRAIN*PREP	4	1.85

Chapter 10

EXERCISES

1.

Source	df	SS	*F* Value
TRT	2	2.766	1.30
EXP	1	1.580	
TRT*EXP	2	2.130	
TOTAL	29		

3. (a)

Source	df	SS	*F* Value
REP	3	1.981	1.62
YEAR	2	29.835	36.70
VAR	2	6.231	7.66
NIT	1	0.0003	0.00
YEAR*VAR	4	33.531	20.62
VAR*NIT	2	15.667	19.2
YEAR*NIT	2	22.926	28.20
YEAR*VAR*NIT	4	18.796	11.56
ERROR	51	20.72	

(b)

Source	df	SS	*F* Value
YEAR	2	29.835	44.07
REP	3	1.981	
YEAR*REP (A)	6	2.031	
VAR	2	6.231	1.82
NIT	1	0.0003	0.00
VAR*NIT	2	15.667	4.59
ERROR (B)	55	93.951	

5.

Source	df	SS	*F* Value
REP	2	246.542	
SALT	3	4340.917	6.16
REP*SALT (A)	6	1408.458	
DAY	3	8147.417	352.32
DAY*SALT	9	3457.583	49.84
ERROR (B)	24	185.000	

7. (a)

Source	df	SS	*F* Value
REP	4	1.033	
LIGHT	2	1.947	6.16
REP*LIGHT (A)	8	1.264	
LEAF	4	0.885	9.90
LIGHT*LEAF	8	0.302	1.69
ERROR (B)	48	1.073	

(b) $\hat{\mu}_{y|x} = 1.781 + 0.013\text{LIGHT} - 0.0001\text{LIGHT}^2 + 0.267\text{LEAF} - 0.047\text{LEAF}^2 + 0.0005\text{LIGHT*LEAF}.$

(c) $F = 3.33.$

9. SURFACE:

Source	df	SS	*F* Value
TRT(Shade)	3	73.541	16.61
COLOR	1	0.003	0.00
TRT*COLOR	3	3.176	0.72

RECTAL:

Source	df	SS	*F* Value
TRT(Shade)	3	1.253	1.79
COLOR	1	0.062	0.26
TRT*COLOR	3	0.521	0.75

11. SAT: $F = 2.03$
BTY: $F = 2.74$
FNC: $F = 6.77$
INT: $F = 1.01$
DIG: $F = 8.81$
CST: $F = 12.54$
FSH: $F = 4.96$

13. WEIGHT: $F = 16.54$
LENGTH: $F = 8.31$
RELWT: $F = 27.91$

Chapter 11

EXERCISES

1.

Source	df	SS	*F* Value
STAGE	2	289.82	12.59
WWT	1	394.08	34.24

3. $\hat{\mu}_{y|x} = -2.75 + 0.52\text{CONC}.$

5. Unweighted: $\hat{\mu}_{y|x} = -0.360 + 0.012\text{VENT}.$
Weighted: $\hat{\mu}_{y|x} = -0.322 + 0.0115\text{VENT}.$
LOG: $\hat{\mu}_{y|x} = -4.048 + 0.057\text{VENT}.$

7. (a) ANOVA:

Source	df	SS	*F* Value
PAVE	2	216.774	43.67
TREAD	2	203.676	41.03
PAVE*TREAD	4	22.154	2.23

DUMMY VARIABLE:

Source	df	SS	*F* Value
PAVE	2	233.584	47.06
TREAD	2	212.463	42.80
PAVE*TREAD	4	6.699	0.67

(b)

Source	df	SS	*F* Value
PAVE	2	232.818	52.31
TREAD	1	219.062	98.44

(c) $\hat{\mu}_{y|x} = 26.194 + 28.660\text{FRICT} + 1.374\text{TREAD}.$

9. (a)

Source	df	SS	*F* Value
MEDIUM	2	3137.392	50.97
TIME	3	1514.468	16.40
MEDIUM*TIME	6	514.574	2.79

11. (a)

Source	df	SS	*F* Value
SUBJ	39	59.664	1.22
BTY	1	38.738	31.01
FNC	1	3.951	3.16
INT	1	83.250	66.64
DIG	1	0.333	0.27
CST	1	9.682	7.75
FSH	1	30.543	24.45

13.

Source	df	SS	*F* Value
SIZE	1	913381.32	3881.24
TYPE	1	85.55	0.36
SIZE*TYPE	1	461.12	1.96

Chapter 12

EXERCISES

1. (1) (0.67, 0.75)
(2) $X^2 = 3.23$

3. $X^2 = 0.275$

5. $X^2 = 4.21$

7. $X^2 = 3.306$

9. $X_2^2 = 7.66$

11. (a) $X^2 = 26.25$

13. $X^2 = 11.217$, $X^2 = 0.731$

Chapter 13

EXERCISES

1. $T(+) = 9$

3. $T = 81$

5. (a) $T(+) = 2.0$

7. $T^* = 0.486$

9. $H = 22.68$

11. $T^* = 1.448$

INDEX

D

E

F

N

O

P

Q

T

U

V

W

Y